Handbook of Research on Progressive Trends in Wireless Communications and Networking

M.A. Matin
Institut Teknologi Brunei, Brunei Darussalam

A volume in the Advances in Wireless
Technologies and Telecommunication
(AWTT) Book Series

Information Science
REFERENCE
An Imprint of IGI Global

Managing Director:	Lindsay Johnston
Book Production Manager:	Jennifer Yoder
Development Editor:	Austin DeMarco
Assistant Acquisitions Editor:	Kayla Wolfe
Typesetter:	John Crodian
Cover Design:	Jason Mull

Published in the United States of America by
Information Science Reference (an imprint of IGI Global)
701 E. Chocolate Avenue
Hershey PA 17033
Tel: 717-533-8845
Fax: 717-533-8661
E-mail: cust@igi-global.com
Web site: http://www.igi-global.com

Library of Congress Cataloging-in-Publication Data

Handbook of research on progressive trends in wireless communications and networking / M.A. Matin, editor.
    pages cm
  Includes bibliographical references and index.
    ISBN 978-1-4666-5170-8 (hardcover) -- ISBN 978-1-4666-5171-5 (ebook) -- ISBN 978-1-4666-5173-9 (print & perpetual access) 1. Wireless communication systems. 2. Mobile communication systems. 3. Sensor networks. I. Matin, Mohammad A., 1977-
  TK5103.2.H3369 2014
  621.382--dc23
                                2013044828

This book is published in the IGI Global book series Advances in Wireless Technologies and Telecommunication (AWTT) (ISSN: 2327-3305; eISSN: 2327-3313)

British Cataloguing in Publication Data
A Cataloguing in Publication record for this book is available from the British Library.

All work contributed to this book is new, previously-unpublished material. The views expressed in this book are those of the authors, but not necessarily of the publisher.

For electronic access to this publication, please contact: eresources@igi-global.com.

Advances in Wireless Technologies and Telecommunication (AWTT) Book Series

ISSN: 2327-3305
EISSN: 2327-3313

MISSION

The wireless computing industry is constantly evolving, redesigning the ways in which individuals share information. Wireless technology and telecommunication remain one of the most important technologies in business organizations. The utilization of these technologies has enhanced business efficiency by enabling dynamic resources in all aspects of society.

The **Advances in Wireless Technologies and Telecommunication Book Series** aims to provide researchers and academic communities with quality research on the concepts and developments in the wireless technology fields. Developers, engineers, students, research strategists, and IT managers will find this series useful to gain insight into next generation wireless technologies and telecommunication.

COVERAGE

- Cellular Networks
- Digital Communication
- Global Telecommunications
- Grid Communications
- Mobile Technology
- Mobile Web Services
- Network Management
- Virtual Network Operations
- Wireless Broadband
- Wireless Sensor Networks

IGI Global is currently accepting manuscripts for publication within this series. To submit a proposal for a volume in this series, please contact our Acquisition Editors at Acquisitions@igi-global.com or visit: http://www.igi-global.com/publish/.

Titles in this Series

For a list of additional titles in this series, please visit: www.igi-global.com

Handbook of Research on Progressive Trends in Wireless Communications and Networking
M.A. Matin (Institut Teknologi Brunei, Brunei Darussalam)
Information Science Reference • copyright 2014 • 592pp • H/C (ISBN: 9781466651708) • US $380.00 (our price)

Broadband Wireless Access Networks for 4G Theory, Application, and Experimentation
Raul Aquino Santos (University of Colima, Mexico) Victor Rangel Licea (National Autonomous University of Mexico, Mexico) and Arthur Edwards-Block (University of Colima, Mexico)
Information Science Reference • copyright 2014 • 289pp • H/C (ISBN: 9781466648883) • US $180.00 (our price)

Multidisciplinary Perspectives on Telecommunications, Wireless Systems, and Mobile Computing
Wen-Chen Hu (University of North Dakota, USA)
Information Science Reference • copyright 2014 • 274pp • H/C (ISBN: 9781466647152) • US $175.00 (our price)

Mobile Networks and Cloud Computing Convergence for Progressive Services and Applications
Joel J.P.C. Rodrigues (Instituto de Telecomunicações, University of Beira Interior, Portugal) Kai Lin (Dalian University of Technology, China) and Jaime Lloret (Polytechnic University of Valencia, Spain)
Information Science Reference • copyright 2014 • 408pp • H/C (ISBN: 9781466647817) • US $180.00 (our price)

Research and Design Innovations for Mobile User Experience
Kerem Rızvanoğlu (Galatasaray University, Turkey) and Görkem Çetin (Turkcell, Turkey)
Information Science Reference • copyright 2014 • 377pp • H/C (ISBN: 9781466644465) • US $190.00 (our price)

Cognitive Radio Technology Applications for Wireless and Mobile Ad Hoc Networks
Natarajan Meghanathan (Jackson State University, USA) and Yenumula B. Reddy (Grambling State University, USA)
Information Science Reference • copyright 2013 • 370pp • H/C (ISBN: 9781466642218) • US $190.00 (our price)

Evolution of Cognitive Networks and Self-Adaptive Communication Systems
Thomas D. Lagkas (University of Western Macedonia, Greece) Panagiotis Sarigiannidis (University of Western Macedonia, Greece) Malamati Louta (University of Western Macedonia, Greece) and Periklis Chatzimisios (Alexander TEI of Thessaloniki, Greece)
Information Science Reference • copyright 2013 • 438pp • H/C (ISBN: 9781466641891) • US $195.00 (our price)

Tools for Mobile Multimedia Programming and Development
D. Tjondronegoro (Queensland University of Technology, Australia)
Information Science Reference • copyright 2013 • 357pp • H/C (ISBN: 9781466640542) • US $190.00 (our price)

DISSEMINATOR OF KNOWLEDGE

www.igi-global.com

701 E. Chocolate Ave., Hershey, PA 17033
Order online at www.igi-global.com or call 717-533-8845 x100
To place a standing order for titles released in this series, contact: cust@igi-global.com
Mon-Fri 8:00 am - 5:00 pm (est) or fax 24 hours a day 717-533-8661

Table of Contents

Detailed Table of Contents

Chapter 1
M. A. Matin, Institut Teknologi Brunei, Brunei Darussalam

There has been a tremendous growth of mobile communications markets all over the world, as they provide ubiquitous communication access to citizens. Wireless technologies are the core of mobile communications. They fundamentally revolutionize data networking, telecommunication, and make integrated networks to increase capacity and coverage. This has made the network portable because of affordable digital modulation, adaptive modulation, information compression, wireless access, multiplexing, and so on. It supports exciting applications such as sensor networks, smart homes, telemedicine, video conferencing and distance learning, cognitive radio networks, automation, and so on. This chapter provides an overview of the evolution of wireless and mobile communications from 2G to 4G.

Chapter 2
Siyi Wang, University of South Australia, Australia
Weisi Guo, University of Warwick, UK

It has been widely recognised that the exchange of information is one of the underpinning factors for economic growth in developing and developed nations. One of the fastest growing areas of information transfer is the mobile data sector. In 2012, global mobile data traffic grew by 70%. There is an urgent need to improve the wireless capacity of cellular networks in order to match this growth. One of the key issues faced by mobile operators is the fall in Average Revenue Per User (ARPU) and the growing Operational Expenditure (OPEX) due to capacity growth and rising energy prices. The challenge is therefore how to grow the wireless capacity in a way that minimizes the OPEX and thus improves the ARPU. Furthermore, there is growing focus on the environmental impact of Information and Communication Technology (ICT) sectors. There are tangible, financial, and environmental motivations for reducing the energy expenditure of wireless networks whilst growing its capacity. This chapter examines recent research in the area of future wireless network architectures and deployments. This is done in the context of improving capacity in a sustainable way. That is to say, what is the lowest-cost and -energy method of achieving certain capacity targets? The authors of this chapter were researchers in the world's first green wireless communications project—Mobile VCE Green Radio (2007-2012).

Virtual Private Network (VPN) services are widely used in the present corporate world to securely inter-connect geographically distributed private network segments through unsecure public networks. Among various VPN techniques, Internet Protocol (IP)-based VPN services are dominating due to the ubiquitous use of IP-based provider networks and the Internet. Over last few decades, the usage of cellular/mobile networks has increased enormously due to the rapid increment of the number of mobile subscribers and the evolvement of telecommunication technologies. Furthermore, cellular network-based broadband services are able to provide the same set of network services as wired Internet services. Thus, mobile broadband services are also becoming popular among corporate customers. Hence, the usage of mobile broadband services in corporate networks demands to implement various broadband services on top of mobile networks, including VPN services. On the other hand, the all- IP-based mobile network archi-tecture, which is proposed for beyond-LTE (Long Term Evolution) networks, is fuel to adapt IP-based VPN services in to cellular networks. This chapter is focused on identifying high-level use cases and scenarios where IP-based VPN services can be implemented on top of cellular networks. Furthermore, the authors predict the future involvement of IP-based VPNs in beyond-LTE cellular networks.

#### Chapter 4

A major challenge in the context of LTE networks is a cost-effective network operation, which can be done by carefully controlling the network Operational Expenses (OPEX). Therefore, to minimize OPEX costs while optimizing network performance, Self-Organizing Network (SON) principles were pro-posed. These networks are the main focus of this chapter, which highlights the state of art and provides a comprehensive investigation of current research efforts in the field of SONs. A major contribution of the chapter is the handling of SON use cases, going through their challenges, solutions. and open research questions. The chapter also presents efforts to provide coordination frameworks between SON use cases and routines. An additional essential contribution of the chapter is the description of SON activities within 3GPP.

#### Chapter 5

The Long Term Evolution (LTE) femtocell has promised to improve indoor coverage and enhance data rate capacity. Due to the special characteristic of the femtocell, it introduces several challenges in terms of mobility and interference management. This chapter focuses on mobility prediction in a wireless net-work in order to enhance handover performance. The mobility prediction technique via Markov Chains and a user's mobility history is proposed as a technique to predict user movement in the deployment of the LTE femtocell. Simulations have been developed to evaluate the relationship between prediction accuracy and the amount of non-random data, as well as the relationship between the prediction accuracy and the duration of the simulation. The result shows that the prediction is more accurate if the user moves in regular mode, which is directly proportional to the amount of non-random data. Moreover, the prediction accuracy is maintained at 0.7 when the number of weeks is larger than 50.

Chapter 6

Miroslav Škorić, IEEE Section, Austria & National Institute of Amateur Radio, India

In modern (amateur radio) wireless communications, we use computers. Depending on particular situations, such as employers' and personal preferences, users can adopt more or less proprietary operating systems and related end-user programs. In emerging and developing societies, the usage of proprietary software can be costly. Not only that, contrary to so-called "open" software, the "closed" software is not able to motivate its users to upgrade those programs regularly, not only because of high prices and restricted licensing policies, but also because of its nature, which is the "closedness" of program codes, where the end-users are not allowed to change programmed software, and so assist companies in improving features of their software products. Therefore, the authors help prospective newcomers in the amateur wireless communications to become familiar with the "open" software and, as well, to encourage them in implementing many "free" software solutions at home or work.

Chapter 7

Peter J. Hawrylak, University of Tulsa, USA
Steven Reed, University of Tulsa, USA
Matthew Butler, University of Tulsa, USA
John Hale, University of Tulsa, USA

Access to resources, both physical and cyber, must be controlled to maintain security. The increasingly connected nature of our world makes access control a paramount issue. The expansion of the Internet of Things into everyday life has created numerous opportunities to share information and resources with other people and other devices. The Internet of Things will contain numerous wireless devices. The level of access each user (human or device) is given must be controlled. Most conventional access control schemes are rigid in that they do not account for environmental context. This solution is not sufficient for the Internet of Things. What is needed is a more granular control of access rights and a gradual degradation or expansion of access based on observed facts. This chapter presents an access control system termed the Access of Things, which employs a gradual degradation of privilege philosophy. The Access of Things concept is applicable to the dynamic security environment present in the Internet of Things.

Chapter 8

Noman Islam, Technology Promotion International, Pakistan
Zubair A. Shaikh, National University of Computer and Emerging Sciences, Pakistan

Ad hoc networks enable network creation on the fly without support of any predefined infrastructure. The spontaneous erection of networks in anytime and anywhere fashion enables development of various novel applications based on ad hoc networks. However, ad hoc networks present several new challenges. Different research proposals have came forward to resolve these challenges. This chapter provides a survey of current issues, solutions, and research trends in wireless ad hoc networks. Even though various surveys are already available on the topic, rapid developments in recent years call for an updated account. The chapter has been organized as follows. In the first part of the chapter, various ad hoc network issues arising at different layers of TCP/IP protocol stack are presented. An overview of research proposals to address each of these issues is also provided. The second part of the chapter investigates various emerging models of ad hoc networks, discusses their distinctive properties, and highlights various research issues arising due to these properties. The authors specifically provide discussion on ad hoc grids, ad hoc clouds, wireless mesh networks, and cognitive radio ad hoc networks. The chapter ends with a presenting summary of the current research on ad hoc networks, ignored research areas and directions for further research.

Chapter 9

This chapter presents three algorithms to determine stable connected dominating sets (CDS) for wireless mobile ad hoc networks (MANETs) whose topology changes dynamically with time. The three stability-based CDS algorithms are (1) Minimum Velocity (MinV)-based algorithm, which prefers to include a slow moving node as part of the CDS as long as it covers one uncovered neighbor node; (2) Node Stability Index (NSI)-based algorithm, which characterizes the stability of a node as the sum of the predicted expiration times of the links (LET) with its uncovered neighbor nodes, the nodes preferred for inclusion to the CDS in the decreasing order of their NSI values; (3) Strong Neighborhood (SN)-based algorithm, which prefers to include nodes that cover the maximum number of uncovered neighbors within its strong neighborhood (region identified by the Threshold Neighborhood Ratio and the fixed transmission range of the nodes). The three CDS algorithms have been designed to capture the node size—lifetime tradeoff at various levels. In addition to presenting a detailed description of the three stability-based CDS algorithms with illustrative examples, the authors present an exhaustive simulation study of these algorithms and compare their performance with respect to several metrics vis-à-vis an unstable maximum density-based MaxD-CDS algorithm that serves as the benchmark for the minimum CDS Node Size.

Chapter 10

In this chapter, the authors present an improvement to the interactions between MAC (Medium Access Control) and TCP (Transmission Control Protocol) protocols for better performance in MANET. This improvement is called IB-MAC (Improvement of Backoff algorithm of MAC protocol) and proposes a new backoff algorithm. The principle idea is to make dynamic the maximal limit of the backoff interval according to the number of nodes and their mobility. IB-MAC reduces the number of collisions between nodes. It is also able to distinguish between packet losses due to collisions and those due to nodes' mobility. The evaluation of IB-MAC solution and the study of its incidences on MANET performance are done with TCP New Reno transport protocol. The authors varied the network conditions such as the network density and the mobility of nodes. Obtained results are satisfactory, and they showed that IB-MAC can outperform not only MAC standard, but also similar techniques that have been proposed in the literature like MAC-LDA and MAC-WCCP.

#### Chapter 11

This chapter gives an overview of research works on secure routing protocol and also describes a Novel Secure Routing Protocol RSRP proposed by the authors. The routes, which are free from any malicious node and which belong to the set of disjoint routes between a source destination pair, are considered as probable routes. Shamir's secret sharing principle is applied on those probable routes to obtain secure routes. Finally, the most trustworthy and stable route is selected among those secure routes using some criteria of the nodes present in a route (e.g., battery power, mobility, and trust value). In addition, complexity of key generation is reduced to a large extent by using RSA-CRT instead of RSA. In turn, the routing becomes less expensive and highly secure and robust. Performance of this routing protocol is then compared with non-secure routing protocols (AODV and DSR), secure routing scheme using secret sharing, security routing protocol using ZRP and SEAD, depending on basic characteristics of these protocols. All such comparisons show that RSRP shows better performance in terms of computational cost, end-to-end delay, and packet dropping in presence of malicious nodes in the MANET.

#### Chapter 12

Wireless Sensor Networks constitute one of the highest developing and most promising fields in modern data communication networks. The benefits of such networks are expected to be of great importance, since applicability is possible in multiple significant areas. The research community's interest is mainly attracted by the usability of sensor networks in healthcare services and environmental monitoring. This overview presents the latest developments in the field of sensor-based systems, focusing on systems designed for healthcare and environmental monitoring. The technical aspects and the structure of sensor networks are discussed, while representative examples of implemented systems are also provided. This chapter sets the starting point for the development of an integrated cooperative architecture capable of widely providing distributed services based on sensed data.

The purpose of this chapter is the study of the clustering process in Wireless Sensor Networks (WSN), starting with clarifying why there are different clustering protocols for WSN by stating and briefly describing some of the variate features in their design; these features can represent questions the clustering protocol designer asks before the design, and their brief description can be considered probabilities for these questions' answers to represent design options for the designer. The designer can choose the best answer to each design question or, in better words, the best design options that will make its protocol different from the others and make the resultant clustered network satisfies some requirements for improving the overall performance of the network. The chapter also mentions some of these requirements. The chapter then gives illustrative examples for these design variations and requirements by studying them on three well-known clustering protocols: Low-Energy Adaptive Clustering Hierarchy (LEACH), Energy-Efficient Clustering Scheme (EECS), and Hybrid, Energy-Efficient, Distributed clustering approach for ad-hoc sensor networks (HEED).

#### Chapter 14

In the last few decades, the Wireless Sensor Network (WSN) paradigm has received huge interest from the industry and academia. Wireless sensor networking is used in various fields like weather monitoring, wildfire detection/monitoring, battlefield surveillance, security systems, military applications, etc. Moreover, various networking and technical issues still need to be addressed for successful deployment of WSN, especially power management. In this chapter, the various methods of saving energy in sensor nodes and a method by which energy can be saved are discussed with emphasis on various energy saving protocols and techniques, and the improvement in the Performance of Clustered WSN by using Multi-tier Clustering. By using a two-tier architecture in the clustering and operation of sensor nodes, an increase in the network lifetime of the WSN is gained. Since this clustering approach has better results in term of energy savings and organizing the network, the main objective of this chapter is to describe power management techniques, two-tier architecture, clustering approaches, and network models to save the energy of a sensor network.

The need for reliable data delivery at the transport layer for video transmission over IEEE 802.15.4 Wireless Sensor Networks (WSNs) has attracted great attention from the research community due to the applicability of multimedia transmission for many applications. The IEEE 802.15.4 standard is designed to transmit data within a network at a low rate and a short distance. However, the characteristics of WSNs such as dense deployment, limited processing ability, memory, and power supply provide unique challenges to transport protocol designers. Additionally, multimedia applications add further challenges such as requiring large bandwidth, large memory, and high data rate. This chapter discusses the challenges and evaluates the feasibility of transmitting data over an IEEE 802.15.4 network for different transport protocols. The analysis result highlights the comparison of standard transport protocols, namely User Datagram Protocol (UDP), Transport Control Protocol (TCP), and Stream Control Transmission Protocol (SCTP). The performance metrics are analyzed in terms of the packet delivery ratio, energy consumption, and end-to-end delay. Based on the study and analysis that has been done, the standard transport protocol can be modified and improved for multimedia data transmission in WSN. As a conclusion, SCTP shows significant improvement up to 18.635% and 40.19% for delivery ratio compared to TCP and UDP, respectively.

#### Chapter 16

Wireless Sensor Networks (WSNs) have been attracting increasing interest lately from the research community and industry. The main reason for such interest is the fact that WSNs are considered a promising means of low power and low cost communication that can be easily deployed. Nowadays, the advanced protocol design in WSNs has enhanced their capability to transfer video in the wireless medium. In this chapter, a comprehensive study of Medium Access Control (MAC) and MPEG-4 video transmission is presented. Various classifications of MAC protocols are explained such as random access, schedule access, and hybrid access. In addition, a hybrid MAC layer protocol design is proposed, which combines Carrier Sense Multiple Access (CSMA) and unsynchronized Time Division Multiple Access (TDMA) protocols using a token approach protocol. The main objective of this chapters is to present the design of a MAC layer that can support video transfer between nodes at low power consumption and achieve the level of quality of service (QoS) required by video applications.

The authors propose a graph intersection-based benchmarking algorithm to determine the sequence of longest-living stable data gathering trees for wireless mobile sensor networks whose topology changes dynamically with time due to the random movement of the sensor nodes. Referred to as the Maximum Stability-based Data Gathering (Max.Stable-DG) algorithm, the algorithm assumes the availability of complete knowledge of future topology changes and is based on the following greedy principle coupled with the idea of graph intersections: Whenever a new data gathering tree is required at time instant t corresponding to a round of data aggregation, choose the longest-living data gathering tree from time t. The above strategy is repeated for subsequent rounds over the lifetime of the sensor network to obtain the sequence of longest-living stable data gathering trees spanning all the live sensor nodes in the network such that the number of tree discoveries is the global minimum. In addition to theoretically proving the correctness of the Max.Stable-DG algorithm (that it yields the lower bound for the number of discoveries for any network-wide communication topology like spanning trees), the authors also conduct exhaustive simulations to evaluate the performance of the Max.Stable-DG trees and compare to that of the minimum-distance spanning tree-based data gathering trees with respect to metrics such as tree lifetime, delay per round, node lifetime and network lifetime, under both sufficient-energy and energy-constrained scenarios.

This chapter considers low-complexity detection in hybrid Direct-Sequence Time-Hopping (DS-TH) Ultrawide Bandwidth (UWB) systems. A range of Minimum Mean-Square Error (MMSE) assisted Multiuser Detection (MUD) schemes are comparatively investigated with emphasis on the low-complexity adaptive MMSE-MUDs, which are free from channel estimation. In this contribution, three types of adaptive MUDs are considered, which are derived based on the principles of Least Mean-Square (LMS), Normalized Least Mean-Square (NLMS), and Recursive Least-Square (RLS), respectively. The authors study comparatively the achievable Bit Error-Rate (BER) performance of these adaptive MUDs and of the ideal MMSE-MUD, which requires ideal knowledge about the UWB channels and the signature sequences of all active users. Both the advantages and disadvantages of the various adaptive MUDs are analyzed when communicating over indoor UWB channels modeled by the Saleh-Valenzuela (S-V) channel model. Furthermore, the complexity of the adaptive MUDs is analyzed and compared with that of the single-user RAKE receiver and also with that of the ideal MMSE-MUD. The study and simulation results show that the considered adaptive MUDs constitute feasible detection techniques for deployment in practical UWB systems. It can be shown that, with the aid of a training sequence of reasonable length, an adaptive MUD is capable of achieving a similar BER performance as the ideal MMSE-MUD while requiring a complexity that is even lower than that of a corresponding RAKE receiver.

Directional antennas have gained immense popularity among researchers working in the area of wireless networks. These antennas help in enhancing the performance of wireless networks through increased spatial reuse, extended communication range, energy efficiency, reduced latency, and communication reliability. Traditional Medium Access Control (MAC) protocols such as IEEE 802.11 are designed based on use of omnidirectional antennas. Therefore, suitable design changes are required to exploit the benefits of directional antennas in wireless networks. Though directional antennas provide many benefits to enhance network performance, their inclusion in the network also results in certain challenges in network operation. Deafness is one such problem that occurs among nodes using directional antennas. This chapter concentrates on the problem of deafness, which is introduced due to the use of directional antennas in wireless ad-hoc, sensor, and mesh networks. Many researchers have provided numerous solutions to deal with the problem of deafness in these networks. In this chapter, the authors first explain the problem of deafness and then present an extensive survey of solutions available in the literature to deal with the problem of deafness in wireless ad-hoc, sensor, and mesh networks. The survey is accompanied by a critical analysis and comparison of available solutions. Drawbacks of available solutions are discussed and future research directions are presented.

J. G. Joshi, Government Polytechnic, India
Shyam S. Pattnaik, National Institute of Technical Teachers Training and Research, India

This chapter presents metamaterial-based wearable, that is, textile-based, antennas for Wi-Fi, WLAN, ISM, BAN and public safety band applications, which have been designed, fabricated, and tested. Textile substrates like polyester and polypropylene are used to design and fabricate these antennas. The metamaterial inclusions are directly used to load the different microstrip patch antennas on the same substrate, which significantly enhances the gain and bandwidth with considerable size reduction. The microstrip patch antenna generates sub-wavelength resonances under loading condition due to the modifications of its resonant modes. The DNG and SNG metamaterials are used to load the microstrip patch antennas for size reduction by generating the sub-wavelength resonances. The simulated and measured results are found to be in good agreement for all the presented wearable antennas. The bending effect on antenna performance due to human body movements is also presented in this chapter.

Preface

The Wireless Communication Network is developing at an accelerated pace, which provides ubiquitous communication access to people, enabling real-time multimedia communications and supporting exciting applications such as sensor networks, smart homes, telemedicine, video conferencing and distance learning, cognitive radio networks, automation, and so on. It brings fundamental changes to data networking and telecommunication, and is making integrated networks to increase capacity and coverage. This book highlights the current research issues and trends in wireless communications and networking. Moreover, this book includes some chapters on the fundamentals of wireless communications, which puts the reader in a good place to be able to understand more advanced research and make a contribution in this field for themselves.

Chapter 1 presents the evolution of wireless and mobile technologies and analyzes the market trends of mobile communication in addition to the amount of traffic projected for future mobile communications. This chapter also addresses different network architectures, including 4G networks.

Chapter 2 informs the reader on the latest developments in power-efficient and cost-effective cellular architectures, as well as guides promising research directions in cellular wireless networks.

Chapter 3 is focused on identifying the high level use cases and scenarios where IP-based VPN services can be implemented on top of the cellular network. Furthermore, the authors predict the future involvement of IP-based VPNs in beyond-LTE cellular networks.

Chapter 4 presents current efforts to provide coordination frameworks between various self-organizing network (SON) use cases and lists the major research challenges and open issues to be focused on. An additional important contribution of the chapter is the description of SON activities within 3GPP.

Chapter 5 focuses on mobility prediction in wireless networks in order to enhance the handover performance. The mobility prediction technique via Markov Chain and user's mobility history is proposed as a technique to predict user movement in deployment of LTE femtocells.

Chapter 6 helps prospective newcomers in amateur wireless communications to become familiar with the 'open' software and, as well, to encourage them in implementing many 'free' software solutions at home or work.

Chapter 7 presents an access control system termed the Access of Things, which employs a gradual degradation of privilege philosophy. The Access of Things concept is applicable to the dynamic security environment present in the Internet of Things.

Chapter 8 provides the emerging trends in wireless ad hoc networks. It highlights various issues pertaining to the implementation of ad hoc networks and the research efforts that have been done to resolve these issues, as well as gives future research direction.

Chapter 9 presents a detailed description of the three stability-based connected dominating sets (CDS) algorithms with illustrative examples for wireless mobile ad hoc networks (MANETs) whose topology changes dynamically with time.

Chapter 10 presents an improvement in the interactions between Medium Access Control (MAC) and Transmission Control Protocol (TCP) for better performance in MANET, which is called the Improvement of Backoff algorithm of MAC protocol (IB-MAC) and proposes a new backoff algorithm.

Chapter 11 gives an overview of current available research works on secure routing protocol and proposes a novel Secure Routing.

Chapter 12 presents the latest developments in the field of sensor-based networks, focusing on cooperative healthcare and environmental monitoring.

Chapter 13 provides an overview of three well-known clustering protocols: LEACH, EECS and HEED.

Chapter 14 emphasizes various power saving protocols, techniques and the improvement in the Performance of Clustered WSN by using Multi-tier Clustering.

Chapter 15 deals with the challenges and evaluation of performance in transmitting data over an IEEE 802.15.4 network for different transport protocols: User Datagram Protocol (UDP), Transport Control Protocol (TCP) and Stream Control Transmission Protocol (SCTP).

Chapter 16 presents a hybrid MAC layer design that can support video application with low power consumption and achieve better quality of service (QoS) as required by video application.

Chapter 17 proposes a graph intersection-based benchmarking algorithm to determine the sequence of longest-living stable data-gathering trees for wireless mobile sensor networks whose topology changes dynamically with time due to the random movement of the sensor nodes.

Chapter 18 presents adaptive multiuser detection (MUD) techniques that constitute feasible detection techniques for deployment in practical UWB systems. With the aid of a training sequence of reasonable length, an adaptive MUD is capable of achieving a similar BER performance as the ideal MMSE-MUD.

Chapter 19 concentrates on the problem of deafness, which is introduced due to the use of directional antennas in wireless ad-hoc, sensor, and mesh networks, and drawbacks of available solutions and future research directions.

Chapter 20 presents metamaterial-based wearable antennas that are useful for WLAN, BAN, and Wi-Fi applications, is fabricated and tested. The bending effect on the antenna performance due to the movement of the human body is also presented in this chapter.

This book serves as a comprehensive reference for graduate and undergraduate senior students who seek to learn about the latest developments in wireless communication and networks.

Mohammad A. Matin
Institut Teknologi Brunei, Brunei Darussalam

Chapter 1
Evolution of Wireless and Mobile Communications

M. A. Matin
Institut Teknologi Brunei, Brunei Darussalam

ABSTRACT

There has been a tremendous growth of mobile communications markets all over the world, as they provide ubiquitous communication access to citizens. Wireless technologies are the core of mobile communications. They fundamentally revolutionize data networking, telecommunication, and make integrated networks to increase capacity and coverage. This has made the network portable because of affordable digital modulation, adaptive modulation, information compression, wireless access, multiplexing, and so on. It supports exciting applications such as sensor networks, smart homes, telemedicine, video conferencing and distance learning, cognitive radio networks, automation, and so on. This chapter provides an overview of the evolution of wireless and mobile communications from 2G to 4G.

1. INTRODUCTION

The world is undergoing a major wireless revolution both in terms of wireless and mobile technology that provides ubiquitous communication access to citizens (Matin, 2012). The rapid worldwide growth of cellular communication is based on the second-generation (2G) Global System for Mobile Communications (GSM) standard which uses generalized minimum shift keying modulation, block coding, and time-division multiple access (TDMA) to achieve circuit switched bit rates 16 kb/s, and packet data rates 100 kb/s (GSM, 1998). The 2G mobile communications support new or extended services, as well as consumer needs and technology advances in access, and network areas. The other 2G U.S. digital cellular standard known as code-division multiple access (CDMA) developed by Qualcomm, Inc. and standardized

DOI: 10.4018/978-1-4666-5170-8.ch001

by the Telecommunications Industry Association (TIA) as an Interim Standard (IS-95), uses spread spectrum modulation, convolutional coding, and CDMA to achieve roughly similar bit rates as GSM (CDMA Cellular Standard) . The IS-95 can support up to 64 users that are orthogonally coded and simultaneously transmitted on each 1.25 MHz channel. New data-centric standards have been developed to superimpose on existing 2G technologies to provide high data rate transmission that are required to support modern Internet applications.. These new standards represent 2.5G and allow existing 2G equipment to be modified and enhanced with new base station software upgrades to support higher data rate transmissions for Web browsing, email traffic, mobile commerce and location based mobile services. Both GSM and CDMA migrated to the "3G" standards called wideband CDMA known as the Universal Mobile Telecommunication System (UMTS) in Europe and CDMA-2000 in the United States. The most important changing step of GSM towards UMTS is GPRS. GPRS introduces Packet Switching into the GSM core network and allows direct access to packet data networks. This enables packet data transmission well ahead of 64 kbit/s limit of integrated service digital network (ISDN) through the GSM core network. GPRS prepares and optimizes the core network for high data rate packet switching transmission, as UMTS does with UTRAN over the RAN. Thus, GPRS is a precondition for the UMTS introduction. These 3G standards use wideband spread spectrum, adaptive modulation, convolutional coding, and CDMA to achieve this peak service bit rates of up to 2 Mb/s (ITU, 2009). Cellular networks are growing fast from their telecom roots to become more Internet protocol (IP)-based network as in Fourth generation (4G). 4G has started around 2005 which enables a wide range of services including computing and multimedia applications ranging from navigation to mobile video streaming. The emergence of such technologies and the increasing growth of subscriber demand have triggered the researcher and industries to move on to the 4G network and it is expected that 4G will further converge into the future mobile Internet protocols over the next decade.

2. WIRELESS TECHNOLOGIES

Wireless technologies have been a crucial part of communication in the last few decades which enables multimedia communications between people and devices which is far apart from each other. It brings fundamental revolutionizes to data networking, telecommunication, and makes integrated networks. Wireless technologies have made the network portable because of digital modulation, adaptive modulation, information compression, wireless access and multiplexing. It supports exciting applications such as sensor networks, smart homes, telemedicine, and automation. The early wireless technology is primarily used in the military, emergency services, and law enforcement organizations. As the society moves forward to the information centricity, the need to have information accessible at anytime and anywhere takes on a new dimension. With the rapid growth of mobile telephony and networks, the vision of a mobile information society (introduced by Nokia) is slowly becoming a reality. It is common to see people are communicating with each other via their mobile phones and devices. With today's networks and coverage, it is possible for a user to have connectivity almost everywhere. The growth in commercial wireless networks occurred primarily in the late 1980s and 1990s.

Figure 1 shows a summary roadmap for wireless technology (Raychaudhuri & Mandayam, 2012), identifying some revolution during the period 2000–2010. The diagram is structured into four levels: radio hardware platforms, wireless physical layer technologies, network protocols and software, and mobile systems/applications. In the hardware platform level, it is observed that there has been a proliferation of new radio equipment during this period including 3G, 4G, WiFi, Bluetooth, open mobile handsets, software-defined radio, and most recently, open virtualized access points and base stations. In terms of the radio physical layer, it can be seen that cellular radio link speed has increased from about 2 Mb/s with early 3G systems in the year 2000 to 100 Mb/s with 4G (LTE and WiMax) systems using multiple-input–multiple-output (MIMO) radio technology. Similarly, short-range WiFi radio speeds have increased from 11 Mb/s 802.11b in the year 2000 to 300 Mb/s with 802.11n (Raychaudhuri et al., 2012). The huge competition in the wireless industry and the mass acceptance of wireless devices have caused costs associated with the terminals and air time to come down significantly in the last 10 years.

Wireless local area networking (WLAN) products conform to the 802.11 standards upon which wireless networking devices are built, have expanded a well-built status in a different kind of markets during the last decade. It can set up to offer wireless connection within a limited exposure area which can be a hospital, a university, airport, health care provider or a gas plant. The adopted

802.11 specification for WLAN started out with direct sequence spreading (DSS), quadrature phase shift keying (QPSK) modulation, and carrier sense multiple access with collision avoidance (CSMA/CA) MAC (in the 802.11b standard) at 1 Mb/s, later adding the option of higher order adaptive quadrature amplitude modulation (QAM) without spreading to achieve up to 11 Mb/s.

As wireless data rate increases and worldwide standards begin to converge, WLAN standards and high speed cellular (i.e., 4G cellular including WiMAX and LTE, and 802.11a,g,n) have migrated to orthogonal frequency division multiplexing (OFDM) modulation, which offers higher spectral efficiency and performance (Chang, 1996). A fourth-generation cellular system based on LTE-Advanced-3GPP Long Term Evolution is using OFDM. It also utilize a different MAC protocol based on dynamic FDMA/TDMA in which time-frequency slots are allocated to each subscriber on a frame-by-frame basis. LTE is a series of upgrades to the existing UMTS technology. Both LTE and WiMax use OFDMA with FDMA/TDMA to achieve basic service bit rates in the range of 10–20 Mb/s. With the addition of multiple

Figure 1. Wireless technology roadmap (Raychaudhuri et al., 2012)

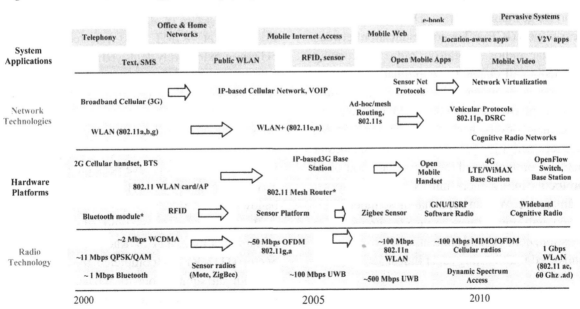

antennas, MIMO signal processing (Foschini & Gans, 1998) and wider band channels, it becomes possible to increase peak bit rates to the range of 100 Mb/s in both LTE and WiMax systems (Li, Li, Lee, il Lee, Mazzarese, Clerckx & Li, 2010). OFDM/MIMO technologies have also been used in WLAN to achieve significant higher bit rates. 802.11n has a peak bit rate of 300 Mb/s using OFDM with higher order adaptive modulation and MIMO along with multiple-channel channel bonding techniques (Nee, Jones, Awater, Zelst, Gardner & Steele, 2006). Note that the faster 802.11 standards based on OFDM differ from wide-area cellular radios in the sense that they continue to use CSMA/CA at the MAC layer in order to maintain compatibility with previous versions of the standard and to avoid implementation complexity.

With the rapid growth of wearable computers such as PDAs, cell phones, smart cards and position location devices, Personal Area Networks (PANs) can provide the connection between each other for remote retrieval and monitoring of surroundings information. WiFi technologies which begin as local area networks with limited indoor coverage have also been extended to incorporate ad hoc and mesh networking protocols for wide-area coverage. There is also an emerging 802.11p/dedicated short range communication (DSRC) standard for peer-to-peer (P2P) ad hoc communication between vehicular radios. Cognitive radio networking protocols will enable the coordination between different technologies sharing the same white space band over the next decade. It can be noted here that convergence of cellular and Internet services will drive further integration of cellular network protocols and the next-generation of IP protocols into a more unified mobile Internet architecture. We can hope that these mobile networks will provide new service features such as location, context, and content-aware routing and enhanced multicasting capabilities.

3. MOBILE NETWORK EVOLUTION

1G refers to analog cellular technologies which became available in 1980s. 2G systems were first introduced in the early 1990s, and evolved from the first generation of analog cellular systems. GSM (Global system for mobile communication), IS-136, PDC, cdmaOne are the different standards of 2G cellular technologies. GSM is a worldwide accepted standard developed by the European Telecommunications Standards Institute (ETSI) for digital cellular communication. The GSM standard was developed as a replacement for first generation (1G) analog cellular networks, and described a digital, circuit switched network that was optimized for full duplex voice communication. Even with small user data rates, 2G standards are able to support limited Internet browsing and sophisticated short messaging capabilities using a circuit switched approach. This was improved over time to include data communications, first with circuit switched transport, then packet data transport via GPRS (General Packet Radio Services) and EDGE (Enhanced Data rates for GSM Evolution or EGPRS). Further development or improvements is done using 3G standard (UMTS) and fourth generation (4G) LTE Advanced standards. Since 2000, the world has seen the introduction of the first series of standards derived from the IMT concept — IMT 2000 (referred to as 3G). 3G is now extensively deployed and being rapidly enhanced as it promises unparalleled wireless access in ways that have never been possible before. Multi-megabit Internet accesses, communications using VoIP are some of the examples of 3G technologies. 4G enables IP based network with a objective to achieve peak data rates up to 1 Gb/s with up to 100 MHz supported spectrum bandwidth.

To produce global standards for mobile communications, ITU's Radiocommunication Sector (ITU–R) has completed the assessment of candidate submissions for the next generation global mobile broadband technology in 2011.

Coordination among these proposals has resulted in the selection of two technologies i.e. LTE-Advanced and Wireless MAN-Advanced. These radio interface standards were submitted to the Radiocommunication Assembly, held in Geneva on 16–20 January, 2012 for final endorsement by the ITU Member States and have been agreed.

4. MARKET TRENDS OF MOBILE COMMUNICATIONS

The Global mobile Suppliers Association (GSA) has reinforced the promotion of GSM/EDGE, WCDMA-HSPA/HSPA+, and 4G/LTE systems worldwide for the successful launching of mobile broadband, enhanced multimedia, and voice services. GSA confirms that 412 operators are investing in LTE technology in 125 countries across the world. 357 operators have made firm deployment commitments in 113 countries, of

which 156 operators have launched commercial services in 67 countries. A further 55 operators are engaged in trials and studies in 12 additional countries. For many operators, LTE represents a significant shift from legacy mobile systems as the first all Internet technology. Figure 2 shows commercial launch of LTE networks as of March, 2013.

5. TRAFFIC ESTIMATION

If mobile communications are used for voice communications, the number of mobile subscribers and traffic volume will reach it saturated point. However, this also provides data communications between non-human objects as well as people. Moreover, the amount of mobile communications traffic increases due to the development of new applications. In February 1999 ' i-mode service" was introduced in Japan. This service covers a wide

Figure 2. Commercial LTE network launches (www.gsacom.com)

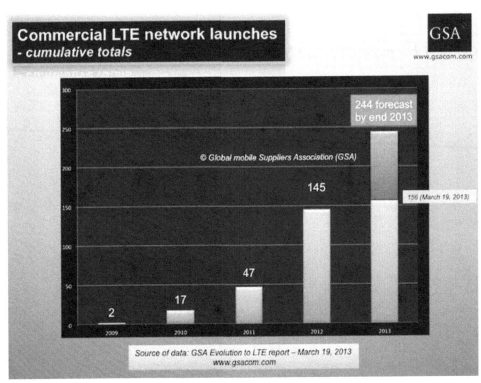

5

range of applications, such as enabling customers to access Web sites, exchange mail, and buy tickets (Enoki, 1999). Since the service is introduced, the number of i-mode subscribers has dramatically increased. At the end of March 2001 the number of mobile Internet subscribers was 3.5 million in Japan (a total of three operator groups).

As of end of 2011 shown in Figure 3, the total number of mobile subscriptions has reached 6 billion and is expected to reach around 9 billion within 2017. This means that the demand for mobile broadband services continues to rise and we need the right technology to be appeared at the right time to provide fast, high responsive mobile data services. The number of mobile broadband subscriptions reached close to 1 billion, and is predicted to reach 5 billion in 2017. PC and tablet mobile subscriptions are increasing and are expected to grow around 650 million in 2017 from around 200 million in 2011.

It can be seen from Figure 4 that there are some similarities in short term estimation, but variations in the future forecasts. Some discrepancies are related to different assumptions assumed in each forecasts. Based on existing mobile broadband growth and new trends, several new traffic estimation, is depicted in Figure 4.

Moreover, when looking at the traffic forecasts over the next decade, one source (UMTS Forum) anticipates total worldwide mobile traffic of more than 127 Exabytes (EB) in 2020 that represents a 33 times increased compared with 2010 figures. The same source also anticipates that Asia will represent 34.3% of total world mobile traffic while Europe and The Americas (including North, Central and South America) represent 22% and 21.4%, respectively as depicted in Figure 5.

If we observe into the deeper for example future ahead of 2020, the same source anticipates global mobile traffic would be 350 EB in 2025 (worldwide) which represents a 174% increase compared to 2020. However, it should be noted that the 2025 forecasts are given in order to show mobile traffic trends, but that the model used was designed for 2010-2020. Figure 6 shows a comparison of actual and forecast traffic. This information is just a reference at this stage.

Figure 3. Global mobile traffic (2010-2017). Source: Ericson.

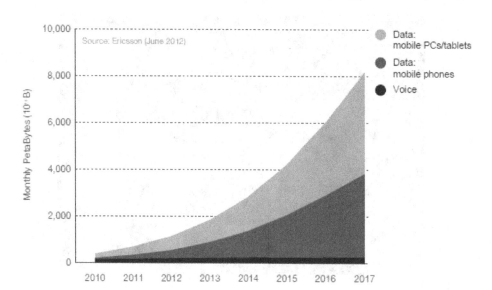

Figure 4. Mobile global data traffic estimates from 2011 to 2015 based on multiple sources

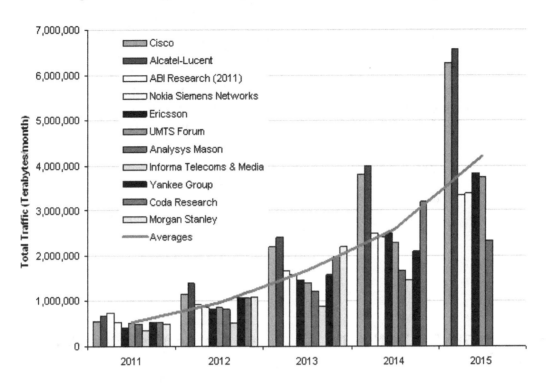

Figure 5. Regional traffic forecasts for 2020. Source: IDATE.

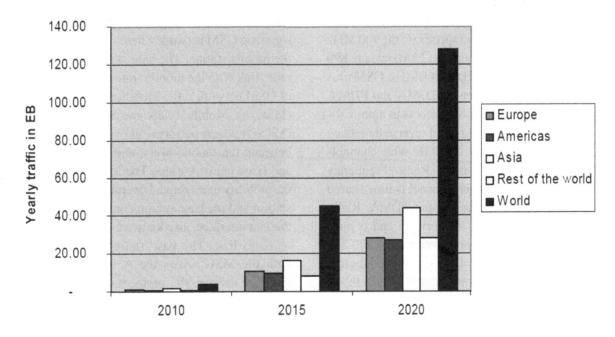

Figure 6. Comparison of ITU-R M.2072 with current data. Source: ITU-R M. 2243 Report.

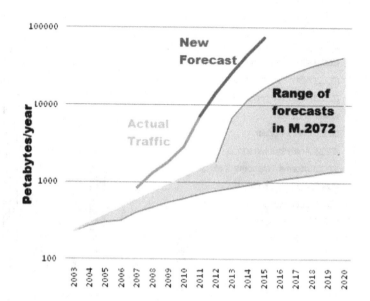

6. SECOND GENERATION: GSM NETWORK

Second generation mobile networks operate in a number of different bands, with most 2G GSM networks operating in the 900 MHz or 1800 MHz bands. The primary band specific to the 900 MHz band includes two sub bands of 25 MHz each, 890 to 915 MHz and 935 MHz to 960 MHz. GSM uses FDD and a combination of TDMA and FHMA schemes to provide multiple access to mobile users. The available forward and reverse frequency bands are divided into 200 KHz wide channels called ARFCNs (Absolute Radio Frequency Channel Numbers). Each channel is time shared between eight subscribers using TDMA. Radio transmissions on both the forward and reverse link are made at a channel data rate of 270.833 kbps using binary BT=0.3 GMSK modulation. Each time slot has time duration of 576.92 µs and a single GSM TDMA frame spans 4.615 ms.

6.1 Architecture of the GSM Network

A GSM network consists of three major interconnected subsystems that interact between themselves and with the users through certain network interfaces. Figure 7 shows the layout of a generic GSM network where subscriber carries the mobile station. The base station controls the radio link with the mobile station. The main part of GSM network is the Mobile Switching Center (MSC) or Mobile Telephone Switching Office (MTSO) which performs the switching of calls between the mobile users, and between mobile and fixed network users. The MSC also handles the mobility management operations. The mobile station and the base station communicate across the Um interface, also known as the air interface or radio link. The base station communicates with the MSC across the A interface. The A interface uses an SS7 protocol called Signalling Correction Control Part (SCCP) which supports communication between the MSC and the BSS, as well as network messages between the individual subscribers and the MSC.

Figure 7. General architecture of a GSM network

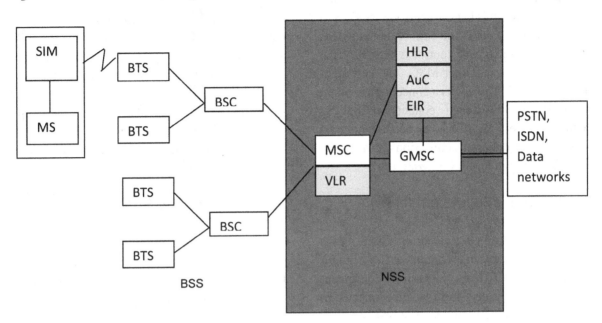

6.1.1 Mobile Station (MS)

The mobile station (MS) consists of the mobile equipment (ME) and a smart card called the subscriber identity module (SIM). The SIM is a detachable smart card which contains the user's subscription information and phone book. This allows the user to retain his or her information after switching handsets and the user is able to receive calls at that terminal, make calls from that terminal, and receive other subscribed services. The SIM card contains the international mobile subscriber identity (IMSI), of 15 digit long number, which identifies the ME but can also be shorter. IMSI introduces the subscriber to the system, a secret key for authentication, and for other information. The IMEI and the IMSI are independent, thereby allowing personal mobility. The SIM card may be protected against unauthorized use by a password or personal identity number.

6.1.2 Base Station Subsystem (BSS)

The Base Station Subsystem is composed of two parts namely the Base Transceiver Station (BTS) and the Base Station Controller (BSC). Each BSC typically controls up to several hundred BTS. Some of the BTSs may be co-located at the BSC, and others may be remotely distributed and physically connected to the BSC by microwave links. The BSS is responsible for handling traffic and signalling between a mobile phone and the network switching subsystem.

The base transceiver station houses the radio transceiver that handles the radio-link protocols with the mobile station. In a large urban area, there will potentially be a large number of BTSs deployed, thus the requirements for a BTS are ruggedness, reliability, portability, and minimum cost.

The BSC is the connection between the mobile station and the MSC. It handles allocation of radio channels, receives measurements from the mobile phones, and controls handovers from one BTS to another BTS (except in the case of an inter-BSC handover in which case control is in part of the responsibility of the other MSC. The key function of the BSC is to act as a concentrator where many different low capacity connections to BTSs (with relatively low utilization) become reduced to a smaller number of connections towards the MSC with a high level of utilization.

6.1.3 Network Subsystem (NSS)

MSC is the central component of a NSS which contains the switching equipment for routing mobile phone calls. It also contains the equipment for controlling the cell sites that are connected to the MSC. The other components of NSS are three different databases called home location register (HLR), visitor location register (VLR), and the authentication centre (AUC). The NSS provides all the functionality needed to handle a mobile subscriber, such as registration, authentication, location updating, handovers, and call routing to a roaming subscriber. The MSC provides the connection to the fixed networks such as the PSTN or ISDN). Signalling between functional entities in the network subsystem uses Signalling System Number 7 (SS7), used for trunk signalling in ISDN and widely used in current public networks.

HLR and VLR, together with the MSC, provide the call-routing facilities and roaming capabilities of GSM. The HLR is a database used for storing and managing subscriber information and location information for each subscriber who resides in the same area as the MSC. The VLR is a database which stores the IMSI and customer information for each roaming subscriber who is visiting the coverage area of particular MSC. E Once a roaming mobile is logged in the VLR, the MSC sends the essential information to the visiting subscriber's HLR so that calls to the roaming mobile can be appropriately routed over the PSTN by the roaming user's HLR

The Equipment Identity Register (EIR) is a database that contains a list of all valid mobile equipment on the network, where each mobile station is identified by its IMEI. An IMEI becomes invalid if it has been reported stolen or is not type approved. The Authentication Center (AuC) is a protected database that stores a copy of the secret key in each subscriber's SIM card, which is used for authentication and encryption over the radio channel.

6.2 GSM Channel Structure

GSM logical channels are classified into two groups called Traffic Channels (TCHs) and Control Channels (CCHs). Different logical channels are mapped under physical channels (slots) depending on the kind of information transmitted (user data and control signalling). Digitized speech is sent on TCHs, and control signalling and synchronizing commands are sent on CCHs.

6.2.1 Traffic Channels (TCHs)

A TCH is used to carry either encoded speech or user data traffic in both up and down directions in a point to point communication. There are two types of TCHs – Full Rate TCH (gross rate of 22.82 Kbps) and Half Rate TCH (half of full rate channels) which are differentiated by their traffic rates. TCHs are defined using a 26-frame multiframe, or group of 26 TDMA frames. The length of a 26-frame multiframe is 120 ms, which is the total length of a burst period. (120 ms divided by 26 frames, each having 8 burst periods). Out of the 26 frames, 24 are used for traffic, 1 is used for the Slow Associated Control Channel (SACCH) and 1 is kept standby. TCHs for the uplink and downlink are separated in time by 3 burst periods, so that the mobile station does not have to transmit and receive simultaneously, thus simplifying the electronics

6.2.2 Control Channels (CCHs)

Logical Control Channels (LCCs) are of three types:

- Broadcast Control Channel (BCCH)
- Common Control Channel (CCCH)
- Dedicated Control Channel (DCCH)

The functions of different control channels are given in Table 1.

7. THIRD GENERATION: UNIVERSAL MOBILE TELECOMMUNICATIONS SYSTEM (UMTS) NETWORK

7.1 Network Architecture

A UMTS network shown in Figure 8 consists of three interconnected domains such as Core Network (CN), UMTS Terrestrial Radio Access Network (UTRAN) and User Equipment (UE) that interact between themselves through certain network interface. All equipment in each domain has to be modified for UMTS operation and services. The basic CN architecture for UMTS is based on GSM network with GPRS and the main function of the CN is to provide switching, routing and transit for user traffic. CN also contains the databases and network management functions. The UTRAN provides the air interface access method for User Equipment (UE). Base Station in UMTS is referred as Node-B and control equipment for Node-B's is called Radio Network Controller (RNC).

7.1.1 Core Network (CN)

Core network is similar to other backbone network and provides the connection to transit networks, Internet or PSTN/ISDN networks. The Core Network is divided in circuit switched (CS) and packet switched (PS) domains. Some of the CS elements are MSC, VLR and Gateway MSC (GMSC). PS

Table 1. Logical control channels

Broadcast Control Channel (BCCH)	Frequency Correction Channel (FCCH)	FCCH is used to Allow an MS to Accurately Tune to a BS. The FCCH Carries Information for the Frequency Correction of MS Downlink.
	Synchronization channel (SCH)	**SCH** is used to provide TDMA frame oriented synchronization data to a MS. When a mobile recovers both **FCCH** and **SCH** signals, the synchronization is said to be complete.
Common Control Channel (CCCH)	Paging channel (PCH),	PCH is used to search (page) the MS in the downlink direction,
	Random access channel (RACH)	RACH is used by MS to request of an **SDCCH** either as a page response from MS or call origination/ registration from the MS. This is uplink channel and operates in point-point mode (MS to BTS).This uses slotted ALOHA protocol.
	Access grant channel (AGCH)	AGCH is a downlink channel used to assign a MS to a specific SDCCH or a TCH. AGCH operates in point-to-point mode.
Dedicated Control Channel (DCCH)	Standalone DCCH (SDCCH)	**SDCCH** is used for system signalling during idle periods and call setup before allocating a TCH
	Associated Control Channel (ACCH)	**ACCH** is a **DCCH** whose allocation is linked to the allocation of a CCH. A FACCH or burst stealing is a DCCH obtained by pre-emptive dynamic multiplexing on a TCH.

Figure 8. Basic architecture of a UMTS network

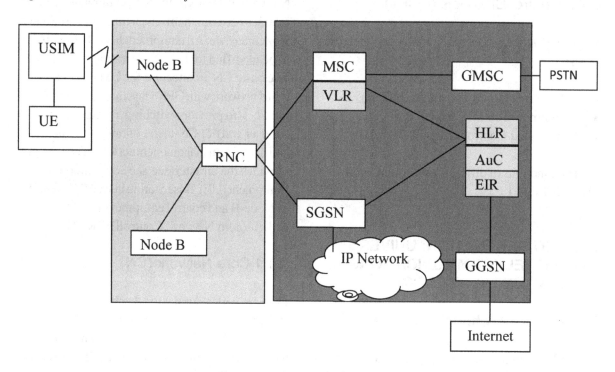

elements are Serving GPRS Support Node (SGSN) and Gateway GPRS Support Node (GGSN). Some network elements, like EIR, HLR, VLR and AUC are shared by both CS and PS domains.

In packet switched domain the SGSN/GGSN nodes provide, amongst other things, support for packet switched services towards mobile stations, including mobility management, access control and control of packet data protocol contexts. In addition, the GGSN provides interworking with external packet-switched networks such as the public Internet.

The Asynchronous Transfer Mode (ATM) is designated for UMTS core transmission. ATM Adaptation Layer type 2 (AAL2) handles circuit switched connection and packet connection while protocol AAL5 is designed for data delivery.

The architecture of the CN may change when new services and features are introduced. Number Portability Database (NPDB) will be used to enable user to change the network while keeping their old phone number. Gateway Location Register

(GLR) may be used to optimize the subscriber handling between network boundaries. MSC, VLR and SGSN can merge to become a UMTS MSC.

7.1.2 UMTS Terrestrial Radio Access Network (UTRAN)

UTRAN provides CDMA or WCDMA access and radio resource management functionalities to mobile stations for a cellular system. UMTS WCDMA is a Direct Sequence CDMA system where user data is multiplied with quasi-random bits derived from WCDMA Spreading codes. In UMTS, in addition to channelization, codes are used for synchronization and scrambling. WCDMA has two basic modes of operation: Frequency Division Duplex (FDD) and Time Division Duplex (TDD).

UTRAN is comprised with Node-B and RNC. The functions of Node-B include air interface transmission or reception, modulation and demodulation, CDMA Physical Channel coding, error handing, closed loop power control. From traffic point of view, RAN contains a large number of flows. Presently, this traffic is mainly composed of voice traffic and smaller contribution of data traffic. Basically, the RAN traffic is heterogeneous and is composed of small volume traffic like, compressed voice packets, and signalling traffic. Because of this, the Iub and Iur connections are in general 2 Mbps.

7.1.3 User Equipment (UE)

The UMTS standard does not restrict the function of the UE in anyhow. Terminals work as an air interface counterpart for Node-B and have many different types of identities. Most of these UMTS identity types are taken directly from GSM specifications. UMTS IC card has same physical characteristics as GSM SIM card. It has several functions:

- Support one or more user profile on the USIM
- Update USIM specific information over the air
- Security functions
- User authentication
- Optional inclusion of payment methods
- Optional secure downloading of new applications

7.2 Modes of Operation of UMTS Mobile Station

UMTS mobile station can operate in one of the three modes of operation such as PS/CS mode, PS mode and CS mode. In PS/CS mode of operation, The MS is attached to both the PS domain and CS domain, and the MS is capable of simultaneously operating PS services and CS services. In PS mode, the MS is attached to the PS domain only and may only operate services of the PS domain. However, this does not prevent CS-like services

to be offered over the PS domain (like VoIP). The other mode of operation is CS where the MS is attached to the CS domain only and may only operate services of the CS domain.

8. FOURTH GENERATION (4G): LTE-ADVANCED (LTE-A)

The Long Term Evolution (LTE) is an IP-based wireless technology that will drive a major network transformation as the traditional circuit-based applications and services migrate to an all-IP environment. LTE will open the door to new converged multimedia services; however, introducing complex voice and multimedia applications into a wireless network is not a trivial task. LTE of the 3G services are turning towards the next development that is of LTE-Advanced (LTE-A). LTE-A is 4G technology named IMT Advanced which is being developed under the auspices of 3GPP. LTE -A is not the only candidate for 4G technology. WiMAX is also there, offering very high data rates and high levels of mobility. The 4G objective is to meet challenges by targeting peak data rates up to 1 Gb/s with up to 100 MHz supported spectrum bandwidth.

There are a number of base technologies such as MIMO and OFDM that enable LTE Advanced to achieve the high data throughput rates. Along with these, there are a number of other techniques and technologies such as coordinated multipoint (CoMP) transmission, advanced heterogeneous networks (HetNets), relay and cross layer optimization that will be employed to enhance cell capacity and coverage drastically. To meet the various deployment needs, 3GPP LTE-Advanced aims to support intraband contiguous, intraband non-contiguous, and interband carrier aggregation. In addition, the design of 3GPP LTE Advanced carrier aggregation further considers backward compatibility, specification impact, implementation complexity, and so on.

8.1 Architecture

The overall LTE architecture shown in Figure 9 is a simplified access network. It comprises of two networks: the E-UTRAN and the Evolved Packet Core (EPC). The access network, E-UTRAN is characterized with a network of Evolved Nodes B (eNode B) which support OFDMA and advanced antenna techniques. Each eNodeB has an IP address, and is part of the all-IP network. The eNode B in LTE collaborate to perform functions such as handover and interference mitigation. LTE's packet domain is called the Evolved Packet Core (EPC). All IP packet core network enables the deployment and efficient delivery of packet-oriented multimedia services, through the IP Multimedia Subsystem (IMS).

8.1.1 Evoveld UMTS Terrestrial Radio Access Network (E-UTRAN)

The Evolved UMTS Terrestrial Radio Access Network (EUTRAN) implements the LTE access network as a network of eNodesB . The main difference between UTRAN and E-UTRAN is the absence of a centralized radio network controller. The eNodes B is responsible for a lot of functions including radio resource management, IP header compression, user data encryption, the scheduling and allocation of uplink and downlink radio resources, coordinating handover with the neighbouring eNodesB.

8.1.2 Evolved Packet Core (EPC) Network

Evolved Packet Core (EPC), is the core network of the LTE system. The User Equipment (UE) is connected to the EPC over E-UTRAN (LTE access network). the EPC is comprised of four network elements: the Serving Gateway (Serving GW), the PDN Gateway (PDN GW), the MME and the HSS. The EPC is connected to the external networks, which can include the IP Multimedia Core Network Subsystem (IMS). The HSS (for Home Subscriber Server) is a database that contains user-related and subscriber-related information. It also provides support functions in mobility management, call and session setup, user authentication and access authorization.

Figure 9. LTE architecture

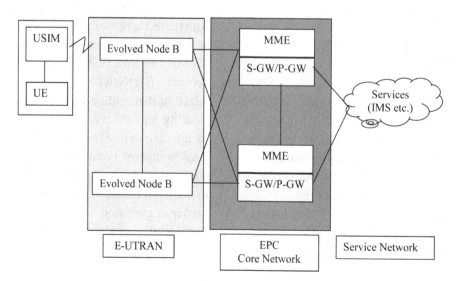

8.1.3 IP Multimedia System (IMS)

3GPP has developed a complete service network system for mobile networks, called the IP Multimedia Subsystem (IMS). Its functions include charging, billing and bandwidth management.

8.2 UMTS vs. LTE

UMTS and LTE networks are similar in some aspects. The fundamental difference between UMTS and LTE is shown in Figure 10.

The main components in 3G UMTS include Node B, RNC and SGSN/GGSN. 3G networks enable network operators to offer users a wide range of advance services which includes voice calls, video calls, broadband wireless services with High Speed Packet Access (HSPA) data transmission capabilities able to deliver speeds up to 14.4 Mbps on the downlink and 5.8 Mbps

on the uplink. The fundamental difference between 3G and LTE is the functionality of the RNC is now distributed to the eNodeB. In contrast to 3G, the new 4G framework to be established has to accomplish new levels of user experience and multi-service capacity. The difference between 3G and 4G is highlighted in Table 2.

9. FIFTH GENERATION (5G): FUTURE MOBILE CARRIER NETWORK

5G (5th generation mobile networks) is the next major phase of mobile telecommunications standards after the current 4G/IMT-Advanced standards. The growth of data traffic requires the urgent introduction of new 5G advanced technologies that maximize the use of the limited available radio spectrum for high data transmission. Though 4G

Figure 10. LTE vs. UMTS architecture

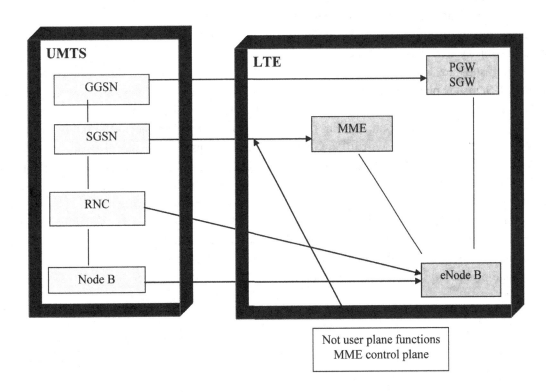

Table 2. Comparison between 3G and 4G

Features	3G	4G
Goal	Provide multimedia multi-rate mobile communications anytime and anywhere	Extension on the 3G goal to provide a wider range of new and improved multimedia services.
Network Architecture	Wide area cell-based	Integration of wireless LAN, wide area and fixed wire systems to appear as a single seamless network.
Speed	Up to 2 Mbps	Up to 100 Mbps for mobile access
Frequency band	1.8-2.4 GHz	2-8 GHz
Switching technique	Circuit and packet switching	IP packet switched network
Access technologies	W-CDMA, 1×RTT, Edge	OFDM and MC-CDMA
Forward Error correction	Uses turbo codes for error correction	Uses concatenated codes for error correction
Component Design	Optimized antenna design, multi-band adapters	Smarter antennas, software multiband and wideband radios
Protocols	Air link protocols including IP 5.0	All IP (IP 6.0)

technology hasn't really reached most consumers, it isn't stopping the race for setting the standards for 5G speeds.

Samsung announced the world's first adaptive array transceiver technology operating in the millimeter-wave Ka bands for cellular communications. The technology will purportedly be the core component of the 5G mobile communication system. However, there are no defined set of standards on what comes after 4G even though Samsung plans on commercializing their new service by 2020. According to South Korea's Yonhap News Agency, Samsung has successfully tested its 5G network with 1.056 Gbps speed (potentially up to 10 Gbps), to a distance of up to 2 kilometers(Yonhap, 2012; Smith, 2013).

If 5G appears, the major difference from a user's point of view between 4G and 5G techniques must be something else than increased maximum throughput; for example less energy consumption, lower outage probability (better coverage), high bit rates in larger portions of the coverage area, cheaper or no traffic fees due to low infrastructure deployment costs, or higher aggregate capacity for many simultaneous users (i.e. higher system level spectral efficiency).

10. CONCLUSION

This chapter presented the evolution of wireless and mobile technologies and analyzed the market trends of mobile communication in addition to the amount of traffic projected for future mobile communications. This chapter also addressed different network architectures including 4G networks that comprise an IP-based core network and radio access networks. Furthermore, it summarizes the difference between 3G and 4G networks and provides a brief introduction of future mobile carrier network.

REFERENCES

Chang, R. W. (1996). Synthesis band-limited orthogonal signals for multichannel data transmission. *The Bell System Technical Journal, 45*(10), 1775–1796. doi:10.1002/j.1538-7305.1966.tb02435.x

Enoki, K. (1999). Concept of i-mode service: New communication infrastructure in the 21st century. *NTT DoCoMo Technical Journal, 1*(1), 4–9.

Foschini, G. J., & Gans, M. J. (1998). On limits of wireless communications in a fading environment when using multiple antennas. *Wireless Personal Communications, 6*(3), 311–335. doi:10.1023/A:1008889222784

GSM. (1998). *Global system for mobile communications: Technical specifications*. Retrieved from http://www.etsi.org/deliver/etsi_gts/07/0705/05.05.00_60/gsmts_0705v050500p.pdf

ITU. (2009). *International telecommunication union, what really is a third generation (3G) mobile technology*. Retrieved from http://www.itu.int/ITU-D/imt-2000/ DocumentsIMT2000/What_really_3G.pdf

Li, Q., Li, G., Lee, W., Il Lee, M., Mazzarese, D., Clerckx, B., & Li, Z. (2010). MIMO techniques in WiMAX and LTE: A feature overview. *IEEE Communications Magazine, 48*(5), 86–92. doi:10.1109/MCOM.2010.5458368

Matin, M. A. (2012). *Developments in wireless network prototyping, design, and deployment: Future generations*. Hershey, PA: IGI Global. doi:10.4018/978-1-4666-1797-1

Nee, R., Jones, V. K., Awater, G., Zelst, A., Gardner, J., & Steele, G. (2006). The 802.11n MIMO-OFDM standard for wireless LAN and beyond. *Wireless Personal Communications, 37*(3-4), 445–453. doi:10.1007/s11277-006-9073-2

Raychaudhuri, D., & Mandayam, N. B. (2012). Frontiers of wireless and mobile communications. *Proceedings of the IEEE, 100*(4), 824–840. doi:10.1109/JPROC.2011.2182095

Smith, J. (2013). *Samsung wants to offer 5G in 2020, speeds of 1Gbps reached in tests*. Retrieved from http://www.pocket-lint.com/news/120967-samsung-wants-to-offer-5g-in-2020-speeds-of-1gbps-reached-in-tests

Yonhap. (2012). *Samsung to offer 5G service by 2020*. Retrieved from http://english.yonhapnews.co.kr/news/2013/05/12/0200000000AEN20130512000900320.HTML

Chapter 2
Sustainable Growth for Cellular Wireless Networks

Siyi Wang
University of South Australia, Australia

Weisi Guo
University of Warwick, UK

ABSTRACT

It has been widely recognised that the exchange of information is one of the underpinning factors for economic growth in developing and developed nations. One of the fastest growing areas of information transfer is the mobile data sector. In 2012, global mobile data traffic grew by 70%. There is an urgent need to improve the wireless capacity of cellular networks in order to match this growth. One of the key issues faced by mobile operators is the fall in Average Revenue Per User (ARPU) and the growing Operational Expenditure (OPEX) due to capacity growth and rising energy prices. The challenge is therefore how to grow the wireless capacity in a way that minimizes the OPEX and thus improves the ARPU. Furthermore, there is growing focus on the environmental impact of Information and Communication Technology (ICT) sectors. There are tangible, financial, and environmental motivations for reducing the energy expenditure of wireless networks whilst growing its capacity. This chapter examines recent research in the area of future wireless network architectures and deployments. This is done in the context of improving capacity in a sustainable way. That is to say, what is the lowest-cost and -energy method of achieving certain capacity targets? The authors of this chapter were researchers in the world's first green wireless communications project—Mobile VCE Green Radio (2007-2012).

1. INTRODUCTION

It has been widely recognised that the exchange of information is one of the underpinning factors for economic growth in developing and developed nations. One of the fastest growing areas of information transfer is the mobile data sector. In 2012, the global traffic for mobile data alone grew by 70%, most of the content being video based. One of the fastest emerging growing data transmissions is machine-2-machine (M2M) communications. There is therefore an urgent need to improve the

DOI: 10.4018/978-1-4666-5170-8.ch002

wireless capacity of cellular networks in order to match this growth in data demand.

One of the key issues faced by mobile operators is the fall in average revenue per user (ARPU) and the growing operational expenditure (OPEX) due to capacity growth and rising energy prices. The challenge is therefore how to grow the wireless capacity in a way that minimises the OPEX, and thus improves the ARPU. Furthermore, there is growing focus on the environmental impact of information and communication technologies (ICT) sectors. There are tangible financial and environmental motivations for reducing the energy expenditure of wireless networks, whilst growing its capacity (Fehske, Fettweis, Malmodin, & Biczok, 2011).

This chapter will examine recent research in the area of future wireless network architectures and deployments. This is done in the context of improving capacity in a sustainable way. That is to say, what is the lowest cost and energy method of achieving certain capacity targets? The authors of this book chapter were researchers in the world's first green wireless communications project Mobile VCE Green Radio (2007-2012) ("Mobile Virtual Centre of Excellence (VCE) - Green Radio Programme," n.d.).

Currently, the energy story as shown in Figure 1 is as follows:

- 0.5% of the world's total energy is consumed by wireless communications, equivalent to 650 TWh of energy per year (35 2000 MW power plants).
- Over 90% of this energy is consumed in the outdoor cellular network, of which 75% is consumed by base-stations.

In terms of digital connectivity, approximately 70% of the developed world and less than 20% of the developing world is digitally connected (Rinaldi & Veca, 2007). Yet, the volume of data communication has increased by more than a factor of 10 over the past 5 years. To foster economic growth and reduce the wealth and knowledge gap: a low energy solution that can increase connectivity and meet the growing data demand must be found.

The chapter will be organised to follow a logical research methodology of proposing metrics for measurement, models for characterization, and technologies as hypothesis to be tested. They are as follows:

1. **Background**: Of cellular network challenges and deployment evolution directions
2. **Metrics:** For measuring average capacity quality-of-service (QoS), energy expenditure, operational cost Expenditure of a multi-cell radio-access-network (RAN). Theoretical bounds for maximum energy

Figure 1. Energy consumption of a) ICT and b) Wireless communications as of 2008-2010. A single UK cellular network typically consumes 40MW. c) operational expenditure (OPEX) of typical 3G cellular network.

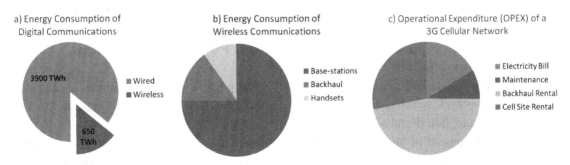

and cost savings achievable for different technology categories are also presented

3. **Modelling Techniques for Multi-Cell RAN:** Monte-Carlo simulation and stochastic geometry methods
4. **Heterogeneous Network Deployment Optimization:** Cell densities, transmit power levels, frequency planning, and cell location. Indoor network deployment is also considered in the context of: cellular vs. Wi-Fi, access-point location planning, indoor-outdoor interference mitigation
5. **Future Work and Current Research Trends:** Dynamic base stations and self-organisation

The objective of this chapter is to inform the reader on the latest developments between 2007–2012 on energy- and cost-efficient cellular architectures, as well as show promising research directions going forwards.

2. BACKGROUND

The Information and Communications Technology (ICT) infrastructure is recognized as a key enabler to the growth of the global economy. With increased data transfer, there is an unprecedented growth in the associated energy consumption, carbon emissions and operational cost of ICT. In order for operators to increase competitiveness, the challenge is how to satisfy the growing data demand, whilst reducing the energy consumption and costs.

The volume of data exchanged in ICT has increased by a factor of 10 over the past 5 years and the associated energy consumption by 20% (Fettweis & Zimmermann, 2008). One of the most challenging aspects of the ICT infrastructure is the wireless access network, which constitutes 14% of ICT energy consumption (Figure 1a). The global cellular network consists of over 3 million cells and 1.5 billion sub- scribers. Roughly 70% of the wireless ICT energy consumption is consumed by

the outdoor network (Figure 1b), which includes 60TWh of electricity (20 million households). The utility bill is over $10 billion and 40MT of CO_2 is directly attributed, with a further 500MT indirectly attributed. Many operators are also pledging to reduce carbon emissions (Sustainability Report 2010-2011, 2011).

Currently, many operators are considering upgrading their 3rd Generation High-Speed-Packet-Access (HSPA) network with the 4th Generation Long-Term-Evolution (LTE) network. This is primarily due to the higher spectral efficiency and increased bandwidth of 4G LTE, which amounts to a higher throughput rate (Holma & Toskala, 2009; Ericsson, 2007). However, it is unclear what the most cost and energy efficient deployment of 4G LTE is. Given an existing reference network deployment (3GPP, 2010), the key research questions being addressed in recent years are:

*   Fundamental saving bounds for energy consumption and operating costs
*   Deployment architecture that achieves the same capacity growth, but with lower energy consumption and operating costs
*   Transmission Schemes that achieves the same capacity growth, but with lower energy consumption

The next section discusses the commonly used metrics for measuring the performance of a cellular network in terms of throughput, quality-of-service, energy consumption and cost expenditure.

3. PERFORMANCE METRICS

3.1. Power Consumption

The power consumption model of a base station largely depends on its peak transmit power, which is related to the intended coverage area size. Generically speaking, the power consumption model for any cell size consists of 3 distinctive parts:

- **Radio-Head (RH):** Power amplifier and transceivers
- **Over-Head (OH):** Signal processing, base-band and cooling elements
- **Back-Haul (BH):** Back-haul transceivers

The amalgamated power consumption of the aforementioned 3 elements yields the total power consumption as:

$$P_{\text{total}} = P_{\text{tx}} + P_{\text{OH}} + P_{\text{BH}}, \tag{1}$$

where the RH element is a function of the transmit power (P_{tx}) and the power amplifier efficiency (μ). Irrespective of whether the cell is transmitting or not, the overhead and backhaul elements generally expend power.

The values presented in Table 1 are amalgamated from the European Community's Energy Aware Radio and Network Technologies (EARTH) project (Auer et al., 2011).

3.2. Power Consumption

The energy efficiency of a single communication link with capacity C can be compared to its spectral efficiency, in that it is defined as (Chen, Zhang, Xu, & Li, 2011):

$$EE = \frac{C}{P_{\text{total}}} \quad \text{bits/s/W}, \tag{2}$$

which can be interpreted as the number of bits transmitted per Joule of energy. Alternative forms of the EE metric also exist, whereby only the transmit power is considered and the ratio is inverted (Badic, O'Farrell, Loskot, & He, n. d.).

3.3. Energy Saving Bounds

The energy efficiency metric is an effective metric that compares the link-level efficiency of trans-

mission. However, it does not take into account the multi-user dynamics of a cellular network, especially with respect to how traffic demand is scheduled and satisfied.

A more effective metric takes into account the traffic demand and the cell throughput. Figure 2 compares a reference and a test network. The total energy consumed by a network is:

$$E_{\text{total}} = P_{\text{RH}}T_{\text{RH}} + (P_{\text{OH}} + P_{\text{BH}})T_{\text{total}}$$

In order to compare two systems, a useful metric is the Energy Reduction Gain (ERG) is defined as the ratio between energy saved and the reference energy consumption:

$$ERG = \frac{\text{total,ref.} - \text{total,test}}{\text{total,ref.}}. \tag{3}$$

In order to improve the performance of a cellular network, the network can be split into two

Table 1. Power consumption for macro to pico cells (2010)

Power Consumption					
Cell Radius, m	Macro	Micro	Micro	Pico	Pico
	>1000	600-1000	400-600	200-400	<200
Max. Transmit Power,	40W	20W	10W	6W	1W
Efficiency, μ	0.25	0.25	0.21	0.18	0.1
RH Power, P_{RH}	160W	80W	48W	33W	10W
Overhead Power, P_{OH} POH	84W	68W	52W	40W	30W
Backhaul Power, P_{RH}	50W	40W	20W	20W	10W
Total Power, P_{total}	294W	188W	120W	93W	50W
Sleep Mode Power	42W	34W	26W	20W	15W

Figure 2. Energy model comparing two systems with a constant operational time T_{OP} and different RAN power consumption values for radio transmission P_{RF} and overhead P_{OH}

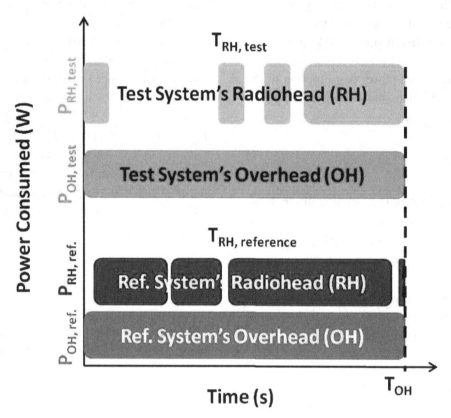

categories. The maximum energy saving owing to either category are as follows:

- **Radio Resource Management:** Improve the transmission efficiency of links can reduce the trans- mission duration and associated radio-head energy consumption. The ERG upper-bound of this is approximately 40%, depending on the cell technology and size (Guo, Wang, O'Farrell, & Fletcher, 2013).
- **Deployment:** whereby additional hardware is added to cells or additional cells are added. The energy saving comes from achieving the same capacity with lower power consuming infrastructure. The ERG upper-bound of this can be shown to theoretically achieve 100% for very large capacity gains

That is to say, capacity improving techniques on existing cells can only hope to achieve a 40% ERG by decreasing the load part of energy consumption, whereas changing the deployment can achieve up to 100% ERG.

3.4. Cost Expenditure

This section introduces the economic cost of deploying and running a cellular network, which can be generally broken down into the following categories (Lang, Redana, & Raaf, 2009):

- **Capital and Implementation Expenditure (CAPEX includes IMPEX):** One off insertion costs that include: planning, equipment and installation costs.
- **Operational Expenditure (OPEX):** Maintenance and operational costs that

occur over a period of time. The initial CAPEX-IMPEX per cell ($\wp$) can be broken down into that owing to cells and the core-network:

$$\text{CAPEX} = \wp_{\text{cell}} + \wp_{\text{deploy}} + \wp_{\text{core}}, \qquad (4)$$

where $\wp_{\text{cell}}$ and $\wp_{\text{deploy}}$ are the equipment and the insertion building cost of a cell-site respectively; and $\wp_{\text{core}}$ is the core network cost on a per cell basis.

The annual OPEX per cell includes the costs associated with marketing and billing, electricity bills, site leasing costs, backhaul rental, hardware and software maintenance:

$$\text{OPEX} = \wp_{\text{rent}} + \wp_{\text{BH}} + E_{\text{total}}\wp_{\text{bill}} + \eta\text{CAPEX}, \qquad (5)$$

where $\wp_{\text{rent}}$, $\wp_{\text{BH}}$, and $\wp_{\text{bill}}$ are the cell site rental, backhaul rental, and electricity utility costs respectively. The annual operational costs attributed to marketing and upgrades can be represented as a function of the initial CAPEX-IMPEX costs, whereby the parameter η is the factor by which a percentage of the CAPEX is used to maintain the RAN. The values for the OPEX and CAPEX are given in Table 2.

Usually the CAPEX-IMPEX amount needed is raised by paying a loan at an annual interest rate i over Y years. The annual total cost of a RAN with N_{cell} cells, is therefore the sum of the CAPEX repayment and OPEX costs:

$$\wp_{\text{total}} = \text{CAPEX}\,\frac{i(1+i)^Y}{(1+i)^Y - 1} + \text{OPEX}. \qquad (6)$$

The global energy consumption of cellular networks (3.5 million cells) is 60TWh and costs over \$200 billion (Sustainability Report 2010-2011, 2011).

4. MODELLING TECHNIQUES

4.1. Monte-Carlo Simulators

In this section, details of Monte-Carlo simulators are provided, with a special focus on how they should be structured and implemented. This simulator has been conceived to simulate uplink and downlink multiple access strategies in radio-access-networks (RAN) by taking into account several inter-linked elements. Typically the simulator simulates either or both the physical (PHY) and the MAC layer of the RAN. A list of typical features include, but are not limited to: cell location, user mobility, radio resource optimisation, interference, frequency reuse techniques, adaptive modulation and coding (AMC), Multiple Input Multiple Output (MIMO), and other aspects that are relevant for industrial and scientific research.

More advanced simulators encompasses both the Evolved Universal Terrestrial Radio Access (E-UTRAN), the air interface of 3GPP's LTE upgrade path for mobile networks, and the Evolved Packet System (EPS), the Internet Protocol (IP)-based network architecture as shown in Figure 3.

1. **Basics:** Three kinds of network nodes are typically modelled in modern simulators: the User Equipment (UE), the evolved Node B (eNB), and the Home Node B (HNB). This simulator provides a support for radio resource allocation in a time-frequency domain. We focus primarily on the latest cellular network, the OFDMA based LTE, which allows different timing granularities (3GPP, 2006). Radio resource allocation is performed every Transmission Time Interval (TTI), each one lasting 1 ms. Each TTI consists of two slots of 0.5 ms corresponding to 14 Orthogonal Frequency-division Multiplexing (OFDM) symbols in the default configuration with short cyclic prefix (CP) (3GPP, 2009). In the frequency domain, the whole bandwidth is divided into 180 kHz

Table 2. Annual Capex and Opex cost values for a network

Shared Parameters		
Parameter	**Symbol**	**Value**
Interest Rate	i	5%
Loan Duration	Y	15 years
Electricity Price	$\wp_{bill}$	$0.05-0.5 (mean: $0.2)
Carbon Emission Ratio	Ω	0.52-0.64 kg/kWh
CAPEX-IMPEX (per cell)		
Core Network	$\wp_{core}$	$5k
Macro-cell Equipment	$\wp_{macro}$	$50k
Micro-cell Equipment	$\wp_{micro}$	$20k
Pico-cell Equipment	$\wp_{pico}$	$5k
Macro Build Insertion Cost	$\wp_{deploy}$	$120k
Micro Build Insertion Cost Pico	$\wp_{deploy}$	$15k
Build Insertion Cost	$\wp_{deploy}$	$3k
OPEX (per cell)		
Maintenance Cost Ratio	η	0.04
Backhaul Rental	$\wp_{BH}$	$10k
Macro-cell Site Rental	$\wp_{rent}$	$6-10k
Micro-cell Site Rental	$\wp_{rent}$	$1-6k
Pico-cell Site Rental	$\wp_{rent}$	$1k
Number of Backhaul	N_{BH}	1-4
Energy Consumption (per cell)		
Macro Cell Operational Energy	E_{macro}	17.5MWh
Micro Cell Operational Energy	E_{micro}	4-12MWh
Pico Cell Operational Energy	E_{pico}	1MWh

sub-channels, each of which contains 12 consecutive 15 kHz sub-carriers. A Physical Resource Block (PRB) is defined as a time/frequency radio resource spanning over 0.5 ms time slot in the time domain and one sub-channel in the frequency domain. This is also the smallest resource element which can be assigned to a UE for data transmission. Therefore, one TTI is made up of 100 PRBs in the frequency domain and 14 OFDM symbols in the time domain. The first three OFDM symbols of every TTI are reserved for transmission of related downlink control channels. The number of PRBs depends on the system bandwidth since the sub-channel dimension is fixed. A system bandwidth of 20 MHz with 100 resource blocks is considered in this chapter unless otherwise specified.

Physical layer (PHY) aspects are managed for all UEs, eNBs and HNBs. This is to say, physical parameters and radio channel models proposed in (3GPP, 2006) are connected to each device. Information such as channel quality, perceived interference level, available bandwidth, list of available PRBs and frequency reuse pattern are stored and in matrix form for future reference.

2. **MAC-Scheduling:** The purpose of the scheduler is to determine which users to transmit data (time domain scheduler) and on which set of resource blocks (frequency domain scheduler). For simplicity in this simulator, the time domain scheduler treats each user fairly and accepts all the connected users to the frequency scheduler. The frequency scheduler implemented in each femtocell AP manages and allocates network resources using a Round-Robin (RR) policy. This scheduler assigns resource blocks to the users in a closed loop circular manner regardless the users Channel State Information (CSI) or SINR. It has the property of allocating resource blocks fairly to all the users without using power control. At any time slot, a user will be allocated as many RBs that are sufficient to transmit the offered load bits and then the next user will do the same. Once all the RBs have been assigned, any users left will be discarded. For the purpose of fairness, users who have the least RBs in the last TTI will be prioritised in the next TTI.

3. **MAC-QoS:** The instantaneous user data rate in each TTI is calculated by the multiplication of the number of bits per resource element obtained from the relative link adaptation table and the number of resource elements that a user has been assigned. Overall QoS requirements are set in two respects: 1). a minimum target data rate for individual users in the network; 2) a threshold on the percentage of users that can achieve the target data rate. From the network service point of view, a technology (LTE-femtocells or 802.11n) with a specific network configuration can be said to satisfy the network QoS requirement only if the percentage of users that achieve the targeted data rate is larger than the percentage threshold, under any given network topology. In this chapter, the user QoS achieved for the network service is defined as the highest 95%-ile threshold of the user data rate, thus if the user QoS achieves 2 Mbit/s then at least 95% of users achieve a minimum data rate of 2 Mbit/s.

4. **PHY- Link Adaptation:** Link adaptation, or Adaptive Coding and Modulation (ACM), is a term used in wireless communications to denote the matching of the modulation, coding and other signal and protocol parameters to the conditions on the radio link (e.g. the pathloss, the interference due to signals coming from other transmitters, the sensitivity of the receiver, the available transmitter power margin, etc.). The process of link adaptation is a dynamic one and the signal and protocol parameters change as the radio link conditions change. The simulator supports link adaptation by changing the Modulation and Coding Scheme (MCS) based on the channel quality (i.e., SINR). The specific look-up tables can be generated from PHY simulator such as the Vienna link level LTE simulator (Mehlfu"hrer, Wrulich, Ikuno, Bosanska, & Rupp, 2009) under WINNER II multipath model (Kyö'sti et al., n.d.).

Figure 3. EUTRAN architecture as part of a LTE network

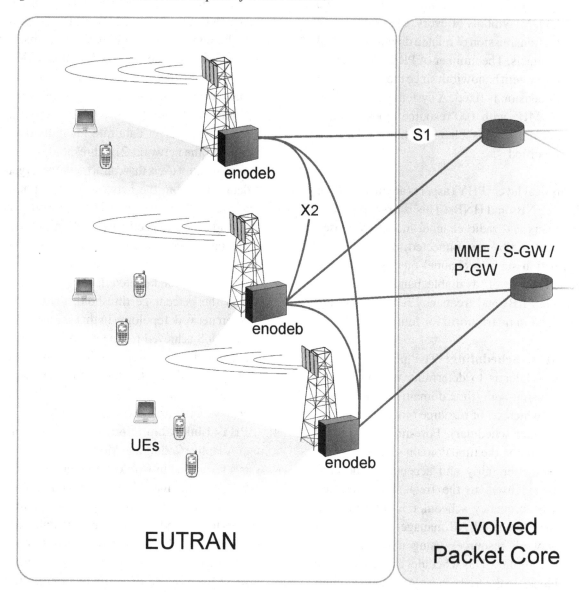

4.2. Stochastic Geometry

4.2.1 Background

The chapter now outlines an alternative theoretical approach to modelling the stochastic communication channels in a RAN. Two key stochastic effects are:

1. Movement of UEs relative to the BSs
2. Multipath fading of the channels

There exists a historical open issue on how to analytically model the performance of the heterogeneous cellular network accurately. The mathematical intractability mainly arises from the difficulty of precisely modelling adjacent interference by taking into account the spatial locations of the MBSs and the stochastic character of the wireless channel (J. G. Andrews, Ganti, Haenggi, Jindal, & Weber, 2010). An alternative and popular approach of conducting such analysis through costly, time-consuming and often proprietary system-level simulators (Baccelli, Klein, Lebourges, & Zuyev, 1997). However, this usual method seldom provides insightful information on system design and the results of optimisation vary from case to case in terms of the dependency of the system parameters. With the introduction of new infrastructure elements to the conventional single tier cellular network, e.g., femto/pico BSs, fixed/mobile relays, cognitive radios, and distributed antennas, the downside of excessive reliance of simulations is even more exacerbated in the future cellular network, which is becoming more heterogeneous (Heath & Kountouris, 2012; Andrews, Claussen, Dohler, Rangan, & Reed, 2012). Therefore, accurate and tractable models for downlink capacity and coverage, considering full network interference is needed in order not to impede the development of techniques to combat the other-cell interference.

4.2.2 Existing Common Analytical Models

To tackle the interference, communications researchers usually abstract tractable but overly simple spatial models of the BSs' locations for performance analysis. In particular, there are four abstraction models widely used by information theorists in this community.

The Wyner model (Wyner, 1994): Only accurate if there is a very large amount of interference averaging over the space because this model is normally one-dimensional and assumes a unit gain from each BS to the associated user and an equal gain that is less than one to the two users in the two neighbouring cells. It is highly inaccurate due to the fact that other-cell interference was modelled as a constant factor of the aggregate interference. For evolving cellular systems, such as LTE and Worldwide Interpretability for Microwave Access (WiMAX) using OFDMA, the Wyner model and related mean-value methods are highly inaccurate as the SINR values over a cell vary dramatically over irregular cell deployment. However, some

of the latest discussions were still conducted via this model to evaluate the capacity of multi-cell systems.

1-D single interference cell model (Tarokh, Chae, Hwang, & Heath Jr., 2009): At least correct in the two cell case, where SINR does vary for different user position and possibly fading. Nevertheless, most sources of interference in the network of this approach is still neglected and therefore it is highly idealised. A recent work of such a model for the purposes of BS cooperation to reduce interference is given in (Gesbert et al., 2010). That such a simplified approach to other-cell interference modelling was still considered state-of-the-art for analysis speaks to the difficulty in finding more realistic tractable approaches.

2-D regular hexagonal lattice or square grid model (MacDonald, 1979): Tractable analysis can some- times be achieved for a fixed user with only a small number of interfering BSs, for example by considering the "worst-case" user location-the cell edge-and finding the SINR which is still a random variable in terms of shadowing and/or fading. The performance metrics such as average user rate and network outage probability can be determined only for this "worst-case" scenario (Rappaport & Others, 1996; Goldsmith, 2005). These are very pessimistic results which do not provide much insight to the performance of most users in the system. Moreover, such a model still requires either intensive numerical simulations or multi-fold Monte Carlo integrations done by computers. Tractable expressions for the SINR in general for a random user location in the cell and the probability of outage/coverage over the entire cell are still unavailable in this model and it usually provides information for specific BSs deployments, and typically fails to provide useful guidance for more random, unplanned, emerging underlaid femtocells and pico-cells heterogeneous cellular networks. In a nutshell, this 2-D model is still highly idealised and may be increasingly inaccurate for the heterogeneous deployments common in urban and suburban areas where cell radii vary considerably.

Homogeneous Spatial Poisson Point Process (SPPP) model (Haenggi & Ganti, 2009; Haenggi, Andrews, Baccelli, Dousse, & Franceschetti, 2009a; Baccelli & Blaszczyszyn, 2010b, 2010a): Motivated by the aforementioned considerations, a new abstraction model is currently emerging and gaining popularity, according to which an additional source of randomness is introduced that the positions of the BSs are modelled as points of a Poisson Point Process (PPP) compared to that they are placed deterministically on a regular grid. Powerful tools from applied probability, such as stochastic geometry, are leveraged to develop tractable integrals and tractable mathematical frameworks for several key performance metrics (e.g., network coverage and average rate). Such an approach for BS modelling has been considered as early as 1997 (Baccelli & Zuyev, 1996).

Subsequently, a similar shotgun-based, PPP-based abstraction model was proposed in (Brown, 2000), and it was shown that, compared with the traditional hexagonal grid model, the shotgun approach provides upper performance bounds, but the key metrics of coverage (SINR distribution) and average network rate have not been determined. More recently, the SPPP model has been widely accepted for the analysis of spatial and opportunistic Aloha protocol (Baccelli, Miihlethaler, & Blaszczyszyn, 2009), and for the characterisation of the SINR of single cellular networks. In particular, a comprehensive framework to compute coverage and average rate of single-tier deployments is provided in (Andrews, Baccelli, & Ganti, 2010). Until then, the random-based abstraction model for the positions of the BSs had not received the attention it deserved. It has been demonstrated that the SPPP model is as accurate as regular grid models, but it has the main advantage of being more analytically tractable. An example of the wireless network modelled as a PPP can be seen in Figure 4.

It can be noted that the homogeneous (or stationary) SPPP model providing an advantage to use stochastic geometry characteristics will continue being the main tool to study and analyse large networks of essentially randomly deployed nodes. This is more true for the femtocells underlaid heterogeneous network where FAPs are formed due to the end-user deployments which are regarded as falling to random deployment category. A comprehensive study based on real BSs deployments obtained from the open source project OpenCellID has revealed that the SPPP model can indeed be used for accurate coverage analysis in major cities worldwide (Lee, Shih, & Chen, 2012). There are many other researchers currently using the SPPP-based abstraction model to study one- and multi-tier cellular networks and interference in a vast literature corresponding to stochastic geometric modelling and analysis of systems with randomly deployed nodes (Dhillon, Ganti, Baccelli, & Andrews, 2011).

There have been three independent efforts to applying stochastic geometry to study heterogeneous networks: downlink coverage and average network rate in a single-tier cellular network with general fading and minimum distance cell association policy, downlink coverage and average network rate in an open-access multi-tier co-channel cellular network with general fading and maximum average received power cell association policy, uplink interference and power control modelling in a single-tier cellular network with Rayleigh fading. This chapter will incorporate all the modelling details available in the literature for studying the downlink performance of heterogeneous cellular network, introduce a few additional important features, and obtain the complete characterisation of the downlink SINR, and the outage probability (i.e., difference between 1 and coverage probability) in both open- and closed-access multi-tier co-channel/non-co-channel heterogeneous network, for the original case where the mobile user is tagged to the cell that provides the maximum long-term averaged received power to the user.

Figure 4. Network topology of non-uniform SPPP two-tier network model in an 20 km ×20 km area with: Macro-BSs, UEs, and Femto- Access-Points (FAPs)

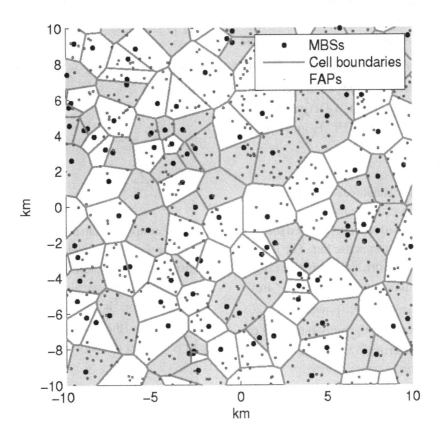

Stochastic Geometry Model: The central conceit to stochastic geometry is to model a random spatial distribution of communication links, each subject to a fading distribution. Generically, the model considers a collection of BSs modelled by SPPP Φ_i of intensity $\lambda_i (i = 1...k)$ in the Euclidean plane, respectively. Then, a heterogeneous cellular deployment can be modelled as a k-tier network where each tier models the BSs of a particular class and the k SPPPs are assumed to be spatially independent. The mobile users are also arranged to some independent SPPP Φ_u of intensity λ_u.

Without loss of generality, the analysis of the model is focused on a typical mobile user located at the origin. The multi-path fading between the typical mobile user and a BS is assumed to be independent and identically distributed (i.i.d.) Rayleigh fading and denoted by $H_i \sim \exp(1)$. The standard path loss propagation model is applied with the path loss exponent $\alpha_i > 2$. Therefore, the downlink received SINR assuming the user connects to l^{th} BS in an ith tier is calculated as below ignoring antenna gain (A) and log-normal shadowing (S):

$$\gamma_{il} = \frac{P_{ti} \mathrm{PL}_C h_{il} \mathbf{d}_{il}^{-\alpha_i}}{\sum_{k=1}^{k} \sum_{\substack{j \in \Phi_k \\ \backslash BS_{il}}} P_{tk} \mathrm{PL}_C h_{kj} \mathbf{d}_{kj}^{-\alpha_k} + \sigma^2} = \frac{P_{ti} \mathrm{PL}_C h_{il} \mathbf{d}_{il}^{-\alpha_i}}{I_{kj} + \sigma^2},$$

(7)

where d_{il} and d_{kj} are the distance between the typical mobile user and its associated home BS and j^{th} interfering BSs in the kth tier, respectively. h_{il} and h_{kj} follow the defined exponential distribution. P_{ti} is the transmitted power of BS or FAP in the ith tier. PLC is the path loss constant, which is typical equal to 10^{-4}. Define the new random variable $G_i = P_{ti}PL_CH_i$ and hence $G_i \sim \exp(\beta_i)$ where $\beta_i = \dfrac{1}{P_{ti}PL_C}$.

Open Access: The chapter defines open access as that the typical user can connect to the BS in any tier while the closed access is defined the user can only be tagged to a specific tier. The typical user in open access network is defined to be tagged to the ith tier only if the received power P_{ri} in i^{th} tier is greater than any received power P_{rk} ($k \neq i$) in other tiers. Therefore, the probability that the typical user is tagged in the ith tier can be expressed as the following mathematical form

$$p_{Ti} = \mathbb{E}_{R_i}\left\{\mathbb{P}[P_{ri}(r_i) > \max_{k,k\neq i} P_{rk}(r_k)]\right\},$$

$$2\lambda_i\pi\int_0^{+\infty} r_i \exp\left[-\pi\sum_{k=1}^{K}\lambda_k\left(\frac{P_{tk}}{P_{ti}}\right)^{\frac{2}{\alpha_k}} r_i^{\frac{2\alpha_i}{\alpha_k}}\right] dr_i, \quad (8)$$

where f_{R_i} is the probability density function (pdf) of the random variable R_i denoted as the distance between the typical user and the nearest BS in i^{th} tier. The pdf and cumulative distribution function (CDF) for a random variable X are denoted, unless otherwise specified, as $f_X(x)$ and $F_X(x)$. The derivation of f_{R_i} is based on the fact that no other BS within the tier is located closer than r_i which is the distance separating the typical mobile user and its associated closest home BS. Therefore there is no BS at a distance from the typical user smaller than r_i. This probability can be derived by employing the 2-D Poisson process defined as

$$\mathbb{P}[N(D) = m] = \frac{(\lambda\,|D|)^m e^{-\lambda|D|}}{m!} \quad \text{when} \quad m = 0$$

and $D = \pi r^2$, and thus $R_i \sim \text{Rayleigh}\left(\dfrac{1}{\sqrt{2\lambda_i\pi}}\right)$.

Setting $\alpha_i = \alpha_k = \alpha$, p_{Ti} yields a more elegant form as

$$p_{Ti}(\alpha) = \frac{\lambda_i}{\displaystyle\sum_{k=1}^{K}\lambda_k\left(\frac{P_{tk}}{P_{ti}}\right)^{\frac{2}{\alpha}}}. \quad (9)$$

Equation (1.9) confirms the intuitive result that a user prefers to connect to a tier with higher BS density and transmit power.

The average typical user's data rate of the overall network is taken over both the SPPP Φ_i for all tiers and the exponential fading distribution and is defined as follows

$$C = \sum_{i=1}^{K} C_i p_{Ti}, \quad (10)$$

where C_i is the mean achievable user data rate in the ith tier and is averaged over both the distribution of D_i and the exponential fading distribution

$$C_i = B_{eff}\int_0^{+\infty}\int_0^{+\infty} e^{-\beta\gamma_{eff}(2^\zeta-1)\sigma^2 d_{il}^{\alpha_i}}\prod_{k=1}^{K} e^{-\lambda_k\pi\left(\frac{P_{tk}}{P_{ti}}\right)^{\frac{2}{\alpha_k}}d_{il}^{\alpha_k}\mathcal{A}(\zeta,\alpha_k)} d\zeta$$

$$\cdot\frac{2\lambda_i\pi d_{il}\exp\left[-\pi\sum_{k=1}^{K}\lambda_k\left(\frac{P_{tk}}{P_{ti}}\right)^{\frac{2}{\alpha_k}}d_{il}^{\frac{2\alpha_i}{\alpha_k}}\right]}{p_{Ti}} dd_{il}$$

$$\quad (11)$$

where $\mathcal{A}(\zeta,\alpha_k) = \displaystyle\int_{[\gamma_{eff}(2^\zeta-1)]^{-\frac{2}{\alpha_k}}}^{+\infty}\frac{[\gamma_{eff}(2^\zeta-1)]^{\frac{2}{\alpha_k}}}{1+u^{\frac{\alpha_k}{2}}} du$.

In an interference-limited network, setting $\sigma^2 = 0$ and all path loss exponents to α, C_i can be further simplified as

$$\mathsf{C}_{\mathrm{i}}(\alpha)\,\big|_{\sigma^2=0} = \int_0^{+\infty} \frac{B_{\mathrm{eff}}}{1+\mathcal{A}(\zeta,\alpha)}\,\mathrm{d}\zeta. \tag{12}$$

Plugging Equation (12) in Equation (10), $\mathsf{C}_{\mathrm{i}}(\alpha)\,\big|_{\sigma^2=0}$ can be obtained as

$$\begin{aligned}
\mathsf{C}(\alpha)\,\big|_{\sigma^2=0} &= \int_0^{+\infty} \frac{B_{\mathrm{eff}}}{1+\mathcal{A}(\zeta,\alpha)}\,\mathrm{d}\zeta \sum_{k=1}^{K} p_{\mathrm{Ti}},\\
&= \int_0^{+\infty} \frac{B_{\mathrm{eff}}}{1+\mathcal{A}(\zeta,\alpha)}\,\mathrm{d}\zeta.
\end{aligned} \tag{13}$$

For a special case when $\alpha = 4$, the mean achievable user data rate of the overall open access LTE heterogeneous network can be computed as

$$\begin{aligned}
\mathsf{C}(4)\,\big|_{\sigma^2=0} &= \int_0^{+\infty} \frac{B_{\mathrm{eff}}}{1+\mathcal{A}(\zeta,4)}\,\mathrm{d}\zeta,\\
&= 0.56\int_0^{+\infty} \frac{1}{1+\sqrt{\gamma_{\mathrm{eff}}(2^\zeta-1)}\arctan\sqrt{\gamma_{\mathrm{eff}}(2^\zeta-1)}}\,\mathrm{d}\zeta,\\
&\approx 0.92\,\mathrm{bit/s/Hz}.
\end{aligned} \tag{14}$$

This suggests that the mean downlink spectral efficiency in an LTE heterogeneous cellular network with Rayleigh fading and no AWGN power is predicted to be 0.92 bit/s/Hz. It reveals that the average interference-limited network data rate is not affected by BS density, BS transmit power, or even the number of tiers. This means that adding BSs or raising the power increases interference and desired signal power by the same amount, and they eventually offset each other. Therefore, the network sum data rate increases in direct proportion to the total number of BSs.

The last metric of performance characterisation to be studied is the outage probability (i.e., difference between 1 and coverage probability). The probability of coverage averaging the whole plane in an open access interference-limited heterogeneous network is given by (assume $\alpha_{\mathrm{i}} = \alpha_{\mathrm{k}} = \alpha$)

$$\overline{\mathsf{P}}_{\mathrm{cov}} = \left(1 + \xi^{\frac{2}{\alpha}}\int_{\xi^{-\frac{2}{\alpha}}}^{+\infty} \frac{1}{1+\mathsf{u}^{\frac{\alpha}{2}}}\,\mathrm{du}\right)^{-1}, \tag{15}$$

and the probability of outage averaging the whole plane is given as

$$\overline{\mathsf{P}}_{\mathrm{out}} = 1 - \left(1 + \xi^{\frac{2}{\alpha}}\int_{\xi^{-\frac{2}{\alpha}}}^{+\infty} \frac{1}{1+\mathsf{u}^{\frac{\alpha}{2}}}\,\mathrm{du}\right)^{-1}. \tag{16}$$

Not surprisingly, the probability of outage and coverage is not function of the BS density, transmit power or the number of tiers in an open access heterogeneous cellular network.

4.4.2.1 Closed Access

In this subsection, expressions of metrics derived in open access network will be provided in closed access scenario accordingly. Since the user can only be associated to a specific tier. The probability that the typical user is tagged in the specific i^{th} tier is 1 and to other tier will be 0 by definition. It should be noted that the pdf of random variable D_{i} turns out to be the same as the one of R_{i} in closed access network. The mean achievable user data rate in the ith in closed access scenario can be expressed as

$$\begin{aligned}
\mathsf{C}_{\mathrm{i}}^{\mathrm{CA}} = B_{\mathrm{eff}}\int_0^{+\infty}\int_0^{+\infty} e^{-\beta\gamma_{\mathrm{eff}}(2^\zeta-1)\sigma^2 \mathsf{d}_{\mathrm{ii}}^{\alpha_{\mathrm{i}}}} \prod_{k=1}^{K} e^{-\lambda_k\pi\left(\frac{P_{\mathrm{tk}}}{P_{\mathrm{ti}}}\right)^{\frac{2}{\alpha_k}} \mathsf{d}_{\mathrm{ii}}^{\frac{2\alpha_{\mathrm{i}}}{\alpha_k}} \mathcal{B}(\zeta,\alpha_k)}\,\mathrm{d}\zeta\\
\cdot 2\lambda_{\mathrm{i}}\pi\mathsf{d}_{\mathrm{ii}}\,e^{-\lambda\pi\mathsf{d}_{\mathrm{ii}}^2}\,\mathrm{dd}_{\mathrm{ii}},
\end{aligned} \tag{17}$$

where $\mathcal{B}(\zeta,\alpha_{\mathrm{k}}) = \dfrac{2\pi[\gamma_{\mathrm{eff}}(2^\zeta-1)]^{\frac{2}{\alpha_{\mathrm{k}}}}}{\alpha_{\mathrm{k}}\sin\left(\dfrac{2\pi}{\alpha_{\mathrm{k}}}\right)}$. In an interference-limited network, setting $\sigma^2 = 0$ and all path loss exponents to α, C_i^{CA} can be further simplified as

$$C_i^{CA}(4)\big|_{\sigma^2=0} = B_{eff}\int_0^{+\infty}\left[1+\frac{\pi}{2}\sum_{k=1}^{K}\frac{\lambda_k}{\lambda_1}\sqrt{\frac{P_{tk}}{P_{ti}}}\gamma_{eff}(2^\varsigma-1)\right]^{-1}d\zeta.$$

(18)

For a special case when $\alpha = 4$, the mean achievable user data rate of the ith tier in a closed access LTE heterogeneous network can be computed as

$$C_i^{CA}(4)\big|_{\sigma^2=0} = B_{eff}\int_0^{+\infty}\left[1+\frac{\pi}{2}\sum_{k=1}^{K}\frac{\lambda_k}{\lambda_1}\sqrt{\frac{P_{tk}}{P_{ti}}}\gamma_{eff}(2^\varsigma-1)\right]^{-1}d\zeta.$$

(19)

It is also an intuitive result that the higher BS density and transmit power provide a higher mean achievable user data rate. Figure 5 depicts average spectra efficiency versus different FAP density to macro-cell BS density ratio in a two-tier interference-limited LTE heterogeneous network. The power of MBS to FAP is set to 400. The crossover point occurs at the density ratio of around 20, which means the closed access network of any tier provides the same network performance in terms of the average data rate when the number of FAPs is 20 times more than MBS.

Recall Equation (16), the ith tier probability of outage averaging the whole plane in an interference-limited heterogeneous network (assume $\alpha_i = \alpha_k = \alpha$) can be achieved and yields

$$\overline{P}_{iout}^{CA} = 1 - \overline{P}_{icov}^{CA},$$

$$= 1 - \left(1+\frac{\pi}{2}\sum_{k=1}^{K}\frac{\lambda_k}{\lambda_1}\sqrt{\frac{P_{tk}}{P_{ti}}}\xi\right)^{-1}.$$

(20)

Figure 6 shows the probability of outage versus SNR threshold comparison between open access and closed access policy in a two-tier inference-limited heterogeneous network, with

three different FAP to MBS. It is expected that the closed access model to gives pessimistic results compared to a open access deployment due to the strong interference generated by nearby BS.

4.4.2.2 Non-Co-channel Deployment

In this sub-section, a brief discussion on the average user data rate and the probability of outage in non-co-channel deployment with open and closed access policy will be given. The term "non-co-channel" means there is no existence of inter-tier interference. The mean achievable user data rate of the ith tier network is equivalent to the result of one-tier network and therefore the mean achievable user data rate of a K-tier open access interference-limited network with same path loss exponent is the same as Equation (13). Moreover, the mean achievable user data rate of the ith tier network is also the result of average spectral efficiency of any ith network with closed access policy. It is no wonder that the probability of outage in open access network and probability of outage in any i^{th} tier closed network are identical to Equation (16). That is to say, while using more radio resource in non-co-channel deployment, this scenario does not provide a more significant enhancement on the mean achievable user data rate nor does it offer a lower probability of outage in a whole.

5. HETEROGENEOUS NETWORK DEPLOYMENT

5.1. Background

Whilst OFDMA based Long-Term-Evolution (LTE) can significantly boost spectral efficiency compared to the existing High-Speed-Packet-Access (HSPA) network, it remains unclear what the lowest energy deployment is. The standards define a reference and enhanced deployment of LTE micro-cells. Two competing architectures

Figure 5. Average LTE spectral efficiency versus the ratio of FAP density to the macro-cell BS density comparison between open access and closed access policy in a two-tier interference-limited heterogeneous network

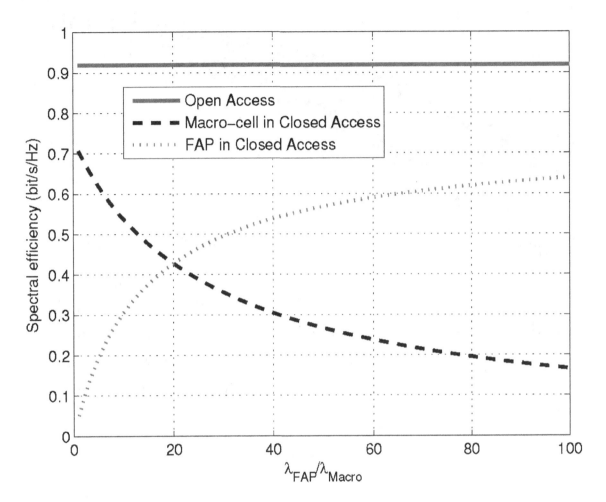

have emerged from recent studies and they can be classified as:

- **Simple Small-Nets:** A dense deployment of low power pico-cells that dramatically increase network capacity by spectrum re-use, whilst consuming a low level of power. The cells tend to be single-sector cell-sites with a single omni-directional antenna. A large macro-overlay is required to serve a small percentage of high mobility users.
- **Wireless Relays:** A sparse deployment of high power macro-cells that dramatically increase network capacity by using radio

techniques such as MIMO and relaying. The cells tend to be richly sectorized with multiple directional antennas. The cell-edge of the large macro-cells employ low power wireless relays to improve the edge quality-of-service.

In addition to deciding which architecture is better suited to delivering sustainable capacity growth, optimizing the cellular network's cell locations is one of the most fundamental problems of network design. It has been shown that whilst unarticulated cell deployments can lead to localized improvements, there is a significant

Figure 6. Probability of outage vs. SNR threshold comparison between open access and closed access policy in a two-tier inference-limited heterogeneous network, with FAP to MBS ratio 10, 30 and 50

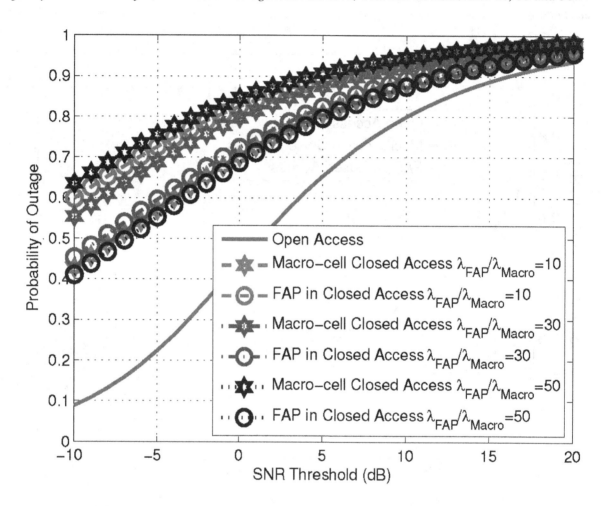

risk posed to network-wide performance due to the additional interference.

A key challenge to such a HetNet is how to mitigate the excessive interference in areas that traditionally would have good coverage, but now suffer degradation due to the additional inference created by nearby cells (Lopez-Perez et al., 2011). In Figure 7, the mean received signal-to-interference-plus-noise ratio (SINR) in an example HetNet is shown. The HetNet consists of a sectorized macro base station (BS) with 12 LPNs deployed within its coverage area. It can be seen that the SINR is high near the LPNs, but rapidly falls to a level that is below the original

macro-BS serving SINR in regions surrounding the LPN coverage areas. This is due to the excessive cross-tier interference.

Cell-site planning has traditionally targeted coverage percentage and traffic density. The latter is difficult to characterize, especially given its dynamic nature and the shifting trends in usage patterns and social mobility. Nonetheless, a great deal of traffic information is inferred and forecasted from:

- **Demographic Data:** The residential and business population distribution based on demographic census;

- **Traffic Data:** Vehicular data based on public transport and private vehicle movement patterns;
- **Fixed Line Data:** Based on correlation with fixed line telephony records, given that most mobile data traffic occurs indoors.

On a macro- and statistical-scale, the stochastic framework introduced in (Haenggi, Andrews, Baccelli, Dousse, & Franceschetti, 2009b) can calculate the LPN density as a function of the transmit powers, statistical pathloss exponent, and noise level.

For radio planning on a micro scale, Monte-Carlo simulations are employed along with detailed urban terrain maps and ray-traced pathloss models. This is recognized as an NP-hard problem. Given a set of possible cell-site or LPN locations, iterative techniques are usually used to scan the optimal locations for cells. Optimization methods such as integer programming, simulated annealing, and multi-era genetic programming algorithms are employed to search for optimal solutions. Meta-heuristic methods such as Tabu search can accelerate the process by ignoring previous negative search results (within a certain iteration period) that are stored in a memory. The ultimate deliverable goal is to make the search complexity linearly proportional to the number of BSs and user equipments (UEs) considered.

To give an idea of the scale and complexity of the challenge, a typical developed urban metropolis has approximately 2 BS sites per square kilometre per operator. This equates to approximately 100 BSs per city, incorporating over 300 macro-cell sectors. In order to deploy LPNs in a HetNet, investigations carried out by the industry have shown that the typical number of LPNs required

Figure 7. HetNet with a macro-BS and randomly deployed Femto-cells. Femto-cells improve local signal strength but severely degrade the surrounding-area signal strength due to excessive interference.

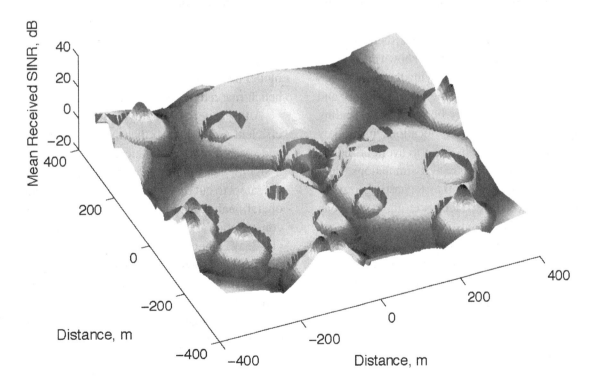

to boost indoor coverage to outdoor levels, ranges from 30 to 100 per BS, yielding a lower-bound of 60 cells per square kilometer and 3000 cells per operator in a city. The resulting radio planning complexity for the HetNet is extremely high, primarily because:

- **Cell Densification:** 30 to 100 fold increase in cells;
- **Coverage Resolution:** 100 fold increase (from 20m to 2m) in coverage resolution for LPNs and indoor areas, and at least a 3 fold increase in coverage height resolution;
- **Indoor-Outdoor Pathloss Complexity:** Unknown increase in computation time;

which lead to at least a 10000 fold increase in the computation time for radio coverage analysis or prediction. This would increase deployment planning and more importantly system optimization times to unfeasible levels. There is therefore a temptation to deploy LPNs without articulated radio planning and rely on signal processing techniques to improve performance. The danger with this approach is that in the absence of effective interference mitigation techniques, there might be zones of intense interference as shown in Figure 7.

The complexity of deploying femto-cells and relays is how to predict their performance, and how can the complexity be reduced by finding approximate deployment locations using key network parameters. In order to avoid or reduce the complexity of protracted simulations, analytical methods such as the stochastic geometry model proposed in (Haenggi et al., 2009b) can be used. Whilst stochastic geometry offers offer network-wide mean performance bounds that relate to node density and other parameters, the challenge of how to plan each specific BS of a HetNet remains open.

5.2. Reducing Search Algorithm Complexity

Reducing algorithm complexity in cell deployment is a concept that attempts to deterministically find the optimal location of a new cell, subject to knowledge about the locations of existing cells, users and the propagation environment. This is in contrast to random deployment or optimization using brute-force search methods in simulations. In order to gain insight of small cell deployment location under the coverage of a single macro-BS, the latest development in network performance modelling has included the effects of interference from the co-channel transmission of a dominant neighbouring cell, and the capacity saturation of realistic transmission schemes.

Whilst some of the deployment solutions are known to experienced radio-planning engineers, the availability of the deployment location in closed-form as a function of transmit power, transmission scheme and pathloss parameters, is novel and significantly beneficial. The work has been applied to outdoor wireless relays (Guo & O'Farrell, 2012), and access-points (APs) for indoor areas (Wang, Guo, & O'Farrell, 2012; Guo & Wang, 2012). The reduced complexity deployment model has been validated against an industrially bench-marked multi-cell system simulator.

A key benefit of deploying a more spectrally efficient network is so that the carbon footprint and expenditures are reduced. There is already a significant commitment from major wireless operators to cut their carbon footprint and reduce operational expenditures. Whilst the total power consumption of a small cell is typically small (10-25W), there are already hundreds of millions of LPNs across the world, and this figure is set to grow rapidly. Therefore it is important to consider their ecological and economical impact. Using bench-marked system simulation tools, it was found that the network-wide spectral- and transmit energy-efficiency improvement achieved by the proposed automated deployment over the random deployment is approximately 20-50%, depending on the environment (Guo & O'Farrell, 2012; Wang et al., 2012; Guo & Wang, 2012). This leads to a carbon footprint reduction of 7-16% and a small operational expenditure (OPEX) saving of 5-12% (Guo et al., 2013). Furthermore, as a result of deploying LPNs more efficiently, it can be argued that fewer LPNs need to be deployed to achieve the same mean network performance than the reference system (random deployment). In that case, both the energy and cost savings are more profound and can reach 40-50%.

Investigations conducted under the MVCE Green Radio program have shown that for an urban environment, the lowest energy architecture depends on the percentage of traffic that is of high mobility. For a low percentage (less than 8%), the lowest energy architecture has a macro-cell overlay that handles the high-mobility users and multiple femto-cells to handle the majority of the traffic load. For a percentage greater than 8%, the lowest energy architecture is the relaying HetNet solution with comprises of several macro-cells that employ co-frequency wireless relays at the cell-edge. The achievable energy reduction from deploying either solutions is approximately 55% when compared with the reference homogeneous LTE reference deployment, and 80% when compared with a reference HSPA reference deployment.

One of the key challenges with deployment optimization generally is that the optimal capacity location of a cell, may not be available for practical and economic reasons. In that case, each node should be equipped with certain self-optimization features such that sub-optimal placement does not exacerbate the network performance. Another reality is that there is a complex balancing act between profit margins from capacity improvements and those from savings made to site rental costs.

6. DYNAMIC AND SMART BASE STATIONS

6.1 Background

Mobile traffic is very dynamic in both the temporal and spatial domains, but the network coverage and capacity is very static. By adapting the coverage and capacity in accordance to the traffic environment, significant energy can be saved (Niu, Wu, Gong, & Yang, 2010). A key consideration of base station (BS) design is to maximise spectrum reuse. This is achieved by sectorisation, so that a BS consists of multiple sectors with directional antennas. The azimuth beamwidth of the antennas pre-dominantly depends on the number of sectors. Generally speaking, the amount of bandwidth and the power consumption at the BS scales linearly with the number of sectors (Auer et al., 2011). Typically, a network consisting of multiple BSs is deployed to meet an offered peak traffic rate. In the temporal domain, the traffic intensity typically varies by 4 to 6 folds during the course of a day. In the spatial domain, the traffic intensity between BSs can vary by up to 10 folds.

6.2 Sleep Mode and Coverage Compensation

The motivation for sleep mode implementation is that reducing the transmit power of a cell can only save a certain amount of radio-head energy, whereas putting an entire cell in sleep mode can save most, if not all, of its total energy consumption. The challenge of multiple cell coordination has been considered in (Niu et al., 2010), and current research has focused on resolving contention between cells that can both enter sleep mode, as well as how to compensate for the lost coverage of a cell in sleep mode.

Current state-of-the-art research integrates dynamic antenna-beam tilting and cooperative transmission to efficiently tackle the aforementioned challenges (Guo & O'Farrell, 2013). The benefit of this approach is that it is able to expand the cell coverage with greater spectral efficiency and achieve this with a low-complexity distributed algorithm. This is accomplished by switching off low load cells and compensating for the coverage loss by expanding the neighbouring cells through antenna beam tilting. The multi-cell coordination is resolved by using either a centralized controller or a distributed self-organizing-network (SON) algorithm. The analysis in (Guo & O'Farrell, 2013) demonstrates that a distributed algorithm is able to exploit flexibility and performance uncertainty through reinforced learning. The energy saving benefit is up to 50% compared to a reference deployment and 44% compared with alternative state-of-the-art dynamic base-station techniques.

6.3 Robotic Base Station Antennas

To date, adaptive antennas for cellular networks have included designs for dual-band operation where the frequency band of operation can be electronically switched and designs which can be switched between omni-directional and sectorized radiation patterns. The designs mostly comprise of high-gain collinear antennas surrounded by an active cylindrical frequency selective surface (FSS). By electronically controlling the state of the FSS a directive radiation pattern can be obtained that can be swept in the entire azimuth plane, or if the FSS is switched off omni-directional coverage can be provided. The disadvantage of this design and similar designs is that that resultant BS is capacity-limited as all users share the same RF feed regardless of the configuration.

One promising concept that can efficiently scale energy consumption with the traffic load, is to allow the BS to dynamically switch between:

- **High Capacity:** High Energy: multiple directional sectors, during periods of high traffic

- **Low Capacity:** Low Energy: single omni-directional sector, during periods of low traffic

Given that the offered traffic varies in time and space, a scalable radio-access-network (RAN) should consist of multiple BSs that have different number of sectors, that can change in accordance with the traffic environment. Each BS only needs to be aware of its own traffic load and does not need knowledge on the state of neighbouring BSs. Furthermore, no additional infrastructure or network-wide controller is required in the network. The antenna design itself does not save energy, but it enables the BSs to dynamically sleep and wake its sectors, as shown in Figure 8.

By incorporating such novel antenna designs and combining it with a self-organization framework using machine-learning, it has been demonstrated how a cellular network can dynamically adjust its capacity and energy consumption in accordance to the traffic load (Guo, Rigelsford, Ford, & O'Farrell, 2012). In comparison with the conventional BS design, the energy savings achieved reach a peak of 75% and a mean of 38% under a realistic 3G data traffic profile. On a system level, the dynamic BS shows better scalability than the current BSs in terms of reducing energy consumption in accordance to the offered traffic, as well as not sacrificing the coverage pattern of the network significantly. On an antenna level, the proposed design has the potential to yield improved coverage, higher diversity and lower transmit power; when compared to existing dynamic antenna designs.

7. SUMMARY

In this chapter, we introduced various methods of modelling a wireless network from a system perspective. Performance metrics of interest typically resolve around mean capacity and the

Figure 8. Dynamic BS's operational modes (top) and mean received SINR plots (bottom) across different traffic loads: a) 6 sectors, b) 3 sectors, c) 1 sector

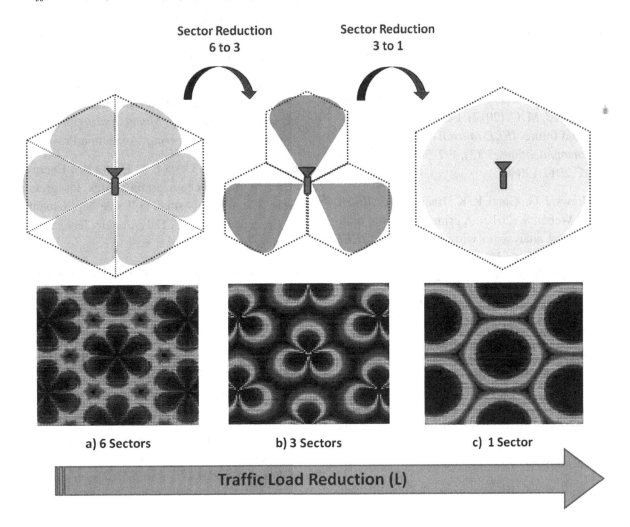

quality-of-service provided to the end users. Recently the energy and cost expenditure has also drawn attention and we have linked them with the system capacity and traffic load. In terms of modelling, researchers and the industry primarily use Monte-Carlo simulation and stochastic geometry theoretical frameworks. The latter was discussed in detail for open and closed access heterogeneous networks (HetNets).

A more detailed examination of HetNets and dynamic networks and the impact they have on performance metrics was then discussed. The chapter showed how careful network planning was needed to avoid excessive interference caused by Femto-cells in a HetNet configuration. It is shown that it can be very beneficial in terms of energy and cost expenditure to vary the network cell operations in accordance to the traffic load to both reduce expenditures and minimise co-channel interference.

REFERENCES

Andrews, J., Baccelli, F., & Ganti, R. (2010). A tractable approach to coverage and rate in cellular networks. *IEEE Transactions on Communications*, (99): 1–13.

Andrews, J. G., Claussen, H., Dohler, M., Rangan, S., & Reed, M. C. (2012). Femtocells: Past, present, and future. *IEEE Journal on Selected Areas in Communications*, *30*(3), 497–508. doi:10.1109/JSAC.2012.120401

Andrews, J. G., Ganti, R. K., Haenggi, M., Jindal, N., & Weber, S. (2010). A primer on spatial modelling and analysis in wireless networks. *IEEE Communications Magazine*, *48*(11), 156–163. doi:10.1109/MCOM.2010.5621983

Auer, G., Giannini, V., Godor, I., Skillermark, P., Olsson, M., Imran, M., & Blume, O. (2011). Cellular energy efficiency evaluation framework. In *Proceedings of IEEE Vehicular Technology Conference* (VTC Spring). IEEE.

Baccelli, F., & Blaszczyszyn, B. (2010a). Stochastic geometry and wireless networks: Applications. *Foundations and Trends in Networking*, *4*(1-2), 1–312. doi:10.1561/1300000026

Baccelli, F., & Blaszczyszyn, B. (2010b). Stochastic geometry and wireless networks: Theory. *Foundations and Trends in Networking*, *3*(3-4), 249–449. doi:10.1561/1300000006

Baccelli, F., Klein, M., Lebourges, M., & Zuyev, S. (1997). Stochastic geometry and architecture of communication networks. *Telecommunication Systems*, *7*(1-3), 209–227. doi:10.1023/A:1019172312328

Baccelli, F., Miihlethaler, P., & Blaszczyszyn, B. (2009). Stochastic analysis of spatial and opportunistic Aloha. *IEEE Journal on Selected Areas in Communications*, *27*(7), 1105–1119. doi:10.1109/JSAC.2009.090908

Baccelli, F., & Zuyev, S. (1996). Stochastic geometry models of mobile communication networks. In *Frontiers in queueing: Models and applications in science and engineering*. Academic Press.

Badic, O'Farrell, Loskot, & He. (n.d.). Energy efficient radio access architectures for green radio: Large versus small cell size deployment. In *Proceedings of IEEE Vehicular Technology Conference*. IEEE.

Brown, T. X. (2000). Cellular performance bounds via shotgun cellular systems. *IEEE Journal on Selected Areas in Communications*, *18*(11), 2443–2455. doi:10.1109/49.895048

Chen, Y., Zhang, S., Xu, S., & Li, G. (2011). Fundamental trade-offs on green wireless networks. *IEEE Communications Magazine*, 49.

Dhillon, H. S., Ganti, R. K., Baccelli, F., & Andrews, J. G. (2011). Coverage and ergodic rate in K-tier downlink heterogeneous cellular networks. In *Proceedings of 49th Annual Allerton Conference on Communication, Control, and Computing* (Allerton), (pp. 1627–1632). Allerton.

Ericsson. (2007). *Summary of downlink performance evaluation* (Technical Report). 3GPP. TSG RAN R1-072444.

Fehske, A. J., Fettweis, G., Malmodin, J., & Biczok, G. (2011). The global footprint of mobile communications: The ecological and economic perspective. *IEEE Communications Magazine*, 49.

Fettweis, G., & Zimmermann, E. (2008). ICT energy consumption - Trends and challenges. In *Proceedings of the IEEE Wireless Personal Multimedia Communications*. IEEE.

Gesbert, D., Hanly, S., Huang, H., Shamai Shitz, S., Simeone, O., & Yu, W. (2010). Multi-cell MIMO cooperative networks: A new look at interference. *IEEE Journal on Selected Areas in Communications*, 28(9), 1380–1408. doi:10.1109/JSAC.2010.101202

Goldsmith, A. (2005). *Wireless communications*. Cambridge, UK: Cambridge University Press. doi:10.1017/CBO9780511841224

3. GPP. (2006). *Physical layer aspects for evolved universal terrestrial radio access (UTRA)* (Tech. Rep. No. TR 25.814 V7.1.0). 3rd Generation Partnership Project.

GPP. (2009). *Physical channels and modulation* (Tech. Rep. No. TS 36.211 V8.8.0 Release 8). 3rd Generation Partnership Project.

GPP. (2010). Further advancements for E-UTRA physical layer aspects (Rel.9) (Technical Report). 3GPP. TR36.814v9.

Guo, W., & O'Farrell, T. (2012). Relay deployment in cellular networks: Planning and optimization. *IEEE Journal on Selected Areas in Communications*, 31.

Guo, W., & O'Farrell, T. (2013). Dynamic cell expansion with self-organizing cooperation. *IEEE Journal on Selected Areas in Communications*, 31.

Guo, W., Rigelsford, J., Ford, K., & O'Farrell, T. (2012). Dynamic basestation antenna design for low energy networks. *Progress in Electromagnetics Research*, C, 31.

Guo, W., & Wang, S. (2012, November). Interference-aware self-deploying femto-cell. *IEEE Wireless Communications Letters*.

Guo, W., Wang, S., O'Farrell, T., & Fletcher, S. (2013). Energy consumption of 4G cellular networks: A London case study. In *Proceedings of IEEE Vehicular Technology Conference* (VTC). IEEE.

Haenggi, M., Andrews, J. G., Baccelli, F., Dousse, O., & Franceschetti, M. (2009a). Stochastic geometry and random graphs for the analysis and design of wireless networks. *IEEE Journal on Selected Areas in Communications*, 27(7), 1029–1046. doi:10.1109/JSAC.2009.090902

Haenggi, M., Andrews, J. G., Baccelli, F., Dousse, O., & Franceschetti, M. (2009b). Stochastic geometry and random graphs for the analysis and design of wireless networks. *IEEE Journal on Selected Areas in Communications*, 28, 1029–1046. doi:10.1109/JSAC.2009.090902

Haenggi, M., & Ganti, R. K. (2009). *Interference in large wireless networks*. Now Publishers Inc.

Heath, R. W., & Kountouris, M. (2012). Modeling heterogeneous network interference. In *Proceedings of Information Theory and Applications Workshop* (ITA), (pp. 17–22). ITA.

Holma, H., & Toskala, A. (2009). *LTE for UMTS: OFDMA and SC-FDMA based radio access*. Chichester, UK: Wiley. doi:10.1002/9780470745489

Kyösti, P., Meinilä, J., Hentilä, L., Zhao, X., Jämsä, T., & Schneider, C. (n.d.). *WINNER II channel models* (d1. 1.2 v1. 1). no. IST-4-027756 WINNER II, D, 1.

Lang, E., Redana, S., & Raaf, B. (2009). Business impact of relay deployment for coverage extension in 3GPP LTE-advanced. In *Proceedings of IEEE International Conference on Communications: Communications Workshops*. IEEE.

Lee, C.-H., Shih, C.-Y., & Chen, Y.-S. (2012). Stochastic geometry based models for modelling cellular networks in urban areas. *Wireless Networks*, 1–10.

Lopez-Perez, D., Guvenc, I., de la Roche, G., Kountouris, M., Quek, T., & Zhang, J. (2011). Enhanced intercell interference coordination challenges in heterogeneous networks. *IEEE Transactions on Wireless Communications, 18*, 22–30. doi:10.1109/MWC.2011.5876497

MacDonald, V. H. (1979). The cellular concept. *The Bell System Technical Journal, 58*(1), 15–41. doi:10.1002/j.1538-7305.1979.tb02209.x

Mehlführer, C., Wrulich, M., Ikuno, J. C., Bosanska, D., & Rupp, M. (2009). Simulating the long term evolution physical layer. In *Proceedings of the 17th European Signal Processing Conference* (EUSIPCO 2009), (pp. 1471–1478). EUSIPCO.

Mobile Virtual Centre of Excellence (VCE). (n.d.). *Green radio programme*. Retrieved from http://mobilevce.com/

Niu, Z., Wu, Y., Gong, J., & Yang, Z. (2010, November). Cell zooming for cost-efficient green cellular networks. *IEEE Communications Magazine*, 74–79. doi:10.1109/MCOM.2010.5621970

Rappaport, T. S. et al. (1996). *Wireless communications: Principles and practice* (Vol. 2). Upper Saddle River, NJ: Prentice Hall.

Rinaldi, R., & Veca, G. (2007). The hydrogen for base radio stations. In *Proceedings of 29th International Telecommunication Energy Conference* (INTELEC) (pp. 288-292). Rome, Italy: INTELEC.

Sustainability Report 2010-2011. (2011). Vodafone Group Plc.

Tarokh, V., Chae, C.-B., Hwang, I., & Heath, R. W. Jr. (2009). *Interference aware-coordinated beamforming system in a two-cell environment*. Academic Press.

Wang, S., Guo, W., & O'Farrell, T. (2012, May). Low energy indoor network: Deployment optimisation. *EURASIP Journal on Wireless Communications and Networking*. doi:10.1186/1687-1499-2012-193

Wyner, A. D. (1994). Shannon-theoretic approach to a Gaussian cellular multiple-access channel. *IEEE Transactions on Information Theory, 40*(6), 1713–1727. doi:10.1109/18.340450

KEY TERMS AND DEFINITION

4G: In telecommunication systems, 4G is the fourth generation of mobile phone and mobile communication technology standards.

Cellular Network: A radio network distributed over land areas called cells, each served by at least one fixed-location transceiver, known as a cell site or base station.

Femtocell: In telecommunications, a femtocell is a small, low-power cellular base station, typically designed for use in a home or small business.

Heterogeneous Network: A network connecting computers and other devices with different operating systems and/or protocols.

LTE: Long Term Evolution, marketed as 4G LTE, is a standard for wireless communication of high-speed data for mobile phones and data terminals.

Monte-Carlo: Monte-Carlo methods are computational algorithms that rely on repeated random sampling to obtain numerical results; i.e., by running simulations many times over in order to calculate those same probabilities heuristically just like actually playing and recording your results in a real casino situation.

Network Capacity: The tightest upper bound on the rate of information bits that can be reliably transmitted within the network.

Power Consumption: Energy consumption that uses electric energy

QoS: The quality of service ("QoS") refers to the networks that allow the transport of traffic with special requirements.

Relay: A station that relays messages between various points, so as to facilitate communications between units.

Sleep Mode: A low power mode.

Spectral Efficiency: The information rate that can be transmitted over a given bandwidth in a specific communication system.

Stochastic Geometry: The study of random spatial patterns with an emphasis on spatial point processes in this chapter.

Chapter 3
IP–Based Virtual Private Network Implementations in Future Cellular Networks

Madhusanka Liyanage
University of Oulu, Finland

Mika Ylianttila
University of Oulu, Finland

Andrei Gurtov
Aalto University, Finland

ABSTRACT

Virtual Private Network (VPN) services are widely used in the present corporate world to securely inter-connect geographically distributed private network segments through unsecure public networks. Among various VPN techniques, Internet Protocol (IP)-based VPN services are dominating due to the ubiquitous use of IP-based provider networks and the Internet. Over last few decades, the usage of cellular/mobile networks has increased enormously due to the rapid increment of the number of mobile subscribers and the evolvement of telecommunication technologies. Furthermore, cellular network-based broadband services are able to provide the same set of network services as wired Internet services. Thus, mobile broadband services are also becoming popular among corporate customers. Hence, the usage of mobile broadband services in corporate networks demands to implement various broadband services on top of mobile networks, including VPN services. On the other hand, the all- IP-based mobile network archi-tecture, which is proposed for beyond-LTE (Long Term Evolution) networks, is fuel to adapt IP-based VPN services in to cellular networks. This chapter is focused on identifying high-level use cases and scenarios where IP-based VPN services can be implemented on top of cellular networks. Furthermore, the authors predict the future involvement of IP-based VPNs in beyond-LTE cellular networks.

DOI: 10.4018/978-1-4666-5170-8.ch003

INTRODUCTION

Global marketing strategies increase abilities of a firm to conduct its business in various locations across the world. However, the secure communication among these sites is also mandatory to perform a smooth operation of the organization. Many firms use advance communication services such as VPN services to interconnect these geographically distributed branches to headquarters. A VPN service is the first choice of many organizations since it is the most prominent communication methodology to provide a secure inter-site connectivity.

The notion of VPN or Virtual Network (VN) services has been around for last four decades which is almost the same as the life span of data networks. The usage of VPN services is constantly improving due to various factors. Primarily, the implementation cost of VPNs is drastically decreasing with the use of low cost network equipments and communication devices. Furthermore, the competition between different network service providers causes to reduce subscription fee for a VPN service. On the other hand, the technological advancement of VPN technologies in terms of enhanced security, high speed connectivity and high reliability are motivating many organizations to use VPN based services.

The usage of mobile network based broadband services has drastically increased over the past few years. The number of mobile subscribers is increasing rapidly and the total mobile broadband traffic volume is growing faster than the fixed Internet traffic. Furthermore, the steady development in telecommunication techniques causes to provide almost the same level of broadband services as fixed Internet in terms of bandwidth, reliability and Quality of Service (QoS). The recent surveys showed that the number of worldwide mobile broadband subscribers has already exceeded the number of fixed broadband subscribers (Cisco, 2010). Moreover, a telecommunication network is able to provide anytime anywhere broadband connectivity regardless of the mobility pattern

or the location of the subscriber. This is the most prominent advantage of a mobile broadband service. For instance, many organizations often have "road warriors" who equip with portable computing devices such as laptops, smart phones and various tablet devices. These road warriors need to work from anywhere without being physically present in the office. The integration of virtual networks and mobile broadband services is a promising solution to provide efficient and secure connectivity for these road warriors. Furthermore, there are large numbers of mobile network operators than fixed network service operators. Thus, the competition among the mobile network operators is very high and it drastically decreases the mobile broadband charges. These facts fuel corporate customers to choose mobile broadband over wired network services.

The LTE specification introduces all-IP network architecture and beyond LTE networks will operate on top of IP infrastructures. Thus, we focus only on IP VPNs in this chapter. We present high level use cases and scenarios of IP based VPN services which are implemented on top of cellular networks.

VIRTUAL PRIVATE NETWORK

A virtual network is a communication network which contains virtual network links. In other terms, it is a collection of virtual links which are established on top of a physical network. These virtual links are implemented by using methods of network virtualization and they are transparent to end users.

There are two commonly used methods of network virtualization; namely protocol based virtual networks and virtual device based virtual networks (Metz, 2003). However, protocol based virtual networks are easy to implement and globally ubiquitous than virtual device based virtual networks. For instance, VPNs, VLANs (Virtual Local Area Networks), VPLSs (Virtual Private

LAN Services) are the widely used protocol based virtual network implementations (Rosenbaum et al., 2003).

A virtual network which extends a private network across public networks is called Virtual Private Network (VPN). It allows a remote host or site to be a part of the private network with all the functionality by communicating data across a shared or a public network.

These are several building blocks of a VPN.

- **Customer (C) Device:** C device is a legacy device which is owned by the customer. Customer devices are not aware of the presence of a VPN.
- **Customer Edge Equipment (CE):** CEs are located at the edge of the customer's network. It might be a host equipment or a router or a switch which is located at the customer's premise. Furthermore, CE is the interface device between the customer and provider networks.
- **Provider Edge Equipment (PE):** PEs are located at the edge of the provider network and connects to customer sites through CEs. Hence, PEs aware the existence of VPNs and contain all the VPN intelligence.
- **The Core Network:** The core network is the backbone network of the VPN. It belongs to the network service provider. This core network can be operated based on several network protocols, such as IPv4, IPv6, MPLS (Multiprotocol Label Switching).
- **Provider Device (P):** A P device is located inside the core network. They are not directly interfacing to any customer network. Hence, the P devices are not VPN-aware and maintaining any VPN states. Its principal role is to support the traffic routing and aggregation functions of the core network.

Figure 1 Illustrates a topology of a simple VPN.

History of Virtual Private Networks

The history of VPNs spans as long as the history of the data communication networks (Metz, 2003). The first generation of VPNs was based on X.25 carriers. In early 1970's, VPNs were consisted of privately operated network devices which are connected by using the dial-up or dedicated leased lines over an operator network. In the late 1970s, the development of X.25 introduced the virtual connection concept for data networks. The connection-oriented networks were capable to logically separate customer communication channels for virtual connections. Then, operators were capable of multiplexing various virtual connections which are interconnecting users from the same customer cooperation, through a single switched network infrastructure. TCP/IP protocol was introduced to data networks in the 1980s (Rosenbaum et al., 2003). In the early 1980s, X.25 carriers began to offer VPN services to early adopters of TCP/IP protocol stack.

The second generation of VPNs was the virtual connection oriented VPNs based on frame relay and ATM (Asynchronous Transfer Mode) switching. The advancement of frame relay and ATM switching technologies were allowed to provide high speed virtual connection based services. In the 1990s, the connectivity speed of virtual network connections grew up to 155 Mbps (Knight & Lewis, 2004).

The third generation of VPNs was IP based VPNs. During the mid-1990s, the traditional connection oriented VPNs services such as ATM and frame relay were in the decline stage of the product lifecycle. Simultaneously, Internet began to popular in communication networks. Thus, network service providers began to build new network infrastructures to support IP based services. These migrations fuel the adaptation of IP based virtual networks as well. IP based VPNs were attracted by various customers since they provide significant economic advantages and better service functionalities than the connection oriented VPN

Figure 1. A topology of a simple VPN

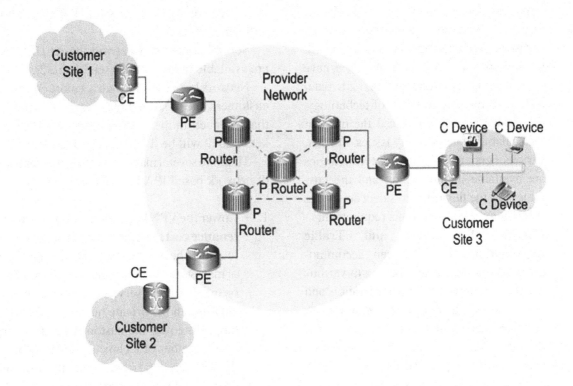

services. Meantime, service providers started to offer IP VPNs as a Value Added Service (VAS) for customers. It was the initiation of the provider provisioned VPN concept.

The fourth generation of VPNs aims to extend IP based Layer-2 VPN (L2VPN) solutions over WAN (Wide Area Network) connectivity through an IP or MPLS network. Especially, Virtual Private LAN Service (VPLS) and Virtual Private Wire Service (VPWS) are becoming popular as a cost effective alternative to Layer-3 VPN (L3VPN) solutions.

IP-BASED VIRTUAL PRIVATE NETWORK (IP VPN)

IP VPN is a virtual network which is it deployed over a shared IP based network infrastructure. It connects customer sites by using IP VPN tunnels. These IP VPN tunnels offer the packet based VPN connectivity. It forwards the customer data, packet by packet basis. A separate tunnel header is imposed on the VPN data packet at the source site and disposed it at the destination site. It allows to opaquely forwarding the VPN data packets through the provider network or the Internet (Carug & Clercq, 2004).

IP VPNs are widely used in corporate organizations since they provide various benefits to the customer (Harris, 2002).

- **Global Ubiquitous Availability:** IP based networks are widely available all over the globe. All most all the network operators provide IP based network connectivity for customers. Literally, the IP VPN services can be implement in anywhere in the world.

- **Any-to-Any Connectivity:** IP VPNs allow establishing a full mesh of VPN tunnels over the provider network. Hence, the customer can achieve any to any connectivity.
- **Secure, Scalable, Flexible, Robust Network Architecture:** IP based network technologies such as MPLS and IPsec, have been developed thought out the last two decades and matured in terms of technology. They have already addressed the most of the scalability and security issues. Further, various versions of IP based technologies are available in the market and the customer has the flexibility to select any of the technology according to his requirements.
- **Classes of Service and Traffic Aggregation:** IP VPNs can accommodate and optimize traffic belong to various classes. It allows traffic prioritization and extra bandwidth allocations according to requirements of different traffic classes. On the other hand, it is possible to integrate various traffic types and delivers them in a single network infrastructure due to the packet based VPN connectivity.
- **Economic Advantages:** IP based VPNs have lower implementation and maintenance cost than traditional virtual connection based VPN services.

IP VPNs can be categorized in to two main classes (Rosenbaum et al., 2003).

1. Network based IP VPN
2. Premises based IP VPN

Network Based IP VPN

The service provider is responsible for all operational and management tasks of a network based IP VPN. Hence, the customer does not need to consider about the implementation aspects on the VPN service over the public network. The customer and the provider exchange several Service Level Agreements (SLAs) to define the required service level such as bandwidth, the number of connected sites, traffic priorities and the classes of service quality (Venkateswaran, 2001; Callon & Suzuki, 2005).

Several versions of network based IP VPNs are available in the market (see Figure 2).

Network based IP VPNs attract more corporate customers due to several reasons. It is estimated that revenues for the network based IP VPN market in 2009 will be double in 2014 (XO, 2010).

There are several market drivers and advantages of network based IP VPNs (Daniel, 2004).

1. **Lower the VPN implementation and maintenance cost:** The customer's capital cost is low since he is not responsible to implement or manage the VPN devices. He can purchase on the shelf VPN services without changing anything in his private network infrastructure. Moreover, the customer needs a very small number of technical staff to manage the VPN since the routing, maintenance and operational functions are outsourced to the provider. On the other hand, the customer has opportunity to tailor-made his VPN service at any time just by exchanging few SLAs with the provider.
2. **Converged network services:** In contrast to other VPN services, network based IP VPNs can transport traffic belong to various applications such as VoIP, data, video and multimedia applications, in a single VPN services.
3. **Traffic prioritization:** Network based IP VPNs support traffic prioritization which allows the provider to tailor the bandwidth according to the requirement of each customer traffic type.
4. **Global ubiquitous availability:** Network based IP VPNs are widely available all over the globe. Many network operators provide IP VPNs as an additional broadband service for their customers.

Figure 2. Network-based IP VPN

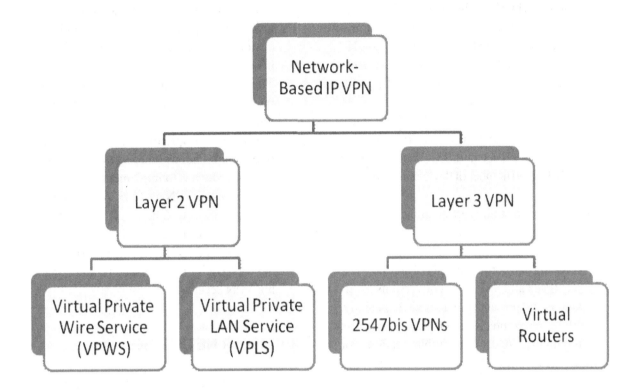

The market restraints of network based IP VPNs

1. **Sunken investments on ATM & Frame Relay networks:** ATM & frame relay networks are still dominating the access network space. Many of the network operators still use ATM & frame relay networks. Thus, they are reluctant to deploy IP based infrastructures due to the sunken investment to deploy these old technologies. It delays the process of adaptation of new network based IP VPN technologies.

Premises Based IP VPN

The customer is responsible for all the operational and management tasks of a premises based IP VPN. The network operator only provides the connectivity among the customer sites and he is not aware the existence of premises based VPNs. Hence, the customer must have expertise in the implementation of VPN service over the public network. Furthermore, premises based IP VPNs require to install intelligent devices at the customer premises which can provision VPN services over the provider backbone or the Internet. Firewalls or VPN tunnel termination devices are installed at the edge of the customer network. Usually, premises based IP VPNs are developed based on IPsec (Internet Security Protocol) or SSL (Secure Sockets Layer) protocols (Cisco, 2004).

There are several market drivers and advantages of premises based IP VPNs (Rosenbaum, 2003).

1. **Advanced security features:** Both IPsec and SSL protocols ensure the confidentiality, integrity, and authenticity of the VPN traffic. Furthermore, it is possible to integrate additional security mechanisms such as firewalling services, intrusion detection, URL (Uniform Resource Locator) filtering and virus scanning to premises based IP

VPNs. Hence, they provide the security features which are far beyond the features of a traditional IPSec or SSL connection.

2. **The global connectivity:** Premises based IP VPNs required only the broadband connectivity from the network operator. Literally, a simple Internet connection is enough to implement a premises based IP VPN service.

3. **Simple implementation:** The implementation of both IPsec and SSL is quite simple and a limited number of devices are required.

The market restraints of premises-based IP VPNs

1. **Lacking of interoperability:** There are various IPsec and SSL protocol versions available in the present IP networks. Especially, IPsec has imprecise standards. Thus, it is required to select compatible versions for implementation of a premises based IP VPN service.

2. **Lacking of scalability:** Both IPsec and SSL VPNs do not utilize any auto discovery mechanism. Also, it is not possible to utilize the provider's auto discovery mechanism in premises based IP VPNs. Thus, the manual configuration is still required. Hence, the large scale premises based IP VPNs are highly complex to implement and maintain.

3. **The issue of lawful interception:** Authorize personals have right to intercept the communication data to investigate any communication that may threat to national security or support potential terrorist attacks. This process is known as the lawful interception. However, the lawful interception is not possible at the provider network for premises based IP VPN traffic since the customer data is encrypted. Therefore, some countries and service providers prevent the transfer of encrypted premises based IP VPNs data over their networks and gateways.

4. **Problems in QoS management:** Most of the QoS management schemes are based on layer 3 header attributes. The encryption of customer data prevents the access of layer 3 header information at the provider network. Thus, it is not possible to use QoS management mechanisms at the provider network for premises based IP VPNs.

5. **Higher VPN implementation and maintenance cost:** The customer has a higher capital cost since he has to implement and manage devices in the VPN network. Moreover, the customer needs technical staff to manage the VPN. On the other hand, the customer has to change his infrastructure including the VPN devices, each time he wants to tailor-made his VPN service.

CELLULAR NETWORKS

A cellular network or a mobile network is a radio network which provides wireless network services for mobile users. The radio network is consisted of cells. A cell is a small geographical area which is served by at least one base station. A base station is a fixed location transceiver which provides wireless connectivity for mobile users. These cells jointly provide the radio coverage over a wide geographical area. The cellular network allows a mobile user to make or receive network services from almost anywhere within its radio coverage. Furthermore, the user is receiving uninterrupted network services even while on the move within the radio coverage.

Initially, cellular networks were designed to provide only voice call services. However, these cellular networks have experienced rapid technological advancement during past three decades. Today, millions of people around the world are using mobile networks for various network services including voice call, video call, broadband access, multimedia, VPN and mobile cloud services. The cellular network technologies have passed four generations.

- **The first generation (1G):** The first generation of cellular networks was operated based on analog telecommunication technologies. The first analog telecommunication standards were introduced in the 1980s. 1G telecommunication networks provided the connectivity speed up to 56 Kbps.
- **The second generation (2G):** The second generation of cellular networks was operated based on digital telecommunication technologies. In 1991, the first 2G cellular telecom networks were commercially launched on the GSM (Global System for Mobile Communications) standard in Finland. Digital TDMA (Time Division Multiple Access) and CDMA (Code Division Multiple Access) technologies are used in the 2G telecommunication networks to increase the network capacity.
- **The third generation (3G):** The first 3G cellular telecom networks were commercially launched in South Korea in January 2002. 3G telecommunication networks provide the data transfer at a minimum speed of 200 Kbps. However, many 3G telecommunication networks provide higher speed than the minimum data transfer speed of a 3G service. These various 3G releases often denoted as 3.5G and 3.75G. They provide the mobile broadband access at a speed of several Mbps. Furthermore, 3G introduces various network applications such as wireless voice telephony, mobile Internet access, fixed wireless Internet access, video calls and mobile TV.
- **The fourth generation (4G):** The first 4G cellular telecom networks were commercially launched in the USA in 2008. A 4G system provides mobile ultra-broadband Internet access to laptops with USB (Universal Serial Bus) wireless modems, to smartphones, and to other mobile devices. The advance version of 4G networks is introduced as LTE in 2009. LTE cellular networks offer conceivable applications including amended mobile Web access, IP telephony, gaming services, high definition mobile TV, video conferencing, 3D television and cloud computing.

Long Term Evolution (LTE)

Long Term Evolution (LTE) is a wireless communication technology that allows operators to provide higher data rate and broadband connectivity than the previous 2G/3G network technologies. Even an earliest LTE network can offer very fast data speeds up to 100 Mbps in the downlink and 50 Mbps in the uplink. LTE is belonging to the fourth generation mobile networks which is marketed as 4G LTE. The overall objective for LTE is to provide an extremely high performance radio access technology that offers full vehicular speed mobility and that can readily coexist with earlier mobile networks. The first commercial LTE networks were launched by TeliaSonera in Norway and Sweden in December 2009. Since then, the LTE networks are becoming popular all around the globe. There were 119 commercial LTE networks in over 50 countries by November 2012.

Integration of IP VPNs in LTE Networks

The implementation of IP VPNs in cellular networks is not motivated until the availability of the IP based LTE mobile architectures. Especially, the architectural differences between the data network and the cellular networks prevent IP VPN implementations in earlier 2G/3G architectures. Initially, cellular networks provide only the network connectivity for IP VPNs. Mobile network operators did not participate in the service provisioning of VPNs. Later, mobile network operators were interested to offer provider provisioned IP VPN services by actively participating in VPN service provisioning functionalities. They generated additional revenue by providing the IP VPNs as a Value Added Service (VAS) and it is the main motivation factor for them to offer provider provisioned VPN services (Shneyderman et al., 2000).

However, the deployment of IP VPNs in cellular network environment is not matured enough as wired IP VPN architectures. Only a very few number of mobile VPN architectures are proposed so far. Many IP VPN architectures for wired networks have been developed during past few decades. Many of these wired IP VPN architectures are properly implemented and keep using for a long period of time. However, it is not feasible to directly implement the existing wired IP VPN architectures in cellular networks. The availability of resources, network topology, data transmission media and behavior of end users are completely different in a cellular network environment. Thus, a cautious consideration on these factors is required while implementing IP VPNs in cellular networks. Several mobile IP VPN architectures are proposed by considering these factors (Shneyderman & Casati, 2003; Benenati et al., 2002; Feder et al., 2003). In this chapter, we investigate the feasibility of such proposals and possible applications based on these mobile IP VPN architectures. Furthermore, we present the adaptation efforts of existing wired IP VPN architectures into cellular networks.

APPLICATIONS OF IP VPNS IN CELLULAR NETWORKS

IP VPNs can offer various services by integrating them with cellular networks. Since the LTE architecture is proposed an all-IP mobile architecture, this chapter mainly focuses only on LTE and beyond LTE cellular networks. On the other hand, it is the future of present telecommunication networks. Main applications of IP VPNs in LTE cellular network are as follows (see also Figure 3):

1. Service provisioning of conventional IP VPN services as a value added service.
2. IP VPN based backhaul traffic transportation architectures for LTE networks.
3. Securing the LTE backhaul traffic by using IP VPN techniques.
4. IP VPN based traffic architectures for Mobile Virtual Network Operators (MVNO).

Service Provisioning of Conventional IP VPN Services as a Value Added Services

The mobile Internet usage has increased during the past few years. Hence, service providers are eager to provide various services in addition to the traditional voice communication. The conventional VPN services over telecommunication networks or Mobile VPNs (MVPNs) are also newly embraced as a VAS by mobile network operators (Shneyderman & Casati, 2003; Benenati et al., 2002; Feder et al., 2003). They expand their role in VPN context from a simple connectivity provider to a VPN service provider.

The MVPN concept has fundamentally changed the method of communication and network access of a traditional VPN service. Especially, many of enterprises have road warriors who equipped with mobile devices. MVPNs provide a location independent connectivity for these road warriors to access the customer's private network. Hence, the remote workers and road warriors can reach the private network regardless of the location and the time. Furthermore, MVPNs provide the seamless mobility for users. For instant, a customer can use the VPN services even while he is travelling in a high speed train or a bus. Moreover, mobile Internet tariffs are decreasing day by day. Hence, MVPNs are becoming an affordable and cost effective solution for users.

The architectural framework for MVPNs is defined in several research articles and standards. Furthermore, various forms of MVPNs are currently in (Shneyderman & Casati, 2003; Benenati et al., 2002; Feder et al., 2003). It is possible to

Figure 3. The implementation concept of IP VPNs in LTE cellular networks

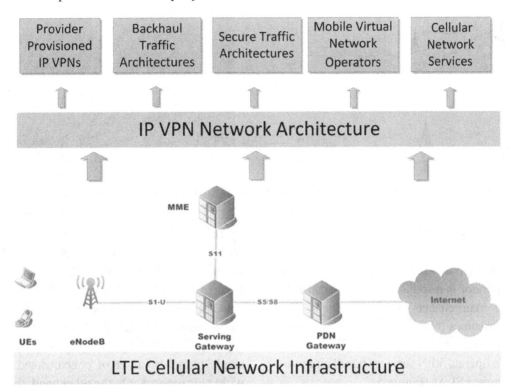

categorize these VPN implementations over the telecommunication networks in to three categories based on the tunnel establishment mechanism; namely, end to end tunnel mode, network based tunnel mode and chained tunnel mode.

End to End (E2E) Tunnel Mode MVPNs

The E2E tunnel mode is considered as the most widely used VPN mode over the globe. In the E2E tunnel mode VPN scenario, remote user establishes a direct VPN connection to the customer private network by using the telecommunication network. The user can connect either a single device such as PDA (Personal Digital Assistant), laptop, smart phone or a complete customer network site by using a mobile communication capable CE device. The most of these VPN tunnels are IPsec tunnels. The administrative staff of the customer network is responsible to implement the security mechanism such as user authentication functions, access control mechanisms, firewalls and confidentiality protection schemes for these VPNs. Hence, E2E tunnel mode VPNs can be categorized as premises based VPN.

Figure 4 illustrates an end-to-end tunnel mode VPN over a LTE network.

The operation of an E2E tunnel mode VPN is as followed. First, the user device set up a dial up access connection over the telecommunication network to gain the Internet access. Then, it built a VPN tunnel (IPsec tunnel) with the VPN gateway which is residing at the customer private network by using this dial up connection. Hence, the part of the tunnel is built across the telecommunication network and the rest over the Internet. The mobile operator has no knowledge on the presence of these VPN tunnels. However, the tunnel establishment is temporal and it exists as long as the mobile user is connected to the Internet.

Figure 4. The topology of an end to end tunnel mode MVPN

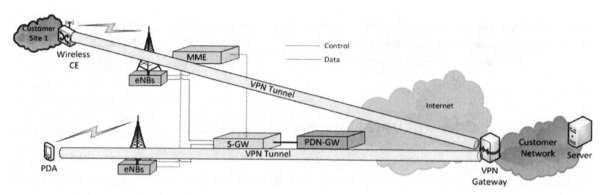

Network-Based Tunnel Mode

In the network-based tunnel mode scenario, the mobile operator is responsible for the establishment and maintenance of VPN tunnels. The customer assumes that the mobile operator has a network infrastructure with the intelligence and features to operate VPN services based on mutually exchanged SLAs (Figure 5).

In this scenario, VPN tunnels terminate at the entry point of the mobile network. Several tunnel establishment mechanisms such as IPsec, GRE (Generic Routing Encapsulation), MPLS can be used to establish these VPN tunnels. The operator and the customer exchange the VPN information and SLAs to establish the VPN service. Since the operator is responsible for the access control and the denial of unauthorized users, the customer has to trust the operator to provide the confidentiality of the private data.

Chained Tunnel VPN

The third type of VPN mode is the chained tunnel VPN. It establishes a VPN connectivity between the remote mobile user and the VPN gateway of customer network by using a set of concatenated tunnels (Figure 6).

There are many forms of chained tunnel VPNs. Similar to the network based VPN, operator builds a VPN tunnel between the border GW of the operator network and the customer network. This tunnel can be built by using various establishment mechanisms such as IPsec, GRE, MPLS. Then, the operator builds VPN tunnels within his mobile network to interconnect the remote mobile users. These tunnels have different forms and different expendabilities. The provider can deploy either a single tunnel or a set of concatenated tunnels within his network. On the other hand, these tunnels can be extended up to the mobile user device or only up to the base stations/eNodeBs. Furthermore, the operator can build some tunnels with foreign operators to facilitate the secure transmission for roaming users. These VPN tunnels within the operator network can be built by using different tunneling protocols such as GTP (GPRS Tunneling Protocol), IPsec, HIP (Host Identity Protocol).

Traffic Transportation Architecture from LTE networks

The virtual network based traffic transportation is also a major use case of virtual networks in telecommunication network domain. Two scenarios have been identified. Namely, transport models to transmit the traditional traffic over IP\Ethernet based networks during the migration phase and general backhaul traffic transportation architectures for LTE networks.

Figure 5. The topology of a network based tunnel MVPN

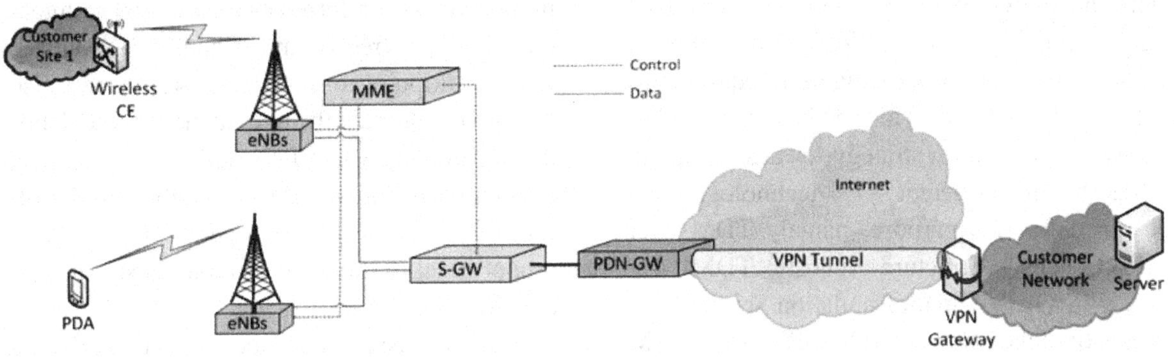

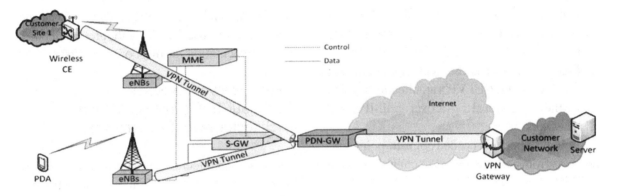

Figure 6. The topology of chained tunnel MVPN

Virtual Network Based Transport Model to Transport the Traditional Traffic During the Migration Phase

The truly affordable broadband bandwidth from GSM (Global System for Mobile Communications), UMTS (Universal Mobile Telecommunications System) and HSDPA (High-Speed Downlink Packet Access) is saturating due to the rapid increment of the mobile broadband usage. Hence, all telecommunication operators have to migrate for the LTE and LTE-A (LTE Advanced) architectures to satisfy future demands. Ethernet based networks provide higher bandwidth and scalability which are necessary features to accommodate future evolvements. The LTE architecture proposes a new Ethernet based all-IP network infrastructure. Hence, operators have to build a

new Ethernet based network infrastructures at the backhaul level (Ronai & Officer, 2009; Dahlman et al., 2011). However, the present GSM and UMTS networks are operating based on TDM (Time-division multiplexing) or ATM technologies and operators are not ready to abandon these network structure. Several market researchers predict that GSM networks will remain at least until 2020 (Croy, 2011).

On the other hand, it is not feasible to manage multiple backhaul traffic technologies such as TDM, ATM and Ethernet over a single operator network due to high operating costs and compatibility issues. Thus, operators are trying to converge all backhaul traffic into single traffic model.

Several solutions are proposed to achieve this goal. One possible solution is to tunnel the traditional TDM or ATM services over a virtual network which overlays on an Ethernet based network. Here, all the GSM (TDM) and UMTS (ATM) traffics transport over an IP-based infrastructure without changing the existing radio equipments at the access network. This can be considered as a part of a long-term strategy towards an all-IP infrastructure. Different VPN technologies are designed for this purpose, namely TDM over MPLS such as Structure Agnostic TDM over Packet (SAToP), Circuit Emulation Service over Packet Switched Network (CESoPSN) and ATM over MPLS such as ATM Pseudowire Emulation Edge-to-Edge (Alvarez et al., 2011; Cisco, 2010). Both L2VPN and L3VPN architectures are feasible to implement. However, the architectural choices will depend on the protocol stack of intermediate nodes in the underlay operator network.

Figure 7 illustrates the deployment and the protocol stack of L2/ L3 VPN architectures which tunnel the conventional traffic over an all-IP based LTE network.

Initially, operators needed point-to-point connections from cell site routers to controllers such as RNC (Radio Network Controller), BSC (Base Station Controllers). Hence, the E-line (Ethernet pseudowire) is preferred as the L2VPN architecture. Later, E-tree (point-to-multipoint) and E-LAN (multipoint-to-multipoint) services are also acquired to provide the redundancy and scalability. However, these L2VPN solutions have complex operational features and less flexibility on controlling. The complexity of operation is directly proportional to the level of redundancy support and the scalability.

Due to these reasons, operators are motivated to deploy L3VPNs such as MPLS based 2457bis VPNs. These L3VPNs provide various advantages including higher scalability, simpler integration, less operational complexity, better resilience and redundancy. Hence, L3VPN is the preferred choice of implementation for most of operators.

Virtual Network Based Backhaul Traffic Transportation Architectures for LTE Network

The LTE architecture proposed a whole new architecture for telecommunication networks.

Figure 7. The protocol stack of the L2/ L3 VPN architectures which tunnel the conventional traffic over an all-IP-based LTE networks

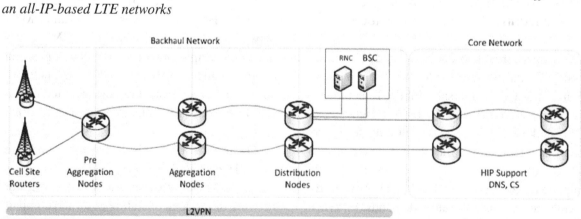

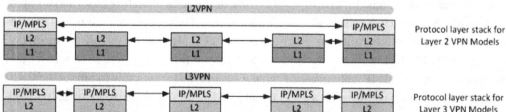

Especially, it proposed to use an all-IP based flat backhaul network architecture. Also it defined several new interfaces such as S1, X2 to enhance the performance of the traffic transportation. Furthermore, the flat architecture concept proposes to distribute the centralized controlling entities across the network to optimize the controlling functions.

On the other hand, the traffic architecture is well defined in LTE architecture. It proposed several traffic types namely, S1-U for the user traffic between eNodeBs and the S-GW (Service Gateway), S1-C for the control traffic between the eNodeBs and the MME (Mobility Management Entity), X2-u and X2-c for the traffic between eNodeBs, OSS (Operations Support System) traffic to provide fault, configuration, and performance management and the network synchronization traffic (Alvarez et al., 2011). However, these traffic classes have different operational, controlling and QoS requirements. It is challenging to satisfy these individual requirements in a shared network infrastructure. A VPN based traffic architecture is an ideal solution to provide different level of services for these different traffic types.

Several VPN based traffic architectures are proposed for LTE backhaul networks (Liyanage & Gurtov, 2012; Cisco 2010; Alvarez et al., 2011; UTStarcom, 2009). These architectures can be categorized in to two types based on traffic separation options for VPNs. In the first scenario, the backhaul traffic is divided into two VPNs. One VPN is carrying the traffic between the eNodeBs and core network elements. Several interfaces such as S1-U and S1-C, belong to this VPN. The other VPN is carrying the traffic between eNodeBs. Several interfaces such as X1-U, X1-C belong to later VPN. In second scenario, each backhaul traffic class is divided into a separate VPN. In other terms, each interface has a separate VPN. The first scenario is preferred by the operators due the lesser operational complexity (Cisco, 2010). In contrast, VPN based traffic architectures again categorize in to two types based on operational network layer. It depends on the underlay network infrastructure. Generally, intermediate routers in LTE backhaul networks support either layer 2 or layer 3 protocols. Therefore, it is possible to implement these VPN-based traffic architectures as L2VPNs or L3VPNs.

Layer 2 VPN Model for LTE/ EPC Deployments

The L2VPN traffic architecture is proposed for LTE networks which have layer 2 network nodes. Figure 8 illustrates the deployment and the proto-

Figure 8. The deployment and the protocol stack of L2VPN traffic architecture for a LTE backhaul network

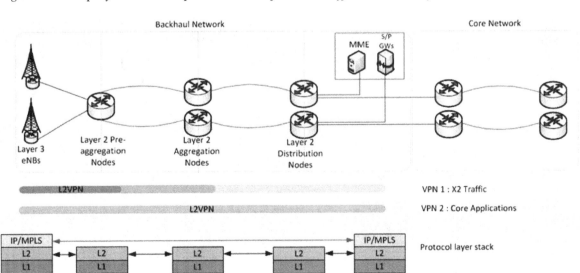

col stack of L2VPN traffic architecture for a LTE backhaul network.

There are two VPNs in Figure 8. VPN1 transmits the X2 traffic and VPN2 transmits the traffic for core applications. VPN1 requires multipoint to multipoint connectivity as eNodeBs are mesh connected in LTE architecture. Hence, VPLS is the preferred VPN technology for VPN1. Furthermore, this VPN can spans until several nodes such as pre-aggregation nodes, aggregation nodes, distribution nodes or control elements. VPN2 requires point to point connectivity to transfer the traffic for core network elements. Hence, VPWS is the preferred VPN technology for VPN2.

The L2VPN model has several advantages. L2VPN architectures have a low implementation cost. Furthermore, they can support multiple layer 3 protocols as the implementation is independent of layer 3 protocols. However, L2VPN model has several disadvantages as well. L2VPNs can results for larger broadcast domains at the backhaul network and it may cause to DoS (Denial of Service) and DDoS (Distributed DoS) attacks. Hence, L2VPNs are lacking of scalability. Also, it is not possible to implement higher layer security mechanisms such as IPsec in L2VPNs as IPsec

need layer 3 equipments at end nodes of tunnels. Moreover, eNodeB authentication mechanisms such as 802.1x have some incompatibility issues with layer 2 network attributes. The layer 2 attributes based traffic separation is also not efficient. Due to these reasons, L2VPN techniques are not mature and globally ubiquitous as L3 techniques.

Layer 3 MPLS VPN Model for LTE/EPC Deployments

L3VPN traffic architecture is proposed for LTE networks which have layer 3 network nodes. Figure 9 illustrates the deployment and the protocol stack of L3VPN traffic architecture for a LTE backhaul network.

VPN1 transmits the X2 traffic and VPN2 transmits the traffic for core applications. MPLS based 2547bis VPN is the preferred VPN architecture for both VPNs. It is possible to use the same VPN technology for both VPNs as MPLS VPNs provides both point to point and multipoint to multipoint connectivity.

L3VPN model has several advantages. L3VPN architecture is flexible and it can be modified with a minimum effort. For instance, the upper layer

Figure 9. The deployment and the protocol stack of L3VPN traffic architecture for a LTE backhaul network

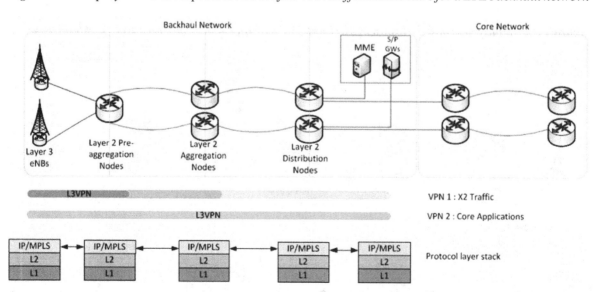

security mechanisms such as IPsec tunnels can be implemented without modifying the intermediate nodes. Furthermore, the use of a single VPN technology for all VPN instances is reducing the operational cost and the complexity of the architecture. Moreover, the traffic separation and QoS management are managed based on L3 attributes. L3VPN technologies are efficient, matured and globally ubiquitous than L2 techniques. However, there are some issues which are associated with L3VPNs. The L3VPN model cannot be used for a LTE backhaul network with layer 2 nodes. Besides, MPLS based L3VPNs have higher implementation cost than L2VPN technologies.

Virtual Network based Secure Architectures for LTE Network

LTE networks confront several security threats which do not exist in 2G/3G networks. These treats can be originated in different segments such as customer nodes, backhaul network, customer provider interface network and core network (Liyanage & Gurtov, 2012; Alvarez et al., 2011; Alvarez et al., 2012). Hence, 3GPP has defined several security requirements such as user authentication and authorization, payload encryption, privacy protection and IP based attack protection to prevent these security threats (Alvarez et al., 2011; Alvarez et al., 2012). It is necessary to implement segment specific security functions for each segments of the transport network to protect the LTE network. VPN based traffic architectures are promising solution to provide the required level of security for the LTE backhaul network segment.

Three main reasons have identified for security threats in the LTE backhaul network (Alvarez et al., 2011; Alvarez et al., 2012).

First, the LTE architecture proposed all-IP network architecture. Hence, the LTE backhaul network also consists of the IP-based entities such as MME, SGW, eNodeBs and IP based interfaces such as X2, S1. Thus, the backhaul network is now vulnerable to IP based attacks. Also, it is

possible to destruct important core elements and gateways directly by compromising nodes in the access network.

Second, the LTE backhaul network has hundreds or thousands of end nodes (e.g. eNodeBs). Hence, an attacker has multiple places to mount an attack to the backhaul network. On the other hand, the flat architecture concept distributes the control functionality over the backhaul nodes. Therefore, an attack on a single node can cause a significant damage in the network.

Third, all the traffic which are transferred through the access and the aggregation networks, are unencrypted in LTE networks. In the previous mobile network architectures such as GSM and UMTS, these traffic were encrypted by radio network layer protocols. However, these radio network layer protocol based encryptions are terminated at the eNodeBs in the LTE architecture. Hence, the backhaul traffic is vulnerable to common security threats such as eavesdropping and man in middle attacks.

Several VPN based secure traffic architectures are proposed to avoid these security threats. Most of these architectures are based on IPsec VPNs (Liyanage & Gurtov, 2012; Alvarez et al., 2011; Alvarez et al., 2012). Figure 10 illustrates a deployment and the protocol stack of IPsec based L3VPN architectures for LTE backhaul network (Liyanage & Gurtov, 2012).

Two modes of IPsec tunnels can be used to develop secure traffic architectures, namely IPsec tunnel mode and IPsec BEET (Bound End-to-End Tunnel) mode (Liyanage & Gurtov, 2012). Both of these architectures have similar properties as previously described L3VPN traffic transportation architecture. The IPsec tunnel mode VPN architecture is built by using Internet Key exchange version 2 (IKEv2) Mobility and Multihoming Protocol (MOBIKE) and the IPsec BEET mode VPN architecture is built by using Host Identity Protocol (HIP). Both architectures fulfill the 3GPP security requirements for the LTE backhaul network and provide additional features such as load

Figure 10. The protocol stack of IPsec-based L3VPN architectures for LTE backhaul network

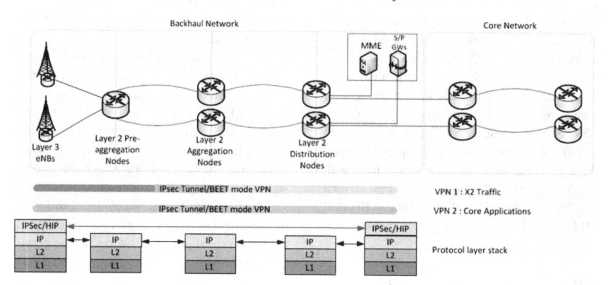

Mobile Virtual Network Operator (MVNO)

A MVNO is a mobile network service provider who operates his network over a leased telecommunication network. A MVNO obtains bulk access to network services such as radio spectrum, backhaul nodes, network controllers at wholesale rates based on the business agreement with a physical mobile network operator. Then, MVNO uses these leased resources to provide his own operator network. From a customer point of view, there is no difference between the services provided by a MVNO versus a traditional operator. Hence, the MVNO concept is becoming popular across the globe. The first MVNO was created by Tele2 in Denmark, and subsequently rolled out in several European markets. There are 633 active MVNO operations worldwide by October 2012 (MVNO, 2012).

The deployment of the MVNO concept in various countries varied based on different factors. In some countries, the regulatory intervention is the key reason to adapt the MVNO concept for mobile networks. These regulatory interventions are forced to mobile network operators to offer wholesale access to their network to ensure robust competition and provide benefits to the consumer. However, the most of mobile network operators adapt the MVNO concept to sell their excess capacity at wholesale rates to other entities in an effort to bring an incremental revenue which otherwise be regarded as unused network capacity.

The different types of MVNOs are present around the globe. They are categorized based on their participation for each function of the value chain. There are three main types (see also Figure 11) (Balon & Liau, 2012).

1. **Branded Reseller:** The branded reseller is the lightest MVNO business model. The MVNO just operates its brand and distribution channels. The host mobile network operator operates all other functionalities of the network. The branded reseller model requires the lowest investment for a new MVNO. Therefore it is the fastest MVNO type to implement.

2. **Full-MVNO:** The full-MVNO is the most complete MVNO model. Here, the host mobile network operator just provides the

Figure 11. The different types of MVNOs

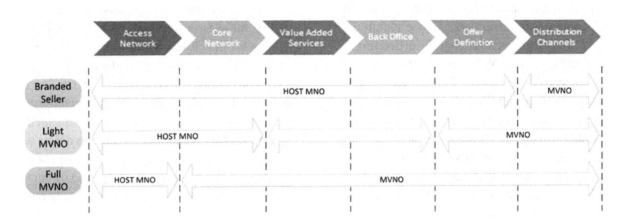

access network infrastructure. The MVNO operates the rest of functionalities of the network. The full-MVNO model is typically adopted by telecom players who wish to obtain all most all the control functionalities of their network.

3. **Light-MVNO:** The light-MVNO is an intermediate type in between a branded reseller and a full-MVNO. Light-MVNO allows the MVNO to take control over some of the back-office processes and control functionalities while having the full control of the marketing and sales areas. The MVNO can adapt his responsibilities according to his requirement.

Virtual private networks play a special role in a MVNO context. A MVNO itself can be considered as a special kind of VPN customer of the host network operator. A large scale host operator can simultaneously support several MVNOs by selling his network resources. Thus, a secure virtual network architecture is a promising solution to provide traffic separation, bandwidth allocation and privacy protection for these MVNOs. The existing VPN architectures are proposed only for traffic separation. However, primary operators need VPN architectures that not only separate the traffic but also slice the resources at network elements.

On the other hand, a MVNO also provides the traditional VPN services for his customers. Technically, it builds VPNs over another VPN. However, this dual VPN encapsulation wastes the scarce radio bandwidth and increases the operational cost. Hence, it is interesting to discover the novel VPN architectures to avoid these dual VPN encapsulations.

FUTURE TRENDS OF IP VIRTUAL PRIVATE NETWORKS

The future of IP VPNs in the telecommunication domain is not easy to predict due to the influence of numerous factors. Especially, VPN architectures and services in telecommunication networks are not stably deployed and widely used to predict a steady future. However, we identified several applications which may be interested in the future.

The Convergence Different Wireless Technologies

The LTE architecture proposes to integrate different wireless packet data technologies and systems such as Wireless Local Area Network (WLAN), GPRS, UMTS, CDMA (Code Division Multiple Access) with LTE networks. For instance, the convergence of WLAN with LTE systems provides better broadband services. WLAN has a far superior throughput rates than telecommunication networks. Hence, mobile user can achieve a higher throughput by offloading the user traffic to WLAN. On the other hand, WLAN has a cheap connection fee. Furthermore, WLAN equipments are significantly less expensive than telecommunication nodes and they are easy to install and maintain. These factors cause to achieve a significant economic advantage for operators. However, existing VPN architectures have to consider the impact of such a network convergence and modify them accordingly.

Mobile Cloud Services

Cloud computing and cloud based services are becoming popular among present communication networks. The main concept of a cloud service is to outsource the computing and storage functions to the service provider. Cloud service users can use mainframe resources which are located at the various places in a network (Sosinsky, 2010). The integration of the cloud computing with mobile devices or mobile cloud computing is accomplished due to the tremendous development of the mobile broadband services. The deployment of mobile cloud computing provides different advantages such as higher scalability, lower overall costs, higher computation power and more storage facility to mobile users. Furthermore, cloud base security is also achievable as a mobile cloud service. Here, the mobile user offloads complex security functions from the device to a virtualized cloud infrastructure. However, mobile cloud computing is still new to the telecommunication domain. There are many scalability and compatibility issues which need to be tackled before worldwide implementations.

The security is considered as a key requirement of a mobile cloud service. Cloud service providers can secure data communication to some extent. However, customers cannot rely only on service providers to provide the privacy, integrity and confidentiality for their private data. VPN based security architectures are promising solutions to solve these issues. While service providers interconnect the customer network with the rest of the world, well advanced and matured VPN security precautions protect the private data from third party users. However, VPN architectures to secure mobile cloud services are not implemented yet. On the other hand, it is possible to define a generalized virtual network platform on top of a telecommunication network. It will provide secure communication not only for mobile clouds but also other network services such as mobile P2P (Peer-to-Peer), multicast services.

Software Defined Networking (SDN) based Network Virtualization

As a result of increasing demand in data and the increase in the number of base stations over the telecommunication networks, the backhaul network will face congestion, in a manner similar to data center networks. In order to overcome this problem, Software Defined Networking (SDN) is a very promising solution. Initially, SDN has been defined only for fixed networks (McKeown, 2009). SDN separates the control plane from the data plane in network switches and routers.

Although, the widely discussed marketing applications in SDN is consolidated data centers, the basics of SDN could be extended to any application. These extensions also mean that SDN based network virtualization is applicable in telecommunication network. Several recent research projects (MEVICO, 2010; SIGMONA, 2013) are focused on these aspects. Hence, it is important to study the implementation aspect of IP VPNs in SDN based telecommunication network.

Person-to-Machine Services

Initially, telecommunication networks are used only for person-to-person services such as voice call and text message services. However, the recent evolution of telecommunication networks inspires to provide person-to-machine services such as dynamic traffic condition reports, interactive voice response systems, interactive short message systems, online navigation systems. These person-to-machine services are well integrated with the present human life style. It is expecting that the service usage will be increased in the future. Present, virtual networks are playing a vital role in person-to-machine services in wired Internet environment to avoid the unauthorized access, enforce service exclusivity and protect against IP based attacks. Some of these VPN architectures might be applicable up to certain extend in the telecommunication networks. However, it will be interesting to implement novel IP VPN architectures to protect the person-to-machine services over the telecommunication networks.

CONCLUSION

Virtual private networks are widely used in wide area networks to securely interconnect geographically distributed customer sites through public networks. On the other hand, the tremendous development of the telecommunication technologies causes to increase the number of mobile subscribers and provide almost the same level of broadband services as wired networks. Hence, corporate customers are motivated to use telecommunication networks to obtain the broadband access and it fuels implementations of the IP VPNs on top of the cellular networks.

This chapter contains an extensive survey on these IP VPN implementations in the cellular network domain. In the beginning, we described the history, the evolvement and the classification of VPNs. We identified four main applications of virtual networks in the telecommunication domain; namely mobile VPN services, VPN based traffic transportation models for LTE networks, secure backhaul traffic transportation of telecommunication network and Mobile Virtual Network

Operators (MVNO). Finally, we discussed the future trends of virtual networks in telecommunication network context; namely, mobile cloud services, mobile person-to-machine services and Software Defined Networking (SDN) based network virtualization. Hence, we can conclude that the integration of IP VPN technologies with telecommunication networks is a highly demanded requirement and it is a timely context to conduct extensive research works.

REFERENCES

Balon, M., & Liau, B. (2012). Mobile virtual network operator. In *Proceedings of XVth International Telecommunications Network Strategy and Planning Symposium (NETWORKS)* (pp. 1-6). IEEE.

Benenati, D., Feder, P. M., Lee, N. Y., Martin-Leon, S., & Shapira, R. (2002). A seamless mobile VPN data solution for CDMA2000,* UMTS, and WLAN users. *Bell Labs Technical Journal, 7*(2), 143–165. doi:10.1002/bltj.10010

Callon, R., & Suzuki, M. (2005). *A framework for layer 3 provider-provisioned virtual private networks (ppvpns). Request for Comments (RFC) 4110*. IETF.

Carugi, M., & De Clercq, J. (2004). Virtual private network services: Scenarios, requirements and architectural constructs from a standardization perspective. *IEEE Communications Magazine, 42*(6), 116–122. doi:10.1109/MCOM.2004.1304246

Cisco. (2004). *Comparing MPLS-based VPNs, IPSec-based VPNs, and a combined approach from Cisco Systems* (Technical Report). Cisco Cooperation.

Alvarez, M. A., Jounay, F., Major, T., & Volpato, P. (2011). *LTE backhauling deployment scenarios* (Technical Report). Next Generation Mobile Networks Alliance.

Alvarez, M. A., Jounay, F., & Volpato, P. (2013). *Security in LTE backhauling* (Technical Report). Next Generation Mobile Networks Alliance.

Cisco. (2010). *Architectural considerations for backhaul of 2G/3G and long term evolution networks* (Technical Report). Cisco Cooperation.

Croy, P. (2011). *LTE backhual requirements, reality check (Technical Report)*. Aviat Networks Inc.

Dahlman, E., Parkvall, S., & Skold, J. (2011). *4G: LTE/LTE-advanced for mobile broadband: LTE/ LTE-advanced for mobile broadband.* Academic Press.

Daniel, A. (2004). IP virtual private networks– A service provider perspective. *IEEE Communications*, *151*(1), 62–70. doi:10.1049/ip-com:20040133

Feder, P. M., Lee, N. Y., & Martin-Leon, S. (2003). A seamless mobile VPN data solution for UMTS and WLAN users. In *Proceedings of 4th International Conference on 3G Mobile Communication Technologies,* (pp. 210-216). IET.

Harris, S. (2002). *IP VPNs: An overview for network executives* (Technical Report). Retrieved from http://www.onsiteaustin.com/whitepapers/VPN20justfication.pdf

Knight, P., & Lewis, C. (2004). Layer 2 and 3 virtual private networks: Taxonomy, technology, and standardization efforts. *IEEE Communications Magazine*, *42*(6), 124–131. doi:10.1109/MCOM.2004.1304248

Liyanage, M., & Gurtov, A. (2012). Secured VPN models for LTE backhaul networks. In *Proceedings of the Vehicular Technology Conference* (pp. 1-5). IEEE.

McKeown, N. (2009). Software-defined networking. In *Proceedings of INFOCOM*. IEEE.

Metz, C. (2003). The latest in virtual private networks: Part I. *IEEE Internet Computing*, 7(1), 87–91. doi:10.1109/MIC.2003.1167346

MEVICO. (2010). *Mobile networks evolution for individual communications experience (MEVICO)*. Retrieved from http://www.celtic-initiative.org/Projects/Celtic-projects/Call7/MEVICO/mevico-default.asp

MVNO. (2012). *The MVNO directory*. Retrieved from http://www.mvnodirectory.com/overview.html

Ronai, A., & Officer, C. M. (2009). *LTE ready mobile backhaul (Technical Report)*. Ceragon Networks Ltd.

Rosenbaum, G., Lau, W., & Jha, S. (2003). Recent directions in virtual private network solutions. In *Proceedings of the 11th IEEE International Conference on Networks* (pp. 217–223). IEEE.

Rosenbaum, G., Lau, W., & Jha, S. (2003). An analysis of virtual private network solutions. In *Proceedings of the 28th Annual IEEE International Conference on Local Computer Networks* (pp. 395–404). IEEE.

Shneyderman, A., Bagasrawala, A., & Casati, A. (2000). *Mobile VPNs for next generation GPRS and UMTS networks* (White Paper). Lucent Technologies.

Shneyderman, A., & Casati, A. (2003). *Mobile VPN: Delivering advanced services in next generation wireless systems.* Hoboken, NJ: John Wiley & Sons.

SIGMONA. (2013). *SDN concept in generalized mobile network architectures (SIGMONA)*. Retrieved from http://www.celticinitiative.org/Projects/Celtic-Plus-Projects/2012/SIGMONA/sigmona-default.asp

Sosinsky, B. (2010). *Cloud computing bible*. Hoboken, NJ: Wiley Publications.

UTStarcom. (2009). *3G/LTE mobile backhaul network MPLS-TP based solution* (Technical Report). UTStarcom Inc.

Venkateswaran, R. (2001). Virtual private networks. *IEEE Potentials*, *20*(1), 11–15. doi:10.1109/45.913204

XO. (2010). *A business guide to MPLS IP VPN migration: Five critical factors (Technical Report)*. XO Communications Inc.

ADDITIONAL READING

Alvarez, M. A., Jounay, F., Major, T., & Volpato, P. (2011). *LTE backhauling deployment scenarios. Technical Report*. Next Generation Mobile Networks Alliance.

Alvarez, M. A., Jounay, F., & Volpato, P. (2013). *Security in LTE backhauling. Technical Report*. Next Generation Mobile Networks Alliance.

Aspas, J. P., & Arroyo, F. B. (2002). Design of a Mobile VPN able to Support a Large Number of Users. In *proceeding of 2nd European Conference on Universal Multiservice Networks, 2002, ECUMN 2002 (pp. 219-222)*. IEEE.

Chen, J. C., Liang, J. C., Wang, S. T., Pan, S. Y., Chen, Y. S., & Chen, Y. Y. (2006). Fast handoff in mobile virtual private networks. In *Proceedings of the 2006 International Symposium on World of Wireless, Mobile and Multimedia Networks (pp. 548-552)*. IEEE.

Cisco, (2010). Architectural Considerations for Backhaul of 2G/3G and Long Term Evolution Networks. *Technical Report*, Cisco Cooperation.

Feder, P. M., Lee, N. Y., & Martin-Leon, S. (2003). A seamless mobile VPN data solution for UMTS and WLAN users. In *Proceedings of 4th International Conference on 3G Mobile Communication Technologies, 2003, 3G 2003 (pp. 210-216)*. IET.

Gurtov, A. (2008). *Host identity protocol (HIP), towards the secure mobile Internet*. Wiley Publications. doi:10.1002/9780470772898

Kim, K., Byun, H., & Lee, M. (2007). Route Optimization Mechanism for the Mobile VPN users in Foreign Networks. In *Proceedings of the 9th International Conference on Advanced Communication Technology*, (Vol. 3, pp. 1969-1973). IEEE.

Liu, Z. H., Chen, J. C., & Chen, T. C. (2009). Design and analysis of SIP-based mobile VPN for real-time applications. *IEEE Transactions on Wireless Communications*, *8*(11), 5650–5661. doi:10.1109/TWC.2009.090076

Liyanage, M., & Gurtov, A. (2012). Secured VPN Models for LTE Backhaul Networks. In *Proceedings of the Vehicular Technology Conference (VTC Fall), 2012 (pp. 1-5)*. IEEE.

Ntantogian, C., & Xenakis, C. (2007). A security protocol for mutual authentication and mobile VPN deployment in B3G networks. In *proceeding of 18th International Symposium on Personal, Indoor and Mobile Radio Communications, 2007, PIMRC 2007 (pp. 1-5)*. IEEE.

Petrescu, A., & Olivereau, A. (2009). Mobile VPN and V2V NEMO for public transportation. In *Proceedings of 9th International Conference on Intelligent Transport Systems Telecommunications (ITST 2009) (pp. 63-68)*. IEEE.

Ronai, A., & Officer, C. M. (2009). *LTE Ready Mobile Backhaul. Technical Report*. Ceragon Networks Ltd.

Shneyderman, A., Bagasrawala, A., & Casati, A. (2000). Mobile VPNs for Next Generation GPRS and UMTS Networks. *White Paper*. Lucent Technologies.

Shneyderman, A., & Casati, A. (2003). *Mobile VPN: Delivering Advanced Services in Next Generation Wireless Systems*. John Wiley & Sons.

Sosinsky, B. (2010). *Cloud computing bible*. Wiley Publications.

Xenakis, C., Gazis, E., & Merakos, L. (2002). Secure VPN deployment in GPRS mobile network. In *Proceedings of European Wireless* (pp. 293-300).

Xenakis, C., & Merakos, L. (2004). Security in third generation mobile networks. *Computer Communications, 27*(7), 638–650. doi:10.1016/j.comcom.2003.12.004

Xenakis, C., Ntantogian, C., & Stavrakakis, I. (2008). A network-assisted mobile VPN for securing users data in UMTS. *Computer Communications, 31*(14), 3315–3327. doi:10.1016/j.comcom.2008.05.018

Ye, R., Feng, Y., Yu, S., & Song, C. (2005). A Lightweight WTLS-based Mobile VPN scheme. *Microelectronics & Computer, 22*(4), 126–133.

Yu, J., & Liu, C. (2006). Performance Analysis of Mobile VPN Architecture. In *proceeding of 4th Annual Conference on Telecommunications and Information Technology*. Las Vegas, USA.

KEY TERMS AND DEFINITIONS

Cellular Network: A radio network that provides wireless network services for mobile users.

LTE (Long Term Evaluation): A telecommunication standard for wireless communication of high speed data for mobile phones and data terminals.

Mobile: A device, user, technology, or service that has the ability to move or be moved.

MVPN (Mobile Virtual Private Network): A virtualization mechanism that allows mobile devices to access network resources on their home/private network, when they connect via other public mobile networks.

Telecommunication: Long distance communication by using technological means such as electrical signals or electromagnetic waves.

VPN (Virtual Private Network): A virtualization mechanism to extend a private network across a public network, such as the Internet.

Chapter 4
Self–Organization Activities in LTE–Advanced Networks

Ali Diab
Al-Baath University, Syria

Andreas Mitschele-Thiel
Ilmenau University of Technology, Germany

ABSTRACT

A major challenge in the context of LTE networks is a cost-effective network operation, which can be done by carefully controlling the network Operational Expenses (OPEX). Therefore, to minimize OPEX costs while optimizing network performance, Self-Organizing Network (SON) principles were proposed. These networks are the main focus of this chapter, which highlights the state of art and provides a comprehensive investigation of current research efforts in the field of SONs. A major contribution of the chapter is the handling of SON use cases, going through their challenges, solutions. and open research questions. The chapter also presents efforts to provide coordination frameworks between SON use cases and routines. An additional essential contribution of the chapter is the description of SON activities within 3GPP.

1. INTRODUCTION

The satisfaction of the always increasing demands of broadband applications and new services, the better support of mobility, the less costs, etc. are the main drivers standing behind the standardization and development of 4th Generation (4G) mobile communication networks, termed Long Term Evolution (LTE), which aim at providing incredible customer experience (Dahlman, Parkvall, & Skoeld, 2011). Researchers expect that the number of LTE subscribers will reach the base of 3G/UMTS networks (1.087 million subscribers) by the year 2015 (Garza, Ashai, Monturus & Syputa, 2010).

DOI: 10.4018/978-1-4666-5170-8.ch004

A major challenge in LTE networks is a cost-effective provision of new services. Sure, this necessitates carefully controlling the Capital Expenditures (CAPEX) associated with the infrastructure and also the Operational Expenses (OPEX) resulting from operating this infrastructure. Cost-effectiveness gets more crucial when one considers the additional cost resulting from the expected affordable operation of overlaid multi-standard networks (2G, 3G, 4G, etc.). While CAPEX costs remain an important issue for network operators and are normally affected by factors outside their control (the cost of new technologies for instance), OPEX costs become more significant part in the cost structure. 3G mobile communication technologies have shown that the contribution of network-related OPEX costs to the total cost associated with the network is ~30%, see (Motorola, 2009). So, operators need to keep their OPEX costs minimized, while enhancing the performance of their networks (Ramiro & Hamied, 2012). Note that OPEX costs strongly relate to the solutions implemented in the infrastructure. This implies that OPEX costs reduction requires implementing adequate solutions that, for sure, will be best if self-organized.

The release being standardized by the 3rd Generation Project Partnership (3GPP) (3GPP, 2013) with self-organization capabilities is termed LTE-Advanced (Dahlman, Parkvall, & Skoeld, 2011). This standard aims at minimizing OPEX costs, while maximizing resource usage, network capacity, etc. To further contribute to Self-Organizing Networks (SONs), the Next Generation Mobile Networks (NGMN) alliance was also constructed. This alliance has summarized SONs requirements in a number of operation use cases, see Ramiro and Hamied (2012) and Lehser (2008).

This chapter focuses on the principles and activities of SONs. It provides in section 2 a thorough overview of SONs functionalities. Following that, the chapter handles in section 3 the main SON activities done in the scope of 3GPP going through the activities in releases 9, 10, 11 and even beyond.

Section 4 investigates coordination frameworks that coordinate between various SON activities to guarantee systems' stability and reliability in addition to performance optimization. Furthermore, the chapter provides in section 5 various SON use cases going through their goals, proposed solutions and research challenges. Finally, the chapter summarizes with the main results.

2. SON FUNCTIONALITIES

There are different schemes proposed to categorize SON functionalities, see NGMN (2006), NGMN (2008), and NGMN (2008a). Throughout this chapter, we will use the wide-accepted categorization provided in Ramiro and Hamied (2012), which states that SON functionalities are categorized into five categories, namely self-planning, self-deployment, self-optimization, self-healing and SON enablers. Figure 1 provides a summarized view of these categories and their corresponding routines.

Self-planning functionalities group the functions that cover the derivation of settings for new network nodes, the selection of site locations and hardware configuration including radio and transport parameters. Radio parameters include an initial selection of handover settings, Random Access Channel (RACH) settings, Paging Channel (PCH) settings, the configuration of new evolved NodeBs (eNBs) with suitable Physical Cell Identities (PCIs), the determination of an initial Neighbor Cell Relation (NCR) list, etc. Transport parameters enable starting tunnels between new network nodes and other network entities, so that further configurations can be received. Transport parameters include IP addresses, Virtual Local Area Network (VLAN) identifiers, QoS settings, etc. Self-planning functionalities also take care of determining which nodes must update their databases, white and black lists, etc. Note that the functions related to

Figure 1. Summarized view of SON functionalities

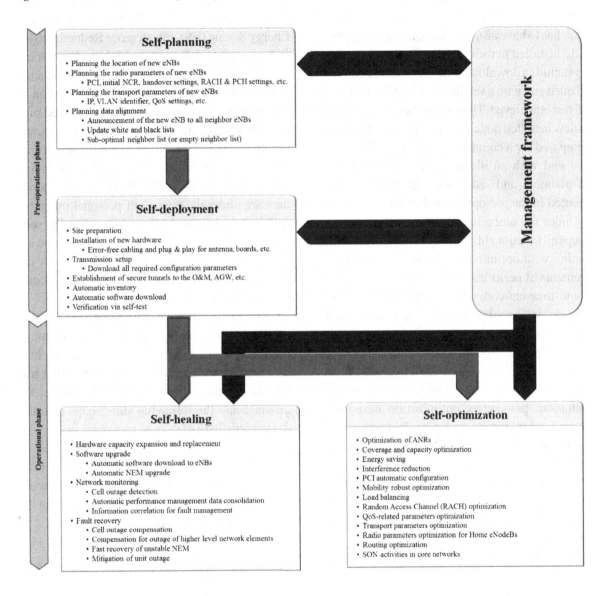

the site acquisition and preparation are excluded from this category.

Self-deployment functionalities include all procedures required to bring new network nodes into commercial operation. This includes the preparation, installation, authentication, verification and status reports of new network nodes. The installation of new network nodes covers the site preparation and hardware installation including error-free cabling and plug and play behavior

for all components, e.g. antenna, boards, etc. After the hardware installation is accomplished, secure tunnels must be established to a specific configuration server which sends all configuration parameters required to authenticate the new node and to establish tunnels to the Operation and Maintenance (O&M) center, Access Gateway (AGW), etc. Following that, the new node is authenticated and authorized. This enables it to establish secure tunnels to the O&M center and the AGW. All installed hardware units must be autonomously delivered to an inventory system, which has to have an up-to-date picture of the currently installed network nodes. Self-deployment may include a download of new software packages and must end with a verification process, within a self-test is achieved. The self-test verifies whether the new installed node is in the expected state and is prepared for a commercial operation. This test must end with an illustrative report. Note that self-planning and self-deployment routines are executed in the pre-operational phase.

Under the umbrella of *self-optimization* one groups all functionalities done to keep the network running with optimized performance. The measurements of performance indicators are utilized to auto-tune network settings, so that the overall network performance remains optimized. The measurements and traces themselves are achieved by User Equipment (UE), eNBs, the O&M center, etc. Self-optimization is a continuous closed-loop process that encompasses periodic performance evaluation, parameters optimization based on the evaluation results and a re-deployment of the optimized parameters. The combination between the current system's configurations and the achieved measurements we mentioned are input data to the optimization processes in charge. These processes result in selecting optimized configuration parameters that are re-deployed again in the network. Note that the optimization itself is a trade-off between various performance indicators, e.g. quality, coverage, capacity, security, etc., to achieve common well-defined goals. Note also that network equipment of different vendors may implement different optimization methods and different performance figures, which complicates self-optimization routines. Therefore, it is essential for operators in multi-vendor environments to develop adaptation layers to harmonize the performance measures of various vendors, so that a common optimization platform can be used. The routines of self-optimization cover, in principle, the use cases defined by the NGMN alliance,

namely the Automatic Neighbor Relation (ANR), Coverage and Capacity Optimization (CCO), Energy Saving (ES), Interference Reduction (IR), PCI automatic configuration, Mobility Robustness Optimization (MRO), Mobility Load Balancing (MLB) and RACH optimization. In addition to these use cases, other topics must be covered such as QoS-related, transport and radio parameters optimization. Beside the mentioned topics, efforts must be done to optimize the scenarios containing Home eNBs (HeNBs) (FemtoForum, 2011) since they are under the control of personal operators and switched on and off arbitrary. Finally, self-optimization functionalities are executed during the operational phase, which is defined as the phase when the radio frequency interface is commercially active.

Self-healing routines are responsible of keeping the network running and preventing disruptive problems from arising. These routines include the upgrade/replacement of software and hardware. Concrete, self-healing functionalities are grouped into the following sub-groups (Ramiro & Hamied, 2012):

- Hardware capacity extension and replacement: these routines take care of autonomous system surveillance by proposing the convenience of hardware expansions.
- Software upgrade: the routines of this sub-group should allow the upgrade of existing software with minimal operator intervention or even without. The routines belonging to this sub-group include automatic software download to eNBs and automatic Network Element Manager (NEM) upgrade.
- Network monitoring: the system must be capable of measuring and analyzing the Radio Access Network (RAN) performance to enable taking decisions concerning further developments. Multi-vendor scenarios and call tracing must be supported without extra complexity. The

following routines belong to this sub-group: cell/service outage detection, automatic performance management data consolidation and information correlation for fault management.

- Fault recovery: the routines belonging to this sub-group have to autonomously recover from failures when happening. Example routines include Cell Outage Compensation (COC), compensation for outage of higher level network elements, fast recovery of unstable NEM and mitigation of unit outage.

SON enablers cover all actions that facilitate the execution as well as availability of other SON routines. Note that SON enablers themselves are not SON routines. They include a standardized open northbound interface (the interfaces between NEM and other management tools), a real-time performance management via Inf-N, a real-time Key Performance Indicators (KPIs) reporting, subscribers and equipment traces, etc.

Note that the routines in the five categories mentioned do not work sequentially, see the figure. First, self-planning activities are executed followed by self-deployment routines. Thereafter, self-optimization and also self-healing routines run in a cyclic-like behavior. A management framework works in the background and manages the system as a whole. This framework must coordinate between different SON actions, more details follow in section 4.

For more clarifications, we will discuss the cyclic-behavior of self-optimization functionalities, see Figure 2. First, continuous measurements of certain performance figures are achieved. The results of the measurements and traces are periodically summarized in measurement reports and sent as input to the optimization processes in charge. These processes judge the reported figures against the required/accepted thresholds. The optimization processes decide based on some polices employed how to tune the system parameters, so that the

performance figures remain realizing the pre-defined goals. Note that there is a trade-off between various performance figures to achieve the global goal. After taking the optimization decisions, the new tuned parameters are re-deployed in the system again. The described process is repeated in a cycle so that the performance remains optimized.

3. 3GPP SON ACTIVITIES

3GPP is a standardization body created in December 1998 (3GPP, 1998) within the scope of the International Telecommunication Union (ITU) (ITU, 2013) with the goal: development of 3G and 4G specifications based on evolved Global System for Mobile communication (GSM) standards as a starting point. 3GPP was originally responsible of producing Technical Specifications (TSs) as well as Technical Reports (TRs) for global 3G cellular systems. The responsibility of 3GPP was extended later to cover TSs and TRs for the maintenance and development of GSM (3GPP, 2011). 3GPP is currently responsible of supporting the evolution of 3G by developing TSs as well as TRs for LTE and LTE-Advanced systems.

The SON standardization processing was started in 2008 and is continued since then. The work on SONs is distributed among Radio Access Network Working Group 2 (RAN WG2 (RAN2)), RAN WG3 (RAN3) and technical Specification group Service and system Aspects WG5 (SA5) in collaboration sometimes with other groups, e.g. SA2 for instance. Note that 3GPP does not discuss which SON algorithms should be implemented. Rather, it focuses on architecture, use cases and associated performance requirements. RAN2 is responsible of one use case, namely the Minimization of Drive Tests (MDT). The use cases covered by RAN3 are gathered in the 3GPP TR36.902 (3GPP, 2013). These use cases are mainly the use cases defined by the NGMN alliance. The work in the scope of RAN3 reached different stages. Some use cases are considered

Figure 2. The cyclic-behavior of self-optimization functionalities

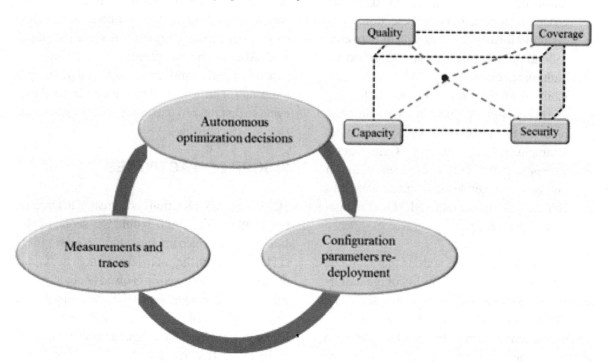

closed, e.g. RACH optimization and ANR, while others are still being enhanced, e.g. MLB, MRO, ES, etc. Other use cases are suspended or leaved aside to be handled later, see Hamalainen, Sanneck, and Sartori (2012).

The work in the scope of SA5 was guided by the operator use cases. The first step was the definition of the configuration parameters necessary for the ANR function in the Network Resource Model (NRM) and the definition of how self-configuration (3GPP TS32.50x series) and automatic software management (3GPP TS32.53x series) work in the Integration Reference Point (IRP) interface. Next, Automatic Radio Configuration Function (ARCF) was considered as a part of self-configuration processes. The motivation behind this is the fact that the characteristics of radio network environment should be considered while self-configuring network elements. Sure, this will enhance network performance and positively affect further self-optimization tasks. Basic management principles for self-optimization were

agreed on in 2009 as a part of SA5. Concrete, the principles included target- and policy-controlled optimization (3GPP TS32.52x series). The targets were defined for HandOver Optimization (HOO) and Load Balancing Optimization (LBO) and included new measurements to monitor targets achievement. Later, more targets are defined in 2010 and 2011 in the scope of SA5, e.g. for RACH optimization. Extended trace functionalities were enhanced as well to support automated Drive Tests (DT). The issue of conflicting targets was also tackled within SA5.

3.1. SON Activities Up to Release 9

Since release 8 features related to SON started to be added to progressively support the use cases we mentioned. Release 8 started with specifying the concepts and requirements of SONs and continued with developing self-eNB establishment procedures and ANR management routines. Release 9 continued, following that, by adding self-optimization and self-healing functionalities. The following provides a summary of SON activities up to release 9. Note that the final specifications of release 9 have been taken as a reference point, see 3GPP (2010d), 3GPP (2010e), 3GPP (2009d), and 3GPP (2012).

- **Self-configuration**: To put an eNB in an operational state with minimal human intervention, a framework has been proposed in 3GPP (2010f), 3GPP (2010g), and 3GPP (2010h). The self-configuration procedure is triggered after the self-test process and includes:
  - Transfer of configuration data to the eNB.
  - Allocation of a new IP address.
  - Provision of basic information about the transport network.
  - Announcement of the eNB's key characteristics to the O&M center.
  - Interconnection between the eNB and the O&M center in charge of self-configuration.
  - Interconnection between the eNB and the O&M center in charge of normal operation.
  - Automatic software download.
  - Automatic configuration of initial transport and radio parameters.
  - Establishment of S1 links.
  - Establishment of planned X2 links.
  - Announcement of the eNB to the inventory system.
  - Execution of self-test process.
  - Generation of status reports.
  - Announcement of the new eNB to network entities above the Itf-N
  - Software installation and activation.
- **Self-optimization**: The standardized SON routines are the following:
  - **ANR**: The construction of neighbors includes intra- as well as inter-frequency LTE neighbors, 2G neighbors and 3G neighbors (3GPP, 2009; 3GPP, 2013d). There are also management functionalities to control basic settings, for white and black lists for instance (3GPP, 2010i).
  - **MLB:** The set of procedures required are standardized (3GPP, 2010q; 3GPP, 2013d; 3GPP, 2010a; 3GPP, 2010j; 3GPP, 2010c).
  - **MRO:** The focus is on providing seamless handovers (3GPP, 2010q; 3GPP, 2013d; 3GPP, 2010a).
  - **RACH optimization:** 3GPP investigates the routines required to optimize the access via this channel (3GPP, 2010q; 3GPP, 2013d; 3GPP, 2010a; 3GPP, 2010c).
  - **IR:** 3GPP standardizes this use case including the definition of specific information elements to be exchanged through X2 interface (3GPP, 2010q).
  - **CCO:** 3GPP sets the requirements and takes initial decisions concerning this use case. It considers the provision of optimal coverage and capacity as an essential objective (3GPP, 2010a). Some generic scenarios to be investigated are described in 3GPP (2010a), while some configuration parameters for this use case are provided in 3GPP (2010i).

In addition to the above listed use cases, some management capabilities are also proposed for the use cases MLB and MRO. Concrete, they allow the O&M center to activate/deactivate the use cases, set targets with different priorities, collect performance measurements, etc. (3GPP, 2010b; 3Gpp, 2010k).

- **Self-healing**: the term self-healing has been introduced in release 9 (3GPP, 2009e). In general, a workflow for self-healing has been provided in addition to various recovery actions for different types of problems. Concrete, three use cases for self-healing have been identified, namely

- ○ Self-recovery of network elements software.
- ○ Self-healing of board faults.
- ○ Self-healing of cell outage.
- ○ Self-testing routines.

- **Operation, Administration and Maintenance center (OAM) aspects**: 3GPP handles in this context issues related to the OAM architecture for SONs, management issues, standardized interfaces between the OAM center and other entities in the network. The following were focused on:

  - ○ SON self- management.
    - Design of coordination functionalities (turning the automatic functions on/off, coordination among different targets in MRO use case, etc.).
  - ○ Design and standardization of OAM interfaces for HeNBs.
  - ○ Self-optimization management. (3GPP, 2012)

3.2. SON Activities in Release 10

The most important SON-related activities in release 10 are the following (3GPP, 2012; 3GPP, 2010l).

- **Self-optimization**: the following tasks have been accomplished.
  - ○ CCO enhancement
    - Focus was on the detection of coverage as well as capacity problems.
    - Work on management aspects.
  - ○ MRO and MLB enhancements
    - Inter-RAT, load and capacity information enhancements.
    - Multi-RAT environments.

- Consideration of HeNBs and macro-cells.
- Support of unsuccessful (re-) establishments.
- Connection failures in inter-RAT environment.
- Measurements achieved by the UE after connection failure.
- Ping-pong handoffs in idle mode (inter- and intra-RAT) as well as active mode (inter-RAT only).
- Handovers to wrong cells (intra-LTE) without connection failures.

- ○ ES use case
  - ES for inter-RAT.
  - Intra-LTE solutions enhancements (in combination with coverage optimization).
  - Principles about ES management are provided in (3GPP, 2010m).
  - Configurable load thresholds in cells and their neighborhood for starting/leaving ES state.
  - Possibilities to define cells not affected by ES routines.
  - Work on management aspects (3GPP, 2011a).
  - Solutions to enable ES within UMTS networks and preliminary evaluations (3GPP, 2009f).

- ○ ANR for 3G
  - Investigation of the methods that enable updating neighborhood relation lists in 2G, 3G and LTE systems.

- ○ RACH optimization enhancements
  - Based on access or access delay probability.
  - Work on management aspects.

- ○ COC
  - Design of object models to manage CCO.

- Possibilities to define cells not affected by COC routines.
  - Interference Control (IC)
    - Work on management aspects.
- **Self-healing**:
  - Requirements specification.
  - Definition of inputs and outputs for self-healing routines.
  - O&M support. (3GPP, 2010n)
- **OAM aspects**:
  - SON Self- management (continuation of the work started in release 9).
    - Completing the management aspects of IR, CCO and RACH optimization.
    - Addressing extra coordination functionalities.
  - Self-healing management: 3GPP aimed in this context at documenting self-healing OAM requirements and defining inputs and outputs from/ for self-healing entities in addition to their locations in the management architecture as well as the associated algorithms standardization degree.
  - OAM aspects of ES in Radio Networks
    - Defining ES management OAM requirements for the following scenarios: eNB overlaid, carrier restricted and capacity limited network.
    - Selecting existing/new measurements that can be used for the assessment of the impact of ES actions corresponding to the above mentioned scenarios.

In addition to the above mentioned tasks, the MDT in E-UTRAN and in UTRAN was a major focus in release 10 since RAN2 has worked on specifying solutions for MDT with the coverage optimization as a priority use case. MDT-related activities are to be captured in (3GPP, 2013a).

Furthermore, a study on integration of device management information with Itf-N was completed (3GPP, 2010o). Another important issue discussed is the coordination between different SON use cases. The work on coordination aspects also remained in newer releases. The most important topic started to be discussed in release 10 is the handle of competing self-X functionalities. This topic remained open in release 10 and shifted to further releases.

3.3. SON Activities in Release 11

The work on SON in release 11 (3GPP, 2013b) focuses on two trends. The first handles the coordination between various SON use cases and their management aspects, while the second provides further self-optimization enhancements (3GPP, 2012). The SON activities in this release can be summarized as follows:

- **SON management**: The goals are:
  - Defining management aspects for the following SON use cases (UTRAN context):
    - ANR (inter-UTRAN, UTRAN ANR (UTRAN to GERAN), UTRAN ANR (UTRAN to E-UTRAN), E- UTRAN ANR (E-UTRAN to GERAN), E-UTRAN ANR (E-UTRAN to UTRAN) and E-UTRAN ANR (E-UTRAN to CDMA2000)).
    - CCO use case.
  - Specification of UTRAN SON management solutions.
  - Specification of SON management solutions capturing the common SON management part of E-UTRAN and UTRAN.
  - SON coordination management: the objectives is to design coordination solutions for the following aspects:

- Coordination between configuration management via Itf-N and configuration parameters change induced by SON routines below Itf-N
- Coordination between various SON use cases below Itf-N based on case by case approach. The coordination comprises several typical scenarios, namely: coordination between configuration changes triggered by centralized and distributed SON functions, coordination between configuration changes triggered by distributed SON functions, and Coordination between SON use cases.
  - ES Management (ESM): this item aims at supporting inter-RAT ESM while considering the results of TR 32.834 (3GPP, 2012a).
- **Self-optimization enhancements**: Following activities are of interest for this release.
  - Inter-RAT ANR enhancements.
  - Inter-RAT MRO enhancements.
  - Inter-RAT ping-pong scenarios: as stated in TR 36.902 (3GPP, 2011b), there are some mobility scenarios that do not produce connection failures, e.g. inter-RAT ping-pongs. However, they produce unnecessary signaling. Following scenarios are to be considered.
    - Inter-RAT failure resulting from the deployment of LTE over broader 2G/3G coverage.
    - Support of connection failures resolution for heterogeneous deployment.
    - Inter-RAT ping-pong resolution.

3.4. SON Activities in Release 12

SON activities in this release (3GPP, 2013c) focus on OAM aspects and on the provision of a study about SON use cases in LTE-HRPD (High Rate Packet Data in 3GPP2) inter-RAT, see 3GPP, (2012).

- **OAM aspects**: Following items have been addressed:
  - Enhanced network management for centralized CCO use case.
  - Multi-vendor plug and play eNB interconnection.
- **SON use cases in LTE-HRPD:** This study has to include the following use cases with possible solutions.
  - Mobility load balancing between LTE and HRPD.
  - ANR between LTE and HRPD.
  - MRO between LTE and HRPD.

4. SON COORDINATION FRAMEWORKS

So far, most efforts to develop SON solutions have focused on stand-alone solutions. However, with increasing employment of SON functions, increasing dependencies and complex relations between them arise. This may result in conflicts and may transit the whole system into an unstable state. Such conflicts may arise, for instance, when two independent SON use cases (e.g. CCO and IR use cases) aim at different goals when optimizing a configuration parameter (e.g. antenna tilt). Another source for such conflicts may be the different impacts of specific configuration parameters on various SON use cases. In other words, the modification of a configuration parameter may positively affect a SON use case and, at the same time, negatively impact another, see Schmelz et al. (2011). Keep also in mind that there may be specific combinations of a set of configuration parameters with interdependencies, which also produce complex relations among them and the applied SON use cases. Thus, a coordination between SON functionalities, use cases, etc. becomes essential to avoid and/or resolve conflicts and dependencies between them, and consequently, to guarantee systems' stability and reliability (Jansen et al., 2009).

4.1. Conflict Types

There are two types of the conflicts that result from the existence of multiple SON functions, namely control parameters conflicts and dependencies-based conflicts (Schmelz et al., 2011), see figure 3. Control parameters conflicts arise when two or more SON use cases request a change of a configuration parameter with different values. This type of conflicts can be further divided into directionality and magnitude conflicts. Directionality control parameters conflicts appear when a SON use case requests an increase of a configuration parameter, while another use case asks for a decrease of the parameter. An example scenario can be observed for instance when the MLB use case requests a decrease in the handover offset to optimize the load distribution among cells, and simultaneously, the MRO use case wants to

increase the handover hysteresis to reduce ping-pong effects. The magnitude conflicts arise when a SON use case requires a large increase (decrease) in the value of a configuration parameter, while another SON use case wants to cautiously allow a gradual increment (decrement) of the configuration parameter value and simultaneously monitor the resulting effects.

Dependencies-based conflicts arise when a performance metric used as an input for a SON use case is affected by another SON use case applied to optimize this performance metric or other metrics. This type of conflicts also depends on the relative timescale of the participated SON use cases. Concrete, when a SON use case operates at a short timescale and affects a specific performance metric, while another SON use case also affects the same performance metric, however, operates at a long timescale. In this case, the system may reach a stable state as a result of the application of the short timescale SON use case. Note that this may prohibit noting the interdependency between the both mentioned use cases.

Finally, we have to mention that the interdependencies between co-existing SON use cases depend on various factors, e.g. the choice of control parameters, SON use cases timescales, targeted performance metrics, input measurements, etc.

Figure 3. Conflict types

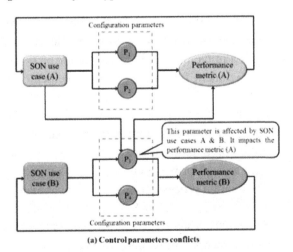

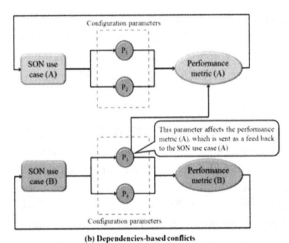

(a) Control parameters conflicts   (b) Dependencies-based conflicts

So, interdependent SON use cases can either be co-designed (thus, the need for coordination is minimized) or independently designed (coordination framework is then essential).

4.2. SOCRATES Coordination Concept

To guarantee a joint and harmonized operation of individual SON use cases towards common goals specified normally via operator's high-level object, a coordination framework has been proposed in Bandh (2011) as a part of the Self-Optimisation and self-ConfiguRATion in wirelEss networkS (SOCRATES) FP7 research project (SOCRATES, 2008). SOCRATES introduces two types of harmonization to control the behavior of various SON use cases towards high-level objectives, namely heading and tailing harmonization. The heading harmonization is comparable to a design-time SON use cases co-design. It is applied to avoid conflicts among the use cases at design time. The co-design can also be seen as conflicts avoidance by appropriate setting of policies. The tailing harmonization is a run-time conflict resolution mechanism. Any configuration change request is evaluated via tailing harmonization against the impacts on the pre-defined common goals. Heading and tailing harmonization interact complementary with each other. In other words, the more harmonization through the policies is performed (heading harmonization), the less tailing harmonization is required. Note that decisions concerning both types of harmonization are made based on policies. Moreover, SOCRATES uses a wide definition of policies different from the common definition of technical policies, see SOCRATES (2008). One can consider that SOCRATES policies are a kind of abstract high-level non-technical policies.

SOCRATES determined an architecture with multiple building blocks for the proposed coordination concept. The building blocks include a policy function, an alignment function, guard function and auto-gnostic function, see Figure 4.

The policy function is used to refine operator policies. The tasks of the alignment function are distributed among two sub-functions, activation and arbitration. The arbitration sub-function is used for conflict resolution and detection. The activation sub-function has two tasks. First, it activates the required SON use case. Second, it reacts on guard function undesired behavior. The guard function assumes that undesired behavior cannot be avoided when a large number of SON functions are employed. The detection of this undesired behavior based on inputs from operator policies is the main task of the guard function. Whenever such behavior is detected, the guard function notifies the alignment function. The auto-gnostic function provides all data the coordination framework and other SON functions require. Concrete, once a specific SON function requires some data, the auto-gnostic function collects, pr-processes and provides the data in the requested format.

The developers of the coordination framework state that it is hard to derive a coordination process that shows which tasks are executed in which functional entity of the SON coordination framework when a configuration change request arises. Concerning the implementation of the proposed architecture, the guard function was not implemented for simplicity reasons. Furthermore, the auto-gnostic is integrated directly into other SON-functions. The same is also applied to the policies required for the alignment function. Therefore, the policy function is not needed and also not implemented. In fact, a small fraction of the alignment function is only implemented in the coordination framework. The developers also state that there are many functions and information replicated among the functional building blocks of the coordination framework. Thus, the same tasks are performed multiple times during the coordination process. Some questions remain to be answered, e.g. how SON-function instances are actually triggered within SOCRATES, how the framework will be implemented, etc.

Figure 4. The architecture of SOCRATES coordination framework

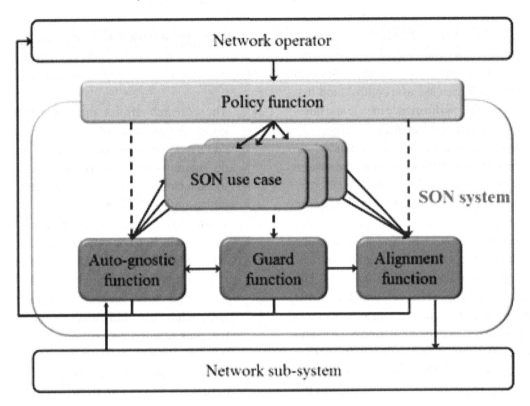

4.3. Decision Tree-Based Coordination Concept

The basic idea of the coordination framework with policy-based decision making process proposed in Romeikat et al. (2010) and Bandh, Romeikat, and Sanneck (2011) is to construct a decision tree for each use case. The coordination function operational knowledge is used to define decision trees for applied SON use cases. Sure, the decision tree of each use case considers all other existing SON use cases that have impacts on it. In this way, the performance of the whole system remains optimized as a whole and the system remains stable. The authors have shown in Bandh, Sanneck, and Romeikat (2011) that the application of their framework on the coverage and capacity optimization use case results in a full coverage, whereas the neglect of coordination among SON use cases contributes to the

appearance of coverage holes. A main drawback of the proposed framework is the complexity of required policies and policies-trees since the relations and interdependencies among various use cases must be expressed in the policies, which will massively complicate them when the number of SON use cases considerably increases. Moreover, there are no ways to capture the interdependencies not known to the system, i.e. emerging ones.

4.4. Control Theory-Based Coordination Framework

An interesting new trend in developing coordination frameworks is presented in Combes, Altman, and Altman (2013). The authors tried to provide a generic mathematical model for the interaction between multiple SON use cases operating in parallel. Based on the generic mathematical model developed, the authors proposed a coordination framework. The mathematical model of the proposed framework has been derived based on control theory (Boyd, El Ghaoui, Feron, & Balakrishnan, 1994) and stochastic approximations (Kushner & Yin, 2003). The basic idea is to model each SON use case as a control loop that measures a specific performance metric and tunes a certain configuration parameter accordingly. Each SON use case is modeled as a stand-alone control loop, which enables modeling the system by means of an ordinary differential equation.

The authors say that the main feature to be hold is the stability of the system, which enables a scalable deployment of SON systems. So, the authors studied the stability by means of the Lyapunov approach used by control theorists. The authors studied many scenarios and found that even for very simple scenarios (one eNB and two SON mechanisms, one for admission control and one for resource allocation), instability occurs with high probability. The coordination mechanism proposed corresponds, in principle, to the concepts of controllability and state-feedback synthesis. The authors interpret the coordination as a transformation of the performance indicator the SON use case i monitors from the performance indicator of the SON use case itself (f_i) to a linear combination of the performance indicators monitored by all applied SON use cases (Combes, 2013). This assumption presents the main drawback of the model since the relation between the measured performance indicators is, in general, not linear. In addition, SON systems adapt their behavior based on measurement reports, which may be corrupted by additive noise. The authors proposed to cope with this case the application of stochastic approximation theorems (Boyd, El Ghaoui, Feron, & Balakrishnan, 1994). These drawbacks, however, do not speak against the great contribution of the proposal since it opens a new thinking way in developing SON coordination frameworks.

4.5. Open Issues

As shown, there exist few works dealing with coordination frameworks. Therefore, a wide range of open issues exist, examples include:

- The research efforts mentioned focused on the design of coordination frameworks, and even from a structural view. There are no assumptions concerning which SON mechanism has to be implemented in which entity. Furthermore, there exist no studies to evaluate where the coordination frameworks must be implemented.
- The majority of coordination frameworks rely on high-level policies defined by networks operators. There are no assumptions and studies describing how to automate the mapping of high-level policies to low-level policies specific to each SON use case. Another open issue in this concern is the definition of the relations between the policies of various SON use cases. The control theory-based coordination framework assumes a linear system and a linear relation between the performance indicators of all SON use cases. As known, existing SON systems are not linear.

5. SON USE CASES

This section presents a main contribution of the chapter, since it provides a comprehensive research-oriented discussion of SON use cases. Concrete, the paragraph investigates each use case in detail, discusses its research challenges and highlights the state of art along with a discussion of the pros and cons of well-known self-organized mechanisms developed to face the challenges of the use case.

5.1. Automatic Neighbor Relation (ANR)

ANR is in charge of relieving the manual management of neighbor relations (3GPP, 2013d), which are captured in a Neighbor Relation Table (NRT). This table contains an entry for each neighbor eNB including the Target Cell Identifier (TCI) and some other attributes. TCI identifies each neighbor cell by means of its E-UTRAN Global cell Identifier (GID) and the PCI of the cell. The standard ANR function is implemented in each eNB and works in a distributed manner. It carries out the following tasks:

- Adds/removes neighbor relations entries.
- Provides management functions to the O&M center to enable updating neighbor relations.
- Notifies the O&M center about changes in the NRT.

The ANR function in intra-RAT operates as follows: each UE receives from the serving eNB a list with neighbor PCIs and their cell individual offsets. Once the UE notes that the signal of a given cell gets stronger than the signal of the current one plus an offset, the UE reports the PCI of the given cell and the associated measurement report. This is done without caring whether the detected PCI is present in the list received from the serving eNB or not. In case the PCI detected is not known to the serving eNB, the corresponding UE is instructed to do measurements to provide the GID of the cell, the Tracking Area Code (TAC) and all available Public Land Mobile Network (PLMN) IDs. The UE accomplishes the required measurements and reports the results to the serving eNB. The serving eNB actualizes in this case its NRT and installs a new X2 connection towards the new detected neighbor eNB. In a multi-technology environment, the definition of non E-UTRAN neighbors is under the responsibility of the network operator. Currently, 3GPP works on extending this item to

enable automating the construction of neighbor relations in such environments, see (Ramiro & Hamied, 2012).

In the world of mobile communication networks, one of the early-developed approaches to construct neighbor relations was designed for GSM, D-AMPS, and PDC (Olofsson, Magnusson, & Almgren, 1996; Magnusson & Olofsson, 1997). The basic idea is to provide users in each cell with a list of neighbor cells. Then, each user measures the cells currently not present in the provided list and reports to the serving cell. A detailed investigation of the performance of this approach is provided in Gustås, Magnusson, Oom, and Storm (2002). In WCDMA systems, a Detected Set Reporting (DSR) is specified, with which mobiles will be capable of measuring cells not present in the current neighbor lists (3GPP, 2009a; 3GPP, 2005). A method to self-optimize NCRs in UTRA FDD network is provided in Soldani and Ore (2007). This method uses the DSR measurements and is not directly applicable to LTE systems since UEs are not capable of providing the globally unique cells identifiers without clear instructions to do this. Another mechanism for constructing NCRs is proposed in Baliosian and Stadler (2007). The proposal relies on a centralized schema, where each base station intersects the set of mobiles residing in its area with other sets of mobiles in the service area of all other base stations. A further approach for NCRs proposes to approximate the service area of the cells and, consequently, computes the overlapping areas, which simplifies extracting neighbor relations (Parodi, Kylvaejae, Alford, Li, & Pradas, 2007). In contrast to all mentioned approaches, the mechanism proposed in Amirijoo et al. (2008) proposes to utilize measurement reports sent from UEs to update the NCRs in each eNB, which conforms to the specification of LTE-Advanced networks. Another mechanism proposed in Feng and Seidel (2008). This mechanism relies on the OAM center to manage NRTs based on measurements done by eNBs and UEs.

5.1.1. Open Issues

A major challenge is the capturing of fast changing neighbor relations due to the dynamic nature of LTE networks. These updates have sometimes short lifetimes. However, they may strongly impact the network and may result in unstable behavior.

5.2. Coverage and Capacity Optimization (CCO)

The CCO use case aims at optimizing the network coverage with guaranteed service continuity and maximized capacity along with reduced interferences as well as delays. Proper solutions for CCO considerably reduce the costs of DTs and enhance the cell edges' performance. The traditional technique used is the adjustment of antenna settings. This can be achieved by applying autonomous closed-loop schemes or open-loop self-optimization techniques if the Remote Electrical Tilt (RET) is available (Ramiro & Hamied, 2012). Assuming the RET is available, the trade-offs are similar to those observed in UMTS networks, which are:

- Decreasing the Electrical Tilt (ET) (antenna up-tilting) results in expanding the cell coverage area. However, problems may be created/aggravated, e.g. overshooting, cell umbrella, etc.
- Increasing the ET (antenna down-tilting) reduces the cell coverage footprint. This is useful to recover from overshooting problems, correct cell umbrella problems, offload traffic and reduce the interference the cell produces to its neighbors.

A centralized planning tool to solve the problem of CCO is proposed in Feng and Seidel (2008). First, measurements are collected from eNBs. These measurements are passed then to the CCO unit, which detects coverage and also capacity problems. After detecting such problems, the radio configuration control unit adjusts radio related parameters, e.g. the downlink transmission power, reference signal power offset and antenna tilt. These changes are applied to the LTE system. Following that, measurements are collected again and so on. Another approach for CCO is proposed in Luketic, Simunic, and Blajic (2011). The developers distinguish between three types of coverage and capacity problems, namely E-UTRAN coverage holes with 2G/3G coverage, E-UTRAN coverage holes with no other radio coverage and E-UTRAN coverage holes with isolated island cell coverage. The third type of problems is, in fact, the type handled in the proposal. Isolated island cells appear when the actual coverage is smaller than the planned coverage due to some reasons. The basic idea is to automatically adjust the antenna tilt based on network traffic and the UE's location. Concrete, an initial antenna tilt value is applied and the signal strength of each user is measured taking into account his location inside the cell. The results are compared with the results of other previous values of tilts. Based on the comparison, the best value of the antenna tilt is selected.

Another CCO solution is proposed in Combes, Altman, and Altman (2010). The solution depends on α-fair schedulers including Proportional Fair (PF), Max ThroughPut (MTP) and Max-Min Fair (MMF) schedulers. The basic idea is to dynamically adjust the packet scheduling strategy using different KPIs to find the optimal α, which results in an optimized coverage and capacity. The approach presented in Naseer ul Islam and Mitschele-Thiel (2012) uses a dynamic antenna down-tilt adaptation to provide enhanced coverage and capacity. Fuzzy Q-Learning strategies are used to solve the problem of CCO. These strategies provide independent optimization process by presenting learning speed and the convergence to optimal setting. The proposal investigates two different Fuzzy Q-Learning strategies, namely stable and dynamic strategies. While stable strategies allow one cell to carry out an action at a time, dynamic strategies enable many cells to achieve many actions simultaneously. Based on the investigation results, the authors proposed a hybrid strategy to benefit from the advantages of both strategies.

5.2.1. Open Issues

As noted, there are considerable efforts done to solve the CCO problem in a self-optimized manner. An important trend in these efforts depends on learning techniques. There are, however, many challenges in this concern such as which learning technique is better for which situation and for which type of coverage and capacity problems? What is the impact of learning process on the performance and how the learning speed can be increased?, etc.

5.3. Energy Saving (ES)

The high cost of power has motivated network operators to switch off capacity of booster cells when extra capacity is not needed, e.g. in shopping centers, companies etc. during closing days. The capacity offered by the network must be as closed as possible to the traffic demands to maintain the required network performance and, simultaneously, cutting down OPEX costs. There is, however, an essential issue, namely the coverage must be provided at all times (Holma & Toskala, 2011). It is not allowed to create coverage holes when running ES approaches, which can be achieved based on many solutions, namely a transmission power adaptation, multi-antenna scheme adaptation and switching off cells.

Note that switching off cells can be simply implemented in a distributed manner since each cell knows when there are no active UEs in its range. Such ES actions does not negatively affect handovers since UEs will not report switched-off cells as possible handover candidates. The major problem is to define the moment, at which the cell has to wake up. This is done either by periodic waking up and sensing the present traffic, or by

depending on the cells that provide the extra coverage on behalf of the switched-off cells, which are instructed to wake up when necessary. This is, in fact, the most used solution.

Two ES approaches are described in Xu, Sun, Li, Lim, and He (2009). The approaches focus on ES for HeNBs using adaptive transmission methods. As known, HeNBs are assumed to send simultaneously even if they do not serve any UE. UEs are also assumed to move to the macro-cells that contain HeNBs in their coverage area before moving to these HeNBs. The first ES approach controls the messages sent from HeNBs in relation to the UEs they serve. The basic idea of the second approach is to introduce different transmission states for each HeNB based on the need to exchange information and the type of the information. Another two schemes are presented in del Apio et al. (2011), namely a selective cells disconnection and a power reduction scheme. The selective cells disconnection scheme chooses some cells to be switched off, while maintaining a coverage gaps-free network. This approach reduces the energy consumption, however, negatively affects the available throughput. The power reduction scheme gradually reduces the power of all overlapped eNBs. Based on the simulation results handled in del Apio et al. (2011), the power reduction scheme is considerably better than the selective cells disconnection mechanism.

5.3.1. Open Issues

Note that the method responsible of switching off cells is not specified in the specification. Therefore, it is possible to relay on a gradual power-down. How fast the power should be switched down relates to the number of UEs remaining in the cell. Sure, incoming handovers while switching off may be rejected. This necessitates considering a new reason for handover failure, namely "switch off ongoing." In general, major efforts are necessary in this context.

5.4. Interference Reduction (IR)

As known, the orthogonality of the Orthogonal Frequency Division Multiplexing (OFDM) applied in LTE enables avoiding the intra-cell interference and the near-far problem typical for CDMA-based systems (Holma & Toskala, 2004). However, LTE systems remain sensitive to the Inter-Cell Interference (ICI), which makes the ICI coordination together with RF optimization essential mechanisms to minimize the interference in LTE systems and to improve network performance (Ramiro & Hamied, 2012). ICI reduction can also be achieved by applying ES mechanisms, see the previous section and Feng and Seidel (2008).

The ICI level a UE experiences depends on the amount of power received from neighbor cells and the set of resources assigned to the UE. LTE systems support coordinated dynamic adjustments (Holma & Toskala, 2009) of power as well as frequency resources. The coordination is achieved via signaling exchanged between eNBs over X2 interfaces. This adjustment is accomplished slowly to keep the network operated around the desired operation status and to guarantee system's stability. In addition, each eNB is authorized to carry out autonomous actions based on per Transmission Time Interval (TTI) to enable adapting to fast channel variations depending on packet scheduling and link adaptation policies.

Based on Ramiro and Hamied (2012), ICI coordination mechanisms can be classified into reactive and proactive mechanisms. Reactive schemes monitor the experienced interference and do transmit power adjustments or packet scheduling-related actions to reduce the interference when exceeding certain limits. Proactive schemes work cooperatively with neighbor eNBs since they depend on a distribution of future scheduling plans among these neighbors. The aim is to take coordinated actions beforehand to hold the interference under control.

Various scheduling and frequency allocation mechanisms are applied to minimize ICI, e.g. Soft Frequency Re-use (SFR), Partial Frequency Re-use (PFR), Fractional Frequency Re-use (FFR), etc. see Kim, Ryu, Cho, and Park (2011). In ETSI (2010), another mechanism for ICI reduction is proposed. This mechanism manages the radio resources, while taking overload, resources priority and transmission power into account. The scheme proposed in Wunder, Kasparick, Stolyar, and Viswanathan (2010) minimizes the ICI by applying Soft Fractional Frequency Re-use (SFFR) and adjusting transmission power on a per-beam base. The Virtual Sub-band Algorithm (VSA) uses a new technique to minimize ICI (Dababneh, 2013). VSA scheme is prediction-based, in which all beams are always switched-on. This makes the transmission power and the channels known and also predictable.

5.4.1. Open Issues

The coordination of ICI is not a simple task since the power and the activities of neighbor cells must be monitored and hold under control, so that ICI is kept at a minimum. Although major efforts are accomplished, there is a need for more work in this context.

5.5. Physical Cell-ID (PCI) Automatic Configuration

As known, LTE cells are identified within measurement reports either based on their broadcasted PCIs or GIDs (Diab & Mitschele-Thiel, 2012). PCIs are reference signal sequences comparable to 3G scrambling codes (Bandh, Carle, & Sanneck, 2009). They are used as regional unique identifiers on the physical level and serve as time and frequency references since the structure of the channels of any cell depends on its PCI (3GPP, 2009b). PCIs themselves are given by synchronization channels and need less than 5 ms to be obtained and decoded by UEs. For GIDs, they are globally unique, not reference signals and longer than PCIs (Golaup, Mustapha, & Patanapongpibul, 2009; Kwak, Lee, Kim, Saxena, & Shin, 2008). UEs must read the system information to obtain any GID. This implies a long measurement gab (may reach 160 ms (Lee, Jeong, Saxena, & Shin, 2009)) and consumes the battery power of UEs. Moreover, UEs cannot exchange data with serving cells during measurement gabs without multiple receiver capabilities. Therefore, the use of GIDs is restricted to when it is absolutely necessary. This also means that it is essential to assign PCIs properly, so that no PCIs conflicts arise. This is, however, challenging due to the limited range of PCIs available (only 504 values) (3GPP, 2009b). Thus, the available PCIs should be handled carefully, so that PCIs are re-used without causing conflicts. The addressed challenge gets more crucial when the available space of PCIs is further divided into smaller ranges. This is achieved for multiple reasons, e.g. a part of the PCIs may be reserved for different types of cells (Lee, Jeong, Saxena, & Shin, 2009), newly switched on eNBs (Nokia, 2008), etc. Moreover, network operators may introduce several artificial sub-groups of PCIs for specific reasons. Remember also that several properties of the channels used in any cell depend on its PCI. This further constraints the PCI allocation process, see 3GPP (2008).

The configuration of an eNB with a PCI is defined by the NGMN alliance as a part of the auto-configuration process achieved in the pre-operational phase. An adequate scheme for automatic PCI configuration has to fulfill the following requirements (Ahmed et al., 2010):

- Collision-free PCIs assignment, i.e. there is no eNB assigned a PCI used by a neighbor of it.
- Confusion-free PCIs assignment, this implies that any given eNB must not have two or more neighbor eNBs configured with the same PCI.

- Coping with dynamically changing network topologies
- Quick PCI assignment process.
- Minimized overhead.
- Security and scalability.

In brief, the assignment of conflict-free PCIs is not just a matter of only selecting adequate PCIs during the pre-operational phase. In fact, this task is a continuous process achieved also during the operational phase. The reasons are the PCIs re-configurations resulting from dynamically changing topologies. Keep also in mind that the recover from conflicts may trigger consecutive conflicts, which may lead to network instability. Note also that it is unlikely that PCIs re-configurations could be done "at any time." Such actions must be judged against the impacts they may have on ongoing applications and services (3GPP, 2008a).

Concerning existing schemes for PCIs assignment, there exist two categories of schemes, namely centralized and distributed (Diab & Mitschele-Thiel, 2012; 3GPP, 2009c). Centralized schemes use a central entity with global knowledge to assign conflict-free PCIs to eNBs. On the contrary, distributed schemes do not use a central entity. The assignment of PCIs is done based on local information gathered by the eNBs themselves (Amirijoo et al., 2008). The following provides more details.

Let us now give more insight into centralized approaches. They deploy a central entity with global knowledge. This entity implements all functions related to PCI auto-configuration or provides all information necessary to select a suitable PCI (Liu, Li, Zhang, & Lu, 2010). Example schemes include a standard scheme defined in LTE specification (3GPP, 2009c) and the centralized Graphic Coloring (GC) approach (Bandh, Carle, & Sanneck, 2009). Concerning the standard scheme defined in LTE specification, it uses the Operation, Administration and Maintenance center (OAM) as a central entity. The OAM chooses during the pre-operational phase a specific PCI and signals it to the eNB under configuration. The given eNB configures itself then using this PCI. No implementation details are specified.

The centralized GC scheme maps the PCI assignment problem to the well-known graphic coloring problem. The approach uses the OAM as a central entity. It considers eNBs as vertices and X2 interfaces as edges between them. So as to guarantee conflicts freeness, a graph of 2-hops neighbors is considered. After coloring the graph, the color of each vertex is translated to a PCI to be used for the configuration of radio parameters. To economize PCIs, the centralized GC scheme assigns the eNB under configuration an adequate PCI selected from the PCIs assigned to its 3-hops neighbors.

Let us discuss now distributed schemes. The PCI assignment process is achieved based on local knowledge, i.e. there is no central entity. These schemes gain increasing interest since improved scalability, better coping with dynamic changing topologies, etc. are expected. Examples include the distributed scheme defined in the LTE specification (3GPP, 2009c), distributed GC approaches (Ahmed et al., 2010), the NS solution (Nokia, 2008), and the Stable GC (SGC) scheme (Diab & Mitschele-Thiel, 2013). The standard distributed scheme proposed for LTE uses the OAM as well. However, the tasks of the OAM are restricted to the transmission of a PCIs list, from which the eNB under configuration selects a PCI different from those reported by neighbor eNBs over X2 interfaces, obtained from UEs measurement reports or acquired through other implementation-dependent methods. Again, no implementation details are discussed.

Distributed GC approaches follow similar principles as centralized GC schemes. They differ from them in the scope of knowledge available since they utilize local knowledge including 1-hop and 2-hops neighbors. Two types of distributed GC algorithms, namely those using real-valued interference pricing (Neel & Reed, 2006; Babadi & Tarokh, 2008) and binary interference pricing of conflicts (Zhang, Wang, Xing, & Wittenburg,

2005), are investigated in Ahmed et al. (2010). The results of the study showed that in terms of the attempts to find a global optimum, the binary interference pricing algorithms show better performance than those depending on a real-valued pricing, which converge quickly to a local optimum.

The NS approach reserves a range of PCIs, termed temporary PCIs, for newly switched on eNBs. Once a new eNB is switched on, it randomly selects a temporal PCI and operates with it for a limited duration (the duration of the configuration phase). During this phase, the newly switched on eNB tries to detect the eNBs locating in its vicinity. Thereafter, the eNB randomly selects a PCI used neither by its 1-hop nor by its 2-hops neighbors. After that, the new eNB boots itself up and notifies its neighbors of the new PCI.

The SGC scheme aims at improving the performance, minimizing the impact of NCRs inaccuracy and maximizing the PCIs usage rate. For these purposes, the SGC scheme combines between the basic ideas of GC approaches and the NS solution. Concrete, the SGC approach divides the range of PCIs into two sub-ranges, namely a temporal and a stable sub-range. The handling of these sub-ranges is similar to what the NS solution does. The assignment of stable PCIs is done based on the graph coloring principles. The economizing of the PCIs is done by following the principles followed by the enhanced GC approach.

5.5.1. Open Issues

From the above induced discussion, one notes that there exist many challenges to face. The most important one is to guarantee the stability of the system under all circumstances. Another important challenge is the minimization of the impact of network dynamics on the performance of PCI assignment approaches. This necessitates the consideration of this dynamic when designing PCIs assignment approaches. A careful re-use of PCIs is also a main challenge since the more the PCIs re-use ratio, the denser the network that can be served.

5.6. Mobility Robustness Optimization (MRO)

The main goal of MRO use case is to minimize handover failures, which in turn improves the network resource utilization. The configuration parameters to be optimized can be the handover hysteresis, Time-to-Trigger (TTT), cell-specific offset, frequency-specific offset and/or cell re-selection parameters (3GPP, 2010a). These parameters affect the following events (Ramiro & Hamied, 2012) (For more details, readers are directed to 3GPP (2010b) and 3GPP (2010c)):

- **A3:** A neighbor cell becomes better than the serving one plus an offset.
- **A4:** A neighbor cell becomes better than a given threshold.
- **A5:** The serving cell becomes worse than a certain threshold and a neighbor cell becomes better than another threshold.
- **B1:** A neighbor cell of different RAT becomes better than a threshold.
- **B2:** The serving cell becomes worse than a threshold and a neighbor cell of different RAT becomes better than another threshold.

Handover-related failures are grouped in three groups, namely too late handovers, too early handovers and handovers to a wrong cell. To optimize the network performance, the reasons for the Radio Link Failure (RLF) must be analyzed and, consequently, the MRO solution executed. If the optimization process must be achieved collaboratively with neighbor cells, the solutions of MLB use case might be applied here, see the next section. Sure, these solutions must consider a handoff optimization as a special case for handling, see Ramiro and Hamied (2012).

The MRO scheme proposed in Kitagawa, Komine, Yamamoto, and Konishi (2011) observes the handoff failures occurred and derives the root cause of them before doing any changes in handoff parameters. A main property of this scheme is that it adjusts the parameters without adding UE mobility estimation functions. The studies achieved in Kitagawa, Komine, Yamamoto, and Konishi, (2011) showed that the algorithm is robust against changes in UEs mobility. The User Speed Mobility Robust Optimization (US-MRO) algorithm (Wei, 2010) optimizes the mobility of UEs by considering the relationship between the UE speed and hysteresis parameters. In fact, the proposed scheme assigns different hysteresis values to different UEs' speeds. Note that there exist three levels for speeds in LTE networks, namely normal, medium and high. So, a hysteresis value is assigned to each speed level and reported to UEs which choose the most appropriate handover parameters based on their speeds. A policy-driven inter-RAT MRO scheme is proposed in Awada, Wegmann, Rose, Viering, and Klein (2011). This scheme uses handoff technology-depended thresholds, TTT and filter coefficient as inter-RAT mobility configuration parameters. As known, the cell edge problem does not exist with inter-RAT mobility since source and target cells operate at different frequencies. In addition, the area where UEs sense good signal strength for both source and target cells is large. The basic idea is to tune based on policies and the type of handoff error occurred mobility configuration parameters, so that the performance is optimized. The studies achieved in Awada, Wegmann, Rose, Viering, and Klein (2011) have shown that the proposal enhances the performance and improves user experiences.

5.6.1. Open Issues

As one can note, there are a wide range of open issues to study, e.g. how to change the mobility configuration parameters in a cooperative manner?, which configuration parameter is the best for which type of handoff error?, what is the relation between the mobility configuration parameters and the load, traffic characteristics, services as well as the speed of UEs?, etc.

5.7. Mobility Load Balancing (MLB)

The demand for resources in upcoming broadband applications increases quickly which makes resource shortage in cellular networks a common issue. One possible solution to overcome the resource shortage is to balance the load in the network. Based on Ruiz-Avil´es, Luna-Ram´ırez, Toril, and Ruiz (2012) and Wei and Peng (2011), load balancing can be done either by adjusting eNB's physical parameters, e.g. data or pilot transmission power (Kojima & Mizoe, 1984), antenna radiation pattern (Saraydar & Yener, 2001), etc., or by changing parameters in Radio Resource Management (RRM) processes, e.g. cell re-selection (Papaoulakis, Nikitopoulos, & Kyriazakos, 2003), handover (Wille, Pedraza, Toril, Ferrer, & Escobar, 2003), etc. Adjusting the physical parameters is rarely used since it necessitates considerable maintenance actions and may create coverage holes unless the applied changes in the physical parameters are synchronized with neighboring cells (Ruiz-Avil´es, Luna-Ram´ırez, Toril, & Ruiz, 2012). Therefore, the adjustment of RRM parameters is the preferred technique to realize load balancing. In fact, most existing load balancing algorithms relay on altering handover parameters since changing the cell re-selection parameters is effective only during call set-up (Kwan, Arnott, Paterson, Trivisonno, & Kubota, 2010; Mu˜noz, Barco, De la Bandera, Toril, & Luna-Ram´ırez, 2011). Thus, the shift of some UEs residing at the borders of adjacent cells or those locating in overlapping cells from more to less congested cells balances the load and, consequently, improves the performance. Note, however, that the traffic is time-varying and, in principle, unpredictable, which makes static as well as pre-fixed network planning not suitable to adapt to varying load. Therefore, to utilize resources efficiently and remain competitive at reduced cost (from market side), development of self-organized MLB

The MLB algorithm presented in Feng and Seidel (2008) optimizes the cell re-selection and handover parameters to balance the load taking into account that the number of handovers must be also minimized. Each eNB measures the load in its cell and exchanges the load information with adjacent cells. When the cell gets congested, it distributes a part of its load to neighboring cells by tuning re-selection and handover parameters in the cell and its neighbors. A new MLB algorithm with penalized handovers is proposed in Hu, Zhang, Zheng, Yang, and Wu (2010). This algorithm takes into account the average delay of the system and the average number of handovers. The authors state that balancing the load by shifting a set of UEs to a neighbor cell with spare resources will, on one side, result in improved network capacity, enhanced queue backlogs and data rates for UEs. On the other side, this will considerably increase the number of handovers events. As known, handovers are costly and not preferred to occur frequently. Therefore, the proposal assumes that only one UE (not a set of UEs) is allowed to hand over to another cell at the same time. The developed approach assumes that the channel conditions, power allocation and interference level can be known by UEs and eNBs. A penalty (between 0 & 1) is assigned to a pair of a UE and a cell if it is required to assign the UE to the cell. The handover event takes place in relation to (1—penalty), not only to the best data rate and service. So, the problem of selecting a set of handovers is formulated as a Maximum Weighted Matching (MWM) problem, which is further solved by a greedy distributed algorithm. This provides low complexity and system overhead. The optimization question is a trade-off between the number of handovers and the average queue backlog. The larger is the penalty, the less the number of handovers shifted to another cells and the larger the queue backlog.

A capacity-based MLB algorithm is proposed in Wei and Peng (2011). The algorithm aims at shifting the maximum number of UEs from overloaded cells to neighboring ones with a minimum number of rejections. The proposal handles the case where two congested cells transfer UEs to a cell at the same time. For this purpose, the load balancing algorithm considers the load in neighboring cells in addition to the current cell. This reduces the collisions produced by the MLB scheme when many neighboring cells require load balancing to a cell at the same time. The simulation results presented in (Wei & Peng, 2011) showed that the proposal results in an improved throughput, enhanced user experience and eliminated ping-pong handoffs. Another capacity-based MLB algorithm is proposed in Lv, Li, Zhang and Liu (2010). This algorithm also considers the load status in the current and neighboring cells, while dynamically adjusting the handoff margin. Different from the algorithm proposed in Wei and Peng, (2011), which deals with the throughput, this proposal handles the Signal to Interference and Noise Ratio (SINR). The results of the simulations induced in Lv, Li, Zhang, and Liu (2010) show that the adjustment of the handover margin improves the user satisfaction and minimizes the number of unsatisfied users.

Examples of MLB mechanisms that apply heuristic rules to solve the load balancing optimization problem are proposed in Lobinger, Stefanski, Jansen, and Balan (2010) and Mu̅noz, Barco, De la Bandera, Toril, and Luna-Ram´ırez (2011). They use a diffusive load sharing algorithm and differ from each other in the performance indicator to be balanced. While the proposal presented in Lobinger, Stefanski, Jansen, and Balan (2010) considers the cell average load as the performance indicator of interest, the proposal described in Mu̅noz, Barco, De la Bandera, Toril, and Luna-Ram´ırez (2011) considers the call blocking rate. The latter has a better performance according to the studies achieved in Toril and Wille (2008) since higher system's stability results without need for any hardware upgrade in legacy equipment.

5.7.1. Open Issues

From the discussion induced above one notes that although there are considerable efforts done to optimize the load balancing routines in a self-organized manner, these routines are far from being optimized. The major problem lies in the complexity of LTE-advanced networks and the high dynamic grad they have. Questions like how many UEs should be shifted to which cells?, which UEs can be shifted?, How the load balancing actions relate to the services being running?, which performance indicators should be focused on in which network situation?, etc. still need research efforts to be answered.

5.8. Random Access Channel (RACH) Optimization

The random access procedure is performed in various situations, e.g. initial access of a UE in idle mode, connection re-establishment after a radio link failure, etc. The main goal of the procedure is to establish an uplink time synchronization (Amirijoo, Frenger, Gunnarsson, Moe, & Zetterberg, 2013). Once a UE wants to establish a connection with an eNB, it scans the carrier frequencies aiming at determining the best suitable cell to associate with. After this cell is defined, the UE reads the cell's broadcast information to obtain cell-specific random access procedure details (Dahlman, Parkvall, Skoeld, & Beming, 2007). Optimized random access performance is crucial to obtain the intended coverage with low-delay communication, while preventing excessive interference to other UEs and simultaneously balancing the radio resource allocation between random access and data services, see Amirijoo, Frenger, Gunnarsson, Moe, and Zetterberg (2013). The random access procedure is also crucial for call setup delays, sessions resuming delays, etc. In principle, there are two challenges to face. First, the uplink interference that may result from random access procedure must be minimized. Second, an optimized balancing of the radio resource allocation between random access and data services, i.e. the number of allocated access opportunities, must be guaranteed. The parameters to be optimized are: RACH configuration parameters, RACH preambles parameters and RACH transmission power control parameters (Kottkamp, Roessler, Schlienz, & Schuetz, 2011).

A solution applied is to use a set of standard configuration parameters for the random access procedure in all eNBs. The set of parameters are selected based on extensive simulations, experiments, etc. One clearly notes that such solution will result in a sub-optimal performance since the radio conditions are not the same at all eNBs and they are time-dependent. Another approach is to choose cell-specific set of parameters that satisfy specific requirements. This way of thinking will improve the performance since each cell has its group of parameters sets for selected network conditions. However, there are no possibilities to be dynamically adaptable to all changes in the radio environment of the cell. Amirijoo, Frenger, Gunnarsson, Moe, and Zetterberg (2013) provided another approach to self-optimize the performance of random access procedure. The approach depends on tuning RACH power control parameters, so that the miss probability (the probability to miss a response for a preamble) of transmitted preambles relate to given requirements. So as to enable the RACH optimization process to autonomously work, UEs must report the number of transmitted preambles to eNBs. This enables the derivation of miss probability. Note that this parameter is further used to control the network access delay. This approach is, for sure, better than the approaches discussed above since it improves the performance of random access procedure and reduces the uplink interference. However, the collisions and contentions are not directly affected. Moreover, no statements have been done concerning the ideal number of slots to be reserved for the RACH.

5.8.1. Open Issues

From the above discussion one notes that there are many open issues to be worked on. The major challenge is to provide a solution that controls and minimizes the preamble transmission power, minimizes the collisions and contentions and defines the ideal number of required access slots. All these must be dynamically controlled based on current network conditions.

6. CONCLUSION

The chapter has provided a comprehensive study of SON principles and activities. One of the main contributions of the chapter is the thorough study of SON use cases going through the research challenges and the state of art. The coordination between SON activities has been studied as well. Another major contribution is the highlight of the main SON activities done in the scope of 3GPP going through the activities in releases 9, 10, 11 and even beyond. In brief, there are major efforts done to construct SON solutions, however, they are far from being optimal. Considerable work is necessary, especially for the coordination among SON use cases.

REFERENCES

3.^{d} Generation Project Partnership (3GPP). (1998). *The 3rd generation partnership project agreement.* Retrieved June 07, 2013, from http://www.3gpp. org/ftp/Inbox/2008_Web_files/3gppagre.pdf

3rd Generation Project Partnership (3GPP). (2005). *Requirements for support of radio resource management (FDD).* TS 25.133—V3.22.0. Retrieved June 07, 2013, from http://www.arib.or.jp/english/ html/overview/doc/STD-T63v9_50/5_Appendix/ R99/25/25133-3m0.pdf

3rd Generation Project Partnership (3GPP). (2008a). *Evolved universal terrestrial radio access (E-UTRA), LTE physical layer—General description (release 8).* TS 36.211—V8.3.0. Retrieved June 07, 2013, from http://www.etsi.org/deliver/ etsi_ts/136200_136299/136211/08.07.00_60/ ts_136211v080700p.pdf

3rd Generation Project Partnership (3GPP). (2008b). *Automatic physical cell ID assignment.* 3GPP. TSG-SA5 (Telecom Management). S5-081185.

3rd Generation Project Partnership (3GPP). (2009a). *Automatic neighbour relation (ANR) management, concepts and requirements.* 3GPP. TS 32.511—Version 9.0.0, release 9. Retrieved June 07, 2013, from http://www.3gpp.org/ftp/Specs/archive/32_series/32.511/32511-900.zip

3rd Generation Project Partnership (3GPP). (2009b). Technical specification group radio access network, requirements for evolved UTRA (E-UTRA) and evolved UTRAN (E-UTRAN). TR 25.913, Release 7.

3rd Generation Project Partnership (3GPP). (2009c). *Evolved universal terrestrial radio access, physical channels and modulation (release 8).* TS 36.211—V8.6.0. Retrieved June 07, 2013, from http://www.3gpp.org/ftp/specs/archive/36_series/36.211/36211-860.zip

3rd Generation Project Partnership (3GPP). (2009d). *Self-organizing networks (SON), concepts and requirements.* 3GPP. TS 32.500—Version 9.0.0, Release 9. Retrieved June 03, 2013, from http://www.3gpp.org/ftp/Specs/archive/32_series/32.500/32500-900.zip

3rd Generation Project Partnership (3GPP). (2009e). *Study on self-healing.* 3GPP. TR 32.823—Version 9.0.0, Release 9. Retrieved June 07, 2013, from http://www.3gpp.org/ftp/Specs/archive/32_series/32.823/32823-900.zip

3rd Generation Project Partnership (3GPP). (2009f). *SID on study on solutions for energy saving within UTRA NodeB.* RP-091439, Release 9. Retrieved June 03, 2013, from http://www.3gpp.org/ftp/tsg_ran/TSG_RAN/TSGR_46/Docs/RP-091439.zip

3rd Generation Project Partnership (3GPP). (2010a). *Self-organizing networks (SON) policy network resource model (NRM) integration reference point (IRP), requirements.* 3GPP. TS 32.521—Version 9.0.0, Release 9. Retrieved June 07, 2013, from http://www.3gpp.org/ftp/Specs/archive/32_series/32.521/32521-900.zip

3rd Generation Project Partnership (3GPP). (2010b). *Self-configuring and self-optimizing network (SON) use cases and solutions.* 3GPP. TS 36.902—Version 9.2.0, Release 9, 15. Retrieved June 07, 2013, from http://www.3gpp.org/ftp/Specs/archive/36_series/36.902/36902-920.zip

3rd Generation Project Partnership (3GPP). (2010c). *Self-organizing networks (SON) policy network resource model (NRM) integration reference point (IRP), information service (IS).* 3GPP. TS 32.522—Version 9.1.0, Release 9, 8. Retrieved June 07, 2013, from http://www.3gpp.org/ftp/Specs/archive/32_series/32.522/32522-910.zip

3rd Generation Project Partnership (3GPP). (2010d). *Radio resource control (RRC), protocol specification.* 3GPP. TS 36.331—Version 9.3.0, Release 9. Retrieved June 07, 2013, from http://www.3gpp.org/ftp/Specs/archive/36_series/36.331/36331-930.zip

3rd Generation Project Partnership (3GPP). (2010e). *Overview of 3GPP.* Release 8. Version 0.1.0. Retrieved June 07, 2013, from http://www.3gpp.org/ftp/Information/WORK_PLAN/Description_Releases/Previous_versions/Rel-08_description_20100421.zip

3rd Generation Project Partnership (3GPP). (2010f). *Overview of 3GPP.* Release 9. Version 0.1.0. Retrieved June 07, 2013, from http://www.3gpp.org/ftp/Information/WORK_PLAN/Description_Releases/Previous_versions/Rel-09_description_20100621.zip

3rd Generation Project Partnership (3GPP). (2010g). *Self-configuration of network elements, concepts and requirements*. 3GPP. TS 32.501—Version 9.1.0, Release 9. Retrieved June 07, 2013, from http://www.3gpp.org/ftp/Specs/archive/32_series/32.501/32501-910.zip

3rd Generation Project Partnership (3GPP). (2010h). *Self-configuration of network elements integration reference point (IRP), information service (IS)*. 3GPP. TS 32.502—Version 9.2.0, Release 9. Retrieved June 07, 2013, from http://www.3gpp.org/ftp/Specs/archive/32_series/32.502/32502-920.zip

3rd Generation Project Partnership (3GPP). (2010i). Self-configuration of network elements integration reference point (IRP), common object request broker architecture (CORBA) solution set (SS). 3GPP. TS 32.503—Version 9.1.0, Release 9. Retrieved June 07, 2013, from http://www.3gpp.org/ftp/Specs/archive/32_series/32.503/32503-910.zip

3rd Generation Project Partnership (3GPP). (2010j). Evolved universal terrestrial radio access network (E-UTRAN) network resource model (NRM) integration reference point (IRP), information service (IS). 3GPP. TS 32.762—Version 9.4.0, Release 9. Retrieved June 07, 2013, from http://www.3gpp.org/ftp/Specs/archive/32_series/32.762/32762-940.zip

3rd Generation Project Partnership (3GPP). (2010k). *S1 application protocol (S1AP)*. 3GPP. TS 36.413—Version 9.3.0, Release 9. Retrieved June 07, 2013, from http://www.3gpp.org/ftp/Specs/archive/36_series/36.413/36413-930.zip

3rd Generation Project Partnership (3GPP). (2010l). Self-organizing networks (SON), policy network resource model (NRM) integration reference point (IRP), common object request broker architecture (CORBA) solution set (SS). 3GPP. TS 32.523—Version 9.0.0, Release 9. Retrieved June 07, 2013, from http://www.3gpp.org/ftp/Specs/archive/32_series/32.523/32523-900.zip

3rd Generation Project Partnership (3GPP). (2010m). *Study on energy savings management (ESM)*. 3GPP. TR 32.826—V 10.0.0. Retrieved June 03, 2013, from http://www.quintillion.co.jp/3GPP/Specs/32826-a00.pdf

3rd Generation Project Partnership (3GPP). (2010n). *Overview of 3GPP*. Release 10. Version 0.0.7. Retrieved June 03, 2013, from http://www.3gpp.org/ftp/Information/WORK_PLAN/Description_Releases/Previous_versions/Rel-10_description_20100621.zip

3rd Generation Project Partnership (3GPP). (2010o). *Self-healing concepts and requirements*. 3GPP. TS 32.541—Version 1.4.0, Release 10. Retrieved June 03, 2013, from http://www.3gpp.org/ftp/Specs/archive/32_series/32.541/32541-140.zip

3rd Generation Project Partnership (3GPP). (2010p). *Integration of device management information with Itf-N*. 3GPP. TR 32.827—Version 10.1.0, Release 10. Retrieved June 03, 2013, from http://www.3gpp.org/ftp/Specs/archive/32_series/32.827/32827-a10.zip

3rd Generation Project Partnership (3GPP). (2010q). *X2 application protocol (X2AP)*. 3GPP. TS 36.423—Version 9.3.0, Release 9. Retrieved June 07, 2013, from http://www.3gpp.org/ftp/Specs/archive/36_series/36.423/36423-930.zip

3rd Generation Project Partnership (3GPP). (2011a). *About 3GPP*. Retrieved June 07, 2013, from http://www.3gpp.org/About-3GPP

3rd Generation Project Partnership (3GPP). (2011b). *Energy saving management (ESM), concepts and requirements*. 3GPP. TS 32.551—V 10.1.0. Retrieved June 03, 2013, from http://www.3gpp.org/ftp/Specs/archive/32_series/32.551/

3rd Generation Project Partnership (3GPP). (2011c). Evolved universal terrestrial radio access network (E-UTRAN), self-configuring and self-optimizing network (SON) use cases and solutions. TR 36.902. Retrieved June 03, 2013, from http://www.3gpp.org/ftp/Specs/archive/36_series/36.902/

3rd Generation Project Partnership (3GPP). (2012a). *3GPP work items on self-organizing networks*—Version 0.0.9. Retrieved June 02, 2013, from http://www.3gpp.org/ftp/Information/WORK_PLAN/Description_Releases/SON_20120924.zip

3rd Generation Project Partnership (3GPP). (2012b). *Study on operations, administration and maintenance (OAM) aspects of inter-radio-access-technology (RAT) energy saving.* TR 32.834. Retrieved June 03, 2013, from http://www.3gpp.org/ftp/Specs/archive/32_series/32.834/

3rd Generation Project Partnership (3GPP). (2013a). Universal terrestrial radio access (UTRA) and evolved universal terrestrial radio access (E-UTRA), radio measurement collection for minimization of drive tests (MDT), overall description, stage 2. TS 37.320. Retrieved June 03, 2013, from http://www.3gpp.org/ftp/Specs/archive/37_series/37.320/

3rd Generation Project Partnership (3GPP). (2013b). *Release 11.* Retrieved June 03, 2013, from http://www.3gpp.org/ftp/Information/WORK_PLAN/Description_Releases/

3rd Generation Project Partnership (3GPP). (2013c). *Release 12.* Retrieved June 03, 2013, from http://www.3gpp.org/ftp/Information/WORK_PLAN/Description_Releases/

3rd Generation Project Partnership (3GPP). (2013d). Evolved universal terrestrial radio access (E-UTRA) and evolved universal terrestrial radio access network (E-UTRAN), overall description, stage 2 (release 11). Technical Specification. TS 36.300—V11.6.0. Retrieved June 07, 2013, from http://www.3gpp.org/ftp/Specs/archive/36_series/36.300/36300-b60.zip

3rd Generation Project Partnership (3GPP) Official Website. (2013). Retrieved June 07, 2013, from http://www.3gpp.org/

Ahmed, F., Tirkkonen, O., Peltomäki, M., Koljonen, J. M., Yu, C. H., & Alava, M. (2010). Distributed graph coloring for self-organization in LTE networks. *Journal of Electrical and Computer Engineering.* doi:10.1155/2010/402831

Amirijoo, M., Frenger, P., Gunnarsson, F., Kallin, H., Moe, J., & Zetterberg, K. (2008). Neighbor cell relation list and physical cell identity self-organization in LTE. In *Proceeding of IEEE International Conference on Communications Workshops (ICC'08).* IEEE.

Amirijoo, M., Frenger, P., Gunnarsson, F., Moe, J., & Zetterberg, K. (2013). On self-optimization of the random access procedure in 3G long term evolution. In *Proceedings of IFIP/IEEE International Symposium on Integrated Network Management (IM)-Workshops.* New York: IEEE.

Awada, A., Wegmann, B., Rose, D., Viering, I., & Klein, A. (2011). Towards self-organizing mobility robustness optimization in inter-RAT scenario. In *Proceeding of the 73rd IEEE Vehicular Technology Conference.* IEEE.

Babadi, B., & Tarokh, V. (2008). A distributed asynchronous algorithm for spectrum sharing in wireless ad hoc networks. In *Proceedings of the 42nd Annual Conference on Information Sciences and Systems (CISS '08).* CISS.

Baliosian, J., & Stadler, R. (2007). Decentralized configuration of neighboring cells for radio access networks. In *Proceeding of the 1st IEEE Workshop on Autonomic Wireless Access (in Conjunction with IEEE WoWMoM)*. Helsinki, Finland: IEEE.

Bandh, T. (2011). *The SOCRATES SON-function coordination concept* (Technical report). Retrieved June 07, 2013, from http://www.net.in.tum.de/fileadmin/bibtex/publications/papers/socrates.pdf

Bandh, T., Carle, G., & Sanneck, H. (2009). Graph coloring based physical-cell-ID assignment for LTE networks. In *Proceeding of the ACM International Wireless Communications and Mobile Computing Conference. (IWCMC '09)*. Leipzig, Germany: ACM.

Bandh, T., Romeikat, R., & Sanneck, H. (2011). Policy-based coordination and management of SON functions. In *Proceeding of IM 2011*. IM.

Bandh, T., Sanneck, H., & Romeikat, R. (2011). An experimental system for SON function co-ordination. In *Proceeding of the 73rd Vehicular Technology Conference (VTC Spring)*. IEEE.

Boyd, S., El Ghaoui, L., Feron, E., & Balakrishnan, V. (1994). *Linear matrix inequalities in system and control theory*. Philadelphia, PA: SIAM. doi:10.1137/1.9781611970777

Combes, R. (2013). *Mécanismes auto-organisants dans les réseaux sans fil*. (Dissertation thesis). Universite Pierre et Marie Curi, Paris, France.

Combes, R., Altman, Z., & Altman, E. (2010). On the use of packet scheduling in self-optimization processes: Application to coverage-capacity optimization. In *Proceeding of the 8th International Symposium on Modeling and Optimization in Mobile, Ad Hoc and Wireless Networks (WiOpt`10)*. IEEE.

Combes, R., Altman, Z., & Altman, E. (2013). Coordination of autonomic functionalities in communications networks. In *Proceeding of the 11th International Symposium on Modeling and Optimization in Mobile, Ad Hoc, and Wireless Networks (WiOpt'13)*. IEEE.

Dababneh, D. (2013). *LTE network planning and traffic generation*. (Master Thesis). Ottawa-Carleton Institute for Electrical and Computer Engineering (OCIECE), Carleton University, Ottawa, Canada.

Dahlman, E., Parkvall, S., Skoeld, J., & Beming. (2007). *3G evolution—HSPA and LTE for mobile broadband* (2nd ed.). Academic Press.

Dahlman, E., Parkvall, S., & Skoeld, J. (2011). *4G—LTE/LTE-advanced for mobile broadband*. Academic Press.

del Apio, M., Mino, E., Cucala, L., Moreno, O., Berberana, I., & Torrecilla, E. (2011). Energy efficiency and performance in mobile networks deployments with femtocells. In *Proceeding of the 22nd IEEE International Symposium on Personal Indoor and Mobile Radio Communications (PIMRC'11)*. Toronto, Canada: IEEE.

Diab, A., & Mitschele-Thiel, A. (2012). Comparative evaluation of distributed physical cell identity assignment schemes for LTE-advanced systems. In *Proceeding of the 7th Performance Monitoring, Measurement and Evaluation of Heterogeneous Wireless and Wired Networks Workshop*. Paphos, Cyprus: IEEE.

Diab, A., & Mitschele-Thiel, A. (2013). Development of distributed and self-organized physical cell identity assignment schemes for LTE-advanced systems. In *Proceedings of the 16th ACM International Conference on Modeling, Analysis and Simulation of Wireless and Mobile Systems*. Barcelona, Spain: ACM.

ETSI Technical Report. (2010). *LTE, evolved universal terrestrial radio access network (E-UTRAN), self-configuring and self-optimizing network use cases and solutions.* ETSI TR 136 902—V9.2.0. Retrieved June 07, 2013, from http://www.etsi.org/deliver/etsi_tr/136900_136999/136902/09.02.00_60/tr_136902v090200p.pdf

FemtoForum. (2011). *HeNB (LTE-Femto) network architecture.* Retrieved June 07, 2013, from http://www.smallcellforum.org/

Feng, S., & Seidel, E. (2008). *Self-organizing networks (SON) in 3GPP long term evolution: Novel mobile radio (NOMOR) research center.* Retrieved June 07, 2013, from http://www.nomor.de/uploads/gc/TQ/gcTQfDWApo9osPfQwQoBzw/SelfOrganisingNetworksInLTE_2008-05.pdf

Garza, C., Ashai, B. H., Monturus, E., & Syputa, R. (2010). [*LTE operator commitments: Deployment scenarios and growth opportunities.* MARAVEDIS Wireless Market Research & Analysis.]. *Top (Madrid)*, 25.

Golaup, A., Mustapha, M., & Patanapongpibul, L. (2009). Femtocell access control strategy in UMTS and LTE. *IEEE Communications Magazine, 47*(9). doi:10.1109/MCOM.2009.5277464

Gustås, P., Magnusson, P., Oom, J., & Storm, N. (2002). Real-time performance monitoring and optimization of cellular systems. *Ericsson Review, 1.*

Hamalainen, S., Sanneck, H., & Sartori, C. (Eds.). (2012). *LTE self-organising networks (SON), network management automation for operational efficiency.* Hoboken, NJ: John Wiley & Sons, Ltd.

Holma, H., & Toskala, A. (Eds.). (2004). *WCDMA for UMTS: Radio access for third generation mobile communications.* Hoboken, NJ: John Wiley & Sons, Ltd.

Holma, H., & Toskala, A. (Eds.). (2009). *LTE for UMTS: OFDMA and SC-FDMA based radio access.* Hoboken, NJ: John Wiley & Sons, Ltd. doi:10.1002/9780470745489

Holma, H., & Toskala, A. (Eds.). (2011). *LTE for UMTS evolution to LTE-advanced* (2nd ed.). Hoboken, NJ: John Wiley & Sons Inc. doi:10.1002/9781119992943

Hu, H., Zhang, J., Zheng, X., Yang, Y., & Wu, P. (2010). Self-configuration and self-optimization for LTE networks. *IEEE Communications Magazine, 48*(2). doi:10.1109/MCOM.2010.5402670

International Telecommunication Union (ITU) Official Website. (n.d.). Retrieved June 07, 2013, from http://www.itu.int/en/Pages/default.aspx

Jansen, T., Amirijoo, M., Tuerke, U., Jorguseski, L., Zetterberg, K., & Nascimento, R. ... Balan, I. (2009). Embedding multiple self-organisation functionalities in future radio access networks. In *Proceedings of VTC Spring 2009.* IEEE.

Kim, J., Ryu, B., Cho, K., & Park, N. (2011). Interference control technology for heterogeneous networks. In *Proceeding of the 5th International Conference on Mobile Ubiquitous Computing, Systems, Services and Technologies.* Lisbon, Portugal: IEEE.

Kitagawa, K., Komine, T., Yamamoto, T., & Konishi, S. (2011). A handover optimization algorithm with mobility robustness for LTE systems. In *Proceeding of the 22nd IEEE International Symposium on Personal Indoor and Mobile Radio Communications (PIMRC`11).* IEEE.

Kojima, J., & Mizoe, K. (1984). *Radio Mobile communication system wherein probability of loss of calls is reduced without a surplus of base station equipment.* U.S Patent 4435840 (1984). Washington, DC: US Patent Office.

Kottkamp, M., Roessler, A., Schlienz, J., & Schuetz, J. (2011). *LTE release 9—Technology introduction*. White paper.

Kushner, H. J., & Yin, G. G. (2003). *Stochastic approximation and recursive algorithms and applications* (2nd ed.). Springer Stochastic Modeling and Applied Probability.

Kwak, H., Lee, P., Kim, Y., Saxena, N., & Shin, J. (2008). Mobility management survey for home-eNB based 3GPP LTE systems. *Journal of Information Processing Systems*, 4(4). doi:10.3745/JIPS.2008.4.4.145

Kwan, R., Arnott, R., Paterson, R., Trivisonno, R., & Kubota, M. (2010). On mobility load balancing for LTE systems. In = *Proceeding of the 72nd IEEE Vehicular Technology Conference (VTC)*. Ottawa, Canada: IEEE.

Lee, P., Jeong, J., Saxena, N., & Shin, J. (2009). Dynamic reservation scheme of physical cell identity for 3GPP LTE femtocell systems. *Journal of Information Processing Systems*, 5(4). doi:10.3745/JIPS.2009.5.4.207

Lehser, F. (Ed.). (2008). *Next generation mobile networks—Recommendation on SON and O&M requirements (Technical Report)*. NGMN Alliance.

Liu, Y., Li, W., Zhang, H., & Lu, W. (2010). Graph based automatic centralized PCI assignment in LTE. In *Proceeding of the IEEE Symposium on Computers and Communications (ISCC'10)*. Riccione, Italy: IEEE.

Lobinger, A., Stefanski, S., Jansen, T., & Balan, I. (2010). Load balancing in downlink LTE self-optimizing networks. In *Proceeding of 71st IEEE Vehicular Technology Conference (VTC'10)*. Taipei, Taiwan: IEEE.

Luketic, I., Simunic, D., & Blajic, T. (2011). Optimization of coverage and capacity of self-organizing network in LTE. In *Proceeding of the 34th International Convention MIPRO*. MIPRO.

Lv, W., Li, W., Zhang, H., & Liu, Y. (2010). Distributed mobility load balancing with RRM in LTE. In *Proceeding of the 3rd IEEE International Conference on Broadband Network and Multimedia Technology*. IEEE.

Magnusson, S., & Olofsson, H. (1997). Dynamic neighbor cell list planning in a micro cellular network. In *Proceeding of the IEEE International Conference on Universal Personal Communications*. San Diego, CA: IEEE.

Motorola. (2009). *LTE operations and maintenance strategy—Using self-organizing networks to reduce OPEX*. White paper.

Mu~noz, P., Barco, R., De la Bandera, I., Toril, M., & Luna-Ram´ırez, S. (2011). Optimization of a fuzzy logic controller for handover-based load balancing. In *Proceeding of 73rd IEEE Vehicular Technology Conference (VTC)*. Budapest: IEEE.

Naseer ul Islam, M., & Mitschele-Thiel, A. (2012). Reinforcement learning strategies for self-organized coverage and capacity optimization. In *Proceeding of the IEEE Conference on Wireless Communications and Networking*. IEEE.

Neel, J. O., & Reed, J. H. (2006). Performance of distributed dynamic frequency selection schemes for interference reducing networks. In *Proceedings of the IEEE Military Communications Conference (MILCOM '06)*. IEEE.

Next Generation Mobile Networks (NGMN) Alliance. (2006). *Next generation mobile networks beyond HSPA & EVDO*. White Paper, Version 3.0. Retrieved June 07, 2013, from http://www.ngmn.org/uploads/media/Next_Generation_Mobile_Networks_Beyond_HSPA_EVDO_Web.pdf

Next Generation Mobile Networks (NGMN) Alliance. (2008a). *NGMN recommendation on SON and O&M requirements*. Requirement Specification, Version 1.23. Retrieved June 07, 2013, from http://www.ngmn.org/uploads/media/NGMN_Recommendation_on_SON_and_O_M_Requirements.pdf

Next Generation Mobile Networks (NGMN) Alliance. (2008b). *NGMN use cases related to self-organizing network, overall description.* Deliverable, Version 2.02. Retrieved June 07, 2013, from http://www.ngmn.org/uploads/media/ NGMN_Use_Cases_related_to_Self_Organising_Network__Overall_Description.pdf

Nokia Siemens Networks and Nokia. (2008). *SON use case: Cell Phy ID automated configuration.* R3-080376. Technical Report.

Olofsson, H., Magnusson, S., & Almgren, M. (1996). A concept for dynamic neighbor cell list planning in a cellular system. In *Proceedings of IEEE Personal, Indoor and Mobile Radio Communications.* Taipei, Taiwan: IEEE.

Papaoulakis, N., Nikitopoulos, D., & Kyriazakos, S. (2003). Practical radio resource management techniques for increased mobile network performance. In *Proceedings of the 12th IST Mobile and Wireless Communications Summit.* IST.

Parodi, F. Kylvaejae, Alford, M. G., Li, J., & Pradas, J. (2007). An automatic procedure for neighbor cell list definition in cellular networks. In *Proceeding of the 1st IEEE Workshop on Autonomic Wireless Access (in Conjunction with IEEE WoWMoM).* Helsinki, Finland: IEEE.

Ramiro, J., & Hamied, K. (2012). *Self-organizing networks (SON) self-planning—Self-optimization and self-healing for GSM, UMTS and LTE.* Hoboken, NJ: John Wiley & Sons.

Romeikat, R., Bauer, B., Bandh, T., Carle, G., Sanneck, H., & Schmelz, L.-C. (2010). Policy-driven workflows for mobile network management automation. In *Proceeding of the ACM (IWCMC'10).* ACM.

Ruiz-Avil'es, J. M., Luna-Ram'ırez, S., Toril, M., & Ruiz, F. (2012). Traffic steering by self-tuning controllers in enterprise LTE femtocells. *EURASIP Journal on Wireless Communications and Networking.* Retrieved June 07, 2013, from http:// jwcn.eurasipjournals.com/content/2012/1/337

Saraydar, C., & Yener, A. (2001). Adaptive cell sectorization for CDMA systems. *IEEE Journal on Selected Areas in Communications, 19*(6), 1041–1051. doi:10.1109/49.926360

Schmelz, L. C., Amirijoo, M., Eisenblaetter, A., Litjensm, R., Neuland, M., & Turk, J. (2011). A coordination framework for self-organisation in LTE networks. In *Proceedings of IEEE International Symposium on Integrated Network Management.* Dublin, Ireland: IEEE.

Self-Optimization and self-ConfiguRATion in wirelEss networkS (SOCRATES) Project. (2008). Retrieved June 07, 2013, from http://www.fp7-socrates.eu, http://www.fp7-socrates.org

Soldani, D., & Ore, I. (2007). Self-optimizing neighbor cell lists for UTRA FDD networks using detected set reporting. In *Proceeding of IEEE Vehicular Technology Conference.* IEEE.

Toril, M., & Wille, V. (2008). Optimization of handover parameters for traffic sharing in GERAN. *Wireless Personal Communications, 47*(3), 315–336. doi:10.1007/s11277-008-9467-4

Wei, Y., & Peng, M. (2011). A mobility load balancing optimization method for hybrid architecture in self organizing network. In *Proceeding of the IET International Conference on Communication Technology and Application.* IET.

Wei, Z. (2010). Mobility robustness optimization based on UE mobility for LTE system. In *Proceeding of the International Conference on Wireless Communications and Signal Processing (WCSP`10)*. WCSP.

Wille, V., Pedraza, S., Toril, M., Ferrer, R., & Escobar, J. (2003). Trial results from adaptive handover boundary modification in GERAN. *Electronics Letters, 39*(4), 405–407. doi:10.1049/el:20030244

Wunder, G., Kasparick, M., Stolyar, A., & Viswanathan, H. (2010). Self-organizing distributed inter-cell beam coordination in cellular networks with best effort traffic. In *Proceeding of the 8th International Symposium on Modeling and Optimization in Mobile, Ad Hoc and Wireless Networks (WiOpt'10)*. Avignon, France: WiOpt.

Xu, L., Sun, C., Li, X., Lim, C., & He, H. (2009). The methods to implement self optimization in LTE system. In *Proceeding of the International Conference on Communications Technology and Applications*. Alexandria, Egypt: IEEE.

Zhang, W., Wang, G., Xing, Z., & Wittenburg, L. (2005). Distributed stochastic search and distributed breakout: Properties, comparison and applications to constraint optimization problems in sensor networks. *Journal of Artificial Intelligence, 161*(1-2).

KEY TERMS AND DEFINITIONS

3GPP SON Activities: The SON activities done in the scope of the 3rd Generation Project Partnership (3GPP).

4G Networks: 4th Generation mobile communication networks, termed also Long Term Evolution (LTE).

LTE-Advanced: The release being standardized by the 3rd Generation Project Partnership (3GPP) with self-organization capabilities.

Self-Organization: The ability of a system to arrange and organize itself spontaneously under appropriate circumstances in a purposeful (non-random) manner without any help of external agencies.

Self-Organizing Networks (SON): Networks with the capability to configure, organize, and optimize their performance while minimizing network Operational Expenses (OPEX).

SON Coordination Framework: Frameworks that coordinate between various SON activities to guarantee systems' stability and reliability in addition to performance optimization.

SON Use Cases: The use cases defined by the Next Generation Mobile Networks (NGMN) alliance to benefit from self-organization capabilities.

Chapter 5
Mobility Prediction in Long Term Evolution (LTE) Femtocell Network

Nurul 'Ain Amirrudin
Universiti Teknologi Malaysia, Malaysia

N. N. N. Abd Malik
Universiti Teknologi Malaysia, Malaysia

Sharifah H. S. Ariffin
Universiti Teknologi Malaysia, Malaysia

N. Effiyana Ghazali
Universiti Teknologi Malaysia, Malaysia

ABSTRACT

The Long Term Evolution (LTE) femtocell has promised to improve indoor coverage and enhance data rate capacity. Due to the special characteristic of the femtocell, it introduces several challenges in terms of mobility and interference management. This chapter focuses on mobility prediction in a wireless network in order to enhance handover performance. The mobility prediction technique via Markov Chains and a user's mobility history is proposed as a technique to predict user movement in the deployment of the LTE femtocell. Simulations have been developed to evaluate the relationship between prediction accuracy and the amount of non-random data, as well as the relationship between the prediction accuracy and the duration of the simulation. The result shows that the prediction is more accurate if the user moves in regular mode, which is directly proportional to the amount of non-random data. Moreover, the prediction accuracy is maintained at 0.7 when the number of weeks is larger than 50.

INTRODUCTION

Introduction of Long Term Evolution

Long Term Evolution (LTE) is introduced due to the emergence of new application such as Multimedia Online Gaming, Web 2.0 (i.e. Facebook, myspace) and high growth of consumer technology such as a laptop, tablet and smart phone. The LTE is the latest standard in a mobile network technology tree that was introduced by Third Generation Partnership Project (3GPP). The LTE whose radio access called as Evolved UMTS Terrestrial Radio Access Network (E-UTRAN) is expected to improve end-user throughput, reduce user plane latency, and improve users experience with full

DOI: 10.4018/978-1-4666-5170-8.ch005

mobility. The LTE is provided for Internet Protocol (IP) based traffic with end-to-end Quality of Service (QoS) to meet a requirement of high peak data rate. Some of LTE requirements are as follows (Motorola, 2007):

1. Peak data rate of 100Mbps for downlink and 50Mbps for uplink (for 20MHz spectrum).
2. Scalable channel bandwidths of 1.4, 3, 5, 10, 15, and 20MHz in both downlink and uplink.
3. Mobility supports for up to 500kmph but optimized for low speeds from 0 to 15kmph.
4. Control plane latency is less than 100ms.
5. User plane latency is less than 5ms.
6. Can serve for more than 200 users per cell (for 5MHz spectrum).
7. Cell coverage from 5km to 100km with slight degradation after 30km.

The LTE has been developed based on orthogonal frequency-division multiplexing (OFDM) waveform for downlink and single-carrier FDM (SC-FDM) waveform for uplink, together with higher-order multiple-input multiple-output (MIMO) spatial processing techniques. In addition to the LTE, the 3GPP defines IP-based, flat core network architecture where this architecture is a part of System Architecture Evolution (SAE) that consists of Evolved Packet Core (EPC) and E-UTRAN. The EPC consists of Mobility Management Entity (MME), Serving Gateway (S-GW), and Packet Data Network Gateway (PDN GW). Some of the functions of the MME are; to control the signal between user equipment (UE) and core network (CN), to establish and release radio bearer services, responsible for paging and tracking the UE in idle mode, and to check the authorization of the UE to camp on the service provider's Public Land Mobile Network (PLMN). The SGW is responsible to route and forward user data packets, and acts as mobility anchors for user plane during handovers. The PDN GW provides connectivity to the UE to external packet data

networks and allocates IP addresses for the UE and quality of service (QoS) enforcement. The E-UTRAN consists of evolved Node B (eNB) where the terminology of base station in the LTE. The eNBs are interconnected with each other by means of X2 interface and S1 interface to the EPC.

The LTE can be used in both paired frequency division duplexing (FDD) and unpaired time division duplexing (TDD) spectrum (Ericsson, 2009). The FDD requires two separate carriers where one for downlink and another for uplink. Downlink and uplink share a single carrier, but separate in time domain for TDD. In general, FDD is more efficient but more complex where it is difficult to make antenna bandwidths broad enough to cover both sets of spectrum. For TDD, it is only required/ needed a single carrier and no spectrum-wasteful guard bands or channel separation.

Implementation of the LTE Femtocell

Femtocell technology had attracted many industries in late 2007 and early 2008. As per survey on wireless usage, more than 50% of all voice calls and more than 70% of data traffic originates indoors (Chandrasekhar, 2008). Therefore it is significant for network provider to provide a good coverage in indoor environments. Femtocell is an effective solution to provide better coverage in the indoor environment as well as can improve macrocell reliability. Femtocell is a low-power wireless access points with small cell coverage and operates in a spectrum licensed to connect standard mobile devices to a mobile operator's network using residential digital subscriber line (DSL) or cable broadband connections (see Figure 1). Femtocells have been devised for offering broadband services indoor (i.e. home and offices) and outdoor scenarios with a very limited geographical coverage (Capozzi, Piro, Grieco, Boggia, & Camarda, 2012). It is also called as a home based station which also known as home evolved Node B (HeNB) in 3GPP LTE. Femtocell

Figure 1. The basic structure of femtocell

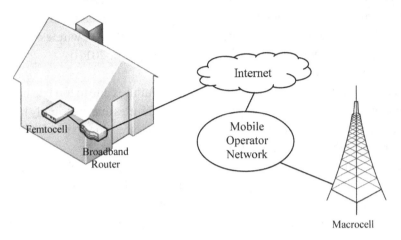

requires low power which is between 13 to 20dBm with coverage from 15 to 50 meters (Wu, 2011).

There are many benefits from the deployment of the femtocells to mobile users and network providers (Qutqut & Hassanein, 2012):

1. Mobile users
   a. Improve indoor coverage because femtocell is close to the users.
   b. Improve data rate capacity because femtocell utilizes the user's high data rate broadband connection as its backhaul.
   c. Reduce indoor cost charge.
   d. Reduce power consumption for the UEs due to the lower transmit power of the femtocell if compare to the macrocell.
   e. Able to offer new services such as a home gateway, location based services.
   f. Simple deployment since femtocell works as a 'plug-and-play' device.
2. Network provider
   a. Reduce capital expenditures since no new expensive macrocell are needed.
   b. Lower operational expenditures because no new cell site, cell site backhaul, and maintenance are needed.
   c. Increase mobile usage indoors due to the low-cost fare. Therefore it will increase Average Revenue per User (ARPU).
   d. Reduce customers churn rate because customers will be potentially satisfied with the offered services through femtocells.

There are three access control modes in femtocell: open, hybrid, and closed. For the open access mode, all users can access to the femtocell. For example, an operator deploys this femtocell to provide a good coverage in an area where there is a coverage hole. For the closed access mode, only users under closed subscriber group (CSG) can access the femtocell. For the hybrid access mode, basically it is similar to closed access mode but open to the non CSG if there is any available bandwidth. In order to differentiate the femtocell access modes, CSG identity (ID) and CSG indicator are introduced. CSG identity is a unique numeric identifier that broadcasts in system information (SI) by CSG and hybrid cell which is used by the UE during handover process. On the other hand, CSG indicator is presented with a value of TRUE for CSG cell and absent for hybrid and open cell (3GPP, 2011a) (Golaup, Mustapha, Patanapongpibul, & Group, 2009) (Lin, Zhang, Chen, & Zhang, 2011).

There are three potential deployment scenarios for LTE femtocell namely domestic femtocell, enterprise femtocell, and outdoor femtocell (Network, 2011). For the domestic femtocell, femtocell is installed in a residential house where generally the coverage area is about 25 meters. It is operating mainly in closed and hybrid access modes, and can serve between 4 and 8 users simultaneously. For the enterprise femtocell, it is installed in a large building such as office and shopping complex. It can serve a larger number of users than the domestic femtocell ranges between 32 and 64 users. For the outdoor femtocell, it emphasizes on improving a limited area due to the building penetration loss. This scenario mainly is installed by the network provider and operates in open access mode.

Mobility Management in LTE Femtocell

The LTE is expected to provide fast and seamless handover from one cell to another cell where user is unaware of the handover. This handover can be achieved by reducing handover latency, minimize packet loss, and minimize loss of communication state. Handover in LTE is based on network controlled-UE assisted, and only support hard handover where old radio links in the UE are removed before a new radio links are established. The eNBs make a handover decision based on a measurement report send by the UE. Several parameters are used to make the handover decision such as the signal to interference ratio (SIR), received signal strength (RSS), distance from a base station and velocity (Hussein, Ali, Varahram, & Sali, 2011). Handover parameters are important in order to have a successful handover because incorrect handover parameter setting may affect the users experience and waste network resources due to a ping-pong effect, handover failures, and radio link failures (RLF). The ping-pong effect or also referred as unnecessary handover has happened when a call is handover to a new cell and handed

back to the source cell in less than a critical time (T_{crit}) (Jansen, Balan, Turk, Moerman, & Kurner, 2010). The RLF in wireless network may happen due to (3GPP, 2011b):

1. Failures due to too late handover triggering.
2. Failures due to too early handover triggering.
3. Failures due to handover to a wrong cell.

Considering the femtocell deployment, femtocell can be deployed in anywhere since it is installed by consumer without any centralized coordination. Femtocell may deploy in coverage of macrocell, therefore there are three possible handover scenarios may happen in femtocell deployment:

1. Inbound handover; means handover from macrocell to femtocell,
2. Outbound handover; means handover from femtocell to macrocell,
3. Inter femtocell handover; means handover between femtocells.

Due to the characteristic of femtocell (i.e. small cell coverage, deployed in an unplanned manner), a large number of femtocells may deploy in a single macrocell. Therefore, it will create a large neighbour cell list (NCL) and interference problem (Zhang, Wen, Wang, Zheng, & Sun, 2010). Also, femtocell may deploy as CSG access mode whereby not all users can connect to the femtocell. These aspects make the inbound handover more challenges than other scenarios. For the outbound handover, whenever the users move out from femtocell's coverage, macrocell signal strength may be stronger than the femtocell. Hence the outbound handover is not complex compare to the inbound handover. The inter femtocell handover may happen in a scenario where femtocells are adjacent to each other. Since there is a direct connection between femtocells which is X2 interface, handover between them shall not be so complex at all. During the handover procedure, resource allocation takes a lot of time and becomes the main factor of handover latency. One of the most effective techniques to reduce delay in the resource allocation and finally reduce the handover latency is by predicting a next location of the users. This chapter will provide an overview of handover in LTE femtocell as well as the handover principle in conventional networks without femtocell. Related work on mobility prediction is presented and finally a proposed work on mobility prediction via Chains Markov and user mobility history is discussed.

BACKGROUND

Conventional Handover in Wireless Network

Handover management is an important aspect to be considered in any wireless network technology to support mobility and maintain user's quality of service (QoS). Handover enables the network to maintain mobile's connection when move from one coverage cell to another. Handover is a process of changing a channel (frequency, time slot, spreading code, or combination of them) associates with a current connection while a call is in progress (Zeng & Agrawal, 2002). Handover is divided into two broad categories: hard and soft handover. The soft handover or also called as 'make before break' allows the mobile station to communicate and exchange data with multiple interfaces simultaneously during the handover. On the other hand, the hard handover or 'break before make' results in disconnecting from an old access point when signal strength is below a threshold before connecting to a new access point (Atiquzzaman & Reaz, 2005).

Basically a handover procedure is divided into three stages: preparation (initiation), execution, and completion (Hussein et al., 2011). The preparation stage is started when the mobile station sends a handover request message to a source base station. The signals from the source and neighbour base stations are measured continuously, and once the neighbour's signal is higher than the source base station, the handover request message is triggered. Then the source base station sends the handover request message to a target base station and a handover decision is performed if the target base station can provide the mobile station with required resource. Some of the parameters that used for the handover decision are based on signal strength, velocity, and SIR (Zeng & Agrawal, 2002; Quang, 2012):

1. **Based on signal strength:**
   a. **Relative Signal Strength (RSS) with threshold:** The handover is requested once the RSS of the source base station is lower than a threshold value and the RSS of the target base station is stronger than the source base station. This method helps the network to limit the handover when the signal from the source base station is strong enough to serve the mobile station. However, the appropriate threshold value is required because connection may drop if the threshold value is too low.
   b. **Relative Signal Strength (RSS) with hysteresis:** The handover is requested when the target base station is sufficiently stronger (by a hysteresis margin) than the source base station. This method helps to mitigate the ping-pong effect, but still needs to find the appropriate hysteresis margin. If the value is too high, the current signal may fall to a very low value and the connection may drop. Contrary, if the value is too low, the handover may happen unnecessarily, while the signal strength of the source base station is enough to maintain the connection.
   c. **Relative Signal Strength (RSS) with hysteresis and threshold:** This method

combines both the threshold value and hysteresis margin to reduce a number of handover. The handover is requested if RSS of the source base station lower than the threshold value and the RSS of the target base station is stronger than the source base station by the hysteresis value.

2. **User's velocity:** If the user moves fast, a probability of call drop may be high due to excessive delay during the handover. Means, the handover process is not complete but the user has moved to another base station. For low speed user, the user spends much time in handover area, therefore the threshold value is assigned higher than for the high speed user.

3. **SIR:** Signal to interference ratio (SIR) is a measure of the communication quality. The handover is triggered if the SIR of the source base station is lower than the threshold and the SIR of the target base station is better.

Handover may affect many aspects on the wireless network such as user's QoS and capacity of the network. If the value of parameters for the handover decision is wrongly chosen, the connection may be dropped or may increase the capacity of the network. During the handover process, it consumes network resources to reroute the call to the new base station, therefore minimizing the number of handover will reduce a signalling overhead. Some of handover requirements for the wireless networks are as follows (Quang, 2012):

1. The latency of handover must be low.
2. The total number of handover should be minimal.
3. The effect of handover on QoS should be minimal.
4. The additional signalling during the handover process should be minimized.

Handover Procedure in LTE Femtocell

Handover in LTE is network controlled-UE assisted handover where the handover decision is made by the source base station based on the measurement from the UE. Handover procedure in LTE consists of three parts (Ulvan, Bestak, & Ulvan, 2010; 3GPP, 2013):

1. Handover preparation
   a. Measurement control/report where the source eNB configures and triggers the UE measurement procedure and the UE sends the measurement report to the source eNB.
   b. The source eNB performs the handover decision based on the measurement report and Radio Resource Management (RRM) information.
   c. The source eNB issues the handover request message to the target eNB with necessary information.
   d. The target eNB performs the admission control depends on the QoS to increase the successfulness of the handover. If the resource can be granted, the target eNB prepares the handover with L1/L2 and sends the handover request acknowledgement to the source eNB.
   e. The source eNB sends the handover command to the UE.
2. Handover execution
   a. The UE detach from the source eNB and synchronize to the target eNB. There is a short interruption in service between the time that the UE receives the handover command from the source eNB and the time that the target eNB receives the handover confirmation from the UE (Incorporated, 2010).
3. Handover completion
   a. The handover has complete when serving gateway (S-GW) has switched

downlink data path to the target eNB and the source eNB release radio and C-plane related resources associated with the UE context. Then the target eNB can transmit the downlink packet data.

Handover procedure discusses above is for a normal base station which is macrocell or eNB. Handover decision is not involving the MME and S-GW if there is an X2 interface between eNBs. However, since there is no direct interface between eNB and HeNB, the MME and S-GW are involved in handover decision for the handover from eNB to HeNB. Handover to HeNB is different from the normal handover procedure in three aspects (3GPP, 2013):

1. Proximity estimation; in case the UE is able to determine that it is near the CSG cell who's the CSG ID is in the UE's whitelist, the UE may provide the proximity indication to the eNB. In response, the source eNB configures the UE to perform relevant measurements of the CSG cell (Seidel & Saad, 2010). This proximity indication can allow the UE to continuously make measurements and read the System Information (SI) with lots of CSG cells in cases of large scale HeNB deployments (Zhang et al., 2010).
2. Physical Cell Identity (PCI) confusion; due to the small size of the HeNB, there can be multiple HeNBs within the coverage of the source eNB that has the same PCI, and cause PCI confusion. This situation can cause the source eNB unable to determine the correct target cell for the handover. This PCI confusion is solved by the Cell Global Identity (CGI) of the targets HeNB reported by the UE.
3. Access control; the UE reports the membership status based on the CSG ID of the target cell either open, close, or hybrid, and then verified by the network.

For handover to HeNB, when source eNB receives handover request from the UE, it sends handover request to the MME including CSG ID and access mode. The access control is performed by the MME instead of target cell as normal procedure. If the access control succeeds, the MME sends handover request to the target HeNB through HeNB gateway (HeNB GW), if available. If the access control fails, the MME ends the handover procedure by replying with Handover Preparation Failure message. Further process is similar to normal procedure as discussed earlier. The additional aspects need to be considered on the femtocell's handover and the special characteristic of the femtocell require an enhancement on handover algorithm in order to achieve seamless and fast handover in the femtocell deployment. One of the techniques to enhance handover performance in femtocell deployment is by predicting the target base station in advance.

Mobility Prediction in the Wireless Network

Whenever the users are moving, they will experience a number of handover during a call to make sure a continuation of the call. The number of handover is depending on the users' speed and cell's coverage. The higher the user's speed, the more frequent handover is. During handover procedure, resource allocation is performed to assign any available resources in a new cell. This activity takes a lot of time and increase the handover latency. By knowing user's movement, the next base station where the user will connect can be predicted in advance. Therefore the resource can be reserved prior to the actual handover and finally reduce the handover latency. In order to achieve seamless and fast handover, continuous resource reservation is required (Doss, Jennings, & Shenoy, 2004). Moreover, mobility prediction can reduce the number of control packets needed to reconstruct routes and thus minimizes overhead (Su, Lee, & Gerla, 2000). Because of that reasons, the mobility prediction plays an important role to enhance the handover performance.

The important role of the mobility prediction in order to enhance the handover performance has attracted attention of many researchers to investigate and enhance mobility prediction technique in the wireless networks. Many mobility prediction schemes have been proposed such as:

1. Prediction based on user's movement history.
2. Prediction based on Markov Model.
3. Prediction based on user's direction.
4. Prediction based on signal strength.

Prediction Based on User's Mobility History

The prediction based on the user's movement history has been proposed in Ge and Lu (2009). The proposed technique requires the network to recognize user who frequently visit a cell and then track and record movement information about the user. This information is needed to find a route of the user. Based on the information of the user's movement history and current location of the user, the network is able to search the route of the user and predict a set of potential handover. A signal strength of the user is considered and put as candidate handover once the signal strength is higher than a certain threshold. The proposed technique has shown a decreasing of the number of handover and lowers a ping-pong rate in the LTE system, however it can only be implemented for regular users.

The authors Wang and Prabhala (2012) have considered a spatio temporal technique to predict the next location of the user where it considers the user's location and time as well. The prediction is based on the current location rather than past sequence. The user's mobility history is collected and then frequently visit place and aggregate total time spent in place is computed. After that, periodicity analysis for the top frequent place is performed. Data file of the user's mobility history consists of user ID, current place ID, start time for a visit and end time of a visit. Based on the data file, the most frequent place and at what time can be detected, and thus the next place of the user can be predicted. The result shows that prediction accuracy is acceptable which around 55%.

Another proposed technique that relies on the user's mobility history is presented in Daoui, M'zoughi, and Lalam (2008). The authors have proposed the mobility prediction by modelling a mobile user's behaviour based on ant colony optimization (ACO). The technique is inspired by ants' behaviour when they search for food where the ants will follow the most attractive route to find its food. During searching for food, every ant will deposit a quantity of pheromone. After that, the ant that just comes will follow the route with higher pheromone intensity. Based on the ACO, the proposed technique requires the network to store the user's mobility history in order to predict the next base station of the user. The data of user's mobility history shall consist of Mobile ID, period, source cell, destination cell, and date. This proposed technique can apply for a new user where if they enter the system and the base station does not have its history, the base station can use other's history that has a same mobility profile. The advantage of this proposed technique is it can be applied to all users whether the user is a regular user or new user.

Since the user's mobility history can give useful information regarding the user's movement and can assist the prediction of the next location, the network needs to maintain the history data properly so that the data can assist efficiently without burdening the network itself. The larger the user's history, the more accurate the prediction is. However, the database of user's history becomes larger and the process of prediction may become longer. Therefore, a proper technique to process the database of the user's mobility history has been studied. In Duong and Tran (2012), the authors have applied a spatiotemporal data mining technique to discover frequent mobility patterns in order to predict the next location of a mobile node. A predefined timestamps have been introduced so that the data mining process becomes easy. A log file of the user's history contains of time and location. Based on the log file, a transactional database is developed where it shows a sequence transition of the user. A time constraint between locations is introduced in order to avoid missing data and make significant mobility sequences. If a time gap is larger than the maximal time gap, a new transaction is created. Then a mobility pattern is defined where it shows the frequent location that the user commonly visit. The next location of the user can be predicted based on the mobility pattern.

Prediction Based on Markov Model

In Ulvan, Ulvan, and Bestak (2009), the authors have proposed a prediction technique that relies on Markov Chains which is a mathematical system that undergoes transitions from one state to another. The proposed technique requires the network to track a position of the user. It is assumed that the user is able to send its location periodically to the source base station, and at the same time the source base station is able to maintain a database of roads within its coverage. Main parameter of this proposed technique is a transition probability matrix, where in this paper the value of the transition probability matrix is assumed. Based on simulation results, it is shown a user's direction after several movements and a form of the user's movement (i.e. linear, reside, random, and patterned) can be predicted as well. Therefore we can predict the user's movement by using the Markov Chains.

An enhancement of the Markov Chains standard is proposed in Gambs, Killijian, and Del Prado Cortez (2012) where the next state depends on n-previous location instead of current location only. This extended Markov Chains mobility is called as n-MMC. Standard Markov Chains is memoryless where the prediction of the next location only depends on the current location. The authors claim that the limitation of the standard Markov Chains that forget the previous locations visited can give a negative impact on prediction accuracy. There are two steps to discover the user's mobility where first use a clustering algorithm to discover points of interests (POIs), and then completes the transition between those POIs. In the standard Markov Chains, if there are three POIs, the transition probability matrix becomes 3 x 3. However, since it considers the n-previous location, the transition probability matrix becomes 3^n x 3, where n is the number of previous locations. For example, if we consider two previous locations, the transition probability matrix becomes 9 x 3. Result shows that the prediction accuracy is higher when considering two previous locations. However, the transition probability matrix may become more complex if consider many POIs or base stations such as the femtocell deployment.

Another type of the Markov Chains is Hidden Markov Chains, which is the standard Markov Chains with additional input of observation. Parameter that considers in the standard Markov Chains is the transition probability matrix, while in the Hidden Markov Chains the additional parameter call observation matrix is considered. The observation matrix stores a probability of observation that influences the transitions. In Fazio and Marano (2012), the authors have proposed a prediction algorithm based on Hidden Markov Chains process. The Hidden Markov Chains is used by each coverage cell to forward passive reservation message to the predicted neighbouring cell. By using the Hidden Markov Chains process, a handover direction can be predicted. The input of the observation matrix derives from training observation that consists of the set of handover direction. Result shows that the prediction accuracy is directly proportional to the number of observations.

Prediction Based on User's Direction

Another technique of the mobility prediction is based on the user direction as presented in Sadiq and Bakar (2011). The authors have proposed the prediction technique based on the user's direction and signal strength called as mobility and signal strength-aware handover decision (MSSHD) approach. These parameters are denoted as an input to a fuzzy inference system to predict the handover decision. Received signal strength indicator (RSSI) shall be higher than RSSI threshold that has been set in advance. A direction of the user is calculated based on an angle of direction between the mobile station and the target base station. If the angle is less than the angle threshold, means the mobile station is moving towards the base station and handover can be triggered. This proposed technique displays a decreasing of latency in Layer 2 (L2).

Other proposed technique that relies on the user's direction has been presented in Kim and Kim (2013). The authors have proposed handover optimization technique by predicting the target base station based on the user moving direction. The proposed algorithm is using conventional RSS based handover decision to trigger the handover request. If the target base station is not the predicted cell, the handover process is delayed. If the target cell is the predicted cell, an assurance level is checked and the handover is executed if the assurance is satisfying the threshold level. The predicted cell is defined by calculating a variance of relative distance from a mobile station to each of candidate target base stations. A signal loss is considered as well. This technique is mainly proposed to postpone the handover as much as possible in order to minimize the number of handover and finally enhance the handover performance.

MAIN FOCUS

Mobility Issues in Femtocell Deployment

Due to the special characteristic of femtocell (i.e. the cell coverage is much smaller than macrocell, can be placed at any location and the implementation of the access control), the mobility management in femtocell deployment become more challenges. The most critical issue in femtocell handover is on unnecessary handover where this result of the network overhead and degradation of the user's QoS. Moreover, maintenance of the NCL to individual femtocells gives an impact to the handover performance in femtocell deployment. This result will increase of scanning interval and increase packet jitter.

The unnecessary handover happens when a call is handover to a new base station within a short time. The ping-pong effect may happen from the unnecessary handover if the call is handover back to a previous base station and this may cause a network overhead and degradation of the QoS. This scenario happens especially for small coverage area like femtocell and also for high speed user where they will stay in femtocell coverage area within a short time. During the conventional handover procedure, a handover margin (HM) and time-to-trigger (TTT) are introduced in order to mitigate the unnecessary handover and reduce the ping-pong effect. This technique has been proved in Kim and Lee (2010) where it shows a drop number of the unnecessary handover when introduced the TTT. However, it does not consider on throughput. Zetterberg et al. (2010) show that the low TTT values give high throughput but increase the number of handover. Therefore, HM and TTT itself could not give a good result of handover performance and another approach is required to enhance the handover performance in femtocell deployment.

Another issue in the handover femtocell is maintenance of the NCL of femtocell. The NCL contains a list of base station ID that adjacent to the base station. During the handover procedure, the NCL is broadcast by the source base station to the UE to assist the UE with a scanning process. Consider the femtocell deployment, there may have a large number of femtocell in a single macrocell, therefore the NCL becomes large. This resulted in increased packet jitter and end-to-end delay. In other hands, if the NCL is not complete due to large shadowing for example, the UE might handover to undesirable base station or even fail to handover (Woo, Kang, & Choi, 2010).

Moreover, resources in femtocell need to be considered in order to achieve a seamless handover. Enough resource is a factor to serve the users seamlessly where the users may experience a drop call if there is not enough resource. Femtocell can serve a small number of users which are maximum 16 users simultaneously (depends on vendor) (Forum, 2011). Hence the resource is much smaller than macrocell. Therefore some users may not connect to the femtocell due to the limitation of the resources, and it may cause the users loses the connection and finally face a bad experience during a call. Based on these factors, the mobility prediction is proposed in order to enhance the handover performance where it is one of the main objectives of the LTE.

Proposed of Prediction Technique via Markov Chains

In this chapter, a mobility prediction via Markov Chains is proposed as a technique to assist the handover procedure in LTE femtocell. As discussed earlier, the mobility prediction can enhance the handover performance by reducing handover latency, packet jitter and end-to-end delay. Markov Chains is a mathematical system that undergoes transitions from one state to another. It is a random process usually characterized as memoryless; which is the next state depends only on the current state and not on any previous states. However, there is some modification of the Markov Chains

theory where the next state depends on the previous state as well. It has been presented in Gambs et al. (2012) and it shows that prediction accuracy is higher when the system considers two previous states. Nevertheless, this technique doesn't suit for a system consists of large number of base station, because the Markov Chains might become large and complex. Therefore this proposed work will use the standard Markov Chains where the next state only depends on the current state. Markov Chains theory is a very powerful tool to analyze stochastic systems over time, and is regularly used to model an impressively diverse range of practical systems (Izquierdo & Izquierdo, 2009). Markov Chains is a transition system composed of:

1. A set of state $S = \{s_1, \ldots, s_n\}$; in which each state corresponds to a base station (or a set of base stations).
2. A set of transitions n; in which each transition represents a movement from the state.

Main parameter in Markov Chains is transition probability matrix P where it represents a probability of transition between the states. For instance, if the number of state S is k, therefore the transition probability matrix P becomes:

$$P = \begin{bmatrix} p_{11} & \cdots & p_{1k} \\ \vdots & \ddots & \vdots \\ p_{k1} & \cdots & p_{kk} \end{bmatrix}$$

In the matrix P above, p_{11} is a value of transition probability that the user moves from state 1 to state 1 and p_{k1} denotes the transition probability from state k to state 1. Summation of all the transition probabilities in any row is equal to 1 (Mathivaruni & Vaidehi, 2008):

$$\sum_{i=1}^{j=7} p_{i,j} = 1$$

The values of transition probability matrix P are derived from a diagram called state Markov Chains diagram. Figure 2 shows a relationship between the state Markov Chains diagram and the transition probability matrix P. A and B represents a name of states, means there are two states here or two base stations in our scenario. Every row represents a source state or a source base station, while column represents a destination state or a destination base station. Value of α and $(1 - \alpha)$ represent the transition probability that user moves from state A to state A and from state A to state B, respectively. The value of the transition probability matrix P shall not in negative value. Summation of any row is equal to 1, e.g. the summation of α and $(1 - \alpha)$ is equal to 1. From the transition probability matrix P, the most frequent base station that the users visit can be detected easily.

Besides the transition probability matrix P, another parameter needs to be considered in Markov Chains is initial distribution matrix p. The value of initial distribution matrix p can be derived from user's velocity, distance, or initial state. This value is introduced so that the mobility prediction can be calculated precisely. In this project, the user's initial state is chosen as the value of an initial distribution matrix p. This is because the user's movement starting their initial state is to be determined. As per example above

where the number of states is two, therefore the initial distribution matrix p becomes 1 x 2:

$$p = \begin{bmatrix} a & b \end{bmatrix} \tag{1}$$

The initial distribution matrix p only has one row, and number of columns depends on the number of state or base station. Then a position of the user after one movement can be predicted as pP. Therefore the position of the user after n movement can be derived as:

$$p_n = pP_{n-1}{}^n \tag{2}$$

where,

p = initial distribution,
P_{n-1} = current transition probability matrix,
n = number of state transition.

User's Mobility History as an Input of Markov Chains

Equation (2) is used to predict the user's movement in this project. Based on Amirrudin, Ariffin, Malik, and Ghazali (2013), the main parameter that influences the prediction of the user's movement is the transition probability matrix. In this project, the value of the transition probability matrix is derived from user's mobility history. An

Figure 2. Relationship between Markov Chains diagram and transition probability matrix

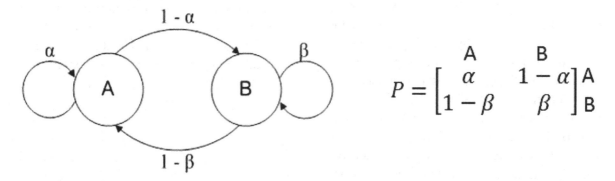

analysis in Azevedo, Bezerra, Campos, and De Moraes (2009) shows that people walking through the same places at different time and with a high spatial regularity. Means, the user often follows a same route or in other words the regular movement. This fact has attracted much research on the prediction based on the user's mobility history recently. The user's mobility history provides much useful information such as the most frequent place that the users visit, common routes the user drives on, and time they spend on certain places (Kirmse, Udeshi, Bellver, & Shuma, 2011). However, the user's mobility history consumes much memory, energy and bandwidth, especially for a base station that many users visit. Therefore the user's mobility history should manage properly to avoid any consumption but at the same time can provide all information needed. Hence a method to extract and analyze the user's mobility history data is important. This activity calls as data mining which is a process of analyzing data from different perspectives and summarizing it into useful information (Palace, 1996).

In the data mining process, firstly a log file is created as per format in Figure 3. A structure of the log file is inspired by Daoui et al. (2008) but simplified so that the log file becomes less complicated. Whenever the user enters a location, a prediction system is launched where the network keeps a required data and stores it in the log file. The descriptions of the log file are as follows:

1. **User ID:** Represent a unique mobile identifier which can be Pin code, Mac address or other.
2. **Time:** Represent a date and time where the user connected to the particular base station.
3. **Location:** Represent a base station ID where the user connected at a particular time.

Formal Definitions of Mobility Pattern

In this subsection, a formal definition of model mobility patterns is discussed. Assume that the user moves around a space freely without any rules. Let c be the base station ID to which the user connected at time t, we define a point as $p = (c, t)$.

Definition 1: Let C and T as a set of base station ID and time respectively. The ordered pairs $p = (c, t)$, where $c \in C$ and $t \in T$, is called a point. Denote P to be the set of all points $P = C \times T = \{(c, t) \mid c \in C \text{ and } t \in T\}$.

Two points $p_1 = (c_1, t_1)$ and $p_2 = (c_2, t_2)$ are considered. The point p_1 is equivalent to the p_2 if and only if $c_1 = c_2$, and $t_1 = t_2$. If $t_1 < t_2$, the point p_1 is defined as earlier than point p_2, means the user connects to the point p_1 before connecting to the point p_2.

Definition 2: A trajectory of the user is defined as a finite sequence of the points p (i.e. p_1, p_2, ..., p_k) in C x T space where point $p_i = (c_i, t_i)$ for $1 \le i \le k$.

The value of the point p must be unique in the trajectory for each user (e.g. $p_1 \ne p_2$), where the user cannot connect to the same base station at the same time. The value of time t shall be unique in the trajectory, however the value of base station ID c can be same. Means, the user can connect to the same base station but at different time.

Definition 3: Time interval t_i is an interval time between the two sequence points p (i.e. p_i and p_{i-1}). Time gap t_g is a max interval time that the user connects to a particular base

Figure 3. A structure of log file

User ID	Time	Location

station. If the time interval $t_i \geq$ time gap t_g, we assume that the user has stop moving or reach its destination.

The time gap t_g is introduced in order to identify a last destination of the user. This is to make sure a transactional database is more accurate. The value of time gap t_g should choose properly because it may affect an accuracy of the mobility prediction.

Discovering Mobility Pattern

From the log file, a transactional database is created to identify a relationship between the source and the destination base station for each user. From the transactional database, most frequent base station that the user always visits can be detected. Figure 4 shows a step of creating the transactional database from the log file. Assume there are two users UE1 and UE2 that move around four base stations (i.e. BS1, BS2, BS3, and BS4). The time gap t_g is set as 5 hours. Consider the UE1, a source-destination table is created in order to identify the trajectory of the user. At time 1/3/2013 7:37, the time interval t_i is more than the time gap t_g, therefore the value of the destination base station is same as the source destination. This represents the user is reaching its destination. Notes that the source-destination table is catered per user. From the source-destination table, the transactional database is created. The database shows the relationship between the source and the destination base station in terms of total number of the transition. First column in the transactional database is a list of the source base station, and first row represents a list of the destination base station. For example, the value 1 in the transactional database (refer to the third column and second row) is a frequency of the user moves to BS2 from BS1 at time T. Grand total is total amount that the user moving from the particular base station. For example, value 4 (refer to the last column and second row) represents a total time the user moves from BS1 to other base station.

Once the transactional database is created, the transition probability matrix is computed by dividing a transition value to the grand total for each particular source base station. For example in

Figure 4. Step of creating the transactional database from the log file

A Log File

User ID	Time	Location
<user ID>	<D/M/YYYY> <h:mm>	<BS ID>
UE1	1/3/2013 7:00	BS1
UE1	1/3/2013 7:10	BS2
UE2	1/3/2013 7:22	BS2
UE2	1/3/2013 7:28	BS3
UE1	1/3/2013 7:30	BS3
UE1	1/3/2013 7:37	BS4
UE1	2/3/2013 9:00	BS1
UE1	2/3/2013 9:25	BS3
UE2	2/3/2013 9:27	BS2
UE1	2/3/2013 9:33	BS2
UE2	2/3/2013 9:40	BS3
UE1	3/3/2013 8:05	BS2
UE1	3/3/2013 8:31	BS4
UE1	3/3/2013 8:37	BS1
UE1	4/3/2013 13:55	BS4
UE1	4/3/2013 14:03	BS3
UE1	4/3/2013 14:13	BS2
UE1	4/3/2013 14:24	BS1

Source-Destination Table

Source	Destination
BS1	BS2
BS2	BS3
BS3	BS4
BS4	BS4
BS1	BS3
BS3	BS2
BS2	BS2
BS2	BS4
BS4	BS1
BS1	BS1
BS4	BS3
BS3	BS2
BS2	BS1
BS1	BS1

Per user

Destination base station

Transactional Database

Base Station	BS1	BS2	BS3	BS4	Grand Total
BS1	2	1	1	0	4
BS2	1	1	1	1	4
BS3	0	2	0	1	3
BS4	1	0	1	1	3

Source base station

Destination base station is same as source base station due to high time interval

Figure 5, a value of ½ (refer table transition probability matrix, second column and second row) is derived by dividing a value of 2 to the total value of 4 from the transactional database. To verify the value of the transition probability matrix, the total value for each row shall equal to 1.

Prediction of the User's Movement

Figure 5 shows a final step of the data mining process on discovering the mobility pattern. From the transition probability matrix the most frequent base station ID can be identified. By using the user's initial state as the value of initial distribution matrix, and the transition probability matrix, Equation (2) can be executed to predict the user's movement. Predicted result shall be checked to mitigate an error prediction. In an ideal case, the handover only occurs between adjacent cells. Therefore if the predicted base station is not adjacent to the previous base station, the predicted base station is removed from the prediction result. This consider as an error prediction. Then, the value of the transition probability matrix is checked for each predicted base station. If the value of the transition probability matrix is null, the predicted base station is removed. The algorithm of the proposed mobility prediction is shown in Figure 6. The efficiency of the proposed mobility prediction technique is evaluated on the basis of prediction accuracy where it derives from the ratio of the number of correct predictions to the total number of predictions (Daoui et al., 2008; Ali-Yahiya, 2011).

$$Prediction\, accuracy = \frac{number\, of\, correct\, predictions}{number\, of\, predictions}$$

(3)

Experimental Evaluation

The evaluation has considered a scenario in urban area where 29 HeNBs are installed along the roads with a T-junction as shown in Figure 7. The HeNBs are installed within the eNB coverage area. The HeNB has a radius of 10m, therefore the coverage area of the simulation is 120m x 200m. The user is moving around the space in two types of movement (i.e. random and non-random movement). There are two datasets of the user's mobility history. First dataset is meant to identify an effect of the number of random data to the mobility prediction. The data of user's movement history are stored in the log file for 20 weeks where a number of state transitions are 20. Means, the user move for 20 transitions every day for 20 weeks. Therefore, the total number of trajectories is 140, and total number of transitions becomes 2800. The number of non-random movements have varied from 0% to 100% in 10% incremental steps to find how significant the non-random movement to the prediction accuracy. For the value of 0%, means the movement is purely random, and for the value of 100%, means the movement is purely regular. For the value of 30%, the number of trajectories is 98 and 42 for random and non-random movement, respectively.

Second dataset is to find an effect of duration of the simulation of the prediction accuracy. The data of user's movement history are stored in the

Figure 5. Step in calculating the transition probability matrix

| Transactional Database | | | | | | | Transition Probability Matrix | | | | |
Base Station	BS1	BS2	BS3	BS4	Grand Total			Base Station	BS1	BS2	BS3	BS4
BS1	2	1	1	0	4		BS1	1/2	1/4	1/4	0
BS2	1	1	1	1	4		BS2	1/4	1/4	1/4	1/4
BS3	0	2	0	1	3		BS3	0	2/3	0	1/3
BS4	1	0	1	1	3		BS4	1/3	0	1/3	1/3

Figure 6. An algorithm of proposed mobility prediction

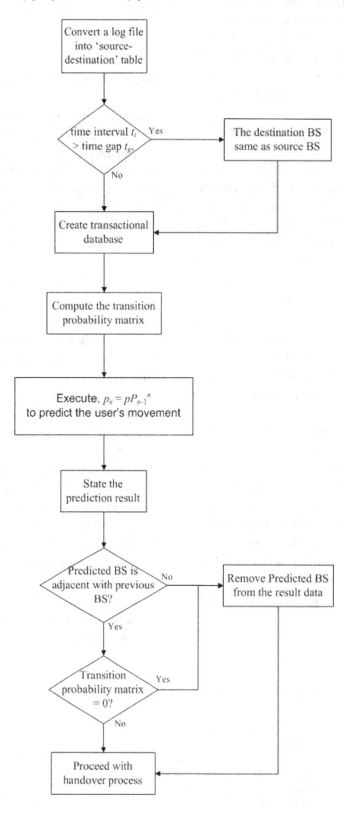

Figure 7. Scenario of the evaluation

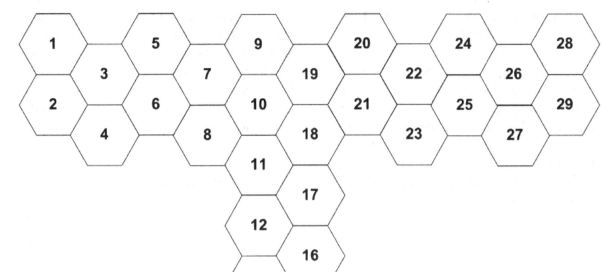

log file up to 56 weeks with the number of state transition is 20. The number of weeks varies from 2 weeks to 56 weeks with increments of 2 weeks. The user is moved in two types of movement which are in non-random movement for weekdays and random movement for weekends. Therefore, percentages of trajectories are 71.4% and 28.6% for non-random and random movement, respectively. The scenarios considered the access control of open access mode where all users can attach to the HeNB. It is assumed that all HeNBs are switched on and have sufficient resources to serve the UE. Therefore, the UE will attach to the HeNBs in the simulation instead of the eNB.

Experimental Analysis and Discussion

The performance of the proposed mobility prediction is evaluated by calculating the prediction accuracy. The prediction accuracy is derived from the ratio of the number of correct predictions to the total number of predictions. First of all, the transactional database is created from the log file of the user. From the transactional database, the transition probability matrix is computed. By using the Markov Chains Equation (2), the user's movement can be predicted via Matlab. Based on the predicted result, the prediction accuracy is defined.

For the first dataset, the prediction accuracy is defined and plotted as per Figure 8 (a). The proposed mobility prediction technique is compared with another technique that also uses the user's mobility history which is from Ge and Lu (2009) and Duong and Tran (2012). From the graph, the prediction accuracy increases slightly as the percentage of non-random data increase from 0% to 10%. When the percentage of non-random data is larger than 10%, the prediction accuracy is strongly increased. It is shown that the prediction accuracy is increased when the user moves more regularly. The accuracy is very low when the user is purely moving in random movement. This is because there is much probability of routes that the user may go. As the user moves more regularly, the probability of routes has decreased and finally increases the accuracy of the prediction. It also shows that our proposed technique is much better than others when the percentage of non-random data is larger than 50%. The reason is because other techniques do not consider a scenario where the user has reached their destination. Moreover, Duong and Tran (2012) also consider the next two transitions instead of the next transitions only.

For the second dataset, the prediction accuracy when the number of simulation's duration are varied is shown in Figure 8 (b). From the graph, the prediction accuracy decreases strongly from 0.75 to 0.67 as the number of weeks increase from week 2 to week 4. The prediction accuracy is increased from 0.67 to 0.7 from week 4 to week 12, and decreases after week 12. The value of prediction accuracy is acceptable which is between 0.67 and 0.74 with a mean value of 0.7. The prediction accuracy is maintained at the same value from week 50 to week 56. This value is defined as a stable state for the user. The num-

ber of weeks where the prediction accuracy is unchanging is put as a threshold week. The threshold week for the scenario is 54, it means that the network can remove the older history and store a data for 54 weeks only, since the value of prediction accuracy is not changed. Therefore, it can minimize the burden of the network due to the large amount of data history.

FUTURE RESEARCH

Due to the emergence of new application such as Multimedia Online Gaming, Mobile TV, 'e' application, online banking, Web 2.0 (i.e. Facebook, twitter, etc.), there is a high demand on high peak data rate especially in indoor environments. Femtocell has been developed in order to extend coverage and increase system capacity in indoor environments. The special characteristics of femtocell introduce many challenges such as interference management and mobility management. In terms of interference, it happens due to the fact that the femtocells are installed in an ad-hoc manner, or independent of the structure of cellular network (Elleithy & Rao, 2011). Interference

Figure 8. Result of the prediction accuracy

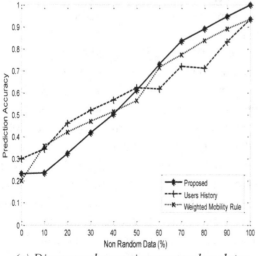

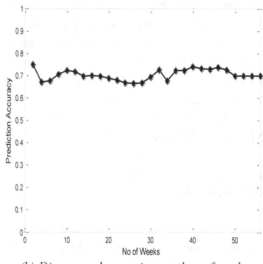

(a) Diagram when varies non-random data   *(b) Diagram when varies number of weeks*

scenarios in femtocell deployment may happen between femtocells and also with macrocell at the same frequency. The user may experience a degradation of data rate if there is high interference. Therefore, a method of reducing interference in femtocell deployment is necessary so that the femtocell may be able to operate efficiently and can be implemented as fault-free devices.

CONCLUSION

Seamless and fast handover is one of the LTE goals in order to support mobility and maintain user's quality of service (QoS). Mobility prediction has been proved to enhance the handover performance by performing resource allocation in advance and reducing the unnecessary handover. Based on the results, it shows that the user's movement can be predicted by using Markov Chains technique. It also shows that the input of the Markov Chains can be extracted from the user's mobility history. The prediction is increasing when the user moves to more regularly. Moreover, the duration of data history is effect on certain duration. The prediction accuracy is maintained at the same value when it reaches certain duration. Therefore, the network can take initiative action by store the data history with certain duration. This action can minimize the burden of the network due to the large amount of data history. The data mining process of the user's mobility history is the main steps to discover the mobility pattern. The weakness of the data mining may affect the prediction result and reduce the prediction accuracy of the user's movement.

REFERENCES

Ali-Yahiya, T. (2011). *Understanding LTE and its performance*. Academic Press. doi:10.1007/978-1-4419-6457-1

Amirrudin, N. A., Ariffin, S. H. S., Malik, N. N. N. A., & Ghazali, N. E. (2013). Mobility prediction via Markov model in LTE femtocell. *International Journal of Computers and Applications*, *65*(18), 40–44.

Atiquzzaman, M., & Reaz, A. (2005). Survey and classification of transport layer mobility management schemes. In Proceedings of Personal, Indoor and Mobile Radio Communications, 2005 (Vol. 6151, pp. 2109–2115). IEEE.

Azevedo, T. S., Bezerra, R. L., Campos, C. A. V., & De Moraes, L. F. M. (2009). An analysis of human mobility using real traces. In *Proceedings of 2009 IEEE Wireless Communications and Networking Conference.* IEEE. doi:10.1109/WCNC.2009.4917569

Capozzi, F., Piro, G., Grieco, L. A., Boggia, G., & Camarda, P. (2012). On accurate simulations of LTE femtocells using an open source simulator. *EURASIP Journal on Wireless Communications and Networking*, (1): 1–13. doi: doi:10.1186/1687-1499-2012-328

Chandrasekhar, V. (2008). Femtocell networks: A survey. *IEEE Communications Magazine*, 1–23.

Daoui, M., M'zoughi, A., & Lalam, M. (2008). Mobility prediction based on an ant system. *Computer Communications*, *31*, 3090–3097. doi:10.1016/j.comcom.2008.04.009

Doss, R., Jennings, A., & Shenoy, N. (2004). *Mobility prediction for seamless mobility in wireless networks*. Melbourne, Australia.

Duong, T., & Tran, D. (2012). An effective approach for mobility prediction in wireless network based on temporal weighted mobility rule. *International Journal of Computer Science and Telecommunications*, *3*(2).

Elleithy, K., & Rao, V. (2011). Femto cells: Current status and future directions. *International Journal of Next-Generation Networks*, *3*(1), 1–9. doi:10.5121/ijngn.2011.3101

Ericsson. (2009). *LTE - An introduction*. Ericsson.

Fazio, P., & Marano, S. (2012). A new Markov-based mobility prediction scheme for wireless networks with mobile hosts. In Proceedings of Performance Evaluation of Computer and Telecommunication Systems (SPECTS). SPECTS.

Forum, F. (2011, December). Femtocell market status. *Informa Telecoms & Media*.

Gambs, S., Killijian, M.-O., & Del Prado Cortez, M. N. (2012). Next place prediction using mobility Markov chains. In *Proceedings of the First Workshop on Measurement, Privacy, and Mobility*. doi:10.1145/2181196.2181199

Ge, H., & Lu, Z. (2009). A history-based handover prediction for LTE systems. In *Proceeding of 4th International Conference on Ubi-Media Computing (U-Media)*. U-Media.

Golaup, A., Mustapha, M., Patanapongpibul, L. B., & Group, V. (2009, September). Femtocell access control strategy in UMTS and LTE. *IEEE Communications Magazine*, 117–123. doi:10.1109/MCOM.2009.5277464

GPP. (2011a). *TR25.367 v 10.0.0: Mobility procedures for home node B (HNB), overall description*. 3GPP.

GPP. (2011b). TR36.902 v 9.3.1: Self-configuring and self-optimizing networks (SON) use cases and solutions. 3GPP.

GPP. (2013). TS36.300 v 11.5.0: Technical specification group radio access network, evolved universal terrestrial radio access (E-UTRA) and evolved universal terrestrial radio access network (E-UTRAN), overall description, stage 2. 3GPP.

Hussein, Y., Ali, B., Varahram, P., & Sali, A. (2011). Enhanced handover mechanism in long term evolution (LTE) networks. *Scientific Research and Essays*, 6(24), 5138–5152. doi: doi:10.5897/SRE11.480

Izquierdo, L., & Izquierdo, S. (2009). Techniques to understand computer simulations: Markov chain analysis. *Journal of Artificial Societies and Social Simulation*, 12(1).

Jansen, T., Balan, I., Turk, J., Moerman, I., & Kurner, T. (2010). Handover parameter optimization in LTE self-organizing networks. In *Proceedings of 2010 IEEE 72nd Vehicular Technology Conference - Fall*. IEEE. doi:10.1109/VETECF.2010.5594245

Kim, J., & Lee, T. (2010). Handover in UMTS networks with hybrid access femtocells. In Proceedings of Advanced Communication Technology (ICACT). ICACT.

Kim, T., & Kim, J. (2013). Handover optimization with user mobility prediction for femtocell-based wireless networks. *International Journal of Engineering and Technology*, 5(2), 1829–1837.

Kirmse, A., Udeshi, T., Bellver, P., & Shuma, J. (2011). Extracting patterns from location history. In *Proceedings of the 19th ACM SIGSPATIAL International Conference on Advances in Geographic Information Systems*, (pp. 397–400). ACM.

Lin, P., Zhang, J., Chen, Y., & Zhang, Q. (2011, June). Macro-femto heterogeneous network deployment and management: From business models to technical solutions. *IEEE Wireless Communications*, 64–70.

Mathivaruni, R., & Vaidehi, V. (2008). An activity based mobility prediction strategy using markov modeling for wireless networks. In *Proceedings of the World Congress on Engineering and Computer Science 2008 WCECS 2008* (pp. 379–384). WCECS.

Motorola. (2007). *Long term evolution (LTE): A Technical overview* (Technical White Paper). Motorola.

Network, N. S. (2011). *Improving 4G coverage and capacity indoors and at hotspots with LTE femtocells.* Author.

Palace, B. (1996). *Data mining: What is data mining?* Retrieved from http://www.anderson.ucla.edu/faculty/jason.frand/teacher/technologies/palace/datamining.htm

Quang, B. V. (2012). A survey on handoffs—Lessons for 60 GHz based wireless systems. *IEEE Communications Surveys & Tutorials, 14*(1), 64–86. doi:10.1109/SURV.2011.101310.00005

Qutqut, M., & Hassanein, H. (2012). *Mobility management in wireless broadband femtocells.* Academic Press.

Sadiq, A., & Bakar, K. (2011). Mobility and signal strength-aware handover decision in mobile IPv6 based wireless LAN. In *Proceedings of the International MultiConference of Engineers and Computer Scientists.* IEEE.

Seidel, E., & Saad, E. (2010). LTE home node Bs and its enhancements in release 9. *Nomor Research,* 1–5.

Su, W., Lee, S.-J., & Gerla, M. (2000). Mobility prediction in wireless networks. In Proceedings 21st Century Military Communications Architectures and Technologies for Information Superiority (Cat. No.00CH37155), (pp. 491–495). doi: doi:10.1109/MILCOM.2000.905001

Ulvan, A., Bestak, R., & Ulvan, M. (2010). The study of handover procedure in LTE-based femtocell network. In *Proceeding of 3rd Joint IFIP on Wireless and Mobile Networking Conference (WMNC).* IFIP.

Wang, J., & Prabhala, B. (2012). Periodicity based next place prediction. In *Proceedings of Nokia Mobile Data Challenge 2012 Workshop.* Nokia.

Woo, S., Kang, D., & Choi, S. (2010). Automatic neighbouring BS list generation scheme for femtocell network. In *Proceedings of 2010 Second International Conference on Ubiquitous and Future Networks (ICUFN),* (pp. 251–255). ICUFN. doi:10.1109/ICUFN.2010.5547200

Wu, S.-J. (2011). A new handover strategy between femtocell and macrocell for LTE-based network. In *Proceedings of 2011 Fourth International Conference on Ubi-Media Computing* (pp. 203–208). IEEE. doi:10.1109/U-MEDIA.2011.58

Zeng, Q., & Agrawal, D. (2002). Handoff in wireless mobile networks. In *Handbook of wireless networks and mobile computing* (pp. 1–26). Academic Press. doi:10.1002/0471224561.ch1

Zetterberg, K., Ab, E., Scully, N., Turk, J., Jorguseski, L., & Pais, A. (2010). Controllability for of home eNodeBs. In *Proceedings of Joint Workshop COST 2100 SWG 3.1 & FP7-ICT-SOCRATES.* COST.

Zhang, H., Wen, X., Wang, B., Zheng, W., & Sun, Y. (2010). A novel handover mechanism between femtocell and macrocell for LTE Based networks. In *Proceedings of 2010 Second International Conference on Communication Software and Networks,* (pp. 228–231). IEEE. doi:10.1109/ICCSN.2010.91

ADDITIONAL READING

Aad, I., & Niemi, V. (2010, November). NRC data collection and the privacy by design principles. In Proceedings of the International Workshop on Sensing for App. Phones PhoneSense2010.

Butun, I., Cagatay Talay, A., Turgay Altilar, D., Khalid, M., & Sankar, R. (2010, April). Impact of mobility prediction on the performance of cognitive radio networks. In Wireless Telecommunications Symposium (WTS), 2010 (pp. 1-5). IEEE.

Dahlman, E., Parkvall, S., & Skold, J. (2011). *4G: LTE/LTE-Advanced for Mobile Broadband: LTE/LTE-Advanced for Mobile Broadband*. Academic Press.

Damnjanovic, A., Montojo, J., Wei, Y., Ji, T., Luo, T., Vajapeyam, M., & Malladi, D. (2011). A survey on 3GPP. heterogeneous networks. *Wireless Communications, IEEE, 18*(3), 10–21. doi:10.1109/MWC.2011.5876496

De La Roche, G., Valcarce, A., López-Pérez, D., & Zhang, J. (2010). Access control mechanisms for femtocells. *Communications Magazine, IEEE, 48*(1), 33–39. doi:10.1109/MCOM.2010.5394027

Eagle, N., & Pentland, A. (2006). Reality mining: sensing complex social systems. *Personal and Ubiquitous Computing, 10*(4), 255–268. doi:10.1007/s00779-005-0046-3

Etter, V., Kafsi, M., & Kazemi, E. (2012). Been there, done that: What your mobility traces reveal about your behaviour. In Nokia Mobile Data Challenge 2012 Workshop. p. Dedicated task (Vol. 2, No. 3).

François, J., & Leduc, G. (2005). Mobility Prediction's Influence on QoS in Wireless Networks: A Study on a Call Admission Algorithm. *3rd International Symposium on Modeling and Optimization in Mobile, Ad Hoc, and Wireless Networks.*

Gambs, S., Killijian, M. O., & del Prado Cortez, M. N. (2010, November). Show me how you move and I will tell you who you are. In Proceedings of the 3rd ACM SIGSPATIAL International Workshop on Security and Privacy in GIS and LBS (pp. 34-41). ACM. Giannotti, F., & Trasarti, R. Mobility, Data Mining and Privacy: The GeoPKDD Paradigm.

Hasan, S. F., Siddique, N. H., & Chakraborty, S. (2009, May). Femtocell versus WiFi-A survey and comparison of architecture and performance. In Wireless Communication, Vehicular Technology, Information Theory and Aerospace & Electronic Systems Technology, 2009. Wireless VITAE 2009. 1st International Conference on (pp. 916-920). IEEE.

Huang, Q., Chan, S., & Zukerman, M. (2007). Improving handoff QoS with or without mobility prediction. *Electronics Letters, 43*(9), 534–535. doi:10.1049/el:20070626

Katsaros, D., & Manolopoulos, Y. (2009). Prediction in wireless networks by Markov chains. *Wireless Communications, IEEE, 16*(2), 56–64. doi:10.1109/MWC.2009.4907561

Khandekar, A., Bhushan, N., Tingfang, J., & Vanghi, V. (2010, April). LTE-advanced: Heterogeneous networks. In Wireless Conference (EW), 2010 European (pp. 978-982). IEEE.

Kiukkonen, N., Blom, J., Dousse, O., Gatica-Perez, D., & Laurila, J. (2010). Towards rich mobile phone datasets: Lausanne data collection campaign. Proc. ICPS, Berlin.

Knisely, D., Yoshizawa, T., & Favichia, F. (2009). Standardization of femtocells in 3GPP. *Communications Magazine, IEEE, 47*(9), 68–75. doi:10.1109/MCOM.2009.5277458

Laurila, J. K., Gatica-Perez, D., Aad, I., Blom, J., Bornet, O., Do, T. M. T., & Miettinen, M. (2012, June). The mobile data challenge: Big data for mobile computing research. In Proceedings of the Workshop on the Nokia Mobile Data Challenge. In Conjunction with the 10th International Conference on Pervasive Computing (pp. 1-8).

Mathew, W., Raposo, R., & Martins, B. (2012, September). Predicting future locations with hidden markov models. In Proceedings of the 2012 ACM Conference on Ubiquitous Computing (pp. 911-918). ACM.

Moore, H., & Bhat, S. (2007). *MATLAB for Engineers*. Pearson Prentice Hall.

Neruda, M., Vrana, J., & Bestak, R. (2009, June). Femtocells in 3G mobile networks. In Systems, Signals and Image Processing, 2009. IWSSIP 2009. 16th International Conference on (pp. 1-4). IEEE.

Qiu, Q. L., Chen, J., Zhang, Q. F., & Pan, X. Z. (2009, December). LTE/SAE Model and its Implementation in NS 2. In Mobile Ad-hoc and Sensor Networks, 2009. MSN'09. 5th International Conference on (pp. 299-303). IEEE.

Shaffer, J., Siewiorek, D. P., & Smailagic, A. (2005, October). Analysis of movement and mobility of wireless network users. In Wearable Computers, 2005. Proceedings. Ninth IEEE International Symposium on (pp. 60-67). IEEE.

Stamp, M. (2004). *A revealing introduction to hidden Markov models*. Department of Computer Science San Jose State University.

TS22.220. (2011). Service requirements for Home Node B (HNB) and Home eNode B (HeNB).

TR36.839. (2011). Mobility Enhancements in Heterogeneous Networks.

Vivier, G., Kamoun, M., & Becvar, Z. de MARINIS, E., Lostanlen, Y., & Widiawan, A. (2010, June). Femtocells for next-G wireless systems: the FREEDOM approach. In Future Network and Mobile Summit, 2010 (pp. 1-9). IEEE.

Zhang, J., & De la Roche, G. (2010). *Femtocells: Technologies and Deployment*. New York: Wiley. doi:10.1002/9780470686812

KEY TERMS AND DEFINITIONS

CSG Identity: A unique numeric identifier that broadcast by a CSG/Hybrid cell and used by the UE to check the accessibility of the CSG/Hybrid cell.

Destination Base Station: A base station that the UE will attach soon.

Inbound Handover: Handover from the macrocell to the femtocell.

Inter Femtocell Handover: Handover between the femtocells.

Neighbor Cell List: List of all base stations in the UE's neighborhood.

Outbound Handover: Handover from the femtocell to the acrocell.

PCI: Physical cell identity that use to identify the cell.

Random Data: A data where the user move randomly.

Source Base Station: A base station that serves the UE currently.

Trajectory: A path that the UE uses when moving.

Chapter 6
Software in Amateur "Packet Radio" Communications and Networking

Miroslav Škorić
IEEE Section, Austria & National Institute of Amateur Radio, India

ABSTRACT

In modern (amateur radio) wireless communications, we use computers. Depending on particular situations, such as employers' and personal preferences, users can adopt more or less proprietary operating systems and related end-user programs. In emerging and developing societies, the usage of proprietary software can be costly. Not only that, contrary to so-called "open" software, the "closed" software is not able to motivate its users to upgrade those programs regularly, not only because of high prices and restricted licensing policies, but also because of its nature, which is the "closedness" of program codes, where the end-users are not allowed to change programmed software, and so assist companies in improving features of their software products. Therefore, the authors help prospective newcomers in the amateur wireless communications to become familiar with the "open" software and, as well, to encourage them in implementing many "free" software solutions at home or work.

INTRODUCTION

Most of the time, somebody of us spends money on purchasing new computer hardware. Regardless the type of consumers we are—individual citizens, or schools and companies where we study or work—ICT markets constantly push us and other prospective buyers into replacing our technical equipment in order to, among the others, adopt a newer operating system (OS), or user programs or computing procedures. Every so often, the new versions of our programs are no more capable to run on existing hardware. Even though our computing machines might not be obsolete enough so that they had to be replaced with brand new models, in many occasions software trends succeed initiating new purchases. For example, popular operating systems produced by

DOI: 10.4018/978-1-4666-5170-8.ch006

Microsoft™ are well known as 'hardware-hungry' environments, which means that any new member of the Windows OS family requires better (read: newer generation of) computer hardware. But that is not all. Some producers of computer programs for Microsoft Windows operating systems do not preserve backward compatibilities with previous versions of that OS, hence the new program features often require installing newer versions of Windows that, in turn, requires new hardware, and so on. Sometimes the differences between various versions of an operating system and related end-user programs are so significant that it is not easy to recognize where the 'upgrade' actually is: Is it in the users program itself, or in a specific characteristics of an OS, or it is maybe in a new hardware specification. The final question might be: Is it all worth the upgrade and would not be better if a user could continue with his or her existing hardware and software solutions—without investments in new equipment and without having losses in productivity during periods of those mostly time-consuming upgrades. In the first part of this chapter we are going to elaborate recent advances in some amateur radio software for Microsoft-based 'packet-radio' node installations, where computer hardware and the operating system are seemingly going toward the end of their lifecycle (considered as a hardware + software combination). After that, we will discuss some feasible alternatives for non-Microsoft environments that, in turn, ensure the prolonged usage of the same hardware, and in addition to save the customers from spending money on purchasing new licenses for operating system upgrades.

Background

Amateur radio is an old hobby. Having in mind that most of its practitioners and wireless enthusiasts have been always keen on spending money for *real* technical equipment, such as radio transmitters, antennas, power supplies, and grounding installations, it is for sure that they mostly practiced to save

finances by avoiding unnecessary expenditures in non-primary but expensive purchases. An example of such accessory are computers dedicated for supporting amateur radio activities, such as logging wireless correspondence or calculating distances and contest results. In most cases, average computers of older generations, such as Amiga™, Atari™, Commodore™, as well as the first generations of Intel™-based machines, satisfied all those needs. Particularly, whenever such computers have *not* been used for more complex operations at either home or work, everything is fine. In addition, the mentioned computer brands and types also satisfy basic needs in digital amateur radio communications, such as 'packet radio'. Although that simple computer-related amateur radio mode is often *wrongly* considered as a low-cost replacement for commercial wireless services—particularly in urban city zones, it is a valuable learners' tool for prospective wireless communicators and educational institutions in rural areas and remote locations with no other types of wireless connectivity (Skoric, 2009). By implementing 'packet radio' at its premises, a school can establish an AMUNET, the experimenting wireless network for their students, parents, and teachers.

To start exploring amateur wireless communications and networks, you do not need to obtain brand new computers and newest versions of MS Windows (though nobody is going to stop you from doing that). Instead, you can build your own version of the operating system based on *Linux*, and save a lot of money. To be precise, by rebuilding the core of Linux, its *kernel*, you are able to "eliminate the device drivers that you don't need … [and] … reduce the amount of memory used by the kernel itself" (Welsh & Kaufman, 1995, p. 162). As a result, you will have your system software to be perfectly 'tailored' to satisfy your computer hardware's expectations. As Komarinski and Collett (1998) state "Linux gives you things that Windows … can only dream about … [such as] … source code for the entire kernel … [or] … full configurability of the operating system … [or]

... ability to turn features of the system on and off without rebooting" (p. 3). In the same work, Komarinski and Collett (1998) emphasize that Linux bases its huge pool of software from twenty five years of experience in UNIX programming.

However, early developers of Linux and others who advocated using that then new intriguing OS, realized that immediate and complete replacing *MS DOS*, *PC DOS*, *DR DOS*, Windows, or *OS/2* with Linux was not feasible in many cases, and that existed obstacles for prospective users of Linux to implement the new environment. Among the other, at many homes and businesses there is a variety of end-user programs that were written for Microsoft OS family and many of those programs do not have appropriate Linux counterparts. For example, buyers of portable camcorders often receive proprietary software for managing photos and videos with their devices. Most of the time, camera vendors dispatch their software exclusively for the Microsoft product family. From our personal experience, and despite the best efforts, we did not find (yet) any suitable alternative for combined video/photo managing in a Linux environment—featuring the same characteristics and abilities our camera program provides in Windows. In at least one case, the camera manufacturer did not express interest or plan for releasing Linux-oriented software in the near future. Having that reality in mind, we could have remained depending to Windows for at least that particular end-user application. In that case, full switching to Linux would not be possible just like that, and that fact, in turn, might not motivate newcomers to enter the world of free and open source solutions.

Luckily, Linux programmers offered possibilities for installing Linux on previously Windows-equipped computers, so that it became possible to use both operating systems interchangeably. The approach that we practice is to reserve enough free space on a computer hard disk for Linux installations, which means not to install Windows by using the whole space on the disk. (Nevertheless, if you happen to have your Windows-equipped machine

installed in the latter way, there are options for shrinking Windows partitions, so to acquire free space for adding Linux.) Depending on users' needs and preferences, one of possible approaches can be to split the space on the hard disk into, say, two halves and put each operating system onto one half. Although Bill Wohler, the first creator of the on-line manual *Linux+WindowsNT mini-HOWTO*, advocated installing "a minimal Linux" at first, and then to add a Microsoft product, the second maintainer of the document practiced rather the opposite installation—where Windows was intended to be the first one operating system, and Linux as the second one (Skoric, 2010a). In that case *and* with newer Linux distributions, the installer *should* recognize that Windows is already there and according to that it would suggest the most optimal process in organizing the remaining space on the disk. However, there is always a possibility that a user has an older Linux distribution package where the installer might not be 'clever enough' to recognize an existing OS and act accordingly. In that case, the implementers of Linux OS might search for some external software utilities, such as *Partition Magic*™ utility by Power Quest™. It is a graphical tool that is able to 'see' all partitions on all hard disks a user may have on his or her computer. The best thing is that by using that utility the user can make some changes with existing computer partitions but *not* to destroy data on them.

It is important to install a program routine that will give you an option what operating system to activate at each computer boot. That routine is known as a 'Linux loader' (or *LI*nux *LO*ader, or just *LILO* in short), although other options such as *Grub* are also available. While being computer traditionalists, we are more oriented to LILO (Skoric, 2010b). That program is capable to boot various operating systems and allows a user to select at startup time which one OS to boot (Welsh & Kaufman, 1995). In many installations, LILO comes with a red-orange introductory screen, displaying so-called *LILO boot menu*, see Figure 1.

Figure 1. LILO boot menu with two options

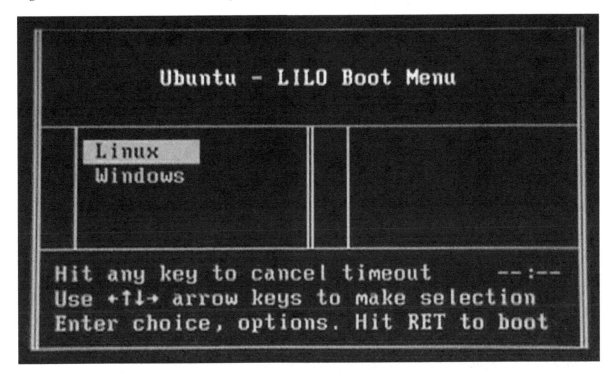

The photography of the screen in Figure 1 represents a menu provided by LILO that was included with Linux distribution *Ubuntu 10.04.4 LTS*, and the version of the LILO program was *1:22.8-8ubuntu1*. Software updates and changes in LILO versions appear relatively rarely, and until recently the very same boot screen was produced with other Linux distributions, such as *Debian 6.0.7*. When we noticed that there was a possibility to update LILO from Debian package, from so-called 'stable' version named *1:22.8-10*, to a newer one named *1:23.2-4*, we decided to upgrade.

As it can be seen in Figure 2, the program's basic appearance slightly changed. (For example, there is a new heading LILO 23, as well as some changes in the text located at the bottom of the menu screen.) However, the main and structural change occurred in the file */etc/lilo.conf* (both versions of that parameter file are provided in Appendix A and Appendix B). As you can see in those appendices, the program parameter specifying the boot device (the line starting with *boot=*) changed its value from, an example, *boot=/dev/sda* to *boot=/dev/disk/by-id/ata-Maxtor_6E040L0_E1728VJN*, and the parameter specifying the device that should be mounted as root (the line starting with *root=*) changed its value from *root=/dev/sda5* to *root="UUID=83c6882b-88a5-433b-9712-0ffa1663bb57."* It seems that the former notation style used for describing hard disk partitions was not precise enough, so that the notation change introduced more precise approach. If we compare the remaining content of those two files, we can also notice that it is possible to define more than two operating systems. (In fact, we did that in our second example, by adding the entry for a 'LinuxOLD' kernel-image, which is useful because we wanted to have more than one Linux kernel installed).

An example of three computers running LILO is shown in Figure 3. Depending on preferred choice, all three machines can be booted into Windows (left half in Figure 4), or into Linux

Figure 2. LILO boot menu with three options

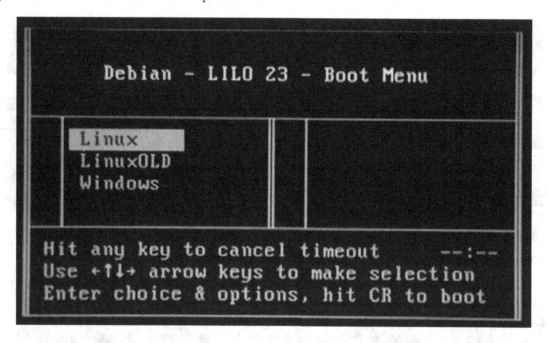

Figure 3. Three computers in a LAN displaying LILO menu

(right half in Figure 4, where two computers on the left run Debian Linux, while the third machine on the right runs Ubuntu Linux). We made those choices for the sake of simplicity, which means that other combinations were possible and readers are encouraged to experiment with various operating systems whenever possible.

DISCUSSING SOFTWARE SOLUTIONS

Adopting Linux Operating System

In opposite to buyers of operating systems made by a popular software company in Redmond, WA, the average user of Linux is going to have his or her hands wet by performing various administrative tasks, such as system maintenance or the software upgrades. But it gives a lot of joy. As others also realized:

Running a Linux system is not unlike riding and taking care of a motorcycle. Many motorcycle hobbyists prefer caring for their own equipment—routinely cleaning the points, replacing worn-out parts, and so forth. Linux gives you the opportunity to experience the same kind of "hands-on" *maintenance with a complex operating system. (Welsh & Kaufman, 1995, p. 109)*

Tranter (1997) states that Linux offers more than DOS or Windows when it comes to implement amateur radio. For example, Linux has native support in its kernel for AX.25 protocol, which leads to the tighter integration between the amateur radio applications and Linux-based networking interfaces, device drivers, and utilities. With that operating system, it is easier to have multiple packet radio interfaces with more than one simultaneous connection over each interface. In addition, a computer running Linux can act as a router to forward packet radio traffic to the Internet or to the local area network (Tranter, 1997).

It is usually suggested to add Linux to an existing (or to a newly installed) Windows platform. Depending on a particular situation, a Microsoft product can spread on the whole hard disk, which is common for older installations of an OS or it can be installed on a specific, dedicated portion of the hard disk, which is common for newer installations where a second OS had been also planned. Whatever the particular situation is, a prospective experimenter has to allow enough free space for Linux partitions. The details on installing Linux alongside with Windows are beyond the theme

Figure 4. The same computers booted into Windows (left), or into Linux (right)

of this book, and one of possible sources on that segment might be (Skoric, 2010a). Nevertheless, LILO should be taken into consideration as a possible choice for a boot loader during the Linux installation, and if everything goes well after the final installation step that requires a system restart, the user should be presented with welcome menus displayed in Figures 1-3. Before going further with amateur radio program deployment, the user should check if either OS is capable to boot and shutdown properly, so to get the screens depicted in Figure 4.

Our research with operating systems for amateur radio practitioners included *Microsoft Windows XP™* product family, as shown in the left half of Figure 4. According to the personal experience that OS proved as a reliable solution for the owners of computers of the same or similar generation as of Windows XP. That means it served well for users whose machines were produced around the beginning of the new millennia. (Of course, *XP* works quite fine on newer equipments too.) But, one of the issues with all Microsoft products is the end of their planned lifespan, which results in stopping technical support, releasing important security patches, etc. Practically, it means that there are no more improvements for a particular product after a specific date. In case of Windows XP, the end of support has been scheduled for April 2014, which means there will not be security patches and other assistance from Microsoft after that date. Among predictable consequences, millions of computers still using that OS will become seriously vulnerable in the near future. Although that reality might not be frightening for isolated systems (those without Internet connectivity and with restricted file exchange by using uncontrolled *BYOD*s /stands for *Bring Your Own Device*/ such as USB flash memories or other portable data carriers, it is the right time to plan the future usage of existing hardware that support our wireless operations.

The inevitable end of the XP's lifespan should be another motivating factor for its users to add Linux to their Windows-equipped machines, and take some practice with the new environment *before* Microsoft ceases technical support for its now 14-year-old OS. Making a decision to replace Windows XP with a Linux OS, would be an ideal solution for those who do not want to waste their existing computers because of increased hardware requirements mandated by newer operating systems made by Microsoft.

However, that does not mean that different working environments, including Microsoft operating systems, should not coexist in our wireless and other communicating activities. Furthermore, there are many opportunities for those who plan to continue operating under Windows or even older solutions. Among the others, Skoric (2012) offered some examples in using Microsoft Windows-related amateur radio-relay software *BPQ* in its 32-bit versions (also known as *BPQ32*). The author experimented in a closed LAN, and provided tests that simulated real wireless networks. Besides Windows-based computers, there was an old Microsoft DOS™-equipped computer in the same LAN; hence it was possible to describe various DOS-to-Windows interactions, including functional collaboration between different amateur radio programs in mixed computing environments. For example, there existed (and still exist) many versions of BPQ for MS DOS or PC DOS™ operating systems, as well as a variety of other DOS-based amateur radio applications for outdated computers, such as *XRouter*, *JNOS*, or *DosFBB*. The results of Skoric's work proved that amateur wireless enthusiasts can freely implement software that best fit to their home or work equipment and personal budgets. An important advice for newcomers in the amateur wireless communications and networking is that it is much better to put extra money in radio devices and antenna equipment, rather than into computers.

In the other work, Skoric (2013) discussed security issues and solutions in the amateur radio networks, where he mentioned some Linux-related amateur radio programs with characteristics similar to their Windows-related counterparts. For example, the program *FPAC* for Linux-based wireless radio-relays can replace BPQ32. Similarly, the radio mailbox software *LinFBB* can replace *WinFBB*, etc. However, the author did not provide detailed installation procedures and most suitable approaches for different user situations; neither had he covered specific considerations related to appropriate program choices.

Installing Node Software BPQ32 in Microsoft Windows XP

BPQ has been continually improved for more than two decades now, thankfully to the enthusiasm and tirelessness of its author John Wiseman, G8BPQ. For the purpose of this chapter, we decided to

upgrade an older instance of the program, BPQ32 version 4.10n (dated July 2010), which we used more-less successfully in the past (Skoric, 2013). Because of the significant gap between the used version and the newer ones, one of which was 5.2.9.1 (dated September 2012, please also note the change in numbering convention, see Figure 5), we had to take an intermediate step that was suggested by Ron Stordahl, AE5E, (Stordahl, 2012).

In fact, besides the changed naming, the version 5.2.1.3 (dated October 2011) was the last one that supported both the old method of separate configuration files, and the new method of a single combined file. (In other words, the old method had placed working parameters into several files, which meant that main program parameters were placed in *BPQCFG.TXT* that, in

Figure 5. The new look of BPQ32 console

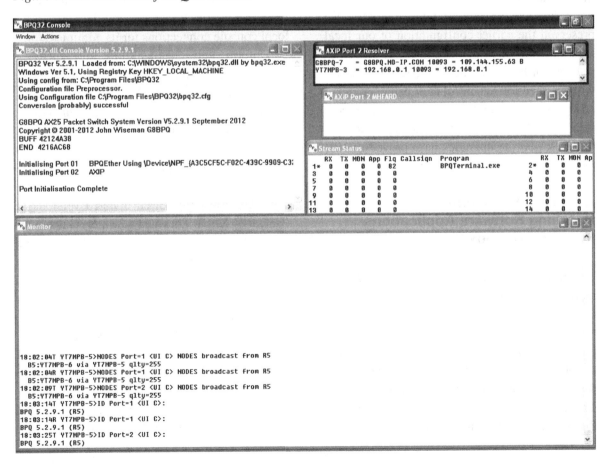

turn, had to be re-compiled by the external program *BPQCFG.EXE* each time when a user changed values of one or more parameters. In addition, there were other files, among which *BPQAXIP. CFG* and *BPQETHER.CFG* that handled parameters for *AXIP* and *Ethernet* configuration, respectfully. In opposite to that, the new method combined almost all configuring statements into *BPQ32. CFG* and it was not needed to perform subsequent re-compilations anymore.)

Newer versions of BPQ32 included even more upgrades and new features, such as the capability to transfer node location data to a centralized Web server (in case the node computer is connected to the Internet). That allows a new BPQ-based wireless node to be shown on a global map that is accessible on the Website http://nodemap.g8bpq. net:81/. An example of a new configuration file for an 'intermediate' version (v. 5.2.1.3) is given in Appendix C.

During the process of node upgrades, we preserved the earlier configuration of our local area network, which included so-called *BPQether* protocol for each of the three BPQ32-based nodes, with an intention to control their behavior and to spot eventual program malfunctions. The process went smoothly with all three machines, and tests that followed the upgrades have proved that all installations had been successful.

A Historical Example of Installing Software Applications for Red Hat Linux 6.2

In Figure 6 we can see an instance of *XFBB* radio mailbox software that was installed as a Linux GUI *application.*

Some time ago it was quite usual to install that type of amateur radio software in a relatively simple way—almost not much complicated than the way we handle Windows-based programs. That means a Linux user was directed to extract the content of a compressed archive file, and by

using a pair of simple commands to install the whole program, including its graphical representation on the screen. More details on doing that can be found in an on-line operating manual for FBB (Skoric, 2010c).

In those times, some software installers would even add a new program icon on the desktop, similarly to the programs for Microsoft products; otherwise it was needed to execute a specific command within the Linux terminal, such as *xterm* (a kind of a 'command prompt' in Linux) - in order to start the new program. Because of the obvious similarity with deploying Windows-based amateur radio applications, that approach was probably one of the most proper ways for motivating experimenters to switch from Microsoft environments to non-Microsoft alternatives. As shown in Figure 6, the application invokes three windows (from the top-left, clockwise): the main configuration and administration, the window for monitoring radio frequency, and the window for local console connections for its administrator.

Unfortunately but thankfully to maturity of XFBB's more popular 'older brother' *WinFBB*, which is the version for Windows-based machines, as well as thankfully to the strong inclination in many Linux programmers against using Linux GUI, XFBB has never reached the visual quality in its graphical presentation—compared to vivid and colorful appearance of WinFBB. The same fate followed newer, updated versions of the early XFBB, nowadays widely known as *LinFBB*, which have been probably produced with a single intention to run in background, as a Linux *daemon*. Kirch (1995) defined a daemon as "a program that opens a certain port and waits for incoming connections" (p. 125). In our case, it means that LinFBB would wait silently (or not, if we configured it to announce its presence every few minutes) for the radio amateurs to connect to the server and manipulate their emails. Nevertheless, the operators of LinFBB servers have at least two options. The first one is to access the Linux

Figure 6. An earlier version of FBB mailbox software for Linux

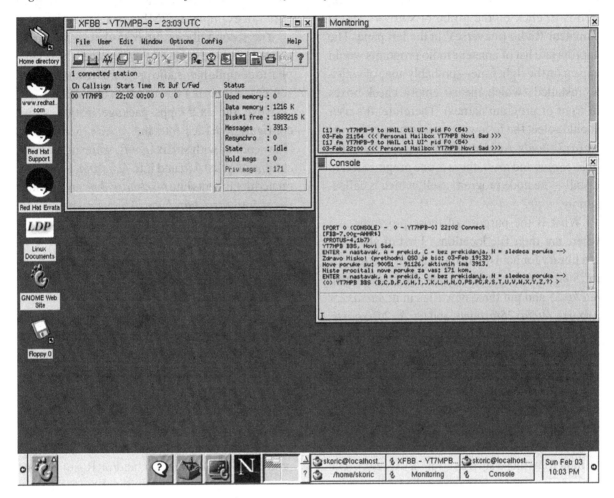

machine remotely from a Windows-based computer that must be equipped with *xfbbW*, which is a Windows client program for LinFBB. The second option is to compile *xfbbX*, a Linux GUI client program, and run it on the same machine where the LinFBB server is active. (To make the second option possible, a user has to install Linux by including the graphical user interface.) However, and to make things worse, even the newer iterations of xfbbX lack several features that XFBB had in the past, such as the absence of a message editor for properly formatting texts in replying mails, or the option for using computer mouse for some administrative tasks, etc.

Installing 'Prefabricated' FBB Software in Ubuntu Linux 10.04.4 LTS

The easiest way for beginners to do that type of node installation is to open an included utility called *Synaptic Package Manager* and check if the 'All' group is selected within its left pane (although it should be selected by default). Then, the user should search for and select *libax25* and *libax25-dev* from the right pane, in order for those programs to be prepared for installation. After those programs are selected, the user should click on 'Apply' to install them. That is the first phase in installing amateur radio software. The second step is to click on the group of software called 'Amateur Radio (universe)' in the left pane. The appropriate list of amateur radio programs would appear on the right pane—probably none of which as installed (which means empty check-boxes in front of program names). Therefore, the user should select the following programs: *ax25-apps* and *ax25-tools*. After installing all those, so-called *libax*, *apps*, and *tools*, there is one program left to install—the node program itself, which is called, simply, *node*.

What is the purpose of those programs and where are their file locations? First of all, libax is a library for ham radio applications that use the ax25 protocol and it adds a new directory named */etc/ax25* and put three new files in it: */etc/ax25/axports*, */etc/ax25/nrports*, and */etc/ax25/rsports*. It also adds some new libraries into */usr/lib*, new documents into */usr/share/doc*, and manual pages into */usr/share/man*. The main purpose of *axports* is to define amateur packet-radio 'ports' within the Linux OS (although the term 'port' here does not mean physical computer connectors i.e. radio ports such as *COM1*, *COM2*, etc., or in Linux terminology those such as *ttyS0*, *ttyS1*, etc.). The second file, *nrports*, defines so-called *netrom* ports i.e. packet-radio ports that use netrom protocol rather than the basic amateur radio one - *AX.25* (that *axports* uses). Finally, there is the third file, *rsports*, which handles ports associated with so-called *Rose* protocol. The user should edit axports and nrports manually, while rsports will be edited automatically by Rose-related programs. To help you learning basic configuration details of axports and nrports, you should consult newly-installed manual pages by using commands: *man axports*, and *man nrports* respectively. Keep in mind that in order to have things working properly, you have to assign different amateur radio identifiers, aka *callsigns*, to each axport and nrport you add to your Linux system. Actually, you can use the same callsign for all ports, but with adding different suf-

fixes, aka *SSIDs* (substation identifiers) for each port: For example, YT7MPB-5, YT7MPB-6, etc.

The second library, libax25-dev, is primarily needed for those 'sysops' (system operators) who plan to compile ham radio programs from software sources, so it is not discussed in this section.

Next, the ax25-apps package installs some configuration files into the */etc/ax25* directory [files ending with suffix *.conf*], some user applications into */usr/bin* and into */usr/sbin*, as well as related documents into */usr/share/doc*, and manual pages into */usr/share/man*. The user should edit *.conf* files, although we did that with */etc/ax25/ax25ipd.conf* only.

Finally, ax25-tools install more configuration files, executables, and docs.

Have in mind that the majority of *.conf* files (if not all) come with their authors' sample data filled-in, so it is important to edit the files and fill in appropriate parameters related to the particular system operator identification and the newly-planned amateur radio node. For example, the file */etc/ax25/ax25d.conf* should be edited in order to have appropriate programs responding to incoming requests from users using different protocols (AX.25, netrom, and/or Rose). In our case, we configured incoming AX.25 requests to the callsign YT7MPB-5 to be responded by the command */usr/sbin/node* (see Appendix D). Besides the .conf files, there are some other files to be checked, such as for example */etc/ax25/nrbroadcast*. That file is responsible for how a node is going to periodically announce its presence to the surrounding node network, and by using which particular port to perform that activity. Related to the frequently mentioned AX.25 protocol, it is a derivation from the X.25 protocol, mainly used in public data networks, where sophisticated hardware such as a *Packet Assembler/Disassembler* has been required (Kirch, 1995).

After libax, apps, and tools have been configured, the next phase is to set parameters for node program. That is done by editing */etc/ax25/node.conf* file. Terry Dawson, VK2KTJ, was among the first ham radio enthusiasts who provided written information on configuring the node software, as well as how to install and configure native AX.25 support in Linux OS (Dawson, 1996).

For the simplicity of this study, and to help beginners to start quickly, we did not want do go deeply into parameter tweaking. Instead, we rather gave a basic setup and let you a freedom to make additional fine-tuning by yourself.

To check if the configuration phases have been done properly, you should type several command by using terminal application (i.e. a command prompt mode). Later, after all commands have been proven to give expected outputs, it is possible to insert those commands into the file */etc/*

rc.local so they can be executed automatically at each Linux boot. (Of course, only in those situations where the users want their nodes to be active immediately after they switch on their computers, see Appendix D.) The smooth interaction between a Linux node and BPQ32 is shown in Figure 7.

A user of 'prefabricated' node software (*LinuxNode*) should be warned that there have been discussions reporting on security issues with the program. It is unknown if those issues are resolved. As a protective measure—especially if the same computer has Internet connectivity, a user may implement alternatives such as Unode, Uronode, and so on.

Figure 7. A new Linux node R5:YT7MPB-5 communicates the Windows node R3:YT7MPB-3

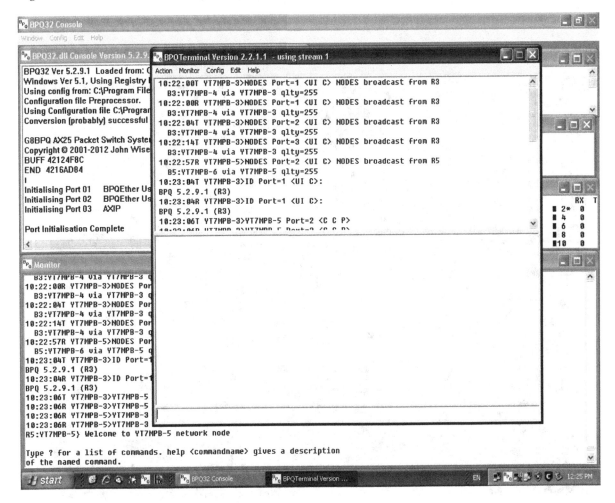

Although a user may want his or her computer to serve as a node only, it is always possible to add some mailbox (a.k.a. BBS) software. On the amateur radio software market, there is more than one opportunity for Linux users. However, the most popular mailbox server program is FBB, and it is already listed in Synaptic Package Manager. With Ubuntu 10.04.4 it comes in version *7.04j-8.2* that is actually not a new one, but for the beginners' first exercises it would fit. The installation process goes the same way as with the other amateur radio programs mentioned in this section.

After installing and configuring FBB on your Linux system, you may want to test its functionality. One possible way to perform that is to open Ubuntu's *Terminal* (i.e. the 'command line'), such as the one in Figure 8:

In this example, the command was *sudo fbb*, followed by the operator's password. In a couple of seconds, the mailbox was started and reachable from the network, as it can be seen in Figure 9.

In the same time, it is possible to perform some operations from within Linux desktop, such as checking the node users, and monitoring the mailbox (and node) frequency, see Figure 10. What does not exist after the program installation is a graphical interface, and that fact probably suggests that GUI-oriented users should link their Linux mailbox servers to other, Windows-based machines, equipped with *xfbbW* clients.

When shutting down such a Linux computer, it would broadcast an informative message to the radio channel, informing other network participants of its incoming unavailability, such as described in Figure 11.

Figure 8. Starting newly-installed LinFBB from the command line

Figure 9. Accessing LinFBB from the amateur radio network

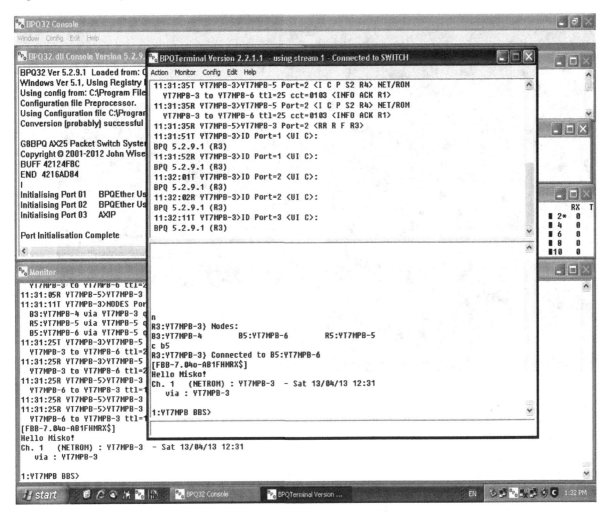

Installing 'Compiled' FBB Software in Debian Linux 6.0.7

Ubuntu Linux is mainly based on Debian Linux distribution. Within the Linux community, it is usual to consider Debian as an 'older brother' or, as some say, a 'father' of Ubuntu. However, despite many similarities between those two products, there are a plenty of differences. For the purpose of this study, we want to warn the readers that Debian is also considered as a more 'serious' distribution, which means it never includes fully-new versions of end-user software. Instead of that, Debian developers practice longtime tests over new versions of their software, which means they do not suggest using programs that they have not fully tested. As a consequence, many inexperienced users of Debian distribution tend to see it as an overformal, too-traditional and old-fashioned product. That is far from truth, and instead of that one can be sure that there will be less program bugs with Linux systems based on Debian distributions, and such systems will run more stable in a long run—compared to Linux distributions that tend to involve newer versions so often. And last, but not least, is that our experiments with older hardware and slower machines

Figure 10. Checking node users (left), and monitoring the frequency (right)

proved that Debian OS works faster and more reliable than Ubuntu.

To make a difference with the previous section—where we implemented an easier approach with installing 'prefabricated' amateur radio software onto an Ubuntu system, in this section we are going to use software *sources* and *compile* them on a Debian-based computer. For the sake of simplicity, the term of compiling can be understood as a way of preparing and fine-tuning software for a particular piece of hardware. That approach is supposed to dispatch a computer program that should better fit to a particular hardware and system software combined.

Having in mind that the actual software maintainer of LinFBB, Bernard Pidoux, F6BVP, suggested implementing his updated ('patched') versions of basic AX.25 programs (Pidoux, 2012), we downloaded the following software packages from his Website:

```
libax25-0.0.12-rc2.patched_f6bvp.tar.
bz2
ax25-apps-0.0.8-rc2.patched_f6bvp.
tar.bz2
ax25-tools-0.0.10-rc2.patched_f6bvp.
tar.bz2
```

Now we are going to perform some very similar procedures with all three archives. At first, we will make the following steps with the *libax* package, by typing the following commands within Debian's root terminal:

```
mkdir /usr/local/src/ax25
cp libax25-0.0.12-rc2.patched_f6bvp.
tar.bz2 /usr/local/src/ax25
cd /usr/local/src/ax25
tar xf libax25-0.0.12-rc2.patched_
f6bvp.tar.bz2
cd libax25-0.0.12-rc2.patched_f6bvp
./configure
make
```

Figure 11. Switching off a Linux-based amateur radio node and mailbox

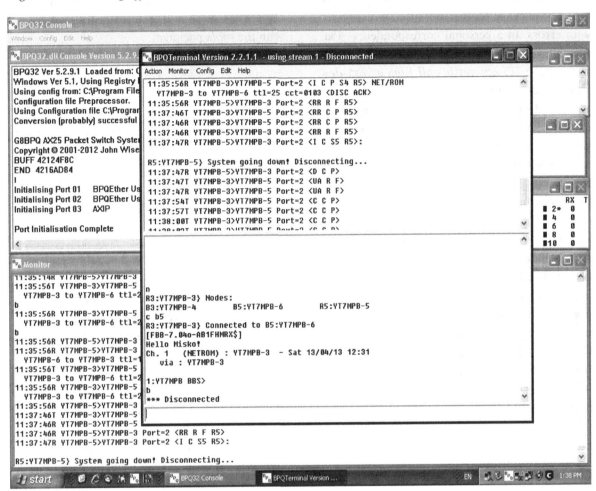

```
make install
make installconf
```

The procedure above creates samples of new configuration files and places them into */usr/local/etc/ax25* directory (files named axports, rsports, and nrports), include files into */usr/local/include*, documentation man pages into */usr/local/man*, and AX.25 libraries into the */usr/local/lib/* directory.

In his documentation, Bernard suggested to add a line with */usr/local/lib* at the end of */etc/ld.so.conf* file and run the command */sbin/ldconfig* (although he added that the change in *ld.so.conf* and commanding *ldconfig* could be performed automatically by running a specific script described

later in his manual). Nevertheless, we followed the former approach.

The second phase is to perform the similar procedure but this time with *apps* package and compile AX.25 applications. Therefore, we are going to execute the following commands in the terminal:

```
cp ax25-apps-0.0.8-rc2.patched_f6bvp.
tar.bz2 /usr/local/src/ax25
cd /usr/local/src/ax25
tar xf ax25-apps-0.0.8-rc2.patched_
f6bvp.tar.bz2
cd ax25-apps-0.0.8-rc2.patched_f6bvp
./configure
make
```

Notice: When we tried the command *make* with the *apps* package for the first time, it responded with errors. In a personal correspondence, Patrick NE4PO, suggested to perform two Linux commands: (1) *apt-get install libncurses5-dev*, and (2) *apt-get build-dep ax25-apps*. It was obvious that our Debian installation did not have *libncurses5-dev* package at that time. Even though we installed it, a repeating *make* resulted again with some errors. On the other hand, performing the step in the second suggestion - (2) seemed to require from a user to install several dozens of megabytes in additional programs, including somewhat irritating warning of various software dependencies - including libax that we already had installed in our machine. To overcome that unpleasant situation, we installed the package *libax25-dev*, and then tried to update/upgrade our Linux system as much as possible. When the OS reported that no more updates were available, we performed a system shutdown and rebooted it subsequently. After that, the command 'make' responded successfully.

```
make install
make installconf
```

The steps with the *apps* package in the previous procedure produces new configuration sample files and places them into */usr/local/etc/ax25* directory (files named *ax25ipd.conf, ax25mond. conf*, and *ax25rtd.conf*), additional documentation man pages are placed into */usr/local/man*, and application tools in */usr/local/sbin* and */usr/ local/bin* directories.

As Bernard also suggested, we downloaded *rc.init.script* from his Website and typed the following:

```
chmod a+x rc.init.script
./rc.init.script
```

Both commands ran flawlessly.

The third phase is to compile and install AX.25 tools. The user should perform the following steps with the *tools* package:

```
cp ax25-tools-0.0.10-rc2.patched_
f6bvp.tar.bz2 /usr/local/src/ax25
cd /usr/local/src/ax25
tar xf ax25-tools-0.0.10-rc2.patched_
f6bvp.tar.bz2
cd ax25-tools-0.0.10-rc2.patched_
f6bvp
./configure
make
make install
make installconf
```

The procedure above produces new configuration sample files and positions them into */usr/local/ etc/ax25* directory (files named *ax25d.conf, ax25. profile, ax25spawn, nrbroadcast.conf, rip98d. conf, rxecho.conf*, and *ttylinkd.conf*), additional documentation man pages into */usr/local/man* and application tools in */usr/local/sbin* and */usr/ local/bin* directories.

The next important step is to configure parameter files. To enable AX.25 functionality, as well as to explore possibilities of testing the newly-installed packet-radio system in the local area network, we will partly edit */usr/local/etc/ax25/ axports*, by inserting proper callsign and SSID, as well as the file */usr/local/etc/ax25/ax25ipd.conf*, by making it looking similar to the corresponding *ax25ipd.conf* we have on the other Linux-based computer in the same LAN.

The activities mentioned up to now will result with basic AX.25 programs being installed on our Debian-based system.

In order to complete the installation of an amateur radio-relay (a 'node'), we will follow Bernard's suggestion to try another node-related software—a French product named *FPAC*. As with the other program sources mentioned in this section, we downloaded *fpac-3.27.18.tar.*

bz2 from Bernard's Website and performed the following steps:

```
cp fpac-3.27.18.tar.bz2 /usr/local/
src/ax25
cd /usr/local/src/ax25
tar xf fpac-3.27.18.tar.bz2
cd fpac-3.27.18
./configure
make
make install
make installconf
```

The procedure above creates new subdirectories and configuration files for *FPAC* and places them into */usr/local/var/ax25* and */usr/local/etc/ax25*, while FPAC application programs are added in */usr/local/sbin*.

By the way, *rc.init.script* suggested a pair of simple changes in parameters within *fpac.sh*—related to the sysop's radio identification (a.k.a *callsign*), so according to that you can type the following:

```
cd /usr/local/sbin
nano fpac.sh
```

In order to simplify learning process, it is suggested to download sample configuration files such as *fpac.conf*, *fpac.nodes* and *fpac.routes* from Bernard's Website, and use them to inform you on the proper configuration of a new node. In that manner, we can adopt necessary changes within the following files:

```
/usr/local/etc/ax25/fpac.conf
/usr/local/etc/ax25/fpac.nodes
/usr/local/etc/ax25/fpac.routes
```

The functionality of a newly-installed FPAC node can be now tested by executing the command:

```
/usr/local/sbin/fpac.sh -start
```

In a couple of seconds the node would broadcast to the network and could be immediately accessible by surrounding radio amateurs and nearby wireless network nodes. You will recognize that there are other node-related parameter files to be edited, such as *fpac.hello* that is displayed when a user connects to the node, or *fpac.info* that is displayed when a user run the I(nfo) command after he or she is connected to the node.

After that, it might be suitable to make an AX.25 startup script. Once again, it is possible to download the script from Bernard's site and to place it into the */etc/rc.d/* directory. In our operating system configuration, that directory did not exist, so we did the following:

```
mkdir /etc/rc.d
cp rc.ax25 /etc/rc.d/rc.ax25
chmod a+x /etc/rc.d/rc.ax25
```

Notice: Before trying to execute that script (*/etc/rc.d/rc.ax25*), you should familiarize yourself with its content and function, because it is user-specific. That means, if not edited properly this script would try to access processes and devices that you might not have planned and/or configured for your particular wireless node, and that would result in returning one or more error messages. To avoid that annoying situation, we suggest beginners not to execute the script as it is, but instead to activate script commands manually—one by one, in a terminal window (by preceding every command with a '*sudo*'), and learn the output of each command. If and when the user is sure that all parts of the script satisfy his or her requirements, the script might be executed in whole—at first manually and later automatically.

For some users it might sound as interesting to initialize basic AX.25 programs and activate their FPAC wireless nodes each time the computers start. In order to enable that opportunity, the script */etc/rc.d/rc.ax25* should be invoked by the other script, which is */etc/rc.d/rc.local*. (In our installation, it was */etc/rc.local*.) Therefore you could add the following lines into your *rc.local* script:

```
/etc/rc.d/rc.ax25
/usr/local/sbin/fpac.sh -start
```

However, that approach did not prove as effective in our Debian-based installation, which means that it might be distribution-dependent. Nevertheless, if the automatic start of the wireless node is not possible due to whatever reason, or just because it is not wanted, there is a set of commands to start the node manually. In the list that follows this paragraph, you can see the commands for activating FPAC node on a Debian-based computer in a way where the node serves as a simulated radio-relay station in between two other wireless nodes. That is an excellent way to learn the amateur radio technology by simulating processes in a computer network.

```
/sbin/ifconfig bpq0 hw ax25 yt7mpb-15
/sbin/ifconfig bpq0 192.168.0.1 net-
mask 255.255.255.0 up
/sbin/ifconfig bpq1 hw ax25 yt7mpb-14
/sbin/ifconfig bpq1 192.168.2.1 net-
mask 255.255.255.0 up
/usr/local/sbin/nrattach -i
192.168.0.1 netrom
/usr/local/sbin/nrattach -i
192.168.0.1 netbbs
/usr/local/sbin/nrattach -i
192.168.2.1 netrom
/usr/local/sbin/nrattach -i
192.168.2.1 netbbs
/usr/local/sbin/ax25d
/usr/local/sbin/netromd -i
/usr/local/sbin/fpac.sh -start
```

The above set of commands is suitable for those users who want to experiment with wireless nodes only. For users who want to add a wireless mailbox (a.k.a. *BBS*), it is needed to continue with software installations by compiling the mailbox

server program—LinFBB. That procedure has been explained in other sources, such as in (Skoric, 2010c), but for those researchers who want to have a graphical user interface of the mailbox program, the question is how to compile an appropriate GUI front-end of LinFBB. The procedure may include the following steps with X11 (graphical) parts of LinFBB software sources (we used the source package *xd705c09-src.tar.bz2*):

```
cd /usr/local/src/ax25/fbbsrc.705c9/
src/X11
make
```

In our experiment with Debian installations the command 'make' returned some errors, and the graphical front-end was not created. Obviously, our system needed some additional software for producing GUI applications and therefore we decided to add *libxpm-dev* package. After installing that package, the command 'make' was successful, as follows:

```
make
gcc -Wall -Wstrict-prototypes -O2 -g
-funsigned-char  -D__LINUX__ -DPROTO-
TYPES -I../../include -I/usr/X11R6/
LessTif/Motif2.0/include -I/usr/
X11R6/include  -I/usr/include/ne-
tax25 -o xfbb xfbb.o ../arbre.o ../
autobin.o ../balise.o ../bidexms.o
../conf.o ../console.o ../date.o ../
devio.o ../dos_1.o ../dos_dir.o ../
driver.o ../drv_aea.o ../drv_ded.o
../drv_hst.o ../drv_kam.o ../drv_
mod.o ../drv_pop.o ../drv_sock.o ../
drv_tcp.o ../edit.o ../ems.o ../
error.o ../exec_pg.o ../fbb_conf.o
../fortify.o ../forward.o ../
fwdovl1.o ../fwdovl2.o ../fwdovl3.o
../fwdovl4.o ../fwdovl5.o ../
fwdovl6.o ../fwdovl7.o ../fwdutil.o
../gesfic.o ../ibm.o ../info.o ../
init.o ../init_srv.o ../init_tnc.o
```

```
../initfwd.o ../initport.o ../k_
tasks.o ../kernel.o ../lzhuf.o ../
maint_fw.o ../mbl_edit.o ../mbl_
expo.o ../mbl_impo.o ../mbl_kill.o
../mbl_lc.o ../mbl_list.o ../mbl_
log.o ../mbl_menu.o ../mbl_opt.o ../
mbl_prn.o ../mbl_read.o ../mbl_rev.o
../mbl_stat.o ../mbl_sys.o ../mbl_
user.o ../mblutil.o ../md5c.o ../
modem.o ../nomenc.o ../nouvfwd.o ../
pac_crc.o ../pacsat.o ../qraloc.o
../redist.o ../rx25.o ../serv.o ../
serveur.o ../statis.o ../themes.o
../tnc.o ../tncio.o ../trait.o ../
trajec.o ../trajovl.o ../variable.o
../warning.o ../watchdog.o ../
waveplay.o ../wp.o ../wp_mess.o
../wpserv.o ../xfwd.o ../xmodem.o
../yapp.o xfbbabtd.o xfbbcnsl.o
xfbbedtm.o xfbbedtu.o xfbblcnx.o
xfbbmain.o xfbbpndd.o xeditor.o -L/
usr/local/lib -lm -lax25 -L/usr/
X11R6/lib -L/usr/X11R6/LessTif/Mo-
tif2.0/lib -lXm -lXt -lXpm -lXext
-lX11
gcc -Wall -Wstrict-prototypes -O2 -g
-funsigned-char  -D__LINUX__ -DPROTO-
TYPES -I../../include -I/usr/X11R6/
LessTif/Motif2.0/include -I/usr/
X11R6/include  -I/usr/include/netax25
-c -o xfbbXabtd.o xfbbXabtd.c
gcc -Wall -Wstrict-prototypes -O2 -g
-funsigned-char  -D__LINUX__ -DPROTO-
TYPES -I../../include -I/usr/X11R6/
LessTif/Motif2.0/include -I/usr/
X11R6/include  -I/usr/include/netax25
-c -o xfbbXcnsl.o xfbbXcnsl.c
gcc -Wall -Wstrict-prototypes -O2 -g
-funsigned-char  -D__LINUX__ -DPROTO-
TYPES -I../../include -I/usr/X11R6/
LessTif/Motif2.0/include -I/usr/
X11R6/include  -I/usr/include/netax25
-c -o xfbbX.o xfbbX.c
gcc -Wall -Wstrict-prototypes -O2 -g
-funsigned-char  -D__LINUX__ -DPROTO-
TYPES -I../../include -I/usr/X11R6/
LessTif/Motif2.0/include -I/usr/
X11R6/include  -I/usr/include/ne-
tax25 -o xfbbX_cl xfbbX.o ../md5c.o
xfbbXabtd.o xfbbXcnsl.o -L/usr/local/
lib -lm -lax25 -L/usr/X11R6/lib -L/
usr/X11R6/LessTif/Motif2.0/lib -lXm
-lXt -lXpm -lXext -lX11
root@localhost:/usr/local/src/ax25/
fbbsrc.705c9/src/X11#
```

The final step is to test if we created the graphical version of LinFBB properly. In our experiment we used the command *./xfbb* (within the same directory */usr/local/src/ax25/fbbsrc.705c9/src/X11/*) to start the GUI interface of the server. The result is shown in Figure 12.

As it can be seen in Figure 12, the graphical version of the program has still not much improved, if not slightly degraded in the visual quality—compared to some earlier versions, including the one we had tested several years ago with Linux distribution Mandrake 9.1 (see Appendix G).

Installing 'Compiled' FBB Software in Ubuntu Linux 10.04.4 LTS

In the following section we are going to perform a very similar installation of Linux amateur radio software, but this time we will work with different operating system, and will use different versions of some programs. In that manner, our procedures will actually replicate some steps we did in the previous section, related to Debian 6.0.7, but we will add some instructions that are specific for our platform that is Ubuntu 10.04.4. Once again, we will start with the same versions of basic AX.25 packages:

```
libax25-0.0.12-rc2.patched_f6bvp.tar.
bz2
ax25-apps-0.0.8-rc2.patched_f6bvp.
tar.bz2
```

Figure 12. Compiled LinFBB mailbox software in Debian 6.0.7

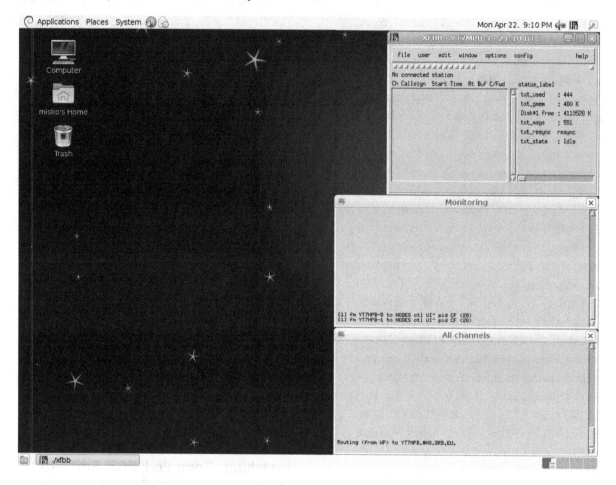

```
ax25-tools-0.0.10-rc2.patched_f6bvp.
tar.bz2
```

We are going to perform similar procedures with all three archives. At first, we will make the following steps with the *libax* package:

(We already had */usr/local/src/ax25* directory on our Ubuntu file system)

```
sudo cp libax25-0.0.12-rc2.patched_
f6bvp.tar.bz2 /usr/local/src/ax25
```

(Note added 'sudo' in front of the command 'cp' because in Ubuntu we did not have a "root terminal" for administrative tasks; what follows is asking for [*sudo*] password of a user)

```
cd /usr/local/src/ax25
sudo tar xf libax25-0.0.12-rc2.
patched_f6bvp.tar.bz2
```

(Note again the added word 'sudo' in front of a particular command)

```
cd libax25-0.0.12-rc2.patched_f6bvp
sudo ./configure
sudo make
sudo make install
sudo make installconf
```

The procedure above creates samples of new configuration files and places them into */usr/local/ etc/ax25* directory (files named *axports*, *rsports*, and *nrports*), include files into */usr/local/include*, documentation man pages into */usr/local/man* and AX.25 libraries into the */usr/local/lib/* directory.

Once again, as Bernard suggested in his documentation, we added a line with */usr/local/lib* at the end of */etc/ld.so.conf* file and ran the command */sbin/ldconfig* after that.

The second phase is to perform the similar procedure but this time with *apps* package and compile AX.25 applications. Therefore, we are going to execute the following commands in the Linux terminal:

```
sudo cp ax25-apps-0.0.8-rc2.patched_
f6bvp.tar.bz2 /usr/local/src/ax25
cd /usr/local/src/ax25
sudo tar xf ax25-apps-0.0.8-rc2.
patched_f6bvp.tar.bz2
cd ax25-apps-0.0.8-rc2.patched_f6bvp
sudo ./configure
sudo make
```

(Note that the command 'make' with *apps* package did not respond with errors here. By checking with Ubuntu's Synaptic Package Manager, we noticed that we had *libncurses5-dev* (version *5.7+20090803-2ubuntu3*) in our Ubuntu installation. That is probably the reason why the command 'make' responded successfully and without any objection.

```
sudo make install
sudo make installconf
```

The procedure above produces new configuration sample files and places them into */usr/local/etc/ax25* directory (files named *ax25ipd.conf*, *ax25mond.conf*, and *ax25rtd.conf*), additional documentation man pages are placed into */usr/local/man* and application tools in */usr/local/sbin* and */usr/local/bin* directories.

The third phase is to compile and install AX.25 tools. The user should perform the following steps with *tools* package:

```
sudo cp ax25-tools-0.0.10-rc2.
patched_f6bvp.tar.bz2 /usr/local/src/
ax25
cd /usr/local/src/ax25
sudo tar xf ax25-tools-0.0.10-rc2.
patched_f6bvp.tar.bz2
cd ax25-tools-0.0.10-rc2.patched_
f6bvp
sudo ./configure
sudo make
sudo make install
sudo make installconf
```

The procedure above produces new configuration sample files and positions them into */usr/local/etc/ax25* directory (files named *ax25d.conf*, *ax25.profile*, *axspawn.conf*, *nrbroadcast*, *rip98d.conf*, *rxecho.conf*, and *ttylinkd.conf*), additional documentation man pages into */usr/local/man* and application tools in */usr/local/sbin* and */usr/local/bin* directories.

As it was with Debian, the next important step is to configure appropriate parameter files with Ubuntu. To enable AX.25 functionality, as well as to explore possibilities of testing the newly-installed packet-radio system in the local area network, we will partly edit */usr/local/etc/ax25/axports*, by inserting proper callsigns and SSIDs, as well as the file */usr/local/etc/ax25/ax25ipd.conf*, by making it looking similar to the corresponding *ax25ipd.conf* we have on the other Linux-based computer in the same LAN.

The activities mentioned up to now will result with basic AX.25 programs being installed on our Ubuntu-based computer.

In order to complete the installation of the wireless node, we will install FPAC software. The procedure for an Ubuntu distribution follows:

```
sudo cp fpac-3.27.18.tar.bz2 /usr/lo-
cal/src/ax25
cd /usr/local/src/ax25
sudo tar xf fpac-3.27.18.tar.bz2
cd fpac-3.27.18
```

```
sudo ./configure
sudo make
sudo make install
sudo make installconf
```

The procedure above creates new subdirectories and configuration files for FPAC and places them into */usr/local/var/ax25* and */usr/local/etc/ax25*, while FPAC application programs will be added in */usr/local/sbin*.

By the way, *rc.init.script* suggests some changes within *fpac.sh*—related to the sysop's radio identification - callsign, so according to that we can do the following:

```
cd /usr/local/sbin
sudo nano fpac.sh
```

Once again, in order to simplify the learning process, it is suggested to download sample configuration files *fpac.conf*, *fpac.nodes* and *fpac.routes* from Bernard's Website, and use them to teach yourself on the proper configuration of a new node. In that manner, we can adopt necessary changes within the following files:

```
/usr/local/etc/ax25/fpac.conf
/usr/local/etc/ax25/fpac.nodes
/usr/local/etc/ax25/fpac.routes
```

The functionality of a newly-installed FPAC node can be now tested by executing the command:

```
/usr/local/sbin/fpac.sh -start
```

In a couple of seconds the node would broadcast to the network and could be immediately accessible by surrounding radio amateurs and nearby nodes. You will recognize that there are other node-related parameter files to be edited, such as *fpac.hello* that is displayed when a user connects to the node, or *fpac.info* that is displayed when a user run the *I(nfo)* command after he or she is connected to the node.

After that, it might be suitable to make an AX.25 startup script. Once again, it is possible to download the script from Bernard's site and to place it into the */etc/rc.d/* directory. In our operating system configuration, that directory did not exist, so we did the following:

```
sudo mkdir /etc/rc.d
sudo cp rc.ax25 /etc/rc.d/rc.ax25
sudo chmod a+x /etc/rc.d/rc.ax25
```

Notice: Before trying to execute that script (*/etc/rc.d/rc.ax25*), you should familiarize yourself with its content and function, because it is user-specific. That means, if not edited properly this script would try to access processes and devices that you might not have planned and/or configured for your particular wireless node, what would result in returning one or more error messages. Because of that, we suggest beginners not to use the script just like that, but instead to activate script commands manually, one by one, in a terminal window (by preceding any command with a 'sudo'), and learn the output of each command. If and when the user is sure that all parts of the script satisfy his or her requirements, the script might be executed in whole—at first manually and later automatically.

For those users who might find it as interesting to initialize their AX.25 programs and activate their FPAC nodes each time the computers start, the script /etc/rc.d/rc.ax25 and other related programs should be invoked by another script, which is */etc/rc.d/rc.local* (in some Linux installations it is */etc/rc.local*). Therefore the user can add additional lines into *rc.local* such as in this example:

```
/etc/rc.d/rc.ax25
/usr/local/sbin/fpac.sh -start
```

As we saw in the previous section, although that approach did not work in our Debian-based installation, what probably meant that this approach was distribution-dependent, it worked well in Ubuntu. However, if the automatic start of the wireless node is not possible due to any reason, or if it is not wanted, here we provide a sample set of commands to start the node in *Ubuntu v.10.04.4* manually:

```
/sbin/ifconfig bpq0 hw ax25 yt7mpb-15
/sbin/ifconfig bpq0 192.168.0.2 net-
mask 255.255.255.0 up
/usr/local/sbin/nrattach -i
192.168.0.2 netrom
/usr/local/sbin/nrattach -i
192.168.0.2 netbbs
/usr/local/sbin/ax25d
/usr/local/sbin/netromd -i
/usr/local/sbin/fpac.sh -start
```

The above set of commands is suitable for users who want to experiment with the nodes only. For those of you who want to add an amateur radio wireless BBS, it is needed to continue with software installations by compiling the mailbox server program—LinFBB. Although the procedure for compiling that software can be found in other sources, such as in Skoric (2010c), we will provide some details here because of the differences between operating system environments and the versions of the software that were described elsewhere in literature—particularly having in mind those researchers who want looking to a graphical user interface of the mailbox program within the newer Linux environments. Therefore we will use the new source package *xd705e-src. tar.bz2* and perform the following steps:

```
sudo cp xd705e-src.tar.bz2 /usr/lo-
cal/src/ax25
cd /usr/local/src/ax25
sudo tar xf xd705e-src.tar.bz2
cd fbbsrc.705e/src
sudo make
cd ..
sudo ./install_sh
```

(The command *sudo ./install_sh* will start the FBB installation utility that creates directory configuration tree, copy new binary files or replaces binary files from older program versions /if they exists/, etc. By the way, the existing configuration files of an older program version are not removed or replaced, so after upgrading the software, the parameters are fully preserved and the mailbox will be functional immediately.)

When the installation is finished, the server can be tested by issuing the following commands:

```
cd /usr/local/sbin
sudo ./fbb
```

(Interestingly, during LinFBB's compilation process, we did not perform a shutdown of the FPAC node running *bellow* the mailbox server. Instead, the node remained active 'on the air', so as soon as we started the newly-compiled server it became accessible from the packet network, by using the same FPAC node.)

Finally, what remains is to compile the graphical front-end for visually-appealing users of LinFBB:

```
cd /usr/local/src/ax25/fbbsrc.705e/
src/X11
sudo make
```

Similarly to Debian 6.0.7, it seemed that our Ubuntu 10.04.4 LTS distribution did not include *libxpm-dev* package in its default installation. Besides that, the Ubuntu installation program did not include software for producing GUI applications, such as *LessTif* development libraries, etc. Nevertheless, after installing required software packages, the command 'make' performs successfully and without reporting any error.

The final step is to test if the graphical version of LinFBB has been properly created for our Ubuntu environment. For activating the mailbox server in GUI interface, you can use the command *sudo ./xfbb* (within directory */usr/local/src/ax25/ fbbsrc.705e/src/X11/*). The output is shown in Figure 13.

Similarly to the visual appearance in Debian (described in Figure 12), it is obvious that the quality of xfbbX GUI client is not the top-priority task for LinFBB developers anymore. (Be aware that we had to decrease resolution of our 22'' monitor from its standard 1920 x 1080 pixels to 1152 x 864 pixels—just to make the GUI client's letters in Figure 13 to be recognizable.) As mentioned earlier, some practitioners recommend implementing xfbbW, a Microsoft Windows-based client, instead of xfbbX but that solution requires another computer in the LAN, what cost-savvy enthusiasts may find unsuitable for their budgets.

Upgrading Ubuntu 10.04.4 LTS with 'compiled' FBB → to Ubuntu 12.04.2 LTS

According to Ubuntu's naming convention practice, the version number tells its release date. In that manner, the version 10.04.4 was launched in April 2010—hence the numbers 10.04. (The trailing number *.4* means that the particular system configuration includes so-called 4th 'release point' (it is something similar to a 'service pack' in MS Windows terminology.) It was obvious that during the work on this study, this configuration—despite being fully 'patched'—was some three years old. That fact would not be a problem for us, if Ubuntu developers' team had

Figure 13. Compiled LinFBB mailbox software in Ubuntu 10.04.4 LTS

not announced that the support for this version would cease in April 2013. Furthermore, some time in between—precisely somewhere in April 2012, they launched a new long-term-support (LTS) release: *Ubuntu 12.04 LTS*, which should be a logical upgrade for the users who preferred LTS versions. At first, we tested the new release on a spare computer, but did not make a decision to upgrade our 'production' system because the Ubuntu team had suggested the users to wait for at least the first release point—12.04.1, which was supposed to fix eventual defects to be found in the initial 12.04 distribution. Therefore we have rather chosen to wait until April 2013—in order to see what would happen on the scene. In addition, during the period April-May 2013 we noticed that the version 10.04.4 continued to receive security fixes and occasional distribution updates. However, having in mind the timeliness of this study and after careful preparations (such as performing backup of important data, etc) we decided to upgrade. To make that decision easier, our *Update Manager* has notified us about the newest release point at the time—*Ubuntu 12.04.2*, which was available since February 2013.

Be aware, though, that software upgrading should be cleverly thought and performed if and when is really needed:

When should you upgrade? In general, you should consider upgrading a portion of your system only when you have a demonstrated need to upgrade. For example, if you hear of a new release of some application that fixes important bugs (that is, those bugs which actually affect your personal use of the application), you might want to consider upgrading that application. If the new version of the program provides new features that you might find useful, or has a performance boost over your present version, it's also a good idea to upgrade. However, upgrading just for the sake of having the newest version of a particular program is probably silly. (Welsh & Kaufman, 1995, p. 153)

We strongly agree with the statement above, as well as with opinions of other authors—particularly when they warn less experienced computing enthusiasts about precautions that have to be respected before upgrading or modifying their Linux system:

When you upgrade your entire distribution, you will definitely want to back up at least the / (root) and /usr file systems, and probably your whole machine, as well. It doesn't happen often, but it is possible for a critical library or program not to upgrade properly, crippling your machine. You should probably also back up your /etc directory and /home directories. (Komarinski & Collett, 1998, p. 306)

Due to relatively slow Internet connectivity at our testing location, we decided to obtain a copy of Ubuntu *'alternate'* CD, to perform the process off-line. (Be aware, though, that Ubuntu developers suggest performing network-upgrades in order to fetch most new packages from their Web repository.) Nevertheless, and despite our best efforts, the off-line approach failed, so we ended up by opting for the network-based procedure. As expected, it took several hours for the whole process to complete (see Appendix H), but after rebooting the system everything seemed to run properly. In order to check FBB's behavior, we started the software by using commands *cd / usr/local/src/ax25/fbbsrc.705e/src/X11* and *sudo ./xfbb* (as expected, the new terminal program 'remembered' commands we used earlier with 10.04.4 so we just needed to scroll down the past and rerun the same syntax). The result is shown in Figure 14.

Installing 'Compiled' JNOS software in Debian Linux 6.0.7

JNOS is another type of email server software that runs on most popular platforms (DOS, Windows, and Linux). Its actual programmer, Maiko Lange-

Figure 14. LinFBB mailbox software in Ubuntu 12.04.2 LTS

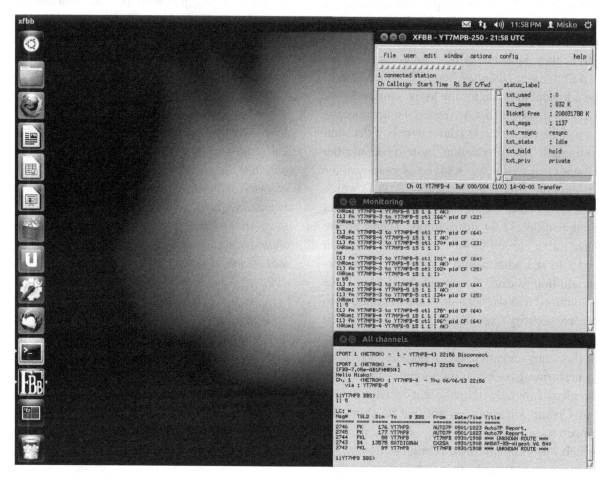

laar, VE4KLM, announced his primary intention for continual upgrading Linux versions of that software, although updated versions for DOS- and Windows-based systems are also available, although a little bit less frequently. Therefore, in the following sections we are going to see Linux-based installations of JNOS.

For the purpose of this chapter segment, we used another Debian 6.0.7-based computer, which was equipped with a VHF amateur radio station connected via TNC2 modem on *ttyS0* (serial port *COM1* in Linux terminology). In fact, our goal was to test the program's capability to transmit radio signals immediately after the software deployment and activation.

Of course, before that we needed to compile the program from the software source. For that purpose we used the source package *jnos20j.tar.gz* that was available for download at Maiko's Website. What follows is the procedure we performed with that package.

At first, we copied the source package file to Debian's source repository, unpacked the compressed file, and extracted the content into the source folder named *jnos2*:

```
cp jnos20j.tar.gz /usr/local/src
cd /usr/local/src
gunzip jnos20j.tar.gz
tar xvf jnos20j.tar
```

Next, we relocated to the new folder, and used *make* to produce (compile) a sub-folder *jnos*:

```
cd jnos2
make defconfig
make
```

After repositioning to that folder, we copied the source of the installer program to Debian's source repository, and used *make* again to produce (compile) the installer program. Mister Lange-laar periodically updates the software installer, so that it was possible to download the newer version in a package named *installerv2.1.tar.gz* that we unpacked and extracted into a new folder (*installerv2.1*).

```
cd jnos
cp jnosinstaller.c /usr/local/src
cd /usr/local/src
make jnosinstaller
./jnosinstaller
cd jnos
cp installerv2.1.tar.gz /usr/local/
src
cd /usr/local/src
gunzip installerv2.1.tar.gz
tar xvf installerv2.1.tar
cd installerv2.1
```

After repositioning to that folder, we activated the installer itself by using the command:

```
./jnosinstaller
cd installerv2.1
./jnosinstaller
cd ..
cp jnosinstaller /usr/local/src/in-
stallerv2.1
cd installerv2.1
./jnosinstaller
cd installerv2.1
cp jnos /usr/local/src/installerv2.1
```

```
cd ..
./jnosinstaller
```

As you can see, we performed some additional commands and file copying operations, listed above, to produce other files and folders. (Be aware that *your* command lists might slightly differ from ours, depending on particular Linux deployments. Nevertheless, you can use the same commands as above, and as long as you do that within your local */usr/local/src/* folder *only*, you should not expect any damage to your Linux system.) Finally, we can test our newly-compiled JNOS program, by performing the following steps:

```
cd /usr/local/jnos
./jnos -d /usr/local/jnos
```

The result should be the same or very similar to what you can see in the upper left part of Figure 15, i.e. within the window titled *'Terminal (as superuser)'*. As mentioned before, if you have a radio station connected to your Linux machine, as soon as the content of that window appears, the server will send its initial 'radio beacon' in order to announce its presence to the network participants.

After we ensure that the program executes, we have various possibilities to continue testing its basic features. The simplest way we can recommend to beginners is to open another terminal screen (that one does not need to be for *superusers* (that is, system administrators), so instead it might be a terminal for regular users who perform non-administrative tasks). As depicted in the bottom right part of Figure 15, we entered the command telnet *<IP address of our JNOS server>*, and the server responded with usual login and password prompts.

Some remarks related to telnet connectivity: During our initial tests we were not capable to reach JNOS server by using telnet command until we temporarily deactivated the firewall on that computer. That means the firewall configuration

Figure 15. Compiled JNOS mailbox software in Debian 6.0.7

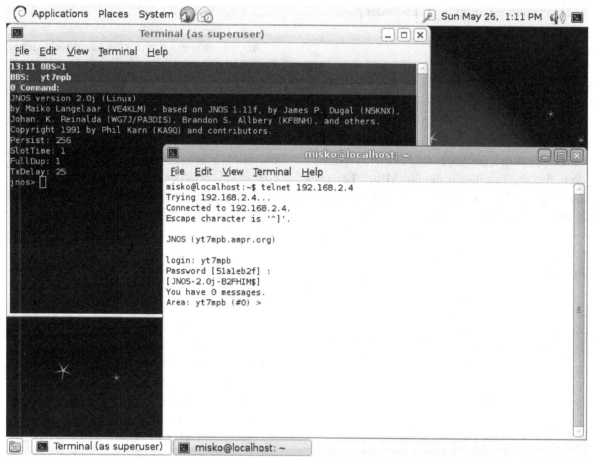

did not allow us to use telnet *locally* on the same computer. To make things more difficult, the firewall also did not allow us to access the server by using telnet command *remotely* from other machines in the local area network. Nevertheless, it did allow incoming command *ping* from the same LAN to check if JNOS was running, but interestingly after deactivating the firewall, even the 'pings' were not responding anymore. Please notice that our JNOS machine resided within the DMZ (abbr. 'demilitarized zone') of the LAN because it was expected from the server to handle telnet connections, both incoming and outgoing (means from the Internet to that computer and vice versa) *and* in the same time to ensure computer's security regarding its exposure to the unknown cyber world.

Therefore, we had to inspect and fine-tune the firewall parameters. First of all, we noticed that a new network interface *tun0* was reported by *ifconfig* command only when JNOS software was active and running. However, regardless *tun0* was present in the interface list or not, it obviously had to be defined within the firewall configuration. Therefore and in order to satisfy that requirement, we added some new entries to the firewall parameters, as can be seen in Appendix I. At first, we defined the new interface tun0 and assigned it to the firewall zone *loc* (the local, safe zone at that particular computer). Secondly, we defined two pairs of new entries for firewall rules: The first pair of new rules enabled sending 'pings' (icmp) and establishing connections with JNOS by using telnet command (tcp telnet) *locally*—on the same computer (direction fw → loc). The second pair of new rules enabled sending 'pings' and establishing telnet connections *remotely*—from the other machines in the LAN (direction net → loc).

After introducing those new parameters, there were no more requests for repeated deactivation of the firewall on that JNOS computer. Having in mind that tweaking the firewall is a sensitive task that might result with unwanted consequences, such as in having an unnecessary 'open system', we avoided the common mistake of changing/adding more than one parameter at a time. Instead, after each change we restarted the firewall, checked for eventual warning messages, and tested the new environment. Although that might seem as a too slow procedure, we are pretty sure now that we did not make mistakes that could render the server unprotected.

Upgrading Debian Linux 6.0.7 with 'Compiled' JNOS Software → to Debian 7.0

During the work on this study, Debian developers' community proudly announced a major operating system upgrade, which was Debian 7.0—a new 'stable' release, which included many changes such as program improvements, security updates, and so on. Therefore we concluded that it would be a good opportunity to test our present operating system's capability to upgrade itself to a newer version, but in the same time to preserve functionalities of previously installed programs, such as JNOS.

Detailed instructions for performing that particular distribution upgrade can be easily found elsewhere on the Internet, for example on the Website http://www.debian.org/releases/wheezy/i386/release-notes/ch-upgrading.en.html where it was suggested to do the whole process manually, rather than by using *software managers* such as *Synaptics* or so on. In order to do perform the suggested path, we needed to make some changes within the file */etc/apt/sources.list* that handles

data on software repositories (both Internet repositories and the local ones, such as hard disks, CDs or DVDs, etc.) Related to that and according to the instructions, we added to the mentioned file a new line:

```
deb http://mirrors.kernel.org/debian
wheezy main contrib
```

To allow the system to check for available new software packages, we performed the command *apt-get update*, which provided the following output:

```
root@localhost:/home/misko# apt-get
update
Get:1 http://mirrors.kernel.org
wheezy Release.gpg [1,672 B]
Get:2 http://mirrors.kernel.org/
debian/wheezy/contrib Translation-en
[34.8 kB]
Ign http://mirrors.kernel.org/debian/
wheezy/contrib Translation-en_US
Get:3 http://mirrors.kernel.org/debi-
an/wheezy/main Translation-en [3,851
kB]
Ign http://mirrors.kernel.org/debian/
wheezy/main Translation-en_US
Get:4 http://mirrors.kernel.org
wheezy Release [159 kB]
Get:5 http://mirrors.kernel.org
wheezy/main i386 Packages [5,864 kB]
Ign http://mirrors.kernel.org wheezy/
main i386 Packages
Get:6 http://mirrors.kernel.org
wheezy/contrib i386 Packages [42.3
kB]
Get:7 http://mirrors.kernel.org
wheezy/main i386 Packages [7,775 kB]
Fetched 11.9 MB in 8min 48s (22.4 kB
/s)
Reading package lists... Done
root@localhost:/home/misko#
```

To test if the available space on the hard disk would be enough for the upgrade, we performed the next command:

```
apt-get -o APT::Get::Trivial-
Only=true dist-upgrade

Reading package lists... Done
Building dependency tree
Reading state information... Done
Calculating upgrade... Done
The following packages will be RE-
MOVED:
  at-spi capplets-data defoma desk-
bar-applet gcj-4.4-base gcj-4.4-jre
  gcj-4.4-jre-headless gcj-4.4-
jre-lib girl.0-clutter-1.0 girl.0-
freedesktop
...
The following NEW packages will be
installed:
  accountsservice acl aisleriot ant
ant-optional apg aptdaemon-data
  aptitude-common argyll at-spi2-core
caribou caribou-antler colord cpp-4.7
...
The following packages will be up-
graded:
  acpi acpi-support-base acpid ad-
duser alacarte alsa-base alsa-utils
anacron
  apache2.2-bin app-install-data apt
apt-listchanges apt-offline apt-utils
...
1048 upgraded, 637 newly installed,
58 to remove and 0 not upgraded.
Need to get 1,081 MB of archives.
After this operation, 718 MB of addi-
tional disk space will be used.
```

As seen in the previous step's report above, many software packages should have been changed—either upgraded i.e. replaced from their earlier versions, or be added as completely new

components, as well as some of existing packages to be removed from the system. In regard to that, Debian developers included a useful warning in their instructions, related to a possibility that the full operating system upgrade sometimes gets interrupted or otherwise wrongly done, which could lead to a unusable system. To prevent that possibility (or just to lower the chances to face the problem), the programmers suggested an intermediate step—the so-called 'minimal system upgrade', by using the command apt-get upgrade.

We followed that direction and a part of the (pretty long!) result is shown here:

```
Reading package lists... Done
Building dependency tree
Reading state information... Done
The following packages have been kept
back:
  alacarte alsa-base alsa-utils
apache2.2-bin apt apt-offline apt-
utils
  aptdaemon aptitude aspell at-spi
baobab base-files bash bind9-host
...
The following packages will be up-
graded:
  acpi acpi-support-base acpid ad-
duser anacron app-install-data
  apt-listchanges apt-xapian-index
aptoncd aspell-en at autoconf auto-
make
...
277 upgraded, 0 newly installed, 0 to
remove and 773 not upgraded.
Need to get 167 MB of archives.
After this operation, 14.2 MB disk
space will be freed.
Do you want to continue [Y/n]?
```

If we compare the final lines from two last commands' outputs, we can see that the 'minimal system upgrade' only updates some of the existing programs, but it does not add anything new, neither has it removed anything from the user's computer. It is obvious that such approach seems to be safer. Therefore, we responded by 'y' to the question above, and let the machine to download new packages from Debian's Website. (Be aware that we intentionally did not want to use the same machine, or any other, for downloading so-called *ISO* files for producing a CD or DVD installation disk that is needed for an 'offline' installation, where 'offline' means the absence of Internet connectivity.)

After some time, the computer reported that the upgrade was finished and asked for rebooting the computer. During the system reboot, we noticed some minor changes, however the machine reported it was still in its 6.0.7 version. Nevertheless, the next step was to perform the main, longer part of the operating system upgrade, by using the command apt-get dist-upgrade.

Here you have the main part of a much longer output of that command:

```
Reading package lists... Done
Building dependency tree
Reading state information... Done
Calculating upgrade... Done
The following packages will be RE-
MOVED:
  at-spi capplets-data defoma desk-
bar-applet gcj-4.4-base gcj-4.4-jre
  gcj-4.4-jre-headless gcj-4.4-
jre-lib girl.0-clutter-1.0 girl.0-
freedesktop
...
The following NEW packages will be
installed:
  accountsservice acl aisleriot apg
aptdaemon-data aptitude-common argyll
  at-spi2-core caribou caribou-antler
colord cpp-4.7 crda cryptsetup-bin
...
The following packages will be up-
```

```
graded:
  alacarte alsa-base alsa-utils
apache2.2-bin apt apt-offline apt-
utils
  aptdaemon aptitude aspell baobab
base-files bash bind9-host binfmt-
support
...

771 upgraded, 628 newly installed, 58
to remove and 0 not upgraded.
Need to get 910 MB of archives.
After this operation, 724 MB of addi-
tional disk space will be used.
Do you want to continue [Y/n]?
```

If we now compare the final few lines in the command's output above, and the output of the command apt-get -o APT::Get::Trivial-Only=true dist-upgrade, which we used before to test the potential distribution upgrade, we can now see that what left was to update the remaining programs, to add many ones, as well as to remove some software that Debian 7.0 had excluded from the distribution. It is obvious that this step is going to be longer than the previous one. However, we responded by 'y' to the question above, and let the machine to do the job. It was a long night, and in the morning we noticed that during the process of 'fetching' files from Debian repositories, some network disconnections occurred. Because of that, it was needed to instruct the machine to 'pull' remaining files and install.

When the process was finished and the computer rebooted again, it appeared that the root file system was full. Therefore, we had to check the kernel status, by using the command dpkg --list | grep linux-image that responded with the following:

```
rc  linux-image-2.6-686
3.2+46            i386    Linux for
modern PCs (dummy package)
```

```
ii  linux-image-2.6.32-5-686
2.6.32-48squeeze3   i386    Linux
2.6.32 for modern PCs
ii  linux-image-3.2.0-4-486
3.2.41-2            i386    Linux 3.2
for older PCs
ri  linux-image-3.2.0-4-686-pae
3.2.41-2            i386    Linux 3.2
for modern PCs
```

According to the output above, the second kernel from the top seemed to be a 'relict of the past' because it contained '*Squeeze*' (a code name of Debian 6.0.7) in the file name description. Therefore we decided to remove the kernel named linux-image-2.6.32-5-686 in order to free

some 79.9 MB disk space. After this operation, the machine did not complain about lack of space in the root file system.

Finally, to check if the OS upgrade might have brought any negative impact to the functionality of the installed ham radio software, we replicated the appearance of JNOS, very similar to what we have previously described in Figure 15. The result is shown in Figure 16.

By comparing Figures 15 and 16, it is obvious that there are no functional differences in running JNOS server and accessing the mailbox—before and after the upgrading operation. To be precise, the only visible difference is the visual presentation of the command menus within the user's desktop, but that is a difference between the op-

Figure 16. JNOS mailbox software in Debian 7.0

erating system versions, and not of the user applications.

For the purpose of this study, and to simplify explanations, we avoided details related to fine-tuning the newly-upgraded Debian 7.0 operating system. For example, an average user might probably face to challenges such as the change in how the new distribution handles program screens (i.e. its 'windows'), or how and where the new environment handles the program icons a user might have been familiar in the previous versions, or so on. Of course, such things should in no case discourage prospective experimenters from testing new working environments. However, one has to be careful when and where to perform such structural tests. The best solution is to have a secondary machine in the experimenter's lab—in order to avoid potential failures in 'production' equipment.

After having confirmed that JNOS was accessible from within the newly-upgraded operating system, we extended our test by including a Windows-based computer in the LAN, similarly to the experiment shown in Figure 9. This time, we opted to upgrade the communicating software on the secondary computer too, so that it was the version *6.0.1.1* of BPQ32. Furthermore, if you carefully look at the upper left part of Figure 9, and compare it with its counterpart of Figure 17; you may notice that the former configuration consisted of three ports, where port *01* and port *02* were of so-called *BPQEther* type, while port *03* was of *AXIP* type. In fact, for the purpose of the first experiment we needed BPQEther protocol in the LAN because we configured LinFBB 7.04j in Ubuntu 10.04.4 to use the same protocol—so that the two systems could communicate each other. In opposite to that, the latter configuration (Figure 17) had only one port—port 01 of the AXIP type because the same protocol was initially set up in JNOS.

Despite the differences in those two protocols we were using with BPQ32 for Windows, the final outputs were almost the same. For those who want to learn more about particular details of BPQ32 parameters, we recommend instructions provided by John Wiseman, G8BPQ (Wiseman 2011a, 2011b). Nevertheless, the latter tests have proved several things: a) at first, we confirmed uninterrupted network functionalities of the Linux-based computer after it had been upgraded; b) secondly, we confirmed that the initial setup of JNOS server software remained intact and fully operable after upgrading Linux operating system 'bellow' the program; c) next, we confirmed the functionality of BPQ32 after it had been upgraded from version 5.2.9.1 to 6.0.1.1; and d) we proved that Linux-based wireless software successfully correspondent with its MS Windows-based partners.

In addition, it is worth to mention that by establishing more ports and implementing different protocols on those ports, the 'sysop' can provide a 'gateway' for his or her users. For example, there might be a gateway between the amateur radio (wireless) traffic and the Internet or LAN (wired) traffic. In case of implementing JNOS on Linux, the gateway could provide radio amateurs the opportunity to use their radio stations to send emails where the recipients will use the Internet to read and respond, and vice versa. Furthermore, JNOS was developed on the original KA9Q project, a brainchild of Phil Karn, KA9Q that some two decades ago wanted to enable TCP/IP traffic to travel over the amateur radio frequencies (i.e. TCP/IP over AX.25).

FUTURE RESEARCH DIRECTIONS

When it comes to the amateur radio (wireless) software development, it has to be said that it is a continual process. We can say that many authors of our communication programs are fully dedicated for improving capabilities of their products—on behalf of the broad ham radio community, so the software changes come very quickly. Somehow a consequence of that is that our work on this scholarly publication—describing software solutions—cannot ever compete with vivid dynamics

Figure 17. Accessing JNOS from the amateur radio network

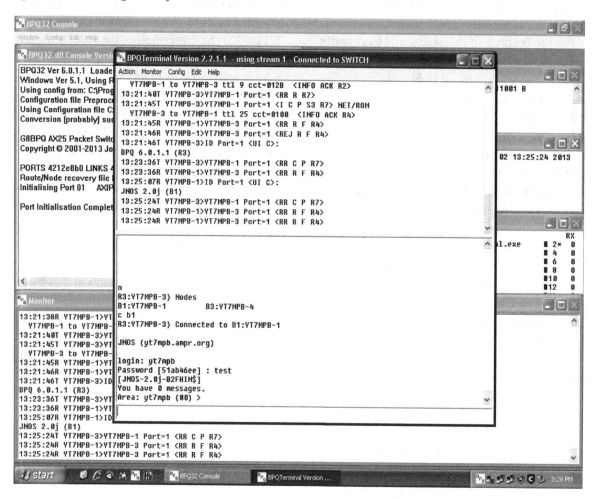

in releasing new programs for wireless communications. Therefore, possible future research shall include testing the newer versions of the software we presented in this chapter, as well as to incorporate experiments with other distributions of Linux OS. Related to that, the purpose of switching to improved versions of operating systems and user applications is not only to avoid inevitable program defects ('bugs') in older and otherwise outdated releases, but also to implement new practices in wireless communications, and explore new horizons as well. In that manner, newer versions of BPQ32 wireless node and accompanied mail server could be a future research target for MS Windows users, while newer releases of LinFBB

or JNOS would remain an exploring challenge for Linux community.

Linux comes in different flavors (i.e. distributions). In this chapter we described some amateur radio solutions in Debian, Ubuntu, Red Hat, and Mandrake distributions. However, there are many others. One of them is Centos, which is favorable by David Ranch, KI6ZHD. In his on-line manual, David explored many different topics from his personal experience with that particular OS (Ranch, 2013). What might differentiate Debian or Ubuntu from Centos is a fact that the latter seemingly does not support AX.25 protocol in its default kernel versions (Centos 5, Centos 6). In addition, Centos neither supports some USB features that are needed to connect newer computers and modems for amateur radio usage. In that case, a user of that OS has to 'patch' and recompile the kernel in order to enable mentioned capabilities, and after that to continue with installations and procedures described in this chapter. Having that in mind, it seems that additional efforts have to be invested within the Linux developers' community, in order to standardize approaches to the amateur wireless operations with that OS. Nevertheless, Steve Blanche, VK2KFJ encourages prospective newcomers to investigate a plenty of opportunities in both Linux platforms and appropriate end-user amateur radio applications for that operating system (Blanche, 2013). Bill Leonard, KF8GR also provided an interesting insight to the amateur radio opportunities with Linux platforms (Leonard, 2005).

When we discuss possible further experiments with JNOS, there are many interesting areas to be explored. Some are security features of the program because, as it can be seen in Figures 15, 16, and 17, JNOS can be compiled to perform user authentication with basic level of security (either without asking for passwords, or to provide them in plain text), and with elevated level of protection (by implementing MD5 algorithm for enciphering passwords). Next, there are possibilities for linking amateur wireless access to Linux computers with different programs, processes, and databases existing within the same machines, such as librarian databases of scholarly publications, or something similar. As Jeff Tranter also pointed out:

Once up and running, you can let users [to] telnet into your Linux system via packet radio, offer them a Unix shell or one of several BBS programs, or even let them surf your system with a Web browser. Thanks to Linux's multiuser and multitasking capability, this occurs without affecting the normal use of the system. (Tranter, 1997, p. 1)

In addition to program innovations, correlated with cleaning the codes for software bugs

and adding new features, we can also expect that new installation procedures i.e. program 'installers' will appear on the software scene. In some cases, improved installation routines will support beginners in wireless communications in acquiring computing skills more quickly, while in other cases the traditional, manual methods of deploying our programs will continue to be used. To fulfill the gap in availability of reliable Linux versions of Windows-specific software, we expect more breakthroughs in adapting the latter to run in Windows emulators for Linux, such as *Wine* or various *virtual machine* applications. At the moment, these emulators have yet to cure some shortcomings to achieve full functionality, although some Windows-based programs emulate in Linux at acceptable level.

Related to possible experiments 'in the field', we consider going out to the nearest university campus and other public places where wireless *Wi-Fi* access to the Internet is available. By using portable computers, equipped with both amateur radio and Internet connectivity, it would be a challenge to compare differences in capacities of both media. Having in mind that there exist alternative ways for 'storing-and-forwarding' the amateur radio messaging system—in cases where reliable radio links are not available, exploring the tradeoffs between the speed of transfer and the coverage areas might also lead to interesting results. Furthermore, by exploring different wireless media—combined and collaborating, we will provoke new ways in ICT convergence. That is particularly important in case of emergency when urgent messages have to be promptly dispatched and received; regardless the wireless path is built on amateur radio or on commercial technology. The importance of being ready, trained and prepared for natural or artificial disasters, has been recognized by many groups, including the ARRL Michigan Section who introduced a new mantra for 2013: "Learn the ins and outs of JNOS management." As they underscored, their ultimate goal is to move bits from point-A to point-B by whatever means they can, including the use of the Internet where terrestrial RF paths may not be available (Nugent, 2010). Ron Hashiro, AH6RH, pointed out the significance of JNOS software in emergency situations: "This program was highlighted in the ARRL Emergency Communications course in 2000 as a means of passing messages (Internet email) during emergencies using the digital modes of amateur radio" (Hashiro, 2012).

Alongside the traditional concepts in computer size and shape, such as desktops and laptops, there appear new emerging trends, such as *RaspBerry Pi* project of a credit-card-sized single-board computer (The Raspberry Pi Foundation, 2012). Besides that, there are other hardware solutions such as industrial-type computers that have been made for deployment in harsh environments, which requires specific attributes such as being dust- and water-proof. Such constructions might be (and some of them have already been) a challenging playground for amateur radio enthusiasts, and especially for their hardware and software laboratories in innovative educational institutions. Therefore we predict that AMUNETs—the amateur radio university networks will appear at academic institutions, colleges, and elementary schools.

Related to other hardware implementations in the amateur wireless networks, it should be noted that specific radio modems are used for converting digital signals produced by computers, into analog signals being transmitted and received by amateur radio stations. For readers who plan to use ISA or PCI cards as radio modems for their personal computers, also known as SCC and DRSI boards, Dave Calder, N4ZKF, has written detailed instructions on software parameters required by FPAC program in the version for DOS-based computers (Calder, 2000). Instead of radio modems, it is possible to use inexpensive soundcards. In modern PC-compatible computer configurations, the soundcards have already been installed by the factories. By consulting Ranch (2013) and other

sources of information, any user can start with amateur wireless communications by interfacing computers and radio stations without separate modems. For example, Dr. Thomas Sailer, HB9JNX authored software solutions for using 'soundcard modem' in both Linux and Windows environment (Sailer, 1997a, 1997b). In addition, Karl Norton, KA1FSB provided detailed description on how to configure a sound modem for JNOS or AX.25 utilities (Norton, 2009).

When it comes to environmental movements such as those belonging to various 'green computing' areas, we can expect additional efforts in exploring solar and wind resources for producing electricity to drive amateur radio appliances and networks outside urban metropolitan zones.

CONCLUSION

Amateur radio offer many opportunities for handling computer hardware and software in wireless communications. In this chapter we wanted to make it clear that amateur wireless communicators have freedom in choosing their preferred working environments and tools. However, our goal was to promote and intensify the usage of open software (open source, etc.) in wireless communications and networking, as well as to motivate researchers and practitioners to improve and update existing software applications in the amateur radio communications. As described in the chapter's sections, there are no requirements for making any 'final choice' for good, so anyone can easily upgrade or switch to another operating system, as well as to another end-user application. Besides software solutions we examined in this study and those described in literature, there are always plenty of exciting possibilities for prospective wireless communicators.

It is worth to mention that by practicing amateur radio digital communications, we are capable to acquire specific knowledge and skills that can be used in other aspects of our lives and in some cases even to lead to a remarkable profession career. As Alan Sieg, WB5RMG pointed out, the experience with JNOS and TCP/IP networking has led him to working for NASA (Sieg, 2009).

For examples described in this work, we used the following equipment:

- **Server YT7MPB-1:**
  - Computer AMD Athlon™ CPU clock 1.10 GHz, 512 MB RAM
  - Operating systems MS Windows XP SP3, Linux Debian 6.0.7, and Linux Debian 7.0.0
  - Network node software BPQ32 5.2.9.1 for Windows
  - E-mail server software JNOS 2.0j for Linux.
- **Server YT7MPB-3:**
  - Computer Intel Celeron™ CPU clock 400 MHz, 224 MB RAM
  - Operating systems MS Windows XP SP3, Linux Debian 6.0.7
  - Network node software BPQ32 5.2.9.1 for Windows, BPQ32 6.0.1.1 for Windows, FPAC 3.27.18 for Linux
  - E-mail server software WinFBB 7.00i for Windows, LinFBB 7.05c9 for Linux.
- **Server YT7MPB-5:**
  - Computer Intel Pentium™ Dual-Core CPU clock 3.06 GHz, 3 GB RAM
  - Operating systems MS Windows XP SP3, Linux Ubuntu 10.04.4 LTS, Linux Ubuntu 12.04.2 LTS
  - Network node software BPQ32 5.2.9.1 for Windows, Node 0.3.2-7.1 for Linux
  - E-mail server software WinFBB 7.00i for Windows, LinFBB 7.04j-8.2 for Linux, LinFBB 7.05e for Linux.

As mentioned, if a user possesses a fast and reliable Internet connectivity, the network-based upgrade of the operating system is not only recommended by the Ubuntu team, but also it proves to be efficient and successful—though that type of upgrade takes more time than that of a CD/DVD-based upgrade. Actually, you have a tradeoff between the time consumed for upgrading operations and the quality of the updated product. Secondly, a user should take in consideration a fact that—regardless it was Debian Linux or Ubuntu Linux (according to our tests), the upgrades were just what the name says: the upgrades of the operating systems, but not the 'fresh' installations. For example, by comparing a 'fresh' installation of Debian 7.0 with an upgrade from version 6.0.7 to version 7.0, it was obvious that the latter operation was *not* a complete installation because we realized later that some expected new features were missing after performing the upgrade. On the other side, the upgrade procedure has successfully preserved the most of the original system- and user-related settings. However, there is a high probability that some residues and software 'orphans' from the previous versions may survived the upgrade operation. Related to that, it is recommended for your communicating computers to be additionally 'cleaned' against the software residues and 'orphaned' files after the OS upgrade.

A user should always respect various warning messages that may occasionally appear on the screen during the upgrade procedure. Some of those warnings may be related to whether to keep previous software settings such as firewall parameters, or to replace them with 'fresh' parameter files. If a user opts for a 'fresh' (default) configuration, then he or she will have to correct the parameters according the particular situation.

An issue we noticed with both Linux distributions was the absence of system fonts in some warning messages that appeared on the screen during the OS upgrade. In fact, some of those messages contained unintelligible small squares (such as □ or ☐) in place of alphabet characters, so it was not easy to understand the meanings of some warnings, and make decisions on most adequate response.

By the way, when compared capabilities of Linux and Windows in the LAN, we spotted another advantage of Linux: If it is deployed at the gateway computer that serves as a distributing point or a router that shares Internet connectivity within the LAN, it successfully shares the Web resource with all subnets in the network (which means regardless of different IP addresses that may exist in the same LAN). In opposite to that, Windows XP Prof. allowed us to redirect Internet traffic to only one subnet at a time, which means that the other subnet(s) remained restricted to that resource. During the work on this chapter, we were not in a position to double-check that feature with other flavors of MS Windows, so we are not able now to claim if it is positive or negative software characteristic.

Finally, some analysts predict that Microsoft will inevitably lose its dominance as the OS platform in the near future (Schofield, 2012). For example, Cearley (2012) found that "the implications for IT is that the era of PC dominance with Windows as the single platform will be replaced with a post-PC era where Windows is just one of a variety of environments IT will need to support." Similarly, Noyes (2013) stated that "there's no better time to embrace a more free, open, and desktop-friendly alternative."

ACKNOWLEDGMENT

Teachers are not the only ones whom we often find as responsible for discovering new talents in worlds of engineering and technology. We strongly believe that the process of learning starts within the family. Therefore, parents should master their abilities for recognizing young potentials at homes. This book chapter is dedicated to the late persons, Mr. Sava Škorić and Mrs. Radmila Škorić, who spotted the exploring nature of their son during his early age. As brave parents, they did not punish him too much for his tireless breaking and destroying various home appliances—while inspecting the secrets of hidden mechanisms *living* inside. Instead, they supported their kid in materializing his imagination and in searching for new horizons. Furthermore, Mrs. Škorić gave her son a nickname "Teslich" (means "little Tesla" in Serbian). Credits also go to all mentioned, known and unknown amateur radio enthusiasts who unselfishly and continually have been donating their time, efforts and funds in developing and improving wireless amateur radio technologies.

REFERENCES

Blanche, S. (2013). *Linux for amateur radio applications*. Retrieved July 28, 2013, from http://www.qsl.net/vk2kfj/linux.html

Calder, D. (2000). *FPAC sysop manual*. Retrieved May 24, 2013, from http://www.n4zkf.com/SYS_FPAC.htm

Cearley, D. (2012). Gartner identifies the top 10 strategic technology trends for 2013. *Gartner Press Release*. Retrieved August 11, 2013, http://www.gartner.com/newsroom/id/2209615

Dawson, T. (1996). *Linux AX25-HOWTO, amateur radio*. Retrieved May 24, 2013, from http://www.linuxdocs.org/HOWTOs/AX25-HOWTO.html

Hashiro, R. (2012). *JNOS, amateur radio and mobile IP email/BBS*. Retrieved July 20, 2013, from http://www.qsl.net/ah6rh/am-radio/packet/jnos.html

Kirch, O. (1995). *Linux network administrator's guide*. Sebastopol, CA: O'Reilly & Associates.

Komarinski, M., & Collett, C. (1998). *Linux system administration handbook*. Upper Saddle River, NJ: Prentice Hall.

Leonard, B. (2005). *KF8GR linux ham home page*. Retrieved July 28, 2013, from http://www.qsl.net/kf8gr/

Norton, K. (2009). *Using a soundmodem on packet radio*. Retrieved July 27, 2013, from http://mysite.verizon.net/ka1fsb/sndmodem.html

Noyes, K. (2013). Make 2013 the year you switch to linux. *Network World News*. Retrieved August 11, 2013, from http://www.networkworld.com/news/2013/010213-make-2013-the-year-you-265433.html

Nugent, J. (2010). *The ARRL Michigan section digital radio group (DRG)*. Retrieved July 21, 2013, from http://www.mi-drg.org/

Pidoux, B. (2012). *Linux FPAC mini-HOWTO*. Retrieved April 20, 2013, from http://rose.fpac.free.fr/MINI-HOWTO/

Ranch, D. (2013). *Hampacketizing centos-6 and centos-5*. Retrieved July 21, 2013, from http://www.trinityos.com/HAM/CentosDigitalModes/hampacketizing-centos.html

Raspberry Pi Foundation. (2011-2013). Why raspberry pi? *FAQs | Raspberry Pi*. Retrieved May 24, 2013, from http://www.raspberrypi.org/faqs

Sailer, T. (1997). Using a PC and a soundcard for popular amateur digital modes. In *Proceedings of 16th ARRL and TAPR Digital Communications Conference*. Newington, CT: American Radio Relay League.

Sailer, T. (1997). *PacketBlaster 97 - Soundkarten-PR mit aktuellen betriebssystemen*. Paper presented at 13 Internationale Packet-Radio-Tagung. Darmstadt, Germany.

Schofield, A. (2012). 2013 predictions: Africa. *ICT and the Global Community*. Retrieved August 11, 2013, from http://www.idgconnect.com/blog-abstract/704/adrian-schofield-africa-2013-predictions-africa

Sieg, A. (2009). *JNOS—15 years later | WB5RMG: RadioActive blog*. Retrieved July 27, 2013, from http://wb5rmg.wordpress.com/2009/11/28/jnos-15-years-later/

Skoric, M. (2009). Amateur radio in education. In H. Song & T. Kidd (Eds.), *Handbook of research on human performance and instructional technology* (pp. 223-245). Hershey, PA: Information Science Reference (IGI-Global).

Skoric, M. (2012). Simulation in amateur packet radio networks. In *Simulation in computer network design and modeling: Use and analysis* (pp. 216–256). Hershey, PA: IGI Global. doi:10.4018/978-1-4666-0191-8.ch011

Skoric, M. (2013). Security in amateur packet radio networks. In *Wireless networks and security: Issues, challenges and research trends* (pp. 1–47). Berlin, Germany: Springer. doi:10.1007/978-3-642-36169-2_1

Skoric, M. (2000-2010a). Linux+WindowsNT mini-HOWTO. *The Linux Documentation Project*. Retrieved March 24, 2013, from http://tldp.org/HOWTO/Linux+WinNT.html

Skoric, M. (2000-2010b). LILO mini-HOWTO. *The Linux Documentation Project*. Retrieved March 24, 2013, from http://tldp.org/HOWTO/LILO.html

Skoric, M. (2000-2010c). FBB packet radio BBS mini-HOWTO. *The Linux Documentation Project*. Retrieved March 24, 2013, from http://tldp.org/HOWTO/FBB.html

Stordahl, R. (2012). BPQ32_5.2.9.1_20120925 including BPQAPRS is now available. *BPQ32 Yahoo Group*. Retrieved May 24, 2013, from http://groups.yahoo.com/group/BPQ32/message/9075

Tranter, J. (1997). Packet radio under linux. *Linux Journal*. Retrieved July 21, 2013, from http://www.linuxjournal.com/article/2218

Welsh, M., & Kaufman, L. (1995). *Running linux*. Sebastopol, CA: O'Reilly & Associates.

Wiseman, J. (2011). *BPQAXIP configuration*. Retrieved June 02, 2013, from http://www.cantab.net/users/john.wiseman/Documents/BPQAXIP%20Configuration.htm

Wiseman, J. (2011). *BPQETHER ethernet driver for BPQ32 switch*. Retrieved April 06, 2013, from http://www.cantab.net/users/john.wiseman/Documents/BPQ%20Ethernet.htm

ADDITIONAL READING

Blystone, K., & Watson, M. (1995). *Alternative Information Highways: Networking School BBSs*. Denton, USA: Texas Center for Educational Technology

Constanza, T. (2012). Youths love tech, but not necessarily tech career. *Silicon Republic*. Retrieved June 16, 2012, from http://www.siliconrepublic.com/careers/item/27746-youths-love-tech-but-not/

Corley, A. (2010). Hams in Haiti. *IEEE Spectrum*. Retrieved October 17, 2010, from http://spectrum.ieee.org/telecom/wireless/hams-in-haiti/0

Davidoff, M. (1994). *The Satellite Experimenter's Handbook*. Newington, CT, USA: American Radio Relay League.

Diggens, M. (1990). Enhancing distance education through radio-computer communication. In R. Atkinson and C. McBeath (Eds.), *Open Learning and New Technology: Conference proceedings* (pp. 113-116). Perth: Australian Society for Educational Technology

Dowie, P. (2002). *Presentation: XROUTER Network Infrastructure Software - Paula G8PZT*. Retrieved September 28, 2010, from http://www.g8pzt.pwp.blueyonder.co.uk/fourpak/2002_03.htm

Edwards, J. (2009). Want to bone up on wireless tech? Try ham radio. *Computerworld*. Retrieved October 17, 2010, from http://www.computerworld.com/s/article/9139771/Want_to_bone_up_on_wireless_tech_Try_ham_radio

Erhardt, W. (2010). *Bill's Amateur Radio Page*. Retrieved January 16, 2011, from http://www.k7mt.com/AmateurRadio.htm

Ford, S. (1995). *Your Packet Companion*. Newington, CT, USA: American Radio Relay League.

Ford, S. (1995). *Your HF Digital Companion*. Newington, CT, USA: American Radio Relay League.

Hill, J. (2002). Amateur Radio—A Powerful Voice in Education. *QST, 86*(12), 52–54.

Hudspeth, D., & Plumlee, R. C. (1994). *BBS Uses in Education. Denton, USA: Texas Center for Educational Technology*

Jones, G. (1996). *Packet Radio: What? Why? How?* Tuckson, AZ, USA: Tuckson Amateur Packet Radio.

Kasal, M. (2010). *Experimental Satellites Laboratory*. Retrieved January 16, 2011, from http://www.urel.feec.vutbr.cz/esl/

Langelaar, M. (2009). *JNOS 2.0 - DOS Install (The Easy Way)*. Retrieved June 10, 2012, from http://www.langelaar.net/projects/jnos2/documents/install/dos/

Langelaar, M. (2009). *JNOS 2.0—Linux Install (The Easy Way)*. Retrieved May 29, 2013, from http://www.langelaar.net/projects/jnos2/documents/install/linux/

Lucas, L. W., Jones, J. G., & Moore, D. L. (1992). *Packet Radio: An Educator's Alternative to Costly Telecommunications*. Denton, USA: Texas Center for Educational Technology

Lucas, L. (1997). *Wide Area Networking Guide for Texas School Districts*. Denton, USA: Texas Center for Educational Technology

Martin, J. (2006). *Linux - Jnos Setup and Configuration HOW-TO*. Retrieved October 17, 2010, from http://www.kf8kk.com/packet/jnos-linux/linux-jnos-setup-9.htm

Martin, J. (2006). *JNOS Operators Guide*. Retrieved September 13, 2010, 2010, from http://www.nyc-arecs.org/JNOS_OpGuide.pdf

Martin, J. (2006). *Whetting your feet with Jnos*. Retrieved October 17, 2010, from http://legitimate.org/iook/packet/jnos/whetting/whetting.htm

McCosker, R. (2010). *Creating a XRouter Remote Node*. Retrieved September 28, 2010, from http://vk2dot.dyndns.org/XRouter/XRouter.htm

McDonough, J. (2007). *The Michigan Digital Network*. Retrieved September 7, 2010, 2010, from http://packet.mi-nts.org/257/MIdigital.pdf

McLarnon, B. (2008). *Packet Radio Technology: An Overview*. Retrieved October 17, 2010, from http://www.friends-partners.org/glosas/Tampere_Conference/Reference_Materials/Packet_Radio_Technology.html

Moxon, L. (1993). *HF Antennas for All Locations*. Potters Bar, Great Britain: Radio Society of Great Britain.

Przybylski, J. (1997). *Home Web SP1LOP*. Retrieved January 16, 2011, from http://www.sp1lop.ampr.org/

Przybylski, J. (2009). *All Poland Packet Info Server*. Retrieved January 16, 2011, from http://www.packet.poland.ampr.org/

Skoric, M. (2004). The Amateur Radio as a Learning Technology in Developing Countries. In *Proceedings of the 4th IEEE International Conference on Advanced Learning Technologies* (pp. 1029-1033). Los Alamitos, CA USA: IEEE Computer Society

Skoric, M. (2005). The perspectives of the Amateur University Networks—AMUNETs. *WSEAS Transactions on Communications, 4*, 834–845.

Skoric, M. (2006). The New Amateur Radio University Network—AMUNET (Part 2). In *Proceedings of the 10th WSEAS International Conference on Computers* (pp. 45-50). Athens, Greece: World Scientific and Engineering Academy and Society

Skoric, M. (2007). Summer schools on the amateur radio computing. In *Proceedings of the 12th annual SIGCSE conference on Innovation and Technology in Computer Science Education* (pp. 346-346). New York, NY USA: Association for Computing Machinery

Skoric, M. (2008). The New Amateur Radio University Network—AMUNET (Part 3). In *Proceedings of the 12th WSEAS International Conference on Computers: New Aspects of Computers* (pp. 432-439). Athens, Greece: World Scientific and Engineering Academy and Society

Skoric, M. (2009). The New Amateur Radio University Network—AMUNET (Part 4). In *Recent Advances in Computers: Proceedings of the 13th WSEAS International Conference on Computers* (pp. 323-328). Athens: WSEAS Press.

Sumner, D. (Ed.). (2003). *22nd ARRL and TAPR Digital Communications Conference*. Newington, CT USA: American Radio Relay League

Sumner, D. (Ed.). (2007). *26th ARRL and TAPR Digital Communications Conference*. Newington, CT USA: American Radio Relay League

Wade, I. (1992). *NOSintro—TCP/IP over Packet Radio*. Retrieved September 28, 2010, from http://homepage.ntlworld.com/wadei/nosintro/

Weiss, R. T. (2005). Ham Radio Operator Heads South To Aid Post-Katrina Communications. *Computerworld*. Retrieved October 17, 2010, from http://www.computerworld.com/newsletter/0,4902,104446,00.html?nlid=MW2

Weiss, R. T. (2005). Ham radio volunteers help re-establish communications after Katrina. *Computerworld*. Retrieved October 17, 2010, from http://www.computerworld.com/securitytopics/security/recovery/story/0,10801,104418,00.html

Worcester Polytechnic Institute Wireless Association. (2012). *The history of W1YK*. Retrieved June 09, 2012, from http://users.wpi.edu/~wpiwa/about.html

KEY TERMS AND DEFINITIONS

AMUNET: This acronym stands for the AMateur radio University computer NETwork, which is the proposed name for a wireless network of an amateur radio BBS at a local university, including one or more amateur radio 'digipeaters', and one or more end-user computers in surrounding schools' computer labs, offices or homes.

AX.25: It is the protocol used by radio amateurs in packet radio networking. AX.25 is a derivation (or adaptation) from the X.25 standard protocol used in public data networks.

BBS: An electronic Bulletin Board System, a software that usually operates on a personal computer equipped with one or more amateur radio stations, Internet connections, and telephone lines (a "dial-up mailbox system"), to provide communication between remote users such as electronic mail, conferences, news, chat, file exchange, and database access.

Callsign: In radio communications terminology, the callsign serves as a 'username' that uniquely distinguishes wireless communicators. For example, millions of people worldwide can have a given name Miroslav, or a family name Skoric, but only one person may be assigned and use the callsign YT7MPB.

Compiling; to Compile: Mostly used in Linux terminology: A specific procedure or a process of preparing and fine-tuning computer software for a particular peace of hardware. It is possible to compile not only end-user programs, but also operating system kernels too. If used with Windows: (usually) a process of fine-tuning computer software according to previously formatted parameter configuration lists.

Firewall: A peace of software that is a part of the operating system or can be an external program. It serves to protect a particular computer system it is installed on, or to protect a group of machines in a local area network (LAN).

Gateway: A gateway is a computer that connects two different networks together. The gateway will perform the protocol conversion necessary to go from one network to the other. For example, a gateway could connect a local area network (LAN) of computers in the school to the Internet. In that manner, an amateur radio BBS provides a gateway from the 'air' to the school's LAN, or vice versa.

GUI: The abbreviation for Graphical User Interface. A GUI is mostly found in Linux-related discussions when it comes to Linux programs that require working environment similar to MS Windows. 'To GUI or not to GUI' is an option that a user can choose when installs a Linux platform. In opposite to the GUI, there is a CLI (abbr. Command Line Interface) where only text mode is available.

Kernel: The main, 'core' part of the operating system. The term kernel is mostly used in Linux jargon when adaptations and major upgrades of a Linux-based system are discussed.

Node: In radio communications, a node (i.e. a repeater) is a device that amplifies or regenerates the signal in order to extend the distance of the transmission. Repeaters are available for both analog (voice) and digital (data) signals.

'Digipeater is frequently used abbreviated name for a digital repeater.

Packet Radio: A communication mode between the amateur radio stations where computers control how the radio stations handle the traffic. The computers and attached modems organize information into smaller chunks of it—often referred as 'packets' of data, and route the packets to intended destinations.

Patch; Being Patched: A rather small peace of software which role is to upgrade the main program. Patches are released for fixing security bugs in operating systems, as well as for updating end-user applications. In case of Linux, the patches are often provided in the open source form. That enables system administrators to see what exactly a patch would do if implemented.

Software: A group name for different kinds of computer programs. Widely spoken, the software includes: a) operating systems; b) 'system software', such as device drivers, basic configuration programs, etc.; and c) end-user programs (applications).

SSID: A 'Sub station identifier' in the amateur radio jargon. That is a numbered suffix assigned to a callsign, for example within YT7MPB-**1**, YT7MPB-**2**, etc. The SSID uses to distinguish different occurrences of the same radio station (and/or its operator) in the 'air'. For example, the suffix *-1* can be used for a node, while the suffix *-2* is used for the mailbox server, etc.

Sysop: It is a short name for a system operator. That person runs and maintains an amateur radio BBS or a repeater. Some sources refer to the sysop as 'system administrator'.

APPENDIX A

The following content presents the parameters within */etc/lilo.conf* file that regulates LILO boot loader (with the old approach in naming convention).

```
# Automatically added by lilo postinst script
large-memory
# /etc/lilo.conf - See: `lilo(8)' and `lilo.conf(5)',
# ---------------         `install-mbr(8)', `/usr/share/doc/lilo/',
#                        and `/usr/share/doc/mbr/'.
# +---------------------------------------------------------------+
# |                        !! Reminder !!                         |
# |                                                               |
# | Don't forget to run `lilo' after you make changes to this     |
# | conffile, `/boot/bootmess.txt' (if you have created it), or   |
# | install a new kernel.  The computer will most likely fail to  |
# | boot if a kernel-image post-install script or you don't       |
# | remember to run `lilo'.                                       |
# |                                                               |
# +---------------------------------------------------------------+
# Specifies the boot device.  This is where Lilo installs its boot
# block.  It can be either a partition, or the raw device, in which
# case it installs in the MBR, and will overwrite the current MBR.
#
boot=/dev/sda
# Specifies the device that should be mounted as root. (`/')
#
root=/dev/sda7
# This option may be needed for some software RAID installs.
#
# raid-extra-boot=mbr-only
# Enable map compaction:
# Tries to merge read requests for adjacent sectors into a single
# read request. This drastically reduces load time and keeps the
# map smaller.  Using `compact' is especially recommended when
# booting from a floppy disk.  It is disabled here by default
# because it doesn't always work.
#
# compact
# Installs the specified file as the new boot sector
# You have the choice between: text, bmp, and menu
# Look in lilo.conf(5) manpage for details
#
```

```
install=menu
bitmap=/boot/debianlilo.bmp
# Specifies the location of the map file
#
map=/boot/map
# You can set a password here, and uncomment the `restricted' lines
# in the image definitions below to make it so that a password must
# be typed to boot anything but a default configuration.  If a
# command line is given, other than one specified by an `append'
# statement in `lilo.conf', the password will be required, but a
# standard default boot will not require one.
#
# This will, for instance, prevent anyone with access to the
# console from booting with something like `Linux init=/bin/sh',
# and thus becoming `root' without proper authorization.
#
# Note that if you really need this type of security, you will
# likely also want to use `install-mbr' to reconfigure the MBR
# program, as well as set up your BIOS to disallow booting from
# removable disk or CD-ROM, then put a password on getting into the
# BIOS configuration as well.  Please RTFM `install-mbr(8)'.
#
# password=tatercounter2000
# Specifies the number of deciseconds (0.1 seconds) LILO should
# wait before booting the first image.
#
delay=100
# You can put a customized boot message up if you like.  If you use
# `prompt', and this computer may need to reboot unattended, you
# must specify a `timeout', or it will sit there forever waiting
# for a keypress.  `single-key' goes with the `alias' lines in the
# `image' configurations below.  eg: You can press `1' to boot
# `Linux', `2' to boot `LinuxOLD', if you uncomment the `alias'.
#
# message=/boot/bootmess.txt
        prompt
#        delay=100
        timeout=100
# Specifies the VGA text mode at boot time. (normal, extended, ask, <mode>)
#
# vga=ask
# vga=9
#
```

```
# Kernel command line options that apply to all installed images go
# here.  See: The `boot-prompt-HOWTO' and `kernel-parameters.txt' in
# the Linux kernel `Documentation' directory.
#
# append=""
# If you used a serial console to install Debian, this option should be
# enabled by default.
# serial=
#
# Boot up Linux by default.
#
default=Linux
image=/vmlinuz
label=Linux
read-only
#         restricted
#         alias=1
          initrd=/initrd.img
#image=/vmlinuz.old
#         label=LinuxOLD
#         read-only
#         optional
#         restricted
#         alias=2
#
#         initrd=/initrd.img.old
# If you have another OS on this machine to boot, you can uncomment the
# following lines, changing the device name on the `other' line to
# where your other OS' partition is.
#
# other=/dev/hda4
#         label=HURD
#         restricted
#         alias=3
other=/dev/sda1
        table=/dev/sda
        label=Windows
#         restricted
#         alias=2
```

APPENDIX B

The following content present parameters within */etc/lilo.conf* file that regulates LILO boot loader (with the new approach in naming convention).

```
# Automatically added by lilo postinst script
large-memory
# /etc/lilo.conf - See: `lilo(8)' and `lilo.conf(5)',
# ---------------         `install-mbr(8)', `/usr/share/doc/lilo/',
#                          and `/usr/share/doc/mbr/'.
# +--------------------------------------------------------------+
# |                      !! Reminder !!                          |
# |                                                              |
# | Don't forget to run `lilo' after you make changes to this    |
# | conffile, `/boot/bootmess.txt' (if you have created it), or  |
# | install a new kernel.  The computer will most likely fail to |
# | boot if a kernel-image post-install script or you don't      |
# | remember to run `lilo'.                                      |
# |                                                              |
# +--------------------------------------------------------------+
# Specifies the boot device.  This is where Lilo installs its boot
# block.  It can be either a partition, or the raw device, in which
# case it installs in the MBR, and will overwrite the current MBR.
#
#boot=/dev/sda
boot = /dev/disk/by-id/ata-Maxtor_6E040L0_E1728VJN
# Specifies the device that should be mounted as root. (`/')
#
# root = /dev/sda5
root = "UUID=83c6882b-88a5-433b-9712-0ffa1663bb57"
# This option may be needed for some software RAID installs.
#
# raid-extra-boot=mbr-only
# Enable map compaction:
# Tries to merge read requests for adjacent sectors into a single
# read request. This drastically reduces load time and keeps the
# map smaller.  Using `compact' is especially recommended when
# booting from a floppy disk.  It is disabled here by default
# because it doesn't always work.
#
# compact
# Installs the specified file as the new boot sector
# You have the choice between: text, bmp, and menu
```

```
# Look in lilo.conf(5) manpage for details
#
install=menu
bitmap=/boot/debianlilo.bmp
# Specifies the location of the map file
#
map=/boot/map
# You can set a password here, and uncomment the `restricted' lines
# in the image definitions below to make it so that a password must
# be typed to boot anything but a default configuration.  If a
# command line is given, other than one specified by an `append'
# statement in `lilo.conf', the password will be required, but a
# standard default boot will not require one.
#
# This will, for instance, prevent anyone with access to the
# console from booting with something like `Linux init=/bin/sh',
# and thus becoming `root' without proper authorization.
#
# Note that if you really need this type of security, you will
# likely also want to use `install-mbr' to reconfigure the MBR
# program, as well as set up your BIOS to disallow booting from
# removable disk or CD-ROM, then put a password on getting into the
# BIOS configuration as well.  Please RTFM `install-mbr(8)'.
#
# password=tatercounter2000
# Specifies the number of deciseconds (0.1 seconds) LILO should
# wait before booting the first image.
#
delay=100
# You can put a customized boot message up if you like.  If you use
# `prompt', and this computer may need to reboot unattended, you
# must specify a `timeout', or it will sit there forever waiting
# for a keypress.  `single-key' goes with the `alias' lines in the
# `image' configurations below.  eg: You can press `1' to boot
# `Linux', `2' to boot `LinuxOLD', if you uncomment the `alias'.
#
# message=/boot/bootmess.txt
        prompt
#         delay=100
        timeout=100
# Specifies the VGA text mode at boot time. (normal, extended, ask, <mode>)
#
```

```
#  vga=ask
# vga=9
#
# Kernel command line options that apply to all installed images go
# here.  See: The `boot-prompt-HOWTO' and `kernel-parameters.txt' in
# the Linux kernel `Documentation' directory.
#
# append=""
# If you used a serial console to install Debian, this option should be
# enabled by default.
# serial=
#
# Boot up Linux by default.
#
default=Linux
image=/vmlinuz
        label=Linux
        read-only
#        restricted
#        alias=1
        initrd=/initrd.img
image=/vmlinuz.old
        label=LinuxOLD
        read-only
        optional
#        restricted
#        alias=2
        initrd=/initrd.img.old
# If you have another OS on this machine to boot, you can uncomment the
# following lines, changing the device name on the `other' line to
# where your other OS' partition is.
#
# other=/dev/hda4
#        label=HURD
#        restricted
#        alias=3
other=/dev/sda1
        table=/dev/sda
        label=Windows
#        restricted
#        alias=2
```

APPENDIX C

The following content is a sample parameter file *BPQ32.CFG* that represents a new approach where all configuration statements are combined within a single file:

```
; CONFIGURATION FILE FOR BPQ32: G8BPQ SWITCH SOFTWARE
; The purpose of this configuration is simply to confirm that the basic
; BPQ32 system is operable.  It contains only a LOOPBACK port.
; To perform the basic test, compile this configuration file by executing
; bpqcfg.exe, which will generate bpqcfg.bin. Now execute BPQTerminal.exe,
; and in the lowest window enter:
; C 1 MYNODE v MYCALL
; You should receive the following response:
; MYNODE:MYCALL} Connected to MYNODE
; This is the CTEXT.
; MYNODE:MYCALL} DX CONNECT BYE INFO NODES ROUTES PORTS USERS MHEARD
; If you get the above response the basic test has succeeded.
; \Examples\Minimal
HOSTINTERRUPT=127          ; Interrupt used for BPQ host mode support
EMS=0                       ; Doesn't use EMS RAM
DESQVIEW=0                  ; DesqView unused
NODECALL=YT7MPB-5            ; Node callsign
NODEALIAS=R5          ; Node alias (6 characters max)
;BBSCALL=MYCALL               ; Now defined as APPL1CALL following APPLICA-
TIONS=
;BBSALIAS=MYNODE                ; Now defined as APPL1ALIAS following APPLICA-
TIONS=
IDMSG:                      ; UI broadcast text from NODECALL to fixed dest
ID
BPQ 5.2.1.3 (R5)
***                        ; Denotes end of IDMSG text
BTEXT:                      ; UI broadcast text from BCALL to destination
UNPROTO=
FBB 7.00i (B5)
***                        ; Denotes end of BTEXT text
INFOMSG:                ; The INFO command text follows:
Fwd link to B3:YT7MPB-4 (Only for B5:YT7MPB-6)
QTH Novi Sad - Liman, JN95WF, Serbia
SysOp: Misko YT7MPB
***                        ; Denotes end of INFOMSG text
CTEXT:                      ; The CTEXT text follows:
This is the CTEXT text.
Welcome to R5, test node in Novi Sad.
```

```
Type ? for list of available commands.
***                             ; Denotes end of CTEXT text
FULL_CTEXT=0                     ; 0=send CTEXT to L2 connects to NODEALIAS only
                                 ; 1=send CTEXT to all connectees
; Network System Parameters:
OBSINIT=5                        ; Initial obsolescence set when a node is included
                                 ; in a received nodes broadcast. This value is then
                                 ; decremented by 1 every NODESINTERVAL.
OBSMIN=4                         ; When the obsolescence of a node falls below this
                                 ; value that node's information is not included in
                                 ; a subsequent nodes broadcast.
NODESINTERVAL=15                 ; Nodes broadcast interval in minutes
IDINTERVAL=15                    ; 'IDMSG' UI broadcast interval in minutes, 0=OFF
BTINTERVAL=0                     ; The BTEXT broadcast interval in minutes, 0=OFF
L3TIMETOLIVE=25                  ; Max L3 hops
L4RETRIES=3                      ; Level 4 retry count
L4TIMEOUT=60                     ; Level 4 timeout in seconds s/b > FRACK x RETRIES
L4DELAY=10                       ; Level 4 delayed ack timer in seconds
L4WINDOW=4                       ; Level 4 window size
MAXLINKS=30                      ; Max level 2 links
MAXNODES=120                     ; Max nodes in nodes table
MAXROUTES=35                     ; Max adjacent nodes
MAXCIRCUITS=64                   ; Max L4 circuits
MINQUAL=10                       ; Minimum quality to add to nodes table
BBSQUAL=255                      ; Now defined as APPL1QUAL following APPLICATIONS=
BUFFERS=255                      ; Packet buffers - 255 means allocate as many as
                                 ; possible, normally about 130, depending upon other
                                 ; table sizes.
; TNC default parameters:
PACLEN=128                       ; Max packet size (236 max for net/rom)
; PACLEN is a problem! The ideal size depends on the link(s) over which a
; packet will be sent. For a session involving another node, we have no idea
; what is at the far end. Ideally each node should have the capability to
; combine and then refragment messages to suit each link segment - maybe when
; there are more BPQ nodes about than 'other' ones, I'll do it. When the node
; is accessed directly, things are a bit easier, as we know at least something
; about the link. So, currently there are two PACLEN params, one here and one
; in the PORTS section. This one is used to set the initial value for
; sessions via other nodes and for sessions initiated from here. The other is
; used for incoming direct (Level 2) sessions. In all cases the TNC PACLEN
; command can be used to override the defaults.
TRANSDELAY=1                     ; Transparent node send delay in seconds
; Level 2 Parameters:
```

```
; T1 (FRACK), T2 (RESPTIME) and N2 (RETRIES) are now in the PORTS section
T3=120                          ; Link validation timer in seconds
IDLETIME=720                    ; Idle link shutdown timer in seconds
; Configuration Options:
AUTOSAVE=0                      ; Saves BPQNODES.dat upon program exit
BBS=1                           ; 1 = BBS support included, 0 = No BBS support
NODE=1                          ; Include switch support
HIDENODES=1                     ; If set to 1, nodes beginning with a #
                                ; require a 'N *' command to be displayed.
; The *** LINKED command is intended for use by gateway software, and concern
; has been expressed that it could be misused. It is recommended that it be
; disabled (=N) if unneeded.
ENABLE_LINKED=N                 ; Controls processing of *** LINKED command
                                ; Y = allows unrestricted use
                                ; A = allows use by application program
                                ; N = disabled
; AX25 port definitions:
; The LOOPBACK port simulates a connection by looping input to output.
; To test, start BPQTerminal and enter: 'C 1 MYNODE via MYCALL'
; In this example '1' is the LOOPBACK port number. The LOOPBACK port is
; provided for testing purposes would rarely be included in an established
; system.
;PORT
; PORTNUM=1                     ; Optonal but sets port number if stated
; ID=LOOPBACK                   ; Defines the Loopback port
; TYPE=INTERNAL                 ; Loopback is an internal type
; MAXFRAME=7                    ; Max outstanding frames (1 thru 7)
; FRACK=5000                    ; Level 2 timout in milliseconds
; RESPTIME=1000                 ; Level 2 delayed ack timer in milliseconds
; RETRIES=5                     ; Level 2 maximum retry value
; DIGIFLAG=1                    ; Digipeat: 0=OFF, 1=ALL, 255=UI Only
; UNPROTO=MAIL                  ; DEFAULT UNPROTO ADDR
;ENDPORT
; Ethernet port definition.
PORT
 PORTNUM=1                      ; Optional but sets port number if stated
 ID=BPQEther Link               ; Displayed by PORTS command
 TYPE=EXTERNAL                  ; Calls an external module
 DLLNAME=BPQEther.DLL           ; Uses BPQEther.DLL
 QUALITY=192                    ; Quality factor applied to node broadcasts heard
on
                                ; this port, unless overridden by a locked route
                                ; entry. Setting to 0 stops node broadcasts
```

```
   MINQUAL=0                    ; Entries in the nodes table with qualities greater
or
                               ; equal to MINQUAL will be sent on this port. A value
                               ; of 0 sends everything.
   MAXFRAME=4                   ; Max outstanding frames (1 thru 7)
   FRACK=3000                   ; Level 2 timout in milliseconds
   RESPTIME=1000                 ; Level 2 delayed ack timer in milliseconds
   RETRIES=5                    ; Level 2 maximum retry value
   PACLEN=236                   ; Max = 236
   UNPROTO=NODES
ENDPORT
; AX/IP/UDP port definition.
;PORT
; PORTNUM=1                     ; Optional but sets port number if stated
; ID=AX/IP/UDP                   ; Displayed by PORTS command
; TYPE=EXTERNAL                   ; Calls an external module
; DLLNAME=BPQAXIP.DLL            ; Uses BPQAXIP.DLL
; QUALITY=192                    ; Quality factor applied to node broadcasts heard
on
;                              ; this port, unless overridden by a locked route
;                              ; entry. Setting to 0 stops node broadcasts
; MINQUAL=0                         ; Entries in the nodes table with qualities
greater or
;                              ; equal to MINQUAL will be sent on this port. A
value
;                              ; of 0 sends everything.
; MAXFRAME=1                      ; Max outstanding frames (1 thru 7)
; FRACK=3000                      ; Level 2 timout in milliseconds
; RESPTIME=2000               ; Level 2 delayed ack timer in milliseconds
; RETRIES=15                       ; Level 2 maximum retry value
; PACLEN=64                    ; Max = 236
; UNPROTO=NODES
;ENDPORT
; AX/IP/UDP port definition.
PORT
  PORTNUM=2                     ; Optional but sets port number if stated
  ID=AX/IP/UDP                   ; Displayed by PORTS command
  TYPE=EXTERNAL                   ; Calls an external module
  DLLNAME=BPQAXIP.DLL            ; Uses BPQAXIP.DLL
  QUALITY=192                     ; Quality factor applied to node broadcasts heard
on
                               ; this port, unless overridden by a locked route
                               ; entry. Setting to 0 stops node broadcasts
```

```
    MINQUAL=0                            ; Entries in the nodes table with qualities
greater or
                             ; equal to MINQUAL will be sent on this port. A value
                             ; of 0 sends everything.
    MAXFRAME=1                   ; Max outstanding frames (1 thru 7)
    FRACK=3000                   ; Level 2 timout in milliseconds
    RESPTIME=2000         ; Level 2 delayed ack timer in milliseconds
    RETRIES=15                   ; Level 2 maximum retry value
    PACLEN=64                 ; Max = 236
    UNPROTO=NODES
ENDPORT
ROUTES:                              ; Locked routes (31 maximum)
; CALLSIGN,QUALITY,PORT[,MAXFRAME,FRACK,PACLEN]
; The values in [...] if stated override the port defaults.
; Locked routes tend to be overused and should not be set unless truly needed.
; No routes are specified, as they would be meaningless.
***                              ; Denotes end of locked routes
; Applications:
; Up to 8 applications are supported, which are expressed as positional
; parameters.  Each parameter may be empty, i.e. ',,' or if not empty, up to
; 12 bytes in length. Applications beginning with a * are not shown in the
; valid commands display in response to an entry of '?'.  There are two types
; of applications, local and remote. Local applications are those running on
; the same machine and are associated with an applications mask. A local
; application is entered as a user defined command, such as 'DX', 'BBS' or
; 'MYDX'.  A remote application is in the form 'command/C call(or alias)', for
; example DX/C N5IN.  When a user enters the user defined command 'DX' the
; node issues a 'C N5IN' attempting a connection to the N5IN dxcluster on a
; remote system.
; Associated with each positional parameter is an applications mask:
;    Position: 1,2,3,4,5,6,7,8
;    Decimal Mask: 1,2,4,8,16,32,64,128
;    Hexadecimal Mask: 0x1,0x2,0x4,0x8,0x10,0x20,0x40,0x80
;    Binary Mask: 00000001,00000010,00000100,00001000,
;                    00010000,00100000,01000000,10000000
APPLICATIONS=BBS,*
; In this example no applications are supported.
APPL1CALL=YT7MPB-6
APPL1ALIAS=B5
APPL1QUAL=255
APPL2CALL=YT7MPB-0
APPL2ALIAS=Misko
APPL2QUAL=0
```

```
; APPLnCALL/APPLnALIAS defines 'call's and alias's associated with the
; positional parameters in APPLICATIONS=. N may assume values of [1,8].
; These calls/alias's announce themselves in node broadcasts unless their
; associated APPLnQUAL=0. APPL1CALL,APPL1ALIAS and APPL1QUAL, i.e. for n=1,
; replace the earlier defined BBSCALL/BBSALIAS/BBSQUAL if stated. The 'call's
; may include any of the characters A-Z 0-9, up to 6 characters in length,
; and optionally an SSID. Thus APPL4CALL=ABCDEF as well as APPL4CALL=AA0AAA-13
; are valid. A connection to the APPLnCALL='call' will execute the nth
; positional parameter in the APPLICATIONS= parameter list.
;
; APPLnQUAL defines a multiplier (APPLnQUAL/256) which sets the quality of
; the associated application relative to the node quality.  This setting may
; be used to limit the spread of the associated application information through
; the network to your desired service area.  APPLnQUAL ranges from 0 through
; 255.  If set to 0, nodes broadcasts for the associated application will be
; suppressed.
```

APPENDIX D

The following content includes some examples of Linuxnode configuration files in Ubuntu 10.04.4 LTS. The lines starting with 'hash' sign (#) are either comments or are inactive commands.

```
# /etc/ax25/axports
#
# The format of this file is:
#
# name      callsign   speed      paclen      window     description
#
#1          OH2BNS-1   1200       255         2          144.675 MHz (1200 bps)
#2          OH2BNS-9   38400      255         7          TNOS/Linux  (38400 bps)
bpq0        Y7MPB-15   10000000   255         7          New LAN  (100 Mbps)


# /etc/ax25/nrports
#
# The format of this file is:
#
# name      callsign   alias      paclen      description
#
  netrom    YT7MPB-5   R5         35          Node
  netbbs    YT7MPB-6   B5         235         FBB
```

```
# /etc/ax25/ax25d.conf
#
# ax25d Configuration File.
#
# AX.25 Ports begin with a '['.
#
[YT7MPB-5 via bpq0]
NOCALL    * * * * * *  L
default * * * * * *  -             root  /usr/sbin/node       node
#
#
# NET/ROM Ports begin with a '<'.
#
<netrom>
NOCALL    * * * * * *  L
default * * * * * *  -             root  /usr/sbin/node       node
#


# /etc/ax25/nrbroadcast
#
# The format of this file is:
#
# ax25_name         min_obs         def_qual     worst_qual     verbose
#
# 1                 5               192          100            0
# 2                 5               255          100            1
  bpq0              5               255          100            0


# /etc/ax25/node.conf - LinuxNode configuration file
#
# see node.conf(5)
# Idle timeout (seconds).
#
IdleTimeout                 00
# Timeout when gatewaying (seconds).
#
ConnTimeout                 14400
# Visible hostname. Will be shown at telnet login.
#
HostName        YT7MPB-5
# Node ID.
#
```

```
NodeId              R5:YT7MPB-5}
#NodeId             \033[01;31m***\033[0m
# ReConnect flag.
#
ReConnect           on
# "Local" network.
#
#LocalNet 44.255.0.0/16
#LocalNet 192.168.0.4/16
# Command aliases.
#
# Alias             CAllbook           'telnet %{2:jazz.oh7lzb.ampr.org} 1235%1'
# Alias             CONVers            'telnet %{2:hydra.carleton.ca} 3600 "/n %u
%{1:32768}\n/w *"'
# Alias             CLuster            'c hkiclh'
# Hidden ports.
#
#HiddenPorts        netrom
# External commands
#
# Flags:   1        Run command through pipe
#          2        Reconnect prompt
#
#ExtCmd            PMS       3       root /usr/sbin/pms pms -u \%U -o VK2LID
#ExtCmd            PS        1       nobody     /bin/ps ps ax
#ExtCmd            TPM       1       nobody     /usr/bin/finger
finger tpm
#ExtCmd            Vpaiva    1       nobody     /home/tpm/bin/vpaiva vpaiva
#ExtCmd            NOde      0       root /usr/local/bin/node node
ExtCmd             ECho      1       nobody     /bin/echo echo \%U\%u \%S\%s
\%P\%p \%R\%r \%T\%t \%\% \%0 \%1 \%2 \%3 \%4 \%5 \%6 \%7 \%8 \%9
#ExtCmd            ECho      1       nobody     /bin/echo echo foo\%{1:***}bar
\%{U}\%{0:foo}\%{1:bar}\%{2:huu}\%{3:haa}
ExtCmd             TIme      1       nobody     /bin/echo echo %N Node session
started at %I, current time is \%I.
# Netrom port name. This port is used for outgoing netrom connects.
#
NrPort             netrom
# Logging level
#
LogLevel 3
# The escape character (CTRL-T)
#
```

```
EscapeChar              ^T
# Resolve ip numbers to addresses?
#
ResolveAddrs            off
# Node prompt.
#
NodePrompt              "\n"
#NodePrompt             "\n%s@%h \%i> "
#NodePrompt             "\033[36m%U\033[0m de
\033[01;35m#LNODE\033[0m:\033[01;31mVK2LID-10\033[0m>"
```

APPENDIX E

The following content is an example of the file */etc/rc.local* in Ubuntu 10.04.4 LTS that enables a Linux node to be activated automatically at each Linux boot. Be aware that all lines starting with 'hash' sign (#) are either comments or inactive commands. The remaining, active parts of the file include configuring a BPQ-compatible port bpq0 that successfully correspondents with the Windows-based BPQ32 node shown in Figure 7.

```
#!/bin/sh -e
#
# rc.local
#
# This script is executed at the end of each multiuser runlevel.
# Make sure that the script will "exit 0" on success or any other
# value on error.
#
# In order to enable or disable this script just change the execution
# bits.
#
# By default this script does nothing.
#
# configure the ax25 ethernet port
/sbin/modprobe bpqether
/sbin/ifconfig bpq0 hw ax25 yt7mpb-15
/sbin/ifconfig bpq0 192.168.0.4 netmask 255.255.255.0 up
#
# attach the netrom devices
# "netrom" is a dummy; "netbbs" connects to fbb;
# "netnod" connects to Linuxnode
# ip address via "-i" is necessary to keep from hosing the system
```

```
/sbin/modprobe netrom
/usr/sbin/nrattach -i 192.168.0.4 netrom
/usr/sbin/nrattach -i 192.168.0.4 netbbs
#
# start the daemons
/usr/sbin/ax25d
/usr/sbin/netromd -i
#/usr/sbin/mheardd
#/usr/sbin/ax25ipd
#/usr/sbin/axdigi &
#/var/fbb/xfbb.sh -d &       # we've installed FBB in /var/fbb
exit 0
```

APPENDIX F

The following content is an example of the file */etc/fbb.conf* in Ubuntu 10.04.4 LTS that includes main parameters of LinFBB mailbox server. Be aware that all lines starting with 'hash' sign (#) are either comments or inactive commands. The remaining, active parts of the file include parameters that enabled a remote user to communicate with the server, such as shown in Figure 9.

```
#
# FBB Set-up file
#
# default is /etc/ax25/fbb.conf
#
# may be changed using the $FBBCONF environment variable
#
###################################################################
#
# The following lines are mandatory
#
version = FBB7.04
# Callsign of BBS with hierarchical information
callsign = YT7MPB.#NS.SRB.EU
# SSID of BBS
ssid = -6
# Qra Locator of BBS
qraloc = JN95WF
# Qth of BBS
city = Novi Sad
# First name of SYSOP
```

```
name = Misko
# Callsign of SYSOP
sysop = YT7MPB
###############################################################
#
# Optional lines
#
# Callsign (and route if needed) that will have copy of SYSOP messages
sysmail = YT7MPB
# Line to send WP messages
wpcalls =
# BBS-UP batch or program
upbatch =
# BBS-DW batch or program
dwbatch =
#
# Servers will be searched and run in the "server" directory
#
# REQCFG, REDIST and WP are already built-in
#
#         Name     Filename    Information
server = REQDIR   reqdir      Directory request
###############################################################
#
# The rest of lines overwrites defaults. Here are the default values
#
# Directory of data files
data = /var/ax25/fbb
#data = /mnt/win_d/fbb/system
# Directory of config files
config = /etc/ax25/fbb
#config = /mnt/win_d/fbb/system
# Directory of message files
messages = /var/ax25/fbb/mail
#messages = /mnt/win_d/fbb/mail
# Directory of compressed files
compressed = /var/ax25/fbb/binmail
#compressed = /mnt/win_d/fbb/binmail
# Directory of users
fbbdos = *,*,/var/ax25/fbb/fbbdos,*,*,*,*,*
#fbbdos = *,*,/mnt/win_d/fbb/users,*,*,*,*,*
# Directory of YAPP files
yapp = /var/ax25/fbb/fbbdos/yapp
```

```
#yapp = /mnt/win_d/fbb/users/yapp
# Directory of documentation files
docs = /var/ax25/fbb/docs
#docs = /mnt/win_d/fbb/docs
# Directory of the pg programs
pg = /usr/lib/fbb/pg
#pg = /mnt/win_d/fbb/pg
# Directory of the filter programs
fdir = /usr/lib/fbb/filter
#fdir = /mnt/win_d/fbb/system
# Directory of the server programs
sdir = /usr/lib/fbb/server
# Directory of the tool programs (fbbdos, forward, cron...)
tdir = /usr/lib/fbb/tool
# Path and filename for import file
import = /var/ax25/fbb/mail/mail.in
#import = /mnt/win_d/fbb/mail.in
# Full log
logs = OK
# Test mode
test = NO
# Use (when possible) forward type FBB
fbbfwd = OK 512
# Use (when possible) compressed forward
fbbcomp = OK 3
# Wait for informations (Name, HomeBBS, Qth, ZIP)
askinfo = OK
# First connection mask:
# 0 : Disable
# 1 : Excluded
# 2 : Local
# 4 : Expert
# 8 : Sysop
# 16: BBS
# 32: Pagination
# 64: Guest
# 128: Modem
# 256: See-all-messages
# 512: Unproto list asking is allowed
# 1024: Liste des messages nouveaux.
# 2048:
mask = 32
# Security codes.
```

```
# Users can:
# 1 : Read all messages, including private messages
# 2 : Kill all messages
# 4 : Send SYS command
# 8 : Use remote sysop commands (edit, forward, etc...)
# 16: Edit labels in YAPP, FBBDOS, DOCS
# 32: Can delete files in YAPP, FBBDOS
# 64: Have access to all gateways
# 128: Run DOS commands
# 256: Have access to the entire hard disk
# 512: Have access to commands /A (stop) and /R (Reboot)
# All:    Sysop: Sysop after successful SYS-command:
security = 0 4 1023
# WARNING messages to sysop
# 1 : Less than 1MB in disk
# 2 : Error in system file (FORWARD, BBS, REJECT...)
# 4 : Server error/warning
# 8 : Ping-Pong warning
# 16: Unknown route warning
# 32: Unknown NTS warning
# 64: Message file not found
# 128: Error in proposal
# 256: Message rejected in remote BBS
# 512: Message held in remote BBS
#
warning = 1023
# Time (hour) for housekeeping (cleanup of messages)
housekeeping = 2
# Time-out for normal users / forward
timeout = 10 5
# Download size YAPP / MODEM
maxdownload  = 100 0
# Hours +/- in relation with UTC
#localtime = -4
localtime = 1
# Number of callsigns in mail beacon
beacon = 8
# Number of lines in scroll buffers
#
# User  Console  Monitoring
scroll = 1500 1500 1500
# Text for forward header (Do not change !)
fwdheader = [$c] $$:$R
```

```
# Number of saved BIDs
maxbids = 30000
# Lifetime for bulletins (days)
lifetime = 30
# Zip code of the BBS
zipcode = 21000
# Number of back messages in unproto lists
unprotoes = 500 5 P
#
# End of fbb.conf file
#
```

APPENDIX G

The following screenshot represents a version of LinFBB graphical interface that was compiled and installed in Mandrake Linux 9.1. Besides a couple of icons (bellow the menus *File* and *Window*), the compiled interface introduced some colors, such as yellow and green (Figure 18).

Figure 18. Compiled LinFBB mailbox software in Mandrake 9.1

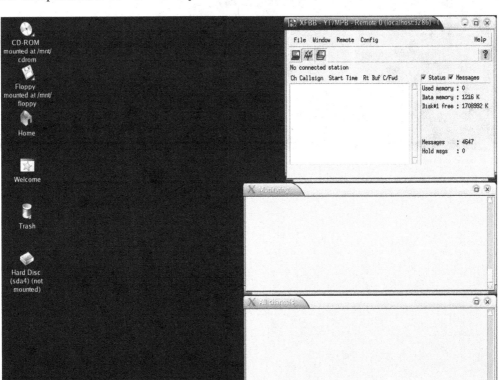

APPENDIX H

The first screenshot displays an intermediate visual content during the operating system upgrade—from Ubuntu 10.04.4 to Ubuntu 12.04.2—where the old menus are still visible on top of the screen (*Applications*, *Places*, and *Window*), while the new background color has been introduced. The second screenshot represents the process flow (Figure 19, Figure 20).

Figure 19. A 'transitional' Ubuntu GUI during the OS upgrade

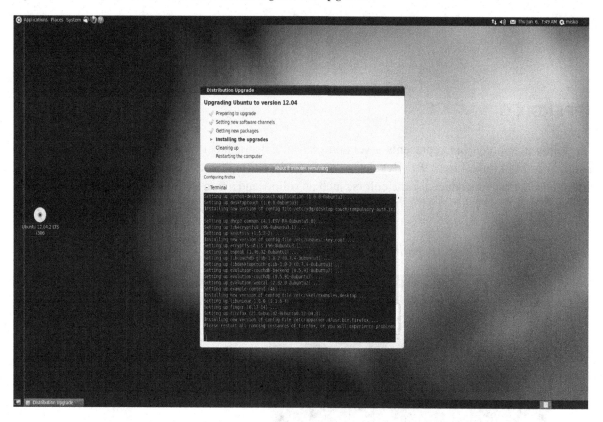

Figure 20. Linux distribution upgrade in action

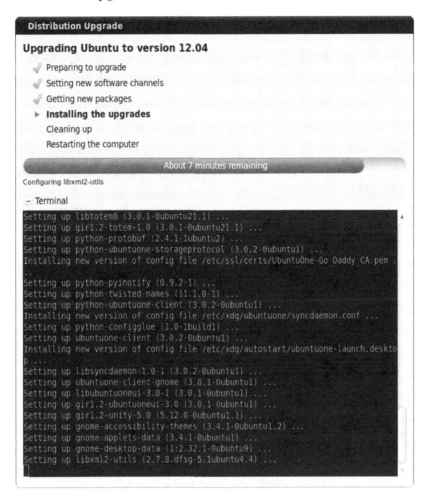

APPENDIX I

The following content describe changes in configuration files of *Shorewall* firewall in Debian 6.0.7. Please note that entries in bold-italic are added in respect to an earlier configuration in order to enable access to the JNOS mailbox from a local area network.

```
#
# For information about entries in this file, type "man shorewall-interfaces"
###############################################################################
#
#ZONE        INTERFACE    BROADCAST     OPTIONS
net          eth0         detect        tcpflags,logmartians,nosmurfs
loc          tun0         -             tcpflags,logmartians,nosmurfs
#
```

```
# For information on entries in this file, type "man shorewall-rules"
############################################################################
#
#ACTION SOURCE DEST PROTO DEST SOURCE   ORIGINAL RATE  USER/  MARK
#                               PORT PORT(S) DEST      LIMIT GROUP
#SECTION ESTABLISHED
#SECTION RELATED
SECTION NEW
ACCEPT net fw icmp 8
ACCEPT net fw tcp ssh,www,https,smtp,pop3,pop3s,imap2,imaps,submission,telnet
ACCEPT net fw udp https
ACCEPT loc fw icmp
ACCEPT fw  loc icmp
ACCEPT fw  loc tcp telnet
ACCEPT net loc icmp
ACCEPT net loc tcp telnet
```

Chapter 7

The Access of Things:
Spatial Access Control for the Internet of Things

Peter J. Hawrylak
University of Tulsa, USA

Matthew Butler
University of Tulsa, USA

Steven Reed
University of Tulsa, USA

John Hale
University of Tulsa, USA

ABSTRACT

Access to resources, both physical and cyber, must be controlled to maintain security. The increasingly connected nature of our world makes access control a paramount issue. The expansion of the Internet of Things into everyday life has created numerous opportunities to share information and resources with other people and other devices. The Internet of Things will contain numerous wireless devices. The level of access each user (human or device) is given must be controlled. Most conventional access control schemes are rigid in that they do not account for environmental context. This solution is not sufficient for the Internet of Things. What is needed is a more granular control of access rights and a gradual degradation or expansion of access based on observed facts. This chapter presents an access control system termed the Access of Things, which employs a gradual degradation of privilege philosophy. The Access of Things concept is applicable to the dynamic security environment present in the Internet of Things.

INTRODUCTION

Today, access to buildings and facilities is often controlled by electronic systems; gone are the days of large key rings. Magnetic stripe technology is one such technology that encodes an identifier linked to a person or employee on their identification card. A scanner at the door reads the identifier and sends it back to a centralized system. The centralized system replies with a message granting access (opening the door) or denying it (keeping the door locked). Wireless versions of this system are available and widely used, with most based on radio frequency identification (RFID)

DOI: 10.4018/978-1-4666-5170-8.ch007

technology. Both systems rely on the person keeping control of their identification card and preventing it from being copied. This has led to binary security policies for access control systems providing either complete access to all authorized areas or no access at all, and does not function well in the case of copied or "cloned" RFID tags or identification cards. The binary security policy often fails to prevent the cloned RFID tag from gaining entry or locks out the legitimate RFID tag preventing the employee from accomplishing their tasks. The wireless nature of RFID makes cloning possible without physical contact with the legitimate user. A security policy and supporting technology is needed to enable access rights to be gradually and gracefully reduced to allow the legitimate user to complete most of their tasks, while preventing significant damage caused by the malicious user. This chapter will present a framework for such a system for RFID and will discuss the integration of other data sources into this system, yielding a truly Internet of Things approach to spatial access control.

The chapter will begin with an overview of the Internet of Things and the need for access control in this environment. Next, current access control systems and policies, the difficulties introduced by the traditional binary access policy, and the requirements for access control within the Internet of Things will be presented. The graceful degradation of privilege policy will then be introduced and the RFID implementation of this policy will be described. Security measures available to RFID technology to prevent cloning will be highlighted. The extension of the graceful degradation of privilege model to the Internet of Things will be presented. This extension is termed "the Access of Things," and provides robust access control policies and options for the Internet of Things. The Access of Things concept integrates information from a variety of sources to achieve multi-point identification and authentication of

the user. The chapter will conclude with a discussion of future research areas, issues facing access control systems, and how the system presented in this chapter begins to address those concerns.

INTERNET OF THINGS CONCEPT

The Internet of Things (IOT) concept envisions an environment where devices automatically connect together to solve problems or better monitor the environment. The problems that can be addressed in the IOT framework are larger than a single device could solve on its own. This may be due to lack of computing power or lack of access to input data. The concept of the IOT is not necessarily one of human-centric applications, but one that will include more machine-to-machine (M2M) applications facilitated by massive M2M networks supported by the IOT's infrastructure. The differentiator between the IOT and a generic Internet capable device is the increased degree of autonomy of the device and reliance on M2M communication. In fact, most IOT applications are based around the M2M communication with the human user being a consumer of information or service rather than the initiator of operations.

The initial idea was to provide every device with an IP address for routing data (traffic) between devices to facilitate M2M communication. While IP is a widely used protocol, it may not be the best protocol for all applications and other protocols are available to supplement IP. This is true for IOT applications because many IOT devices have limited computational and communication resources, especially remote sensor nodes.

IOT applications include smart health, remote healthcare (You, Liu, & Tong, 2011; Revere, Black, & Zalila, 2010; Chen, Gonzalez, Leung, Zhang, & Li, 2010; Wicks, Visich, & Li, 2006), home management, traffic management (Foschini, Taleb, Corradi, & Bottazzi, 2011), smart grid, and industrial control systems. The notion of networks forming, changing, and dissolving on their own raises questions about what resources should be shared. Each actor in the IOT must make this decision on their own and based on their perception of the environment and the application. Spatial access control, including building access control systems, is one example of such a system.

Cooperation and Network Construction Aspects

The IOT will operate over a wide range of wired and wireless networks. Generally, wired networks provide the long-haul backbone connectivity and handle large amounts of data, while wireless networks provide intermediate and the last-mile connection to the user or remote sensor. Some common wireless networking technologies currently used in IOT applications include radio frequency identification (RFID), ZigBee, Bluetooth, and 6LoWPAN (Kushalnagar, Montenegro, & Schumacher, 2007).

RFID systems for retail, inventory, and supply chain applications require a unique identifier for each item. This is in contrast to the typical UPC (Universal Product Code), commonly referred to as a "barcode," which only provides the type of item. RFID needs a unique identifier to allow the reader to distinguish between multiple instances of the same type of item in order to obtain an accurate count. GS1 has produced many specifications for unique identifiers (EPCglobal, 2010) for RFID systems, with the UHF Gen-2 (EPCglobal, 2008) systems being the primary users.

ZigBee is a protocol stack constructed on the PHY and MAC layers provided by IEEE 802.15.4 (IEEE, 2011). Mesh networking is advantageous because it provides a communication infrastructure that is resilient to node failure. ZigBee has been heavily used to provide wireless connectivity between home automation devices. There are two types of network architectures in ZigBee: peer-to-peer and star. The star topology utilizes

a single device, known as the PAN coordinator, that all traffic passes through (IEEE, 2011). The peer-to-peer topology allows any two devices to communicate directly without having to go through the PAN coordinator and can support mesh networking (IEEE, 2011). ZigBee enables significant flexibility in devices connecting (joining) or disconnecting (leaving) the network. However, the protocol is based on a beaconing system that consumes a significant amount of energy over time. This puts strain on the PAN coordinator and those nodes that route a lot of traffic in peer-to-peer topologies. These energy restrictions pose significant problems for ZigBee and other communication protocols for battery powered devices.

6LoWPAN is a protocol focusing on the transmission of small messages and is based on IPv6 and the IEEE 802.15.4 PHY and MAC layers (Kushalnagar, Montenegro, & Schumacher, 2007). It is intended for use in low-functionality devices such as sensor nodes that require long lifetimes and limited or no interaction from humans (e.g., replacing batteries). These devices traditionally send small messages consisting of a few sensor readings. The implementation of the protocol stack in software on an embedded device is a key concern and one of 6LoWPAN's goals was to reduce the software footprint (Mulligan, 2007). Message size in 6LoWPAN is limited 81 octets (1 octet equals 8-bit and is typically identical to a byte) because most messages in the IOT will be small status, informational (e.g. sensor readings), or alarm messages (Kushalnagar, Montenegro, & Schumacher, 2007). The majority of messages transmitted in the IOT will be small, often less than 40 bytes. Most of these messages will be sent within a local network, often to a device's neighbor (single hop messages). 6LoWPAN provides support for efficient transmission of these messages through header compression which leverages information included by headers in the MAC layer of IEEE 802.15.4 and reduces the overhead required by 6LoWPAN on messages (Mulligan, 2007). This allows for more data to be transmitted in each message or allows the minimum message size to be reduced.

These protocols each provide components that are useful for the IOT. RFID systems provide guidance on development and implementation of unique identifiers that can be correlated into device addresses. The development of passive RFID devices with simple computing resources provides a basis for design low-power devices for the IOT. ZigBee provides support for data communication in a dynamic and unpredictable environment where the primary goal is to ensure message delivery. 6LoWPAN provides guidance on implementation of IP using the same MAC and PHY layers (IEEE 802.15.4) as ZigBee but a version of the IP protocol geared to low-power devices.

OVERVIEW OF THE ACCESS CONTROL PROBLEM

Access to resources and places (e.g. buildings) must be limited to authorized users. The access could be physical, virtual, or a combination of the two. Physical access is typically thought of as limiting who can access what rooms or areas, or what resources each person can use or remove from the location. Physical access control is mostly concerned with personal safety (e.g., preventing unauthorized persons from entering a building or floor), preventing theft of property, and workplace safety (e.g., interlock controls on a piece of machinery to prevent operator injury). Virtual access control focuses on protecting computer files (e.g., bank records) and device resources (e.g., processor time). Cyber-physical systems blend physical and cyber components and form the third category of access control problems. Cyber-physical systems have assets from both domains and must support access control policies for both domains. Examples of cyber-physical systems include the electric grid, municipal water

treatment plants, and robotic technology (e.g., surgical robots). Many IOT systems fall into the cyber-physical category with the IOT providing the infrastructure to connect the control system and sensor modules together.

Traditionally, access control has dealt with physically securing locations, such as a building or room, against unauthorized entry, or with securing a computer system against unauthorized use. These two issues are linked in many IOT systems: the first is a physical access control problem with solutions from the physical security domain, while the second is a data access control problem, with solutions primarily from the cyber-security domain, although physical security is an important factor. Cyber-security often relies on physical security at some level to prevent an attacker from gaining physical access to the cyber-asset in question. Access control systems guard resources and places from unauthorized use or entry, as well as provide the means to verify a user's credentials.

Extension of Access Control to the Internet of Things

The IOT will increase the integration of the physical and cyber-security domains because devices will need to determine what resources to provide to the IOT (larger device society) and to whom to provide these resources. This blurs the definition of what the asset is because it may be a collection of cyber and physical items. Access control in the IOT will need to support collecting identity verification input from multiple data sources, while maintaining the assumed quality of service.

OVERVIEW OF EXISTING ACCESS CONTROL SYSTEMS

Physical and cyber access controls systems may be fully characterized by two essential architectural components – their *authorization policy model*, and their *access control enforcement mechanism*

(Bertino, Samarati, & Jajodia, 1993; Sandhu, 1993; Sandhu & Samarati, 1994). A policy model incorporates a language or system for expressing the rules of access by a subject (user or process) to an object (physical or cyber asset or resource). A policy model also commonly allows for the qualification of access type (e.g., "read," "write," or "execute"). The enforcement mechanism of an access control solution integrates authentication services that bind a subject to an identity with system-level controls that accept access requests, and permit or block access to a resource based on the outcome of an access request decision by a policy engine. System-level controls mediating access may exist at the application-level, in an operating system, or on a network.

Conventional access control methods base binary authorization request decisions on simplistic criteria, either providing full privileges or no privileges at all. These systems rely on some method for a user to prove their identity or role, and a backend system that makes access decisions. A wide range of authentication schemes and technologies exist for a user to prove their identity, including passwords, identity (ID) badges with digitized information contained in a magnetic strip or RFID tag, or through biometric identification (e.g., fingerprint or iris scan).

The most prevalent access control solutions are centralized systems that utilize terminals where individuals present their identification credentials. The identification information is transferred to a backend policy engine that determines whether or not access should be granted or what level of access to grant. The policy engine may simply verify that the identity is on a list of authorized subjects or may apply some additional rules to verify an identity or qualify its access to a resource. Examples of such rules include limiting the times when certain individuals can access a building or declining a credit card charge because of suspicious behavior noted by the system. After the decision is made, the appropriate message is sent back to

the physical location to effect the decision (e.g., unlock or not unlock the door). In a cyber-access request, the notion of physical location is often equated to the location of the cyber asset at the present time.

This section contains a brief overview of three of the most common technologies used to identify individuals in an access control system: magnetic stripe, RFID, and biometrics. Each technology offers benefits and drawbacks, which are highlighted in the following subsections. The Access of Things concept requires an architecture that supports taking input from multiple identify verification technologies and other sensors to determine access rights.

Magnetic Stripe Technologies

Magnetic stripe cards utilize a magnetic material to encode information. Traditional embodiments of this technology do not support the ability to write new or to update existing information to the magnet material. As such, they are useful in providing a static identifier that is linked to a person, but cannot be dynamically challenged (e.g., as part of a challenge-response authentication procedure) or updated with new data (e.g., addition or removal of access privileges).

RFID Technologies

RFID technologies remove the need to physically swipe an access card through a magnetic scanner; rather, the user just presents the RFID tag (access badge) within the read range of the RFID reader to gain access. The range at which the RFID tag can be read is dependent on the technology and frequency used. There are three main types of RFID systems: passive, battery-assisted passive (BAP), and active. Passive systems employ RFID tags that have no on-board power source (e.g. a battery) and harvest the energy needed for operation from the RFID reader's RF signal.

Passive RFID tags communicate via backscatter by modulating the signal transmitted by the RFID reader. Backscatter is a very low-power method of communication and does not require the RFID tag to have a dedicated or powered transmitter. The tag modulates the RFID reader's signal by adjusting its antenna impedance to either reflect or absorb the transmitted RF energy. BAP systems utilize BAP tags which communicate using backscatter but have an on-board power source to power sensors or processing elements. However, a BAP tag's on-board power source is not used to power a transmitter or receiver. BAP RFID tags have a longer communication range than passive RFID tags because they do not have to use any of the RF energy harvested from the reader's signal for powering and all of that energy can be used for communication purposes. Active RFID systems employ tags containing an on-board power supply, such as a battery, and communicate using a powered (active) transmitter and receiver. Active RFID tags can communicate at longer distances than BAP RFID tags because the powered receiver can extract weaker signals and the powered transmitter can send stronger signals.

Frequency is another factor in determining the range of RFID systems. Low frequency (LF) systems, typically operating in the 125-135kHz range have a range of 1m. These systems are often used for animal tracking. High frequency (HF) systems operate in the 13.56MHz ISM (industrial, scientific, and medical) band, which is available worldwide, have a range of 1-3m depending on the transmitter power level. HF systems are often used for contactless payment, public transportation fare, and access or identity badges. Ultra-high frequency (UHF) systems operating in the 860-960MHz range, depending on the regulatory region (location), have ranges of 1-6m. These systems are used in asset tracking applications, primarily in the retail industry. This chapter will focus on the access control problem and the HF (13.56MHz) RFID technology.

Biometric Technologies

Biometric technologies (Jain, Hong, & Pankanti, 2000) include fingerprint scanners, iris scanners, hand geometry scanners, and facial recognition. There are concerns about the privacy aspects of these systems because the information (fingerprint, iris print, facial recognition data) would be stored in data repositories and could be used for other activities such as clandestine tracking or in criminal investigations (e.g., universal fingerprint database). There are also health concerns about the transmission of disease through bodily fluids. This is especially true for those systems based on the iris scanner. Facial recognition avoids these health issues and can be easily deployed by using input from a camera. However, the algorithms for facial recognition have met with varying degrees of accuracy, with more accurate algorithms requiring longer processing times.

Motivation to Combine Spatial and Adaptive Access Control Techniques

There are several forms of access control, many of which are not mutually exclusive. The naive solution to user privilege management is to allow full access or no access, termed *Binary Access Control* in this chapter. In a Binary Access Control system, users either have access to a resource, or they are restricted from access to a resource. A Binary Access Control system is simple to implement, understand, and use, but is not without drawbacks. First, if the system does not support automatic modification of user privileges it will not be able to respond quickly to observed events. Second, if a user's access credentials are compromised, few mitigation strategies are available to be employed: either the user retains the same rights, thereby allowing attackers the same access as the legitimate user, or the user is prohibited from all access, preventing the attacker from gaining access, but also inhibiting the legitimate user from completing his or her more safety critical tasks.

Another form of access control is Spatial Access Control. Rather than being concerned with virtual access to data, Spatial Access Control is the practice of managing physical access to areas requiring a security clearance. Areas that require higher levels of clearance are likely to house higher valued resources.

DIFFICULTIES WITH THESE SYSTEMS AND POLICIES

Binary Access Control is often applied for Spatial Access Control purposes, and often times this is all that is necessary. However, in many safety and confidentiality centered environments, this is not the best option. For example, in a hospital, if a doctor is prevented from getting into the operating room, this could be the difference between life and death; or when credit card fraud is detected and the card is deactivated to prevent malicious spending, the legitimate user still needs to be able to make emergency transactions in order to prevent being stranded money-less. In both scenarios it would be better to provide a gradual loss of privilege. For example, preventing the doctor from entering sensitive areas such as the pharmacy, or reducing the spending limit of the credit card in question to enable the real user to obtain taxi fare back to their hotel.

In any given scenario, if the access control system were able to accurately measure the amount of risk involved in permitting or disallowing a given action by one of its users, the system would be better able to achieve the goal of providing a custom tailored access policy in response each request. In order to predict this level of risk and subsequently use it to dynamically make decisions about user privileges, we need a Dynamic Risk Assessment Access Control system, or DRAAC (Butler, 2011).

Implementation of the Graceful Degradation of Privilege System Using RFID

The only way to make an accurate estimation of risk at any given time is to have some related context about the situation. This means a large focus of the DRAAC system is collecting data. Data is not limited to a particular source in any way; almost any information that can be gathered about the environment under supervision can help make better risk assessments. The data gathered is used to make inferences about user behavior or predict future behavior. When an access control decision needs to be made, the system can quickly evaluate whether a user is acting with malicious intent based on data gathered about the user and the state of the system. However, Risk is not directly associated with the user's behavior. Risk is computed as a function of their behavior, represented by a Likelihood of attack value, and the resource they are attempting to access, represented by an Impact value as defined by Equation 1 (Butler, Hawrylak, & Hale, 2011).

$$Risk = Likelihood * Impact \qquad (1)$$

Once the Risk of allowing or disallowing access has been computed, a decision can be made. This may range based on the application ranging from disallowing a user access to contacting the security department.

DRAAC Overview

DRAAC is designed to be modular. The modular architecture of DRAAC is shown in Figure 1. The system used to determine Likelihood values is not dependent on the system for reporting Impact values. Furthermore, the Risk can be calculated in various ways dependent on the scenario, and how this data is purposed to result in access control

Figure 1. DRAAC architecture

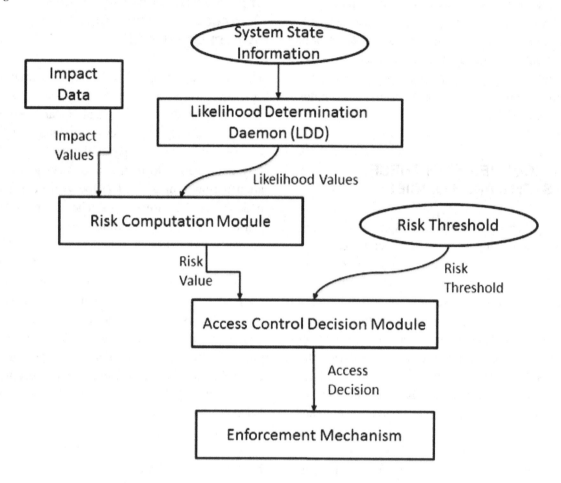

decisions is not dependent on any of the previous steps. One method for determining Risk is given in Equation (2) where Risk is defined by the product of cumulative Likelihood and cumulative Impact (Hartney, 2012).

Risk = (Cumulative Likelihood) * (Cumulative Impact)   (2)

Risk is computed in the "Risk Computation Module" of DRAAC and this module supports the ability to calculate multiple Risk values based on different inputs and/or expressions. This enables risk to be analyzed from different perspectives and may yield a more accurate view of risk.

Notice that the philosophy of DRAAC is not necessarily tied to Spatial Access Control systems at all. A dynamic risk assessment system could be used for virtual access as well, watching user behavior to determine if they are attempting to breach the system. The original implementation of DRAAC was to provide graceful reduction in user access rights on a computer (Butler, 2011).

In our implementation of DRAAC, the Likelihood Determination Daemon (LDD) is written in Python. The LDD is responsible for gathering data as well as operating on the data to make predictions and assertions, all the things involved with deriving the likelihood of an attack. Likelihood is important because it can be used to identify which assets a user is most likely to access. This information can be used to help build a user profile for normal behavior.

Data gathered by the various sensors (e.g., RFID readers) in the system is stored in a simple log file. Items termed "Facts" are derived from these data. A data item could be a Fact or it may be part of a larger Fact that is identified through processing. The Watchdog library is used for cross-platform compatible file-system observation; it is used to watch the log file for new events fired. The identification of suspicious activity is done using the CLIPS expert system. CLIPS has a set of preset "Rules" which operate on dynamically gathered Facts. Rules can generate more Facts, potentially causing more Rules to be applicable and execute, or they can make Python calls to the LDD.

Extension to RFID

The first implementation of DRAAC was used with virtual access control, and the data it collected was SSH login attempts along with geographic location of the attempted login (Butler, 2011). In this implementation, the focus is using reads of RFID tags by RFID readers in different locations to make Spatial Access Control decisions (Butler, Reed, Hawrylak, & Hale, 2013). The RFID reader used was the SecuraKey ET4-AUS-D with a custom client written using Visual C#.

The DRAAC log file is targeted by the RFID reader software where any scan events are written out. Watchdog will catch the changes to the log file and notify the LDD. The LDD will read the newest change to the log and generate a Fact to inject into the CLIPS system. Any applicable CLIPS rules will fire, usually causing a chain reaction of new events: Fact generation, Fact deletion, and more Rule firing. If suspicious activity is identified, the LDD will be notified. Once the LDD has made a decision on the Likelihood of attack, the information is passed on to the Risk Computation Module, which is responsible for Risk computation. In the RFID implementation of DRAAC the LDD monitors the RFID readers located at each door and creates Facts based on their reads. These Facts indicate what doors an individual tries to access. This information can be used to locate the individual which can then be used to help decide if that individual should be granted access to a given area (room).

The syntax for writing Rules for CLIPS is unlike most other programming languages, but the logic is the same. The pattern for a rule is illustrated in Figure 2. The rule starts with a name (line 1 of Figure 2), which allows the rule to be referenced (used) in other rules. Next, the If-clause documents the conditions that must be true (preconditions) for the rule to fire (execute). Finally, the Then-clause contains the actions to be taken when the rule fires.

Figure 3 shows an example of a rule for the RFID implementation of DRAAC that generates a Fact for the first time an individual (ID badge or card) is observed. In English this would be read as: "If the system reads a RFID tag at scanner $s1$, time $t1$, and with card ID $c1$, and the system has never seen the card $c1$ before, Then assert the fact that the system never seen $c1$ before, and assert the fact that now we have seen $c1$."

One thing to take note of is that CLIPS rules use prefix notation. The last part of the If clause (see line 3 of Figure 3) is checking that there is not a fact named "seenBefore" with card ID $c1$ (here card ID and tag ID are used interchangeably). If this Rule successfully fires, the second part of the Then clause is executed (see line 5 of Figure 3) asserting the fact "seenBefore" with card ID $c1$ which prevents this Rule from ever being applicable for this card ID in the future (unless some other rule retracts the relevant Fact). This is the first rule that is evaluated for any RFID tag (ID badge or card) in our implementation of DRAAC.

Figure 2. Syntax of a CLIPS rule

```
1  < Name of rule >
2  < If – conditions >
3  < Then – assertions >
```

Figure 3. Example of a rule in the RFID DRAAC system

```
1   no prior scans
2   If :(rfid – scan(time ?t1)(scannerid ?s1)(cardid ?c1))
3       (not(seenBefore(cardid ?c1)))
4   Then :(assert(noPriorScan(cardid ?c1)))
5       (assert(seenBefore(cardid ?c1)))
```

It is used to identify new ID cards that the system knows nothing about from ID cards which the system does have information about. The "seenBefore" fact is used to track whether or not the system has observed the ID card before or not. The choice of how long the system remembers (e.g., one day) each ID card is important: longer memory requires more space for the log file and processing time, but provides additional information that can be used to make more accurate future decisions. These tradeoffs must be evaluated and balanced to provide adequate security while meeting the transaction time requirements.

The first part of the Then clause (see line 4 in Figure 3) asserts the "noPriorScan" fact which is used to indicate that this is the first time the card ID has been observed. While this may appear to be a contradiction of the "seenBefore" fact, it is not. The "seenBefore" fact only states that the current scan is not the first time the card ID has been observed. The "noPriorScan" fact tells the system that all facts relating to the card ID in question are for the first time that card has been observed and will be used to trigger other rules the next time the card ID is observed. This information is important for physical access control and can be used to help identify cases were an individual tries to enter an interior door (area) without first entering through an exterior door. This can also be used to identify cases of "tailgating" where an individual follows someone else through a checkpoint without carding in themselves. Once the card ID is observed for again, the "noPriorScan" fact is retracted (removed) from the list of Facts maintained by CLIPS.

Extension to the Internet of Things

The IOT will require access control policies for all three domains: physical, cyber, and cyber-physical. Information will be collected through a collection of devices connected over wired and wireless links. However, the majority of these connections will be wireless. The high mobility of powerful wireless computing devices such as a smartphone or table introduce serious complications for access control in the IOT. The ability to identify and collect information from other sources within the IOT will be critical to making access control decisions.

Description of the Graceful Degradation of Privilege Policy

The binary (complete access or no access) access control policy will not provide the required support or quality of service needed for the IOT. The IOT will require multiple levels of access control and support multiple adjustments to access policies in response to observed facts. One such access policy is termed the *graceful degradation of privilege policy* which gradually removes or expands a user's access rights on a given system. The goal of graceful degradation of privilege is to attempt to provide the maximum available use of a device or service for a given level of privilege, while assuring specific security goals and requirements are met. The IOT will support many technologies that are capable of being compromised, cloned, or copied. Ideally, the legitimate user would like to be able to perform as many tasks as they can, provided their resources are protected from malicious users.

The IOT will utilize collaboration between devices and each device will need to determine what resources it shares with the others. The resources that are shared will depend, in large part, on the surroundings and the other devices present in the collaboration. Graceful degradation of privilege provides a means for each device to adjust its level of cooperation based on the observed dynamics of the IOT. For example, a body area network monitoring a person's medical status may want to restrict the information it transmits to non-sensitive information when the person is in a public setting. In this example, the level of information sent would be set by the person (user) and would be based on their desired level of privacy.

Description of the Access Control Problem

The access control decisions made by the DRAAC system are based on the Risk of the access. In this implementation of RFID DRAAC, Risk is based on the Likelihood and Impact quantities as defined by Equation (2). Graceful degradation of privilege is achieved by updating the Likelihood and Impact parameters, or adjusting the Risk threshold for allowing access to a resource based on observed Facts (Butler, 2011). Conceptually, this is a simple and straightforward process, but requires the ability to quantize risk, impact, and likelihood. This quantization requires two steps: first, the information used to derive facts must be identified and classified as affecting risk, impact, or likelihood; and second, methods must be identified and implemented to collect this information.

Quantization of Likelihood and Impact Parameters in Access Control

Cloning or copying of a RFID tag (Halamka, Juels, Stubblefield, & Westhues, 2006) and malicious use of a legitimate RFID tag via a man-in-the-middle attack (Oren & Wool, 2010) are the two main security threats to RFID based access control systems. These attacks allow the malicious user to impersonate the legitimate user to gain access to protected resources. Access control has been widely studied and formal languages have been developed to model access control policies and activities (Bertino & Kirkpatrick, 2011; Kirkpatrick, Damiani, & Bertino, 2011; Ardagna, Cremonini, Damiani, De Capitani di Vimercati, & Samarati, 2006). Geographic location of both users and resources are used in GEO-BRAC to determine access rights in a military setting to restrict access to information to only those that can use that information within a given geographic area (Bertino & Kirkpatrick, 2011). Geographic information can be used to derive additional rules and inferences about the mobility of individuals and to monitor the surrounding environment to make sure that access requests are granted only in secure surroundings (Ardagna, Cremonini, Damiani, De Capitani di Vimercati, & Samarati, 2006). Kirkpatrick, Damiani, and Bertino propose a formal language for describing buildings and floor plans to document the relationship of individual areas to each other for access control decisions (Kirkpatrick, Damiani, & Bertino, 2011). This formal language documents those locations that are contained within other locations and this can be used as a means to derive additional information about the individual's location (Kirkpatrick, Damiani, & Bertino, 2011). Geographic location information is used in the RFID implementation of DRAAC for physical access control to detect cloned tags based on door (room) agency information and the sequence of access requests (Butler, Reed, Hawrylak, & Hale, 2013).

The access control system must identify both the likelihood of a malicious use and the impact of allowing access if the use is malicious. To quantize likelihood it is often necessary to determine the condition of the user's surroundings and the IOT provides significant assistance in obtaining this information. Impact can be determined from the potential follow-on activities resulting from the access and can be documented using an attack graph or attack dependency graph.

Defining and Adjustment of Access Rights in this Context

The IOT provides support to obtain such this information. Scans from RFID readers at access points (e.g., doors) can be used to identify the location of the access request. Cameras, such as the Microsoft Kinect (Kong, 2011), can be used to track a person's location, but also to identify how many individuals are present. This information can be useful in verifying that the location is secure or that the necessary resources are present for the requested action (Bertino & Kirkpatrick, 2011; Ardagna, Cremonini, Damiani, De Capitani di Vimercati, & Samarati, 2006). Location can be determined through processing of wireless signals (Parr, Miesen, & Vossiek, 2013) and because the IOT will employ a large number of wireless devices the accuracy of such systems should be sufficient for access control requirements. Based on this information the likelihood of each possible security breach can be estimated. One method to assist in this quantization is to use attack graphs (Philips & Swiler, 1998) or attack dependency graphs (Louthan, Hardwicke, Hawrylak, & Hale, 2011; Ou, Boyer, & McQueen, 2006) to map the potential security vulnerabilities or threats. While these structures provide information about possible threats, they must be analyzed to obtain actionable information. One option is to identify those vulnerabilities that will result in the greatest impact to the system (Hartney, 2012), and this information can be used to identify the malicious user's next move (Hawrylak, Hartney, Haney, Hamm, & Hale, 2013). This information can be used to identify what resources would be used or gathered (sampled) from the IOT to help verify a user's identity or to counter a malicious user.

Attack graphs and attack dependency graphs are useful tools to quantify impact in addition to assist with determining likelihood. Other techniques are applicable for quantifying impact, including attack surfaces (Manadhata & Wing, 2011) and privilege graphs (Dacier, Deswarte, & Kaâniche, 1996). These techniques can be used to determine what future exploits (attacks) could be carried out if the current access request is from a malicious user.

Regardless of the techniques used to quantify likelihood and impact they must integrate with the heterogeneous technology and network protocols that will be present in the IOT. The time required to compute the likelihood and impact values must not introduce too much delay into the overall interaction. This is especially true for use-cases where the access involves moving objects, such as a high-speed assembly line or automobile. Finally, the computational effort required to quantize these two components must be minimized for energy conservation and to ensure that it can execute on the resources available to the IOT device.

Definition in Relation to the Internet of Things

In the IOT, devices will need to identify their surroundings and form networks to collaborate with other devices. In this context impact is harder to quantify because of the heterogeneous nature of the resulting networks and the high degree of mobility of devices. Mobility complicates impact calculation because it allows a compromised system to move from one network or system to another, and be used to compromise the new system.

Likelihood can be correlated to the user, the requester, and the current surroundings. Here the IOT can help by providing information about the surroundings that the device cannot gather on its own. Further, many wireless protocols are beacon based, including Wi-Fi and ZigBee, enabling a device to identify other wireless devices nearby without having to join the network to determine the composition of the surroundings.

Difference from Current Access Control Policies

The graceful degradation of privilege approach allows access rights to be adjusted on a finer scale than the binary model. It also provides for increased security because in response to a situation where a decision between two access levels cannot be made, the system can provide access rights at the lower level (higher security with the user having fewer access rights) and then gradually increase access rights if no suspicious behavior is observed. In the IOT this is advantageous because collaborations are not suddenly cut off, but go through a graceful degradation of service as conditions change. The graceful *increase* in privilege is also very useful because it limits the damage that can occur in the event that the decision to allow access was incorrect. Compared to the binary method this is very advantageous because the malicious user would have full access rights instead of starting with a limited set of access rights and gradually working their way to full access rights.

The RFID DRAAC system relies on the correct selection of the Impact and Likelihood values to compute the Risk associated with each access request. Accurate information can be difficult to maintain for these two parameters in a traditional physical or computer access situation. The IOT coupled with the significant mobility of devices and the ease of establishing connections using some wireless network makes this more difficult. Each device must maintain data to calculate these values. This may require memory and processing resources that are beyond the ability of the low-power devices that will comprise the bulk of the IOT.

Security Measures Available to RFID Based Systems

There are a number of security measures that can be applied to RFID systems to provide security. Most are geared toward preventing the cloning of

RFID tags or protecting the data stored in the tag from modification. Data can be protected from malicious reading by encryption before being transmitted to the tag. This has the added benefit of protecting the data against eavesdropping. The EPC Gen-2 protocol allows the tag's user-memory to support a lock feature where data can be permanently set to be read only or not readable at all (EPCglobal, 2008).

Other security measures can be inferred from the location of the RFID tag. The tag's location can be determined by a variety of means, including wireless location methods, which are used in many real-time location systems (RTLS). RTLS are not limited to RFID technology but may include Wi-Fi or other proprietary protocols.

RFID Tag Technology to Deter Cloning

Low-cost passive RFID tags offer limited security features and require different approaches to security problems. The security features are limited by the number of gates available in the tag IC (integrated circuit) due to the low price point for the tag (often 5 cents or less), and amount of energy that can be delivered to the tag by the reader. The latter is the primary technical concern because most encryption based security procedures require high-energy (with respect to a passive or battery-less device) computations. In spatial access control authentication of the RFID to prevent cloned RFID tags from being used is the primary concern. Challenge-response and physically unclonable functions are two potential solutions to this problem.

Challenge-Response Architectures

Challenge-response is an authentication technique based on each party knowing a shared secret or parts of a combined secret, such as one key of a key pair. Juels proposed a lightweight challenge-response method using the password exchange procedure used by the tag to valid a KILL command or an ACCESS command (Juels, 2005) termed fulfillment-conditional PIN distribution or FCPD. FCPD is used to authenticate the tag to the reader to defend against the use of cloned tags. By combining the ACCESS and KILL commands the reader and tag can authenticate each other: the reader is authenticated to the tag by providing the correct ACCESS password, and the tag is authenticated to the reader by accepting the correct KILL password (Juels, 2005). One problem with this procedure is the use of the KILL command because it is possible to completely disable the tag (e.g., a single authentication). Juels suggests applying the KILL command when the tag has enough energy to evaluate the KILL password but not enough to complete the disablement process (Juels, 2005). In practice this is difficult to achieve because of the dynamic nature of the RF field that is providing energy to the tag. If the tag harvests too much energy the KILL command will complete and render the tag inoperable.

Physically Unclonable Functions

Physically unclonable functions (PUFs) are based on the manufacturing variations present in every fabrication process (Lim, Lee, Gassend, Suh, van Dijk, & Devadas, 2005; Bolotnyy & Robins, 2007; Devadas, Suh, Paral, Sowell, Ziola, & Khandelwal, 2008). These variations can result in slight, but measurable, delays in data processing operations that can be used to provide an item specific fingerprint to a chip. Such technology can be used to prevent cloning of a RFID tag because it would be impossible to replicate the manufacturing variations present in the original RFID tag. However, the drawback of PUF technology is the large sample space that must be obtained to provide a conclusive fingerprint because often many trails are needed to positively place a given chip in the cloned or un-cloned category.

Range Estimation Methods

There are a number of methods to estimate range from RF signals. The three major methods are time-of-arrival (TOA), angle-of-arrival (AOA), and received signal strength (RSS). TOA is based on the time required for a signal to travel from the transmitted (source) to the receiver (destination). Often, TOA is computed using a round-trip because it is too difficult to maintain sufficient clock synchronization between devices in different locations. TOA accuracy may suffer from queuing and processing delays on both devices because these delays will result in overestimation of distance. RSS uses the strength of the RF signal to determine distance and is often based on the Friis equation for propagation of a RF signal in free space. Inaccuracy is introduced into RSS distances when the RF propagation dynamics differ from the free space approximation and can be problematic inside and near large buildings (e.g., in an urban area). RSS will also suffer from dynamic RF environments and multipath. While TOA and RSS provide distance, they are not able to determine direction. AOA provides the angle or bearing of the RF signal received by the receiver's antenna. Combining AOA with TOA or RSS provides a means to determine the distance and bearing of the device in question.

The Access of Things Concept

The IOT will provide new opportunities for M2M and device-to-device collaboration with many of these machines/devices carried by or implanted in people. Thus, these devices will have significant mobility. Decisions must be made as to what resources a device should share or provide and this will be based on the device, the owner's rules, the device(s) it will collaborate with, and the other devices in the area. Hence, access rights are dependent not only on the identity of those taking part in the collaboration, but also the surroundings. The IOT will blend the concept of physical access control and cyber-access control to form the Access of Things.

The IOT provides significant benefit to the Access of Things because it provides for additional data input, which can be used to verify identity or to provide knowledge about the current surroundings. The blending of sensor inputs, such as GPS and video, can be used to verify a user's location and also determine the number of people present in that location (Ardagna, Cremonini, Damiani, De Capitani di Vimercati, & Samarati, 2006). This type of information must be collected and published in a manner that can be efficiently searched and retrieved by DRAAC systems.

FUTURE RESEARCH AREAS AND CHALLENGES

While DRAAC and other rule-based access control systems provide significant capabilities in securing physical locations and systems against malicious access there are a number of future research areas. The IOT will provide significant amounts of data and methods must be developed to sort, search, and store information relating to access decisions. The current DRAAC and RFID DRAAC implementations are based on a centralized system, but the IOT architecture will require a decentralized system of DRAAC instances to properly manage not only the local device, but also local segments of the IOT. Finally, the ability of DRAAC to provide actionable access control decisions is limited by the accuracy of the Likelihood and Impact values it uses and improving this accuracy will improve the quality of the access decisions.

One challenge is identifying those inputs that will assist with making the access decision for the Access of Things concept. Identification of these inputs is critical because it enables filtering of data to read only those items of interest from the massive amount of data that the IOT will produce.

The access decision must be determined within the specified time constraints for the application (e.g., factory assembly line, or physical access decision) and this places strict requirements on the search operation to identify these inputs. Databases cataloging information about devices, such as location, type, and use policies, must be created. These databases can then be indexed and searched to identify a set of devices in a specific location or that provide information about a specific location. Methods to maintain and expand such a database must be created. A means to efficiently update this information is needed because of device mobility. Procedures to address cases where devices temporarily disconnect from the IOT because they are turned off or set to "airplane mode" must be created to handle updates to the database. Policies to determine when to delete entries completely or to use analytics to estimate the future location of the device (e.g., similar to dead-reckoning in GPS systems when they lose the satellite signal) must be developed.

Another challenge is the design and deployment of a distributed architecture for the Access of Things concept that provides the necessary security while meeting cost requirements for large-scale deployment. The ability to retrofit buildings and other spaces with the system is also a key concern. The RFID DRAAC model works well as a centralized access control mechanism, but the IOT will consist of many small but dynamic collaborative networks. This will require each device to have its own instance of DRAAC to control access to its local resources. These local DRAACs will need to collaborate with each other and share information. To evaluate more complex rules, information must be gathered and stored in a central repository and this will require a central DRAAC module, but the distributed nature of the IOT will most likely cause this central DRAAC to consist of a hierarchy of DRAAC modules. This is similar to the hierarchical nature of the RFID DRAAC architecture described by Butler, Hawrylak, and Hale (Butler, Hawrylak, & Hale, 2011).

RFID DRAAC relies on the correct estimation of Impact and Likelihood. The IOT environment is one in which these parameters are likely to change rapidly and DRAAC must be able to respond to these changes in a timely manner. Recognizing that these parameters have changed is difficult and methods must be developed to collect this information in a computationally efficient manner to not starve the other processes executing on the device. Further, the communication with other devices must be limited to prevent the communication links from being overloaded with traffic just to maintain DRAAC (and prevent useful work from being completed). Efficient and secure manners to distribute updates from the DRAAC hierarchy to the local DRAACs must be developed for the IOT because of the high rate of mobility.

CONCLUSION

This chapter presented a possible model for controlling access to resources, termed Access of Things. The Access of Things is built on a DRAAC, a rule-based access control system, that has been applied to securing workstations (e.g., laptops and PCs) and RFID based physical access control systems. The RFID implementation of DRAAC can be segmented into independent groups to allow access decisions to be made in isolation or collaboratively (Butler, Hawrylak, & Hale, 2011). This architecture is applicable to the Internet of Things (IOT) and is the basis for the Access of Things concept. Its distributed nature allows the system to provide strong access control policies while maintaining quality of service requirements. The DRAAC system is platform agnostic making it an ideal candidate for deployment on IOT devices.

Access decisions in DRAAC rely on collection of accurate data, namely Likelihood and Impact, and the accuracy of these data items is a critical factor determining the effectiveness of the system. The IOT provides data from a variety of sources that can be used to improve the accuracy of these measurements. However, issues such as data overload and identifying the important data items must be addressed.

The Access of Things concept provides a suitable security protection framework for the IOT while maintaining the highest possible quality of service for applications executing within the IOT. The gradual increase of privilege enables the user to be vetted and will limit the damage potential from a malicious user. Likewise, the gradual decrease in privilege enables the application and user to continue to perform some of their tasks while being alerted to potential security threats. Based on this information, the human user can make more informed security decisions.

REFERENCES

Ardagna, C. A., Cremonini, M., Damiani, E., De Capitani di Vimercati, S., & Samarati, P. (2006). Supporting location-based conditions in access control policies. In *Proceedings of the 2006 ACM Symposium on Information, Computer and Communications Security*, (pp. 212-222). ACM.

Bertino, E., & Kirkpatrick, M. S. (2011). Location-based access control systems for mobile users: concepts and research directions. In *Proceedings of the 4th ACM SIGSPATIAL International Workshop on Security and Privacy in GIS and LBS*, (pp. 49-52). ACM.

Bertino, E., Samarati, P., & Jajodia, S. (1993). Authorizations in relational database management systems. In *Proceedings of 1st ACM Conf. on Computer and Commun. Security*, (pp. 130 -139). ACM.

Bolotnyy, L., & Robins, G. (2007). Physically unclonable function-based security and privacy in rfid systems. In *Proceedings of Fifth Annual IEEE International Conference on Pervasive Computing and Communications, 2007,* (pp. 211–220). IEEE.

Butler, M. (2011). *Dynamic risk assessment access control.* (Master's thesis). The University of Tulsa, Tulsa, OK.

Butler, M., Hawrylak, P., & Hale, J. (2011). Graceful privilege reduction in RFID security. In *Proceedings of the Seventh Annual Workshop on Cyber Security and Information Intelligence Research.* IEEE.

Butler, M., Reed, S., Hawrylak, P. J., & Hale, J. (2013). Implementing graceful RFID privilege reduction. In *Proceedings of the Eighth Annual Cyber Security and Information Intelligence Research Workshop.* IEEE.

Chen, M., Gonzalez, S., Leung, V., Zhang, Q., & Li, M. (2010). A 2G-RFID-based e-healthcare system. *IEEE Wireless Communications, 17*(1), 37–43. doi:10.1109/MWC.2010.5416348

Dacier, M., Deswarte, Y., & Kaâniche, M. (1996). *Quantitative assessment of operational security: Models and tools.* LAAS Research Report 96493.

Devadas, S., Suh, E., Paral, S., Sowell, R., Ziola, T., & Khandelwal, V. (2008). Design and implementation of PUF-based unclonable RFID ICS for anti-counterfeiting and security applications. In *Proceedings of 2008 IEEE International Conference on RFID,* (pp. 58–64). IEEE.

EPCglobal. (2008). *EPC™ radio-frequency identity protocols class-1 generation-2 UHF RFID protocol for communications at 860 MHz – 960 MHz version 1.2.0.* EPCglobal Inc.

EPCglobal. (2010). *EPC tag data standard version 1.5.* EPCglobal.

Foschini, L., Taleb, T., Corradi, A., & Bottazzi, D. (2011). M2M-based metropolitan platform for IMS-enabled road traffic management in IoT. *IEEE Communications Magazine, 49*(11), 50–57. doi:10.1109/MCOM.2011.6069709

Halamka, J., Juels, A., Stubblefield, A., & Westhues, J. (2006). The security implications of VeriChip cloning. *Journal of the American Medical Informatics Association, 13*(6), 601–607. doi:10.1197/jamia.M2143 PMID:16929037

Hartney, C. J. (2012). *Security risk metrics: An attack graph-centric approach.* (Unpublished Master's thesis). The University of Tulsa, Tulsa, OK.

Hawrylak, P. J., Hartney, C., Haney, M., Hamm, J., & Hale, J. (2013). Techniques to model and derive a cyber-attacker's intelligence. In B. Igelnik, & J. Zurada (Eds.), *Efficiency and scalability methods for computational intellect* (pp. 162–180). Hershey, PA: Information Science Reference. doi:10.4018/978-1-4666-3942-3.ch008

IEEE. (2011). *IEEE standard for local and metropolitan area networks - Part 15.4: Low-rate wireless personal area networks (LR-WPANs).* IEEE.

Jain, A., Hong, L., & Pankanti, S. (2000). Biometric identification. *Communications of the ACM, 43*(2), 90–98. doi:10.1145/328236.328110

Juels, A. (2005). Strengthening EPC tags against cloning. In *Proceedings of the 4th ACM Workshop on Wireless Security,* (pp. 67-76). ACM.

Kirkpatrick, M. S., Damiani, M. L., & Bertino, E. (2011). Prox-RBAC: A proximity-based spatially aware RBAC. In *Proceedings of the 19th ACM SIGSPATIAL International Conference on Advances in Geographic Information Systems,* (pp. 339-348). ACM.

Kong, L. (2011). *Spatial access control on multi-touch user interface*. (Master's thesis). The University of Tulsa, Tulsa, OK.

Kushalnagar, N., Montenegro, G., & Schumacher, C. (2007). *IPv6 over low-power wireless personal area networks (6LoWPANs), overview, assumptions, problem statement, and goals*. IETF RFC 4919.

Lim, D., Lee, J. W., Gassend, B., Suh, G. E., van Dijk, M., & Devadas, S. (2005). Extracting secret keys from integrated circuits. *IEEE Transactions on Very Large Scale Integration Systems*, *13*(10), 1200–1205. doi:10.1109/TVLSI.2005.859470

Louthan, G., Hardwicke, P., Hawrylak, P., & Hale, J. (2011). Toward hybrid attack dependency graphs. In *Proceedings of the Seventh Annual Workshop on Cyber Security and Information Intelligence Research*. IEEE.

Manadhata, P. K., & Wing, J. M. (2011). An attack surface metric. *IEEE Transactions on Software Engineering*, *37*(3), 371–386. doi:10.1109/TSE.2010.60

Mulligan, G. (2007). The 6LoWPAN architecture. In *Proceedings of the 4th Workshop on Embedded Networked Sensors*, (pp. 78-82). IEEE.

Oren, Y., & Wool, A. (2010). RFID-based electronic voting: What could possibly go wrong? In *Proceedings of 2010 IEEE International Conference on RFID*, (pp. 118-125). IEEE.

Ou, X., Boyer, W. F., & McQueen, M. A. (2006). A scalable approach to attack graph generation. In *Proceedings of the 13th ACM Conference on Computer and Communications Security*, (pp. 336-345). ACM.

Parr, A., Miesen, R., & Vossiek, M. (2013). Inverse SAR approach for localization of moving RFID tags. In *Proceedings of 2013 IEEE International Conference on RFID*, (pp. 104-109). IEEE.

Philips, C., & Swiler, L. (1998). A graph-based system for network-vulnerability analysis. In *Proceedings of the 1998 Workshop on New Security Paradigms*, (pp. 71-79). New York, NY: ACM.

Revere, L., Black, K., & Zalila, F. (2010). RFIDs can improve the patient care supply chain. *Hospital Topics*, *88*(1), 26–31. doi:10.1080/00185860903534315 PMID:20194108

Sandhu, R. (1993). Lattice-based access control models. *Computer*, *26*(11), 9–19. doi:10.1109/2.241422

Sandhu, R., & Samarati, P. (1994). Access control: Principle and practice. *IEEE Communications Magazine*, *32*(9), 40–48. doi:10.1109/35.312842

Wicks, A. M., Visich, J. K., & Li, S. (2006). Radio frequency identification applications in hospital environments. *Hospital Topics*, *84*(3), 3–8. doi:10.3200/HTPS.84.3.3-9 PMID:16913301

You, L., Liu, C., & Tong, S. (2011). Community medical network (CMN), architecture and implementation. *Global Mobile Congress*, 1-6.

KEY TERMS AND DEFINITIONS

Access of Things: The term used to denote the DRAAC based access control system for the Internet of Things.

Attack Graph: A graph based structure representing the state of the system as vertices and actions, such as vulnerabilities or normal system transitions, which can change the state of the system as edges. Attack graphs are similar to finite state machine diagrams that focus on security rather than system state.

CLIPS: CLIPS stands for C Language Integrated Production System and is used to store the Facts and Rules in the DRAAC system. CLIPS is the core of the Access Control Decision Module of DRAAC.

DRACC: DRAAC stands for Dynamic Risk Assessment Access Control system, which is an access control system that provides graceful degradation or elevation privilege in a system. DRAAC is the basis for the Access of Things concept presented in this chapter.

Impact: The potential damage that can be caused by granting an access if that access should have been denied. Impact is primarily a function of the resource that the user is requesting access to, but is also dependent on the user and the current state of the system.

Internet of Things: A term used to denote the massive collection of networked devices. These collaborative networks will autonomously form and collaborate.

Likelihood: The probability that an event will happen or that a malicious user will attempt to exploit a particular vulnerability. Likelihood is used in security and access control systems, such as DRAAC, to identify those actions that the malicious user is likely to perform.

Physically Unclonable Functions: Physically Unclonable Functions or PUFs are variations in the fabrication process of Integrated Circuits (ICs) that cause timing variations in the resulting ICs (chips). These inconsistencies can be used to provide a unique fingerprint for each particular IC.

RFID: RFID stands for Radio Frequency IDentification and represents the use of RFID readers and RFID tags to provide the last-mile connection between a control system and the end devices. In this chapter, RFID is used for identifying individuals for access control purposes and for location determination via radio-frequency (RF) location techniques.

Rule-Based Access Control: A type of access control system that where access requests are evaluated against a specified list of rules. These types of systems often support the collection of "facts" representing the access control system's knowledge about the resources and users it monitors.

Chapter 8
A Study of Research Trends and Issues in Wireless Ad Hoc Networks

Noman Islam
Technology Promotion International, Pakistan

Zubair A. Shaikh
National University of Computer and Emerging Sciences, Pakistan

ABSTRACT

Ad hoc networks enable network creation on the fly without support of any predefined infrastructure. The spontaneous erection of networks in anytime and anywhere fashion enables development of various novel applications based on ad hoc networks. However, ad hoc networks present several new challenges. Different research proposals have came forward to resolve these challenges. This chapter provides a survey of current issues, solutions, and research trends in wireless ad hoc networks. Even though various surveys are already available on the topic, rapid developments in recent years call for an updated account. The chapter has been organized as follows. In the first part of the chapter, various ad hoc network issues arising at different layers of TCP/IP protocol stack are presented. An overview of research proposals to address each of these issues is also provided. The second part of the chapter investigates various emerging models of ad hoc networks, discusses their distinctive properties, and highlights various research issues arising due to these properties. The authors specifically provide discussion on ad hoc grids, ad hoc clouds, wireless mesh networks, and cognitive radio ad hoc networks. The chapter ends with a presenting summary of the current research on ad hoc networks, ignored research areas and directions for further research.

DOI: 10.4018/978-1-4666-5170-8.ch008

INTRODUCTION

During last few years, extensive developments have been observed in the domain of wireless network. Different communication technologies i.e. general packet radio service (GPRS), enhanced data rates for GSM evolution (EDGE) and worldwide interoperability for microwave access (WIMAX) etc. have evolved and newer form of computing devices i.e. personal digital assistant (PDA), tablets and smart phones are appearing in the market. The wireless computing has progressed from 1G to 4G communication networks. During this progression, various modes of wireless networking have emerged. The simplest form of wireless networking is communication among two or more fixed hosts in open air. The conventional television system operates on this mode. Another approach is *wireless networking with access point*. There are different wireless hosts that are allowed to move while the basic infrastructure is supported by set of fixed nodes called base stations or access points. However, this approach doesn't provide the flexibility to be used in emergency situations requiring quick deployment or networking in adversarial surroundings. The evolution of technologies has lead to development a new mode of wireless networking where the nodes arrange themselves on the fly in the form of a network without any infrastructure support. Such networks are called ad hoc networks.

PROPERTIES OF AD HOC NETWORK

Formally, A*d hoc Network G(N,E)* is defined as a collection of nodes $N=\{n_1,n_2,n_3,...\}$ connected by edges (Islam & Shaikh, 2012). The nodes are usually mobile with limited capabilities, links are volatile and insecure, and there are no dedicated nodes for addressing, routing, key management and directory maintenance etc. The nodes are themselves responsible for various network operations i.e. routing, security, addressing and key management etc. It is obvious from these characteristics that network protocols and algorithm that are currently available for wired and infrastructure-less wireless networks are not suitable for ad hoc networks (Islam et al., 2010). For example, a conventional routing algorithm when employed for ad hoc network can suffer from loops, stale routes and other issues due to the very sharp changes in the network. Similarly, the current security solutions are based on availability of authentication servers, certification authority and other security infrastructure, which are not generally available in ad hoc network. Therefore, new solutions are required for addressing various challenges of ad hoc network.

Different research efforts are underway to address various issues of ad hoc networks. In this chapter, we provide an adequate account of these efforts. There are already some surveys available that have summarized the previous researches on ad hoc networks. For example, Dow et al. (2005) and Singh et al. (2012) have provided a quantitative analysis of the number of research proposals appeared during last few years for addressing a particular issue of ad hoc network. Similarly, a summary of various research issues in ad hoc networks have been presented in (Chlamtac et al., 2003; Toh et al., 2005; Ghosekar et al., 2010; K. Al-Omari & Sumari, 2010). However, the focus of this chapter is on research pursued in ad hoc networks during recent years. The major contributions of this chapter are as follows:

- To provide a summary of various research issues in ad hoc networks and the recent approaches adopted to tackle these issues
- To investigate and report on various emerging models of ad hoc networking
- To present a comprehensive overview of issues and corresponding solutions for different ad hoc networking models i.e. ad hoc grids, ad hoc clouds, wireless mesh

networks and cognitive radio ad hoc networks etc.

- To summarize the current state-of-the-art and avenues for further research

RESEARCH ISSUES IN AD HOC NETWORK

We start the discussion with an overview of major research issues in ad hoc networks. As discussed earlier, the issues arising in ad hoc network span across all layers of communication. In addition, cross-layer issues i.e. security, quality of service (QoS) and energy management ad hoc networks etc. demand resolution mechanisms at more than one layer of communication. Different research efforts have been put in to address these issues. Let's discuss the research issues of ad hoc network along with the proposed mechanisms to address them.

Antenna Design

Ad hoc networks are characterized by limited channel capacity and unreliable links. In addition there are presence of noise and interference in the surrounding environment. Therefore, new antenna design techniques are required that can cope with these limitations. One of the approaches to address these problems is to have multiple antenna elements to improve the quality of received signals (Ramanathan, 2001). In *switched diversity antenna* design, the antenna element is changed continuously and the system uses the element with best gain. This solves the problem of fading and multipath effects. Similarly, *diversity combining* is based on correcting the phase error and combining the power to have more gain. To improve the spectral efficiency, directional antenna has also been proposed in recent literature. This approach leads to efficient utilization of spectrum and more power by pointing the signal in a specified direction. Two types of approaches can be used

for directional transmission (Ramanathan et al., 2005). In a *switched beam antenna*, there are fixed beams that can be formed by shifting the phase of antenna by a fixed amount or by switching between various fixed directional antennas. An *steered antenna* can be focused in any direction.

The use of smart antennas for ad hoc networks, calls for addressing multitude of research questions. For instance, at medium access control (MAC) layer, various issues can arise (Bazan & Jaseemuddin, 2012). A node while transmitting is deaf except the direction in which it is transmitting. Hence, it can't respond to RTS/CTS messages. Similarly, head of line blocking occurs, when a node finds the medium busy in the direction in which it is transmitting. It holds the other messages in its queue, even though other messages can be transmitted in a different direction. Various specialized MAC layer protocols have thus been proposed for smart antenna based systems and can be seen in (Bazan & Jaseemuddin, 2012; Lu et al., 2012). Another important question pertains to size and cost of smart antenna. This is dependent on the domain of application. For example, a military network can afford the cost of smart antenna deployments as well as the size of antenna doesn't matter in such applications. However, for small scale applications, further research is required to design cost efficient, small sized antennas.

Energy Management

In ad hoc networks, the nodes have very limited battery power. Algorithms for ad hoc networks should be designed such that they consume minimum energy for their execution. Substantial efforts have been put in to devise energy efficient algorithms for ad hoc networks. One of the techniques is to *control the transmission power* of the nodes. This is done by adjusting the transmission power to an arbitrary value to obtain optimal interference, power savings and improved channel capacity. A good survey on various power control algorithms is provided in (Singh & Kumar, 2010). However, the connectivity of the nodes is affected by the transmission power and requires careful design. Gomez and Campbeel (2007) have analyzed the impact of power control on connectivity, power savings and capacity of the network.

Another approach to energy management is to devise algorithms such that nodes remain in *low energy conservation state* most of the time. A wireless node can be in transmission, receiving, idle listening or sleep state at particular instant of time. In the sleep state, the transceiver of the node is off and consumes a very small amount of energy i.e. 0.740 W (Fotino & De Rango, 2011). Hence, an energy efficient algorithm should keep the nodes in sleeping mode most of the time. There have been various algorithms proposed in literature for power management based on smart utilization of sleep states. Lim et al. (2009) have proposed random cast, an energy efficient scheme over dynamic source routing (DSR) protocol based on IEEE 802.11 power saving mechanisms.

Another approach to energy management is energy efficient routing. An *energy efficient routing algorithm* employs a routing metric that reflects the energy consumption during routing operation. The metrics that can be used are residual battery of the nodes and energy consumed to forward the packet along the route (Hassanein, 2006; Kwon & Shroff, 2006). There are also some routing protocols that attempt to organize the networks in to clusters such that the number of message exchanges can be minimized (Heinzelman et al., 2000; Tariquea et al., 2009). This leads to reduced power consumption by nodes. Finally, there are *multipath routing protocols* that distribute the overall routing process across multiple paths and reduce the power consumptions at a particular node (Shah & Rabaey, 2002; Liu et al., 2009).

MAC Layer Protocols

A MAC layer protocol provides a fair access to the medium for transmission of information. The various MAC schemes proposed in literature are classified as contention free schemes and contention based schemes (Kumar et al., 2006). The *contention free scheme* is based on assignment of time and frequency, and is free from collisions. Examples are Frequency Division Multiple Access (FDMA), Time Division Multiple Access (TDMA) and Code Division Multiple Access (CDMA) protocols etc.

In the *contention base schemes*, the nodes compete for access to the medium. Carrier Sense Multiple Access (CSMA) is the classical example of contention based protocols. In CSMA, the node first senses the medium before any transmission. The node can defer its transmission in case any other message is being transmitted. CSMA protocols however suffer from hidden node and exposed node problems (Kumar et al., 2006). Multiple Access Collision Avoidance (MACA) protocol attempts to solve this problem by transmission of ready to send (RTS) and clear to send (CTS) messages before commencing the transmission. The nodes hearing the RTS messages defer their transmission. MACA-By Invitation (MACA-BI) is a receiver initiated protocol where the receiver requests the sender for transmission by sending ready to receive (RTR) messages (Talucc, 1997). Thus the sequence of message exchanges is RTR-DATA.

IEEE also specifies the distributed coordination function (DCF) and point coordination function (PCF) as access control protocols for wireless network. They are based on RTS-CTS-DATA-ACK sequence of messages for data transmission. A Network Allocation Vector (NAV) is also maintained that represents the expected duration during which the wireless medium will be busy. Nodes update NAV values based on overhearing the transmissions. More details can be seen in (Chen, 1994).

There has been lot of research work done on designing *QoS aware MAC protocols*. Shin et al. (2011) proposed a multichannel MAC protocol that enables QoS over IEEE 802.11 DCF. Kamruzzaman et al. (2010) proposed a QoS based MAC protocol for cognitive radio ad hoc networks. In addition, there have been some specialized *MAC protocols for directional antennas* (Wang et al., 2008). As the directional antenna transmits in a particular direction, these specialized MAC protocols make optimum use of spectrum as well as address the challenges arise due to the directional antennas, as discussed in previous section.

Routing

One of the features of ad hoc network is multi-hopped routing in which every node has to participate in the routing operation. This presents a number of challenges i.e. creation of false routes, routing loops and security attacks on routing protocols etc. Routing protocols proposed for ad hoc networks can be classified as proactive, reactive and hybrid routing protocols. In a *proactive* routing protocol also known as table drive routing protocol, every node periodically disseminate its routing updates to other nodes. Destination Sequenced Distance Vector (DSDV), Global State Routing protocol (GSR), Fisheye State Routing (FSR), Optimized Link State Routing (OLSR) and Cluster Gateway Switch Routing Protocol (CGSR) are examples of some of the proactive routing protocols (Kumar et al., 2010). *DSDV* is based on periodic / event-based dissemination of routes to neighbors. A sequence number is maintained to determine freshness of the routing entry and avoid inconsistencies in routing information. *GSR* is based on exchange of link state information in the form of vectors with neighboring nodes. Based on the information exchanges, global topology of the network is constructed at each node. In *FSR*, the nodes prioritize other nodes based on their distances with other nodes and the routing updates are exchanged with nearer nodes more frequently than distant nodes, thus reducing congestion in network. *OLSR* is an optimization of the link state routing used in wired networks. *CGSR* is a hierarchical protocol based on DSDV

that divides the whole networks into clusters such that routing overhead can be minimized.

The second type of routing protocol presented in literature is *reactive* routing protocol also known as on-demand routing protocol. The routing operation is performed on as need basis. Among the various reactive protocols, the popular ones are Ad hoc On-Demand Distance Vector Routing (AODV), Distance Vector Routing (DSR), Temporally Ordered Routing (TORA), and Cluster Based Routing (CBR) protocols. The *AODV* and *DSR* are based on the broadcasting of a route discovery request (RREQ) to neighbors until the request reaches the destination. The destination sends a reply which reaches back to the source using the same route that is used for request's dissemination. In the process of dissemination, a route is created towards the source. *TORA* is a reactive routing protocol that creates a route towards the destination based on a directed acyclic graph from source to destination. *CBR* protocol is a hierarchical protocol based on formation of clusters among the network nodes, thus minimizing the messages exchanged in the network. The combination of proactive and reactive protocols i.e. *hybrid routing protocols* have also been advocated in some research proposals. Among them are *Zone Routing Protocol (ZRP), Dual-Hybrid Adaptive Routing (DHAR), Adaptive Distance Vector Routing (ADV)* and *Sharp Hybrid Adaptive Routing Protocol (SHARP)* etc. (Kumar et al., 2010).

Some routing protocols exploit position information for routing of packets. They are called *position based routing protocols*. There are three popular approaches to position based routing in ad hoc networks. These are *greedy forwarding, restricted direction flooding* and *hierarchical approaches* (Qabajeh et al., 2009). A greedy forwarding routing protocol doesn't maintain the complete route towards the destination. During forwarding, the packet includes the geographical location of the recipient node. The intermediate nodes propagate the request towards the destination based on some optimization criteria like the next neighbor closest to the destination. An example of such scheme is Most Forward within distance R (MFR) (Takagi & Kleinrock, 1983). In this protocol, a packet is forwarded to the next hop node which has the maximum progress towards the source. A projection line (with respect to next hop node) is defined as the line between sender and receiver and the progress is the distance from the sender to next hop on the projection line. Improved progress Position Based Beacon Less Routing algorithm (I-PBBLR) extends this approach further by incorporating the direction metric to improve the progress definition (Cao & Xie, 2005). In *restricted direction flooding* protocols, the routing request is propagated towards the destination only in forward direction. The receiving nodes check if they are set of nodes towards the destination and if they should forward the packet ahead. In that case, the nodes retransmit the packet. An example of this type of protocol is Distance Routing Effect Algorithm for Mobility (DREAM) that propagates the routing request in a particular sector (Qabajeh et al., 2009). The third strategy is based on forming a hierarchy among the nodes based on location and other parameters. An example of this approach is TERMINODES that maintains a two level hierarchy among the nodes (Qabajeh et al., 2009). If the location of destination node is closer to the source, packets will be routed based on a proactive routing protocol. Greedy routing will be used for routing to distant nodes.

In addition, there are QoS based routing protocols that selects a route based on QoS attributes of the links. Various QoS routing protocols have emerged during recent years. *Core-Extraction Distributed Ad hoc Routing (CEDAR)* is based on establishing a core structure that provides a low over ahead infrastructure for QoS routing (Sivakumar et al., 1999). As the infrastructure comprises stable nodes, the desired QoS of user can be easily maintained. In *Multipath Routing protocol (MRP)*, several alternate paths towards the destination are discovered based on bandwidth and reliability constraints (Qin & Liu, 2009).

Several *extensions over traditional routing protocols* are also proposed to cope with QoS requirements (Gangwar et al., 2012). For example, *Ad hoc On-Demand QoS routing* extends AODV in which a RREQ also includes the bandwidth and delay constraints. Intermediate nodes will only rebroadcast the packet if it meets the specified constraints. Bandwidth reservation is performed while the first packet is sent from the source using this path. A delay is also measured during the route discovery. So, if more than one route exists towards the destination, route with minimum delay is chosen.

Multicasting

Multicasting is defined as the transmission of a message to a set of hosts in the network. The recipients of the message have a group identity. Multicasting plays a pivotal role in data exchanges and collaborative task execution in ad hoc networks, as it reduces the traffic overhead involved when the same data has to be sent to multiple destinations. However, multicasting presents several challenges in ad hoc network due to changing topology, nodes mobility, limited resources and hostile environment. The various schemes proposed in literature for multicasting can be classified as tree, mesh and hybrid multicast routing protocols (Junhai et al., 2008). In the *tree based approach*, a tree like data forwarding path is maintained which is rooted at the source node of the multicast session. Examples are adaptive demand driven multicast routing protocol and associativity based ad hoc multicast routing protocol etc. (Jetcheva & Johnson, 2001; Royer & Toh, 1999). A different approach is *group shared tree approach* where a single tree is used for a group instead of maintaining tree for each node. Each source first forwards the packet to the root of the tree which is then forwarded through the tree to multicast receivers. Examples of protocols in this category are STAMP and MZRP (Canourgues et al., 2006; Zhang & Jacob, 2004) etc. The problem with tree based protocols is that the mobility of the nodes along the tree may cause packet drops.

In a *mesh based approach*, a mesh like structure is maintained among the members. These protocols lead to higher connectivity among the nodes and thus perform reasonably well in mobility conditions. Examples of mesh based multicast protocols can be seen in (Inn & Winston, 2006; Soon et al., 2008). There are some *hybrid protocols* as well that combine the advantages of tree and mesh like structure (Biswas et al., 2004). Thus, these protocols are both robust against mobility and efficient in performing the multicast operation. Interested readers are referred to Junhai et al. (2008) and Badarneh and Kadoch (2009) for detailed surveys on multicast routing protocols.

Addressing

As ad hoc network doesn't have any prior infrastructure, there has to be some mechanism to perform address assignments in a coordinated manner. The problem of addressing becomes complicated with network partitioning or merging, as the addressing algorithms should be smart enough to identify duplicate addresses in this situation. One of the simplest approaches to perform addressing is to maintain complete information about the Internet protocol (IP) addresses allocated to other nodes called *state-full* approach (Nesargi & Prakash, 2002). Any arriving node consults one of the reachable nodes called initiator for an IP address. The initiator assumes an address for the node and confirms it from other nodes by broadcasting that it is assigning this address to a new node. If the positive replies are received from all nodes, the allocation process is successful. In contrast, *state-less approaches* allow configuration of addresses locally (Chen et al., 2009). Duplicate detection techniques are then used to resolve conflicts. *Hybrid schemes* are also possible, where a set of nodes tries to provide a conflict-free address according to its available

knowledge. Duplicates undetected, are resolved afterwards (Li Longjiang, 2009).

Some of the approaches for addressing work by *extending the classical dynamic host configuration protocol (DHCP)* to addressing. In this direction, Ancillotti et al. (2009) proposed an addressing approach called Ad Hoc DHCP. It works by using the conventional DHCP scheme for hybrid ad hoc networks (ad hoc network with some infrastructure support). Any incoming node first discovers the nodes in its surroundings. It then selects a DHCP relay agent from its neighbors. The DHCP relay agent will forward the DCHP request of this node to known DHCP servers, thus a new address is assigned to the node.

An important aspect of addressing is to maintain the *scalability* of addressing process. Hussain et al. (2010) proposed a scalable approach to addressing where a set of servers are evenly placed in the form of grid topology. These servers ensure the identification of duplicate addresses in its vicinity, and thus guarantee balancing of addressing tasks to the servers.

Transport Protocol

A transport protocol provides end-to-end delivery, flow control and reliable transmission of data to other nodes in the network. There are two popular transport protocols used in wired networks. *Transmission Control Protocol (TCP)* is a connection oriented protocol that provides guaranteed delivery of data, while *Unified Datagram Protocol (UDP)* is proposed for real-time multimedia applications.

It has been observed that TCP protocol doesn't provide good performance in ad hoc networks. Several studies have been done on the performance of TCP over ad hoc networks. The problem arises due to the congestion control mechanism used by TCP. The TCP congestion control algorithm is based on additive increase multiplicative decrease strategy. So, whenever any packet's loss is observed (due to timeout or duplicate acknowledgment), TCP assumes it as a congestion and drastically reduces the window size. In wireless ad hoc network, the packet losses are often due to other reasons besides congestion i.e. attenuation, multipath fading, network partitioning and route failures etc. As the window size is tuned based on additive increase multiplicative decrease, it takes time for the window size to regain its original value. Hence, the overall throughput is severely affected.

Several TCP variants have thus been proposed for ad hoc networks. A survey on TCP variants have been provided in (Francis et al., 2012). Among them include approaches that utilize network layer feedback to improve the congestion control algorithm. *TCP-F* works by distinguishing between a route failure and congestion (Chandran et al., 98). By using cross-layer communication, whenever a link failure is detected, the routing agent sends an explicit congestion notification to the sender node. The sender then freezes its state and waits for a route reestablishment notification; the sender then resumes its proceedings. Similar schemes have been proposed in *ATCP* and *TCP Bus* etc. ATP utilizes the network layer information for startup rate estimation, congestion detection, and control and path failure notification (Sundaresan et al., 2005). Changa et al. (2012) proposed a cross-layer approach where physical and routing layers information is exploited by the TCP to discriminate between various types of network events like channel errors, buffer overflows and link layer contention etc. Besides network layer feedback, various proposals use the information available at transport layer to detect route failures (Hanbali et al., 2005). For example, *Fixed RTO* assumes a link failure when a timer expires two times consecutively, thus the RTO is not doubled. Similarly, *TCP-DOOR* identifies the link failure based on the out-of-order delivery events (Wang & Zhang, 2002).

Unlike TCP, there has not been significant research done on UDP for ad hoc networks. Only a few UDP variants have been proposed for real-time data transmission over wireless and ad hoc network (Mao et al., 2006; Yang & Zhang, 2011). However, several studies have been done on analyzing the performance of UDP over ad hoc network. Most of these studies analyzed the behavior when TCP and UDP are competing (Rohner et al., 2005). When TCP and UDP use a common link, UDP suffers from packet loss and interruptions are seen in TCP. Some experiments have also been done on optimum UDP packet sizes for ad hoc network (He et al., 2007).

Security

Ad hoc networks are prone to a large number of security attacks due to environment where they are deployed as well as due to the absence of prior security infrastructure. A number of active and passive threats are possible i.e. include signal jamming, eaves dropping, impersonation and DoS attacks etc. (Wu et al., 2006). Since, these problems span across the whole protocol stack, security in ad hoc network is considered a multilayer issue. Let's discuss the various approaches that can be adopted to counter for security issues. A comprehensive discussion on security in ad hoc network has been provided in (Islam & Shaikh, 2013).

To address the issues of signal jamming, techniques like frequency hopping spread spectrum (FHSS) and direct sequence spread spectrum (DSSS) can be employed (Wu et al., 2006). For resolving the security attacks at routing layer i.e. wormhole attack, black-hole attack and byzantine attack etc., different approaches have been proposed. Among them are techniques based on *cryptography* to ensure the integrity of routing messages that are being exchanged. Examples of such protocols are SAODV, ARAN and Ariadne etc. (Islam & Shaikh, 2013). Other approaches are based on *exploiting the routing header* for identifying any consistencies in the exchanged messages or *multipath routing* to recover a routing message tempered in transit by malicious nodes. In this direction, the approaches by Kurosawa et al. (2007), and Raj and Swadas (2009) utilized the sequence number present in the routing header to identify any consistencies and misbehavior in the network. In multipath routing protocol SPREAD, a message is decomposed in to multiple sub-packets which are retransmitted through various paths (Lou et al, 2009). At the receiving end, the packets are recombined. It is ensured that if a subset of the original message is tempered by intermediate malicious nodes, the original message can even then be reconstructed. To counter for various types of intrusion attacks and viruses etc., techniques based on firewalls, antivirus and intrusion detection have also been proposed in some researches (Wu et al., 2006).

Cooperation

In ad hoc networks, the nodes are dependent on each other for their operations. For example, in multi-hopped routing, nodes are required to relay packets of other nodes. This interdependency requires some cooperation mechanism to ensure proper functioning of the network operations. Issues can arise when the nodes don't cooperate as the actions performed by a node for other users i.e. packet forwarding and resource sharing etc., call for consumption of bandwidth, data storage and power of the node. In addition, there are misbehaving nodes on the network that don't cooperate so that operation of the network can be jeopardized. It has been observed that the presence of the selfish nodes even in small amount, degrades the network performance significantly (Zayani & Zeghlache, 2009).

There are two major approaches proposed in literature in this regard. The approaches belonging to the first category are based on *enforcing cooperation*. The nodes are monitored for their behavior and are punished in case of any misbehavior. Kwon et al. (2010) proposed a reputation based scheme where interaction among nodes is modeled as a Stackelberg game. The nodes reputations are determined based on their cooperation

at each game. Nodes are encouraged to cooperate only with other cooperative nodes. Zayani and Zeghlache (2012) proposed a cooperation enforcement scheme based on the weakest link TV game principle. The nodes try to obtain a longest sequence of successful forwarding of packets. Misbehaving nodes are punished by reducing their utility value drastically.

The other category of schemes proposed for cooperation enforcement is *incentive based* mechanisms. In this direction, Eidenbenz et al. (2008) introduced a virtual currency scheme based. The intermediate nodes receive compensation in form of residual energy level. Zayani and Zeghlache (2009) proposed a fair and secure incentive based scheme where the nodes are rewarded or charged credits. A payment aggregation scheme is proposed to generate a single receipt for multiple packets thus reducing the network traffic. In Secure Incentive Protocol, a destination generates an acknowledgment to increment the credits of all the intermediate nodes (Zhang et al., 2007). This may cause fairness issues, when intermediate nodes don't get rewarded due to the loss of acknowledgments. The payer might have to pay more in case of loss of reward/receive packets or the data packet doesn't reach to the destination.

Data Management

In the last decade, the problem of data management has emerged as an important issue for ad hoc networks. As the devices are getting smaller and the peer to peer connectivity mediums becoming smaller, the amount of data exchanges in the network is becoming huge. It is therefore imperative to have novel data management mechanisms that can analyze the gigantic volume of data to extract meaningful information. Such a system should address the aspects related to acquisition, representation, storage and protection of data etc. The problem arises in data management because of the limited storage, lack of global schema, heterogeneity and spatiotemporal variation in states of data etc. Various data management systems have been proposed in recent years. Among them are MoGATU (Perich et al., 2006), DRIVE (Zhong et al., 2008), CDMAN (Martin & Demeure, 2008) and CHaMeLeoN (Shahram et al., 2006) etc. A survey of data management frameworks is provided in (Islam & Shaikh, 2011).

Testbeds and Simulation

As newer algorithms and approaches are being proposed for ad hoc networks, a deep analysis of them is required before their deployment. Various approaches have been adopted in this regard. One approach is *analytical modeling*, which is not viable because the complex nature of ad hoc network can't be modeled precisely. The second approach is to use *testbeds*. Several testbeds are available for ad hoc networks (Kiess & Mauve, 2007). However, testbed also suffers from different economical, monitoring and implementation issues. The third approach is the *simulation* of the algorithm in a software environment. There has been a number of commercial and open source software proposed for ad hoc network. A summary is provided in Table 1.

Mobility Models

Mobility models are used to describe the movement of real-world entities in to a simulation environment. Mobility models proposed for ad hoc networks can be categorized as *individual mobility* and *group-based* mobility models (Camp et al., 2002). In the former case, the mobility patterns of a node are independent of other nodes; while in the later case, the mobility of a particular node is dependent on other nodes' mobility. Following are some of the most popular mobility models proposed in literature (Cooper & Meghanathan, 2010):

- **Random Walk:** A node randomly chooses a direction and speed, continues moving until a particular destination is reached or a certain time has elapsed
- **Random Way Point:** A node selects a target and speed, moves towards the target, pauses for some time and then reselects a new target
- **Random Direction Walk:** A node walks randomly in a particular direction with random speed, until it reaches the boundary of simulation area, where it pauses for some time and then reselects a new direction

Table 1. A summary of various ad hoc network simulators

Simulator	Free	Open Source	Language	Features	Website
NS-2	√	√	C, TCL	TCP, Routing protocols, Multicasting, Visualization support	http://www.isi.edu/nsnam/ns
OMNET++	√	√	C++	Wireless and mobile network simulation	http://www.omnetpp.org
Glomosim	√	×	Parsec	Scalable Simulation, support for various types of mobility models, visualization, ad hoc networking	http://pcl.cs.ucla.edu/projects/glomosim
SWANS++	√	√	Java	Support for wireless and ad hoc networks, routing protocols, mobility models	http://sourceforge.net/projects/straw
Opnet	×	×	C++	VoIP, TCP, OSPF, ad hoc networks, IPv6, grid computing	http://www.opnet.com

- **Gauss Markov Move:** The node speed and direction is a function of the last move and direction
- **Probabilistic Random Walk:** Three states are defined for movement in x and y axes. In x-axis, the node can move left, right or still. Similarly in y-axis, up, down or still states are possible. The states of the node can also be changed based on some probability
- **Column mobility:** All the nodes move randomly, except a subset of nodes. Among this subset, there is a leader and rest are followers. The followers move behind the leader in a lane
- **City section:** The node movement is constrained to a grid like structure comprising different streets. There is a speed limit for each street and the nodes' movements in a street are within these speed limits. Nodes are place on a particular intersection at the start of simulation. Node then moves to a random intersection with at most one move. The process is repeated until the simulation ends
- **Manhattan model:** Similar to city model, except a probabilistic model is employed to select the subsequent street in which the node can move

Besides these simple mobility models, another approach is to use specialized tools to generate mobility patterns. Various tools have been proposed for this purpose. For example, MOVE is a Java based toolkit that allows the user to quickly generate mobility patterns by using its map editor or importing the maps from other databases (Karnadi et al., 2007).

Standardization

Although lot of research efforts are underway for various issues of ad hoc network as discussed above, there is a need for standardization for its wide range acceptance. Even though there have been various standardization bodies working on the topic. For example, IETF MANET working group is working on standardization of routing protocols (IETF, 2012). Various IEEE wireless standards are available for use in ad hoc networks (Jangra et al., 2010). However, all of these standards are focused on a particular aspect. There is a need for standardization that looks at the global picture. In particular, efforts are required on evolution of architecture, open standards and protocols. These efforts should also have provision for incorporating current efforts and also integrate current wireless and cellular networks.

Table 2 provides a summary of different research issues of ad hoc network. It can be seen that different type of challenges arises when addressing various issues. The different solutions proposed to address challenges are summarized in the table. However, current solutions are still immature and a lot of research areas are still open, as we will see in later sections.

EMERGING MODELS OF AD HOC NETWORKS

During last few years, different emerging modes of ad hoc networks have been proposed. These variants emerged as applications of ad hoc networks in specialized domains or due to their intersection with other technologies. Figure 1 provides five broad categories of ad hoc networks based on their characteristics. A *sensor network* comprises a number of miniscule sensing devices that organize themselves autonomously for monitoring physical environments. These networks are useful for environmental monitoring, body area networks, smart office, smart agriculture and precision farming etc. A *mobile ad hoc network (MANET)* is a type of ad hoc network with mobile nodes. As the nodes are mobile and the network is highly dynamic, links failure and topological changes are common phenomena in such networks. When the

Table 2. A summary of research issues in ad hoc network

Research Issue	Issues/ Challenges	Solutions	Reference
Antenna Design	Channel capacity, multipath effect, interference	Smart Antenna	(Chau et al., 2012)
Energy management	Low powered nodes, coordinated energy management at various layers	Transmission power control Energy efficient routing	(Singh & Kumar, 2010) (Liu et al., 2009)
MAC Layer	Hidden node, exposed node problems, deafness, head of line blocking	Contention free and contention based solutions QoS aware MAC MAC for directional antenna	(Kumar et al., 2006) (Wang et al., 2008) (Kamruzzaman et al., 2010)
Routing	Blackhole attack, worm hole attack, routing loops	Proactive, Reactive, Hybrid routing protocols Position based routing	(Boukerche et al., 2011)
Multicasting	Changing topology of the network, resource limitations	Tree, mesh and hybrid protocols	(Junhai et al., 2008)
Addressing	No addressing server, possible DoS attack	State-less and State-full approaches	(Nesargi & Prakash, 2002) (Chen et al., 2009) (Li Longjiang, 2009)
Transport Protocol	Low throughput due to congestion control	Network layer information utilization	(Francis et al., 2012)
Security	Hostile environment, lack of security infrastructure	Secure routing, intrusion detection	(Lou et al., 2009) (Svecs et al., 2010).
Cooperation	Lack of trust among nodes	Cooperation enforcement Incentive based mechanism	(Kwon et al., 2010) (Zayani & Zeghlache, 2012)
Data Management	Data interoperability, data discovery	Semantic representation, cross-layer information exchange, opportunistic data sharing	(Martin & Demeure, 2008) (Islam & Shaikh, 2011)
Testbeds and simulation	Network complexity, cost of hardware	Test beds Simulator	(Kiess & Mauve, 2007)
Mobility models	Realistic simulation of mobile nodes	Mobility models Patterns generator	(Karnadi et al., 2007)
Standardization	Lack of harmony among standardization bodies	IEEE, IETF	(IETF, 2012)

Figure 1. A classification of various types of ad hoc networks

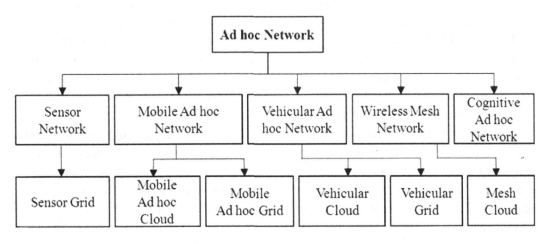

ad hoc network is created among vehicles, it is called *vehicular ad hoc network (VANET)*. These networks are attributed by highly mobile nodes, more processing capabilities and availability of location tracking devices etc. VANET has been used for the management of various transportation problems (Islam, Shaikh et al., 2008). Another type of ad hoc network is *wireless mesh network (WMN)* that is formed among a set of self organizing nodes arranged in a mesh topology. WMN provides more resilience and cost-effective connectivity and has been used for military operations, community networking and telemedicine etc. (Akyildiz et al., 2005). In the last few years, techniques have been proposed to optimally utilize unused licensed spectrum available around a node. *Cognitive Radio Ad hoc Network (CRAHN)* comprises various self-organized nodes that are equipped with cognitive radios that utilize unused portion of spectrum for internal communication.

Grid computing is the concept of aggregating a collection of distributed loosely couple resources. By integrating with ad hoc networks, the concept of *ad hoc grid* has emerged. Similarly, the integration of sensor network and grid computing gives rise to concept of *sensor grid*, where the data collected from sensor networks are passed to a grid for high performance computing. In some of the researches, *vehicular grid* has been proposed as an integration of vehicular network and grid computing. *Cloud computing* is also an emerging discipline where the resources are leased from a collection of tightly coupled and virtualized resources. The merger of ad hoc networks and cloud computing has given rise to the notion of *ad hoc clouds*. Besides, the concept of mesh clouds has also been proposed in some researches.

In addition to the various models discussed above, other novel network models are also possible. Examples are sensor clouds, cognitive vehicular networks, cognitive clouds, cognitive sensor networks, cognitive vehicular network and cognitive wireless mesh networks etc. (Akan et al., 2009; Felice et al., 2012). In the sections

below, the discussion has been restricted to ad hoc grids, ad hoc clouds, wireless mesh networks and cognitive radio ad hoc network. We will discuss briefly these models, their attributes and additional research issues arise during their implementation. The various solutions to address these issues are presented very briefly due to space limitations. However, pointers to corresponding literature are provided to interested readers for further research.

Ad Hoc Grid

Grid Computing is an approach towards accumulation of a number of distributed and loosely coupled resources for coordinated problem solving. The central idea of grid computing is to provide the resources i.e. hardware, software, data and network etc. as a utility similar to conventional electricity grid. By the sharing and optimal utilization of distributed resources, grid computing has enabled large scale computing that was previously possible only with super computers. With the enormity of computing devices around, the potential of integrating grid computing and ad hoc networks has been studied in recent years. This leads to the concept of *ad hoc grid* where various mobile devices organize themselves on the fly in the form of grid.

Due to recent advances, the commonly used appliances in our daily lives i.e. microwave ovens, wrist watches, glasses and air conditioner etc. are expected to behave like computing gadgets. Generally, these resources remain unutilized most of the time. The concept of ad hoc grid is to optimally utilize the capabilities of these devices by self organizing them for different computationally intensive tasks. Ad hoc grids possess all the properties of ad hoc network i.e. self organization, autonomy, distributed operations, multi-hop routing, and mobility of nodes etc. However, they also have some additional properties that set them apart for conventional ad hoc networking. Let's discuss them briefly.

Properties of Ad Hoc Grid

As the grid computing enables resource sharing across organizational boundaries. The different resources available are *heterogeneous* in terms of architecture, capabilities, performance and operating environment etc. The resources can belong to *different administrative domains* or even *geographically distributed*. Ad hoc grids are generally used for solving *large scale* problem. Since, the overall objective of grid computing is to provide resources; *fault tolerance* is an important requirement for any ad hoc grid system. The nodes are required to have better *coordination* and therefore mechanisms for cooperation enforcement is a more stringent requirement. Finally, the consumers of services should provide *transparent access* above the heterogeneous and dynamically changing ad hoc environment.

Research Issues in Ad Hoc Grid

Ad hoc grid presents newer challenges in addition to the one posed by ad hoc network. New *resource discovery algorithms* are required for ad hoc grids that should provide the ease of deployment of ad hoc network coupled with fault tolerance and scalability demands of grid environment. A resource discovery scheme should be energy efficient and takes less time to discover resources. Conventionally, the resource discovery algorithms have been classified as *directory-based* and *directory-less* approaches (Islam et al., 2010). The former utilizes a directory for maintain list of available resources in the network. For instance, Mallah and Quintero (2009) proposed a directory based approach in which a set of backbone nodes are selected for maintaining directories and entertaining discovery requests. Directory-less approach works in decentralized fashion by contacting the hosting nodes for the desired resources. In this direction, Pariselvam and Parvathi (2012) proposed a directory less scheme based on swarm intelligence to perform resource discovery in ad hoc networks.

A similar approach has been proposed by Singh and Chakrabarti (2013), where super nodes are selected based on ant colony optimization. These super nodes are responsible for discovery of resources in the grid. Various cross layer discovery scheme have been proposed in recent literature that merges the discovery process with other protocols. For example, Islam and Shaikh (2013) presented a cross-layer approach to service discovery based on integration of discovery process with routing protocols. A secure cross-layer service discovery SCAODV has also been proposed by (Zhon et al., 2012). Cross-layer solution still remains an unexplored area in the domain of ad hoc grids, however.

Another important research issue in ad hoc grid is the development of *scheduling algorithms* to allocate certain type of resources for the execution of a particular job. For ad hoc grid, scheduling algorithms are required that should not only consider the classical parameters but also consider the attributes reflecting the dynamism in the network. This includes the nodes mobility, failures of links and nodes, and network partitioning etc. There are three types of scheduling schemes proposed in literature. The *centralized approach* is based on a focal entity responsible for scheduling the resources. The *decentralized approach* comprises different schedulers that interact with each other for scheduling of resources. Finally, the *hierarchical scheduling* is based on several low level schedulers and a top-level scheduler for global coordination. According to Bhaskaran and Madheswaran (2010), a decentralized job scheduling is better suited for addressing the fault tolerance and reliability of ad hoc grids. They discussed various challenges in implementing scheduling in ad hoc grids. The authors also proposed a scheme that works in decentralized fashion to perform job scheduling and congestion control in ad hoc network. Among other works, Torkestania (2013) presented a distributed job scheduling algorithm in which the jobs are allocated to a node based on its computational capacity. The schedulers are synchronized by a learning automata. Xu and Yin (2013) have also proposed a task scheduling algorithm based on a mathematical model. The scheme considers the mobility of nodes and resources while performing task scheduling.

Relevant to scheduling, is the development of robust *work flow systems* that can compose complex tasks and schedule its execution in the ad hoc environment (Terracina et al., 2006). The currently available workflow systems can be classified as *non context-aware* and *context-aware* workflows. Among the non-context aware workflow systems is Grid Services Flow Language (GSFL), proposed by Krishnan et al. (2002). GSFL is essentially a non-context-aware workflow description language to represent service providers, activity model, composition model and lifecycle model for implementation of workflows in a grid. However, Abbasi (2013) hypothesized that a context-aware workflow system has more capabilities to adapt for ad hoc changes in the environment. He presented a context-aware approach to design and manage workflows, comprising specialized context-aware activities that can be dynamically loaded based on the context. A context-aware work flow management system for ubiquitous applications has also been presented by Tang et al. (2008).

Security and trust management are essential components of an ad hoc grid system. A security system should provide authorization, protection and establishing the trust among peers to ensure collaboration in a fair manner (Amin et al., 2004). Different security solutions have been proposed for ad hoc grid systems based on trust management. For instance, RETENTION is a reactive trust management scheme based on a mathematical trust model in which malicious nodes are punished for their misbehavior (Bragaa et al., 2013). Singh and Chakrabarti (2013) presented a trust management scheme for mobile grid system. There are super nodes in the network elected based on different parameters i.e. CPU, battery power and bandwidth. For ensuring the trust among nodes, a trust collection mechanism has been adopted. Huraj and

Siládi (2009) have presented a survey of various authorization and authentication shames for ad hoc grids and also presented a security scheme based on trust chains developed using public key infrastructure.

Another important issue in ad hoc gird is the design of *economic models* such that services can be fairly utilized in the network. There should be some utility value associated with a service and a billing mechanism should provide incentives for services provided by a node. A similar work has been cited in (Vazhkudai & Laszewski, 2001), where a basic framework is proposed comprising services for bartering, bidding and trading etc. Li and Li (2012) also presented an ad hoc grid system based on micro economic theory where the producers and consumers are modeled as decision makers who buy and sell resources similar to economic architecture.

New *QoS models and metrics* are required for ensuing QoS in ad hoc grids. In general, the current QoS approaches for grid can't cater with dynamism and the new requirements for collaboration and resource sharing posed by ad hoc grids (Li et al., 2005). Among the few solutions proposed, Li and Li (2012) have presented a scheme for QoS and load balancing. A utility values is associated with the QoS satisfaction and a preference value is computed from the resource point of view. These two values are combined to select a resource for a particular request.

Unlike conventional grid computing system, an ad hoc grid comprises mobile nodes that have limited battery life. Hence, the architecture and solutions proposed for ad hoc grid should be *power efficient*. In this direction, Marinescuand et al. (2003) proposed a basic architecture for ad hoc grid and then presented a model for power consumption for different operations of the network. Shah et al. (2009) have also presented an architecture for establishment of ad hoc grid among resource constrained devices. The architecture takes care of energy limitations, mobility and signal strength issues. There is a configuration profile service running that keeps track of battery power and report to grid job scheduler to switch a job to some other node.

Table 3 summarizes the various research issues related to ad hoc grid as discussed in previous paragraphs. In some of the literature, different sub-types of ad hoc grids systems have also been proposed. This section is concluded with brief discussions on these models.

Sensor Grid

Sensor grid emerged as an integration of sensor network and grid computing system. In a sensor grid, the data collected from sensors is sent to a grid environment for data processing and actuation. Sensor grids have been applied for solving different types of applications. For example, *AgriGrid* is a context-aware sensor grid platform proposed for management of agriculture problems (Aqeel-ur-Rehman & Shaikh, 2008). Similarly, *iHEM* is a sensor grid based approach for in-home management of electrical energy (Erol-Kantarci & Mouftah, 2011). It is based on wireless sensor

based communication among the consumers of electricity and the controllers.

Sensor grid presents several new research issues besides the challenges associated with sensor network and grid computing environment. An overview of various issues is provided in (Tham & Buyya, 2005). One of the core issues is design of *service oriented approaches* for access, distributed processing and management of sensors. SensorML is an effort in this direction that provides XML schema for defining sensor characteristics (OpenGeospatialConsortium, 2013). Efficient and robust *querying* mechanism are also required for sensor grid to fetch data in real-time from the distributed sites by sensors and vice-versa. Novel protocols are required to deal with *security*, *network failures* and *noise issues* that arise due to wireless nature of the network. Finally, mechanisms are required for *addressing*, *scheduling*, *power management* and *QoS provisioning* in sensor networks. Representative solutions can be seen in (Miridakis et al., 2009; Li et al., 2013).

Table 3. Summary of research issues in ad hoc grid

	Research Issue	Challenges / Issues	Representative Approaches
1.	Resource discovery	Limited battery Lack of global schema Heterogeneity	Directory-based Directory-less Cross-layer (Islam & Shaikh, 2013)
2.	Scheduling	New scheduling criteria Node mobility Failures Network partitioning	Centralized scheduling Hierarchical scheduling Distributed scheduling (Bhaskaran & Madheswaran, 2010; Torkestania, 2013; Xu & Yin, 2013)
3.	Workflow	Network dynamism Context management	Context-aware systems (Abbasi, 2013; Tang et al., 2008) Non context-aware system (Krishnan et al., 2002)
4.	Security	Authentication and authorization Trust establishment	(Bragaa et al., 2013) (Singh & Chakrabarti, 2013) (Huraj & Siládi, 2009)
5.	Economic model	Utility assignment Faire utilization of resources Ensuring cooperation	(Vazhkudai & Laszewski, 2001) (Li & Li, 2012)
6.	Power control	Mobility Signal variations	(Marinescuand et al., 2003) (Shah et al., 2009)
7.	Quality of service	New metrics and models considering collaboration and resource sharing	(Li & Li, 2012) (Marinescuand et al., 2003) (Shah et al., 2009)

Vehicular Grid

Vehicular grid is a grid environment established among a number of vehicles travelling on the road that are connected in ad hoc fashion. Different vehicular grid frameworks have been reported in literature. Khorashadi (2009) proposed a vehicular grid framework called *vGrid* in which the vehicles on the road share their information with other vehicles to solve various traffic management problems. The current ITS solutions for traffic management are very expensive and require large amount of time for solving traffic problems. According to Khorashadi (2009), by employing the vehicular grid solution, the traffic problems can be solved in order of seconds. The various types of problems that can be solved using the proposed framework are lane merging, accidental warning and ramp metering etc. In the same direction, Islam et al. (2008) proposed a grid based approach to routing traffic in vehicular ad hoc networks. By establishing a grid between vehicles, a distributed shortest path algorithm is used to find out optimal routes. Another relevant project is *VGITS*, a hybrid system based on data processing and real-time traffic services in a centralized manner, while providing traffic services to drivers in a decentralized manner (Chen et al., 2008).

Beside the various issues of ad hoc grid, vehicular grid systems present several additional challenges. These are mainly related to development of a standards, architecture, security and privacy, routing protocols, proposal for novel applications, and development of testbeds and toolkits etc. for implementation of algorithms and protocols.

Ad Hoc Cloud

Cloud computing is a distributed computing paradigm in which a pool of computing, storage and platforms are delivered as a service over the Internet. Grid and cloud computing have several things in common i.e. utility computing, service oriented architecture and distributed operation etc. However, grid comprises loosely coupled resources spanning across multiple organizational boundaries and cloud is a collection of tightly coupled resources virtualized and available for user as service over Internet. The examples of various cloud computing environments are *Amazon Elastic Cloud Computing, Microsoft Windows Azure* and *Google App Engine* (Islam & Aqeel-ur-Rehman, 2013).

The motivation behind cloud computing is to address resource limitation problems of mobile computing devices. The integration of cloud and mobile computing has been studied recently to address the resource limitations in ad hoc environments. Kovachev et al. (2011) have provided a survey of various models for *mobile cloud computing*. One of the approaches is offloading the task of a mobile device to a cloud. Another approach is *ad hoc cloud* where a set of mobile devices forms a cloud computing environment and provides services to other devices. Ad hoc cloud can be used for different types of applications ranging from cloud robotics, crowd computing and data sharing to image processing etc. (Hu et al., 2012; Fernando et al., 2013).

Properties of Ad Hoc Clouds

The resources in ad hoc clouds are provided as different types of *services* to end users. These services models can be infrastructure, platform, software, network or storage as service (Islam & Aqeel-ur-Rehman, 2013). Unlike ad hoc networks, the resources are *pooled* and provided to consumers as *virtualized* resources. Ad hoc clouds offer a more flexible *on-demand servicing* approach where the resources are *dynamically provisioned* with the increase in demands of users. Contrary to conventional clouds, the nodes in ad hoc cloud are not dedicated, instead every node share their computational resources. Hence, there must be some mechanism for *measurement* of services and users should be provided appropriate incentives for the resources shared. Ad hoc clouds have relatively *more resources* and are therefore useful for computationally intensive tasks. Finally, ad hoc cloud should be *fault tolerant* and provide *transparent access* of services to end users irrespective of failure of certain services in cloud.

Research Issues in Ad Hoc Clouds

There are various research issues associated with implementation of ad hoc cloud as discussed in (Kirby et al., 2010; Fernando et al., 2013). One of the issues is to decide the mechanism to *offload a task* for execution on cloud. Hu et al. (2012) have discussed various issues to be considered while offloading a task for execution. Among the strategies are remote procedure call (RPC) or remote method invocation (RMI) to upload the task, and migrating the virtual machine (Fernando et al., 2013). There are a variety of issues related to *service provisioning* in ad hoc cloud. For example what are the service models supported by the cloud and how the cloud ensures smooth provisioning of these services in failure? The first question is dependent on the application domain. In this direction, Olariu and Yan (2012) have presented a vehicular cloud system in which the services supported are related to traffic management problems. In case of disconnection from mobile cloud, a vicinity node connected to the cloud can be consulted for the service (Dinh et al., 2011). Another relevant issue is the provision of QoS of the services (Zhang & Yan, 2011). To ensure elasticity of services offered on cloud, an application model has been proposed in (Zhang et al., 2009).

Buyya et al. (2009), on his vision on 21st century of computing stressed on the need for marked oriented resource management for cloud computing and presented an architecture for supporting market oriented resource allocation. Hence, an important issue in ad hoc clouds is the availability of an *economic model*. In this direction, Liang et al. (2012) presented a model based on Semi-Markov decision process. The model considers the maxi-

mal system rewards and expense of mobile device to devise an optimal resource allocation policy. *Energy efficiency* is becoming a fundamental issue in cloud computing. Novel mechanisms are required to ensure minimal energy consumption during their operation. Miettinen and Nurminen (2010) presented a model for ad hoc cloud and highlights critical factors responsible for energy consumption in clients in mobile clouds. Some researchers have considered network coding and cooperation among the nodes as mechanisms for energy conservation in mobile clouds (Heide et al., 2012).

Another issue is related to the *interoperability* and heterogeneity of devices that are in used on the cloud (Sanaei et al., 2013). Unfortunately, there are no open standard leading to scalability, deployment and availability of service issues (Chetan et al.). The current Web standards are not useful as they were not designed for mobile devices. HTML5 offers Web-sockets that provides real-time communication and can serve as a useful candidate for ensuring interoperability (Dinh et al., 2011).

There are a number of issues pertaining to *security* in ad hoc clouds. This includes integrity and privacy of data shared on the cloud, authentication of clients that can use these services and establishment of trust among the nodes in the network. Popa et al. (2013) proposed a framework for mobile clouds based on applying security policy depending on the type of data. The security is ensured during communication, and the integrity of the applications are also ensured during their installation and updating process. Discussions on various public key cryptography schemes for mobile cloud have been provided in (Zheng, 2013). In addition, a general comparison of various security schemes for ad hoc clouds has been reported in (Iyer & Durga, 2013).

Table 4 provides a summary of major research issues in ad hoc cloud and representative approaches taken to address them. Besides these approaches, various ad hoc cloud frameworks been proposed in literature to address the various issues. Huerta-Canepa and Lee (2012) presented a preliminary design for creation of ad hoc cloud computing solution. Remote procedure calls are used for offloading of task based on current context. A resource management component performs profiling and monitoring of devices. Harox is another framework that provides an approach towards cloud services on android phones in which various benchmark algorithms of distributed sorting and searching are implemented using Hadoop framework (White, 2012; Marinelli, 2009).

Vehicular Cloud

A sub-discipline of ad hoc cloud is *Vehicular Cloud Computing*, proposed recently in (Olariu & Yan, 2012). The motivation was that vehicles on the street, fuel stations, and parking places of offices, restaurant and shopping plazas etc. have huge underutilized computational resources. These resources include on-board computer, storage, location tracker, communication devices and camera etc. If these resources can be organized in the form of a cloud, they can be optimally utilized for solving various types of traffic problems. The author proposed different types of cloud service models. For example, Internet connectivity available at some nodes can be rented out to other nodes to give rise to concept of *network as a service*. Similarly, nodes can share their extra storage for other nodes that needs large storage for their applications i.e. *storage as a service*. Finally, nodes can cooperate with each other to provide various ITS services. This leads to the concept of *cooperation as a service*.

Let's discuss representative proposals appeared in literature on vehicular cloud. Wang et al. (2011) proposed a vehicular cloud system for improving driving comfort and providing different types of services to cloud. A layered architecture is presented, various types of applications and services are discussed, and research challenges are also highlighted. Hussain et al. (2012) have provided taxonomy of various vehicular cloud models and grouped them in to vehicular clouds, vehicles using clouds and hybrid vehicular clouds. Alazawi et al. (2011) have presented a slightly different approach for disaster management system. It is

Table 4. Summary of research issues in ad hoc cloud

	Research Issue	Challenges / Issues	Representative Approaches
1.	Offloading of task	Heterogeneity Distance	Remote procedure call Remote method invocation Virtual machine migration
2.	Service provisioning	Service models Link failure QoS Dynamic provisioning	Network, storage, coordination as services (Olariu & Yan, 2012) (Dinh et al., 2011) (Zhang & Yan, 2011) (Zhang et al., 2009)
3.	Economic model	Utility assignment Negotiation Payment mechanisms	(Buyya et al., 2009) (Liang et al., 2012)
4.	Energy management	Limited resources Energy aware algorithms	Network coding, node cooperation (Heide et al., 2012)
5.	Interoperability	Heterogeneity Lack of standardization	Websockets (Dinh et al., 2011)
6.	Security	Integrity Privacy Authentication Trust management	(Popa et al., 2013) Cryptography (Zheng, 2013) (Iyer & Durga, 2013)

based on gathering information from multiple sites, which is then processed on a cloud. The point of incident is identified and effective strategies to counter them are devised. This information is then disseminated to vehicles.

Vehicular clouds can play a key role in disaster management, intelligent transportation and similar applications, especially in developing countries. However, several new research challenges needs to be addressed. Among them, some of the challenges are related to information dissemination, standardization, privacy, authentication, trust management and availability of hardware in vehicles etc. (Talebifard & Leung, 2013; Olariu & Yan, 2012; Yan et al., 2012).

Wireless Mesh Networks

Wireless mesh network (WMN) is a self organized network established among a set of stable nodes arranged in mesh topology to provide network access and other types of services to mobile clients. The former set of nodes is called *mesh router* that serves the later type of nodes called *mesh clients*. WMN are generally useful for interconnecting different wireless networks with quick deployments and low upfront costs, and gives rise to several useful applications as discussed earlier.

Properties of Wireless Mesh Network

Sichitiu (2005) and Akyildiz et al. (2005) have discussed various key properties of WMN. One of the important characteristics of WMN is *integration* of different wireless networks. The nodes in WMN are of three different types. There are *mesh routers* that are less mobile and provides services to *mesh clients*. Mesh clients can be stationary or mobile. Some of the mesh nodes serve as *Internet gateways* providing interconnectivity among different networks. This also raises the need for countering the *heterogeneity* among these networks. WMN operates in multi-hopped fashion providing interconnection in short distances. Additionally,

to increase the effective bandwidth, some nodes have *multiple radio interfaces*, allowing them to work on different channels.

Since, WMN comprises a wireless backbone infrastructure with mesh routers, there are *no power management* issues. The network is *scalable* and the load on mesh client side is very minimum. This allows limited capability devices to be part of the network. Most of the operations i.e. routing and configuration etc. are performed by routers and there is a *steady communication* among the mesh routers. Finally, WMN can also be deployed in *preplanned or incremental* fashion in some situations.

Research Issues in WMN

Because of the characteristics of WMN, different new challenges and issues surfaced out at various layers of protocol stack (Akyildiz & Wang, 2005; Pathak & Dutta, 2011). An important research issue is *modeling* of the various parameters of WMN i.e. capacity, data rate and flow assignment etc. (Rawat et al., 2013). The availability of an analytical model provides basis for further analysis and promotes the development of new architecture, applications and protocols for WMN. In this direction, Dely et al. (2010) analyzed the throughput of a multi-radio mesh networks under different conditions. The authors concluded that for low PHY rate, channel separation provides hints for channel capacity. On the contrary, under high PHY rate, channel capacity measurement requires analysis of propagation properties. Research efforts are also required to develop *testbeds and simulators* to validate research proposals. Representative works have also been reported in (Rawat et al., 2013; Lam et al., 2012).

At physical layer, advanced techniques are required to enhance the *transmission rate* of the channel. This includes novel antenna designs techniques i.e. diversity coding, smart antenna and directional antenna etc. The various techniques that can be employed for improving the reliability and data rate of transmission are orthogonal frequency multiple access, code division multiple access and ultra wide band techniques (Sichitiu, 2005).

As the WMN comprises mesh client and routers, new *MAC protocols* are required that can work both for mesh clients and routers. MAC protocols should also be able to work under different wireless technologies and capitalize on multiple communication channels. A simple approach is to use a dedicated channel for controlling. However, the control channel can get saturated leading to non-optimal channel utilization. To address this problem, Lei et al. (2012) proposed a wait-time based approach to multi-channel MAC protocol. When directional antennas are used in WMN, new MAC protocols are required that should counter different problems as discussed in previous section. A survey of MAC protocols for directional antenna is provided in (Bazan & Jaseemuddin, 2012; Lu et al., 2012). *Scalability* is also an important requirement for multiple access protocols for WMN. Unfortunately, most of the MAC and routing protocols for ad hoc network don't cater with the scalability requirements and only a few scalable MAC protocols are currently available (Zhou & Mitchell, 2009). Therefore, designing scalable MAC protocols for WMN is an important area of research.

The *routing protocols* for WMN pose stringent requirements as compared to ad hoc networks. This includes enhanced scalability, fault tolerance, load balancing, power control, and adaptability to work for mobile clients, (Sichitiu, 2005). Unlike the conventional metrics, new metrics should be used to meet these requirements. A routing protocol should combine more than one metric considering the transmission count, round trip time, noise and interference to compute routes. An overview of different routing protocols and metrics for WMN has been provided in (Campista et al., 2008). The authors classified current protocols as those based on reduced link variations, control overhead and network traffic, and opportunistic routing. To enhance the reliability, multi-path

routing scheme has also been proposed in some research (Akyildiz et al., 2005).

The design of *transport layer protocols* also presents several new challenges in mesh networks (Rangwala et al., 2008). It should have the flexibility to operate under different network and protocols running at lower layers. Depending on the network, there can be different flow control and error control parameters. Therefore, transport protocols should be able to cater to these variations. As discussed in previous section, TCP suffers from various issues and a range of TCP variants have been proposed. Similarly, variants of UDP i.e. RCP/ RTCP have also surfaced to provide reliable real-time communication. A survey of various transport protocols for wireless mesh networks have been provided in (Law, 2009).

Like ad hoc networks, WMN also suffers from various security issues. This includes eaves dropping of mesh management and control frames, jamming communication among mesh routers and masquerading as mesh nodes by an intruder etc. (Zdarsky et al., 2010). Among the various approaches proposed to security are mobile client authentication and access control techniques, cryptographic approaches to secure communication, securing mesh routing and securing backbone etc. (Zhang & Fang, 2006; Martignon et al., 2011). The *application layer* of WMN also requires careful design of applications such that they can adapt themselves under the architecture of WMN.

Table 5 provides a summary of discussed research issues in WMN. Besides discussed, there are a number of other research issues that require investigation i.e. cross-layer design, multicasting, network planning and deployment etc. (Akyildiz et al., 2005).

Cognitive Radio Ad Hoc Networks

A *cognitive radio* detects the availability of unused licensed spectrum in its surroundings and dynamically configures itself to operate on that channel. The ISM band proposed for unlicensed used have

Table 5. Summary of research issues in wireless mesh networks

	Research Issue	Challenges / Issues	Representative Approaches
1.	Implementation	Channel modeling Testbeds and simulators	(Dely et al., 2010) (Rawat et al., 2013; Lam et al., 2012)
2.	Transmission rate	New approaches to increase capacity Address noise and interference	New modulation and channel coding techniques Multiple array antenna techniques
3.	MAC protocols	Heterogeneity Multi-channel MAC Scalability	(Lei et al., 2012) (Zhou & Mitchell, 2009).
4.	Routing protocols	New metrics considering fault tolerance and power efficiency	Ad hoc based Controlled flooding Traffic aware Opportunistic routing Multipath routing
5.	Transport protocols	Increased packet loss Heterogeneity in error/ flow control	(Law, 2009)
6.	Security	Mobile client authentication Access control Secure routing	(Zhang & Y.Fang, 2006) (Martignon et al., 2011)
7.	Application layer	Development of new models Architecture	

been crowded due to emergence of plethora of computing devices that compete for access. The solution is to opportunistically capitalize on the licensed spectrums that are not optimally used in most of the cases. A *cognitive radio ad hoc network (CRAHN)* is an ad hoc network created among a set of nodes equipped with cognitive radios. There are two types of users in CRAHN. The *primary user* is the owner of a particular spectrum while the *secondary user* utilizes the unused spectrum of other user.

Properties of CRAHN

CRAHN is differentiated from ad hoc networks by various properties (Akyildiz & Lee, 2009). The nodes in CRAHN are equipped with *cognitive radios* that continuously sense the unused licensed spectrum. These radios should be self-describing and adapts its parameters to operate on different channel. There are *multiple channels* available to a particular user for transmission. The availability of channel for particular user varies continuously depending upon the activities of primary user.

Hence, the routing path between two nodes usually comprises *communication in different spectrum*. The various layers of protocol are usually *coupled with the spectrum sensing mechanism*. For example, an application layer should be able to discriminate between the temporary unavailability of a link due to primary user's activity from other reasons. Additionally, the *neighborhood detection* by means of beacon message requires transmission in all available channels, which is not always feasible in CRAHN.

Research Issues in CRAHN

CRAHN presents several new challenges besides the conventional ad hoc network's challenges and therefore asks for deliberation on several new research issues. Among them, there are a range of research issues related to spectrum management. A cognitive radio user should *sense the spectrum* for unoccupied spectrum and identify spectrum holes. This step involves determining the primary user based on energy of the received signals or observing specific features i.e. modulation type, symbol rate and pilot signals etc. (Cabric et al., 2004). An Eigen value based primary user detection method has been proposed in (Liu et al., 2013). In this scheme, a virtual multi-antenna structure is formed using temporal smoothing technique. A covariance matrix is thus obtained whose maximum and minimum Eigen values are used to detect primary user. However, the problems with local spectrum sensing are channel conditions and time varying nature of channel etc.

Besides performing the spectrum sensing locally, cooperative approaches can be adopted in which different users work together to perform spectrum sensing. In this direction, a collaborative spectrum sensing based on evolutionary game theory has been proposed in (Sasirekha & Bapat, 2012). An adaptation algorithm is proposed such that network utility is maximized even in case of dynamic changes in the network. Cacciapuoti et al. (2012) have also proposed a correlated based spectrum sensing scheme. The proposed distributed approach recommends selection of non-correlated users for cooperation during spectrum sensing. For interested readers, a survey of various spectrum sensing algorithms can be seen in (Yucek & Arslan, 2009).

Once spectrum holes are detected, one of the holes is selected for use by user. Generally, a *spectrum decision approach* is based on characterizing the spectrum, selecting the best spectrum based on this characterization and reconfiguring the communication protocol and hardware according to the radio environment and available QoS (Akyildiz et al., 2009). Most of the approaches proposed for spectrum decision making are cooperative in nature. Lee and Ian F. Akyildiz (2011) have proposed a spectrum decision approach for real-time applications that considers channel dynamics and application requirements for selection. A cooperative weighted decision approach based on SRN tracking has been proposed in (Canberk & Oktug, 2012). The proposal employed a distributed weighted fusion scheme

to combine the decisions of individual users to obtain a cooperative decision.

As there may be multiple users trying to access the spectrum, there should be some coordination to allow *sharing of the spectrum* among multiple secondary users. Game theory has been proposed to provide the equilibrium condition where multiple users compete for the available spectrum (Neel, 2006). A consensus based approach to spectrum sharing has been proposed in (Hu & Ibnkahla, 2012). The scheme performs spectrum sharing based on local information (spectrum or location sensors) and consensus feedback.

For fair utilization of the spectrum among the users, a *MAC protocol* is also required. Unlike conventional ad hoc networks, MAC protocols in CRAHN should be strongly coupled with spectrum sensing process. Among the various protocols for CRAHN, there are protocols based on random access, time slotting and hybrid approaches (Akyildiz et al., 2009). *Random access protocols* are variants of CSMA/CA protocol that are not dependent on time synchronization and can access the medium any time. In *time slotted approaches*, appropriate time slots are provided to user for transmission of data and control frames. This requires global synchronization of nodes at network level. In the *hybrid approaches*, the control frames are transmitted based on time slotting but data can be transmitted randomly without any synchronization. A survey of various MAC protocols along with further details of these protocols can also be seen in (Cormio & Chowdhury, 2009).

Upon detection of primary user, the cognitive radio user must vacate the spectrum. A *spectrum hand off* is thus required that ensures seamless communication by switching the communication to a new spectrum. For this purpose, proactive as well as reactive approaches can be adopted (Feng et al., 2012). In *proactive approach* to hand-off, the future activity is predicted and a new spectrum is determined before the failure of current link. In *reactive approach*, switching to a new spectrum is performed after the links failure has occurred. This approach causes significant delay in current transmission.

A common *control channel* is required to for exchange of control information pertaining to spectrum management. This dedicated channel thus enables the cognitive radio users to seamlessly operate and supports neighborhood discovery, coordination in spectrum sensing and exchange of local measurements etc. (Akyildiz et al., 2009). Among the various choices for control channel include dedicated licensed spectrum, ISM and UWM bands (Marinho & Monteiro, 2012).

Routing in CRAHN is not a trivial task as it must consider different aspects of spectrum management for route computation. Ideally, a route should be selected with the minimum channel switches to transmit a message to the destination. Among the various routing protocols for CRAHN are those based on utilizing spectrum information and middleware based approaches etc. (Chowdhury, 2009; Guan et al., 2010). A survey on various routing protocols for CRAHN has been provided in (Cesana et al., 2010).

New *transport protocols* are required due to fluctuations in channel availability and temporary disconnections. As we discussed earlier, the TCP protocol suffer from issues due to its congestion control algorithm. Hence, the transport protocols for CRAHN should be spectrum aware. It should be able to discriminate between route disruptions due to channel switching and other reasons i.e. node mobility or node/link failure. In CRAHN, as the failure is virtual and the same route is restored after a new channel is selected for transmission. Hence, instead of stopping the transmission the packets should be sent at an optimal route that doesn't overwhelm the intermediate nodes. An example of transport protocol for CRAHN is TP-CRAHN (Chowdhury et al., 2009). It is based on feedback from the physical and link layer to discriminate various types of events in the network. It also adopts a network layer predictive mobility

framework that sends the packet at optimal rate to a sender. TCP-CR also proposed a delayed congestion control scheme based on primary user detection in cognitive radio networks (Yang et al., 2012).

CRAHN suffers from several additional security threats besides the security problems of conventional ad hoc networks. Some of the threats are jamming of control channels, masquerading of primary or cognitive radio users, unauthorized use of spectrum, integrity of cognitive radio messages and the node itself etc. (Baldini et al., 2012). Fadlullah et al. (2013) have presented an intrusion detection system for CRAHN that identifies various types of anomalies based on a learning model. A survey of various security challenges and solutions for cognitive radio networks have also been provided in (Attar et al., 2012).

An important issue is the development of new *applications models* for CRAHN. These application models should be agnostic against events happening at lower layers. For example, an application should be resilient against the temporary disconnection due to spectrum switches. New algorithms are required that focus on QoS at application layers. Yu et al. (2010) proposed an approach towards optimizing application layer QoS parameters using finite state Markov models. Finally, as the experimental evaluation of CRAHN on real environment is very complex, there needs to be *test beds and simulation toolkits* to validate various research proposals. Among the various software platforms and test beds proposed for CRAHN include GNU Radio and SCA (Ishizu, 2006; Gonzalez, 2009) etc. A survey on various implementation choices for CRAHN has been provided in (Chowdhury & Melodia, 2010).

Table 6 summarizes various issues of CRAHN. There have been different variants of CRAHN proposed in literature that are evolved as intersection with other technologies. This section ends the discussion with an overview of them.

Table 6. Summary of research issues in cognitive radio ad hoc network

	Research Issue	Challenges / Issues	Representative Approaches
1.	Spectrum sensing	Sensing accuracy Channel condition i.e. multipath effects, noise and interference etc. Time varying signal properties	Cooperative approaches Non-cooperative approaches
2	Spectrum decision	Power control Time varying signal properties	(Canberk & Oktug, 2012) (Lee &Akyildiz, 2011)
3	Spectrum sharing	Energy management Novel coordination mechanisms	Game theory (Neel, 2006) Consensus based spectrum sharing (Hu & Ibnkahla, 2012)
4.	MAC protocol	Coupling with spectrum sensing	Random access Time slotting Opportunistic access
5.	Hand-off	Delay Connection management	Proactive hand-off Reactive hand-off
6.	Control channel	Control channel contention Jamming attack	Dedicated licensed spectrum ISM band UWM band
7.	Routing	Primary user activity New metrics considering spectrum awareness, channel properties, queuing and switching delay etc. Mechanism for route maintenance	Utilizing spectrum information (Chowdhury, 2009) Middleware based approaches (Guan et al., 2010)
8.	Transport protocol	Spectrum awareness	TP-CRAHN (Chowdhury et al., 2009) TCP-CR (Yang et al., 2012)
9.	Security	Jamming Masquerading Authorization Integrity	(Fadlullah et al., 2013)
10.	Application	Spectrum agnostic Novel Application models QoS	(Yu et al., 2010)
11.	Implementation	Performance Accuracy	GNU Radio SCA

Cognitive Radio Sensor Network (CRSN)

Akan et al. (2009) presented the applications, design and architecture of CRSN. The nodes in sensor networks can be configured in ad hoc style, arranged in the form of clusters or hierarchical, and in some cases, few nodes can be mobile. CRSN are especially useful in places where unlicensed spectrum band (ISM) is not available or there are many users competing for the particular spectrum. This includes applications in the domain of ubiquitous computing and ambient intelligence

etc. CRSN additionally present various research issues i.e. power control, optimal node deployment, spectrum aware grouping of sensors, and analysis of optimal network coverage etc. (Akan et al., 2009; Nguyen & Hwang, 2012).

Cognitive Radio Vehicles (CRV)

Cognitive radio vehicles (CRV) are vehicular networks in which nodes are equipped with radios. They are useful for inter-vehicle communication, public safety and entertainment applications etc. Felice et al. (2012) have provided a summary of different research proposals on cognitive radio vehicles. CRV presents various new issues besides the issues associated with CRAHN. These are related to analyzing impact of high mobility vehicles on spectrum management, security and privacy issues, and development of new testbeds and toolkits for validation of research proposals. Representative solutions can be seen in (Felice et al., 2011; Rawat et al., 2013).

FUTURE RESEARCH DIRECTIONS

We believe that the development on ad hoc networks and related technologies will eventually lead towards pervasive computing environment, a vision perceived by Mark Weiser (1995). Pervasive computing enables the assimilation of information processing in human's life in seamless way. The emergence of 4G wireless networks and corresponding technological developments are already paving the way. Ad hoc networks will play a key role in realization of this vision by enabling spontaneous erection of networks. These networks have the essential ingredients of self-organization, self-management and self-healing required for operations in ubiquitous environment. However, a lot of research efforts are still required before the true realization of pervasive phenomenon. Following paragraphs outline opportunities for further research in ad hoc networks.

For antenna design, current research is being done on design of directional and low dimension antennas. However, the impact of antenna on upper layers has to be analyzed. Research efforts are also required on devising low dimension smart antennas for its usage in practical applications. MAC layer protocols also require several modifications. Future research should be focused on energy efficient, multi-channel and cooperative MAC protocols and medium access control in presence of directional antennas. Since, ad hoc networks are zero-configuration systems; various research efforts are to be done on devising secure

and scalable approaches for address assignments. The various routing protocols must consider the power constraints, the characteristics of channels along the path and other routing metrics to improve the routing operation. At the transport layer, not much work has been done. Protocols are required that utilize network events information from layers below for congestion and flow control. Security is also a challenging research area and research is required on addressing various associated challenges. Existing security solutions can't counter all types of security attacks as almost all of them are developed for a particular type of security threat. Generalized solutions are therefore required that can cope with any type of security threats. Ad hoc networks require cooperation enforcement mechanisms to enable distributed operation of various protocols. In this direction, various efforts are being put in to design error-free incentive based mechanisms to ensure node's cooperation. Data management has recently emerged as a hot topic of research. Several frameworks have recently emerged. However, most of the frameworks are still immature. Future work is required on designing a complete framework encompassing solutions for its various issues i.e. discovery, knowledge management, semantic data representation, and consistency management etc. Simulators and testbeds are vital tools for validation of any research proposal specifically for ad hoc networks. Different proposals have evolved in recent years. The problem with current simulator is that results obtained from a particular simulator are usually not similar when the same algorithm is executed on a different simulator. Research efforts are therefore required on accuracy and reliability of simulators. In addition, research should also be done on designing realistic mobility models that mimic real world situations. Future proposals should also exploit cross layer operations for energy efficient and QoS based MAC, routing and security solutions for ad hoc networks. Finally, the standardization of ad hoc network is an area that also requires attention.

Various models of ad hoc networks are emerging as discussed in previous sections. These new models provide various innovative applications however their potentials are still to be explored. At the same time, several additional challenges require attention from researcher's community. For example, ad hoc grid requires novel resource discovery and scheduling algorithms keeping into consideration dynamically changing QoS requirements of ad hoc environment. In addition, new models for security and economics are required to ensure cooperation among nodes. Similarly, ad hoc cloud requires new approaches for ensuring service provisioning, market oriented management of resources, data interoperability and privacy etc. Research efforts are required in improving data transmission, and addressing the issues arising in multi-channel MAC, routing, transport protocols and security of WMN. For CRAHN, the research is at initial stages and extensive efforts are required in development of techniques that can address spectrum management, mobility management, signaling, routing, transport protocol and security issues. Finally, new application models, implementation tools, analytical models and benchmarks have to be developed to assess the promise of all of these emerging models of ad hoc networks.

CONCLUSION

This chapter discussed the research trends and issues in wireless ad hoc networks. It discussed in detail ad hoc networks and its various types. An overview of various challenges and issues of ad hoc networks have been presented and analysis of research efforts to address these issues has also been provided. We also discussed some of the most recent models of ad hoc network. The last part of this chapter presented several directions for further research.

REFERENCES

Abbasi, A. Z. (2013). *Design of workflows for context-aware applications.* (Unpublished doctoral dissertation). National University of Computer and Emerging Sciences, Karachi, Pakistan.

Akan, O. B., Karli, O. B., & Ergul, O. (2009). Cognitive radio sensor networks. *IEEE Network*, 23(4), 34–40. doi:10.1109/MNET.2009.5191144

Akyildiz, I. F., Lee, W., & Chowdhury, K. R. (2009). CRAHNS: Cognitive radio ad hoc networks. *Ad Hoc Networks*, 7(5), 810–836. doi:10.1016/j.adhoc.2009.01.001

Akyildiz, I. F., Wang, X., & Wang, W. (2005). Wireless mesh networks: A survey. *Computer Networks*, 47, 445–487. doi:10.1016/j.comnet.2004.12.001

Akyildiz, I. F., & Wang, X. D. (2005). A survey on wireless mesh networks. *IEEE Radio Communications*, 43(9), 23–30. doi:10.1109/MCOM.2005.1509968

Al-Omari, S. A. K., & Sumari, P. (2010). An overview of mobile ad hoc networks for the existing protocols and applications. *International Journal on Applications of Graph Theory in Wireless Ad hoc Networks and Sensor Networks*, 2(1), 87-110.

Alazawi, Z., Altowaijri, S., & Mehmood, R. (2011). Intelligent disaster management system based on cloud-enabled vehicular networks. In *Proceedings of 11th International Conference on ITS Telecommunications (ITST)* (pp. 361-368). St. Petersburg, Russia: IEEE.

Amin, K., Laszewski, G. V., & Mikler, A. R. (2004). Toward an architecture for ad hoc grids. In *Proceedings of 12th International Conference on Advanced Computing and Communications* (pp. 1-5). Ahmedabad, India: IEEE.

Ancillotti, E., Bruno, R., Conti, M., & Pinizzott, A. (2009). Dynamic address autoconfiguration in hybrid ad hoc networks. *Pervasive and Mobile Computing, 5*(4), 300–317. doi:10.1016/j.pmcj.2008.09.008

Aqeel-ur-Rehman & Shaikh. Z. A. (2008). Towards design of context-aware sensor grid framework for agriculture. In *Proceedings of Fifth International Conference on Information Technology (ICIT)* (pp. 244-247). Rome, Italy: WASET.

Attar, A., Tang, H., Vasilakos, A. V., Yu, F. R., & Leung, V. C. M. (2012). A survey of security challenges in cognitive radio networks: Solutions and future research directions. *Proceedings of the IEEE, 100*(12), 3172–3186. doi:10.1109/JPROC.2012.2208211

Badarneh, O. S., & Kadoch, M. (2009). Multicast routing protocols in mobile ad hoc networks: A comparative survey and taxonomy. *EURASIP Journal on Wireless Communications and Networking*, (1): 1–52.

Baldini, G., Sturman, T., & Biswas, A. R. (2012). Security aspects in software defined radio and cognitive radio networks: A survey and a way ahead. *IEEE Communications. Surveys & Tutorials, 14*(2), 355–379. doi:10.1109/SURV.2011.032511.00097

Bazan, O., & Jaseemuddin, M. (2012). A survey on MAC protocols for wireless ad hoc networks with beamforming antennas. *IEEE Communications Surveys & Tutorials, 14*(2), 216–239. doi:10.1109/SURV.2011.041311.00099

Bhaskaran, R., & Madheswaran, R. (2010). Performance analysis of congestion control in mobile adhoc grid layer. *International Journal of Computers and Applications, 1*(20), 102–110.

Biswas, J., Barai, M., & Nandy, S. K. (2004). Efficient hybrid multicast routing protocol for ad-hoc wireless networks. In *Proceedings of 29th Annual IEEE International Conference on Local Computer Networks* (pp. 180-187). Tampa, FL: IEEE.

Boukerche, A., Turgut, B., Aydin, N., & Mohammad, Z., Ahmad, Bölöni, L., & Turgut, D. (2011). Routing protocols in ad hoc networks: A survey. *Computer Networks, 55*(13), 3032–3080. doi:10.1016/j.comnet.2011.05.010

Bragaa, R. B., Chavesb, I. A., Oliveiraa, C. T. D., Andradeb, R. M. C., Souzab, J. N. D., Martina, H., & Schulzec, B. (2013). RETENTION: A reactive trust-based mechanism to detect and punish malicious nodes in ad hoc grid environments. *Journal of Network and Computer Applications, 36*(1), 274–283. doi:10.1016/j.jnca.2012.06.002

Buyya, A., Yeo, C. S., & Venugopal, S. (2009). Market-oriented cloud computing: Vision, hype, and reality for delivering it service s as computing utilities. In *Proceedings of 9th IEEE/ACM International Symposium Cluster Computing and the Grid*. Shanghai, China: IEEE.

Cabric, D., Mishra, S. M., & Brodersen. (2004). Implementation issues in spectrum sensing for cognitive radios. In *Proceedings of Asilomar Conference on Signals, Systems & Computers* (pp. 772 - 776). Pacific Grove, CA: IEEE.

Cacciapuoti, A. S., Akyildiz, I. F., & Paura, L. (2012). Correlation-aware user selection for cooperative spectrum sensing in cognitive radio ad hoc networks. *IEEE Journal on Selected Areas in Communications, 30*(2), 297–306. doi:10.1109/JSAC.2012.120208

Camp, T., Boleng, J., & Davies, V. (2002). A survey of mobility models for ad hoc network research. *Wireless Communications and Mobile Computing, 2*(5), 483–502. doi:10.1002/wcm.72

Campista, M. E. M., Esposito, P. M., Moraes, I. M., Costa, L. H. M. K., Duarte, O. C. M. B., & Passos, D. G. et al. (2008). Routing metrics and protocols for wireless mesh networks. *IEEE Network, 22*(1), 6–12. doi:10.1109/MNET.2008.4435897

Canberk, B., & Oktug, S. (2012). A dynamic and weighted spectrum decision mechanism based on SNR tracking in CRAHNS. *Ad Hoc Networks, 10*(6), 752–759. doi:10.1016/j.adhoc.2011.02.006

Canourgues, L., Lephay, J., Soyer, L., & Beylot, A. L. (2006). Stamp: Shared-tree ad-hoc multicast protocol. In *Proceedings of IEEE Military Communications Conference* (pp. 1-7). Washington, DC: IEEE.

Cao, Y., & Xie, S. (2005). A position based beaconless routing algorithm for mobile ad hoc networks. In *Proceedings of International Conference on Communications, Circuits and Systems* (pp. 303-307). Hong Kong, China: IEEE.

Cesana, M., Cuomo, F., & Ekici, E. (2010). Routing in cognitive radio networks: Challenges and solutions. *Ad Hoc Networks, 9*(3), 228–248. doi:10.1016/j.adhoc.2010.06.009

Chakraborty, D., Joshi, A., Yesha, Y., & Finin, T. (2002). GSD: A novel group-based service discovery protocol for MANETs. In *Proceedings of 4th IEEE Conference on Mobile and Wireless Communications Networks (MWCN)* (pp. 140-144). Baltimore, MD: IEEE.

Chakraborty, D., Perich, F., Avancha, S., & Joshi, A. (2001). DReggie: Semantic service discovery for m-commerce applications. In *Proceedings of Workshop on Reliable and Secure Applications in Mobile Environment. In Conjunction with 20th Symposium on Reliable Distributed Systems (SRDS)* (pp. 1-6). New Orleans, LA: STDS.

Chandran, K., Raghunathan, S., Venkatesan, S., & Prakash, R. (1998). A feedback based scheme for improving TCP performance in ad-hoc wireless networks. In *Proceedings of International Conference on Distributed Computing Systems (ICDCS'98)* (pp. 34-39). Amsterdam: IEEE.

Changa, H., Kanb, H., & Hob, M. (2012). Adaptive TCP congestion control and routing schemes using cross-layer information for mobile ad hoc networks. *Computer Communications, 35*(4). PMID:22267882

Chen, K. C. (1994). Medium access protocols of wireless LANs for mobile computing. *IEEE Network, 8*(5), 50–63. doi:10.1109/65.313014

Chen, L., Jiang, C., & Li, J. (2008). VGITS: ITS based on intervehicle communication networks and grid technology. *Journal of Network and Computer Applications, 31*, 285–302. doi:10.1016/j.jnca.2006.11.002

Chen, Y., Fleury, E., & Razafindralambo, T. (2009). Scalable address allocation protocol for mobile ad hoc networks. In *Proceedings of 5th International Conference on Mobile Ad-Hoc and Sensor Networks* (pp. 41-48). WuYi Mountain, China: IEEE.

Chetan, S., Kumar, G., Dinesh, K., & Mathew, K. (n.d.). *Cloud computing for mobile world*. National Institute of Technology. Retrieved from http://chetan.ueuo.com/projects/CCMW.pdf

Chlamtac, I., Conti, M., & Liu, J. J. (2003). Mobile ad hoc networking: Imperatives and challenges. *Ad Hoc Networks, 1*(1), 13–64. doi:10.1016/S1570-8705(03)00013-1

Chowdhury, K., & Felice, M. (2009). Search: A routing protocol for mobile cognitive radio ad-hoc networks. *Computer Communications, 32*(18), 1–6. doi:10.1016/j.comcom.2009.06.011

Chowdhury, K. R., Felice, M. D., & Akyildiz. (2009). TP-CRAHN: A transport protocol for cognitive radio ad-hoc networks. In *Proceedings of IEEE Infocom* (pp. 2482-2490). IEEE.

Chowdhury, K. R., & Melodia, T. (2010). Platforms and testbeds for experimental evaluation of cognitive ad hoc networks. *IEEE Communications Magazine*, *48*(9), 96–104. doi:10.1109/MCOM.2010.5560593

Cooper, N., & Meghanathan, N. (2010). Impact of mobility models on multi-path routing in mobile ad hoc networks. *International Journal of Computer Networks and Communications*, *2*(1), 185–194.

Cormio, C., & Chowdhury, K. R. (2009). A survey on MAC protocols for cognitive radio networks. *Ad Hoc Networks*, *7*(7), 1315–1329. doi:10.1016/j.adhoc.2009.01.002

Dely, P., Castro, M., Soukhakian, S., Moldsvor, A., & Kassler, A. (2010). Practical considerations for channel assignment in wireless mesh networks. In *Proceedings of IEEE GLOBECOM Workshops* (pp. 763-767). Miami, FL: IEEE.

Dinh, H. T., Lee, C., Niyato, D., & Wang, P. (2011). *A survey of mobile cloud computing: Architecture, applications, and approaches*. Wireless Communications and Mobile Computing. doi:10.1002/wcm.1203

Dow, C. R., Lin, P. J., Chen, S. C., Lin, J. H., & Hwang, S. F. (2005). A study of recent research trends and experimental guidelines in mobile ad-hoc network. In *Proceedings of 19th International Conference on Advanced Information Networking and Applications* (pp. 72-77). Tamkang University, Taiwan: IEEE.

Eidenbenz, S., Resta, G., & Santi, P. (2008). The commit protocol for truthful and cost-efficient routing in ad hoc networks with selfish nodes. *IEEE Transactions on Mobile Computing*, *7*(1), 19–33. doi:10.1109/TMC.2007.1069

Erol-Kantarci, M., & Mouftah, H. T. (2011). Wireless sensor networks for cost-efficient residential energy management in the smart grid. *IEEE Transactions on Smart Grid*, *2*(2), 314–325. doi:10.1109/TSG.2011.2114678

Fadlullah, Z. M., Nishiyama, H., Nei Kato, T., & Fouda, M. M. (2013). Intrusion detection system (ids) for combating attacks against cognitive radio networks. *IEEE Network*, *27*(3), 51–56. doi:10.1109/MNET.2013.6523809

Felice, M. D., Doost-Mohammady, R., Chowdhury, K. R., & Bononi, L. (2012). Cognitive vehicular networks: Smart radios for smart vehicles. *IEEE Vehicular Technology Magazine*, *7*(2), 26–33. doi:10.1109/MVT.2012.2190177

Feng, W., Caoa, J., Zhang, C., Zhang, J., & Xin, Q. (2012). Coordination of multi-link spectrum handoff in multi-radio multi-hop cognitive networks. *Journal of Parallel and Distributed Computing*, *72*, 613–625. doi:10.1016/j.jpdc.2011.11.004

Fernando, N., Loke, S. W., & Rahayu, W. (2013). Mobile cloud computing: A survey. *Future Generation Computer Systems*, *29*, 84–106. doi:10.1016/j.future.2012.05.023

Fotino, M., & De Rango, F. (2011). Energy issues and energy aware routing in wireless ad-hoc networks. In X. Wang (Ed.), *Mobile ad-hoc networks: Protocol design* (pp. 156–167). University of Calabria. doi:10.5772/13309

Francis, B., Narasimhan, V., & Nayak, A. (2012). Enhancing TCP congestion control for improved performance in wireless networks. In *Ad-hoc, mobile, and wireless networks* (pp. 472–483). Springer. doi:10.1007/978-3-642-31638-8_36

Gangwar, S., Pal, S., & Kumar, K. (2012). Mobile ad hoc networks: A comparative study of QoS routing protocols. *International Journal of Computer Science Engineering and Technology*, *2*(1), 771–775.

Ghosekar, P., Katkar, G., & Ghorpade, P. (2010). Mobile ad hoc networking: Imperatives and challenges. *International Journal of Computer Application. Special Issue on Mobile Ad-Hoc Networks, 1*(3), 153–158.

Gomez, J., & Campbell, A. T. (2007). Variable-range transmission power control in wireless ad hoc networks. *IEEE Transactions on Mobile Computing, 6*(1), 87–99. doi:10.1109/TMC.2007.250673

Gonzalez, C. R. A. (2009). Open-source SCA-based core framework and rapid development tools enable software-defined radio education and research. *IEEE Communications Magazine, 47*(10), 48–55. doi:10.1109/MCOM.2009.5273808

Guan, Q., Yu, F., Jiang, S., & Wei, G. (2010). Prediction-based topology control and routing in cognitive radio mobile ad hoc networks. *IEEE Transactions on Vehicular Technology, 59*(9), 4443–4452. doi:10.1109/TVT.2010.2069105

Hanbali, A. A., Altman, E., & Nain, P. (2005). A survey of TCP over ad hoc networks. *IEEE Communications Surveys & Tutorials, 7*(3), 22–36. doi:10.1109/COMST.2005.1610548

Hassanein, H. L. J. (2006). Reliable energy aware routing in wireless sensor networks. In *Proceedings of Second IEEE Workshop on Dependability and Security in Sensor Networks and Systems* (pp. 54-56). Columbia, MD: IEEE.

He, W., Ge, Z., & Hu, Y. (2007). Optimizing UDP packet sizes in ad hoc networks. In *Proceedings of International Conference on Wireless Communications, Networking and Mobile Computing* (pp. 1617-1619). Shanghai, China: IEEE.

Heide, J., Fitzek, F. H. P., & Pedersen, M. V. (2012). Green mobile clouds: Network coding and user cooperation for improved energy efficiency. In *Proceedings of IEEE 1st International Conference on Cloud Networking (CLOUDNET)* (pp. 1-8). Paris, France: IEEE.

Heinzelman, W., Chandrakasan, A., & Balakrishnan, H. (2000). Energy-efficient communication protocol for wireless microsensor networks. In *Proceedings of 33rd Annual Hawaii International Conference on System Sciences* (pp. 1-10). Maui, HI: IEEE.

Hu, G., Tay, W. P., & Wen, Y. (2012). Cloud robotics: Architecture, challenges and applications. *IEEE Network, 26*(3), 21–28. doi:10.1109/MNET.2012.6201212

Hu, P., & Ibnkahla, M. (2012). A consensus-based protocol for spectrum sharing fairness in cognitive radio ad hoc and sensor networks. *International Journal of Distributed Sensor Networks, (1)*: 1–12. doi:10.1155/2012/370251

Huerta-Canepa, G., & Lee, D. (2012). A virtual cloud computing provider for mobile devices. In *Proceedings of 1st ACM Workshop on Mobile Cloud Computing & Services: Social Networks and Beyond* (pp. 1-5). San Francisco, CA: ACM.

Huraj, L., & Siládi, V. (2009). Authorization through trust chains in ad hoc grids. In *Proceedings of Euro American Conference on Telematics and Information Systems: New Opportunities to increase Digital Citizenship Prague*. ACM.

Hussain, R., Son, J., Eun, H., & Oh, S. K. H. (2012). Rethinking vehicular communications: Merging vanet with cloud computing. In *Proceedings of IEEE 4th International Conference on Cloud Computing Technology and Science (CloudCom)* (pp. 606-609). Taipei, Taiwan: IEEE.

Hussain, S. R., Saha, S., & Rahman, A. (2010). SAAMAN: Scalable address autoconfiguration in mobile ad hoc networks. *Journal of Network and Systems Management, 19*(3), 394–426. doi:10.1007/s10922-010-9187-4

IBM. (2000). *Salutation service discovery in pervasive computing environments*. Retrieved from http://www-3.ibm.com/pvc/tech/salutation.shtml

IETF. (2012). *IETF MANET working group*. Retrieved on August 2013, from http://www.ietf.org/html.charters/manet-charter.html

Inn, E. R., & Winston, K. G. S. (2006). Distributed steiner-like multicast path setup for mesh-based multicast routing in ad-hoc networks. In *Proceedings of IEEE International Conference on Sensor Networks, Ubiquitous and Trustworthy Computing* (pp. 192-197). Taichung, Taiwan: IEEE.

Ishizu, K. (2006). Adaptive wireless-network testbed for CR technology. In *Proceedings of 1st International Workshop on Wireless Network Testbeds, Experimental Evaluation & Characterization* (pp. 18-25). Los Angeles, CA: ACM.

Islam, N., & Aqeel-ur-Rehman. (2013). A comparative study of major service providers for cloud computing. In *Proceedings of 1st International Conference on Information and Communication Technology Trends (ICICTT)* (pp. 228-232). Federal Urdu University.

Islam, N., Shaikh, N. A., Ali, G., & Shaikh, Z. A., & Aqeel-ur-Rehman. (2010). A network layer service discovery approach for mobile ad hoc network using association rules mining. *Australian Journal of Basic and Applied Sciences, 4*(6), 1305–1315.

Islam, N., & Shaikh, Z. A. (2011). A survey of data management issues & frameworks for mobile ad hoc networks. In *Proceedings of 4th International Conference on Information and Communication Technologies* (pp. 1-5). Karachi: IEEE.

Islam, N., & Shaikh, Z. A. (2012). Towards a robust and scalable semantic service discovery scheme for mobile ad hoc network. *Pakistan Journal of Engineering and Applied Sciences, 10*, 68–88.

Islam, N., & Shaikh, Z.A., Aqeel-ur-Rehman, Siddiqui, M.S. (2013). HANDY: A hybrid association rules mining approach for network layer discovery of services wireless networks. *Wireless Networks*. doi:10.1007/s11276-013-0571-3

Islam, N., & Shaikh, Z. A. (2013). Security issues in mobile ad hoc network. In S. Khan, & A. K. Pathan (Eds.), *Wireless networks and security* (pp. 49–80). Springer. doi:10.1007/978-3-642-36169-2_2

Islam, N., Shaikh, Z. A., & Talpur, S. (2008). Towards a grid based approach to traffic routing in VANET. In *Proceedings of E-INDUS*. Karachi, Pakistan: IIEE.

Iyer, G. N., & Durga, S. (2013). State of the art security mechanisms for mobile cloud environments. *International Journal of Advanced Research in Computer Science and Software Engineering, 3*(4), 470–476.

Jangra, A., & Goel, N., Priyanka, & Bhatia, K. K. (2010). IEEE WLANs standards for mobile ad-hoc networks (MANETs), performance analysis. *Global Journal of Computer Science and Technology, 10*(14), 42–47.

Jayapal, C., & Vembu, S. (2011). Adaptive service discovery protocol for mobile ad hoc networks. *European Journal of Scientific Research, 49*(1), 6–17.

Jetcheva, J. G., & Johnson, D. B. (2001). Adaptive demand-driven multicast routing in multi-hop wireless ad-hoc networks. In *Proceedings of ACM International Symposium on Mobile Ad-hoc Networking and Computing* (pp. 33-44). Long Beach, CA: ACM.

Jun, J., & Sichitiu, M. L. (2003). The nominal capacity of wireless mesh networks. *IEEE Wireless Communications, 10*(5), 8–14. doi:10.1109/MWC.2003.1241089

Junhai, L., Liu, X., & Danxia, Y. (2008). Research on multicast routing protocols for mobile ad-hoc networks. *Computer Networks, 52*(5), 988–997. doi:10.1016/j.comnet.2007.11.016

Kamruzzaman, S. M., Hamdi, M. A., & Abdullah-Al-Wadud, M. (2010). An energy-efficient MAC protocol for QoS provisioning in cognitive radio ad hoc networks. *Radio Engineering, 19*(4), 112–119.

Karnadi, F. K., Mo, Z. H., & Lan, K. (2007). Rapid generation of realistic mobility models for VANET. In *Proceedings of IEEE Wireless Communications and Networking Conference* (pp. 2506-2511). Kowloon: IEEE.

Khorashadi, B. (2009). *Enabling traffic control and data dissemination applications with vGrid - A vehicular ad hoc distributed computing framework*. (Unpublished doctoral dissertation). University of California, Berkeley, CA.

Kiess, W., & Mauve, M. (2007). A survey on real-world implementations of mobile ad-hoc networks. *Ad Hoc Networks, 5*(3), 324–339. doi:10.1016/j.adhoc.2005.12.003

Kirby, G., Dearle, A., Macdonald, A., & Fernandes, A. (2010). An approach to ad hoc cloud computing. *CoRR, 1*, 1–6.

Kovachev, D., Cao, Y., & Klamma, R. (2011). Mobile cloud computing: A comparison of application models. *CoRR, 1*, 1–8.

Krishnan, S., Wagstrom, P., & Laszewski, G. (2002). *GSFL: A workflow framework for grid services*. The Globus Alliance. Retrieved from www.globus.org/cog/papers/gsfl-paper.pdf

Kumar, G. V., Reddyr, Y. V., & Nagendra, M. (2010). Current research work on routing protocols for MANET: A literature survey. *International Journal on Computer Science and Engineering, 2*, 706–713.

Kumar, S., Raghavan, V. S., & Deng, J. (2006). Medium access control protocols for ad hoc wireless networks: A survey. *Ad Hoc Networks, 4*(3), 326–358. doi:10.1016/j.adhoc.2004.10.001

Kurosawa, S., Nakayama, H., Kato, N., Jamalipour, A., & Nemoto, Y. (2007). Detecting blackhole attack on AODV-based mobile ad hoc networks by dynamic learning method. *International Journal of Network Security, 5*(3), 38–346.

Kwon, H., Lee, H., & Cioffi, J. M. (2010). Cooperative strategy by Stackelberg games under energy constraint in multi-hop relay networks. In *Proceedings of 28th IEEE Conference on Global Telecommunications* (pp. 3431-3436). Miami, FL: IEEE.

Kwon, S., & Shroff, N. B. (2006). Energy-efficient interference-based routing for multi-hop wireless networks. [Barcelona, Spain: IEEE.]. *Proceedings - IEEE INFOCOM, 2006*, 1–12.

Lam, J. H., Busan, S. K., Lee, S.-G., & Tan, W. K. (2012). Multi-channel wireless mesh networks test-bed with embedded systems. In *Proceedings of 26th International Conference on Advanced Information Networking and Applications Workshops (WAINA)* (pp. 533-537). Fukuoka, Japan: IEEE.

Law, K. L. E. (2009). Transport protocols for wireless mesh networks. In S. Misra, S. C. Misra, & I. Woungang (Eds.), *Guide to wireless mesh networks* (pp. 255–275). Springer.

Lee, W., & Akyildiz, I. F. (2011). A spectrum decision framework for cognitive radio networks. *IEEE Transactions on Mobile Computing, 10*(2), 161–174. doi:10.1109/TMC.2010.147

Lei, H., Gao, C., Guo, Y., Ren, Z., & Huang, J. (2012). Wait-time-based multi-channel MAC protocol for wireless mesh networks. *Journal of Networks, 7*(8), 1208–1213. doi:10.4304/jnw.7.8.1208-1213

Li, C., & Li, L. (2012a). Design and implementation of economics-based resource management system in ad hoc grid. *Advances in Engineering Software, 45*(1), 281–291. doi:10.1016/j.advengsoft.2011.10.003

Li, C., & Li, L. (2012b). A resource selection scheme for QoS satisfaction and load balancing in ad hoc grid. *The Journal of Supercomputing, 59*(1), 499–525. doi:10.1007/s11227-010-0450-y

Li, C., Li, L., & Luo, Y. (2013). Agent based sensors resource allocation in sensor grid. *Applied Intelligence, 39*(1), 121–131. doi:10.1007/s10489-012-0397-1

Li, Z., Sun, L., & Ifeachor, E. C. (2005). Challenges of mobile ad-hoc grids and their applications in e-healthcare. In *Proceedings of 2nd International Conference on Computational Intelligence in Medicine and Healthcare (CIMED2005)*. Lisbon, Portugal: CIMED.

Li Longjiang, C. Y., & Xiaoming, X. (2009). Cluster-based autoconfiguration for mobile ad hoc networks. *Wireless Personal Communications, 49*(4), 561–573. doi:10.1007/s11277-008-9577-z

Liang, H., Huang, D., & Peng, D. (2012). On economic mobile cloud computing model. In M. Gris, & G. Yang (Eds.), *Mobile computing, applications, and services* (pp. 329–341). Springer. doi:10.1007/978-3-642-29336-8_22

Lim, S., Yu, C., & Das, C. R. (2009). RandomCast: An energy efficient communication scheme for mobile ad hoc networks. *IEEE Transactions on Mobile Computing, 8*(8), 1039–1051. doi:10.1109/TMC.2008.178

Liu, F., Guo, S., & Sun, Y. (2013). Primary user signal detection based on virtual multiple antennas for cognitive radio networks. *Progress in Electromagnetics Research, 42*, 213–227.

Liu, Y., Guo, L., Ma, H., & Jiang, T. (2009). Energy efficient on–demand multipath routing protocol for multi-hop ad hoc networks. In *Proceedings of 10th International Symposium on Spread Spectrum Techniques and Applications*, (pp. 592-597). Bologna, Italy: IEEE.

Lou, W., Liu, W., Zhang, Y., & Fang, Y. (2009). Spread: Improving network security by multipath routing in mobile ad hoc networks. *Wireless Networks, 15*(3), 279–294. doi:10.1007/s11276-007-0039-4

Lu, X., Towsley, D., Lio, P., & Xiong, Z. (2012). An adaptive directional MAC protocol for ad hoc networks using directional antennas. *Science China Information Sciences, 55*(6), 1360–1371. doi:10.1007/s11432-012-4550-6

Mallah, R. A., & Quintero. (2009). A light-weight service discovery protocol for ad hoc networks. *Journal of Computer Science, 5*(4), 330–337. doi:10.3844/jcssp.2009.330.337

Mao, S., Bushmitch, D., Narayanan, S., & Panwar, S. S. (2006). MRTP: A multiflow real-time transport protocol for ad hoc networks. *IEEE Transactions on Multimedia, 8*(2), 356–369. doi:10.1109/TMM.2005.864347

Marinelli, E. E. (2009). *Hyrax: Cloud computing on mobile devices using MapReduce*. (Unpublished Master's thesis). School of Computer Science, Carnegie Mellon University, Pittsburgh, PA.

Marinescuand, D. C., Marinescuand, G. M., & Ji, Y. Boloni, & Siegel, H. J. (2003). Ad hoc grids: Communication and computing in a power constrained environment. In *Proceedings of Workshop on Energy-Efficient Wireless Communications and Networks (EWCN)* (pp. 113-122). Phoenix, AZ: IEEE.

Marinho, J., & Monteiro, E. (2012). Cognitive radio: Survey on communication protocols, spectrum decision issues, and future research directions. *Wireless Networks, 18*(2), 147–164. doi:10.1007/s11276-011-0392-1

Martignon, F., Paris, S., & Capone, A. (2011). DSA-mesh: A distributed security architecture for wireless mesh networks. *Security and Communication Networks, 4*(3), 242–256. doi:10.1002/sec.181

Martin, L., & Demeure, I. (2008). Using structured and segmented data for improving data sharing on MANETs. In *Proceedings of IEEE 19th International Symposium on Personal, Indoor and Mobile Radio Communications (PIMRC 2008)*. Cannes, France: IEEE.

Microsoft. (2000). *Understanding universal plug and play* (White Paper). Retrieved from www.upnp.org/download/UPNP_understandingUPNP.doc

Miettinen, A. P., & Nurminen, J. K. (2010). Energy efficiency of mobile clients in cloud computing. In *Proceedings of 2nd USENIX Conference on Hot Topics in Cloud Computing*. Boston, MA: USENIX Association.

Miridakis, N. I., Giotsas, V., Vergados, D. D., & Douligeris, C. (2009). A novel power-efficient middleware scheme for sensor grid applications. In N. Bartolini, S. Nikoletseas, P. Sinha, V. Cardellini, & A. Mahanti (Eds.), *Quality of service in heterogeneous networks* (pp. 476–492). Springer. doi:10.1007/978-3-642-10625-5_30

Neel, J. (2006). *Analysis and design of cognitive radio networks and distributed radio resource management algorithms*. (Unpublished doctoral dissertation). Virginia Polytechnic Institute and State University, Blacksburg, VA.

Nesargi, S., & Prakash, R. (2002). MANETConf: Configuration of hosts in a mobile ad hoc network. In *Proceedings of IEEE INFOCOM* (pp. 1059–1068). New York: IEEE. doi:10.1109/INFCOM.2002.1019354

Nguyen, K., & Hwang, W. (2012). An efficient power control scheme for spectrum mobility management in cognitive radio sensor networks. In J. J. J. H. Park, Y. Jeong, S. O. Park, & H. Chen (Eds.), *Embedded and multimedia computing technology and service* (pp. 667–676). Springer. doi:10.1007/978-94-007-5076-0_81

Nidd, M. (2001). Service discovery in DeapSpace. *IEEE Personal Communications*, *8*(4), 39–45. doi:10.1109/98.944002

Olariu, S., & Yan, T. H. G. (2012). The next paradigm shift: From vehicular networks to vehicular clouds. In S. Basagni, M. Conti, S. Giordano, & I. Stojmenovic (Eds.), *Mobile ad hoc networking: The cutting edge directions* (pp. 645–700). Hoboken, NJ: Wiley.

Open Geospatial Consortium. (2013). *Sensor model language (sensorml)*. Retrieved on August 2013, from http://vast.nsstc.uah.edu/SensorML

Pariselvam, S., & Parvathi. (2012). Swarm intelligence based service discovery architecture for mobile ad hoc networks. *European Journal of Scientific Research*, *74*(2), 205–216.

Pathak, P. H., & Dutta, R. (2011). A survey of network design problems and joint design approaches in wireless mesh networks. *IEEE Communications Surveys & Tutorials*, *13*(3), 396–428. doi:10.1109/SURV.2011.060710.00062

Perich, F., Joshi, A., & Chirkova, R. (2006). Data management for mobile ad-hoc networks. In W. Kou, & Y. Yesha (Eds.), *Enabling technologies for wireless e-business* (pp. 132–176). Springer. doi:10.1007/978-3-540-30637-5_7

Perich, F., Joshi, A., Finin, T., & Yesha, Y. (2004). On data management in pervasive computing environments. *IEEE Transactions on Knowledge and Data Engineering*, 621–634. doi:10.1109/TKDE.2004.1277823

Popa, D., Cremene, M., Borda, M., & Boudaoud, K. (2013). A security framework for mobile cloud applications. In *Proceedings of 11th Roedunet International Conference (RoEduNet)* (pp. 1-4). Sinaia: IEEE.

Qabajeh, L. K., Kiah, L. M., & Qabajeh, M. M. (2009). A qualitative comparison of position-based routing protocols for ad-hoc networks. *International Journal of Computer Science and Network Security*, *9*(2), 131–140.

Qin, F., & Liu, Y. (2009). Multipath based QoS routing in MANET. *Journal of Networks*, *4*(8), 771–778. doi:10.4304/jnw.4.8.771-778

Raj, P. N., & Swadas, P. B. (2009). DPRAODV: A dynamic learning system against blackhole attack in AODV based MANET. *International Journal of Computer Science Issues*, *2*(3), 1126–1131.

Ramanathan, R. (2001). On the performance of ad hoc networks with beamforming antennas. In *Proceedings of 2nd ACM International Symposium on Mobile Ad Hoc Networking & Computing* (pp. 95-105). Long Beach, CA: ACM.

Ramanathan, R., Redi, J., Santivanez, C., Wiggins, D., & Polit, S. (2005). Ad hoc networking with directional antennas: A complete system solution. *IEEE Journal on Selected Areas in Communications*, *23*(3), 496–506. doi:10.1109/JSAC.2004.842556

Rangwala, S., Jindal, A., Jang, K., Psounis, K., & Govindan, R. (2008). Understanding congestion control in multi-hop wireless mesh networks. In *Proceedings of MobiCom* (pp. 291–302). San Francisco, CA: ACM. doi:10.1145/1409944.1409978

Ratsimor, O., Chakraborty, D., Joshi, A., & Finin, T. (2002). Allia: Alliance-based service discovery for ad-hoc environments. In *Proceedings of International Workshop on Mobile Commerce* (pp. 1-9). Atlanta, GA: ACM.

Rawat, D. B., Zhao, Y., Yan, G., & Song, M. (2013). Crave: Cognitive radio enabled vehicular communications in heterogeneous networks. In *Proceedings of Radio and Wireless Symposium (RWS)* (pp. 190-192). Austin, TX: IEEE.

Rohner, C., Nordström, E., Gunningberg, P., & Tschudin, C. (2005). Interactions between TCP, UDP and routing protocols in wireless multi-hop ad hoc networks. In *Proceedings of 1st IEEE ICPS Workshop on Multi-Hop Ad Hoc Networks: From Theory to Reality* (pp. 1-8). Santorini, Greece: IEEE.

Royer, E. M., & Toh, C. K. (1999). A review of current routing protocols for ad hoc mobile wireless networks. *IEEE Personal Communications*, *6*(2), 46–55. doi:10.1109/98.760423

Sanaei, Z., Abolfazli, S., Gani, A., & Buyya, R. (2013). Heterogeneity in mobile cloud computing: Taxonomy and open challenges. *IEEE Communications Surveys & Tutorials*, (99), 1-24.

Sasirekha, G., & Bapat, J. (2012). Evolutionary game theory-based collaborative sensing model in emergency CRAHNS. *Journal of Electrical and Computer Engineering*, (1): 1–12.

Shah, C. S., Bashir, A. L., Chauhdary, S. H., Jiehui, C., & Park, M.-S. (2009). Mobile ad hoc computational grid for low constraint devices. In *Proceedings of International Conference on Future Computer and Communication* (pp. 416-420). Kuala Lumpur, Malaysia: IEEE.

Shah, R. C., & Rabaey, J. M. (2002). Energy aware routing for low energy ad hoc sensor networks. In *Proceedings of IEEE Wireless Communications and Networking Conference* (pp. 350-355). IEEE.

Shahram, A. H. Krishnamachari, Bar, & Richmond. (2006). Data management techniques for continuous media in ad-hoc networks of wireless devices. In B. Furht (Ed.), Collection of encyclopedia of multimedia (pp. 144-149). Springer.

Shin, K., Yun, S., & Cho, D. (2011). Multi-channel MAC protocol for QoS support in ad-hoc network In *Proceedings of IEEE Consumer Communications and Networking Conference* (pp. 975-976). Las Vegas, NV: IEEE.

Sichitiu, M. L. (2005). Wireless mesh networks: Opportunities and challenges. In *Proceedings of Wireless World Congress* (pp. 1-6). Palo Alto, CA: IEEE.

Singh, A., & Chakrabarti, P. (2013). Ant based resource discovery and mobility aware trust management for mobile grid systems. In *Proceedings of IEEE 3rd International Advance Computing Conference (IACC)* (pp. 637-644). Ghaziabad, India: IEEE.

Singh, S., Dutta, S. C., & Singh, D. K. (2012). A study on recent research trends in MANET. *International Journal of Research and Reviews in Computer Science*, 3(3), 1654.

Singh, V. P., & Kumar, K. (2010). Literature survey on power control algorithms for mobile ad-hoc network. *Wireless Personal Communications*, 60(4), 679–685. doi:10.1007/s11277-010-9967-x

Sistla, A. P., Wolfson, O., & Xu, B. (2005). Opportunistic data dissemination in mobile peer-to-peer networks. In C. B. Medeiros, M. J. Egenhofer, & E. Bertino (Eds.), *Advances in spatial and temporal databases* (pp. 923–923). Springer. doi:10.1007/11535331_20

Sivakumar, R., Sinha, P., & Bharghavan, V. (1999). CEDAR: A core-extraction distributed ad hoc routing algorithm. *IEEE Journal on Selected Areas in Communications*, 17(8), 1454–1465. doi:10.1109/49.779926

Soon, Y. O., Park, J. S., & Gerla, M. (2008). E-ODMRP: Enhanced ODMRP with motion adaptive refresh. *Journal of Parallel and Distributed Computing*, 68(8), 130–134.

Sun Microsystems. (2001). *JINI technology core platform specification, version 1.2*. Retrieved from www-csag.ucsd.edu/teaching/cse291s03/Readings/core1_2.pdf

Sundaresan, K., Anantharaman, V., & Hsieh, H. (2005). ATP: A reliable transport protocol for ad hoc networks. *IEEE Transactions on Mobile Computing*, 4(6), 588–603. doi:10.1109/TMC.2005.81

Svecs, I., Sarkar, T., Basu, S., & Wong, J. S. (2010). XIDR: A dynamic framework utilizing cross-layer intrusion detection for effective response deployment. In *Proceedings of 34th Annual IEEE Computer Software and Applications Conference Workshops* (pp. 287-292). Seoul, Korea: IEEE.

Takagi, H., & Kleinrock, L. (1983). Optimal transmission ranges for randomly distributed packet radio terminals. *IEEE Transactions on Communications*, 32(3), 246–257. doi:10.1109/TCOM.1984.1096061

Talebifard, P., & Leung, V. C. M. (2013). Towards a content-centric approach to crowd-sensing in vehicular cloud. *Journal of Systems Architecture*. Retrieved from http://www.sciencedirect.com/science/article/pii/S1383762113001501

Talucc, F. T., & Bari, D. P. D. (1997). MACA-BI (MACA by invitation), a wireless MAC protocol for high speed ad hoc networking. In *Proceedings of International Conference on Universal Personal Communications* (pp. 913-917). San Diego, CA: IEEE.

Tang, F., Guo, M., Dong, M., Li, M., & Guan, H. (2008). Towards context-aware workflow management for ubiquitous computing. In *Proceedings of International Conference on Embedded Software and Systems* (pp. 221-228). Las Vegas, NV: IEEE.

Tariquea, M., Vitae, A., & Tepe, K. E. (2009). Minimum energy hierarchical dynamic source routing for mobile adhocnetworks. *Ad Hoc Networks*, 7(6), 1125–1135. doi:10.1016/j.adhoc.2008.10.002

Terracina, A., Beco, S., Kirkham, T., Gallop, J., Johnson, I., Randal, D. M., & Ritchie, B. (2006). Orchestration and workflow in a mobile grid environment. In *Proceedings of Fifth International Conference on Grid and Cooperative Computing Workshops* (pp. 251-258). Hunan, China: IEEE.

Tham, C., & Buyya, R. (2005). Sensorgrid: Integrating sensor networks and grid computing. *CSI Communications, 29*(1), 24–29.

Toh, C. K., Mahonen, P., & Uusitalo, M. (2005). Standardization efforts & future research issues for wireless sensors & mobile ad hoc networks. *IEICE Transactions on Communications, 88*(9), 3500. doi:10.1093/ietcom/e88-b.9.3500

Torkestania, J. A. (2013). A new distributed job scheduling algorithm for grid systems. *Cybernetics and Systems: An International Journal, 44*(1), 77–93. doi:10.1080/01969722.2012.744556

Vazhkudai, S., & Laszewski, G. V. (2001). A greedy grid - The grid economic engine directive. In *Proceedings of First International Workshop on Internet Computing and E-Commerce* (pp. 1806-1815). San Francisco, CA: IEEE.

Wang, F., & Zhang, Y. (2002). Improving TCP performance over mobile ad hoc networks with out-of-order detection and response. In *Proceedings of MOBIHOC* (pp. 217-225). Lausanne, Switzerland: ACM.

Wang, J., Cho, J., Lee, S., & Ma, T. (2011). Real time services for future cloud computing enabled vehicle networks. In *Proceedings of International Conference on Wireless Communications and Signal Processing (WCSP)* (pp. 1-5). Nanjing, China: IEEE.

Wang, J., Zhai, H., Li, P., Fang, Y., & Wu, D. (2008). Directional medium access control for ad hoc networks. *Wireless Networks, 15*(8), 1059–1073. doi:10.1007/s11276-008-0102-9

Weiser, M. (1995). The computer for the 21st century. *Scientific American, 272*(3), 78–89.

White, T. (2012). *Hadoop: The definitive guide.* Sebastopol, CA: O'Reilly Media.

Wu, B., Chen, J., Wu, J., & Cardei, M. (2006). A survey of attacks and countermeasures in mobile ad hoc networks. In Y. Xiao, X. Shen, & D.-Z. Du (Eds.), *Wireless mobile network security* (pp. 103–135). Springer.

Xu, B., Ouksel, A., & Wolfson, O. (2004). Opportunistic resource exchange in inter-vehicle ad-hoc networks. In *Proceedings of IEEE International Conference on Mobile Data Management* (pp. 4-12). Brisbane, Australia: IEEE.

Xu, B., & Wolfson, O. (2005). Data management in mobile peer-to-peer networks. In W. S. Ng, B.-C. Ooi, A. M. Ouksel, & C. Sartori (Eds.), *Databases, information systems, and peer-to-peer computing* (pp. 1–15). Springer. doi:10.1007/978-3-540-31838-5_1

Xu, Y. Q., & Yin, M. (2013). A mobility-aware task scheduling model in mobile grid. *Applied Mechanics and Materials, 336-338*, 1786–1791. doi:10.4028/www.scientific.net/AMM.336-338.1786

Yan, G., Rawat, D. B., & Bista, B. B. (2012). Towards secure vehicular clouds. In *Proceedings of Sixth International Conference on Complex, Intelligent and Software Intensive Systems (CISIS)* (pp. 370-375). Palermo, CA: IEEE.

Yang, B., & Zhang, Z. (2011). An improved UDP-Lite protocol for 3D model transmission over wireless network. In *Proceedings of International Conference on Informatics, Cybernetics, and Computer Engineering (ICCE2011)* (pp. 351-357). Melbourne, Australia: Springer.

Yang, H., Cho, S., & Park, C. Y. (2012). Improving performance of remote TCP in cognitive radio networks. *Transactions on Internet and Information Systems (Seoul)*, 6(9), 2323–2338.

Yu, F., Sun, B., Krishnamurthy, V., & Ali, S. (2010). Application layer QoS optimization for multimedia transmission over cognitive radio networks. *Wireless Networks*, 17(2), 371–383. doi:10.1007/s11276-010-0285-8

Yucek, T., & Arslan, H. (2009). A survey of spectrum sensing algorithms for cognitive radio applications. *IEEE Communications Surveys & Tutorials*, 11(1), 116–130. doi:10.1109/SURV.2009.090109

Zayani, M., & Zeghlache, D. (2009). FESCIM: Fair, efficient, and secure cooperation incentive mechanism for hybrid ad hoc networks. *Transactions on Mobile Computing*, 11(5), 753–766.

Zayani, M., & Zeghlache, D. (2012). Cooperation enforcement for packet forwarding optimization in multi-hop ad-hoc networks. In *Proceedings of IEEE Wireless Communications and Networking Conference* (pp. 3150-3163). Paris, France: IEEE Press.

Zdarsky, F. A., Robitzsch, S., & Banchs, A. (2010). Security analysis of wireless mesh backhauls for mobile networks. *Journal of Network and Computer Applications*, 34(2), 432–442. doi:10.1016/j.jnca.2010.03.029

Zhang, P., & Yan, Z. (2011). A QoS-aware system for mobile cloud computing. In *Proceedings of IEEE International Conference on Cloud Computing and Intelligence Systems (CCIS)* (pp. 518-522). Beijing, China: IEEE.

Zhang, X., & Jacob, L. (2004). MZRP: An extension of the zone routing protocol for multicasting in MANETs. *Journal of Information Science and Engineering*, 20(3), 535–551.

Zhang, X., Schiffman, J., & Gibbs, S. (2009). Securing elastic applications on mobile devices for cloud computing. In *Proceedings of ACM Workshop on Cloud Computing Security* (pp. 127-134). Chicago: ACM.

Zhang, Y., & Fang. (2006). ARSA: An attack-resilient security architecture for multihop wireless mesh network. *IEEE Journal on Selected Areas in Communications*, 24(10), 1916–1928. doi:10.1109/JSAC.2006.877223

Zhang, Y., Lou, W., & Fang. (2007). A secure incentive protocol for mobile ad hoc networks. *ACM Wireless Networks*, 13(5), 569-582.

Zheng, Y. (2013). Public key cryptography for mobile cloud. In C. Boyd, & L. Simpson (Eds.), *Information security and privacy*. Berlin: Springer. doi:10.1007/978-3-642-39059-3_30

Zhon, J., Geng, Weng, & Li. (2012). A cross-layers service discovery protocol for MANET. *Journal of Computer Information Systems*, 8(12), 5085–6092.

Zhong, T., Xu, B., & Wolfson, O. (2008). Disseminating real-time traffic information in vehicular ad-hoc networks. In *Proceedings of Intelligent Vehicles Symposium* (pp. 1056-1067). Eindhoven, The Netherlands: IEEE.

Zhou, J., & Mitchell, K. (2009). A scalable delay based analytical framework for CSMA/CA wireless mesh networks. *Computer Networks*.

ADDITIONAL READING

Akyildiz, I. F., Lee, W., & Chowdhury, K. R. (2009). CRAHNS: Cognitive radio ad hoc networks. *Ad Hoc Networks*, 7(5), 810–836. doi:10.1016/j.adhoc.2009.01.001

Amin, K., Laszewski, G. V., & Mikler, A. R. (2004). Toward an architecture for ad hoc grids. In *12th International Conference on Advanced Computing and Communications* (pp. 1-5). Ahmedabad, India.

Badarneh, O. S., & Kadoch, M. (2009). Multicast routing protocols in mobile ad hoc networks: A comparative survey and taxonomy. *EURASIP Journal on Wireless Communications and Networking*, (1): 1–52.

Bazan, O., & Jaseemuddin, M. (2012). A survey on MAC protocols for wireless ad hoc networks with beamforming antennas. *IEEE Communications Surveys & Tutorials*, *14*(2), 216–239. doi:10.1109/SURV.2011.041311.00099

Campista, M. E. M., Esposito, P. M., Moraes, I. M., Costa, L. H. M. K., Duarte, O. C. M. B., & Passos, D. G. et al. (2008). Routing metrics and protocols for wireless mesh networks. *IEEE Network*, *22*(1), 6–12. doi:10.1109/MNET.2008.4435897

Chlamtac, I., Conti, M., & Liu, J. J.-N. (2003). Mobile ad hoc networking: Imperatives and challenges. *Ad Hoc Networks*, *1*(1), 13–64. doi:10.1016/S1570-8705(03)00013-1

Chowdhury, K. R., & Melodia, T. (2010). Platforms and testbeds for experimental evaluation of cognitive ad hoc networks. *IEEE Communications Magazine*, *48*(9), 96–104. doi:10.1109/MCOM.2010.5560593

Cordeiro, C. D. M., & Agrawal, D. P. (2011). *Ad Hoc and Sensor Networks: Theory and Applications*. World Scientific Publishing. doi:10.1142/8066

Junhai, L., Liu, X., & Danxia, Y. (2008). Research on multicast routing protocols for mobile ad-hoc networks. *Computer Networks*, *52*(5), 988–997. doi:10.1016/j.comnet.2007.11.016

Khorashadi, B. (2009). *Enabling traffic control and data dissemination applications with vGrid - a vehicular ad hoc distributed computing framework*. (Unpublished doctoral dissertation). University of California.

Kumar, S. A. (2010). Classification and review of security schemes in mobile computing. *Wireless Sensor Network*, *2*(6), 419–440. doi:10.4236/wsn.2010.26054

Marinho, J., & Monteiro, E. (2012). Cognitive radio: Survey on communication protocols, spectrum decision issues, and future research directions. *Wireless Networks*, *18*(2), 147–164. doi:10.1007/s11276-011-0392-1

Olariu, S., & Yan, T. H. G. (2012). The next paradigm shift: From vehicular networks to vehicular clouds. In S. Basagni, M. Conti, S. Giordano, & I. Stojmenovic (Eds.), *Mobile ad hoc networking: The cutting edge directions* (pp. 645–700). Wiley.

Pathak, P. H., & Dutta, R. (2011). A survey of network design problems and joint design approaches in wireless mesh networks. *IEEE Communications Surveys & Tutorials*, *13*(3), 396–428. doi:10.1109/SURV.2011.060710.00062

Qabajeh, L. K., Kiah, L. M., & Qabajeh, M. M. (2009). A qualitative comparison of position-based routing protocols for ad-hoc networks. *IJCSNS International Journal of Computer Science and Network Security*, *9*(2), 131–140.

Royer, E. M., & Toh, C. K. (1999). A review of current routing protocols for ad hoc mobile wireless networks. *IEEE personal communications, 6*(2), 46-55.

Singh, V. P., & Kumar, K. (2010). Literature survey on power control algorithms for mobile ad-hoc network. *Wireless Personal Communications, 60*(4), 679–685. doi:10.1007/s11277-010-9967-x

Toh, C. K., Mahonen, P., & Uusitalo, M. (2005). Standardization efforts & future research issues for wireless sensors & mobile ad hoc networks. *IEICE Transactions on Communications, 88*(9), 3500. doi:10.1093/ietcom/e88-b.9.3500

Wu, B., Chen, J., Wu, J., & Cardei, M. (2006). A survey on attacks and countermeasures in mobile ad hoc networks. In Y. Xiao, X. Shen, & D.-Z. Du (Eds.), *Wireless mobile network security* (pp. 103–135). Springer.

Yu, F. R. (2011). *Cognitive Radio Mobile Ad Hoc Networks.* Springer. doi:10.1007/978-1-4419-6172-3

Zayani, M., & Zeghlache, D. (2009). FESCIM: Fair, efficient, and secure cooperation incentive mechanism for hybrid ad hoc networks. *Transactions on Mobile Computing, 11*(5), 753–766.

KEY TERMS AND DEFINITIONS

Ad Hoc Network: A network that is formed on the fly without any prior planning and infrastructure.

Cloud Computing: A collection of integrated resources that are dynamically provisioned as services over the Internet.

Cognitive Radio Ad Hoc Network: An ad hoc network where the nodes are equipped with cognitive radios to use unused licensed spectra in their vicinity.

Grid Computing: An integration of loosely coupled resources belonging to different administrative domains for high performance computing.

Mobile Ad Hoc Network: An ad hoc network created among a set of mobile hosts.

Sensor Network: An ad hoc network established among a set of static hosts equipped with sensing capabilities.

Vehicular Ad Hoc Network: An ad hoc network formed between vehicles travelling on the road.

Wireless Mesh Network: A type of network where a set of nodes autonomously organize themselves in the form of mesh topology to provide network access to different clients.

Chapter 9
Algorithms to Determine Stable Connected Dominating Sets for Mobile Ad Hoc Networks

Natarajan Meghanathan
Jackson State University, USA

ABSTRACT

This chapter presents three algorithms to determine stable connected dominating sets (CDS) for wireless mobile ad hoc networks (MANETs) whose topology changes dynamically with time. The three stability-based CDS algorithms are (1) Minimum Velocity (MinV)-based algorithm, which prefers to include a slow moving node as part of the CDS as long as it covers one uncovered neighbor node; (2) Node Stability Index (NSI)-based algorithm, which characterizes the stability of a node as the sum of the predicted expiration times of the links (LET) with its uncovered neighbor nodes, the nodes preferred for inclusion to the CDS in the decreasing order of their NSI values; (3) Strong Neighborhood (SN)-based algorithm, which prefers to include nodes that cover the maximum number of uncovered neighbors within its strong neighborhood (region identified by the Threshold Neighborhood Ratio and the fixed transmission range of the nodes). The three CDS algorithms have been designed to capture the node size—lifetime tradeoff at various levels. In addition to presenting a detailed description of the three stability-based CDS algorithms with illustrative examples, the authors present an exhaustive simulation study of these algorithms and compare their performance with respect to several metrics vis-à-vis an unstable maximum density-based MaxD-CDS algorithm that serves as the benchmark for the minimum CDS Node Size.

DOI: 10.4018/978-1-4666-5170-8.ch009

INTRODUCTION

A mobile ad hoc network (MANET) is a dynamically changing distributed system of arbitrarily moving wireless nodes that operate under resource constraints such as limited battery charge, memory and processing capacity. In addition, the network operates under limited bandwidth, necessitating limited transmission range for the nodes to prevent collisions when packets are transmitted over long-distance. A node within the transmission range of another node is said to be a 'neighbor' of the latter, and the set of nodes within the transmission range of a node are said to constitute the 'neighborhood' of the node. Due to the limited transmission range of the nodes, MANET routes are generally multi-hop in nature, and these routes are discovered through an expensive broadcast query-reply cycle involving all the nodes in the network. Due to the dynamically changing topology, the routes fail over time, leading to the frequent route discoveries.

A connected dominating set (CDS) is considered as an energy-efficient communication topology for network-wide broadcasting vis-à-vis flooding in mobile ad hoc networks (MANETs). Though flooding guarantees delivery of the broadcast message to all the nodes in the network, the number of retransmissions is significantly high as a node receives a copy of the message from each of its neighbors (each node would have to broadcast a message exactly once). A Connected Dominating Set (CDS) of a network graph comprises of a subset of the nodes such that every node in the network is either in the CDS or is a neighbor of a node in the CDS. With a CDS-based broadcast, a message is broadcast only by the CDS nodes (nodes constituting the CDS) and the non-CDS nodes (who are neighbors of the CDS nodes) merely receive the message, once from each of their neighbor CDS nodes. The efficiency of broadcasting depends on the CDS Node Size that directly influences the number of redundant retransmissions.

In a unit-disk graph modeling a MANET, all the nodes in the network are represented as vertices; there exists an edge between any two nodes in the graph only if the two nodes corresponding to the end vertices of the edge are neighbors in the MANET (Kuhn et. al., 2004). We say a network is covered if every node in the network is either in the CDS or is a neighbor of a node in the CDS. As nodes operate with a limited transmission range, a CDS is often not fully connected (i.e., a CDS node is not connected to every other CDS node) and a non-CDS node may not be in the neighborhood of every CDS node. A CDS is considered to be broken if either a CDS node is not reachable to any of the other CDS nodes (directly or through multi-hop paths) or a non-CDS node is not in the neighborhood of at least one CDS node. Due to the dynamically changing topology of MANETs, a CDS may not exist for the entire communication session. In prior research, Meghanathan (2006) as well as Velummylum and Meghanathan (2010) have observed the commonly used minimum node-size based connected dominating set (MCDS) to be quite unstable for MANETs, requiring frequent transitions from one CDS to another. A MCDS is preferred because of its tendency to contain as few nodes as possible to be part of the connected dominating set, minimizing the number of unnecessary retransmissions in network-wide broadcasts.

This book chapter will present three different algorithms to determine stable connected dominating sets (CDS) for MANETs and an exhaustive simulation-based performance comparison study of these algorithms in comparison with the classical MCDS algorithm. The algorithms will be based on the idea of iteratively including nodes into CDS (one node per iteration) until all nodes in the network are covered by at least one node in the CDS. To maintain CDS connectivity, only covered nodes (i.e., a node that is in the neighborhood of a node already in the CDS) will be considered as candidate nodes for inclusion into the CDS. We now describe the underlying principle behind the three algorithms proposed to determine stable connected dominating sets and the associated tradeoffs.

The Minimum Velocity-based CDS (MinV-CDS) algorithm prefers to include a slow-moving node into the CDS (and cover its neighbors) rather than a fast moving node. A MinV-CDS would comprise of nodes that are relatively slow moving and is anticipated to exist for a longer time. However, we anticipate a CDS Node Size-Lifetime tradeoff as the MinV-CDS is constructed without any consideration of minimizing the number of constituent CDS nodes, and a node (with the lowest velocity among all the covered nodes) is included as a MinV-CDS node as long as it has at least one uncovered node in its neighborhood. As node velocity values are modeled as a real number, we do not anticipate any tie to pick up the next node for inclusion into the MinV-CDS (ties are broken by preferring the node with the largest number of uncovered neighbors and if cannot be still resolved, the node with the smaller ID is picked). As a result, the MinV-CDS Node Size is expected to be high.

The Node Stability Index (NSI)-based CDS algorithm prefers to construct a CDS whose constituent nodes have stable links with their neighbors. In this direction, we make use of the model to predict the link expiration time (LET) proposed in an earlier work (Su et. al., 2001). The Node Stability Index (NSI) of a node is the sum of the LETs of the links with its uncovered neighbor nodes (i.e. neighbor nodes that are yet to be covered by a CDS node). We prefer to include a node with the largest NSI into the CDS. Since the NSI is computed based on the links with the uncovered neighbor nodes, we anticipate that nodes with the larger NSI to cover several nodes and thereby balancing the CDS Node Size—Lifetime tradeoff as much as possible. Nevertheless, nodes with fewer neighbors but strong links (with significantly larger LETs) could be still preferred over nodes with more neighbors but weaker links (with low LETs).

The Strong Neighborhood-based CDS (SN-CDS) algorithm introduces a term called the 'Threshold Neighborhood Ratio' ($0 \leq$ TNDR $\leq$ 1) that defines the strong neighborhood of a node. A node j at a physical Euclidean distance of r from node i is said to be in the strong neighborhood of node i if the Edge Distance Ratio $r/R \leq$ TNDR where R is the fixed transmission range of all nodes in the network (i.e., we assume a homogeneous MANET). Note that like MCDS, an SN-CDS also prefers to include nodes with a larger number of uncovered neighbors as part of the CDS. However, the use of TNDR imposes the Strong Neighborhood criteria for defining the unit disk graph (i.e. MANET topology) and limits the physical Euclidean distance between the CDS nodes at the time of formation of the SN-CDS, and this is expected to largely contribute to the stability of the SN-CDS. On the other hand, the Open Neighborhood policy adopted at the time of constructing a MCDS leads to the selection of unstable edges. We anticipate the SN-CDS to incur the smallest the CDS Node Size among the three algorithms; however, since the mobility of the nodes (unlike the case of MinV-CDS or NSI-CDS) is not directly considered as part of the node inclusion criteria into the CDS, we anticipate the SN-CDS Lifetime to be relatively lower compared to the MinV-CDS and NSI-CDS.

RELATED WORK

Very few algorithms have been proposed in the literature to determine a stable connected dominating set for MANETs. Wang et al (2005) proposed a localized algorithm, called the Maximal Independent Set with Multiple Initiators (MCMIS) algorithm to construct stable virtual backbones. The MCMIS algorithm consists of two phases: In the first phase, a forest of multiple dominating trees, each consisting of a subset of the nodes in the network topology, is constructed. The dominating trees, each consisting of a different initiator node, are constructed in parallel. In the second phase, the dominating trees with overlapping branches are interconnected to form a complete virtual backbone. Nodes are ranked according to the tuple (stability, effective degree, ID) and are considered as candidate nodes to be initiators, in decreasing order of importance. Sakai et al (2008) propose a mobility handling algorithm that shortens the recovery time of CDS (i.e., the changes in the CDS membership) in the presence of node mobility as well as maintains the CDS size as small as possible. Sheu et al (2009) propose a link stability driven CDS construction algorithm according to which links are categorized as weak or non-weak links if the beacon signals received through this link are below or above a threshold respectively; and a node with the largest number of non-weak links is preferred for inclusion to the CDS.

Meghanathan (2006) had earlier proposed a centralized algorithm, referred to as *OptCDSTrans*, to determine a sequence of stable static connected dominating sets called the Stable Mobile Connected Dominating Set for MANETs. Algorithm *OptCDSTrans* operates according to a simple greedy principle, described as follows: whenever a new CDS is required at time instant *t*, we choose the longest-living CDS from time *t*. The above strategy when repeated over the duration of the simulation session yields a sequence of long-living stable static connected dominating sets such that the number of CDS transitions (change from one CDS to another) is the global minimum. The sequence of CDS determined by *OptCDSTrans* across the duration of the simulation time will have the longest average time. Under identical conditions and simulation time, the average lifetime of any CDS for MANETs will bounded by the average lifetime per CDS returned by the *OptCDSTrans* algorithm. A comparison on the stability of the MaxD-CDS vis-à-vis the stable CDS determined using the *OptCDSTrans* algorithm is available (Meghanathan, 2006).

NETWORK MODEL, DATA STRUCTURES AND RUN-TIME COMPLEXITY

The network model and data structures employed by all the CDS algorithms discussed in this chapter are listed below. The auxiliary data structures and terminologies specifically used by the individual algorithms are explained in the respective sections.

Network Model

The network model used in this research is described as follows:

- We assume a homogeneous network of wireless nodes, each operating at a fixed transmission range, *R*.
- We use the unit-disk graph model (Kuhn et. al., 2004) according to which there exists a link between any two nodes *i* and *j* at (X_i, Y_i) and (X_j, Y_j) in the network as long as the Euclidean distance $\sqrt{(X_i - X_j)^2 + (Y_i - Y_j)^2}$ is less than or equal to the transmission range per node.
- The set of neighbors of a node *i*, *Neighbors(i)*, comprises of nodes that are connected to vertex *i* in the unit-disk graph model.
- A node learns about its own location through location service schemes such as the Global Positioning System (GPS; Hofmann-Wellenhof et. al., 2004) or any other scheme (e.g. Keiss et. al., 2004).
- A node learns the location and mobility parameters (velocity and direction of movement—measured as the angle subscribed with respect to the positive X-axis) of its neighbor nodes through the beacon messages periodically broadcast by their nodes in the neighborhood.

Data Structures

The data structures used by the CDS algorithms are as follows:

- **CDS-Nodes-List**: This list includes all the nodes that are part of the CDS
- **Covered-Nodes-List**: This list includes all the nodes that are either part of the CDS or is at least a neighbor node of a node in the CDS.
- **Uncovered-Nodes-List**: This list includes all the nodes that are not part of the Covered-Nodes-List.
- **Priority-Queue**: This list includes all the nodes that are in the Covered-Nodes-List (but not in the CDS-Nodes-List) and are considered the candidate nodes for the next node to be selected for inclusion in the CDS-Nodes-List. The order in which the vertices are stored in the Priority-Queue varies with the CDS algorithms as described below. Any tie between the nodes is broken arbitrarily.
  - For the MaxD-CDS algorithm, the *Priority-Queue* stores the covered non-CDS nodes in the decreasing order of the node density (number of uncovered neighbors); the node with the largest number of uncovered neighbors is in the front of the queue.
  - For the MinV-CDS algorithm, the *Priority-Queue* stores the covered non-CDS nodes in the increasing order of the node velocities and have at least one uncovered neighbor node; the node with the lowest velocity and having at least one uncovered neighbor node is in the front of the queue.
  - For the NSI-CDS algorithm, the *Priority-Queue* stores the covered non-CDS nodes in the decreasing order of their NSI values and have at least one uncovered neighbor node;

the node with the NSI value and has at least one uncovered neighbor node is in the front of the queue.
  - For the SN-CDS algorithm, the *Priority-Queue* stores the covered non-CDS nodes in the decreasing order of the number of uncovered neighbors in the strong neighborhood; the node with the largest number of uncovered neighbors in the strong neighborhood is in the front of the queue.

Run-Time Complexity

The run-time complexity of all the CDS algorithms discussed in this chapter is $\Theta((|E|+|V|) * \log|V|)$, where V and E are the set of vertices (number of nodes) and edges in a network graph respectively. This can be attributed to the representation of the *Priority-Queue* as a binary heap of the vertices (of size $|V|$ or less at any point of time). Each enqueue and dequeue operation of the priority-queue takes $\Theta(\log|V|)$ time. As we will be exploring every vertex (and dequeue one vertex per iteration) and edge (visit one or more edges per iteration and dequeue/ enqueue the vertices if their position in the heap needs to be changed) across all the iterations of an algorithm, we would encounter the $\Theta(\log|V|)$ complexity $|E|+|V|$ times. Since $E = O(|V|^2)$ in any graph, we can also simply represent the overall run-time complexity as $\Theta(|V|^2\log|V|)$.

MINIMUM VELOCITY-BASED CDS (MINV-CDS) ALGORITHM

The MinV-CDS (pseudo code in Figure 1) is primarily constructed as follows: The *Start Node* (the node with the minimum velocity and having at least one uncovered neighbor node) is the first node to be added to the *MinV-CDS-Node-List*. As a result of this, all the neighbors of the *Start Node* are said to be covered: removed from the

Figure 1. Pseudo code for the Minimum Velocity-based CDS (MinV-CDS) Algorithm

Input: Snapshot of the Network Graph $G = (V, E)$, where V is the set of vertices and E is the set of edges

Output: *MinV-CDS-Node-List* // contains the list of nodes part of the minimum velocity – based CDS.

Initialization:
MinV-CDS-Node-List = Φ; *Covered-Nodes-List* = Φ; *Priority-Queue* = Φ; *Uncovered-Nodes-List* = V
Begin Construction of MinV-CDS

$$startNode = u \mid \underset{u \in V}{Min}\big(velocity(u)\big)$$

  // Initializing the data structures
  MinV-CDS-Node-List = {*startNode*};   *Priority-Queue* = {*startNode*}
  Covered-Nodes-List = {*startNode*};   *Uncovered-Nodes-List* = *Uncovered-Nodes-List* – {*startNode*}

  // Constructing the MinV-CDS-Node-List
  while (*Uncovered-Nodes-List* $\neq \Phi$ and *Priority-Queue* $\neq \Phi$) **do**
    node s = Dequeue(*Priority-Queue*)
    alreadyCovered = true // to test whether all neighbors of node s have already been covered or not
    for all node $u \in$ Neighbors(s) **do**
      if ($u \in$ *Uncovered-Nodes-List*) **then**
        alreadyCovered = false;   *Uncovered-Nodes-List* = *Uncovered-Nodes-List* – {u}
        Covered-Nodes-List = *Covered-Nodes-List* $\cup$ {u}; *Priority-Queue* = *Priority-Queue* $\cup$ {u}
      end if
    end for
    if (*alreadyCovered* = false) **then**
      MinV-CDS-Node-List = *MinV-CDS-Node-List* $\cup$ {s}
    end if
  end while
  return *MinV-CDS-Node-List*
End Construction of MinV-CDS

Uncovered-Nodes-List and added to the *Covered-Nodes-List* and to the *Priority-Queue*. If both the *Uncovered-Nodes-List* and the *Priority-Queue* are not empty, we dequeue the *Priority-Queue* to extract a node s that has the lowest velocity and is not yet in the *MinV-CDS-Node-List*. If there is at least one neighbor node u of node s that is yet to be covered, all such nodes u are removed from the *Uncovered-Nodes-List* and added to the *Covered-Nodes-List* and to the *Priority-Queue*; node s is also added to the *MinV-CDS-Node-List*. If all neighbors of node s are already covered, then node s is not added to the *MinV-CDS-Node-List*. The above procedure is repeated until the *Uncovered-Nodes-List* becomes empty or the *Priority-Queue* becomes empty. If the *Uncovered-Nodes-List* becomes empty, then all the nodes in the network are covered. If the *Priority-Queue* becomes empty and the *Uncovered-Nodes-List* has at least one node, then the underlying network is considered to be disconnected. During a dequeue operation, if two or more nodes have the same lowest velocity, we choose the node with the larger number of uncovered neighbors. If the tie cannot be still broken, we randomly choose to dequeue one of these candidate nodes.

Figure 2 illustrates an example to demonstrate the working of the MinV-CDS algorithms. In these figures, each circle represents a node. The integer outside the circle represents the node ID and the integer inside the circle represents the number of uncovered neighbors of the corresponding node. The real-number inside the circle represents the velocity (in m/s) for the particular node. The nodes that are part of the CDS have their circles bold. We shade the circles of nodes that are covered, but are not part of the CDS. The circles of nodes that are not yet covered are neither shaded nor made bold. On the 24-node network example considered in Figure 2, it takes 13 iterations for the MinV-CDS algorithm to find the CDS. The MinV-CDS includes 14 nodes and 16 edges. Similar results have also been observed in our simulations. The MinV-CDS includes a rela-

tively larger number of nodes and edges and this helps the former to sustain for a relatively longer lifetime as well as a lower hop count per source-destination path.

NODE STABILITY INDEX-BASED CDS (NSI-CDS) ALGORITHM

The predicted link expiration time (LET) of a link i—j between two nodes i and j, currently at (X_i, Y_i) and (X_j, Y_j), and moving with velocities v_i and v_j in directions θ_i and θ_j (with respect to the positive X-axis) is computed using the formula proposed by Su et al (2001):

Figure 2. Example to illustrate the construction of the Minimum Velocity-based CDS (MinV-CDS)

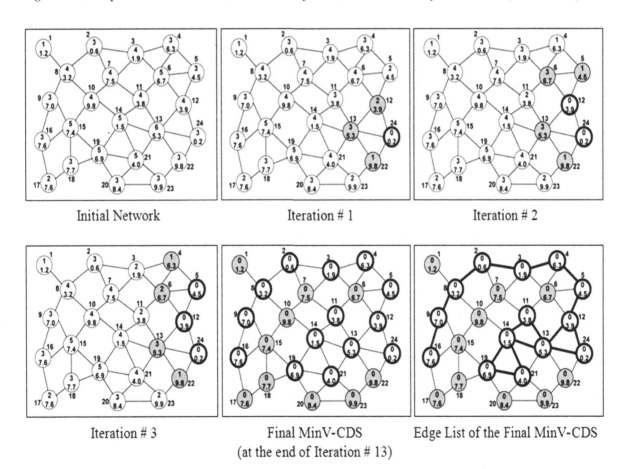

Initial Network   Iteration # 1   Iteration # 2

Iteration # 3   Final MinV-CDS   Edge List of the Final MinV-CDS
                   (at the end of Iteration # 13)

$$LET(i, j) = \frac{-(ab + cd) + \sqrt{(a^2 + c^2)R^2 - (ad - bc)^2}}{a^2 + c^2}$$

(1)

where $a = v_i*\cos\theta_i - v_j*\cos\theta_j$; $b = X_i - X_j$; $c = v_i*\sin\theta_i - v_j*\sin\theta_j$; $d = Y_i - Y_j$

The Node Stability Index (NSI) of a node i during the working of the NSI-CDS algorithm is defined as the sum of the LET of the links with the neighbor nodes j that are not yet covered by a CDS node. The NSI of a node i is represented formally as:

$$NSI(i) = \sum_{\substack{j \in Neighbors(i) \\ j \notin Covered-Nodes-List}} LET(i, j)$$

(2)

The NSI-CDS algorithm (pseudo code in Figure 3) works as follows: For every iteration, the algorithm selects one node from the *Priority-Queue* (the node is also in the *Covered-Nodes-List*) and adds it to the *CDS-Nodes-List*. The criterion to select a covered node from the *Priority-Queue* and include it in the *CDS-Nodes-List* is to give preference for the covered node with the maximum value of the Node Stability Index (NSI). As defined before, the NSI of a node is the sum of the Link Expiration Times (LETs) of the links with the uncovered neighbors of the node (i.e., the neighbor nodes that are not yet in the *Covered-Nodes-List*). Before the first iteration, since none of the nodes are in the *Covered-Nodes-List*, the NSI of a node is simply the sum of the LETs of the links with all of its neighbor nodes. The node with the maximum of such NSI value is the first node to be added to the *Covered-Nodes-List*, *Priority-Queue* and eventually to the *CDS-Nodes-List*. All the uncovered neighbors of the newly included

Figure 3. Pseudo code for the NSI-CDS algorithm

Input: Snapshot of the Network Graph $G = (V, E)$, where V is the set of vertices and E is the set of edges
Output: *CDS-Nodes-List* // contains the list of nodes that are part of the NSI-based CDS
Initialization:
CDS-Nodes-List = Φ; *Covered-Nodes-List* = Φ; *Priority-Queue* = Φ; $\forall i \in V$, $NSI(i) = \sum_{j \in Neighbor(i)} LET(i, j)$

Begin Construction of NSI-CDS

   startNode = $u \mid \underset{u \in V}{Max}(NSI(u))$

   Priority-Queue = {*startNode*}; *Covered-Nodes-List* = {*startNode*}

   while (|*Covered-Nodes-List*| < |V| and *Priority-Queue* $\neq \Phi$) **do**
      node s = Dequeue(*Priority-Queue*) where $s \in$ *Covered-Nodes-List* and $s \notin$ *CDS-Nodes-List*

      CDS-Nodes-List = *CDS-Nodes-List* U {s}
      $\forall v \in$ *Neighbors(s)*;
        if $v \notin$ *Covered-Nodes-List* **then**
          Covered-Nodes-List = *Covered-Nodes-List* U {v}; *Priority-Queue* = *Priority-Queue* U {v}
        end if

      $\forall u \in V$, $NSI(u) = \{ \sum LET(u, v) \mid v \in$ *Neighbors(u)* AND $v \notin$ *Covered-Nodes-List* $\}$

      $\forall u \in$ *Priority-Queue*, **if** ($NSI(u) = 0$) **then** remove node u from *Priority-Queue*

      if (|*Covered-Nodes-List*| < |V| and *Priority-Queue* = Φ) **then**
        return NULL // the network is disconnected and there is no CDS covering all the nodes
      end if

   end while

  return *CDS-Nodes-List*

node to the *CDS-Nodes-List* are now included in the *Covered-Nodes-List* as well as in the *Priority-Queue*. After every iteration, the NSI values of the nodes in the network and the *Priority-Queue* are recomputed based on the updated *Covered-Nodes-List*. Nodes whose NSI value is zero are removed from the *Priority-Queue*. The above procedure is repeated until there is at least one node that is not yet in the *Covered-Nodes-List*; if the underlying network is connected, the *Priority-Queue* will remain non-empty until all nodes are added to the *Covered-Nodes-List* and the algorithm finally returns the *CDS-Nodes-List*. If the *Priority-Queue* gets empty and there is at least one node to be added to the *Covered-Nodes-List*, then it implies the underlying network is disconnected and the algorithm returns NULL.

Figure 4 illustrates an example to demonstrate the working of the NSI-CDS algorithm. In the figure, each circle represents a node. The integer inside the circle represents the node ID; the real-number on an edge represents the LET of the edge. The real-number inside the circle represents the sum of the LETs of the edges with the uncovered neighbor nodes. The nodes that are part of the CDS have their circles bold. We shade the circles of nodes that are covered, but are not part of the CDS. The circles of nodes that are not yet covered are neither shaded nor made bold.

On the 24-node network example considered in Figure 4, it takes 10 iterations for the NSI-CDS algorithm to find the CDS. The NSI-CDS includes 10 nodes and 10 edges. The last iteration of Figure 4 shows the 10 edges (shown in bold thick

Figure 4. Example to Illustrate the Construction of the Node Stability Index-based CDS (NSI-CDS)

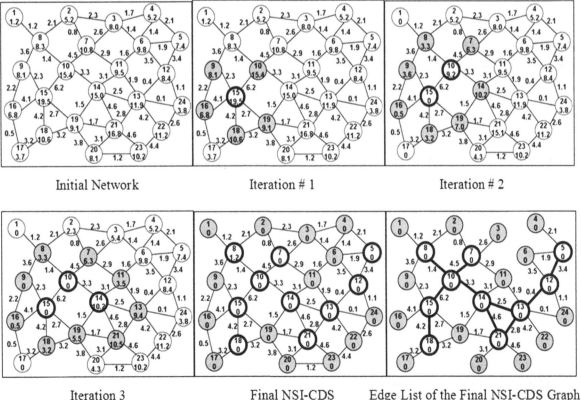

Initial Network	Iteration # 1	Iteration # 2

Iteration 3	Final NSI-CDS (at the end of Iteration # 10)	Edge List of the Final NSI-CDS Graph

lines) that are between the 10 CDS nodes as well as the edges between each non-CDS node with its neighboring CDS nodes (shown in regular lines). The CDS-to-CDS edges and the non-CDS-to-CDS edges constitute the NSI-CDS graph, which is of course a sub graph of the original network graph. The difference between the two graphs is that, in the final NSI-CDS graph, we do not include edges between any two non-CDS nodes.

As we see in the different iterations of this example, the inclusion of a node into the CDS is decided by both the number of uncovered neighbors and the LETs of the links to these uncovered neighbors. The larger the number of uncovered neighbors and/or the larger the LET values of the links to these uncovered neighbors, the greater are the chances of a covered node to be included into the NSI-CDS. It is this characteristic that enables the NSI-CDS to discover a long-living stable CDS with relatively fewer nodes.

STRONG NEIGHBORHOOD-BASED CDS (SN-CDS) ALGORITHM

The SN-CDS algorithm uses the following data structures and auxiliary variables:

- **Open Neighborhood:** Node j belongs to the open neighborhood of node i (denoted ON_i) if the physical Euclidean distance between node i and j is $\leq R$, the transmission range of the nodes in the network. Every node $j \in ON_i$ is simply referred to as a "neighbor" of node i.
- **Edge Distance Ratio (EDR):** The EDR for an edge is the ratio of the physical Euclidean distance between the two constituent end nodes of the edge to that of the fixed transmission range.
- **Threshold Neighbor Distance Ratio (TNDR):** The TNDR is the maximum value of the EDR for an edge i—j, in order for

node i to be considered a "strong" neighbor of node j and vice-versa.

- **Strong Neighborhood:** Node j belongs to the strong neighborhood of node i (denoted SN_i) and vice-versa if the EDR of the edge i—j is $\leq TNDR$. Note that the *strong neighborhood* of a node is a subset of the nodes in its *open neighborhood*.
- **MaxUncoveredStrongNeighbors:** The maximum value for the number of uncovered strong neighbors of any node in the network at the beginning of the algorithm; initialized to $-\infty$ to start with and is computed while determining the strong neighbors of the nodes in the network.
- **UncoveredStrongNeighbors(u):** The set of all strong neighbors of node u that are not yet covered. Initially, $\forall u$, $uncoveredStrongNeighbors(u) = \Phi$.
- **Start Node:** The first node to be included into the SN-CDS. The *Start Node* is the node that has the maximum number of uncovered strong neighbors, to start with. Any tie is broken arbitrarily.

The SN-CDS construction algorithm (pseudo code in Figure 5) primarily works as follows: The algorithm inputs a snapshot of the network (a graph $G = (V, E)$ with vertex set V and edge set E) at the time instant during which we want to determine the SN-CDS. The algorithm also inputs the Threshold Neighborhood Distance Ratio (*TNDR*) and transmission range (R) for the network. When the *Start Node* is added to the *SN-CDS-Node-List*, all of its strong neighbors are said to be covered; these nodes are removed from the *Uncovered-Nodes-List* and added to the *Covered-Nodes-List* and to the *Priority-Queue*. If both the *Uncovered-Nodes-List* and the *Priority-Queue* are not empty, we dequeue the *Priority-Queue* to extract a node s that is not yet in the *SN-CDS-Node-List* and has the largest number of uncovered strong neighbor nodes. All the uncovered strong neighbor nodes of s are now removed from the *Uncovered-*

Nodes-List and added to the *Covered-Nodes-List* as well as to the *Priority-Queue*. The number of uncovered strong neighbors of each node in the network is then updated based on the additional node coverage obtained during the iteration and accordingly, the *Priority-Queue* is re-sorted in the decreasing order of the number of uncovered strong neighbors of the nodes in the queue. The above procedure is repeated for several iterations until the *Uncovered-Nodes-List* becomes empty or the *Priority-Queue* becomes empty.

Note that during a particular iteration, if the node *s* extracted from the *Priority-Queue* has all its strong neighbor nodes already covered, then it implies that all the other nodes, if any, in the

Priority-Queue also have "zero" uncovered strong neighbor nodes. However, we have not yet broken from the while loop (i.e. the *Uncovered-Nodes-List* is not yet empty), indicating that the underlying network based on the strong neighborhood of the nodes is not connected and hence the algorithm returns NULL (i.e. a SN-CDS for the entire network does not exist). Also, even after exiting from the while loop, if the *Priority-Queue* becomes empty and the *Uncovered-Nodes-List* has at least one node, then the underlying network is considered to be disconnected (based on the strong neighborhood of the nodes) and the algorithm returns NULL. If the underlying network is connected based on the strong neighborhood of the

Figure 5. Pseudo code for the SN-CDS construction algorithm

Input: Snapshot of the Network Graph $G = (V, E)$, where V is the set of vertices and E is the set of edges

Output: *CDS-Nodes-List* // contains the list of nodes that are part of the NSI-based CDS

Initialization:

CDS-Nodes-List = Φ; *Covered-Nodes-List* = Φ; *Priority-Queue* = Φ; $\forall i \in V,\ NSI(i) = \sum_{j \in Neighbors(i)} LET(i, j)$

Begin Construction of NSI-CDS

    $startNode = u \mid \underset{u \in V}{Max}(NSI(u))$

    Priority-Queue = {*startNode*}; *Covered-Nodes-List* = {*startNode*}

    while (|*Covered-Nodes-List*| < |*V*| and *Priority-Queue* ≠ Φ) **do**

        node *s* = Dequeue(*Priority-Queue*) where *s* ∈ *Covered-Nodes-List* and *s* ∉ *CDS-Nodes-List*

        CDS-Nodes-List = *CDS-Nodes-List* U {*s*}

        $\forall v \in Neighbors(s)$,

          if *v* ∉ *Covered-Nodes-List* **then**

            Covered-Nodes-List = *Covered-Nodes-List* U {*v*}; *Priority-Queue* = *Priority-Queue* U {*v*}

          end if

        $\forall u \in V,\ NSI(u) = \{ \sum LET(u, v) \mid v \in Neighbors(u)\ AND\ v \notin Covered\text{-}Nodes\text{-}List \}$

        $\forall u \in Priority\text{-}Queue$, **if** (*NSI(u)* = 0) **then** remove node *u* from *Priority-Queue*

        if (|*Covered-Nodes-List*| < |*V*| and *Priority-Queue* = Φ) **then**

          return NULL // the network is disconnected and there is no CDS covering all the nodes

        end if

    end while

return *CDS-Nodes-List*

nodes, then the algorithm does not return NULL and returns the *SN-CDS-Node-List* after all the nodes in the network are included to the *Covered-Nodes-List*.

MAXIMUM DENSITY-BASED (MAXD-CDS) ALGORITHM

The MaxD-CDS algorithm is designed to determine the CDS with the minimum number of constituent nodes. It will be used as the benchmarking algorithm to evaluate the CDS Lifetime—Node Size tradeoff as the stability-based CDS algorithms are anticipated to incur a larger value for CDS Node Size in pursuit of increased CDS lifetime.

The MaxD-CDS algorithm (pseudo code in Figure 6) works as follows: The first node to be included in the *CDS-Node-List* is the node with the maximum number of uncovered neighbors in the open neighborhood of the node (any ties are broken arbitrarily). A CDS member is considered to be "covered," so a CDS member is additionally added to the *Covered-Nodes-List* as it is added to the *CDS-Node-List*. All nodes that are adjacent to a CDS member are also said to be covered, so the uncovered neighbors of a CDS member are also added to the *Covered-Nodes-List* as the member is added to the *CDS-Node-List*. To determine the next node to be added to the *CDS-Node-List*, we must select the node with the largest density amongst the nodes that meet the criteria for inclusion into the CDS. The criteria for inclusion into the MaxD-CDS are same as the criteria for inclusion into the ID-based CDS. Amongst the nodes that meet these criteria for CDS membership inclusion, we select the node with the largest density (i.e., the largest number of uncovered neighbors) to be the next member of the CDS. Ties are broken arbitrarily. This process is repeated until all nodes in the network are included in the *Covered-Nodes-List*. Once all nodes in the network are considered to be "covered," the CDS has been formed and the algorithm returns a list of the members included in the resultant MaxD-CDS (nodes in the *CDS-Node-List*). The primary difference between the

Figure 6. Pseudo code for the algorithm to construct Maximum Density (MaxD)-based CDS

Input: Graph $G = (V, E)$; V – vertex set, E – edge set
Source vertex, s – vertex with the largest number of
uncovered neighbors in V
Auxiliary Variables and Functions: *CDS-Node-List, Covered-Nodes-List, Neighbors*(v) for every v in V
Output: *CDS-Node-List*
Initialization: *Covered-Nodes-List* = $\{s\}$; *CDS-Node-List* = Φ
Begin Construction of *MaxD-CDS*
  while (|*Covered-Nodes-List*| < |V|) **do**
    Select a vertex $r \in$ *Covered-Nodes-List* and $r \notin$ *CDS-Node-List* such that r has the largest number of
    uncovered neighbors that are not in *Covered-Nodes-List*

    CDS-Node-List = *CDS-Node-List* U $\{r\}$

    for all $u \in$ *Neighbors*(r) and $u \notin$ *Covered-Nodes-List*
      Covered-Nodes-List = *Covered-Nodes-List* U $\{u\}$
    end for
  end while
  return *CDS-Node-List*
End Construction of *MaxD-CDS*

MaxD-CDS and SN-CDS algorithms is that the coverage of neighbor nodes is determined with respect to the open neighborhood and strong neighborhood respectively.

Figures 7 and 8 respectively illustrate the construction of the MaxD-CDS and SN-CDS through graph snapshots capturing the initial few iterations and the final iteration of the appropriate CDS construction algorithms. Figure 7 illustrates the unit disk graphs representing the open neighborhood for the execution of the MaxD-CDS algorithm. In the case of SN-CDS (Figure 8), we apply a TNDR of 0.5 on the graph of Figure 7, and derive the graphs of Figure 8 representing the strong neighborhood. The circles represent nodes, and the integer outside the circle for a node rep-

resents the Node ID. The real number on an edge indicates the EDR value for the edge. The integer inside a circle of Figure 7 indicates the number of uncovered neighbors of the node in the open neighborhood. Whereas, the integer inside a circle of Figure 8 indicates the number of uncovered strong neighbors of the node (determined after the application of the TNDR = 0.5 criterion). The CDS nodes are shown in bold circles; the covered (but non-CDS) nodes are shown in shaded circles; the circles representing uncovered nodes are neither made bold nor shaded.

The number of iterations of both the MaxD-CDS and SN-CDS algorithms of Figures 7 and 8 directly correspond to the number of nodes that are part of their corresponding connected domi-

Figure 7. Example to illustrate the construction of the Maximum Density-based CDS (MaxD-CDS)

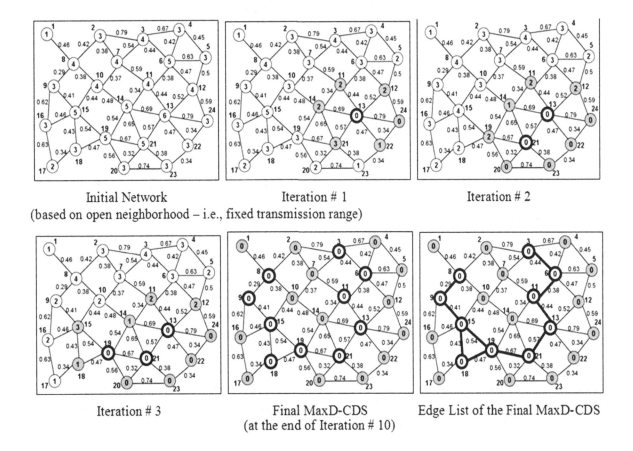

Initial Network
(based on open neighborhood – i.e., fixed transmission range)

Iteration # 1

Iteration # 2

Iteration # 3

Final MaxD-CDS
(at the end of Iteration # 10)

Edge List of the Final MaxD-CDS

Figure 8. Example to Illustrate the Construction of the Strong Neighborhood-based CDS (SN-CDS)

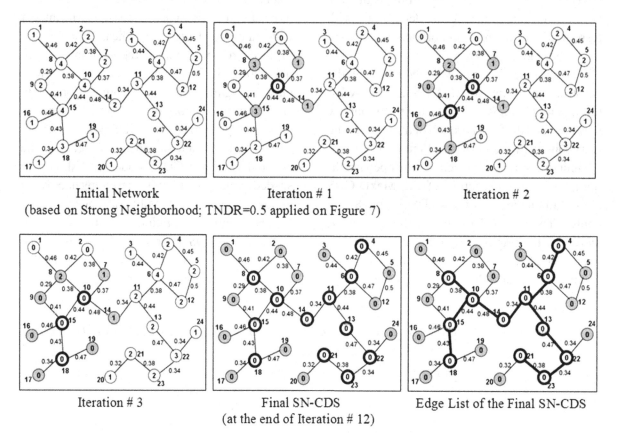

Initial Network
(based on Strong Neighborhood; TNDR=0.5 applied on Figure 7)

Iteration # 1

Iteration # 2

Iteration # 3

Final SN-CDS
(at the end of Iteration # 12)

Edge List of the Final SN-CDS

nating sets. We observe that the SN-CDS has slightly more nodes and edges (12 nodes and 11 edges) compared to the MaxD-CDS (10 nodes and 10 edges). The strength of the SN-CDS lies in the selection of the edges that have an EDR ≤ TNDR. Even though this necessitates slightly more nodes and edges as part of the CDS, we observe that the SN-CDS sustains for a significantly longer lifetime compared to that of the MCDS, as also evidenced in our simulations.

SIMULATIONS

We conduct our simulations in a discrete-event simulator developed in Java. The network dimensions are 1000m x 1000m. The fixed transmission range per node is 250m. We vary the network density by conducting simulations with 50, 100 and 150 nodes to represent networks of low, moderate and high density respectively that also correspond to an average neighborhood size of 10, 20 and 30 nodes for a transmission range of 250m. We use the Random Waypoint model (Bettstetter et. al., 2004) as the mobility model for our simulations. According to this model, each node starts from an arbitrary location, and chooses to move to a randomly chosen destination location (within the network boundary) at a velocity uniform-randomly chosen from the range $[0, ..., v_{max}]$. After reaching the targeted destination location, the node continues to move by randomly choosing another location to move, with a velocity again uniform-randomly chosen from the above range. The new location and velocity values are chosen independent of the mobility history. Each node continues to move like this for the simulation time of 1000 seconds. The movement of one node is independent of the other nodes in the network. The v_{max} values chosen for our simulations are 5, 25 and 50 m/s representing scenarios of low, moderate and high mobility respectively. All of the simulation results presented in Figures 9 through 14 are average values of the performance metrics obtained by running the simulations with 5 different mobility profiles generated for each combination of network density and node mobility.

Simulation Methodology

Our simulation methodology is as follows: We construct snapshots of the network topology for every 0.25 seconds, starting from time 0 to the

simulation time of 1000 seconds. If a CDS is not known at a particular time instant, we run the appropriate CDS construction algorithm on the network snapshot. The CDS determined at a particular time instant is used in the subsequent time instants until the CDS ceases to exist. For a CDS to be considered to exist at a particular time instant, two conditions have to hold good: (1) All the CDS nodes have to stay connected—i.e. reachable from one another directly or through multi-hop paths; and (2) Every non-CDS node should have at least one CDS node as its neighbor. If a CDS ceases to exist at a particular time instant, we again run the appropriate CDS construction algorithm and continue to use the new CDS as explained above. This procedure is continued for the duration of the simulation time.

Figure 9. SN-CDS connectivity vs. threshold neighborhood distance ratio

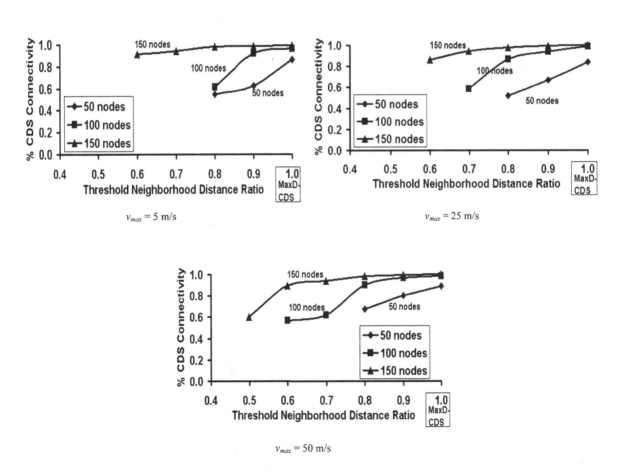

$v_{max} = 5$ m/s

$v_{max} = 25$ m/s

$v_{max} = 50$ m/s

Figure 10. CDS node size

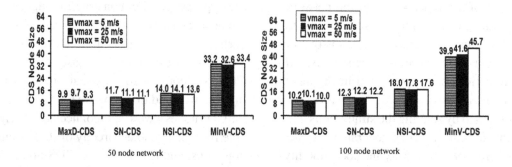

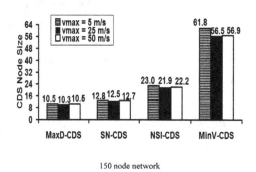

Figure 11. CDS edge size

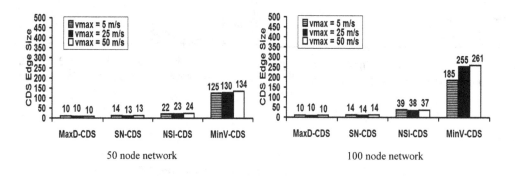

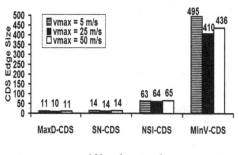

Figure 12. Effective CDS lifetime

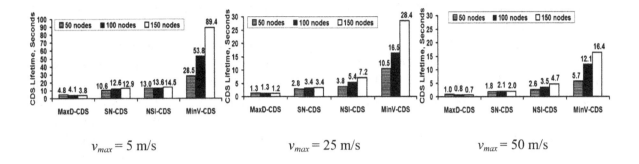

$v_{max} = 5$ m/s $v_{max} = 25$ m/s $v_{max} = 50$ m/s

Figure 13. Hop count per source-destination (s-d) path

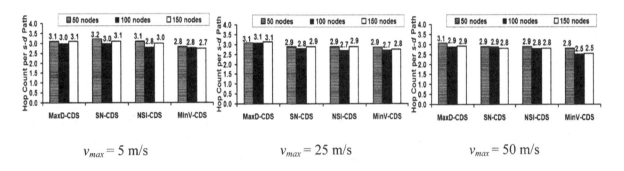

$v_{max} = 5$ m/s $v_{max} = 25$ m/s $v_{max} = 50$ m/s

Figure 14. CDS node size—lifetime tradeoff ratio

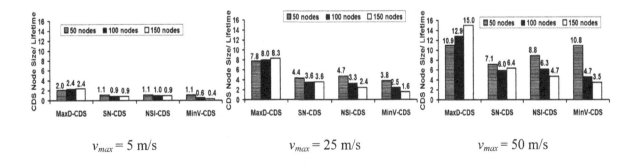

$v_{max} = 5$ m/s $v_{max} = 25$ m/s $v_{max} = 50$ m/s

Connectivity of SN-CDS

We define the connectivity of a CDS as the ratio of the number of time instants there exists a CDS for the underlying network topology divided by the total number of time instants considered over the duration of the simulation time. Since the SN-CDS works on a reduced set of edges (due to the imposition of the strong neighborhood criterion), the connectivity of the SN-CDS for a given unit disk graph (defined according to the fixed transmission range corresponding to the open neighborhood) is heavily dependent on the TNDR value used. The larger the value of TNDR, the larger is the connectivity of SN-CDS, as the number of strong neighbors of a node increases with increase in TNDR. Obviously, if TNDR = 1.0, SN-CDS reverts to MaxD-CDS. Figure 9 illustrates the connectivity of SN-CDS for different network density conditions, observed for v_{max} values of 5 and 50 m/s. As observed in these figures, the connectivity of SN-CDS matches close to that of the MaxD-CDS only when TNDR = 0.9. Hence, in order to be fair to MaxD-CDS and the other stability-CDS algorithms in our comparisons, and also evaluate the performance of SN-CDS in the presence of appreciable CDS connectivity, we will only report the performance of SN-CDS for TNDR = 0.9 for the rest of the metrics evaluated in this section.

Performance Metrics

We measure the following performance metrics in our simulations:

- **Effective CDS Lifetime:** We measure the duration of time each instance of a CDS actually exists, and compute the average of the CDS lifetimes observed across the entire simulation time period (for all the mobility profiles representing a particular combination of network density and node mobility). We refer to these average values of the actually observed CDS lifetimes as the Absolute CDS Lifetime. However, since the connectivity of the different CDSs are less than 1.0 for most of the scenarios evaluated, we introduce a better representative metric called the Effective CDS Lifetime—defined as the product of the Absolute CDS Lifetime and the CDS Connectivity—to effectively capture the stability of a CDS taking into consideration the chances of determining the CDS. For example, if a CDS has an average absolute lifetime of 10 seconds; but its connectivity is only 0.8—it is more prudent to use the effective lifetime of 10*0.8 = 8 seconds as a more realistic measure of the stability (lifetime) of the CDS.

- **CDS Node Size:** This is a time-averaged value for the number of nodes that are part of the CDS used for every time instant over the entire simulation. For example, if there exists a sequence of three CDS of size 30 nodes, 40 nodes and 20 nodes in a network for 6, 10 and 4 seconds respectively, then the average CDS Node Size for a total of 20 seconds is (30*6 + 40*10 + 20*4)/(6 + 10 + 4) = 33.0 and not simply the average of 30, 40 and 20 nodes = 30.

- **CDS Edge Size:** This is a time-averaged value of the number of edges that exist between any two CDS nodes, measured for every time instant a CDS existed during the simulation time period.

- **Hop Count per s-d Path:** The hop count per source-destination (*s-d*) path is the average of the number of hops (edges) on the shortest path between any two nodes on the CDS-induced sub graphs, considered over the entire simulation time and all the 15 *s-d* pairs. Note that the intermediate nodes, if any are needed, for such *s-d* paths must be the CDS nodes. To determine the hop count of the shortest path for every *s-d* pair, we run the Breadth First Search (BFS) algorithm (Cormen et. al., 2009), starting from the source node *s*, on the CDS-induced sub graphs containing edges between any two CDS nodes and between a CDS node and a regular non-CDS node.

- **CDS Node Size / Lifetime Tradeoff Ratio:** This metric captures the tradeoff between CDS Lifetime and Node Size. The CDS algorithm that sustains a lower value for the CDS Node Size/ Lifetime ratio is considered to more effectively balancing the tradeoff.

CDS Node Size

The MaxD-CDS and SN-CDS give preference to nodes with a larger number of uncovered neighbors in the open neighborhood and strong neighborhood respectively. As a result, these two connected dominating sets are formed with a relatively lower number of nodes. The MaxD-CDS, though a heuristic, serves as the optimal benchmark for the minimum CDS Node Size. The SN-CDS incurs a node size that is very close to that of the MaxD-CDS; the slightly larger SN-CDS Node Size is due to restriction of employing a TNDR value to define the neighborhood. The SN-CDS Node Size is about 20% more than the MaxD-CDS Node Size.

The MinV-CDS algorithm gives zero consideration to minimize the number of nodes constituting the CDS. In pursuit of choosing nodes with minimum velocity, it could end up including nodes that as few as one or two uncovered neighbors. This leads to substantially larger MinV-CDS with several constituent nodes. The MinV-CDS Node Size is larger than the MaxD-CDS Node Size by about 225 to 450%, with the difference increasing with increase in network density. As the number of nodes in the network increases, the MinV-CDS incurs relatively more constituent nodes to cover the rest of the nodes in the network.

The NSI-CDS incurs a CDS Node Size that is on the lower end, closer to those incurred by the SN-CDS and MaxD-CDS. Since the NSI value for a node is defined as the sum of the predicted LETs of the links with its neighbors, nodes with multiple uncovered neighbors are likely to get preference for inclusion into NSI-CDS compared to nodes with fewer uncovered neighbors. However, with increase in network density, we could come across situations where a node maintaining long-living stable links with fewer uncovered neighbors ends up having a larger NSI value with a node that has relatively stable links with several uncovered neighbors. The NSI-CDS Node Size is about 40-120% more than that of the MaxD-CDS; the larger differences in magnitudes resulting in networks of 100 and 150 nodes, attributable to the above reasoning of nodes with fewer uncovered neighbors having a larger NSI value.

For a given network density, the CDS Node Size for all the four connected dominating sets is observed to be least impacted due to variations in node mobility. The largest variation in the CDS Node Size is observed for the MinV-CDS due to its dependence on the values of node velocity to decide on the constituent nodes.

CDS Edge Size

The CDS Edge Size is measured by counting the number of edges that may exist between any two CDS nodes in the network. Hence, the larger the CDS Node Size, the larger will be the CDS Edge Size. Also, the larger the CDS Edge Size, the more stable and robust will be the CDS to link failures; the CDS is less likely to be disconnected because of the failure of one or few links. The MaxD-CDS algorithm, in its attempt to minimize the CDS Node Size, chooses CDS nodes that are far away from each other such that each node covers as many uncovered neighbors as possible. As the CDS nodes are more likely to be away from each other, spanning the entire network, the number of edges (Edge Size) between the MaxD-CDS nodes is very low. The SN-CDS incurs about 35% more edges (compared to the MaxD-CDS) and this could be attributed to the 20% increase in the CDS Node Size. The CDS Edge Size incurred with the NSI-based approach is about 100-130% and 260-270% more than that of the CDS Edge Size incurred with the maximum density-based approach for low-density and high-density networks respectively.

On the other hand, since the MinV-CDS algorithm incurs a larger Node Size because of its relative insensitivity to the number of uncovered neighbors of a node, there is a corresponding increase in the number of edges between these CDS nodes. We observe the Edge Size for a MinV-CDS is 12.4 (low network density) to 46.0 (high network density) times larger than that of the Edge Size for a MaxD-CDS. In the case of a MaxD-CDS, for fixed node mobility, as we increase the node density from low to high, there is only at most a 7% increase in the Edge Size. On the other hand, for the MinV-CDS, at fixed node mobility, as we increase the node density from low to high, the Edge Size increases as large as by 400%. This can be attributed to the huge increase (as large as by 190%) in the MinV-CDS Node Size, with increase in network density. The increase in the number of edges and nodes significantly contribute to the increase in the MinV-CDS lifetime as the network density is increased. For a given node density, as we increase the node mobility from low to high, the Edge Size for a MaxD-CDS does not change appreciably, whereas the Edge Size for a MinV-CDS also changes by at most 40%.

Effective CDS Lifetime

The MinV-CDS incurs the largest CDS lifetime, attributed to the relatively larger CDS Node Size and Edge Size. As the constituent nodes of the MinV-CDS are chosen based on the minimum velocity metric, the edges between the CDS nodes are bound to exist for a relatively longer time and the connectivity of the nodes that are part of the MinV-CDS is likely to be maintained for a longer time. On the other hand, the MaxD-CDS algorithm chooses nodes that are far away from each other (but still maintain an edge between them) as part of the CDS. The edges between such nodes are likely to fail sooner, leading to loss of connectivity between the nodes that are part of the MaxD-CDS. We thus observe a tradeoff between the CDS Node Size and the CDS Lifetime.

If we meticulously choose slow-moving nodes to be part of the CDS, the lifetime of the CDS could be significantly improved, at the expense of the Node Size. On the other hand, if we aim to select a CDS with the minimum number of nodes required to cover all the nodes in the network, the lifetime of the CDS would be significantly lower. With respect to the magnitude, the lifetime per MinV-CDS is 6 (low network density) to 25 (high network density) times more than that of the

MaxD-CDS. The relatively high stability of MinV-CDS at high network density can be attributed to the inclusion of a significantly larger number of slow-moving nodes and their associated edges as part of the CDS. The relatively poor stability of MaxD-CDS at high network density can be attributed to the need to cover a larger number of nodes in the network without any significant increase in the number of nodes that are part of the CDS. For both MaxD-CDS and MinV-CDS, for a fixed network density, as we increase node mobility from low (v_{max} = 5 m/s) to moderate (v_{max} = 25 m/s), and from low (v_{max} = 5 m/s) to high (v_{max} = 50 m/s), the lifetime per CDS decreases by a factor of 3.3 and 5.3 respectively.

In terms of absolute values, the NSI-CDS incurs the next largest CDS lifetime; the NSI-CDS lifetime is 16 to 45% of the MinV-CDS lifetime. The difference in magnitude between the lifetimes of the MinV-CDS and NSI-CDS occurs with decrease in node maximum velocity and increase in network density. Nevertheless, the NSI-CDS is still significantly more stable than the MaxD-CDS and the SN-CDS, and it achieves such longer lifetime with a not so significantly high CDS Node Size (unlike the case of MinV-CDS). For a given node mobility, the lifetime of the NSI-CDS is 160-180% and 250-350% more than the lifetime incurred by the MaxD-CDS in low density and high-density networks respectively. Note that the percentage increase in Node Size for NSI-CDS is only about 1/4th of the gain observed in the NSI-CDS lifetime and is at most 45% and 75% more than that of the MaxD-CDS in low-density and high-density networks respectively. Thus, even though we can say that there is a tradeoff between the CDS Lifetime vs. the CDS Node Size and Edge Size, this tradeoff is more favorable towards the NSI-CDS and it significantly gains in the Lifetime metric with a very modest increase in the Node Size. On the other hand, in pursuit of minimizing the number of nodes that are part of the CDS, the MaxD-CDS is quite unstable, especially as the node mobility and/or the network density increases.

Among the three stability-based CDS algorithms, the SN-CDS incurs the lowest lifetime. This could be attributed to its CDS node selection criterion that does not directly consider the stability of the nodes or the links associated with the nodes. It hypothesizes that neighbor nodes that are within the threshold edge distance ratio ($\leq$ TNDR) are likely to stay for a longer time, irrespective of their velocity and/or direction of motion. Such a design criterion helps to the cause of finding a stable CDS only to a certain extent. For a given network density, the difference between the effective lifetime of the SN-CDS and that of the MaxD-CDS decreases with increase in node mobility. This is attributable to the dynamic changes in the network topology with increase in node mobility, and the effectiveness of the strong neighborhood criterion in determining a stable CDS gets marginally diminished in networks of moderate and high mobility. However still, the effective lifetime of the SN-CDS is as large as 250% and 185% more than that of the MaxD-CDS in networks of low mobility and moderate/high mobility respectively. The poor stability of the MaxD-CDS can be attributed to the lack of scope for robustness of the CDS-induced sub graph comprising of only fewer CDS nodes and edges connecting them. In the case of MaxD-CDS, the fact that the distance ratios of the CDS edges are above 0.8, accompanied by the presence of very few CDS edges, contributes to the poor stability of the MaxD-CDS. Even the failure of one or two MaxD-CDS edges could lead to the disintegration of the whole MaxD-CDS. On the other hand, even though the SN-CDS does not involve substantially more edges between the CDS nodes, the SN-CDS edges are part of the strong neighborhood (at least at the time of discovery of the SN-CDS) and hence are likely to exist for a longer time compared to those of the MaxD-CDS edges.

Hop Count per Source-Destination (*S-D*) Path

We determine the average hop count per *s-d* path on a CDS-induced sub graph comprising of all the nodes in the network, and edges between any two CDS nodes and those between a CDS node and non-CDS node, all based on the open neighborhood (even in the case of SN-CDS, since the EDR of the edges is not guaranteed to be less than TNDR on network snapshots other than those in which the SN-CDS was determined). We conduct simulations with randomly selected 15 *s-d* pairs, for each of which we run the BFS algorithm on the CDS-induced sub graphs during every simulation time instant for which there is a CDS available, and then average out the results across all the time instants and *s-d* pairs. We observe the MaxD-CDS, SN-CDS and NSI-CDS to sustain about the same hop count per *s-d* pair.

The MinV-CDS incurs the lowest hop count per source-destination path, mainly attributable to its larger CDS Node Size and Edge Size. It is possible to connect any two nodes (source-destination nodes) with the minimum number of constituent edges that are part of the MinV-CDS. Due to the presence of several edges, there are good chances of connecting the source-destination nodes with a lower hop count path. On the other hand, in the case of a MaxD-CDS, the CDS Edge Size is only one or two units of magnitude more than the CDS Node Size; as a result, there are not many options (routes) to connect a source-destination pair on the MaxD-CDS induced sub graph. The average hop count per *s-d* pair on a MaxD-CDS induced sub graph is 10-15% more than that incurred with a MinV-CDS induced sub graph.

The consequences of having larger hop count per path with a fewer number of nodes per MaxD-CDS are a larger end-to-end delay per data packet and unfairness of node usage (with respect to energy consumption). Nodes that are path of the MaxD-CDS could be relatively heavily used compared to the nodes that are not part of the MaxD-CDS. This could lead to premature failure of critical nodes, mainly nodes lying in the center of the network, resulting in reduction in network connectivity, especially in low-density networks. With MinV-CDS, as multiple nodes are part of the CDS, the packet forwarding load can be distributed across several nodes and this could enhance the fairness of node usage and help to incur a relatively lower end-to-end delay per data packet.

For a given network density, we observe the hop count per *s-d* path to be independent of the node mobility. However, for a given mobility, we observe the hop count per *s-d* path for the CDS algorithms to decrease by about 5-10%, and this could be attributed to the presence of more constituent nodes and edges in the corresponding CDS at higher network density, virtually increasing the number of *s-d* paths from which BFS can choose the shortest one. SN-CDS and MaxD-CDS show the least sensitivity in the hop count per *s-d* path with increase in network density. This could be attributed to the imposition of the urge to choose nodes with the largest number of uncovered neighbors and/or strong neighborhood criterion on the network graphs and the resulting possibility of not a substantial increase in the number of alternate paths between *s-d* pairs.

CDS Node Size: Effective CDS Lifetime Tradeoff Ratio

Since we observe an inverse relationship between the CDS Node Size and the Effective CDS Lifetime, we are interested in identifying which CDS effectively balances the CDS Node Size to Lifetime tradeoff. The ratio has the CDS Node Size in the numerator and the Effective CDS Lifetime in the denominator. Since we prefer to incur a larger CDS lifetime and a lower CDS Node Size, the ratio is ideally expected to be closer to 0. In other words, the CDS that incurs the least

value for the tradeoff ratio is said to be effectively balancing the CDS Node Size and the Lifetime. In this context, we observe (in Figure 14) that the MinV-CDS incurs the lowest tradeoff ratio values for networks of moderate and high density (under all conditions of node mobility) and for low-density networks (under low and moderate node mobility). The MaxD-CDS incurs the largest values for the tradeoff ratio, implying that even though the CDS Node Size is lower, lower values for the CDS Lifetime shoots up the values for the ratio. The SN-CDS and NSI-CDS incur a tradeoff ratio that is relatively lower and much closer to the values observed for the MinV-CDS. Thus, all the three stability-CDS algorithms sustain a CDS Node Size/Effective CDS Lifetime ratio that is lower than that of the MaxD-CDS implying that higher CDS node size incurred by the stability-based algorithms effectively contributes to the larger lifetime of the CDS.

CONCLUSION AND FUTURE WORK

The simulation results confirm our hypothesis that the three stability-based CDS algorithms can be ranked in the following order in decreasing order of CDS lifetime: MinV-CDS, NSI-CDS and SN-CDS and on the other hand, in the following order in increasing order of CDS Node Size: SN-CDS, NSI-CDS and MinV-CDS. Thus, we observe a node size-lifetime tradeoff. However, we observe the MinV-CDS to compensate for the increase in the CDS Node Size by sustaining a significantly longer lifetime. We also observe the larger CDS Node Size and Edge Size of the MinV-CDS to contribute to a lower hop count per *s-d* path, which has the potential of contributing to lower end-to-end delay per path and energy consumption. In comparison to the benchmarking algorithm of minimum node size-based MaxD-CDS, we observe the SN-CDS to be relatively much more stable and incur only a 20% increase in the CDS Node Size. Among the three stability-based CDS algorithms, we consider the NSI-CDS effectively balances the tradeoff between the CDS Node Size and Lifetime, especially in high-density networks where it also incurs a lower hop count per *s-d* pair. In high density networks, the NSI-CDS could incur a lifetime that is about 6-7 times longer than that of the MaxD-CDS and a CDS Node Size that is only about twice that of the MaxD-CDS.

As part of future work, we plan to study the impact of using the CDS-induced sub graphs of the above three connected dominating sets on multicast routing in MANETs and compare the performance metrics (lifetime, hop count per path, number of edges) with respect to those obtained on regular undirected network graphs featuring all possible links according to the unit-disk graph model. We also plan to work on distributed implementations of the NSI-CDS and MinV-CDS algorithms that have been observed to effectively balance the CDS Node Size-Lifetime tradeoff and determine long-living stable CDS that can form a virtual backbone for unicast, multicast and broadcast communication in MANETs.

REFERENCES

Bettstetter, C., Hartenstein, H., & Perez-Costa, X. (2004). Stochastic properties of the random-way point mobility model. *Wireless Networks*, *10*(5), 555–567. doi:10.1023/B:WINE.0000036458.88990.e5

Cormen, T. H., Leiserson, C. E., Rivest, R. L., & Stein, C. (2009). *Introduction to algorithms* (3rd ed.). Cambridge, MA: MIT Press.

Hofmann-Wellenhof, B., Lichtenegger, H., & Collins, J. (2004). *Global positioning system: Theory and practice* (5th ed.). New York: Springer.

Keiss, W., Fuessler, H., & Widmer, J. (2004). Hierarchical location service for mobile ad hoc networks. *ACM SIGMOBILE Mobile Computing and Communications Review*, *8*(4), 47–58. doi:10.1145/1052871.1052875

Kuhn, F., Moscibroda, T., & Wattenhofer, R. (2004). Unit disk graph approximation. In *Proceedings of the Workshop on Foundations of Mobile Computing* (pp. 17-23). Philadelphia, PA: ACM.

Meghanathan, N. (2006). An algorithm to determine the sequence of stable connected dominating sets in mobile ad hoc networks. In *Proceedings of the 2nd Advanced International Conference on Telecommunications*. IARIA.

Sakai, K., Sun, M.-T., & Ku, W.-S. (2008). Maintaining CDS in mobile ad hoc networks. *Lecture Notes in Computer Science*, *5258*, 141–153. doi:10.1007/978-3-540-88582-5_16

Sheu, P.-R., Tsai, H.-Y., Lee, Y.-P., & Cheng, J. Y. (2009). On calculating stable connected dominating sets based on link stability for mobile ad hoc networks. *Tamkang Journal of Science and Engineering*, *12*(4), 417–428.

Su, W., Lee, S.-J., & Gerla, M. (2001). Mobility prediction and routing in ad hoc wireless networks. *International Journal of Network Management*, *11*(1), 3–30. doi:10.1002/nem.386

Velummylum, N., & Meghanathan, N. (2010). On the utilization of ID-based connected dominating sets for mobile ad hoc networks. *International Journal of Advanced Research in Computer Science*, *1*(3), 36–43.

Wang, F., Min, M., Li, Y., & Du, D. (n.d.). On the construction of stable virtual backbones in mobile ad hoc networks. In *Proceedings of the International Performance Computing and Communications Conference*. Phoenix, AZ: IEEE.

ADDITIONAL READING

Agarwal, S., Ahuja, A., Singh, J. P., & Shorey, R. (2000). Route Lifetime Assessment Based Routing Protocol for Mobile Ad hoc Networks. In *Proceedings of the International Conference on Communications* (pp. 1697-1701). New Orleans, LA, USA: IEEE.

Bae, S., Lee, S., Su, W., & Gerla, M. (2000). The Design, Implementation and Performance Evaluation of the On-demand Multicast Routing Protocol in Multi-hop Wireless Networks. *IEEE Network*, 4(1), 70–77. doi:10.1109/65.819173

Bettstetter, C., Hartenstein, H., & Perez-Costa, X. (2004). Stochastic Properties of the Random-Way Point Mobility Model. *Wireless Networks*, 10(5), 555–567. doi:10.1023/B:WINE.0000036458.88990.e5

Bianchi, G. (2000). Performance Analysis of the IEEE 802.11 Distributed Coordination Function. *IEEE Journal on Selected Areas in Communications*, 18(3), 535–547. doi:10.1109/49.840210

Bononi, L., & Felice, M. D. (2006). Performance Analysis of Cross-layered Multi-path Routing and MAC Layer Solutions for Multi-hop Ad Hoc Networks. In *Proceedings of the International Workshop on Mobility Management and Wireless Access* (pp. 190-197). Terromolinos, Spain: ACM.

Broch, J., Maltz, D. A., Johnson, D. B., Hu, Y. C., & Jetcheva, J. (1998). A Performance Comparison of Multi-hop Wireless Ad Hoc Network Routing Protocols. In *Proceedings of the 4th International Mobile Computing and Networking Conference* (pp. 85-97). Dallas, TX, USA: ACM.

Chen, Y. P., & Liestman, A. L. (2002). Approximating Minimum Size Weakly-Connected Dominating Sets for Clustering Mobile Ad hoc Networks. In *Proceedings of the International Symposium on Mobile Ad hoc Networking and Computing*. Lausanne, Switzerland: ACM.

Creixell, W., & Sezaki, K. (2007). Routing Protocol for Ad hoc Mobile Networks using Mobility Prediction. *International Journal of Ad Hoc and Ubiquitous Computing*, 2(3), 149–156. doi:10.1504/IJAHUC.2007.012416

Das, S. K., Manoj, B. S., & Murthy, C. S. R. (2002). Weight-based Multicast Routing Protocol for Ad hoc Wireless Networks. In *Proceedings of the Global Telecommunications Conference* (pp. 117-121). Taipei, Taiwan: IEEE.

Fasolo, E., Zanella, A., & Zorzi, M. (2006). An Effective Broadcast Scheme for Alert Message Propagation in Vehicular Ad hoc Networks. In *Proceedings of the International Conference on Communications, Vol. 9* (pp. 3960-3965). Istanbul, Turkey: IEEE.

Feeney, L. M. (2001). An Energy Consumption Model for Performance Analysis of Routing Protocols for Mobile Ad hoc Networks. *Journal of Mobile Networks and Applications*, 3(6), 239–249. doi:10.1023/A:1011474616255

Gil, H.-R., Yoo, J., & Lee, J.-W. (2003). An On-demand Energy-efficient Routing Algorithm for Wireless Ad hoc Networks. *Lecture Notes in Computer Science*, 2713, 302–311. doi:10.1007/3-540-45036-X_31

Gunes, M., Sorges, U., & Bouazizi, I. (2002). ARA: The Ant Colony based Routing Algorithm for MANETs. In *Proceedings of the ICPP. Workshop on Ad Hoc Networks* (pp. 79-85). Vancouver, Canada: IEEE.

Keiss, W., Fuessler, H., & Widmer, J. (2004). Hierarchical Location Service for Mobile Ad hoc Networks. *ACM Mobile Computing and Communications Review*, 8(4), 47–58. doi:10.1145/1052871.1052875

Lim, G., Shin, K., Lee, S., Yoon, H., & Ma, J. S. (2002). Link Stability and Route Lifetime in Ad hoc Wireless Networks. In *Proceedings of the International Conference on Parallel Processing Workshops* (pp. 116-123), Vancouver, Canada: IEEE.

Lin, X., & Stojmenovic, I. (2003). Location-based Localized Alternate, Disjoint and Multi-path Routing Algorithms for Wireless Networks. *Journal of Parallel and Distributed Computing, 63*, 22–32. doi:10.1016/S0743-7315(02)00037-0

Meghanathan, N. (2007). Impact of Broadcast Route Discovery Strategies on the Performance of Mobile Ad Hoc Network Routing Protocols. In *Proceedings of the International Conference on High Performance Computing, Networking and Communication Systems* (pp. 144-151). Orlando, FL, USA: Promote Research.

Meghanathan, N. (2007). Path Stability based Ranking of Mobile Ad Hoc Network Routing Protocols. *ISAST Transactions Journal on Communications and Networking, 1*(1), 66–73.

Meghanathan, N. (2009). Survey and Taxonomy of Unicast Routing Protocols for Mobile Ad hoc Networks. *The International Journal on Applications of Graph Theory in Wireless Ad hoc Networks and Sensor Networks, 1*(1), 1-21.

Meghanathan, N. (2009). A Location Prediction Based Reactive Routing Protocol to Minimize the Number of Route Discoveries and Hop Count per Path in Mobile Ad hoc Networks. *The Computer Journal, 52*(4), 461–482. doi:10.1093/comjnl/bxn051

Meghanathan, N. (2010). An Algorithm to Determine Minimum Velocity-based Stable Connected Dominating Sets for Ad hoc Networks. *Communications in Computer and Information Science Series, 94*, 206–217. doi:10.1007/978-3-642-14834-7_20

Meghanathan, N., & Sugumar, M. (2009). A Beaconless Minimum Interference Based Routing Protocol for Mobile Ad hoc Networks. *Communications in Computer and Information Science Series, 40*, 58–69. doi:10.1007/978-3-642-03547-0_7

Mueller, S., Tsang, R. P., & Ghosal, D. (2004). Multi-path Routing in Mobile Ad Hoc Networks: Issues and Challenges. *Lecture Notes in Computer Science, 2965*, 209–234. doi:10.1007/978-3-540-24663-3_10

Ni, S., Tseng, Y., Chen, Y., & Sheu, J. (1999). The Broadcast Storm Problem in a Mobile Ad Hoc Network. In *Proceedings of the International Conference on Mobile Computing and Networking* (151-162). Seattle, WA, USA: ACM.

Sheu, P., Tsai, H., Lee, Y., & Cheng, J. (2009). On Calculating Stable Connected Dominating Sets Based on Link Stability for Mobile Ad Hoc Networks. *Tamkang Journal of Science and Engineering, 12*(4), 417–428.

Son, D., Helmy, A., & Krishnamachari, B. (2004). The Effect of Mobility-Induced Location Errors on Geographic Routing in Mobile Ad hoc Sensor Networks: Analysis and Improvement using Mobility Prediction. *IEEE Transactions on Mobile Computing, 3*(3), 233–245. doi:10.1109/TMC.2004.28

Toh, C. K., Guichal, G., & Bunchua, S. (2000). ABAM: On-demand Associatvity-based Multicast Routing for Ad hoc Mobile Networks. In *Proceedings of the 52nd VTS Fall Vehicular Technology Conference, Vol. 3* (pp. 987-993), Boston, MA, USA: IEEE.

Viswanath, K., Obraczka, K., & Tsudik, G. (2006). Exploring Mesh and Tree-based Multicast Routing Protocols for MANETs. *IEEE Transactions on Mobile Computing, 5*(1), 28–42. doi:10.1109/TMC.2006.11

Wang, H., Ma, K., & Yu, N. (2005). Performance Analysis of Multi-path Routing in Wireless Ad Hoc Networks. In *Proceedings of the International Conference on Wireless Communications, Networking and Mobile Computing, Vol. 2* (pp. 723-726). Maui, HI, USA: IEEE.

Ye, Z., Krishnamurthy, S. V., & Tripathi, S. K. (2003). A Framework for Reliable Routing in Mobile Ad Hoc Networks. In *Proceedings of the International Conference on Computer Communications* (pp. 270-280), San Francisco, CA, USA: IEEE.

KEY TERMS AND DEFINITIONS

Algorithm: A sequence of steps to perform a particular task.

Connected Dominating Set (CDS): A subset of nodes in the network (that are reachable from one another either directly or through one or more nodes in the CDS) such that every node in the network is either in the subset or is a neighbor of a node in the subset.

CDS Edge Size: The number of edges connecting any two nodes in the CDS.

CDS Lifetime: The duration of existence of a CDS.

CDS Node Size: The number of nodes constituting the CDS.

Edge Distance Ratio (EDR): The EDR for an edge is the ratio of the physical Euclidean distance between the two constituent end nodes of the edge to that of the fixed transmission range.

Hop Count per *s-d* **Path:** The number of edges on a path, between any two nodes s and d, whose intermediate nodes, if any such exists, are those that are part of the CDS.

Mobile Ad Hoc Network (MANET): A network whose topology changes dynamically with time due to the mobility of the nodes.

Node Stability Index (NSI): The NSI value of a node is the sum of the predicted expiration times of the links with its neighbors.

Open Neighborhood: Node j belongs to the open neighborhood of node i (denoted ON_i) if the physical Euclidean distance between node i and j is $\leq R$, the transmission range of the nodes in the network. Every node $j \in ON_i$ is simply referred to as a "neighbor" of node i.

Stability: A characteristic of communication topologies related to the duration of existence of the topology; the longer the topology exists, the more stable it is.

Strong Neighborhood: Node j belongs to the strong neighborhood of node i (denoted SN_i) and vice-versa if the EDR of the edge i—j is $\leq TNDR$. Note that the *strong neighborhood* of a node is a subset of the nodes in its *open neighborhood*.

Topology: An arrangement of the nodes in the network that can be used for communication between any two nodes in the network.

Threshold Neighbor Distance Ratio (TNDR): The TNDR is the maximum value of the EDR for an edge i—j, in order for node i to be considered a "strong" neighbor of node j and vice-versa.

Transmission Range: The maximum distance between any two nodes such that the signal emanating from one node could directly reach the other node with strength appreciable enough to correctly extract the encoded information.

Chapter 10
Incidence of the Improvement of the Interactions between MAC and Transport Protocols on MANET Performance

Sofiane Hamrioui
UHA University, France & UMMTO University, Algeria, & USTHB University, Algeria

Jaime Lloret
Universidad Politecnica de Valencia, Spain

Mustapha Lalam
UMMTO University, Algeria

Pascal Lorenz
UHA University, France

ABSTRACT

In this chapter, the authors present an improvement to the interactions between MAC (Medium Access Control) and TCP (Transmission Control Protocol) protocols for better performance in MANET. This improvement is called IB-MAC (Improvement of Backoff algorithm of MAC protocol) and proposes a new backoff algorithm. The principle idea is to make dynamic the maximal limit of the backoff interval according to the number of nodes and their mobility. IB-MAC reduces the number of collisions between nodes. It is also able to distinguish between packet losses due to collisions and those due to nodes' mobility. The evaluation of IB-MAC solution and the study of its incidences on MANET performance are done with TCP New Reno transport protocol. The authors varied the network conditions such as the network density and the mobility of nodes. Obtained results are satisfactory, and they showed that IB-MAC can outperform not only MAC standard, but also similar techniques that have been proposed in the literature like MAC-LDA and MAC-WCCP.

DOI: 10.4018/978-1-4666-5170-8.ch010

INTRODUCTION

A MANET (Mobile Ad Hoc Network) (Basagni, Conti, Giordano, & Stojmenovic, 2004) is a complex distributed system which consists of wireless mobile or static nodes. These nodes can freely and dynamically self-organize. In this way they form arbitrary and temporary "Ad hoc" networks topologies, allowing devices to be interconnected wirelessly in areas with no pre-existing infrastructures.

In such network, MAC (Medium Access Control) protocol (Karn, 1990; Bhargavan, Demers, Shenker, & Zhang, 1994; Parsa & Garcia-Luna-Aceves, 1999) must provide an efficient access to the wireless medium and reduces the data interference. Important examples of this protocol include Carrier-Sense Multiple Access CSMA with collision avoidance which uses a random backoff algorithm (IEEE, 1999). CSMA can use a virtual carrier sensing mechanism using Request-To-Send/Clear-To-Send (RTS/CTS) control packets (Mjeku & Gomes, 2008). Both techniques are used in IEEE 802.11 MAC protocol (IEEE, 1999) which is the current standard for wireless networks.

TCP (Transmission Control Protocol) (Holland & Vaidya, 1999; Hanbali, Altman, & Nain, 2005) is the transport protocol used in most IP networks (Kurose and Ross, 2005) and recently in MANETs (Kawadia & Kumar, 2005). It is important to understand the TCP behavior when coupled with IEEE 802.11 MAC protocol in such network.

When the interactions between the MAC and TCP protocols are not taken into account, this can degrade the MANET performance notably, by affecting to the performance of the TCP parameters, i.e. throughput and end-to-end delay (Jiang, Gupta, & Ravishankar, 2003; Nahm, Helmy, & Kuo, 2004; Papanastasiou, Mackenzie, Ould-Khaoua, & Charissis, 2006). In order to adapt the behavior of these protocols for better QoS (Li, 2006), it is very important to study their interactions. In Hamrioui, Bouamra, and Lalam (2007), we

presented a study of the interactions between the MAC and TCP protocols. We showed that TCP performance (notably throughput parameter) is degraded while the number of nodes increases in a MANET using IEEE 802.11 MAC as access control protocol. In Hamrioui and Lalam (2008), we proposed a solution for the problem notified in Hamrioui et al. (2007), but we were just limited to a chain topology and also to the influence of the number of nodes on the TCP performance. In Hamrioui and Lalam (2010) we studied the validity of the solutions proposed previously with several routing and transport protocols and also with different static topologies. The results showed that the proposed solutions are not only influenced by the change of the network topology but also by the routing and the used transport protocols.

Our contribution in this paper is to improve TCP protocol performance by exploiting the backoff algorithm of the MAC protocol. The work done in this chapter is the next step that follows the works (Hamrioui et al., 2007; Hamrioui & Lalam, 2008, 2010). In addition to the number of nodes, we improved our previous solutions by taking into account another parameter which is the mobility of the nodes. We also compared our solution with others solutions which have been proposed in the same context. After a short presentation of MAC and TCP protocols, we will present our IB-MAC (Improvement of the Backoff algorithm of the MAC protocol) and study its incidences on some TCP performance parameters (throughput and end-to-end delay). IB-MAC proposes a dynamic adaptation of the maximal limit of the MAC backoff algorithm. This adaptation includes the number of nodes in the network and their mobility.

Our paper is structured in five sections. Section two gives a short presentation of MAC and TCP protocols. In section three, we present the IB-MAC improvement and in Section four we study its incidences on the TCP performance parameters. We finish our chapter with section five which provides the conclusion and future work.

IEEE 802.11 MAC and TCP Protocols in MANET

IEEE 802.11 MAC protocol defines two different access methods: polling based point coordination function (PCF) and distributed coordination function (DCF) which is used essentially in MANET. DCF access is basically a carrier sense multiple access with collision avoidance (CSMA/CA) mechanism. When a node wants to transmit a frame, it senses the medium. If the medium is busy, it defers this transmission. If the medium is free for a specified time, called the distributed inter-frame space (DIFS), the node can transmit. In order to avoid a collision due to the hidden terminal problem (Jayasuriya, Perreau, Dadej, & Gordon, 2004; Altman & Jimenez, 2003), the node first transmits a Request To Send (RTS) control frame. Then, the destination node responds with a Clear To Send (CTS) control frame. Both RTS and CTS frames include the duration of the transmission that will follow the RTS-CTS exchange. All nearby nodes receiving either the RTS or the CTS frame defer their pending transmissions for this duration. This deferral is referred to as virtual carrier sensing as it "senses" the medium through the exchange of frames at the MAC layer. Once a successful RTS-CTS frame exchange takes place, the data frame is transmitted. The receiving node checks the received data frame, and upon correct receipt, sends an acknowledgement (ACK) frame. If the sending node fails to receive the acknowledgement frame, it assumes that the data frame was lost. It backs-off and attempts a retransmission. After several repeated failures to get the frame across, it simply drops the frame. Thus, although the introduction of RTS-CTS-DATA-ACK frame format makes the transmission more reliable, there is still the possibility of transmission failures. Such failures are more frequent in MANET.

It has been shown that TCP does not work well in a wireless network (Holland & Vaidya, 1999; Kuang, Xiao, & Williamson, 2003). The wireless channel is subject to noise, so the packets losses are frequent. TCP does not distinguish between the packets losses due to the congestion and to the transmission failures at link layer. TCP associates the packet losses to the congestion, and then it starts its congestion control mechanism. Therefore, transmission failures at the MAC layer lead to the congestion control activation by the TCP protocol. Then, the number of packets is reduced (throughput). Several mechanisms have been proposed to address this problem (Bakre & Badrinath, 1995; Brown & Singh, 1997; Bensaou, Wang, & Ko, 2000), but most of them focus on the cellular architecture. The problem is more complex in multi-hop networks such as MANET where there is no base station and each node can act as a router (Gerla, Tang, & Bagrodia, 1999; Gupta & Wormsbecker, 2004).

The performance of TCP throughput and end-to-end delay parameters has been the subject of several evaluations. It has been shown that these parameters degrade when the interactions between MAC and TCP are not taken into account (Holland & Vaidya, 1999; Jayasuriya et al., 2004). In our previous work (Hamrioui et al., 2007), we confirmed these results by studying the effect of the MAC layer when the number of nodes increases. The major source of these effects is the problem of hidden and exposed nodes (Jayasuriya et al., 2004; Altman & Jimenez, 2003). The most important solution, which has been proposed to the hidden node problem, is the use of RTS and CTS frames (Jain, Dubey, Upadhyay, & Charhate, 2008; Marina & Das, 2004). Although the use of RTS/CTS frames is considered as a solution to the hidden node problem, it was shown in Jayasuriya et al. (2004) and Ng, Liew, Sha, and To (2005) that it also leads to further degradation of TCP flow by creating more collisions and introduce an additional overhead. Then these two constraints decrease the TCP performance.

Related Work

Many analyses of TCP protocol performance are done and several solutions on how to improve this performance are proposed (Bakre & Badrinath, 1995; Balakrishnan et al., 1995; Brown & Singh, 1997; Tsaoussidis & Badr, 2000; Zhang & Tsaoussidis, 2001). In this paper, we present the most important of these solutions.

Yuki, Yamamoto, Sugano, Murata, Miyahara, and Hatauchi (2004) have proposed a technique that combines data and ACK packets, and have shown through simulation that this technique can make radio channel utilization more efficient. The technique improved the TCP performance up to 60% and about 10% even when the network load was very high.

Altman and Jimenez (2003) proposed an improvement for TCP performance by delaying 3-4 ACK packets. In their approach the receiver always delays 4 packets (except at the startup) or less if its timeout interval expires. The receiver uses a fixed interval of 100ms and does not react to packets that are out-of-order or filling in a gap in the receiver buffer, as opposed to the RFC 1122.

Kherani and Shorey (2004) suggest significant improvements in TCP performance as the delayed acknowledgement parameter d increases the TCP window size W. The novelty in approach is the analytic modeling of TCP over IEEE 802.11 based networks with a delayed acknowledgement parameter, and, a TCP window size of $W (> 2)$ packets.

Allman (1998) conducted an extensive evaluation on Delayed Acknowledgment (DA) strategies, and they presented a variety of mechanisms to improve the TCP performance in presence of the side-effect of delayed ACKs.

Chandran (1998) proposed TCP-feedback. With this solution, when an intermediate node detects the disruption of a route, it explicitly sends a Route Failure Notification (RFN) to the TCP sender. The source on receiving the RFN, it suspends all packet transmissions and freezes its state. But when a middle node learns of a new route to the destination, it sends a Route Re-establishment Notification (RRN) to the source.

Holland and Vaidya (1999) proposed a similar approach based on ELFN (Explicit Link Failure Notification). When the TCP sender is informed of a link failure, it freezes its state. However, the source continues sending out packets at regular intervals in order to determine if a new route is available.

Liu and Singh (2001) proposed the ATCP protocol. It tries to deal with the problem of high Bit Error Rate (BER) and route failures. The ATCP layer is inserted between the TCP and IP layers. ATCP puts TCP agent into the appropriate state after listening to the network state information provided by ECN (Explicit Congestion Notification) messages and by ICMP "Destination Unreachable" message. On receiving a "Destination Unreachable" message, TCP agent enters a persist state.

Fu, Greenstein, Meng, and Lu (2002) investigated TCP improvements by using multiple end-to-end metrics instead of a single metric. They claim that a single metric may not provide accurate results in all conditions. They used four metrics: inter-packet delay difference at the receiver, short-term throughput, packet out-of-order delivery ratio, and packet loss ratio. These four metrics are cross checked for accurate detection of the network internal state.

Biaz and Vaidya (1998) evaluated three schemes for predicting the reason for packet losses inside wireless networks. They applied simple statistics on observed Round-trip Time (RTT) and/or observed throughput of a TCP connection for deciding whether to increase or decrease the TCP congestion window. The general results were discouraging in that none of the evaluated schemes performed really well.

Liu, Matta, and Crovella (2003) proposed an end-to-end technique for distinguishing between packet losses due to congestion from packet losses in the wireless medium. They designed a Hidden Markov Model (HMM) algorithm to perform the mentioned discrimination taking RTT measurements over the end-to-end channel.

Kim, Toh, and Choi (2001) proposed the TCP-BuS (TCP Buffering capability and Sequence information). Like previous proposals, it uses the network feedback in order to detect route failure events and to take convenient reaction to this event. The novel scheme in this proposal is the introduction of *buffering capability* in mobile nodes. The authors select the source-initiated on-demand ABR (Associativity-Based Routing) routing protocol.

Oliveira and Braun (2005) propose a dynamic adaptive strategy for minimizing the number of ACK packets in transit and mitigating spurious retransmissions. Using this strategy, the receiver adjusts itself to the wireless channel condition by delaying more ACK packets when the channel is in good condition and less otherwise.

Hamadani and Rakocevic (2008) address the problem of TCP intra-flow instability in multi-hop ad hoc networks. They propose a cross layer algorithm called TCP Contention Control that is able to adjust the amount of outstanding data in the network based on the level of contention experienced by the packets as well as the throughput achieved by connections. The main features of this algorithm are its flexibility and adaptation to the network conditions.

Zhai, Chen, and Fang (2007) show that TCP suffers severe performance degradation and unfairness. Realizing that the main reason is the poor interaction between traditional TCP and the MAC layer, they propose a systematic solution named Wireless Congestion Control Protocol (WCCP) to address this problem in both layers. WCCP uses channel busyness ratio to allocate the shared resource and accordingly adjusts the sender's rate so the channel capacity can be fully utilized and fairness is improved.

Lohier, Doudane, and Pujolle (2006) propose to adapt one of the MAC parameters, the Retry Limit (RL), in order to reduce the drop in performance due to the inappropriate triggering of TCP

congestion control mechanisms. Starting from this, a MAC-layer LDA (Loss Differentiation Algorithm) is proposed. This LDA scheme is based on the adaptation of the RL parameter depending on the quality of the 802.11 wireless channels.

In the conclusion, none of the previous approaches has addressed all or even some of the communication parameters involved in the degradation of the TCP performance. Some works have only studied the parameters related to the network traffic, while the others didn't study any of them. In fact, just modeling these parameters and their behavior in an approach to obtain better TCP performance will be a great contribution for MANET performance. Our work focuses precisely on these parameters in order to provide an efficient solution in terms of TCP parameters performance. By exploiting the backoff algorithm of MAC layer, we show that it is possible to ensure better throughput and end-to-end delay parameters performance.

IB-MAC (IMPROVEMENT OF THE BACKOFF ALGORITHM FOR MAC PROTOCOL)

MAC protocol is based on the Backoff algorithm that allows it to determine which node will access to the wireless medium in order to avoid collisions. In the case of station finds a channel busy, the transmission is different according to the Backoff procedure whose principle is as follows. As long as the channel is free for a DCF Inter Frame Space (DIFS) time (after a successful reception) or for an Extended Inter-Frame Space (EIFS) time (after a failed reception), the Backoff time is decreased. This time is calculated as follows:

$$BackoffTime = BackoffCounter * aSlotTime \quad (1)$$

where *aSlotTime* is a time constant and *BackoffCounter* is an integer from uniform distribution in the interval [0, *CW*] and *CW* is the contention window, whose minimum and maximum limits are (CW_{min}, CW_{max}) and are defined in advance.

The CW value is increased in the case of non availability of the channel by using the following formulas:

$$m \leftarrow m + 1$$

$$CW(m) = (CW_{min} + 1)*2^m - 1 \qquad (2)$$

$$CW_{min} <= CW(m) <= CW_{max}$$

where m is the number of retransmissions.

The first parameter used by our IB-MAC solution is the number of nodes in the network. As we have seen through the simulations presented in the previous work, (Hamrioui et al., 2007; Hamrioui & Lalam, 2008), when the number of nodes in the network increases, the performance of TCP deteriorates. The cause of this degradation is the frequent occurrence of collisions between nodes. The more number of nodes, more frequent collisions. These collisions become more frequent with a small backoff interval because the probability of having two or more nodes choosing the same value in a small interval is greater than the probability of having these nodes choosing the same value in a larger interval.

Let I be this interval, S_I its size, and $Pr(i,x)$ the probability that node i chooses the x value in the I interval. The problem then is how to ensure that for any two nodes i and j in the network with $i != j$, we will have:

$$| Pr(i,x) - Pr(j,x) | = y \qquad (3)$$

with $y != 0$

For an important number of nodes in the network, and for a high probability in Equation (**3**), we must have a larger S_I. In order to achieve this we want to make the size of S_I adaptable to the number of nodes in the network. So we intervene on one of the limits of this interval, by proposing the maximum limit: CW_{max}.

Let n be the number of nodes in the network. Then, the first part of the expression of CW_{max} will be:

$$F(n) = log(n) \qquad (4)$$

Log () is used in Equation (4) because we found in Hamrioui et al. (2007) and Hamrioui and Lalam (2008) that the effects of having large number of nodes on the TCP performance are almost the same.

Our solution takes into account all the nodes in the MANET. In our work, we are interested not only to the effective data but also on all kind of data like control packets (information about the status of the nodes, the RTS/CTS packets, etc). Moreover, even when the nodes in the MANET don't transmit, they continually exchange control packets. So we can say that all the nodes belonging to the MANET are continually active. For this reason, our proposed solution IB-MAC takes into account the total number of nodes in the MANET. With our solution, each node must know the variable n. This variable will be constantly updated whenever a new mobile node joins the network or leaves it (for reasons such as low energy, no processor or memory availability, etc). The n value is broadcasted jointly with other information about the MANET by using a piggy backing technique.

Many diffusion algorithms exist (Lehsaini, 2009) and could be implemented with our solution to ensure the availability of the information n for all the nodes in the MANET. Upon receipt of n, all existing nodes in the network (those who are already registered and associated with MANET) will generalize this new value of n on those which they have at their level. When there is a new mobile node in the network, upon receipt of n, it increments n by 1. Then, it is stored it locally and broadcasted to the whole network. When any node leaves the network for any reason, it decrements the value of n and broadcasts it to the others nodes.

Our IB-MAC also takes into account the mobility of nodes because it participates in the degradation of TCP performance (Hamrioui, Lalam, & Lorenz, 2011; Hamrioui, Bouamra, & Lalam, 2008). In fact, node mobility often leads to the breakdown of connectivity between nodes, resulting in TCP packet losses and, then, the degradation of the TCP performance parameters (throughput and end-end-end delay). At the MAC protocol, when packet losses are detected, they are associated to the collisions problem, which is not the case here. Then, the more mobility, the larger backoff interval. It should not happen because these packets are lost due to the rupture of the connectivity, but not due to the collisions. Therefore, we will try to find a compromise between the effect of mobility and the size of the backoff interval.

Mobility is generally characterized by its speed and angle of movement. These two factors determine the degree of the impact of mobility on packet losses. Let node i be in communication with node j, then we present the following notation:

α_i: The angle between the line (i, j) and the movement direction of node i,

W_i: The speed of the mobile node i.

To consider the impact of mobility on the packet losses is equivalent to consider the impact of these two parameters: W_i and α_i. For the speed effect (W_i), as in the case of the number of nodes, we use a logarithmic function because for large speed mobility values the results converge. So this is expressed as follows:

$$H(W_i) = \begin{cases} 1 & \text{if } W_i = 0 \text{ (Without mobility)} \\ Log(W_i) & \text{else} \end{cases} \quad (5)$$

In addition, the direction of the node movement determines the degree of the influence of mobility on packet losses; it is given by $G(W_i, \alpha_i)$ as follows:

$$G(W_i, \alpha) = \begin{cases} 1 & \text{if } W_i = 0 \text{ (without mobility)} \\ 1 & \text{if } -\prod/4 <= \alpha_i <= \prod/4 \quad (6) \\ \sqrt{W_i} & \text{else} \end{cases}$$

The impact of the mobility on the packet losses is given by $K(W_i, \alpha)$ function as follows:

$$K(W_i, \alpha_i) = H(W_i) * G(W_i, \alpha_i) \quad (7)$$

IB-MAC allows each node to get the value of its W_i. The easiest way to do it is to deduce it by knowing the time spent between two geographical points. There are many systems for mobile nodes location such us GPS and power measurement techniques (Elliot, 2005; Doherty, Pister, El ghaoui, 2001). With these systems, each node can know its position at any time, and then it will be able to deduce the distance travelled during an interval of time. With the distance and time we can get the speed W_i.

The same happens to α_i, which represents the angle between the direction of movement and the direction of the communication. With these location systems it is possible to determine the information about node positions and the direction of their movements. With all these information each node is able to determine the value of the angle α_i.

We know that when W_i increases, the packet losses increase too, it increases more when the node is moving in the opposite direction of the communication. This packet losses increase has a negative impact on the backoff interval because they can be associated to the collisions, but it is not the case here (as explained above). To make this impact positive, we must use the inverse of $K(W_i, \alpha_i)$, as follows:

$$M(W_i, \alpha_i) = 1 / K(W_i, \alpha_i) \quad (8)$$

Equation (8) decreases with the increase of $G(W_i, \alpha_i)$ and $H(W_i)$ (when W_i increases). It decreases more when the node is moving in the opposite direction of communication. $M(W_i, \alpha_i)$ provides the impact of the mobility on the packet losses, it is the probability that the cause of these losses is the mobility of nodes.

With (8), we can guarantee that when the mobility of nodes is significant, the adaptation of the backoff algorithm is not important because this mobility may cause higher packet losses. But with weak mobility the same equation makes backoff algorithm be able to get a significant adaptation because in this case the collisions between frames are more probable to be the cause of the packet losses.

We give now the new expression of CW_{max} as follows:

$$CW_{max}(n, W_i, \alpha_i) = CW_{max0} + F(n) * M(W_i, \alpha_i) \ (9)$$

\where CW_{max0} is the initial CW_{max} defined by the MAC protocol (with the IEEE 802.11 version, it is equal to 1024).

After having made the values of CW_{max} adaptive to the number of nodes used and their mobility, IB-MAC (improved version of that given by the Equation (**2**)) becomes:

$$m \leftarrow m + 1$$

$$CW(m) = (CW_{min}(n) + 1) * 2_m - 1$$

$$CW_{min} <= CW(m) <= CW_{max}(n, W_i, \alpha_i) \qquad (10)$$

$$CW_{max}(n, W_i, \alpha_i) = CW_{max0} + (M(W_i, \alpha_i) * F(n))$$

where:

m: the number of retransmissions.
n: the number of the nodes used.
α_i: the angle between the line formed by the mobile node *i* and its corresponding node and the movement direction of this mobile node.
W_i: the speed of mobile node *i*.
CW_{max0}: initial value of CW_{max}.

IB-MAC EVALUATION PERFORMANCE

Simulation Environment

We evaluate our proposed IB-MAC using the simulation environment NS-2 (version 2.34) from Lawrence Berkeley National Laboratory (LBNL) with wireless extension of CMU (Fall & Varadhan, 1998).

The MAC level uses IEEE 802.11b model with the DCF function. The values of its basic parameters are listed in Table 1.

All the nodes communicate through wireless links in half-duplex with an identical bandwidth of 1 Mb/s upstream/downstream. For our simulations, the effective transmission range is of 250 meters and an interference range of 550 meters. All nodes in the area given by this distance to a transmitting node will find the medium busy. Each node has a queue buffer link layer of 50 packets managed with a mode drop-tail (Floyd & Jacobson, 1993). The packet size is 1024 bytes. The scheduling packet transmissions technique is the First In First Out (FIFO) type. The propagation model used is the two-ray ground model (Bullington, 1957).

Our simulations are performed using the reactive routing protocol AODV (Perkins, Royer, & Das, 2006). We used TCP NewReno (Floyd & Henderson, 1999) which is a reactive variant, derived and widely deployed, and whose performance was evaluated under conditions similar to

Table 1. Parameters for IEEE 802.11 MAC

Parameters	Values
Preamble length (bit)	144
RTS length (bit)	160
CTS/ACK length (bit)	112
MAC header (bit)	224
IP header (bit)	160
SIFS (μs)	10
DIFS (μs)	50
Slot time (μs)	20
Contention window	31
Retry limit	7

those conducted here. We chose this because of our previous work (Hamrioui & Lalam, 2010) have shown almost similar results for different routing protocols and TCP versions.

The values, such as the duration of the simulation, the speed of the nodes, and the number of connections have been established in order to obtain interpretable results compared to those published in the literature. The simulations are performed for 1000 seconds in order to analyze the full spectrum of the TCP throughput.

We considered two cases: without and with mobility. In the first case, we used the chain topology. Node 1 sends messages to node *n* (where n is the length of the chain). We just limited our study to the chain topology because in our previous work (Hamrioui & Lalam, 2010) we have studied different topologies and we have found that it influences the communication environment similarly. The distance between two neighboring nodes is 200 meters and each node can communicate only with its nearest neighbor. The interference range of a node is about two times higher than its transmission range (550 meters in our case).

In the case with mobility, we studied a random topology with two cases: weak and strong mobility. In both cases, node 1 sends messages to node n. The mobility model uses the random waypoint model (Hyytiä & Virtamo, 2005). We justify our choice by the fact that the network is not designed for mobility and that this particular model is widely used in the literature. In this model the node mobility is typically random and all nodes are uniformly distributed in simulation space. The nodes move in 2200m*600m area, each one starts its movement from a random location to a random destination. Once the destination is reached, another random destination is targeted after a pause time.

Parameters Evaluation

We have simulated several scenarios with different numbers of nodes n and mobility values. We are interested in two specific parameters. The first

one is the throughput, which is given by the ratio of the received data, taking into account all data sent. The second parameter is the end-to-end delay, which is given by the ratio of the time when the data is received minus the data transmission time divided by the number of data packets received.

Simulations and Results

In these scenarios, we compare our solution (IB-MAC) with MAC standard and two other solutions proposed in the literature. All these solutions use the MAC layer to improve TCP performance in the MANET. The first solution is Wireless Congestion Control Protocol (WCCP) (Zhai et al., 2007) and the second one is MAC-layer LDA (Loss Differentiation Algorithm) (Lohier et al., 2006). The explanation of each solution is given in the related work section.

Scenario 1: Without Mobility

We see in Figure 1, with MAC protocol, that when the number of nodes participating in the network increases, the throughput decreases. This degradation at a given time (from n=100 nodes) begins to be stable. This degradation is due to TCP packet losses, and that becomes more important when the size of the network increases. With the analysis of the file traces for these graphs, we found that RTS and CTS frames, handled at the MAC level, are sensitive to the network size. The more number of nodes, the more losses of these two frames. It has been previously shown that such frame losses in such simulation conditions are mainly given due to the consequences of hidden and exposed nodes. The same result was found in our previous work (Hamrioui et al., 2007; Hamrioui & Lalam, 2008).

But when IB-MAC is used as a MAC protocol, we see that the throughput is better. There is an important improvement of this parameter, even if there is a slight decrease when the number of nodes increases but this decrease is much smaller compared to the first case when the MAC protocol is used. This improvement is due to the

Figure 1. Throughput variation without mobility

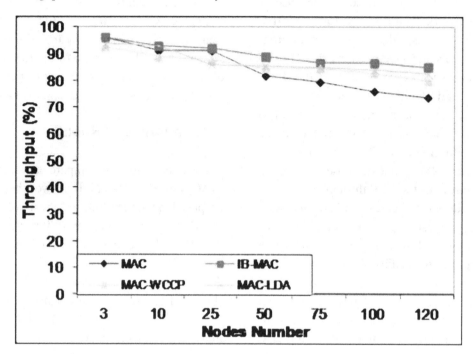

use of the adaptive nature of our IB-MAC solution to the number of nodes in the network. This adaptation reduces the probability of collisions.

Figure 2 shows the evolution of the second studied parameter, which is the end-to-end delay, when the number of nodes increases. With MAC protocol, we can see that this parameter significantly increases with the increase of the number of nodes. It happens because of the frequent loss of TCP packets in the network increases when the number of nodes increases. These losses will be the cause of the frequent start of the congestion avoidance mechanism by the TCP protocol, so that will result in delaying the transmission of TCP packets and the increase of the delay. This delay increase begins to stabilize from n = 110 nodes and provides values below t = 1.2 s approximately.

With IB-MAC, the end-to-end delay is better even if there is a slight increase when the number of nodes increases but this decrease is smaller compared to the first case when the MAC protocol is used.

Based on these results, we can say that our IB-MAC outperforms not only MAC standard, but also similar techniques that have been proposed in the literature. The results of the variation of the throughput and the end-to-end delay parameters are better than those of MAC-LDA, MAC-WCCP and MAC standard. This improvement is due to the dynamic nature of our new IB-MAC algorithm which makes the size of the backoff interval adjustable to the number of nodes in the network. This adjustment reduces the probability of collisions between nodes, thus the number of lost packets is reduced while the throughput and delay are improved.

Scenario 2: Weak Mobility (W = 5 m/s)

For weak mobility, when MAC protocol is used, we found an important degradation of throughput and end-to-end delay parameters compared to the first case (without mobility). To explain this degradation, we analyzed the obtained trace files and we found:

Figure 2. End-To-end delay variation without mobility

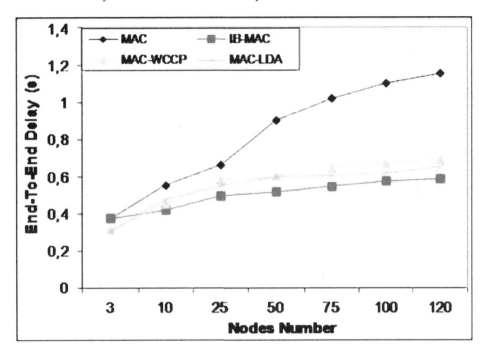

1. RTS/CTS frame losses with the increase of the number of nodes in the network (it is the same than the first case without mobility);
2. TCP packet losses even if there are successful RTS/CTS frame transmissions. In this case, these losses are caused by the route unavailability due the nodes mobility (the used route is outdated, denoted by "NRTE" in the trace file).

We deduce through i) and ii) that the mobility of nodes, although it is weak (here speed W = 5 m/s), participates in the degradation of the throughput and end-to-end delay parameters.

With our IB-MAC solution, we found an important improvement of the throughput and end-to-end delay parameters in comparison to the first case when the MAC protocol is used. Our IB-MAC algorithm makes the size of the backoff interval adjustable to the number of nodes in the network and their mobility. For this reason, even for the case where the nodes are mobiles, the probability of collisions between nodes is reduced, and then

the throughput and end-to-end delay parameters performance is improved.

Figure 3 and Figure 4 show that IB-MAC outperform the others protocols used (MAC-LDA and MAC-WCCP). Although there is a slight difference, our strategy remains better than the other three ones.

In fact, IB-MAC avoids the importunate increase of the backoff interval when nodes are mobile (with the equation (7)). This happens because the probability of having higher packet loss due to nodes' mobility is important. With this control, the nodes don't have to wait uselessly to send (or to re-send) their packets, then it avoids to decrease the TCP parameters performance.

Scenario 3: Strong Mobility (W = 25 m/s)

Figure 5 and Figure 6 shows an important degradation of the throughput and end-to-end delay parameters when MAC protocol is used. In fact,

Figure 3. Throughput variation with weak mobility (speed W=5 m/s)

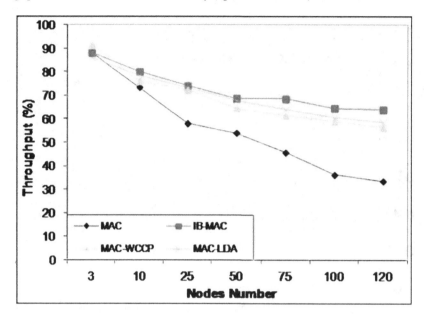

Figure 4. End-to-end delay variation with weak mobility (speed W = 5 m/s)

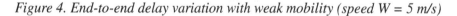

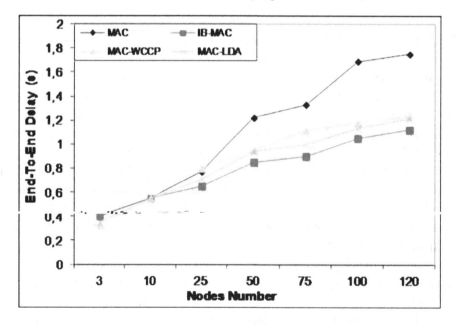

with strong mobility, the number of broken links increases, then the links stability becomes more important. We have done the same analysis as before in order to know the reasons of this degradation. We found that the causes of this degradation are also related to those discussed in i) and ii) in weak mobility case. The amount of data loss is smaller in the case where nodes move at low speeds

(strong mobility), and grows when their mobility increases. When the network has weak mobility (nodes with low speeds), it presents a rather high stability; then link failures are less frequent than the case of high mobility.

For both last cases (weak and strong mobility), IB-MAC presents better performance of throughput and end-to-end delay parameters compared

Figure 5. Throughput variation with strong mobility

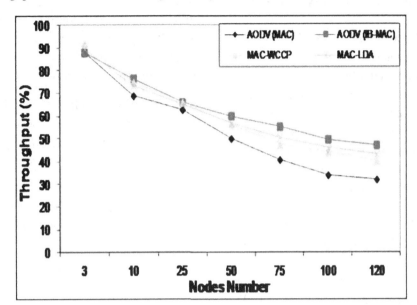

Figure 6. End-to-end delay variation with strong mobility

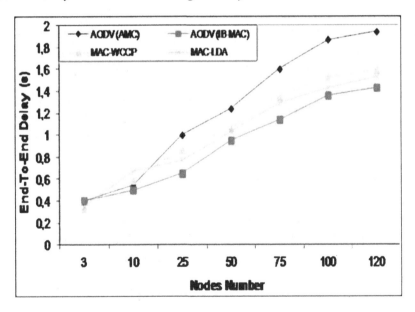

to MAC standard and the other tested solutions (MAC-LDA, MAC-WCCP). In fact, IB-MAC guarantees that when the mobility of nodes is significant, the adaptation of the backoff algorithm is not important because this mobility is more probable to be the cause of many packet losses. But with weak mobility, there is significant adaptation to the backoff algorithm, because in this case the collisions between frames are more probable to be the cause of the packet losses.

Based on these last results, we can say that even in the case of a random topology, where nodes are mobile (a feature specific to MANET networks), the IB-MAC solution improves the TCP performance.

CONCLUSION AND FUTURE WORK

Improving TCP performance over IEEE 802.11 MAC protocol in multi-hop ad-hoc networks is truly a problem on interaction between two layers. In this chapter, we proposed an improvement of TCP protocol performance (throughput and end-to-end delay) in MANET by adapting MAC layer. Our solution, called IB-MAC, proposes a new backoff algorithm which is adjusts the CW_{max} window to the number of nodes used in the network and their mobility.

IB-MAC reduces the number of collisions between nodes and distinguishes between the packets loss due to collisions and those due to the mobility of the nodes. Then, we studied the incidences of IB-MAC on TCP performance. We limited our study on the throughput and end-to-end delay parameters because they have great effects on the TCP and MANET performance. The results are satisfactory and showed that IB-MAC outperforms MAC standard and similar techniques proposed in the literature which are MAC-LDA and MAC-WCCP.

The achieved results in this chapter are encouraging, justifying further investigation on this direction. As a future work, we have to test the number of nodes until our solutions remains valid. The continuation of our work will consist on looking for a complete cross-layer IB-MAC in order to adapt dynamically, and in a coordinated, way the parameters of MAC and TCP protocols for better interactions.

REFERENCES

Allman, M. (1998). On the generation and use of tcp acknowledgements. *ACM Computer Communication Review, 28,* 1114–1118.

Altman, E., & Jimenez, T. (2003). Novel delayed ACK techniques for improving TCP performance in multihop wireless networks. In *Proceeding of IFIP International Federation for Information Processing* (pp. 237-250). Venice, Italy: IFIP.

Bakre, B., & Badrinath, R. (1995). I-TCP: Indirect TCP for mobile hosts. In *Proceeding of the 15th Int. Conf. Distributed Computing Systems* (pp. 136 – 143). Vancouver, Canada: IEEE.

Balakrishnan, H., Seshan, S., Amir, E., & Katz, R. (1995). Improving tcp/ip performance over wireless networks. In *Proceeding of 1st ACM Mobicom* (pp. 2-11). ACM.

Basagni, S., Conti, M., Giordano, S., & Stojmenovic, I. (2004). *Mobile ad hoc networking*. Hoboken, NJ: Wiley-IEEE Press. doi:10.1002/0471656895

Bensaou, B., Wang, Y., & Ko, C. C. (2000). Fair media access in 802.11 based wireless ad-hoc networks. In *Proceeding of Mobihoc* (pp. 99–106). Boston, MA: ACM.

Bhargavan, V., Demers, A., Shenker, S., & Zhang, L. (1994). MACAW, a media access protocol for wireless LANs. In *Proceeding of ACM SIGCOMM* (pp. 212-225). London, UK: ACM.

Biaz, S., & Vaidya, N. H. (1998). Distinguishing congestion losses from wireless transmission losses: A negative result. In *Proceeding of IEEE 7th Int. Conf. on Computer Communications and Networks* (pp. 722-731). IEEE.

Brown, K., & Singh, S. (1997). M-TCP: TCP for mobile cellular networks. *ACM Computer Communications Review, 27,* 19–43. doi:10.1145/269790.269794

Bullington, K. (1957). Radio propagation fundamentals. *The Bell System Technical Journal, 36*(3), 593–625. doi:10.1002/j.1538-7305.1957.tb03855.x

Chandran, K. (1998). A feedback based scheme for improving TCP performance in ad-hoc wireless networks. In *Proceeding International Conference on Distributed Computing Systems* (pp. 472-479). Amsterdam: IEEE.

Doherty, L., Pister, K. S. J., & El Ghaoui, L. (2001). Convex position estimation in wireless sensor networks. In *Proceeding of the IEEE INFOCOM* (pp. 1655-1663). IEEE.

Elliott, D. K. (2005). *Understanding GPS: Principles and applications* (2nd ed.). Artech House Publishers.

Fall, K., & Varadhan, K. (1998). *Notes and documentation*. LBNL. Retrieved from http://www.mash.cs.berkeley.edu/ns

Floyd, S., & Henderson, T. (1999). New reno modification to TCP's fast recovery. *RFC 2582*.

Floyd, S., & Jacobson, V. (1993). Random early detection gateways for congestion avoidance. *IEEE/ACM Transactions on Networking*, *1*, 397–413. doi:10.1109/90.251892

Fu, Z., Greenstein, B., Meng, X., & Lu, S. (2002). Design and implementation of a tcp-friendly transport protocol for ad hoc wireless networks. In *Proceeding of 10th IEEE International Conference on Network Protocosls (ICNP'02)* (pp. 216-225). Paris, France: IEEE.

Gerla, M., Tang, K., & Bagrodia, R. (1999). TCP performance in wireless multihop networks. In *Proceeding of IEEE WMCSA* (pp. 41-50). New Orleans, LA: IEEE.

Gjupta, W. A., & Williamson, C. (2004). Experimental evaluation of TCP performance in multihop wireless ad hoc networks. In *Proceeding of MASCOTS* (pp. 3-11). Volendam, The Netherlands: MASCOTS.

Hamadani, E., & Rakocevic, V. (2008). A cross layer solution to address TCP intra-flow performance degradation in multihop ad hoc networks. *Journal of Internet Engineering*, *2*(1), 146–156.

Hamrioui, S., Bouamra, S., & Lalam, M. (2007). Interactions entre le protocole MAC et le protocole de transport TCP pour l'optimisation des MANET. In *Proceeding of the 1ˢᵗ International Workshop on Mobile Computing & Applications (NOTERE'2007)*. NOTERE.

Hamrioui, S., Bouamra, S., & Lalam, M. (2008). Les effets de la mobilité des nœuds sur la QoS dans un MANET. In *Proceeding of the Maghrebian Conference on Software Engineering and Artificial Intelligence (MCSEAI'08)*. Oran, Algeria: MCSEAI.

Hamrioui, S., & Lalam, M. (2008). Incidence of the improvement of the transport – MAC protocols interactions on MANET performance. In *Proceeding of the 8th Annual International Conference on New Technologies of Distributed Systems (NOTERE 2008)* (pp. 634-648). Lyon, France: NOTERE.

Hamrioui, S., & Lalam, M. (2010). Incidences of the improvement of the MAC-transport and MAC–routing interactions on MANET performance. In *Proceeding of the International Conference on Next Generation Networks and Services*. IEEE.

Hamrioui, S., Lalam, M., & Lorenz, P. (2011). Effets de la mobilité sur les protocoles de transport et de routage dans les MANET. In *Proceeding of Journées Doctorales en Informatique et Réseaux, (JDIR'11)*. Belfort, France: JDIR.

Hanbali, A., Altman, E., & Nain, P. (2005). A survey of TCP over ad hoc networks. *IEEE Communications Surveys & Tutorials*, *7*(3), 22–36. doi:10.1109/COMST.2005.1610548

Holland, G., & Vaidya, N. H. (1999). Analysis of TCP performance over mobile ad hoc networks. In *Proceeding of Annual International Conference on Mobile Computing and Networking (Mobicom'99)* (pp. 207-218). Seattle, WA: ACM.

Hyytia, E., & Vittamo, J. (2005). Random waypoint model in n-dimensional space. *Operations Research Letters*, *33*, 567–571. doi:10.1016/j.orl.2004.11.006

Jain, A., Dubey, A. K., Upadhyay, R., & Charhate, S. V. (2008). Performance evaluation of wireless network in presence of hidden node: A queuing theory approach. In *Proceeding of Second Asia International Conference on Modelling and Simulation* (pp. 225-229). Kuala Lumpur: IEEE.

Jayasuriya, A., Perreau, S., Dadej, A., & Gordon, S. (2004). Hidden vs. exposed terminal problem in ad hoc networks. In *Proceedings of the Australian Telecommunication Networks and Applications Conference* (pp. 52-59). IEEE.

Jiang, R., Gupta, V., & Ravishankar, C. (2003). Interactions between TCP and the IEEE 802.11 MAC protocol. In *Proceeding of DARPA Information Survivability Conference and Exposition* (vol. 1, pp. 273- 282). Washington, DC: DARPA.

Jimenez, T., & Altman, E. (2003). Novel delayed ACK techniques for improving TCP performance in multihop wireless networks. In *Proceeding of Personal Wireless Communications (PWC'03)* (pp. 237-242). Venice, Italy: PWC.

Kam, P. (1990). MACA - A new channel access method for packet radio. In *Proceeding of 9th ARRL/CRRL Amateur Radio Computer Networking Conference* (pp. 134-140). Ontario, Canada: ARRL/CRRL.

Kawadia, V., & Kumar, P. (2005). Experimental investigations into TCP performance over wireless multihop networks. In *Proceeding of SIGCOMM Workshop on Experimental Approaches to Wireless Network Design and Analysis* (pp. 29-34). New York, NY: ACM.

Kherani, A., & Shorey, R. (2004). Throughput analysis of TCP in multi-hop wireless networks with IEEE 802.11 MAC. In *Proceeding of IEEE WCNC'04*, (vol. 1, pp. 237-242). IEEE.

Kim, D., & Toh, C., & Choi. (2001). TCP-BuS: Improving TCP performance in wireless ad hoc networks. *Journal of Communications and Networks*, *3*(2), 175–186.

Kuang, T., Xiao, F., & Williamson, C. (2003). Diagnosing wireless TCP performance problems: A case study. In *Proceeding of SPECTS* (pp. 176-185). SPECTS.

Kurose, J., & Ross, K. (2005). *Computer networking: A top-down approach featuring the internet*. Reading, MA: Addison Wesley.

Lehsaini, M. (2009). *Diffusion et couverture basées sur le clustering dans les réseaux de capteurs: Application à la domotique*. (Thèse de Doctorat Spécialité Informatique). Université de Franche-Comté.

Li, J. (2006). *Quality of service (QoS) provisioning in multihop ad hoc networks. (Doctorate of Philosophy)*. California: Computer Science in the Office of Graduate Studies.

Liu, J., Matta, I., & Crovella, M. (2003). End-to-end inference of loss nature in a hybrid wired/wireless environment. In *Proceeding of WiOpt'03: Modeling and Optimization in Mobile*. Sophia-Antipolis, France: INRIA.

Liu, J., & Singh, S. (2001). ATCP: TCP for mobile ad hoc networks. *IEEE JSAC*, *19*(7), 1300–1315.

Lohier, S., Doudane, Y. G., & Pujolle, G. (2006). MAC-layer adaptation to improve TCP flow performance in 802.11 wireless networks. In *Proceeding of WiMob'06* (pp. 427–433). ACM. doi:10.1109/WIMOB.2006.1696392

Marina, M. K., & Das, S. R. (2004). Impact of caching and MAC overheads on routing performance in ad hoc networks. *Computer Communications Journal, 27*, 239–252. doi:10.1016/j.comcom.2003.08.014

Mjeku, M., & Gomes, N. J. (2008). Analysis of the request to send/clear to send exchange in WLAN over fiber networks. *Journal of Light Ware Technology, 26*(13-16), 2531–2539. doi:10.1109/JLT.2008.927202

NS2. (n.d.). *Network simulator*. Retrieved from http://www.isi.edu/nsnam

Nahm, K., Helmy, A., & Kuo, C.-C. J. (2004). On interactions between MAC and transport layers in 802.11 ad-hoc networks. In *Proceeding of SPIE ITCOM'04*. Philadelphia: SPIE.

Ng, P. C., Liew, S. C., Sha, K. C., & To, W. T. (2005). Experimental study of hidden-node problem in IEEE802.11 wireless networks. In *Proceeding of ACM SIGCOMM'05*. ACM.

Oliveira, R., & Braun, T. A. (2005). Dynamic adaptive acknowledgment strategy for TCP over multihop wireless networks. In *Proceeding of IEEE INFOCOM* (pp. 1863–1874). Miami, FL: IEEE.

Papanastasiou, S., Mackenzie, L. M., Ould-Khaoua, M., & Charissis, V. (2006). On the interaction of TCP and routing protocols in MANETs. In Proceeding of of AICT/ICIW. Guadalupe, French Caribbean: AICT/ICIW.

Parsa, C., & Garcia-Luna-Aceves Tulip, J. J. (1999). A link-level protocol for improving TCP over wireless links. In *Proceeding of IEEE WCNC* (vol. 3, pp. 1253 - 1257). New Orleans, LA: IEEE.

Perkins, C. E., Royer, E. M., & Das, S. R. (2006). *Ad hoc on-demand distance-vector (AODV routing)*. IETF Internet draft (draft-ietf-manet-aodv-o6.txt).

Std, I. E. E. E. 802.11. (1999). Wireless LAN media access control (MAC) and physical layer (PHY) specifications. IEEE.

Toh, C.-K. (1996). A novel distributed routing protocol to support ad-hoc mobile computing. In *Proceeding of IEEE 15th Annual Int'l Phoenix Conf. Comp., & Commun.,* (pp. 480–86). IEEE.

Tsaoussidis, V., & Badr, H. (2000). TCP-probing: Towards an error control schema with energy and throughput performance gains. In *Proceeding of 8th IEEE Conference on Network Protocols*. IEEE.

Yuki, T., Yamamoto, T., Sugano, M., Murata, M., Miyahara, H., & Hatauchi, T. (2004). *Performance improvement of TCP over an ad hoc network by combining of data and ACK packets*. IEICE Transactions on Communications.

Zhai, H., Chen, X., & Fang, Y. (2007). Improving transport layer performance in multihop ad hoc networks by exploiting MAC layer information. *IEEE Transactions on Wireless Communications, 6*(5). doi:10.1109/TWC.2007.360371

Zhang, C., & Tsaoussidis, V. (2001). TCP-probing: Towards an error control schema with energy and throughput performance gains. In *Proceeding of 11th IEEE/ACM NOSSDAV*. New York: IEEE.

KEY TERMS AND DEFINITIONS

End-to-End Delay: Is given by the ratio of the time when the data is received minus the data transmission time divided by the number of data packets received.

IEEE 802.11 Protocol: The IEEE protocol for wireless technology used in a wireless local area network.

Incidences of the Improvement: The impacts which that can have the proposed improvements for the interactions between protocols on the performance of the MANET.

Interactions between Protocols: The exchange of data and information between the protocols belonging to different layers of the OSI model.

MAC Protocols: Is based on the Backoff algorithm that allows it to determine which node will access to the wireless medium in order to avoid collisions.

Mobile Ad Hoc Network (MANET): Is a complex distributed system which that consists of wireless mobile or static nodes. These nodes can freely and dynamically self-organize.

Mobility: The moving of the nodes in the network according to a given speed. It is used as second parameter to improve the interactions between the studied protocols.

Number of Nodes: The number of nodes belonging to the MANET. It is used as first parameter to improve the interactions between the studied protocols.

Performance Evaluation: The goal of this process is to validate the proposed solutions by simulation using the networks simulator (NS).

Routing Protocols: Specifies how routers communicate with each other, disseminating information that enables them to select routes between any two nodes on a computer network.

TCP Protocol: Is the The transport protocol used in most IP networks and recently in MANETs.

Throughput: is given Given by the ratio of the received data, taking into account all data sent.

Transport Protocols: Transport services, which includes the lower-level data link protocol that moves packets from one node to another.

Chapter 11
A Novel Secure Routing Protocol in MANET

Ditipriya Sinha
CIEM, India

Uma Bhattacharya
Bengal Engineering and Science University, India

Rituparna Chaki
University of Calcutta, India

ABSTRACT

This chapter gives an overview of research works on secure routing protocol and also describes a Novel Secure Routing Protocol RSRP proposed by the authors. The routes, which are free from any malicious node and which belong to the set of disjoint routes between a source destination pair, are considered as probable routes. Shamir's secret sharing principle is applied on those probable routes to obtain secure routes. Finally, the most trustworthy and stable route is selected among those secure routes using some criteria of the nodes present in a route (e.g., battery power, mobility, and trust value). In addition, complexity of key generation is reduced to a large extent by using RSA-CRT instead of RSA. In turn, the routing becomes less expensive and highly secure and robust. Performance of this routing protocol is then compared with non-secure routing protocols (AODV and DSR), secure routing scheme using secret sharing, security routing protocol using ZRP and SEAD, depending on basic characteristics of these protocols. All such comparisons show that RSRP shows better performance in terms of computational cost, end-to-end delay, and packet dropping in presence of malicious nodes in the MANET.

1. INTRODUCTION

This chapter gives an overview of secure routing protocols in MANET and describes a robust secure routing protocol (RSRP) proposed in (Sinha et al., 2012). A MANET is a type of ad hoc network that can change locations and configure itself on the fly. Because MANETs are mobile, they use wireless connections to connect to various networks. This can be a standard Wi-Fi connection, or another medium, such as a cellular or satellite transmission. Some MANETs are restricted to a local area of wireless devices (such as a group of laptop computers), while others may be connected

DOI: 10.4018/978-1-4666-5170-8.ch011

to the Internet. Security service requirements of MANET are similar to wired or any infrastructure wireless network. Here, every routing protocol needs secure transmission of data. Authentication, availability, confidentiality, integrity and non-repudiation are five inevitable concepts to provide secure environment in MANETs. Authentication ensures that the communication or transmission of data is done only by the authorized nodes. Without authentication any malicious node can pretend to be a trusted node in the network and can adversely affect the data transfer between the nodes. Availability ensures the survivability of the services even in the presence of the attacks of malicious nodes. Confidentiality ensures that information should be accessible only to the intended party. No other node except sender and receiver node can read the information. This can be possible through data encryption techniques. Encryption and decryption are two important techniques for secure routing in MANETs. Encryption is a technique which converts plain text message in ciphertext. Decryption is the reverse of encryption. Integrity ensures that the transmitted data is not being modified by any other malicious node. Non-repudiation ensures that neither a sender nor a receiver can deny a transmitted message. Non-repudiation helps in detection and isolation of compromised node.

Key generation, encryption and decryption play an important role for providing secure routing in MANETs. However these schemes increase computational overheads for all nodes in the network. The RSA algorithm involves three steps: key generation, encryption and decryption. RSA involves a public key and a private key. The public key can be known to everyone and is used for encrypting messages. Messages encrypted with the public key can only be decrypted using the private key. Chinese remainder theorem (CRT) uses the result about congruence in number theory and its generalizations in abstract algebra. In RSA-CRT, it is a common practice to employ the Chinese Remainder Theorem during decryption which

results in a decryption much faster than modular exponentiation used in RSA.

Secret sharing in MANETs is a challenging issue due to its dynamic nature. Many researchers are involved in solving the secret sharing problem. Shamir's proposal is one of the eminent secret sharing schemes. This scheme uses the concept of Lagrange's Interpolation method, a popular technique for polynomial evaluation. Shamir's scheme divides the data packet into n pieces such that it can be easily reconstructed from any $k = \dfrac{n}{2}$ number of pieces.

Main objective for secure routing is that data should be transmitted in secure and confidential way from source to destination. Trust value, battery power and stability of the nodes are the factors or attributes for determining a reliable, stable and trustworthy path in between a source-destination pair. Absence of any attributes of them may make the path unreliable.

A new security scheme has been proposed in this chapter for MANETs. This paper uses RSA-CRT scheme for its high efficiency in key generation, encryption and decryption of data. For secure route detection a safety key is generated. This safety key is divided into n pieces and propagated through n different available routes in between a source-destination pair. Safetykey can easily be reconstructed from any $k = \dfrac{n}{2}$ pieces. Shamir's secret sharing has been extended to the application of finding secure routes between a source destination pair using Lagrange's Interpolation scheme in the proposed work. Final route amongst those is chosen by using the criteria of a stable and trustworthy path i.e. trust value, battery power and stability of the nodes.

Section 2 contains background of past works related to the area, Section 3 contains scope of the work, Section 4 describes some basic schemes used in secure routing, proposed secure routing protocol is elaborated in section 5, performance of the proposed protocol has been evaluated in section 6, future work is mentioned in section 7 and conclusion has been drawn in section 8.

2. BACKGROUND

Following are the different aspects associated with secure routing protocol in MANET.

2.1 Key Generation for Encryption and Decryption

RSA (Rivest, Shamir, & Adleman, 1978) technique is the example of asymmetric key cryptography. This technique uses two keys, public and private. Source node encrypts message using its public key and destination node decrypts that ciphertext message using its private key. This scheme is very popular technique in cryptography. *Chinese Remainder Theorem* (Shand & Vuillemin, 1993) is simple mathematical result. It helps design of deterministic key pre-distribution using number theory. CRT generates key pool and key chain for key pre distribution. One of the applications of this theorem is encryption and decryption of data. This theorem is used in various secure routing protocols for key generation. The CRT technique improves the throughput rate up to 4 times in the best case. *Fast Decryption Method for RSA Cryptosystem* (Grobler & Penzhorn, 2004) designs new decryption method. It combines RSA with CRT and increases computational speed. This paper shows how CRT decryption gives better performance compared to RSA decryption method. *RSA Cryptosystem Design Based on Chinese Remainder Theorem* (Wu, Hong, & Wu, 2001) proposes a systolic RSA cryptosystem based on a modified Montgomery's algorithm and the Chinese Remainder Theorem (CRT) technique. The CRT technique improves the throughput rate up to 4 times in the best case. This paper adds some control logic to accomplish some modular exponentiation operation. This design consists of modular multiplier and some logic to control the data interleaving and scheduling mechanism. This protocol then adds some partitioning circuit for systolic RSA architecture to get two individual modular multipliers for simultaneous use. The design is suitable for decryption and digital signature. *Provably-Secure Time-Bound Hierarchical Key Assignment Scheme*

(Ateniese & Santis, 2006) is a method to assign time-dependent encryption keys to a set of classes in a partially ordered hierarchy, in such a way that the key of a higher class can be used to derive the keys of all classes lower down in the hierarchy, according to temporal constraints. This paper designs and analyzes time-bound hierarchical key assignment schemes which are provably secure and efficient.

Diffie-Hellman key exchange (Boneh, 1998) scheme is a specific method of exchanging cryptographic keys. It is one of the earliest practical examples of key exchange implemented within the field of cryptography. The Diffie–Hellman key exchange method allows two parties that have no prior knowledge of each other to jointly establish a shared secret key over an insecure communication channel. This key can then be used to encrypt subsequent communications using a symmetric key cipher.

2.2 Secret Sharing and its Applications

Shamir's Secret Sharing (Shamir, 1979) proposes a method for sharing secret. This paper shows how to divide data D into n pieces in such a way that D can easily be re-constructed from any k pieces. This technique enables the construction of robust key management for cryptographic system. This scheme provides most secure key management scheme. Various secure routing protocols use this concept for key management.

Secure Routing Scheme in MANETs using Secret Key Sharing (Amuthan & Baradwaj, 2011) proposes a secret sharing scheme using Shamir's

Secret Sharing method. Here secret is shared to detect malicious nodes in the network. For the key transmission RSA scheme is used in this paper. The key is encrypted using node's public key and then transmitted to them. The original secret key can be reconstructed by applying private key for their corresponding public key encrypted data. This scheme uses RSA modular expansion for decryption whose computational cost is higher than CRT method. *Chinese Remainder Theorem-Based RSA Threshold cryptography based schemes for MANETs using Variable Secret Sharing Scheme* (Sarkar, Kisku et al., 2009) provides a promising secure network. This proposed scheme is based on Chinese Remainder Theorem under the consideration of *Asumth-Bloom secret sharing*. In this protocol key is generated using the concept of RSA key generation. Unlike RSA no public and private key generate in this algorithm. CRT is also used for key generation. Key is shared among network using Asumth-Bloom secret sharing scheme. Computational complexity for key generation is high in this algorithm.

DASR: Distributed Anonymous Secure Routing with Good Scalability for Mobile Ad Hoc Networks (Dang, Xu, Li, & Dang, 2010) proposes a new efficient distributed anonymous secure routing protocol (DASR). There are four major contributions in this paper. Here the anonymous routing and data transmission are achieved by using the dynamic identity pseudonyms of the nodes instead of their real identities based on Incomparable Public Keys. On the other hand in order to achieve the location anonymity perfectly, the routing packets with fixed length are designed to prevent message size attacks. Diffie-Hellman key exchange scheme is used in this scheme to share secret key KSR between the source and destination nodes. No public key operations are involved in this routing protocol for better efficiency. Detailed analysis shows that DASR can achieve strong security and sufficient anonymity, as well as good scalability.

SPREAD: Improving network security by multipath routing in mobile ad hoc networks (Lou, Liu, & Zhang, 2009) presents a complementary mechanism to enhance secure data delivery in a mobile ad hoc network. The basic idea is to transform a secret message into multiple shares, and then deliver the shares via multiple paths to the destination so that even if a certain number of message shares are compromised, the secret message as a whole is not compromised. The fundamental idea of SPREAD is based on two techniques: multipath routing and secret sharing. Suppose a source node has a secret message for a destination node that is multiple hops away. If the source node sends the whole message through a single path, an adversary can intercept it at any one of the intermediate nodes along the path, or it can disrupt the delivery by dropping packets at any one of the nodes along the path. The focus of this paper is to defend against malicious packet dropping by adversaries (compromised nodes) who are on the forwarding paths.

2.3 Secure Group Communication and Secure Routing Protocols

An efficient and attack resistant key agreement scheme for secure group communications in mobile ad hoc networks (Balachandan, Zou, Ramamurthy, & Thukral, 2007) is proposed to reduce problems of secure group communication (SGC) and key management over MANETs. It identifies key features of SGC scheme over MANETs. This paper proposes Chinese Remainder Theorem based DH contributory key agreement protocol. This protocol gives the concept of Group Key (GK). Here GK is generated by the contribution of all members in the group. This algorithm takes care of selection of group members. This protocol is also concerned about problem of leaving the member from the group. It uses the concept of CRT for security of group key.

Secure Group Communication Using Key Graph (Wong, Gouda, & Lam, 2000) proposes routing protocol based upon group communication model. Securing group communication provides integrity, authenticity and confidentiality of message delivered between group members. This protocol introduces key graphs to specify secure groups. This paper maintains three strategies: i) securely distribute re-key messages after joining or leaving of a node of the group ii) specify protocol for joining the group and iii) specify protocol for leaving the group. This paper proposes different hierarchical approach to improve scalability. To implement Secure Group Communication this paper uses hierarchy keys. With a hierarchy key, there are different ways to construct re-key message and securely distribute them to the user. There are three re-keying strategies: user oriented, key oriented and group oriented. This paper presents multiple re-key messages using a single digital signature operation. Each re-key message contains one or more new keys. After each join or leave, a new secure group is formed. This paper gives a novel solution to the scalability problem of group key management.

ECGK: An efficient clustering scheme for group key management in MANETs (Drira, Seba, & Kheddouci, 2010) focuses on group communication confidentiality in the environment of MANETs. Group communication confidentiality prevents non-group members from reading data exchanged within a secure communication session of the group. This confidentiality requires establishing and maintaining a common key between group members. This key, called group key or traffic encryption key (TEK) serves to encrypt/decrypt message exchange within the group. The group key management must meet the following properties:

1. Key secrecy which guarantees that it is computationally infeasible for a passive adversary to discover any group key.
2. Key independence which guarantees that a passive adversary who knows any proper subset of group keys cannot discover any other group key.

2.4 Authentication Based Secure Routing Protocol

Authenticated Routing for Ad-hoc Networks (ARAN) (Djenouril, Mahmoudil, Bouamama, Jones, & Merabti, 2007) proposes cryptographic certificates to prevent and detect most of the security attacks. This protocol introduces authentication, no repudiation and message integrity as part of minimal security policy for the ad hoc environment. ARAN provides preliminary certification process followed by route instantiation process that guarantees end to end authentication. Thus the routing messages are authenticated at end to end and only authorized nodes participate at each hop between source and destination.

A Secure Routing Protocol for Wireless Adhoc Network (Xia, Jia, Li, Ju, & Sha, 2012) presents an on demand secure routing protocol for ad-hoc network based on a distributed authentication mechanism. It is able to tolerate attacks, introduces path failure and provides robust packet delivery. Every node creates a trust table. An optimistic node assigns larger trust value to other nodes. Every node sets up secret key with each trust worthy neighbor by using a two party key establishment protocol. Route request and route reply packets are used for Route Discovery. Attacker isolation phase is implemented using WARNING message. Distributed Authentication model provides protection to data packet communication. Route maintenance phase checks the broken routes in the network using Route Error message. Trust value of a node can be -1(distrust), 0(ignorance), 1(minimal), 2(average), 3(good), and 4(complete), where the number is the trust value. In this protocol, as long as an entity's trust value is ≥ 2, it is assigned a "yes" which means "trustworthy," otherwise, it is assigned a "no," which means "untrustworthy." The trust evaluation process allows a node to decide whether the node is trust worthy or not. This way the routing protocol provides security.

On Securing MANET Routing Protocol against Control Packet (Sun & Yu, 2006) is a secure routing protocol in MANETs. This paper focuses on dropping control packets. Dropping control packet may be beneficial for selfish nodes and malicious nodes. Here each node monitors its successor node. To ensure authentication among two hops asymmetric cryptography is used in this protocol.

2.5 Other Secure Routing Protocols

Trust Based Secure Routing in AODV Routing Protocol (Menaka & Pushpa, 2009) is modified AODV (Perkins, Belding-Royer, & Das, 1999) routing protocol with node trust value. Using this approach, secure route can be established by calculating trust value of each node which is participating in the route establishment process from source to destination. It completely relies only on trust value of the nodes of the concerned route. But in few situations, this method of route establishment is not sufficient to create a secure route.

Security Threats for Cluster Based Ad hoc Networks (Varadharajan, Shankaran, & Hitchens, 2004) proposes security services in cluster-based NTDR ad hoc networks. This protocol describes secure schemes for a mobile node to initiate, join and leave a cluster. This proposed protocol also discusses the secure end-to-end communication and group key management related issues for NTDR networks. Security model makes use of public key systems and involves the use of certificates and certification authorities. All potential participants have access to a public key infrastructure in their own fixed network domains, prior to joining the ad hoc networks. That is, they can be allocated a public key, which can be certified and there is a certification path to the top of their chain that is likely to be publicly recognizable.

Secure Efficient Ad hoc Distance Vector (SEAD) (Hu, Johnson, & Perrig, 2003) is a proactive routing protocol, based on the design of Destination Sequenced Distance Vector routing protocol (DSDV). Nodes maintain distances to destination and keep information about the next hop in the optimal path to a destination. SEAD

provides a robust protocol against attackers trying to create incorrect routing state in other node by modifying the sequence number or the routing metric. But SEAD does not provide a way to prevent an attacker from tampering next hop or destination field in route update. Also, it can't prevent an attacker to use the same metric and sequence number, which are learnt from some recent update message.

I-SEAD: A Secure Routing Protocol for Mobile Ad Hoc Networks (Lin, Lai, Huang, & Chou, 2008) is enhancement of SEAD. In this proposed protocol, called ISEAD, it can let the neighbors check the correctness of the hash value using its TELSA key and reduce the routing overhead. When the start node sends the route request, it chooses a number as a seed randomly. The start node computes the list of values with the seed. Before sending the route request, the start node computes its TESLA key to protect its hash value. Each node can verify the received value after a period time.

Secure Destination-Sequenced Distance-Vector routing protocol (SDSDV) for ad hoc mobile wireless networks (Wang, Chen, & Lin, 2009) is based on the DSDV protocol. Within SDSDV, each node maintains two one-way hash chains about each node in the network. Two additional fields, AL (alteration) field and AC (accumulation) field, are added to each entry of the update packets to carry the hash values. With proper use of the elements of the hash chains, the sequence number and the metric values on a route can be protected from being arbitrarily tampered. In comparison with the secure efficient distance vector (SEAD) protocol previously described in the literature it provides only lower bound protection on the metrics, whereas SDSDV can provide complete protection.

An Encryption Based Dynamic and Secure Routing Protocol for Mobile Ad-hoc Networks (Sehgal & Nath, 2009) proposes an efficient key management mechanisms for enforcing confidentiality, integrity and authentication of messages in ad-hoc networks. This protocol does not place any particular limitations on adversarial entities. It prevents attackers and malicious nodes from tampering with communication process and also prevents large number types of Denial of Service Attack. It uses symmetric key cryptography. The goal of this protocol is to detect selfish and adversarial activities of nodes and mitigate them.

Secure Routing Protocol for Mobile Ad hoc Network (Li & Singhal, 2006) guarantees acquisition of correct topological information in timely manner. This protocol provides accurate connectivity information despite the presence of strong adversaries. This protocol is proven robust against set of attacks that attempt to compromise route discovery. This protocol also prevents from Denial of Service Attack and enhances performance of network.

Secure Data Transmission on Multiple Paths in Mobile Ad Hoc Networks (Xia, Huang, Wang, Cheng, Li, & Znati, 2006) is based on multiple paths for mobile ad hoc networks. The scheme focuses its attention on privacy and robustness in communication. For privacy, a coding scheme using XOR operation is established to strengthen the data confidentiality so that one cannot uncover the information unless he took up all the K paths or special K-1 ones. For robustness, it reduces to a fault tolerant mode which combines backup with checkout.

New Security Algorithm for Mobile Ad hoc Networks Using ZONAL ROUTING PROTOCOL (ZRP) (Varaprasad, Dhanalakshmi, & Rajaram, 2008) designs a secure communication between the mobile nodes. Whenever a source wants to transmit the data packets to the destination it ensures that the source is communicating with real node via the cluster head. The authentication service uses a key management to retrieve the public key, which is trusted by the third party for identification of the destination. The destination also used similar method to authenticate the source. After execution of the key management module, a session key is invoked, this is used by both source

and destination for further confidentially. In this way, all the important messages are transmitted to the destination.

Fault Tolerant and Dynamic File Sharing Ability in Mobile Ad hoc Networks (Pushpalatha, Venkataraman, Khemka, & Rao, 2009) proposes a new method for replication of files within a mobile ad hoc network based on the prediction of the drift of the mobile nodes. The aim of this paper is to create a secure method of dynamically maintaining the number of replicas of files within a system, without saturating the available network resources. This protocol is focused only on a particular issue, node mobility.

Location Aided Secure Scheme in Mobile Ad hoc Networks (Lee, Choi, Choi, & Jung, 2007) proposes a routing scheme that uses geographical position information to reduce intermediate nodes that make a routing path. It gives guarantee of reliability and security of route establishment process. This protocol uses concept of projection line and shadow line to detect the secured route. Here all nodes maintain a safety table. This table concerns about the behavior of neighbor nodes. This table detects all activities of misbehaving nodes. For route detection source node chooses secure path with the help of this safety table.

3. SCOPE OF THE WORK

Following points need to be discussed in this context.

- Securing message transmission normally involves some encryption at source node and decryption at destination node of messages using RSA technique which leads to a large computational overhead.
- Also the presence of the malicious nodes in the network is the cause of packet loss during transmission of messages through those nodes. Identifying malicious nodes and avoiding them in the route of message

passing may thus reduce the number of packets dropped and thus improve the performance in the network.

- Finding secure routes requires generating a Safety Key. This can be constructed using CRT.
- Shamir's Secret Sharing scheme may divide this Safety Key into n (number of available routes) pieces and this key can easily be reconstructed by using pieces out of them. Concept underlying this scheme may be extended successfully to detect all secure routes among all probable routes between a source destination pair.
- Three points to mention are that:
  ○ All nodes in MANETs are subject to loss of battery power during message passing /encrypting/decrypting.
  ○ Stability information about a node may be helpful to establish a stable route because a stable route must consists of a set of stable nodes.
  ○ Trust value of a node will be higher if it takes part in processing of messages successfully from a source to the destination node.

In this context weight of a node may be considered as the summation of trust value, battery power and stability of the node. Amongst all secure routes detected by Shamir's Secret Sharing principle the route having highest average value of nodes may be selected as the final route for message transmission.

Thus motivation of our work is to:

1. Use the less expensive scheme for encryption and decryption of messages.
2. Find set of probable routes without having any malicious nodes to improve the network performance in terms of lesser number of packet drops in the network.

3. Generate Safety Key and detect all secure routes using Shamir's Secret Sharing scheme.
4. Detect finally the most stable and trustworthy paths among those secure paths for message transmission.

4. SOME BASIC SCHEMES USED IN PROPOSED PROTOCOL

The proposed logic uses two keys for message encryption and decryption, RSA technique for encryption of data, CRT technique for decrypting data, safety key to detect the routes without having any malicious node between a source destination pair and Shamir's secret sharing principle to detect secure routes.

4.1 Key Generation

1. Each node i generates two prime numbers p and q so that $N = pq$ and $Ø(N)=(p-1)(q-1)$.
2. Choose e such that e is not divisor of $Ø(N)$ and $1<e< Ø(N)$.
3. d is the modular multiplicative inverse of $e(mod\ Ø(N))$.
   a. $e.d= 1\ mod\ Ø(N)$
4. (e,N) is the public key and (d,N) is the private key of node i.

4.2 Encryption using RSA at Source Node

When source node wants to encrypt message M to ciphertext C for sending it to destination node, it uses public key of destination node using RSA in the way as described below:

$C=M^e\ (mod\ N)$

4.3 Decryption Using RSA at Destination Node

When destination node wants to decrypt cipher text C to plain text M, it uses private key of destination node using RSA in the way as described below:

$$M = C^d \pmod{N}$$

4.4 Decryption Using CRT at Destination Node

When destination node receives the encrypted message it decrypts this encrypted message using CRT following way:

dp=d mod(p-1).
dq=d mod(q-1).
$q_{inv} = q^{-1}$ mod p.
$m_1 = C^{dp}$ mod p.
$m_2 = C^{dq}$ mod q.
$h = (q_{inv} \cdot (m_1 - m_2))$ mod p.
$M = m_2 + h*q$

4.5 Safety Key Generation Using CRT

Safety key (SF) is a key which is used to detect secure routes among all available routes. Source node generates n integers m_1, m_2, m_3, ...m_n, such that gcd $(m_i, m_j)=1$. Then this key is generated by using following equations:

$m = m_1.m_2.....m_n$; $a_i \varepsilon$ I. I = { a_1, a_2,........., a_n }, I is the set of integers.

$z_i = m/m_i$, $y_i \equiv z_i^{-1} \pmod{m_i}$ and $Z \equiv a_1 \pmod{m_1} \equiv a_2 \pmod{m_2}$ .... $\equiv a_n \pmod{m_n}$. Here $\pmod{m_i}$ stands for modular multiplicative inverse operation.

SF = $a_1 y_1 z_1 + a_2 y_2 z_2 + ...+a_n y_n z_n$

Before sending message source node S generates Safetykey SF_s using CRT.

4.6 Shamir's Secret Sharing Principle

The essential idea of Adi Shamir's threshold scheme is that 2 points are sufficient to define a line, 3 points are sufficient to define a parabola, 4 points to define a cubic curve and so forth. That is, it takes k points to define a polynomial of degree (k-1).

Let us suppose we want to use a (k, n) threshold scheme to share our secret S, without loss of generality assumed to be an element in a finite field F of 0<k<=n< P, where P is a prime number.

Let us choose at random k-1 coefficients $a_1, a_2, ..., a_{k-1}$ in F, and let $a_0 = S$ and then build the polynomial

$$F(x) = a_0 + a_1 x + a_2 x^2 + ... + a_{k-1} x^{k-1}$$

Let us construct any n points out of it and for instance set i=1,...,n to retrieve (i,(f(i)). Every participant is given a point (a pair of input to the polynomial and output). Given any subset of k of these pairs, we can find the coefficients of the polynomial using interpolation and the secret is the constant term a_0.

4.7 Secure Route Detection Scheme using Shamir's Secret Sharing

As by the proposed solution source node S divides Safety key SF_s into n parts, where n is the number of available routes from source to destination. Now source node generates a polynomial F(x) of degree floor (n/2)-1=k-1 such that,

$$F(x) = a_0 + a_1 x + a_2 x^2 + ... + a_{k-1} x^{k-1}$$

where a_0, a_1,... a_{k-1} are set of integers.

Source node generates n number of points from this polynomial which are ((x_0, y0); (x_1, y1);..............;(x_{n-1},y_{n-1})) and sends each of these points in encrypted form [section 4.B] through each amongst n different available routes to destination node. Destination node decrypts [section 4. D] those n points and again encrypts [section 4.B] those points and sends to the source node by backtracking in the same route.

Now source node decrypts those n points and takes any k points among them to regenerate the polynomial $F_1(x)$ using Lagrange's Interpolation such that,

$$F_1\left(x\right) = \sum_{r=0}^{k-1} y_r \prod_{i=0, i \neq r}^{k-1} \frac{\left(x - x_i\right)}{\left(x_r - x_i\right)}$$

The first constant part of $F_1(x)$ is called SF_1. If $SF_s = SF_1$, those k points are valid points and the routes used by those k points are also valid and hence secured. Otherwise at least one of the routes used by those k points is not secured. ^nC_k number of combinations are available for computing Safety key. Those combinations generating correct value of Safety key will correspond to the respective secure routes.

5. PROPOSED SECURE ROUTING PROTOCOL

Proposed secure routing protocol is discussed as follows:

5.1 Assumptions

1. Mobility model used here is random waypoint (Laxmi, Jain, & Gaur, 2006).This model restricts movement of the mobile nodes to a rectangle. Each node picks a destination within the rectangle along with a speed. The node travels to the destination at that speed. Upon reaching the destination, the node selects and waits for a uniformly distributed pause time. After waiting, the node picks another destination and another speed, continuing the process. The parameters of this model are the minimum and maximum speed and the maximum pause time.
2. No path loss exists.
3. All nodes are GPS enabled.
4. All nodes have their sequential identity number according to their entry.

5.2 Data Dictionary

W_i: Weight of the node i.

Tv_i: Trust value of node i.

Bp_i: Battery Power of node i.

Bp_{iprev}: Battery power of node i before current processing takes place

Mob_i: Mobility of node i.

a:Energy required for forwarding and receiving each message.

s_t: Total number of messages sent in during the time duration Δt.

r_t: Total number of messages received during the time duration Δt.

u: Number of encryption done by node i during the time duration Δt.

v: Number of decryption done by node i during the time duration Δt.

k: Energy required for each encryption.

l: Energy required for each decryption.

n: Number of available routes in the network during time duration Δt.

hopcount: the number of nodes on a route with same sequence no in the time duration Δt.

$Avgr_{S,D}[j]$: Average weight of jth route in between source destination pair S,D in the time duration Δt.

forward_count$_i$: Number of messages sent by Trust Agent successfully via node i to a neighbor.

tot_forward_message$_i$: Total number of messages sent by the Trust Agent via node i to a neighbor.

Tr: Identity of trust agent.

sus$_{id}$: Identity of suspicious nodes.

S: Identity of source node.

Pk-i: Public key of node i.

Pv-i: Private key of node i.

D: Identity of destination node.

Routetable[ ]: Stores all probable routes between source and destination. Structure of the table: Sequence-no, Source node, List of nodes in the route, Destination node. This table is associated with each node.

Malicious-id[ ]: Stores identities of all malicious nodes. This table is associated with each node.

Secure-route[ ]: Stores all secure routes between source and destination. Structure of this table: Sequence no, Source node, List of nodes in the secure route, Destination node. This table is built within each node when the node becomes a source node.

Suspectlist-Tr[ ]: Trust Agent (Tr) stores identities of suspicious nodes.

Weight-table$_i$[ ]: This table contains weight information of node i. Structure of this table: Node-id, Trust value, Battery Power, Mobility. This table is associated with each node.

Neighbor-key-table$_i$[ ]: This table contains public key information of neighbors of node i. **Structure of this table:** Neighbor node id, public key value of the neighbor node. This table is associated with each node.

Key-table$_i$[ ]: This table contains safety key, public key and private key information of node i. Structure of this table: Node-id, public key value, private key value, safety key value. This table is associated with each node.

5.3 Definitions

Definition 1: Mobility of a node i having n number of neighbors is defined as the summation of difference of distance computed between the node and all its neighbors during time interval Δt, i.e., $mob_i = \Sigma_{s=1..n} dist(i,s)$ where $dist(i,s) = |d^t_{si} - d^{t+\Delta t}_{si}|$, where s being a neighbor of node i d^t_{si} indicates distance between node s and node i at time t. The distance is calculated from the location co-ordinates of the nodes at that instant of time. Stability of the node is defined as the inverse of its mobility.

Definition 2: Battery power consumption of ith node (Bp$_i$) is proportional to energy required for forwarding and receiving number of packets by the node during a specific time interval Δt and also number of encryption and decryption done by node i during the same time interval. BP$_i$ is defined as follows:

At the source and destination node,

$$BP_i = BP_{iprev} - \sum_{t=T}^{T+\Delta t} a*(s_t + r_t) + (u*k + v*l)$$

At any intermediate node,

$$BP_i = BP_{iprev} - \sum_{t=T}^{T+\Delta t} a*(s_t + r_t)$$

Battery power is computed when any processing such as forwarding/receiving or encrypting/decrypting packets takes place.

Definition 3: Trust value Tv$_i$ of a node i is defined as follows:

$$Tv_i = forward\_count_i / tot\_forward\_message_i$$

Definition 4: Weight of a node i (W$_i$) is defined as the summation of its trust value, battery power and stability of the node at that time.

$$W_i = Tv_i + Bp_i + 1/Mob_i$$

5.4 Description of the Proposed Routing Protocol

The proposed routing is a 4-phase one: (i)Assignment of Trust values to all nodes in MANET (ii) Detection of Probable Routes (iii)Detection of Secure Routes (iv)Detection of Final Route.

5.4.1 Assignment of Trust Values to All Nodes in MANET

5.4.1.1 Initialization of Trust Values at the Beginning (Initialize-Trust-MANET())

Assumption: There are a few nodes closely placed in the MANET so that each node is at a distance of one hop to each other and trust values associated with each of them is zero.

Step 1: Arbitrarily, a node is chosen as *Initiator node*. It broadcasts its identity to its neighbors using FIND-NEIGHBOUR message of the format {initiator-id, $Tv_{initiator-id}$ }. Initially, $Tv_{initiator-id} = 0$. Count=0

Step 2: The nodes within one hop distance send an *ACK* message of format {next-hop-id, $Tv_{next-hop-id}$ } to the initiator node with in time out period. Initially $Tv_{next-hop-id}$ value of all next-hop-nodes are zero.

Step 3: All such nodes are added to the neighbor list of the initiator node. Initiator node selects one of them as *Trust Agent (TR) randomly* and sends the neighbor list to all next-hop nodes including the Trust Agent one.

Step 4: *Trust Agent* observes the initiator node for some time interval $t_{mon.}$ During this period *Trust Agent* sends some messages to another node present in the neighbor list using initiator node as the intermediate node. Depending on the successful transmission of a considerable number of messages, *Trust Agent* computes the trust value of the initiator node [Definition 3] and then sends the trust value to the initiator node. If number

of successfully transmitted message $<=30\%$ of the total number of messages transmitted, *Trust Agent* adds the node to Suspect-list [Sus-id] .This is to mention that trust value of a node increases during successful routing of messages for a source destination pair via that node.

Step 5: count = count + 1;

Step 6: If count < 2, Temporary = Tr-id; Tr-id = Initiator-id; Initiator-id = Temporary; Go to step 4. It reverses the role of Initiator node and Trust Agent node and in this way when count = 1, the trust value of the new Initiator node has been computed already.//

Step 7: The next available node in the neighbor list is considered as initiator node and step 4 is executed to compute the trust value of that node. In this way execution of step 4 will continue till the computation of trust value of all nodes in the neighbor list is completed.

Step 8: Stop.

5.4.1.2. Computation of Trust Value for the Newly Entrant Node (New-Entry-Trust())

Step 1: On entry, a new node broadcasts its identity to its neighbors using FIND-NEIGHBOUR {new-id, Tv_{new}} packet where new-id is the identity of new node and Tv_{new} is its trust value which is initialized as 0.

Step 2: The nodes within one hop distance send an ACK message of format {next-hop-id, $Tv_{next-hop-id}$ } to the newly entered node with in time out period where next-hop-id is the identity for this next hop node.

Step 3: New node prepares its list of neighbors associated with the trust value of each node from the ACK messages. It then selects the node as *Trust Agent (TR)* whose trust value is *maximum* and sends its neighbor list to the *Trust Agent*.

Step 4: *Trust Agent* observes new node for some time interval $t_{mon.}$During this period *Trust*

Agent sends some messages to another node in the neighbor list using new node as the intermediate node. *Trust Agent* computes the trust value of the new node [Definition 3] and then sends this value to the new node. If number of successfully transmitted message < 30% of the total number of messages transmitted, the new node is added to Suspect-list [Sus-id] of the trust agent.

Step 5: *Tr* observes suspected node for a particular time duration. If its trust value remains same throughout the period *Trust Agent* declares and broadcasts the node as malicious node in the MANET and enters the identity of the node in its Malicious list. *Trust Agent* sends this information to its neighbors. Neighbor nodes again alert all nodes of its neighbors. All nodes in the network enter new node i in their Malicious list.

Step 6: Stop.

5.4.2 Probable Routes' Detection (Probableroute())

Step 1: Source S broadcasts RREQ messages to all nodes within one hop distance and waits for ACK within Δt duration. The message format is {sequenceno, S, next-hop-id-1, hopcount, D}. Initially value of next-hop-id-1 is 0 and hopcount is 1.

Step 2: When it reaches to next-hop node, if next-hop-id-1 == D, next-hop node replies ACK message to source S. ACK message format is {sequenceno, S, next-hop-id-1, hopcount, D}.

Step 3: If next-hop-id-1 <> D, nexthop node adds its own identity in RREQ packet and again broadcasts it to its neighbors within one hop distance. RREQ packet format is now {sequenceno, S, nexthop-id-1, nexthop-id-2, hopcount,, D}. Hopcount value is increased by 1. next-hop-id-2=0.

Step 4: If next-hop-id-2 = D next-hop node replies ACK message to source S. ACK message

format is {sequenceno, S, next-hop-id-1, next-hop-id-2, hopcount, D }.

Step 5: If next-hop-id-2 <> D, next-hop-id-2 adds its own identity in RREQ message and this process will go on upto n steps (in general) till next-hop-id-n becomes equal to D and hopcount equals n. Finally, nexthop-id-n or D sends ACK message of the format {sequenceno, S, nexthop-id-1, nexthop-id-2,......,nexthop-id-n, n,D} backtracking in the same route traversed already.

Step 6: During execution of step 3- step 5 if any node receives more than one RREQ with same sequenceno, that node simply discards one of them thus ensuring unique routes between S-D pair. Thus each sequenceno remains associated with each unique route between same S-D pair.

Step 7: S accepts all ACK messages during time duration Δt. Source S now checks if any node mentioned in ACK message is present in Malicious-list or not. If any of them is malicious, that ACK message will be discarded, otherwise source node S stores the route obtained from the ACK message in its Routetable {sequenceno, S, nexthop-id-1, nexthop-id-2,......, nexthop-id-n, n, D}.

Step 8: It is obvious that same route may be present in the Routetable with different sequenceno. Source S now searches Routetable to eliminate redundant routes.

Step 9: Stop.

5.4.3 Secure Routes' Detection (Secureroute())

Step 1: Source S generates safety key SF_S [Section 4.4] and stores it in its Key-table$_s$[ ].

Step 2: Source S counts number of available routes from its Routetable whose format is {sequenceno, S, nexthop-id-1, nexthop-id-2,......,nexthop-id-n, n, D }./* n stands for number of available unique routes */

Step 3: S generates a polynomial F(x) of degree

$$k - 1 = \frac{n}{2} - 1$$

such that,

$$F(x) = a_0 + a_1 x + a_2 x^2 + \ldots + a_{k-1} x^{k-1}$$

Step 4: S generates n number of points from F(x) such as $((x_0, y_0); (x_1, y_1); \ldots; (x_{n-1}, y_{n-1}))$

Step 5: S calls RSA and encrypts those n points using[Section 4.4 B]. For encryption S uses public key of destination node D (Pk- D) from its Neighbor-key-table$_s$.

Step 6: S sends each encrypted (x_i, y_i) points with its public key through each unique routes to destination node D. For this reason S sends PointRequestmessage whose format is {sequenceno, S, D, (x_i, y_i)} through n unique routes using its Routetable to D.

Step 7: After receiving those n points D calls CRT and decrypts those n points [Section 4.4 C]. For decryption D uses private key of D (Pv-D) from its Key-table$_D$.

Step 8: Again D encrypts [Section 4.4 B] those n points using public key of S (Pk-S) from PointRequest message and stores value of Pk-S to its Neighbor-key-table$_D$.

Step 9: Now, D sends those encrypted points backtracking in the same unique routes traversed before.

Step 10: S decrypts [Section 4.4 C] each point using its public key (Pk- S).

Step 11: S creates ^nC_k combination of sets where each set contains k out of n number of points.

Step 12: For each set of k points S regenerates the polynomial $F_1(x)$ using Lagrange's Interpolation such that,

$$F_1(x) = \sum_{r=0}^{k-1} y_r \prod_{i=0, i \neq r}^{k-1} \frac{(x - x_i)}{(x_r - x_i)}$$

Step 13: If Constant part of $F_1(x)$ = Constant part of F(x) = SF$_s$, S stores the route information and sequenceno which corresponds to those k valid points to its Secure- route (sequenceno, S, List of Nexthop-nodes in the route, D) table.

Step 14: Go to step 12 until all ^nC_k combination of sets are examined.

Step 15: Stop.

5.4.4 Final Route Detection (Finalroute())

Step 1: S sends WEIGHT_REQ message whose format is {sequenceno, S, List of all node-ids in the route, D} to all nodes on each secure route obtained from its Secure-route[] table.

Step 2: Each node in the secure route sends their weight table [] to S using ACK message whose format is {sequenceno, node-id, Weight-table$_i$, S} with same sequenceno as present in WEIGHT_REQ message.

Step 3: S calculates weight of each node i (W_i) from its Weight-table$_i$ using Definition 4. S then calculates average weight (AVGR$_s$[m]) of all nodes present in m^{th} route which possess the same sequenceno.

Step 4: S selects route j as final route for which average weight is maximum.

Step 5: S uses RSA to encrypt each message using public key Pk-D of destination node D and sends all messages through this j^{th} route to destination node D.

Step 6: After receiving each encrypted message D decrypts this message using its private key Pv-D and sends ACK message to S backtracking in the same route as traversed already. For each node i present in jth route Trust value Tv$_i$ and battery power consumption Bp$_i$ will be incremented and re-computed [Definition 2] respectively in the backtracking period.

Step 7: Step 1 to step 6 are repeated after every Δt time interval.

5.5 Algorithm

Step 1: Initialize-trust-manet() module in executed.

Step 2: For each source destination pair (S-D) S calls **Probableroute()** module which returns all probable disjoint routes between S and D.

Step 3: S calls **Secureroute()** module which returns several secure routes among all available routes.

Step 4: S calls **Finalroute()** module which returns final route through which all packets are sent to D.

Step 5: If any new node i enters in the network **New-entry-trust()** module is executed.

Step 6: After Δt interval S repeats step 2 to 5.

5.6 Case Study

Figure 1 show a network with 9 nodes where trust values of all nodes are initialized. Now source node S wants to send messages to destination node E. At that moment new node T enters into the network. Total case study is divided into following phases.

5.6.1 Computation of Trust Value for Newly Entrant Node T

T broadcasts its identity to its neighbors using FIND-NEIGHBOUR packet. The nodes within one hop distance {S, C, D} send an ACK message to the newly entered node T with in time out period. T prepares its list of neighbors associated with the trust value of each node from the ACK messages. It then selects the node C as *Trust Agent (TR)* whose trust value is maximum and sends its neighbor list to the *Trust Agent C. C* observes new node for some time interval t_{mon}. During this period *Trust Agent* C sends some messages to another node in the neighbor list using new node T as the intermediate node. *Trust Agent* computes the trust value of the new node [Definition 3] and then sends this value to the new node. If number of successfully transmitted message < 30% of the total number of messages transmitted, the new node is added to Suspect-list of the trust agent. C observes T for particular time duration. If its trust value remains same throughout the period *Trust Agent* C declares and broadcasts the node T as malicious node in the MANET and enters the

Figure 1. Network contains 9 nodes and new node T enters into the network

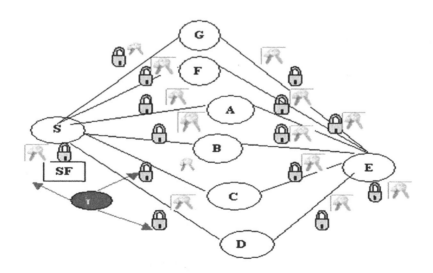

identity of the node T in its Suspect list. **C** sends this information to its neighbors. In the same way all these neighbors sends the information to their neighbors.

5.6.2 Probable Routes' Detection

S broadcasts RREQ messages to all nodes {A, B, C, D, G, F} within one hop distance and waits for ACK message for Δt duration. When it reaches to next-hop nodes A, B, C, D, G and F they add its own identity in RREQ packet and again broadcasts it to its neighbor E within one hop distance. Now the destination E replies ACK message to source S backtracking in the same paths. During this process if any node receives more than one RREQ with same sequenceno, that node simply discards one of them thus ensuring unique routes between S-E pair. Thus each sequenceno remains associated with each unique route between same S-E pair. S accepts all ACK messages during time duration Δt. Source S now checks if any node mentioned in ACK message is present in Malicious-list or not. If any of them is malicious, that ACK message will be discarded, otherwise source node S stores the route obtained from the ACK message in its route-table.

So, route-table associated with source node S contains the value as shown below.

5.6.3 Secure Routes' Detection

Now, S generates safety key SF_S and stores it in its Key-table$_S$. Source S counts number of available routes (n) from its Routetable whose value (n) = 6. Suppose, S generates a number 1234 using CRT and creates a polynomial F(x) such that,

S generates a polynomial F(x) of degree

$$\frac{6}{2} - 1 = 2$$

such that

$$Y = F(x) = 1234 + 166x + 94x^2$$

where S assumes two random numbers (166, 94) to create the polynomial .

S generates 6 number of points from F (x) such as $((x_0, y_0);(x_1,y_1);\ldots\ldots;(x_5, y_5))$.

Those points are as follows:

(1,1496);(2,1942);(3,2578);(4,3402);(5,4414);(6,5614)

S calls RSA and encrypts those 6 points. For encryption S uses public key of destination node E from its Neighbor-key-table$_S$. S sends each encrypted (x_i,y_i) points with its public key through each unique route to destination node E . For this reason S sends PointRequest message through 6 disjoint routes using its Routetable to E. After receiving those 6 points E calls CRT and decrypts those 6 points. For decryption E uses private key of E from its Key-table$_E$. Again E encrypts those 6 points using public key of S from PointRequest message and stores the value of the public key to its Neighbor-key-table$_E$. Now, E sends those encrypted points backtracking in the same unique routes traversed before. S decrypts each point using its public key. S creates $^6C_3$ combination of sets where each set contains 3 out of 6 numbers of points.

For each set of 3 points S regenerates the polynomial $F_1(x)$ using Lagrange's Interpolation such that,

$$F_1\left(x\right) = \sum_{r=0}^{k-1} y_r \prod_{i=0,i \neq r}^{k-1} \frac{\left(x - x_i\right)}{\left(x_r - x_i\right)}$$

Let us consider S takes (2,1942); (4,3402); (5,4414). Values of the variables l_0, l_1, l_2 are computed as follow as a prerequisite for constructing $F_1(x)$.

$$l_0 = [(x-x_1)/(x_0-x_1)].[(x-x_2)/(x_0-x_2)] = 1/6\ x^2 - 1/2\ x + 1$$

$l_1 = [(x-x_0)/(x_1-x_0)] . [(x-x_2)/(x_1-x_2)] = -1/2 \ x^2$
   $+ 3/2 \ x - 5$

$l_2 = [(x-x_0)/(x_2-x_0)] . [(x-x_1)/(x_2-x_1)] = 1/3 \ x^2 - 2x + 4/3$

Therefore, $F_1(x) = y_0 l_0 + y_3 l_1 + y_4 l_2 = 1496 l_0 + 3402 l_1 + 4414 l_2 = 1234 + 166x + 94x^2$

If Constant part of $F_1(x)$ = Constant part of $F(x)$ = SF_s, S stores the route information and sequenceno which corresponds to those k valid points to its Secure-route table and these routes are free of malicious nodes. S again chooses another combination of three points among six available points and checks its validity. S computes ${}^6C_3$ number of sets to detect non secure routes. Lets S->B->E, S->C->E, S->D->E, these three routes are secured according to Figure 2.

5.6.4 Final Route Detection

S sends WEIGHT_REQ message to all nodes {B, C, D} on each secure route obtained from its Secure-route table. Now, B,C,D [Figure 2] in the secure route send their weight table to S using ACK message with same sequenceno as that present in WEIGHT_REQ message sent by S to it. S is informed about weight of each node I (W_i) from its Weight-table$_i$. S then calculates average weight ($AVGR_s[m]$) of all nodes present in m^{th}

route which possess the same sequenceno m . S selects route S->D->E as final route for which average weight is maximum (let) amongst other secure routes (Figure 3).

S uses RSA to encrypt each message using public key of destination node E and sends all messages through this route to destination node E. After receiving each encrypted message E decrypts this message using its private key of E using CRT and sends ACK message to S back-tracking in the same route as traversed already. For each node I in the set {S,D, E} present in the route Trust value Tv_i and battery power consumption Bp_i will increase. This process is repeated after every Δt time interval.

6. PERFORMANCE EVALUATION

Performance of the algorithm RSRP has been evaluated and outlined in the following subsections.

6.1 Performance Metrics

Performance metrics used to evaluate our proposed protocol are packet dropped versus number of malicious nodes, computational cost versus number

Figure 2. Secure routes detection

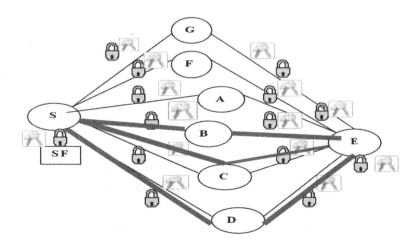

Figure 3. Final route detection amongst all secure routes

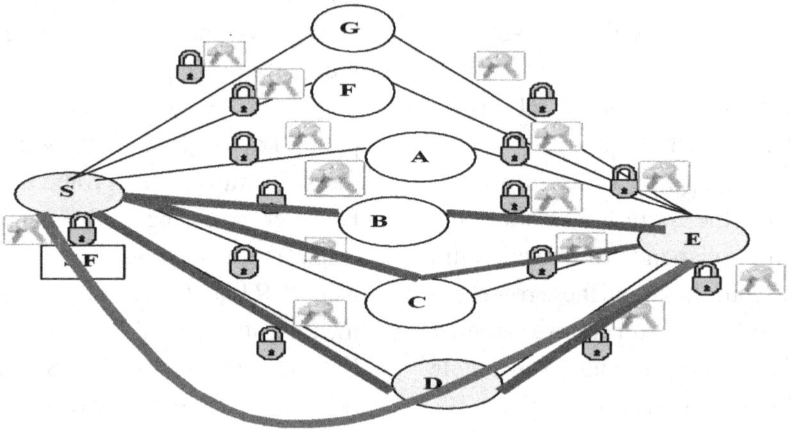

of nodes, end to end delay versus load in terms of number of messages and overhead in terms of number of control packets versus number of nodes.

6.2 Simulation Environment

Performance of proposed protocol is analyzed using simulation techniques. Table 1 shows the simulation setting of the network environment. This protocol is simulated in NS2.29. Simulation environment of this protocol is Fedora 9. We use IEEE 802.11 for wireless LAN as MAC layer. The channel capacity of mobile node is 2 Mbps. In our simulation mobile nodes move in a 600*600 m^2 region for 125 second simulation time. Mobility model is considered here as random waypoint. It is assumed that each node moves independently with the same average speed 10 m/s and pause time is 0-25 second. No path loss is considered. The network size is varied as 10, 20, 30, 40 and 50 nodes. The simulated traffic is constant bit rate (CBR). Table 2 shows the simulation setting of the network environment.

6.3 Results and Analysis

We have compared the performance of proposed protocols with other existing non-secure routing

Table 1. Route-table associated with S

Sequenceno	SID	Intermidiate Node Id	DID
1	S	A	E
2	S	B	E
3	S	C	E
4	S	D	E
5	S	F	E
6	S	G	E

Table 2. Simulation environment

Name	Value
Channel	Wireless
Propagation	Two Way
Network Interface Type	Wireless Phy
Antenna	Omni Antenna
No of nodes	10 to 50
MAC	IEEE 802.11
Simulation Area	600*600 m^2
Timeout period	5.0 sec.

protocols such as AODV (Perkins et al., 2003) and DSR (Johnson et al., 1997) as well as security based routing protocols such as Secure Routing Scheme Using Secret Sharing (Amuthan et al.,

2011), New Security Routing Protocol Using ZRP (Varaprasad et al.,2008). According to basic characteristics of above stated routing protocols, proposed protocol is compared with others.

6.3.1 Number of Packets Dropped vs. Number of Malicious Nodes

This protocol RSRP is compared with two non-secure routing protocols AODV, DSR and also two secure routing protocols i.e. security algorithm using ZRP and Secure Routing Scheme with Secret Sharing which is described in Figure 4. AODV and DSR being non-secure routing protocols, packet dropped in case of RSRP are much less compared to AODV and DSR when number of malicious node increases. Comparison of RSRP with *secure routing scheme with secret sharing* and *new security algorithm using ZRP* shows that its performance is best amongst three.

6.3.2 Computational Cost vs. Number of Nodes

In Figure 5 computational cost of proposed protocol RSRP is compared with another secure routing protocol using secret sharing.

When number of nodes varies from 10 to 40, Computational cost of Secure Routing Scheme using secret sharing which uses RSA modular expansion is always higher than proposed routing protocol using CRT. This is obvious from the following equations describing computational cost in each case.

In RSA decryption computational cost is:

$$3/2*|N||N|^2 = 3/2*|N|^3 \qquad (1)$$

In CRT decryption computational cost is:

$$3/2*|N|/2*|N|^2/2 = 3/8*|N|^3 \qquad (2)$$

6.3.3 End to End Delay vs. Load in MANET

Figure 6 compares end to end delay vs. load amongst DSR (a non-secure routing protocol), New Security Routing Protocol Using ZRP and our proposed protocol RSRP. From the figure it has been observed that overall performance of the proposed protocol is far better than DSR and New Security Routing Protocol Using ZRP. DSR being a non-secure routing protocol is not able to avoid malicious nodes if present in its path which is responsible for increasing end to end delay in comparison to RSRP. Also, performance of New Security Routing Protocol Using ZRP is degraded compared to RSRP. This is because of the fact that our proposed protocol selects the most secure route depending on some parameters such as trust value, battery power and stability of the nodes present in the route which is not the situation in Security Routing Protocol Using ZRP. ZRP requires extra time delay for creating zone and selecting cluster head which takes the responsibility for route selection. Performance of proposed protocol RSRP becomes better than the other for the reasons mentioned above.

6.3.4 Control Packet Overheads vs. Number of Nodes

Here RSRP compares with two secure routing protocols SEAD and New security algorithm for MANET using ZRP. Performance of these three protocols is almost same in terms of control packet overheads vs. number of nodes.

7. FUTURE RESEARCH DIRECTIONS

Security in Mobile Adhoc Network has long been the subject of researchers due to the utmost confidentiality involved in the underlying applications.

Figure 4. Packet dropped vs. number of malicious nodes

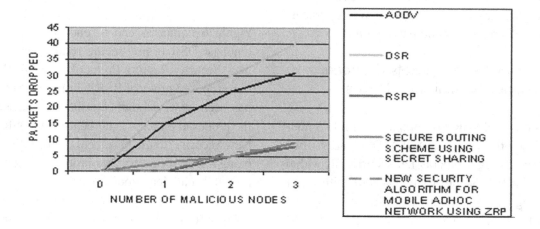

Figure 5. Computational cost vs. number of nodes

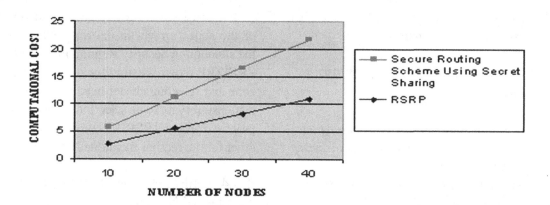

Figure 6. End to end delay vs. load in the network

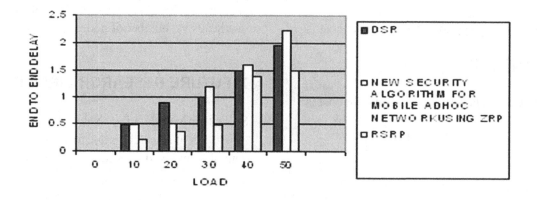

The use of cryptographic techniques along with secure key sharing ensures reliability of data to a large extent. However, as the network becomes heterogeneous and the applications becomes diverse along with the presence of "Internet of Things," these techniques may not be sufficient to provide reliable data delivery. New types of security threats are expected to come up as the number of devices connected to the network becomes too many, and like human beings any other "thing" also becomes ready to communicate with one another. In such a situation, the existing techniques might add to the computational overhead, thus hampering the connectivity process. Lightweight cryptography might provide slight relief, but some serious modification/ alterations will be necessary depending on the amount of network traffic forecast. Our future work may be extended in this direction.

8. CONCLUSION

Ensuring a secure environment in MANET is very challenging. Key generation, encryption and decryption play an important role for providing secure routing in MANETs. The algorithm combines the concepts of RSA, CRT and Shamir's secret sharing for providing secure environment in MANETs. We use combination of RSA and CRT schemes for key generation, encryption and decryption of data. Secure routes not having any malicious nodes in its route are detected using Shamir's secret sharing scheme and finally, secure route having highest weight is selected for message communication. Weight of a route depends on trust value, battery power and stability of all nodes on the route which leads to select a robust, stable, trustworthy and secure route for the messages in MANET.

This protocol RSRP is compared with two non-secure routing protocols AODV and DSR and also two secure routing protocols i.e. security algorithm using ZRP and Secure Routing Scheme with Secret Sharing. As expected it shows that packet dropped in case of RSRP is much less compared to AODV and DSR when number of malicious nodes increases. Comparison with other secure routing protocols shows that RSRP and security algorithm with ZRP shows same and best performance compared to security protocol with secret sharing.

RSRP is now compared with another secure routing scheme with secret sharing in terms of computational cost as number of nodes increases. It shows performance of RSRP is better than the other one since it uses RSA-CRT scheme for encryption and decryption of messages instead of RSA alone.

Then with increase in load in the network, RSRP shows best performance in terms of end to end delay when compared with DSR, a non-secure routing protocol and New Security Routing Protocol Using ZRP. In security protocol with ZRP, as load increases number of zones will also increase leading to highest end-to-end delay amongst three. Under low load, performance of the New Security Routing Protocol Using ZRP is better compared to DSR.

Comparison in terms of Control Packet Overhead vs. Number Of Nodes shows that difference of performance of three secure routing protocols named as RSRP, New security algorithm for MANET using ZRP and SEAD are negligibly small which implies that number of control packets generated in all the cases are almost equal.

All these comparisons lead us to the conclusion that the overall performance of our proposed protocol RSRP is best from different perspectives, when compared to non-secure routing protocols AODV and DSR as well as three secure routing protocols named as New Security Routing Protocol Using ZRP, SEAD, Secure Routing Scheme with secret Sharing whereas overhead in terms of number of control packets of RSRP along with three other secure routing protocols remain same.

REFERENCES

Amuthan, A., & Baradwaj, B. A. (2011). Secure routing scheme in MANETs using secret key sharing. *International Journal of Computers and Applications*, 22(1).

Ateniese, G., Santis, A. D., Ferrara, A. L., & Masucci, B. (2006). Provably-secure time-bound hierarchical key assignment schemes. In *Proceedings of CCS'06*. Alexandria, VA: ACM.

Balachandan, R. K., Zou, X., Ramamurthy, B., & Thukral, A. (2007). An efficient and attack resistant agreement scheme for secure group communications in mobile ad-hoc networks. In *Wireless communication and mobile computing*. Hoboken, NJ: Wiley.

Boneh, D. (1998). The decision diffe hellman problem. In *Proceedings of Third Algorithomic Number Theory Symposium*, (pp. 48-63). Portland, OR: Springer.

Dang, L., Xu, J., Li, H., & Dang, N. (2010). DASR: Distributed anonymous secure routing with good scalability for mobile ad hoc networks. In *Proceedings of 5th IEEE Asia-Pacific Services Computing Conference*, (pp. 454-461). Hangzhou, China: IEEE Computer Society.

Djenouril, D., Mahmoudil, O., Bouamama, M., Jones, D. L., & Merabti, M. (2007). *On securing MANET routing protocol against control packet*. Retrieved from www.researchgate.net/.On_Securing_MANET_Routing_Protocol_AgainstControlPacket

Drira, K., Seba, H., & Kheddouci, H. (2010). ECGK: An efficient clustering scheme for group key management in MANETs. *International Journal of Computer Communications*, 33, 1094–1107. doi:10.1016/j.comcom.2010.02.007

Grobler, T. L., & Penzhorn, W. T. (2004). Fast decryption methods for the RSA cryptosystem. In *Proceedings of 7th AFRICON Conference*. Paris, France: IEEEXPLORE.

Hu, Y. C., Johnson, D. B., & Perrig, A. (2003). SEAD: Secure efficient distance vector routing for mobile wireless adhoc networks. *Ad Hoc Networks Journal*, 1(1), 175–192. doi:10.1016/S1570-8705(03)00019-2

Johnson, D. B., & Maltz, D. A. (1997). Dynamic source routing in ad hoc wireless networks. In T. Imielinski, & H. Korth (Eds.), *Mobile computing* (pp. 153–181). Academic Press.

Laxmi, V., Jain, L., & Gaur, M. S. (2006). Ant colony optimization based routing on NS-2. In *Proceedings of International Conference on Wireless Communication and Sensor Networks*. Allahabad: IEEE

Lee, D., Choi, S., Choi, J., & Jung, J. (2007). Location aided secure routing scheme in mobile ad hoc networks. In *Proceedings of Computational Science and Its Applications* (pp. 131–139). Kuala Lumpur, Malaysia: Springer-Verlag. doi:10.1007/978-3-540-74477-1_13

Li, H., & Singhal, M. (2006). A secure routing protocol for wireless ad hoc networks. In *Proceeding of 39th Hawaii International Conference on System Sciences*. Kauai, HI: IEEE.

Lin, C. H., Lai, W. S., Huang, Y. L., & Chou, M. C. (2008). I-SEAD: A secure routing protocol for mobile ad hoc networks. In *Proceedings of International Conference on Multimedia and Ubiquitous Engineering* (pp. 102-107). Busan, Korea: IEEE.

Lou, W., Liu, W., Zhang, Y., & Fang, Y. (2009). SPREAD: Improving network security by multipath routing in mobile ad hoc networks. *Springer Wireless Network*, 15, 279–294. doi:10.1007/s11276-007-0039-4

Menaka, A., & Pushpa, M. E. (2009). Trust based secure routing in AODV routing protocol. In *Proceedings of the 3rd IEEE International Conference on Internet Multimedia Services Architecture and Applications* (pp. 268-273). Piscataway, NJ: IEEE Xplore.

Perkins, C. E., Belding-Royer, E. M., & Das, S. R. (1999). Ad hoc on-demand distance vector (AODV) routing. In *Proceedings of 2nd Workshop on Mobile Computing and Applications (WMCSA '99)* (pp. 90-100). WMCSA.

Pushpalatha, M., Venkataraman, R., Khemka, R., & Rao, T. R. (2009). Fault tolerant and dynamic file sharing ability in mobile ad hoc networks. In *Proceeding of International Conference on Advances in Computing, Communication and Control* (pp. 474-478). Curan, NY: ACM.

Rivest, R. L., Shamir, A., & Adleman, L. (1978). A method for obtaining digital signatures and public-key cryptosystems. *ACM, 21*(2), 120–126. doi:10.1145/359340.359342

Sanzgiri, K., Dahill, B., Levine, B. N., Shields, C., & Belding-Royer, E. M. (2002). A secure routing protocol for ad hoc networks. In *Proceedings of 10th IEEE International Conference on Network Protocols* (pp.78-79). Paris, France: IEEE.

Sarkar, S., Kisku, B., Misra, S., & Obaidat, M. S. (2009). Chinese remainder theorem-based RSA-threshold cryptography in MANET using verifiable secret sharing scheme. In *Proceedings of IEEE International Conference on Wireless and Mobile Computing, Networking and Communications* (pp.258 – 262). Marrakech: IEEE.

Sehgal, P. K., & Nath, R. (2009). An encryption based dynamic and secure routing protocol for mobile adhoc network. *International Journal of Computer Science and Security, 3*(1).

Shamir, A. (1979). How to share a secret? *Magazine of Communications of the ACM, 22*(11). doi: doi:10.1145/359168.359176

Shand, & Vuillemin. (1993). Fast implementation of RSA cryptography. In *Proceedings of 11th IEEE Symphosium on Computer Arithmetic* (pp. 252-259). IEEE.

Sinha, D., Bhattacharya, U., & Chaki, R. (2012a). A secure routing scheme in MANET with CRT based secret sharing. In *Proceeding of 15th International Conference of Computer and Information Technology* (pp.225-229). Bangladesh: IEEE XPlore.

Sinha, D., Bhattacharya, U., & Chaki, R. (2012b). A CRT based encryption methodology for secure communication in MANET. *International Journal of Computer Applications, 39*(16). DOI number is 10.5120/4904-7406

Sun, H. M., & Wu, M. E. (2005). *An approach towards rebalanced RSA-CRT with short public exponent.* Cryptology ePrint Archive: Report 2005/053. Retrieved from http://eprint.iacr.org/2005/053

Sun, Y., Yu, W., Han, W., & Liu, R. (2006). Information theoretic framework of trust modeling and evaluation for ad hoc networks. *IEEE Journal on Selected Areas in Communications, 24*(2), 305–317. doi:10.1109/JSAC.2005.861389

Varadharajan, V., Shankaran, R., & Hitchens, M. (2004). Security for cluster based ad hoc networks. *Journal of Computer Communications*, (27), 488–501.

Varaprasad, G., Dhanalakshmi, S., & Rajaram, M. (2008). *New security algorithm for mobile adhoc networks using zonal routing protocol.* Ubiquitous Computing and Communication Journal.

Wang, J. W., Chen, H. C., & Lin, Y. P. (2009). A secure DSDV routing protocol for ad hoc mobile networks. In *Proceedings of Fifth International Joint Conference on INC, IMS and IDC* (pp.2079-2084). IEEE.

Wong, C. K., Gouda, M., & Lam, S. S. (2000). Secure group communications using key graphs. *IEEE/ACM Transactions on Networking*, *1*(8).

Wu, C. H., Hong, J. H., & Wu, C. W. (2001). RSA cryptosystem design based on the Chinese remainder theorem. In *Proceedings of Asia and South Pacific Design Automation Conference* (pp. 391-395). Yokohama, Japan: ACM.

Xia, G., Huang, Z. G., Wang, Z., Cheng, X., Li, W., & Znati, T. (2006). Secure data transmission on multiple paths in mobile ad hoc networks. *Lecture Notes in Computer Science*, *4138*, 424–434. doi:10.1007/11814856_41

Xia, H., Jia, Z., Li, X., Ju, L., & Sha, E. H. M. (2012). Trust prediction and trust-based source routing in mobile ad hoc networks. *International Journal of Ad hoc. Networks*, *7*(11), 2096–2114.

KEY TERMS AND DEFINITIONS

CRT: The Chinese Remainder Theorem is a mathematical tool evaluated by Shand and Vuillemin in 1993. It helps design of deterministic key pre-distribution using number theory.

Malicious Node: If trust value of the Suspect node does not increase with time, the node is termed a Malicious node, and the route containing the Malicious node can't be selected as a secure route by the source node for sending its packets.

MANET: A network consisting of mobile nodes with dynamic connectivity pattern.

RSA: A popular cryptography algorithm which was proposed by Rivest, Shamir, and Adleman in 1978. This algorithm uses two keys, public and private. Source node encrypts message using its public key and destination node decrypts that ciphertext message using its private key.

Safety Key: A key value used for guaranteeing the safety of a route.

Suspect Node: The node whose trust value is below the threshold value.

Trust Agent: This type of node monitors newly entrant nodes in a MANET for a specified time interval and assigns trust value to that node and also detects whether the node is malicious or not.

Trust Value: The trust value associated with a node gives an estimation of trusted behavior of the node.

Chapter 12
An Overview of Wireless Sensor Networks:
Towards the Realization of Cooperative Healthcare and Environmental Monitoring

Thomas D. Lagkas
International Faculty of the University of Sheffield, CITY College, Greece

George Eleftherakis
International Faculty of the University of Sheffield, CITY College, Greece

ABSTRACT

Wireless Sensor Networks constitute one of the highest developing and most promising fields in modern data communication networks. The benefits of such networks are expected to be of great importance, since applicability is possible in multiple significant areas. The research community's interest is mainly attracted by the usability of sensor networks in healthcare services and environmental monitoring. This overview presents the latest developments in the field of sensor-based systems, focusing on systems designed for healthcare and environmental monitoring. The technical aspects and the structure of sensor networks are discussed, while representative examples of implemented systems are also provided. This chapter sets the starting point for the development of an integrated cooperative architecture capable of widely providing distributed services based on sensed data.

INTRODUCTION

Lately, the focus of the research community is directed towards the development of flexible and extensive networks, which consist of small devices capable of collecting information regarding the surrounding environment. This procedure is commonly called "sensing" and the corresponding devices "sensors." The realization of sensors has become recently possible and cost efficient, due to the significant advances in the field of wireless communications, hardware design, and computer networking.

DOI: 10.4018/978-1-4666-5170-8.ch012

A Wireless Sensor Network (WSN) involves multiple sensors that are capable of collecting data, perform some processing, and relay them. Sensors are expected to be able to interact with one another and act as intermediate nodes, when information needs to be routed over multiple hops in order to reach the destination. The latter is considered a central entity inside the WSN, often called "sink," where all data are gathered, stored, and processed. The results can then be presented in a comprehensive and useful form, based on the specific case.

Thus, the sensing devices should incorporate components capable of gathering information from the surroundings, convert the measurements into digital form, and forward data using appropriate wireless transmission techniques, medium access control, and routing methodology. The required flexibility in topology is directly related with the need for self-organizing features among different sensors. In specific, the nodes are expected to cooperate and adapt to dynamic conditions, increasing the availability and reliability of the whole system. Taking into account that some popular WSN applications involve deployment in harsh environments, the necessity for self-management becomes even more critical.

The special area that has probably attracted most of the research community's attention regarding WSNs is energy efficiency. Sensors are small wireless devices that usually need to operate on their own energy source for extended periods of time. Hence, power consumption is an issue of great importance that greatly affects the success and usefulness of multiple WSN applications. For this reason, different technologies make significant steps in optimizing energy conservation and ensuring efficient battery charging schemes. According to the specific application, sensors are allowed to transit to low power modes, increase the periodic transmission intervals or even switch on only upon request.

Since the decentralized approach in the operation of WSNs is becoming very popular among the researchers, the ad hoc wireless networks are thought as a promising solution for realizing a WSN. However, ad hoc computer networks are considered to constitute a networking paradigm that is significantly different from the sensor networking concept. Some key differences are listed below (Akyildiz, Su, Sankarasubramaniam, and Cayirci, 2002a, 2002b):

- A WSN may involve far more nodes than a typical ad hoc network.
- The sensing devices are typically placed close to each other to compensate for limited transmission range.
- Each sensor is more likely to fail.
- WSN topology is considered highly dynamic.
- Sensors commonly rely on broadcast transmissions instead of point-to-point links.
- Each node has limited energy, processing power, and memory.
- The involved devices may not carry globally set unique addresses.

Regarding sensors' special characteristics, these are straightly related to their inherent resource constrains (Culler, Estrin, & Srivastava, 2004): The processing speed, the storage capacity, and the communication bandwidth are all limited. From the perspective of a system, WSNs exhibit substantial capabilities, but each device as an independent entity is significantly constrained in resources. Thus, sensors' cooperation and self-organization are fundamental in fulfilling the network requirements.

The demands regarding the operation of WSNs may be significantly challenging; in most cases the sensors are expected to be functional for large time periods using just the energy produced by small size batteries. Since most power saving schemes involve long inactive periods, connectivity typically varies at a great degree, significantly affecting the overall system reliability. For this reason, the deployment pattern, the node density, and the employed communication protocols are of great importance, especially for critical applications and harsh environments. Moreover, a basic prerequisite in WSNs is cost efficiency. The network installation and management requirements should be kept as low as possible, increasing the necessity for efficient self-organization and self-management.

In regards to WSN applications, it is now evident that sensor networks may be proved valuable in multiple different cases and actually applications increase along with the technology advances and the evolvement of users' demands. It is a fact that WSN initial applications were related with military projects. However, through the years sensors have been applicable in various non-military scenarios and nowadays it is considered that their core nature is not military.

Specifically, modern sensor networks can be used in areas such as infrastructure security, environmental-industrial sensing, and health monitoring (Chong & Kumar, 2003). In more detail, a popular WSN application is related to securing critical buildings and facilities. Sensors are able to timely detect possible threats, by collecting video, acoustic, and other types of data. In such cases, the deployment is crucial for providing the required coverage and handling false alarms. Another promising application of sensor networks is related to gathering environmental data. Typical examples involve measurements of temperature, humidity, pH, and air-water pollution. Industrial sensing scenarios in particular may include special devices capable of sensing pressure, vibration, electromagnetic signals or even radiation levels. This case is obviously considered to be critical, since it is straightly related to safety. Another characteristic application of WSNs is traffic control. Sensors are capable of capturing and processing traffic images, leading to key decisions on traffic prioritization and redirection. Lastly, one of the hottest fields where sensors can be proved extremely useful is health monitoring. Real-time or non-real-time collection of vital signals is of high importance to physicians in order to come

up with the correct diagnosis and treatment. This information is also useful to the subjects themselves, since they are given the capability to have continuously updated information on their health status and this way take the required timely action.

The rest of the chapter is organized as follows: The next section discusses the technical aspects of sensor networks, where the communication architecture is presented first and then WSN specific issues are examined. In the following section, we study the application of sensor networks in healthcare. In specific, the required network structure is presented along with some related prototypes. Next, applications of WSNs in environmental monitoring are reviewed and then related research trends are presented and future directions are provided. Finally, the chapter is concluded.

BACKGROUND: WIRELESS SENSOR NETWORKS TECHNICAL ASPECTS

Indisputably, the special nature of the WSNs is accompanied by unique technical aspects that differentiate sensor networks from other types of networks. These aspects actually constitute technical challenges for the realization of successful and efficient WSNs. A typical example of a sensor network that is related with major technical challenges is the case of Industrial WSN (IWSN). These are listed below (Gungor & Hancke, 2009):

- **Resource limitations:** The realization of IWSNs is constrained by energy, memory, and processing resources.
- **Alternating and hostile conditions:** The environmental conditions in industrial deployments are typically harsh and often lead to connection failures and topology changes. Moreover, network performance might be degraded, due to electromagnetic interference, humidity, vibrations, or oth-

er conditions (US Department of Energy, 2004).

- **Quality-of-service (QoS) demands:** Applications of IWSNs may greatly vary having different specifications and QoS requirements. The latter are related with constraints on specific network metrics, such as throughput and delay. For instance, real-time control via sensor feedback in industrial environments requires direct response, hence, very low delays.

- **Data redundancy:** It is a common approach to deploy sensors in close proximity. This way, collected data are typically overlaid and duplicated.

- **Frame errors and varying available bandwidth:** The wireless links under the harsh conditions of IWSNs are extremely unreliable. The exhibited Bit Error Rate (BER) is at high levels, especially when compared to other types of more reliable networks. Thus, the percent of erroneous or lost frame is also high. Furthermore, the link conditions highly vary leading to significant variations of the link capacity.

- **Security:** IWSNs often require increased security provision in order to avert passive or active network attacks. Passive attacks involve eavesdropping on transmissions, which may reveal critical information. Active attacks are usually related with data alternation or service interruption. Both types of attacks may have a great impact on the underlying system and should be prevented by the security aspect of IWSN.

- **Large installation and ad hoc network structure:** IWSNs may include a large number of sensing devices, which operate in a distributed manner. In many cases, there are no preset links and sensors communicate in ad hoc mode without the support and direct coordination of a central network entity.

- **Internetworking:** Network management in IWSNs is expected to provide the capability of remote control. Nowadays, the managers of such networks require complete access over the Internet, in order to be able to control it anytime and from anyplace. This actually means that IWSNs should be integrated to the TCP/IP architecture (Montenegro, Kushalnagar, Hui, & Culler, 2007) using the necessary gateway nodes (Akyildiz, Melodia, & Chowdhury, 2007).

WSN Communication Architecture

Network Design Goals

In order to fulfill the special requirements, WSNs should adopt a suitable communication architecture that corresponds to the primary network design goals. A set of such goals is provided in (Gungor & Hancke, 2009) and the most representative ones are listed below:

- **Flexible architectures and effective protocols:** WSNs are expected to constitute hierarchical and modular networks that exhibit high flexibility, reliability, and robustness.

- **Local processing and data fusion:** In order to keep network load and overhead at low levels, the sensors can pre-process data and transmit only the required information.

- **Energy-efficient design:** The maximization of network lifetime can be only achieved if power saving techniques are employed in all aspects of the network architecture. Protocols in multiple layers of the protocol stack should be designed with energy conservation considerations.

- **Self-organization / adaptive operation:** Sensor networks are characterized by highly dynamic topologies, which may be due to unreliable links, power-down intervals or node mobility. For this reason, the involved devices are expected to self-organize and for instance provide on-demand alternative paths in case of link failures.

- **Synchronization:** In WSNs, a large number of nodes need to cooperate to successfully provide the required service. Cooperation in a highly distributed scheme demands increased synchronization among nodes.
- **Reliability and error tolerance:** According to the application type, sensed data should be delivered reliably through multiple hops, which may cause high packet error rates. Network design is crucial to provide reliable data transfers and efficient error detection/correction techniques.
- **Application-based design:** Each WSN application has specific network design requirements, which significantly differentiate WSN deployments from one another.

System Structure

The typical architecture of a sensor network is scattered as presented in Figure 1. Each device is capable of "sensing" data and relaying information received by adjacent nodes. The destination of all transmissions originated by sensors is a data collecting entity called "sink." The data

packets reach the sink after being routed among interconnected wireless sensors. From that point on, a remotely located task manager node may receive and process or present data.

Sensors are usually deployed very densely, so the sensor field may include hundreds or even thousands of nodes that are located a few feet away from each other. Maintaining network topology in such conditions is a challenging process that can be divided to three phases (Akyildiz, Su, Sankarasubramaniam, and Cayirci, 2002a, 2002b):

- **Pre-deployment and deployment:** Each sensor may be placed one by one or in masses.
- **Post-deployment:** The network topology is prompt to alternations after the initial deployment, due to dynamic existing conditions.
- **Redeploying extra nodes:** At any time the addition of more sensors may be needed to restore or expand the network.

In regards to the architecture of WSNs, it is very important to be scalable. There are two main approaches regarding the nature of the nodes (Gungor & Hancke, 2009). According to the first one (called homogeneous), all devices participating in the network are identical and operate as peer entities. The second approach involves heteroge-

Figure 1. Scattered nodes in WSN

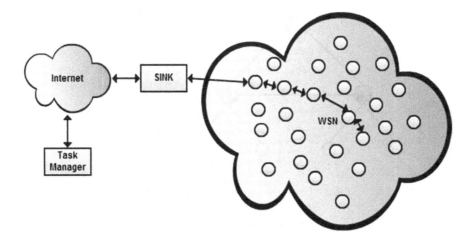

neous sensors, where the simplest ones with great resource limitations perform basic tasks (such as "sensing") and the most advanced ones are responsible for more complicated activities (like data routing). Another typical classification of the parts of a WSN (Qi, Iyengar, & Chakrabarty, 2001) defines as sensor nodes only the devices that produce the original data, the entities which perform advanced tasks are called processing elements, whereas the communications network interconnects the processing elements. Sensors are associated with processing elements creating clusters and data routing takes place over the processing elements (Figure 2).

Some of the most well-known approaches in WSN structure as described in Qi, Iyengar, & Chakrabarty, (2001) are presented here. According to the anarchic committee structure (AC), a mesh network is formed without hierarchy. The dynamic hierarchical cone (DHC) on the other hand provides a type of tree structure. Moreover, hybrid structures that try to combine the advantages of both anarchic and hierarchical networks have been proposed. Specifically, in the flat tree structure, the nodes form complete binary trees with fully connected roots. A similar approach is

adopted in the deBruijn graph (DG), where nodes in the same level also maintain connections. These four types of WSN structure are illustrated in Figure 3.

The latest trends on the WSN topology are towards ad hoc sensor networks. The fundamental concept in such networks is that links are established only when needed (Perkins, 2000). In multiple scenarios, lightweight sensors need to be deployed onsite and create temporary networks. For such dynamic environments, different solutions were proposed to enable connectivity. Authors in Lim, (2001), define three primitive services that need to be implemented in self-organized sensor networks: service lookup, node composition, and dynamic adaptation. Another popular approach considers sensors and the provided services as "resource objects" registered in a centralized "lookup" service repository. Lastly, the basic idea in Estrin, Govindan, Heidemann, and Kumar (1999) is inspired by biological systems and involves directed diffusion of interests for specific type of data.

The establishment of WSN links on demand, which is a feature of self-organized networks, greatly differentiates sensor networks from typi-

Figure 2. A WSN architecture that involves sensors and processing elements

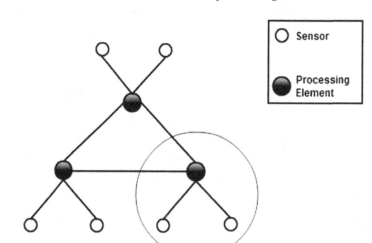

Figure 3. WSN structures: a) AC, b) DHC, c) flat tree, and d) DG

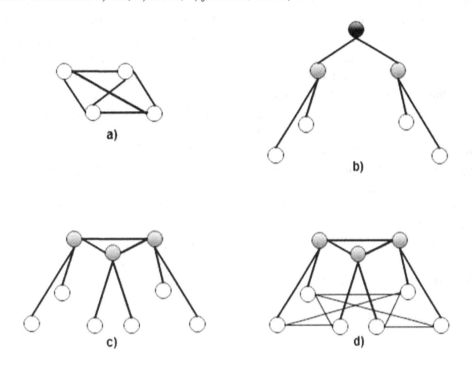

cal networks of stable routes. Each sensor node discovers its neighbors using broadcasting and employ distributed routing algorithms to identify paths to the destination (Culler, Estrin, & Srivastava, 2004). During idle intervals, the nodes scan the wireless medium for a special symbol in the initial fields of a frame in order to synchronize. Network discovery needs to be adaptable to often changes of the network topology.

An additional significant issue that is straightly related to the WSN system structure and the overall communication architecture is definitely the hardware design. The goal here is the development of low-cost and low-power sensors with the following main parts (Gungor & Hancke, 2009): sensing entity, processor, transceiver, and power source. The sensing entity should be able to create a response to alternations of the physical measurement. This response is actually an analog signal that needs to be converted to digital to be processed and forwarded in frames. The processor is responsible for the whole functionality of

the sensor node. A radio unit is implemented in the transceiver, which may typically be in one of the following modes: transmit, receive, idle, and sleep. The latter is related to low energy conservation required by the power saving scheme. Lastly, the power source is critical to the efficiency of a sensor. It usually incorporates a small battery that may be recharged using of the promising energy-harvesting techniques (e.g. solar, air, vibration et. al).

Software is also crucial for the successful operation of a highly distributed network, like a WSN. The related dominant approach is the offering of multiple software components to the users as independent services according to specific needs (Al-Jaroodi & Mohamed, 2012). Lately, a new model that tends to surpass the traditional client-server approach is the Service-Oriented Computing (SOC) (Bichler & Lin, 2006). However, SOC also exhibits major problems mainly related to supporting heterogeneous platforms. For this reason, a middleware system for SOC, called

Service-Oriented Middleware (SOM), seems to be of great importance. Another popular recent approach for multisensory data fusion is the use of software agents (Gascueña & Fernández-Caballero, 2011). A well-known development environment that is adopted in multiple research projects for building agents in sensor networks is the Java Agent DEvelopment Framework (JADE). The created mobile agents are responsible for discovering services and disseminating information. Representative implementations of mobile agents for multisensory data fusion can be found in Biswas, Qi, and Xu (2008) and Karlsson, Backstrom, Kulesza, and Axelsson (2005).

WSN Specific Networking Issues

The network protocol stack that is adopted in WSNs is crucial for the system behavior. As it is already explained, the sensor networks have special requirements that need to be handled with the development of WSN specific mechanisms in the corresponding network layers.

Starting with the physical layer, its functionality includes handling frequency channels, signal modulation, and carrier detection. In order to ensure power consumption minimization, radio transmissions need to be of limited range. This is the main reason behind the dense deployment of sensors in WSNs and leads to the need for multihop communications. Low-power transmissions make high data rate communications extremely challenging. Efficient modulation schemes need to be developed, that enable the transmission of multiple bits per signal symbol, while keeping bit error rate at sufficiently low levels despite the highly unreliable wireless links. A very promising technology for the physical layer of WSNs is Ultra-WideBand (UWB), which is introduced for short-range wireless transmissions at increased data rates. It exhibits low complexity, low cost, and high robustness (Mendes & Rodrigues, 2011). Other candidate technologies are introduced in related standards and are presented later in this subsection.

The role of the data link layer and especially the Medium Access Control (MAC) in self-organized multihop WSNs is twofold. On one hand, the communication links between multiple sensor nodes need to be established, in order to build the hops of the transmission routes. On the other hand, the MAC protocol is responsible for efficient resource sharing among different nodes competing for bandwidth. Existing MAC protocols are of course used as the basis for WSNs' data link layer, however, they cannot be adopted as is. The absence of a central entity that controls transmissions and the dynamic nature of the self-configured systems lead to the necessity for highly distributed schemes. A summary of MAC protocols for sensor networks that base on CSMA and can be used in healthcare systems is presented in Egbogah and Fapojuwo (2011) and provided in Table 1.

As it is already discussed, WSNs are dynamic multihop networks, hence, there is a significant demand for efficient network layer protocols which are able to successfully route data between the sensors and the sink. The main principles when designing a WSN routing protocol according to Akyildiz, Su, Sankarasubramaniam, and Cayirci (2002a, 2002b) are energy efficiency, data-centricity, data aggregation, and location awareness. A typical approach in WSN routing is the division of the network in clusters, where a hierarchical routing process takes place among cluster-heads and members of the same cluster (Mendes & Rodrigues, 2011). One of the most referred cluster-based routing protocols for sensor networks is LEACH (WB, AP, & H, 2002), which is also capable of power saving. Another important feature of modern routing protocols for WSNs is QoS support, which is considered essential for the prioritized forwarding of data based on their differentiation (Hung & Huang, 2010).

As far as the transport layer is concerned, it is actually necessary when WSN access through the Internet is desired, a requirement not placed by a significant number of works. However, remote access to sensors' data is eventually expected to be of great importance. The current transport

Table 1. MAC protocols for sensor networks that base on CSMA and can be used in healthcare systems (Egbogah & Fapojuwo, 2011)

Criteria	S-MAC (Ye, Heidemann, & Estrin, 2004)	Dynamic Sensor MAC (Lin, Qiao, & Wang, 2004)	MS-SMAC (Pham & Jha, 2004)	DS-MAC (Yuan, Bagga, Shen, Balakrishnan, & Benhaddou, 2008)
Main Feature	Message passing	Dynamic duty cycle	Adaptive discovery of neighbors and dynamic duty cycle	Service differentiation via channel prioritization
Energy Efficiency	+ Save energy consumed when idle - Weak for VBR traffic	+ Automatic adjustment of duty cycles	+ Energy efficiency prioritization	(not discussed)
Delay	- Medium sharing unfairness increases latency	+ Latency kept low, thanks to automatic adjustment of duty cycles	- Network lifetime extension at expense of high delay	+ Emergency traffic access channel sooner with low delays
Mobility	- Stations considered fixed	+ Duty cycle adjustment at different update intervals - Different mobility conditions not shown to be efficiently supported	+ Nodes stay connected thanks to the adaptation of the neighbor discovery	(not discussed)

layer protocols may be considered to some degree sufficient Akyildiz, Su, Sankarasubramaniam, and Cayirci, 2002a, 2002b). The fact though that protocols like TCP and UDP employ global addressing makes them difficult to apply to sensors networks, where the nodes usually adopt attribute-based naming. In general, transport layer is considered too complicated, energy demanding, and not straightly necessary for implementation in the WSN layered architecture, hence, some researchers have realized transport layer functionalities, such as congestion control, in lower layer protocols (Mendes & Rodrigues, 2011).

Despite the fact that numerous applications for sensor networks have been proposed, there are very limited introductions of WSN specific application layer protocols in the literature Akyildiz, Su, Sankarasubramaniam, and Cayirci (2002a, 2002b). Three significant tasks for sensor networks that need to be addressed by the corresponding application layer protocols are the following: sensor management, data advertisement, and sensor query – data dissemination. Moreover, multimedia data processing at the application layer of WSNs

is also a field of increasing interest. The comparative analysis of the dominant contemporary video codecs has revealed that VBR traffic transfer over sensor networks is inefficient (Mendes & Rodrigues, 2011).

A very promising modern trend regarding the layered approach in WSNs is the cross-layer design. It is now a common place that there is interdependence between the functions addressed at different layers of a sensor network protocol stack (Kozat, Koutsopoulos, & Tassiulas, 2004). The three first layers affect each other at a great degree. Both physical and data link layers have a high impact on the resource allocation and channel utilization (Gungor & Hancke, 2009). The decisions on the network layer affect the operation of the two lower layers, while the latter undoubtedly affect routing. Moreover, the provision of efficient QoS support, security, and energy conservation is a multilayer issue and cannot be addressed in a single layer (MC & IF, 2009). For these reasons, there are now multiple proposals in the literature of cross-layer schemes for wireless sensor networks (Mendes & Rodrigues, 2011).

The most commonly adopted multilayer wireless networking standards for WSNs come from the IEEE 802.15 workgroup. Specifically, the IEEE 802.15.4 standard was designed for use in wireless personal area networks (IEEE 802.15.4, 2006). It defines four transmission schemes, where the first three operated and the 868/915 MHz frequency band and the last one at the 2450 MHz band. It is considered to be low-rate and it can function in both centralized and distributed architectures, based on the CSMA/CA access technique. Note that the commercial name of the IEEE 802.15.4 standard is now "ZigBee." In order to handle high-rate transmissions, the IEEE 802.15.3 standard was introduced (IEEE 802.15.3, 2003). Its specifications match the demands of multimedia data, such as video traffic. It operates at 2.4 GHz and employs a hybrid CSMA/CA and TDMA scheme. Another popular standard that is also used in WSNs is Bluetooth (Mendes & Rodrigues, 2011). It supports medium-rate transmissions at maximum range of 100 m. Its medium access control is cluster-based (forming piconets) and it is capable of power saving. Lastly, it should be clarified that a popular approach for wirelessly interconnecting sensors is IEEE 802.11 (commercially known as "Wi-Fi") (IEEE 802.11, 2007). It is capable of operating both at the 2.4 GHz and the 5 GHz license-free bands. It supports multiple data rates and different ranges, according to the specific standard version. The IEEE 802.11 standards defines both a centralized and a distributed MAC scheme, however, the CSMA-based distributed MAC scheme is by far the most popular.

WIRELESS SENSOR NETWORKS FOR HEALTHCARE

Overview of Healthcare WSNs

The recent major developments in sensor technology, wireless networking, and computer science have enabled the introduction of new solutions for healthcare provision. Nowadays, small devices can be easily built to take physical measurements. These sensors can now be cost and power efficient. These capabilities along with the sensors' small size allow embedding sensors in different everyday objects, like furniture, watches, or even clothes (Korhonen, Parkka, & Gils, 2003). In-hospital and out-of-hospital monitoring are hot research areas for long time now. Especially the ability of collecting physiological variables' values in real-time and in real-life conditions can be proved extremely useful for applying the appropriate treatment or improving the subject's wellbeing.

A special field of health monitoring that has lately gathered a lot of attention is that of wearable sensor-based systems (Pantelopoulos & Bourbakis, 2010). The increment in healthcare costs and the advances in sensors' technology have greatly affected the development of this specific research area. Different sensing devices of constrained size can be combined in such a health monitoring system. In this manner, the health status of a patient, an elderly person or even an athlete can be monitored seamlessly, that is without affecting the subject's daily activities, and on a constant basis. Among the expected results is high quality health service provision.

These systems actually involve sensor networks that may be formed even on or around a person's body, so they are called Wireless Body Area Networks (WBANs) (Barakah & Ammad-uddin, 2012). The nodes of such networks are characterized of short range and low data rate. The WBAN specifications as presented in Zhen, Li, and Kohno (2009) are provided at Table 2. The IEEE 802.15 Task group 6 is standardizing the current WBAN detailed specifications.

Healthcare System Network Structure

The network structure of a healthcare sensor-based system is examined in Alemdar and Ersoy (2010). According to the specific study, the main components that comprise the system are: the

Table 2. Specifications of a WBAN

Characteristic	Value
Range	2 to 5 meters
Time to start up	Less than 100 ns
Time to setup network	Less than 1 sec per device
Power consumed	About 1mW per Mbps
Density of the network	2 to 4 nodes per m²
End-to-end delay	10 ms
Network size	Maximum of 100 sensors per network

body area network, the personal area network, the gateway, the wide area network, and the monitoring application. A depiction of this architecture is provided in Figure 4. The body area network is the WSN formed by the sensors on the subjects' body. These sensing devices may be electrocardiogram sensors, RFID tags or measure body temperature, blood pressure et al. They come in very small sizes, they can be wearable and they need to be energy efficient. The personal area network consists of environmental sensors, such as audio and video capturing devices, and mobile devices. Moreover, this subsystem can provide location tracking by using the corresponding appliances, like GPS receivers. The gateway is the subsystem that interconnects the body and personal area networks to the wide area network. It can be any device that carries interfaces for the different networks and is capable of routing data from one to the other. For instance, a server computer, a laptop, a tablet or even a smartphone with connectivity to the Internet can be adopted. The nature of the wide area network may vary depending on the scenario. Probably the highest flexibility is provided by the cellular networks that can act as wide area networks interconnecting roaming subjects to the monitoring subsystem. Other possible candidates are the modern wireless broadband access technologies or even the traditional access schemes through the telephone system. The last subsystem in this architecture is the monitoring application, which is the place where all data are

collected and actions are accordingly triggered. It may incorporate machine learning algorithms and alarm systems.

The network structure of the u-healthcare system proposed in Hung and Huang (2010) compared to the one presented above exhibits both similarities and differences. The bottom line of this architecture is the wireless personal area network that consists of the sensors capturing vital information, which is then transmitted using a mobile device with Bluetooth and Wi-Fi interfaces. The access point at the wireless local area network relays data to the corresponding database server, which then forwards data through an Internet router to the database system of the healthcare center.

The general telemedicine architecture that is proposed in Grgić, Zagar, and Krizanović (2011) consists of three main parts: the local sensory area, the communication network area, and the institutional network area. The first one includes medical sensors attached to the patients' body. They support wireless connectivity, typical using ZigBee links. A small mobile device that patient carries collects the information obtained by the sensors and is responsible for forwarding the data destined to the medical institution. The communication network connects the local sensory area to the medical institution using different access technologies, such as 2G/3G, WiMAX (IEEE 802.16, 2004) et al. The data reaching the destination are stored in the corresponding database and are analyzed by the physicians, in order to draw conclusions and possibly take action.

The communication standards that can be employed for wireless connectivity in wearable health monitoring systems are summarized and evaluated in Pantelopoulos and Bourbakis (2010). Long-range data communications between remote devices in the examined system can be supported by a variety of technologies, such as WLAN, GSM, GPRS, and UMTS (Halonen, Romero, & J. Melero, 2003), which offer wide area wireless access. Moreover, the developments in 4G mobile communications are expected to facilitate the col-

Figure 4. Main components of a healthcare sensor-based system

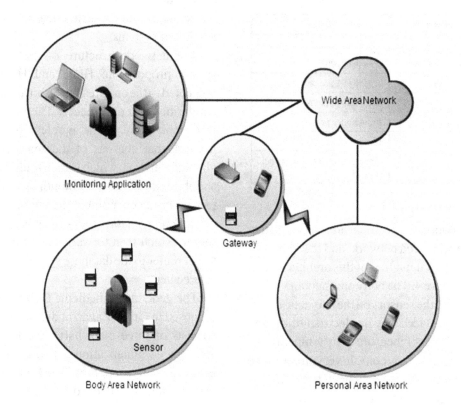

lection of real-time data from remotely located wearable medical sensors. The communication standards that are most commonly used for body area networks are Bluetooth and ZigBee. The main advantage of ZigBee is its power saving capability; however, the latest versions of Bluetooth define efficient energy conservation mechanisms as well. Other candidate schemes include UWB, infrared links, and medical implant communications. Lastly, typical Wi-Fi networks are commonly used for short-range links, despite the fact that they are not specially designed for such applications.

Sensor-Based Healthcare Prototypes

Recently, some prototypes on pervasive health-care monitoring have made their appearance. According to Alemdar and Ersoy (2010), their main focus categories are: monitoring everyday activities, detecting fall and movement, track-

ing location, monitoring medication intake, and monitoring medical status. Regarding health status information routing, only a very few systems use multihop routing, while this is important in extending the area covered by the system. In Table 3, a number of healthcare systems examples are presented, according to the above classification. The corresponding sources can be found in the "ADDITIONAL READING SECTION."

In Hung and Huang (2010), the authors have developed and tested a WSN-based healthcare monitor system, called U-healthcare. Initially, they set up a IPv6 laboratory network including PDA-based sensors for ECG measurements, a system for indoor positioning, physiology signal acquisition, and RFID application. The local area network is divided in two parts: a) servers and Website and b) care service sensor at the demonstration site. The wide area network is connected through a IPv4/IPv6 Stack Router to the Aca-

Table 3. Classification of healthcare systems examples (Alemdar & Ersoy, 2010)

Monitoring Everyday Activities	Detecting Fall and \ Movement	Tracking Location	Monitoring Medication Intake	Monitoring Medical Status
AICO CareNet Disp. Caregiver's Ast. WISP LiveNET	ITALH Act. Mon. & Fall Dr. Fall Detection Smart Phone HCM Smart HCN HipGuard	RFID way finding Ultra Badge ZUPS ALMAS Passive mon.	RFID medic. ctrl. iCabiNET iPackage	MobiHealth CodeBlue AlarmNet LifeGuard Med. Supervision FireLine Baby Glove LISTENse WLAN ECG Mobile ECG PATHS AWARENESS

demica Sinica server, which is then connected to healthcare centers and hospitals. The same structure was later tested in the China Medical University, Beigang Hospital with eight patients and five nursing staff.

Some promising mobile healthcare systems which are based on body area sensor networks are reviewed in Grgić, Zagar, and Krizanović (2011). CodeBlue (Shnayder, Chen, Lorincz, Fulford-Jones, & Welsh, 2005) is a telemedicine system developed in the Harvard University that implements an ad hoc sensor network for medical care purposes. The system includes the hardware infrastructure as well as a scalable software platform. CodeBlue is capable of performing all necessary networking tasks, database management, security provision, and positioning. AMON (Advanced care and alert portable telemedical MONitor) (U, et al., 2004) was the result of a European Union sponsored project for continuous monitoring high-risk patients. It constitutes of a device worn at the wrist, which captures vital signals, analyses them, and sends the results to the remote healthcare center via the cellular network. The Personal Care Connect platform was developed by IBM to provide interfaces for different medical sensors for remote monitoring of patients (M, et al., 2007). Information is forwarded from the sensors to a server that processes, stores, and presents data. It is an open and extendable platform that provides Java-based application programming interfaces.

Smartphones that support Java and Bluetooth can operate as data aggregators. A list of other interesting solutions in sensor-based healthcare monitoring systems examined in Grgić, Zagar, and Krizanović (2011) is the following: Smart Medical Home (Center for Future Health, 2010), Secure Mobile Computing (J & AC, 2010), Medical MoteCare (KF, E, & B, 2009), MEMS Wear (FEH, DG, L, MN, & KL, 2009), MEDIC (WH, AAT, MA, LK, JD, & WJ, 2008).

The authors in Nangalia, Prytherch, and Smith (2010) study a number of examples of healthcare monitoring systems with proven positive outcomes. Lastly, a set of selected wearable sensor-based systems for health monitoring are reviewed in Pantelopoulos and Bourbakis (2010) and the most representative ones are: HealthGear (Microsoft) (Oliver & Flores-Mangas, 2006), MyHeart (EU IST FP6 program) (Habetha, 2006), WEALTHY (EU IST FP5 program) (Paradiso, Loriga, & Taccini, 2005), Human++ (IMEC) (Gyselinckx, Penders, & Vullers, 2007), Lifeshirt (Vivometrics) (Heilman & Porges, 2007), AUBADE (University of Ioannina, Greece) (Katsis, Gianatsas, & Fotiadis, 2006), Bioharness (Zephyr Inc) (05Ze), HeartToGo (University of Pittsburg) (Jin, Oresko, Huang, & Cheng, 2009), Personal Health Monitor (University of Technology, Sydney) (Leijdekkers & Gay, 2008), and MERMOTH (EU IST FP6 program) (Luprano, 2006).

WIRELESS SENSOR NETWORKS FOR ENVIRONMENTAL MONITORING

Overview of Environmental WSNs

WSNs were initially thought to be just new tools for collecting data from the surrounding environment, however, they are now considered to constitute parts of a complete information system (Corke, Wark, Jurdak, Hu, Valencia, & Moore, 2010). This is a multiparty system that involves Internet communication lines, application servers, databases, and presentation software. The productivity of a sensor-based environmental monitoring system mainly depends on the assistance it provides to system users in order to complete their scientific or business tasks. The applications of WSNs may greatly vary and include monitoring the natural environment and agricultural scenarios. In more specific, some examples are related with microclimate monitoring for forests and farms, monitoring of the water-climate, cattle monitoring and control, virtual fencing, and biodiversity monitoring.

Sensor networks are wireless systems that contrarily to other popular applications do not require complicated equipment or fixed infrastructure and are promising to change the perception of men for nature. Authors in Xiaodong, Jun, Ping, and Kaifeng (2011) have worked on the agricultural data collection and studied the related automation system. In the context of the specific project, a greenhouse lab of 85 square meters was built and a control system for greenhouse monitoring via the CAN bus technology was implemented. The system gathered environmental information from a WSN in a greenhouse and forwarded the data to a central control subsystem.

Another promising application related to environmental sensor networks lies on the field of indoor tracking. Outdoor positioning systems are already well established, since most of the implementation challenges are addressed via satellite localization schemes. The use of WSNs in highly populated areas will enable power saving, security enhancement, and improve in response time (Sallabi, Odeh, & Shuaib, 2011).

Examples of Sensor-Based Environmental Monitoring Systems

Lately, a number of environmental WSNs have been implemented. In Corke, Wark, Jurdak, Hu, Valencia, and Moore (2010), the authors present five monitoring systems they have deployed, in order to understand agricultural and natural environments in Australia. The "Cattle Monitoring" system includes a WSN at a research farm, where the project involves different phases. Initially, the positions of cattle had to be recorded as well as soil moisture at several points. Multiple fixed and mobile sensors transmit data to a base station, which then relay data to a remote server through the Internet. The "Fleck1C" series (Corke P.) of nodes were employed, programmed under the TinyOS (Levis, et al., 2005), and adopting the ZTDMA MAC protocol. The second sensor-based project is called "Ground Water Quality Monitoring" and it involves just nine nodes that monitor water table level, salinity, and water extraction rate. The "Fleck3" sensor platform is adopted, the operating system is the TinyOS, and the employed network protocol "MintRoute" (Multihop Routing, 2003). The "Virtual Fencing" system is again related to cattle control. Eighty sensors are included, which are based on the "Fleck3" and the "Fleck Nano" hardware platform and they are programmed under the Fleck Operating System (FOS). The fourth system is about monitoring rainforest ecosystems in the long-run. The employed hardware platform is "Fleck3," there are only eight networked sensors, and the operating system is FOS. Lastly, the "Lake Water Quality Monitoring" project was deployed to measure temperature at various points and depths on a large lake that provides drinking water. The WSN is composed of fifty "Fleck3" sensors that are FOS-based.

REALnet (Albesa, Casas, Penella, & Gasulla, 2007) is an environmental sensor-based testbed implemented at the campus of the UPC University in Spain. The main goal of this system is to monitor environmental variables from ground, water, and air. The built WSN involves temperature and water level sensors, which communicate with a central station. REALnet is characterized as an embryonic WSN for environmental monitoring and was initially started from a project (Laboratori REAL de la UPC) that kicked off in 2001. The final objective is the monitoring of the air temperature, pressure, humidity, ambient light, ground temperature and humidity, and water temperature, level, and conductivity. Moreover, the sensors should be low-cost, energy autonomous, and maintenance-free. In order to achieve these requirements, harvesting environmental energy is needed. Three types of nodes are considered: sensor, router, and coordinator. All nodes employ the ETRX2 module (ETRX-2 Zigbee Module Product Manual) as a transceiver, which is based on EM240 System-on-Chip (EmberZnet Application Developers guide) working in compatibility with ZigBee at 2.4 GHz. The sensors were programmed to measure and send data every 30 seconds (emergency mode) or 2 hours (normal mode). Router nodes were required to be constantly active to forward data between the central node and the sensors. For this reason, two small sized solar cells were connected in parallel to recharge the 3-day autonomous batteries. Lastly, the central node is always on collecting data and connected to a computer via serial port.

The authors in Lu, Qian, Rodriguez, Rivera, and Rodriguez (2007) focus on the design of environmental WSNs, mainly for coastal area acoustic monitoring. The designed framework takes into consideration that WSNs are usually heterogeneous and the fact that the network layer must fulfill the applications' requirements for efficient wireless monitoring. The examined WSN testbed (The WALSAIP Project) was developed at the JBNEER environmental observatory in Puerto Rico. Its main objective was to collect and analyze almost real-time data from different sensors in order to facilitate the studying of large scale natural systems.

FUTURE RESEARCH DIRECTIONS

The latest research trends are toward the optimization of all mechanisms related with WSN operation. All layers of sensor-based networks are now considered hot research fields with promising development perspectives. One of the specific aspects that is generally expected to evolve in WSN is QoS provision. In more detail, modern applications of sensor networks demand efficient differentiation of network traffic, so that the various types of traffic are treated according to their specific features. For example, real-time transmitted data require low latency and jitter. However, the highly unreliable WSN links make QoS provision very challenging. A key factor here is the efficient distribution of the wireless channel resources among multiple nodes. In Lagkas, Angelidis, Stratogiannis, and Tsiropoulos (2010), we present the result of our work toward the development of a bandwidth sharing scheme for sensor networks, which takes into account nodes' queue load and packet priorities. The specific solution focuses on traffic differentiation based on its importance, while addressing the overload buffer problem. The presented analytical and simulation results exhibit a non-linear relationship between the ratio of the bandwidth assigned to different buffers and their loading rates ratio. Moreover, in Lagkas, Stratogiannis, Tsiropoulos, Sarigiannidis, and Louta (2011) we analyze the packet delay values when such a bandwidth allocation scheme is adopted. Our cross-validated results reveal that the specific approach provides efficient bandwidth sharing and minimum packet delay. Furthermore, the cooperative capabilities of a broadcast network (illustrated in Figure 5), like a WSN, are analyzed in Lagkas, Stratogiannis,

Tsiropoulos, and Angelidis (2011), where we provide fairness metrics among traffic flows of different priorities and rates.

Currently, we have started working on the "MedVision" project, which involves the development of a complete, automated, and flexible distributed system for monitoring health status, human activity, and/or environmental variables in different remote locations. Our objective is building a framework that involves different type independent sensors, which would be able to provide autonomous services to any requesting entity. The main goal is the integration of various smart sensors capable of supporting different applications without the limitations imposed by a centralized architecture. For this reason, we are planning to develop advanced software agents, forming an extended overlay network. The respec-

tive agents will be able to advertize and discover services related with the dissemination of sensors' collected data. Regarding the hardware requirements, the latest developments of the system-on-a-chip technology provide the ability of implementing small sized devices of significant processing power, low energy demands, and extensive connectivity. All widely-spread wireless communication schemes for WSNs will be examined for the interconnection of the different mobile devices. Among the latest WSN research issues that are taken into consideration in this project are ad hoc connectivity, network self-configuration, energy efficiency, resource requirements optimization, distributed service provision, and software agents' interaction-behavior.

CONCLUSION

It is now evident that the wireless sensor networks are capable of providing advanced services, which attract the attention of various scientific and commercial fields. On these grounds, the related research attempts have rapidly grown in the last decade and are expected to grow even more in the future, since the tremendous advances in mobile computing have enabled the implementation of smart, autonomous, and self-configured sensor-based systems. This overview presented the most important aspects of WSNs, revealing the latest trends in sensor network design, system structure, and hardware/software platforms. We specially focused on the promising applications of sensor-based networks in healthcare and environmental monitoring, quoting various representative implementation efforts globally. The growing need for accessing different type of data collected by various remote sensors in a completely distributed manner drives us to the development of an integrated cooperative system for the efficient wide dissemination of monitoring information without the constrains imposed by central control.

Figure 5. A cooperative broadcast network (Lagkas T. D., Stratogiannis, Tsiropoulos, & Angelidis, 2011)

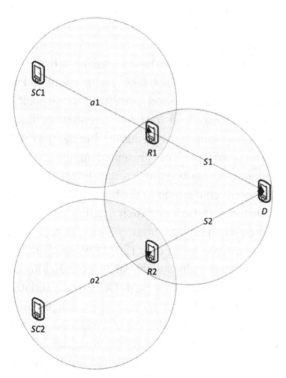

ACKNOWLEDGMENT

This work was supported by the "MedVision" project, which is funded in part by the European Union.

REFERENCES

Akyildiz, I., Su, W., Sankarasubramaniam, Y., & Cayirci, E. (2002a). A survey on sensor networks. *IEEE Communications Magazine, 40*(8), 102–114. doi:10.1109/MCOM.2002.1024422

Akyildiz, I., Su, W., Sankarasubramaniam, Y., & Cayirci, E. (2002b). Wireless sensor networks: A survey. *Computer Networks, 38*(4), 393–422. doi:10.1016/S1389-1286(01)00302-4

Akyildiz, I. F., Melodia, T., & Chowdhury, K. (2007). A survey on wireless multimedia sensor networks. *Computer Networks, 51*(4), 921–960. doi:10.1016/j.comnet.2006.10.002

Al-Jaroodi, J., & Mohamed, N. (2012). Service-oriented middleware: A survey. *Journal of Network and Computer Applications, 35*(1), 211–220. doi:10.1016/j.jnca.2011.07.013

Alabri, H. M., Mukhopadhyay, S. C., Punchihewa, G. A., Suryadevara, N. K., & Huang, Y. M. (2012). Comparison of applying sleep mode function to the smart wireless environmental sensing stations for extending the life time. In *Proceedings of 2012 IEEE International Instrumentation and Measurement Technology Conference* (pp. 2634 –2639). Graz, Austria: IEEE.

Albesa, J., Casas, R., Penella, M. T., & Gasulla, M. (2007). REALnet: An environmental WSN testbed. In *Proceedings of International Conference on Sensor Technologies and Applications, SensorComm 2007* (pp. 502-507). Valencia, Spain: SensorComm.

Alemdar, H., & Ersoy, C. (2010). Wireless sensor networks for healthcare: A survey. *Computer Networks, 54*(15), 2688–2710. doi:10.1016/j. comnet.2010.05.003

Barakah, D. M., & Ammad-Uddin, M. (2012). A survey of challenges and applications of wireless body area network (WBAN) and role of a virtual doctor server in existing architecture. In *Proceedings of 2012 Third International Conference on Intelligent Systems, Modelling and Simulation (ISMS)*. Sabah, Malaysia: ISMS.

Bichler, M., & Lin, K. (2006). Service-oriented computing. *IEEE Computer, 39*(3), 88–90.

Biswas, P. K., Qi, H., & Xu, Y. (2008). Mobile-agent-based collaborative sensor fusion. *Information Fusion, 9*, 399–411. doi:10.1016/j. inffus.2007.09.001

Center for Future Health. (2010, June 1). Retrieved November 20, 2012, from http://www. futurehealth.rochester.edu/smart_home/

Chong, C.-Y., & Kumar, S. (2003). Sensor networks: Evolution, opportunities, and challenges. *Proceedings of the IEEE, 91*(8), 1247–1256. doi:10.1109/JPROC.2003.814918

Corke, P. (n.d.). *CSIRO.* Retrieved November 20, 2012, from http://www.ict.csiro.au/files/AutonomousSystems/Flecks.pdf

Corke, P., Wark, T., Jurdak, R., Hu, W., Valencia, P., & Moore, D. (2010). Environmental wireless sensor networks. *Proceedings of the IEEE, 98*(11), 1903–1917. doi:10.1109/JPROC.2010.2068530

Culler, D., Estrin, D., & Srivastava, M. (2004). Guest editors' introduction: Overview of sensor networks. *Computer, 37*(8), 41–49. doi:10.1109/ MC.2004.93

Egbogah, E. E., & Fapojuwo, A. O. (2011). A survey of system architecture requirements for health care-based wireless sensor networks. *Sensors (Basel, Switzerland)*, *11*(5), 4875–4898. doi:10.3390/s110504875 PMID:22163881

EmberZnet Application Developers Guide. (n.d.). Retrieved November 20, 2012, from http://www.ember.com/

Estrin, D., Govindan, R., Heidemann, J., & Kumar, S. (1999). Next century challenges: Scalable coordination in. In *Proceedings of ACM/IEEE International Conference on Mobile Computing and Networks* (pp. 263-270). Seattle, WA: ACM/IEEE.

ETRX-2 Zigbee Module Product Manual. (n.d.). Retrieved November 20, 2012, from http://www.telegesis.com/

Feh, T., & Dg, G., L, X., Mn, N., & Kl, Y. (2009). MEMSWear - Biomonitoring system for remote vital signs monitoring. *Journal of the Franklin Institute*, *6*, 531–542.

Gascueña, J. M., & Fernández-Caballero, A. (2011). On the use of agent technology in intelligent, multisensory and distributed surveillance. *The Knowledge Engineering Review*, *26*(2), 191–208. doi:10.1017/S0269888911000026

GPP. TS 23.401. (n.d.). *General packet radio service (GPRS) enhancements for evolved universal terrestrial radio access network (E-UTRAN) access (release 8)*. 3GPP.

Grgić, K., Zagar, D., & Krizanović, V. (2011). *Medical applications of wireless sensor networks-current status and future directions*. Official Publication of the Medical Association of Zenica-Doboj Canton Bosnia and Herzegovina.

Gungor, V., & Hancke, G. (2009). Industrial wireless sensor networks: Challenges, design principles, and technical approaches. *IEEE Transactions on Industrial Electronics*, *56*(10), 4258–4265. doi:10.1109/TIE.2009.2015754

Gyselinckx, B., Penders, J., & Vullers, R. (2007). Potential and challenges of body area networks for cardiac monitoring. *Journal of Electrocardiology*, *40*, S165–S168. doi:10.1016/j.jelectrocard.2007.06.016 PMID:17993316

Habetha, J. (2006). The MyHeart project—Fighting cardiovascular diseases by prevention and early diagnosis. In *Proceedings of 28th Ann. Int. IEEE EMBS Conf.* (pp. 6746-6749). IEEE.

Halonen, T., Romero, J., & Melero. (2003). *GSM, GPRS and EDGE performance: Evolution towards 3G/UMTS*. Wiley.

Heilman, K. J., & Porges, S. (2007). Accuracy of the lifeshirt (vivometrics) in the detection of cardiac rhythms. *Biological Psychology*, *3*, 300–305. doi:10.1016/j.biopsycho.2007.04.001 PMID:17540493

Hung, C.-C., & Huang, S.-Y. (2010). On the study of a ubiquitous healthcare network with security and QoS. In *Proceedings of 2010 IET International Conference on Frontier Computing – Theory, Technologies and Applications* (pp. 139-144). Taichung, Taiwan: IET.

IEEE 802.11. (2007). *Part 11: Wireless LAN medium access control (MAC) and physical layer (PHY) specifications*. IEEE Computer Society.

IEEE 802.15.3. (2003). *Part15.3: Wireless medium access control (MAC) and physical layer (PHY) specifications for high rate wireless personal area networks (WPANs)*. IEEE Computer Society.

IEEE 802.15.4. (2006). *Part15.4: Wireless medium access control (MAC) and physical layer (PHY) specifications for low-rate wireless personal area networks (WPANs)*. IEEE Computer Society.

IEEE 802.16. (2004). *Part 16: Air interface for fixed broadband wireless access systems—Standard for local and metropolitan area networks*. IEEE Computer Society.

J, H., & AC, W. (2010, June 15). *A dynamic, context-aware security infrastructure for distributed healthcare applications*. Retrieved from http://www.cs.virginia.edu/~acw/security/doc/Publications/A%20Dynamic,%20Context-Aware%20Security%20Infrastructure%20for%20Distri~1.pdf

Jin, Z., Oresko, J., Huang, S., & Cheng, A. C. (2009). HeartToGo: A personalized medicien technology for cardiovascular disease prevention and detection. In *Proceedings of IEEE/NIH LiSSA, 2009* (pp. 80-83). IEEE.

Karlsson, B., Backstrom, O., Kulesza, W., & Axelsson, L. (2005). Intelligent sensor networks - An agent oriented approach. In *Proceedings of Workshop on Real-World Wireless Sensor Networks*. Stockholm, Sweden: IEEE.

Katsis, C. D., Gianatsas, G., & Fotiadis, D. I. (2006). An integrated telemedicine platform for the assessment of affective physiological states. *Diagnostic Pathology*, *1*, 16. doi:10.1186/1746-1596-1-16 PMID:16879757

KF. N., E, L., & B, L. (2009). Medical MoteCare: A distributed personal healthcare monitoring system. In *Proceedings of International Conference on eHealth Telemedicine, and Social Medicine*. Cancun, Mexico: IEEE.

Korhonen, I., Parkka, J., & Gils, M. V. (2003). Health monitoring in the home of the future. *IEEE Engineering in Medicine and Biology Magazine*, *22*(3), 66–73. doi:10.1109/MEMB.2003.1213628 PMID:12845821

Kozat, U. C., Koutsopoulos, I., & Tassiulas, L. (2004). A framework for cross-layer design of energy-efficient communication with QoS provisioning. [IEEE.]. *Proceedings - IEEE INFOCOM*, *2004*, 1446–1456.

Laboratori REAL de la UPC. (n.d.). Retrieved November 20, 2012, from http://www.upc.edu/mediambient/recerca/lreal1.html

Lagkas, T., Stratogiannis, D. G., Tsiropoulos, G. I., Sarigiannidis, P., & Louta, M. (2011). Load dependent resource allocation in cooperative multiservice wireless networks: Throughput and delay analysis. In *Proceedings of 2011 IEEE Symposium on Computers and Communications (ISCC)* (pp. 185-190). Corfu, Greece: IEEE.

Lagkas, T. D., Angelidis, P., Stratogiannis, D. G., & Tsiropoulos, G. I. (2010). Analysis of queue load effect on channel access prioritization in wireless sensor networks. In *Proceedings of 2010 6th IEEE International Conference on Distributed Computing in Sensor Systems Workshops (DCOSSW)* (pp. 1-6). Santa Barbara, CA: IEEE.

Lagkas, T. D., Stratogiannis, D. G., Tsiropoulos, G. I., & Angelidis, P. (2011). Bandwidth allocation in cooperative wireless networks: Buffer load analysis and fairness evaluation. *Physical Communication*, *4*(3), 227–236. doi:10.1016/j.phycom.2011.03.001

Leijdekkers, P., & Gay, V. (2008). A self-test to detect a heart attack using a mobile phone and wearable sensors. In *Proceedings of 21st IEEE CBMS Int. Symp.* (pp. 93-98). IEEE.

Levis, P., Madden, S., Polastre, J., Szewczyk, R., Whitehouse, K., Woo, A., et al. (2005). TinyOS: An operating system for sensor networks. *Ambient Intell.*, 115-148.

Lim, A. (2001). Distributed services for information dissemination in self-organizing sensor networks. *Journal of the Franklin Institute, 338*(6), 707–727. doi:10.1016/S0016-0032(01)00020-5

Lin, P., Qiao, C., & Wang, X. (2004). Medium access control with a dynamic duty cycle for sensor networks. In *Proceedings of IEEE Wirel. Commun. Net. Conf.* IEEE.

Logan, A. G., McIsaac, W. J., Tisler, A., Irvine, M. J., Saunders, A., & Dunai, A. et al. (2007). Mobile phone–based remote patient monitoring system for management of hypertension in diabetic patients. *American Journal of Hypertension, 20*(9), 942–948. doi:10.1016/j.amjhyper.2007.03.020 PMID:17765133

Lu, K., Qian, Y., Rodriguez, D., Rivera, W., & Rodriguez, M. (2007). Wireless sensor networks for environmental monitoring applications: A design framework. In *Proceedings of IEEE Global Telecommunications Conference, GLOBECOM '07* (pp. 1108 –1112). Washington, DC: IEEE.

Luprano, J. (2006). European projects on smart fabrics, interactive textiles: Sharing opportunities and challenges. In *Proceedings of Workshop Wearable Technol. Intel. Textiles*. Helsinki, Finland: IEEE.

M, B., VM, B., AN, C., MR, E., WF, J., SM, M., et al. (2007). Remote health-care monitoring using Personal Care Connect. *IBM Systems Journal, 1*, 95-114.

MC, V., & IF, A. (2009). Error control in wireless sensor networks: A cross layer analysis. *IEEE/ACM Transactions on Networking, 17*(4), 1186–1199. doi:10.1109/TNET.2008.2009971

Mendes, L. D., & Rodrigues, J. J. (2011). A survey on cross-layer solutions for wireless sensor networks. *Journal of Network and Computer Applications, 34*(2), 523–534. doi:10.1016/j.jnca.2010.11.009

Montenegro, G., Kushalnagar, N., Hui, J., & Culler, D. (2007). *Transmission of IPv6 packets over IEEE 802.15.4 networks*. Internet Engineering Task Force RFC-4944.

Multihop Routing. (2003, September 3). Retrieved November 19, 2012, from http://www.tinyos.net/tinyos-1.x/doc/multihop/multihop_routing.html

Nangalia, V., Prytherch, D. R., & Smith, G. B. (2010). Health technology assessment review: Remote monitoring of vital signs-current status and future challenges. *Critical Care (London, England), 24*(14), 233. doi:10.1186/cc9208

Oliver, N., & Flores-Mangas, F. (2006). *HealthGear: A real-time wearable system for monitoring and analyzing physiological signals*. Microsoft Res. Tech. Rep. MSR-TR-2005-182.

Pantelopoulos, A., & Bourbakis, N. G. (2010). A survey on wearable sensor-based systems for health monitoring and prognosis. *A Survey on Wearable Sensor-Based Systems for Health Monitoring and Prognosis, 40* (1), 1 –12.

Paradiso, R., Loriga, G., & Taccini, N. (2005). A wearable health care system based on knitted integral sensors. *IEEE Transactions on Information Technology in Biomedicine, 9*(3), 337–344. doi:10.1109/TITB.2005.854512 PMID:16167687

Perkins, C. (2000). *Ad hoc networking*. Reading, MA: Addison-Wesley.

Pham, H., & Jha, S. (2004). An adaptive mobility-aware MAC protocol for sensor networks (MS-MAC). In *Proceedings of the IEEE International Conference on Mobile Ad-hoc and Sensor Systems* (pp. 558-560). Fort Lauderdale, FL: IEEE.

Qi, H., Iyengar, S., & Chakrabarty, K. (2001). Distributed sensor networks—A review of recent research. *Journal of the Franklin Institute, 338*(6), 655–668. doi:10.1016/S0016-0032(01)00026-6

Sallabi, F. M., Odeh, A. O., & Shuaib, K. (2011). Performance evaluation of deploying wireless sensor networks in a highly dynamic WLAN and a highly populated indoor environment. In *Proceedings of 2011 7th International Wireless Communications and Mobile Computing Conference (IWCMC)* (pp. 1371 –1376). Istanbul, Turkey: IWCMC.

The WALSAIP Project. (n.d.). Retrieved November 8, 2012, from http://walsaip.uprm.edu/

U, A., JA, W., P, L., G, T., F, D., M, B., et al. (2004). AMON: A wearable multiparameter medical monitoring and alert system. *IEEE Trans Inf Technol Biomed, 4*, 415-27.

US Department of Energy. (2004). *Assessment study on sensors and automation (report)*. Washington, DC: Office of Energy and Renewable Energy, U.S. Department of Energy.

WB, H., AP, C., & H, B. (2002). An application-specific protocol architecture for wireless microsensor networks. *IEEE Transactions on Wireless Communications, 1*(4), 660–670. doi:10.1109/TWC.2002.804190

WH, W., AAT, B., MA, B., LK, A., JD, B., & WJ, K. (2008). MEDIC: Medical embedded device for individualized care. *Artificial Intelligence in Medicine, 42*, 137–152. doi:10.1016/j.artmed.2007.11.006 PMID:18207716

Xiaodong, D., Jun, W., Ping, J., & Kaifeng, G. (2011). Design and implement of wireless measure and control system for greenhouse. In *Proceedings of the 30th Chinese Control Conference (CCC 2011)*. Yantai, China: CCC.

Ye, W., Heidemann, J., & Estrin, D. (2004). Medium access control with coordinated adaptive sleeping for wireless sensor networks. *IEEE/ACM Transactions on Networking, 12*, 493–506. doi:10.1109/TNET.2004.828953

Yuan, X., Bagga, S., Shen, J., Balakrishnan, M., & Benhaddou, D. (2008). DS-MAC: Differential service medium access control design for wireless medical information systems. In *Proceedings of the 30th Annual IEEE Conference on Engineering in Medicine and Biology Society* (pp. 1801-1804). Vancouver, Canada: IEEE.

Zephyr Inc. (n.d.). Retrieved June 15, 2005, from http://www.zephyrtech.co.nz

Zhen, B., Li, H.-B., & Kohno, A. R. (2009). Networking issues in medical implant communications. *International Journal of Multimedia and Ubiquitous Engineering, 4*(1).

ADDITIONAL READING

Baker, C., Armijo, K., Belka, S., Benhabib, M., Bhargava, V., Burkhart, N., et al. (2007). Wireless sensor networks for home health care. *21st International Conference on Advanced Information Networking and Applications Workshops* (pp. 832-837).

Bowser, S., & Woodworth, J. (2004). Wireless multimedia technologies for assisted living. *Second LACCEI International Latin American and Caribbean Conference for Engineering and Technology.*

Chang, Y., Chen, C., Chou, L., & Wang, T. (2008). A novel indoor way finding system based on passive RFID for individuals with cognitive impairments. *Second International Conference on Pervasive Computing Technologies for Healthcare* (pp. 108-111).

Chung, W., Yau, C., & Shin, K. (2007). A cell phone based health monitoring system with self analysis processor using wireless sensor network technology. *29th Annual International Conference of the IEEE EMBS.*

Consolvo, S., Roessler, P., & Shelton, B. (2004). The CareNet display: Lessons learned from an in home evaluation of an ambient display. *6th International Conference on Ubiquitous Computing* (pp. 1-17).

Hansen, T., Eklund, J., Sprinkle, J., Bajcsy, R., & Sastry, S. (2005). Using sensors and a camera phone to detect and verify the fall of elderly persons. *European Medicine, Biology and Engineering Conference.*

Hori, T., & Nishida, Y. (2005). Ultrasonic sensors for the elderly and caregivers in a nursing home. *Seventh International Conference on Enterprise Information Systems* (pp. 110-115).

Iso-ketola, P., Karinsalo, T., & Vanhala, J. (2008). HipGuard: a wearable measurement system for patients recovering from a hip operation. *Second International Conference on Pervasive Computing Technologies for Healthcare.*

Jones, V., Mei, H., Broens, T., Widya, I., & Peuscher, J. (2007). Context aware body area networks for telemedicine. *8th Pacific Rim Conference on Multimedia.*

Kailanto, H., Hyvärinen, E., Hyttinen, J., & Institute, R. (2008). Mobile ecg measurement and analysis system using mobile phone as the base station. *Second International Conference on Pervasive Computing Technologies for Healthcare.*

Konstantas, D., & Herzog, R. (2003). Continuous monitoring of vital constants for mobile users: the MobiHealth approach. *25th Annual International Conference of the IEEE EMBS* (pp. 3728-3731).

Leijdekkers, P., Gay, V., & Lawrence, E. (2007). Smart homecare system for health telemonitoring. *First International Conference on the Digital Society.*

Li, Z., & Zhang, G. (2007). physical activities healthcare system based on wireless sensing technology. *13th IEEE International Conference on Embedded and Real-Time Computing Systems and Applications.*

López-nores, M., Pazos-arias, J., García-duque, J., & Blancofernández, Y. (2008). Monitoring medicine intake in the networked home: the iCabiNET solution. *Second International Conference on Pervasive Computing Technologies for Healthcare.*

Lu, C., & Fu, L. (2009). Robust location-aware activity recognition using wireless sensor network in an attentive home. *IEEE Transactions*, 598-609.

Marco, A., Casas, R., Falco, J., Gracia, H., Artigas, J., & Roy, A. (2008). Location-based services for elderly and disabled people. *Elsevier Computer Communications, 31*, 105–106.

Moh, M., Walker, Z., Hamada, T., & Su, C. (2005). A prototype on RFID and sensor networks for elder healthcare: progress report. *ACM SIGCOMM Workshop on Experimental Approaches to Wireless Netword Design and Analysis* (pp. 70-75).

Montgomery, K., Mundt, C., Thonier, G., Tellier, A., Udoh, U., Barker, V., et al. (2004). Lifeguard – a personal physiological monitor for extreme environments. *26th Annual International Conference of the IEEE Engineering in Medicine and Biology Society.*

Pang, Z., Chen, Q., & Zheng, L. (2009). A pervasive and preventive healthcare solution for medication noncompliance and daily monitoring. *2nd IEE International Symposium on Applied Sciences in Biomedical and Communication Technologies* (pp. 1-6).

Philipose, M., Consolvo, S., Smith, I., Fox, D., Kautz, H., & Patterson, D. (2004). Fast, detailed inference of diverse daily human activities. *Sixth International Conference on Ubiquitous Computing.*

Philipose, M., Smith, J., Jiang, B., Mamishev, A., Sumit, R., & Sundara-Rajan, K. (2005). Battery-free wireless identification and sensing. *Pervasive Computing*, 4(1), 37–45. doi:10.1109/MPRV.2005.7

Purwar, A., Jeong, D., & Chung, W. (2007). Activity monitoring from realtime triaxial accelerometer data using sensor network. *International Conference on Control, Automation and Systems* (pp. 2402-2406).

Shnayder, V., Chen, B., Lorincz, K., Fulford-Jones, T., & Welsh, M. (2005). Sensor networks for medical care. *3rd International Conference on Embedded Networked Sensor Systems.*

Sung, M., Marci, C., & Pentland, A. (2005). Wearable feedback systems for rehabilitation. *Journal of Neuroengineering and Rehabilitation*, 2(1). doi:10.1186/1743-0003-2-17 PMID:15987514

Tabar, A., & Aghajan, H. (2006). Smart home care network using sensor fusion and distributed vision-based reasoning. *4th ACM International Workshop on Video Surveillance and Sensor Networks* (pp. 145-154).

Wang, C., Chiang, C., Lin, P., Chou, Y., Kuo, I., Huang, C., et al. (2008). Development of a fall detecting system for the elderly residents. *2nd IEEE International Conference on Bioinformatics and Biomedical Engineering* (pp. 1359-1362).

Wood, A., Stankovic, J., Virone, G., Selavo, L., He, Z., & Cao, Q. et al. (2008). Context-aware wireless sensor networks for assisted living and residential monitoring. *IEEE Network*, 26–33. doi:10.1109/MNET.2008.4579768

Yan, H., Xu, Y., Gidlund, M., & Nohr, R. (2008). An experimental study on home wireless passive positioning. *Second International Conference on Sensor Technologies and Applications (SENSOR-COMM 2008)* (pp. 223-228).

Zhou, B., Hu, C., Wang, H., & Guo, R. (2007). A wireless sensor network for pervasive medical supervision. *International Conference on Integration Technology* (pp. 740-744).

KEY TERMS AND DEFINITIONS

Cooperative Networking: Networking based on the cooperation among the involved nodes so as to achieve significant improvement in terms of the overall system capacity meeting at the same time the Quality of Service (QoS) requirements.

Environmental Monitoring: The processes and activities that need to take place to characterize and monitor the quality of the environment.

Health Monitoring: Observation of a disease, condition or one or several medical parameters over time.

Wireless Body Area Network (WBAN): A wireless network of wearable computing devices.

Wireless Personal Area Network (WPAN): A wireless network typically involving small devices organized around an individual person.

Wireless Sensor Network (WSN): A group of spatially dispersed and dedicated sensors for monitoring and recording the physical conditions of the environment and organizing the collected data at a central location.

Chapter 13
Clustering in Wireless Sensor Network:
A Study on Three Well-Known Clustering Protocols

Basma M. Mohammad El-Basioni
Electronics Research Institute, Egypt

Hussein S. Eissa
Electronics Research Institute, Egypt

Sherine M. Abd El-Kader
Electronics Research Institute, Egypt

Mohammed M. Zahra
Al-Azhar University, Egypt

ABSTRACT

The purpose of this chapter is the study of the clustering process in Wireless Sensor Networks (WSN), starting with clarifying why there are different clustering protocols for WSN by stating and briefly describing some of the variate features in their design; these features can represent questions the clustering protocol designer asks before the design, and their brief description can be considered probabilities for these questions' answers to represent design options for the designer. The designer can choose the best answer to each design question or, in better words, the best design options that will make its protocol different from the others and make the resultant clustered network satisfies some requirements for improving the overall performance of the network. The chapter also mentions some of these requirements. The chapter then gives illustrative examples for these design variations and requirements by studying them on three well-known clustering protocols: Low-Energy Adaptive Clustering Hierarchy (LEACH), Energy-Efficient Clustering Scheme (EECS), and Hybrid, Energy-Efficient, Distributed clustering approach for ad-hoc sensor networks (HEED).

DOI: 10.4018/978-1-4666-5170-8.ch013

INTRODUCTION

Naturally, grouping sensor nodes into clusters has been widely adopted by the research community (Yang, Zhuang, & Li, 2011; Boregowda, Babu, Puttamadappa, & Mruthyunjaya, 2011; Hasnaoui, Hssane, Ezzat, & Benalla, 2010; Mohammad El-Basioni, Abd El-kader, Eissa, & Zahra, 2012; Abdel Hady, Abd El-kader, Eissa, Salem, & Fahmy, 2012; Abbasi, & Younis, 2007; Katiyar, Chand, & Soni, 2010) to satisfy scalability, achieve high energy efficiency, and prolong network lifetime in large-scale WSN (Akyildiz, 2010; Misra, Woungang, & Misra, 2009; Boukerche, 2009) environments. The corresponding hierarchical routing and data gathering protocols imply cluster-based organization of the sensor nodes in order that data fusion and aggregation are possible, thus leading to significant energy savings. In the hierarchical network structure each cluster has a leader, which is also called the Cluster Head (CH) and usually performs special tasks such as fusion, aggregation, cluster coordination, and transmission to the sink, and several sensor nodes as members.

The cluster formation process eventually leads to a two-level hierarchy where the CH nodes form the higher level and the cluster-member nodes form the lower level. The sensor nodes periodically transmit their data to the corresponding CH nodes. The CH nodes aggregate the data (thus decreasing the total number of relayed packets) and transmit it to the Base Station (BS) either directly or through the intermediate communication with other CH nodes. However, because the CH nodes send all the time data to higher distances than the common (member) nodes and do other additional functions as mentioned before, they naturally spend energy at higher rates (Zhang, Yang, & Chen, 2010). It is essential to rotate the role of CH among nodes especially in homogeneous networks so as not to burden a few nodes with more duties than others; this achieves load-balance and also fault-tolerance.

VARIATION IN THE DESIGN OF CLUSTERING PROTOCOLS

Clustering protocols can differ from each other in many features such as:

1. **The criteria of selecting the nodes that will take the head role:** The head selection criterion may be:
   a. Completely random, for example, node ID or a probability not based on any property of the node itself such as node's residual energy.
   b. Random selection upon a probability based on nodes properties such as residual energy, i.e., as each node compares its CH selection probability against random number, the probability condition randomly may satisfied or not but with a larger chance for the node has the required property to be a CH.
   c. Non-probabilistic CH selection, rather the CH is selected upon its weight which defines its significance or its suitability to become a CH, this weight may depend on a certain metric such as node degree (number of its neighbors which are the nodes in a certain range around it) and residual energy, may depend on more than one metric, or may be a weighted function in more than one metric like:

$$\text{Weight} = w_1 \times \text{metric}_1 + w_2 \times \text{metric}_2 + w_3 \times \text{metric}_3 \qquad (1)$$

The values of the weighting factors of the used metrics may be chosen according to application requirements.

2. **The method by which a node takes the head role:** The CH may appoint itself in the head role, it may be appointed in this role by a centralized authority, or it may nominate

itself in an election, in better words, in a competition with other nodes for selecting CHs. All the nodes in the network may be candidates in this competition or a subset from them; the method and the basis on which this subset is selected also can differ from protocol to another.

3. **The method by which the formed CHs announce about themselves:** The CH may send an advertisement message for all the nodes in the network to advertise them joining its cluster, it may send an advertisement message for nodes in a certain range (*cluster range*) to advertise them joining its cluster, it may send an informative message to a certain set of nodes to inform them that it is their CH, or it may not need to send any announcement message because the nodes already knew about its existence and may the existence of other CHs by the messages exchanged during the head role competition.

4. **The criteria and the method by which a member node chooses and informs its CH:** The protocol designer should answer some questions such as, is cluster overlap allowed, i.e. is it allowed for the node to belong to more than one cluster, if cluster overlap is allowed, what is the degree of overlap; if it is not allowed, what is the criteria upon which the tie condition is broken (the tie condition occurs when the member node located in the range of more than one CH), the node may choose the CH with more residual energy, larger or smaller degree, smaller distance to the sink, larger weight, etc. Is it necessary for the member node to send a join message to inform its CH, or it is not necessary because it is obliged by an informative announcement sent to it from a certain CH to join its cluster; the most or usual case is the member node informing the corresponding CH.

5. **The order of the execution of the two processes, CH selection and cluster for-** mation: May be a choice for the clustering protocol design to make the nodes form clusters first upon some criteria such as the location then they select their CH from themselves. Another and common option is the CH selection process done first then in another period of time each regular node selects from them the best CH to join its cluster. In some protocols, the CH selection process and the join cluster process are performed in parallel in the same time period of protocol operation.

6. **Iterations of clustering process:** The network can be clustered through one iteration or more. If the protocol uses several iterations, it should determine their maximum number and how this number will be controlled.

7. **Centralized or distributed (local) protocol:** A centralized authority uses global information about the network to distribute the roles on nodes, or each node decides on its role only based on its local information.

8. **The method of reclustering or head role rotation for load-balance and fault-tolerance:** The protocol operation can be divided into rounds, each round begins with a clustering phase to recluster the network, or the reclustering process triggered in any time by any node if a certain condition is met, for example by a CH found its energy level below a threshold level.

9. **The clustering is two-level or multi-level**

10. **The intra-cluster communication is single-hop or multi-hop**

11. **The cluster shape is regular or irregular**

12. **The cluster size is equal or non-equal**

13. **Node heterogeneity is supported or not**

14. **Node mobility is supported or not**

15. **Location information is necessary for each node or just the approximate value of its distance from the sink and the other nodes is required**

16. **Are there additional roles or division of roles functions in the clustered network:**

Such as a proxy node for CH or a gateway node in each cluster for communication outside the cluster.

PROPERTIES OF CLUSTERED NETWORKS

For the clustering algorithm to achieve the aimed goals, scalability, energy-efficiency, load-balance, long lifetime, etc., regardless what a combination of design options (mentioned above or not) it uses, there are some needs in the resultant clustered network should be satisfied such as:

1. Uniform CH distribution
2. Optimal CHs selection
3. Optimal count for the generated CHs
4. Little variation of protocol behavior; the instability of the protocol sometimes may lead to undesirable behavior, for example, the number of the generated CHs may be very small in a number of rounds, in another number of rounds it may be very large, in other rounds it may equal zero!
5. It is preferable to the protocol to be adaptive not fixed (clusters count and membership evolve), i.e. its behavior is variable but controllable according to the need, so the variation of its behavior has limits and triggering conditions all over network lifetime
6. Adjustable cluster size aids in balancing the load on CHs which may be unbalanced due to other factors such as transmission to the sink
7. One-node clusters (CH with no members) should be avoided
8. Guarantee for each active node to take an effective role as head node or plain node at the end of each clustering phase
9. The uniform node distribution is preserved for high percentage of network lifetime to approximately maintain the whole area

covered for a long time, in other words, the clustering protocol aids with other design factors in the whole routing protocol to achieve uniform energy dissipation

10. Clustering algorithm should have low overhead in terms of control messages and time convergence

The following sections represent a trial to study the previously mentioned aims for a clustered network using the clustering algorithms of three known routing protocols which are LEACH, EECS, and HEED. From the clustering algorithm, it was concentrated on the shape and the properties of the resulting clustered network along its lifetime, the method by which clusters communicate is not belong to the clustering algorithm but it belongs to the hierarchical routing protocol uses this clustering algorithm which may use other design factors such as coverage algorithm and multi-hop communication among CHs to help the used clustering algorithm in achieving overall good performance of the routing protocol. So in the following evaluation, it is concentrated on the clustering algorithm of the routing protocols and it is assumed that all CHs send directly to the sink, i.e., the routing protocols have the same design except the clustering method, with the knowledge that the actual behavior of the routing protocol will vary if it doesn't use this assumption (for example the clustering method used in one routing protocol may be better than that used in another but the usage of CHs multi-hop communication in this second routing protocol makes its overall behavior better than the first, and accordingly, the overall behavior of the first one may become better if it uses CHs multi-hop communication). Also as a result to the previous assumption the performance metrics related to the routing protocol as whole such as lifetime and delay are not considered.

The evaluation was conducted using OMNeT++ simulator (Varga, 2005). Each clustering protocol is applied to the same node distribution shown in Figure 1 assuming that all nodes begin with equal energies and their initial count equals 100 node.

LEACH CLUSTERING PROTOCOL

LEACH is a distributed clustering protocol, in which, a predefined percentage of nodes are selected as CHs randomly and randomly rotated- with no probability of a CH to become again CH up to

a certain number of rounds- to evenly distribute the load of the head role among nodes.

Each selected CH broadcasts a message to the remaining nodes to offer them joining its cluster; each node selects the closest head, then nodes send data to their CH using a Time Division Multiple Access (TDMA) schedule to aggregate and send it to the sink. This method of cluster formation doesn't consider the remaining energy of the selected CH, its distance from the sink, its distance from the other heads, and its distance from its members. All of these factors result in

Figure 1. The node distribution used to evaluate clustering protocols

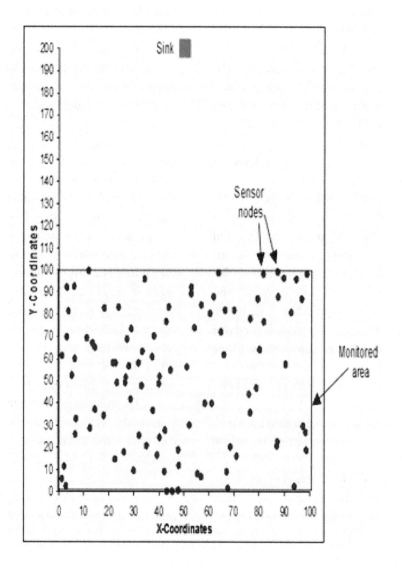

bad CH selection, non-uniform CHs distribution, and instability in generated clusters number and size; this increases the load on CHs as well as on cluster members, which tends to the fast depletion of nodes energy and fast nodes death which indoors fasten the network death.

In addition, sometimes the distance between the CH and its member may be very long, this leads to high energy consumption from the member node, as well as it takes a long time for data to reach the CH which implies widening the time slot of the TDMA schedule, and as a result of the instability in generated clusters number and size, sometimes the whole network formed in one cluster i.e. one CH and all the remaining nodes are its members, this implies the lengthening of the TDMA schedule itself to be enough for all the nodes existing in the network minus one (the alone CH). This long

TDMA schedule with its wide time slots increases the data latency. For protocol details see Heinzelman, Chandrakasan, and Balakrishnan (2000). An example of the clusters formed in LEACH round is shown in Figure 2.

Properties of LEACH Clustering

- Dynamic clustering as it includes regular CH re-election or cluster reorganization procedures
- Distributed
- Adaptive
- Leader election process performed first before clusters formation
- The node appoints itself in the head role if it satisfies the three conditions:

Figure 2. Example of LEACH network

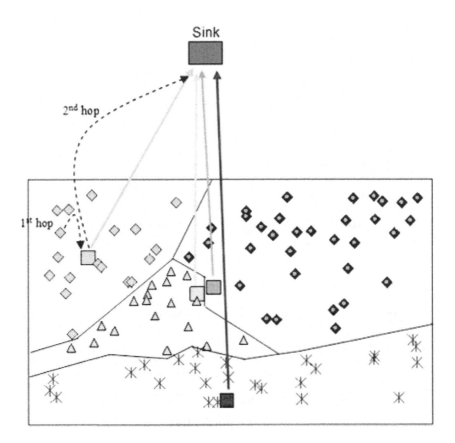

○ **Condition 1:** Residual energy > a specified energy level

○ **Condition 2:** The threshold value (T) > a random number between 0 and 1

○ **Condition 3:** The role of CH has not taken in the current $\frac{1}{P}$ rounds

- No overlap among formed clusters
- The tie condition solved by choosing the nearest CH
- Member nodes informs the CH they selected it
- Irregular cluster shape
- Nodes need no location information
- All nodes are static after deployment
- Doesn't use iterations
- CHs with no members exist (decrease when nodes count increases) see Figure 3

The square CH in Figure 3 has no members because the three CHs (square, triangle, and circle) are very close to each other, and the triangle and circle CHs withheld the nodes close to the square CH from it.

- Average number of members per cluster remains constant with the number of nodes increase
- CHs average percentage remains constant with the increase in nodes number
- Expected maximum number of members per cluster = all nodes − 1, this affects TDMA schedule length
- There may exist rounds without formed CHs, this occurs when all nodes in a round don't satisfy the two conditions, condition 2 and/or condition 3; Actually a different percentage from the specified percentage of CHs is formed, because this percent-

Figure 3. Example of CHs without members in LEACH

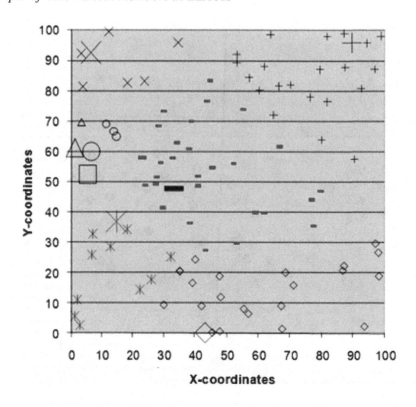

age may be greater than the specified, all nodes may become CHs before the end of the (1/P) round period causes the network to become unclustered for a number of subsequent rounds equals to the number of remaining rounds in that period

- Bad locality, LEACH doesn't satisfy free space model in intra-cluster communication.
- Bad CHs distribution
- Non-optimal CH selection
- Requires synchronization
- There are no non-clustered nodes in case of clusters exist
- Fault tolerant due to dynamic nature
- The CHs are selected randomly with the assumption that in each round a certain percentage of nodes (P) will formed, and it is not allowed to a node to become a CH more than one time each (1/P). The value of the threshold T of a node in a round represents the probability of this node to become a CH in this round and it equals the ratio of the count of the nodes required to be CHs in each round to the count of eligible nodes in this round for CH role which have not been CHs yet in the current (1/P) rounds, i.e. if the number of nodes is 100 and P equals 0.05, then in the first round in the current (1/P) rounds, T will be 5/100= 0.05, in the second round T will be 5/(100-5) =0.052632, in the third round T equals 5/(100-5-5)=0.055556, and so on. As it was seen in the previous example, at the beginning of each (1/P) rounds of protocol operation, T will be equal to P and increases gradually as rounds go until its value reaches one in the last round of the current (1/P) rounds of protocol operation, so that at the last round in each (1/P) round period, all nodes which have not been CHs in this period will be CHs in this round, i.e. it is guaranteed that all nodes in the network will take the head role once during each (1/P) rounds of protocol operations.

- The load is unbalanced among nodes and accordingly as rounds go energy and node distribution becomes non-uniform. Figure 4 shows the energy distribution after 200 round for LEACH network, while Figure 5 shows node distribution after 500 round. As shown in Figure 4, not all nodes drain equal amount of energy, nodes far from the sink consume larger amount of energy and this causes them to die quickly as a result, approximately the lower half of the monitored area gradually becomes uncovered as shown in Figure 5.

- There is a big variation in the generated number of CHs each round as shown in Figure 6. The average number of CHs is approximately equal to 5; this value represents the predefined percentage of CHs which equals 5 in the network setup of the used network example. The count of formed CHs in a round may be 1 as mentioned before; the percentage of one-CH rounds approximately equals 3.4%.

ENERGY-EFFICIENT CLUSTERING SCHEME

Energy-Efficient Clustering Scheme (EECS) is an energy-efficient clustering protocol for periodical data gathering applications in WSNs. A constant number of candidate nodes for CH role are elected with a probability T and compete for the head role according to the node residual energy within radio range $R_{compete}$. The candidate node will be a head if it didn't find another candidate with higher residual energy, otherwise, it will give up competition with the first found candidate with higher residual energy.

The energy expended during data transmission for far away CHs is significant, especially in large scale networks. Thus, for performing load balance

Figure 4. Residual energy distribution in LEACH network as rounds go (each bubble represents a node and the width of each bubble represents the residual energy of the node represented by this bubble)

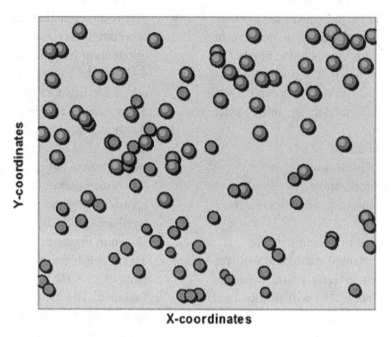

Figure 5. Node distribution in LEACH network as rounds go (black nodes represent nodes still alive while white nodes represent dead nodes)

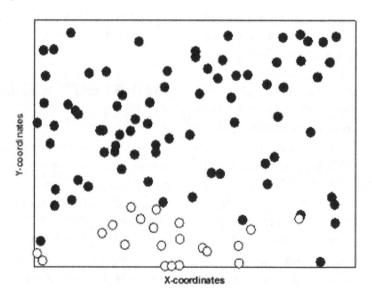

Figure 6. Count of formed CHs in LEACH network

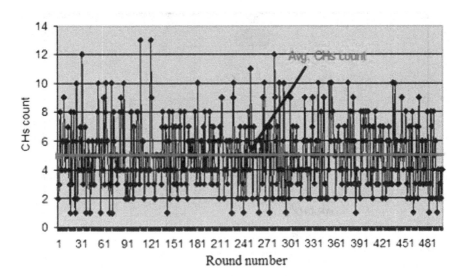

among CHs, the cluster size for each CH should be justified such that, the larger the distance between the CH and the BS is, the smaller member size the CH should accommodate.

So that in EECS the plain nodes choose the CH according to a weighted function introduced to form load balanced cluster, the plain node chooses the CH by considering not only saving its own energy but also balancing the workload of cluster heads, i.e. two distance factors: distance from node to CH and distance from CH to BS, Figure 7 illustrates the weighted function cost for CH selection in EECS and the representation of the distances used in it on the graph according to the used EECS network example . For protocol details see Ye, Li, Chen, and Wu (2005). An example of the EECS clusters is shown in Figure 8.

Properties of EECS Clustering

- Dynamic clustering as it includes regular CH re-election or cluster reorganization procedures
- Distributed
- Adaptive
- Leader election process performed first before clusters formation

- A subset of nodes presents itself as candidates in competition among them for the post of CH. These nodes select themselves to be candidates with a certain probability, but the candidate which win in a competition and takes the head role will be the candidate with larger residual energy in a certain transmission range
- No overlap among formed clusters
- The tie condition solved by choosing the CH with minimum *cost*
- Member nodes informs the CH they selected it
- Irregular cluster shape
- Nodes need no location information
- All nodes are static after deployment
- Doesn't use iterations
- CHs with no members exist as CH 68 in Figure 8. According to (Ye, Li, Chen, & Wu, 2005), if CH2 is closer to the sink than CH1, a node chooses CH2 which is farther than CH1 from it when *cost2 < cost1, i.e.:*

$$\left(w \times f_2 + (1-w) \times g_2 \right) < \left(w \times f_1 + (1-w) \times g_1 \right)$$

$$\Box f_1 > \left(f_2 - \left(\frac{1-w}{w} \right) \times g_c \right) \qquad (2)$$

Figure 7. The weighted function cost for CH selection in EECS

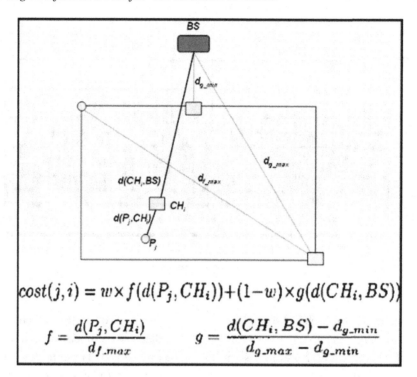

Figure 8. Example of EECS network (T = 0.2, $R_{compete}$ = 26, and w = 0.8)

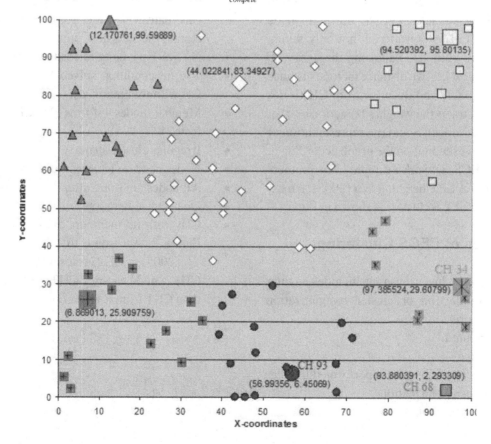

where f_1 and f_2 are normalized functions for the distance between the node and CH1 and between the node and CH2 respectively, g_1 and g_2 are normalized functions for the distance between CH1 and the sink and between CH2 and the sink respectively, w is the weighted factor for the tradeoff between f and g, and $g_c = g_1 - g_2$

This means that, every CH builds a boundary between him and each CH nearer than him to the sink, this boundary form a specific area around this CH, all the nodes, which in this area choose this head as its CH, while all nodes, which is located outside this area may choose the CH located outside this area which is nearer to the sink. As this boundary limits the number of nodes choose the far CH, the load on this CH is reduced. Figure 9 shows the network of Figure 8 illustrated on it the boundaries between CH 68 and each CH. The irregular area which represents the range in which the nodes are allowed to choose CH 68 is equal to the smallest area results from the intersect between two boundaries and contains no other CHs, this area is shaded in Figure 9. By accident the range of CH 68 is void from nodes due to random distribution of nodes, thus usually CH 68 has no members in each round it becomes a CH and there are other CHs near it.

- CHs are elected through local radio communication, but because the CH broadcasts the HEAD_AD_MSG message across the network, EECS may not satisfy free space model in intra-cluster communication.
- The CH distribution is better than CH distribution in LEACH, there are no two CHs in range $R_{compete}$; but, a big portion from the monitored area may be void from CHs as shown in Figure 8, there is any CH in the center area. A cause of this void area is that the candidate node sends the COMPETE_HEAD_MSG without a knowledge about what will be its state, and it receives COMPETE_HEAD_MSGs without any knowledge about the current state of its

senders, as an example, nodes 15, 14, 98, 55, 22, 65, and 49 labeled and redrawn from Figure 8 in Figure 10; all these candidate nodes sent COMPETE_HEAD_MSG at the same time, node 15 gave up competition for node 14 while 14 gave up for 98, similarly, 98 gave up for 55, 55 gave up for 22, 22 gave up for 65, 65 gave up for 49, only 49 became a CH; if 15 knows that 14 will be a plain node and there will not be any CH in its range, it wouldn't give up competition and would become itself a CH, the same for 14, 98, and 55, this case repeated more than once and the result is inexistence of CHs in a considerable portion of the network.

- Non-optimal CH selection, it is true that the CH selected is the candidate with larger residual energy in range $R_{compete}$, but the set of candidate nodes in the competition are selected randomly before the competition
- Requires synchronization
- After clustering, there are no non-clustered nodes
- There are no deceived nodes, each member choose its CH and it is sure that it will be a CH till the end of the round, this is because the CH election procedures and the clustering procedures aren't made in parallel, nodes decide its state in a time period then plain nodes select its CH in another time period.
- Fault tolerant due to dynamic nature
- EECS introduces load-balance among nodes more than that offered by LEACH by reducing the number of member nodes for far CHs. Figure 11 shows the energy distribution after 200 round went in the EECS network, while Figure 12 shows node distribution after 600 round. As shown in Figure 11, approximately nodes drain equal amount of energy, thus as rounds go approximately all the monitored area is covered as shown in Figure 12, the

dead nodes are not concentrated in lower half of the network.

- The average number of CHs is almost equal 5. Figure 13 shows that the actual number of formed CHs varies in each round may be less or more than 5. In the first round the count of CHs is big (for example 18 as shown in the figure) because all nodes or in better words all candidate nodes have equal energy and thus each candidate choose itself. Also it should be noted that due to the problem illustrated in Figure 10, the network may turn into one cluster, all nodes become members for alone CH, as it happens in LEACH, but rarely happen.

HYBRID, ENERGY-EFFICIENT, DISTRIBUTED CLUSTERING PROTOCOL

For prolonging network lifetime, CH selection in the Hybrid, Energy-Efficient, Distributed clustering protocol (HEED) is primarily based on the residual energy of each node, and to increase energy efficiency and further prolong network lifetime, a secondary clustering parameter considers intra-cluster "communication cost" which can be a function of neighbor proximity or cluster density. The node in this algorithm may take one of three states, regular node, tentative CH, or final CH. Tentative CHs are a set of nodes selected randomly in the first iteration and through subsequent iterations with a probability (CH_{prob}) less than 1 based on their residual energy to be in head state. A *tentative_CH* node can become a regular node at a later iteration if it finds a lower cost CH. Final CHs are *tentative_CH* nodes their CH_{prob} reaches 1, while they didn't informed about any other *final_CH* or *tentative_CH* in their ranges or each one of them found that it has the lowest cost among all its compotator CHs.

As it is clear from the pseudo code in Younis and Fahmy (2004), HEED clustering protocol

consists of three phases, Initialize, Repeat, and Finalize phases. The Initialize phase procedures are performed every round once; it includes:

- Neighbor discovery, neighbor discovery is not necessary to be performed every round; HEED authors argue that because in a stationary network, where nodes do not die unexpectedly, the neighbor set of every node does not change very frequently. In addition, HEED distribution of energy consumption extends the lifetime of all the nodes, which adds to the stability of the neighbor set.
- Setting the probability of becoming a CH (CH_{prob}),

$$CH_{prob} = C_{prob} \times \frac{E_{residual}}{E_{max}} \qquad (3)$$

where:

$E_{residual}$ is the estimated current residual energy in the node.

E_{max} is a reference maximum energy (corresponding to a fully charged battery), which is typically identical for all nodes.

C_{prob} is the initial percentage of cluster heads among all nodes.

For terminating the algorithm in reasonable constant number of iterations, CH_{prob} is not allowed to fall below a certain threshold. In our example, the minimum value of CH_{prob} is corresponding to the value of $E_{residual}$ below which the node can't be a head, so the minimum CH_{prob} equals 0.00025 when C_{prob} =0.05, $E_{residual}$ = 0.01, and E_{max} = 2

The Repeat phase consists of a number of steps performed iteratively each until CH_{prob} reaches 1, with the knowledge that each node doubles its CH_{prob} before it goes to the next iteration. In each iteration the node tests three conditions in order, the first is whether the list of CHs is not empty, the second is whether CH_{prob} reaches 1, and the

Figure 9. Example of CHs without members in EECS

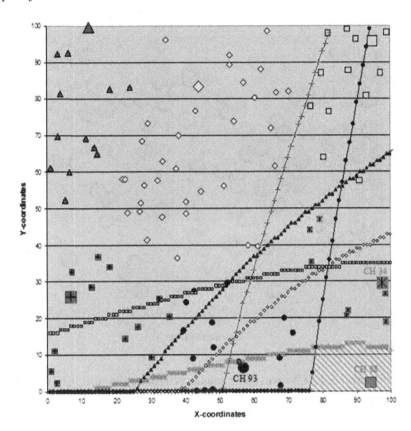

Figure 10. Illustrative example for the cause of large CHs-void area in EECS network

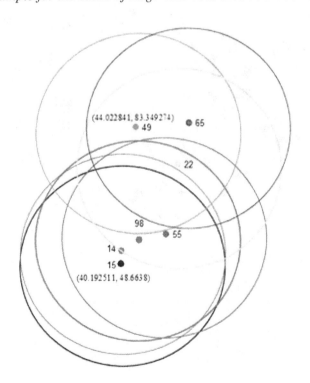

Figure 11. Residual energy distribution in EECS network as rounds go (each bubble represents a node and the width of each bubble represents the residual energy of the node represented by this bubble)

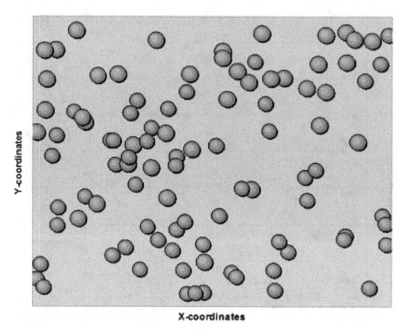

Figure 12. Node distribution in EECS network as rounds go (black nodes represent nodes still alive while white nodes represent dead nodes)

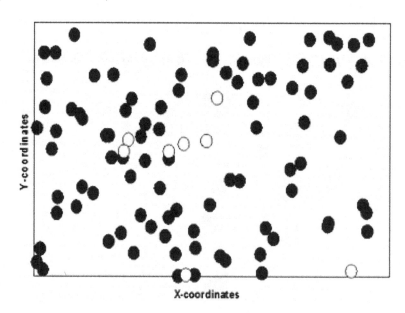

Figure 13. Count of formed CHs in EECS network

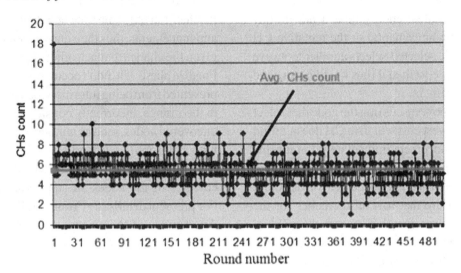

third is whether CH_{prob} greater than or equal to a random number uniformly distributed between 0 and 1.

In the first iteration, all nodes will not meet the first and second condition, so all of them will test the third condition, some of them satisfy it and some do not, the set of nodes that satisfy the third condition in the first iteration are the initial set of tentative CHs. Each tentative CH sends an announcement message with status tentative in its cluster range and of course adds itself to the CHs list; each node receives the announcement message whether it is tentative CH or regular node will add the sender of the message in its CHs list. Thus from iteration 2, the nodes that receives a message will meet the first condition accordingly always they ignore the other conditions, i.e. the tentative CH will be sure that it will not receive *tentative_CH* messages other than that received in the same iteration where it becomes a tentative CH (but it is probable for the regular node to receive tentative messages each round), and regular nodes that received a message will be sure that they will not be a tentative CH as they always ignore the third condition. According to the procedures of the first condition, each tentative CH sees whether it has the lowest cost in its transmission range, if

it has the lowest cost, it will wait until its CH_{prob} reaches 1 then it changes its state to final CH and informs nodes in its cluster range about that.

The term "uncovered node" refers to the node that didn't receive announcement messages from final or tentative CH, the uncovered node has during its clustering phase a chance each iteration to become tentative CH if it satisfies condition 3. If its clustering iterations ended without receiving final or tentative announcements, it will announce itself to be a final CH as condition 2 implies, if its clustering iterations ended while the node received tentative announcements and no final announcements, it considered also uncovered node, or in other words semi-uncovered node (in this case the semi-uncovered node may be tentative CH or regular node) and it will announce itself as a final CH as the procedure number 5 of the Finalize phase in the pseudo code in Younis and Fahmy (2004) implies.

As said before any node sends a tentative CH announcement in iteration will not receive any other tentative CH announcements during the subsequent iterations, but it can receive final CH announcement from semi-uncovered node selected this tentative CH as the lowest cost tentative CH, but it didn't receive its final announcement because this node finishes its iterations faster than it due to its larger residual energy (and it has no any other final announcements) or the tentative CH which the node selected selects another tentative CH with lower cost than it thus it will never send final announcement.

From the above discussion the node that select itself as the lowest cost tentative CH doesn't send the final announcement immediately, but it waits a until its CH_{prob} reaches 1, giving itself a time limit to receive final announcements from other nodes and ensure that it is all the better for the head role in its range with respect to the residual energy and the cost (the cost in the example is the node degree). In the Finalize phase, the nodes that received final announcements select the lowest cost final CH and send it a join message, while the nodes that didn't receive final announcements announce itself as final CH. The Finalize procedures is called when CH_{prob} reaches 1, but an important caution for protocol implementation, the Finalize procedures shouldn't be called immediately when CH_{prob} equals 1, but it must be delayed a period equals to the time each iteration takes because

there may be more than one node their CH_{prob} reached 1 at the same time and they sent final CH announcements, thus their announcements should given the chance to reach other nodes before their Finalize phase. If it didn't occur, some nodes will be prevented from being informed about the final CHs in their range, incorrectly consider themselves as uncovered nodes, send an announcement message with status *final_CH*, and as a result the number of CHs increases especially when the residual energies of the nodes are approximately equal. For protocol details see (Younis & Fahmy, 2004). An example of the clusters formed in HEED is shown in Figure 14.

Properties of HEED Clustering

- Dynamic clustering as it includes regular CH re-election or cluster reorganization procedures
- Distributed
- Adaptive
- Leader election process and clusters formation are performed in the same period of

Figure 14. Example of HEED network

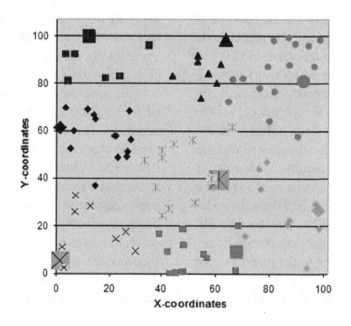

protocol operation, while one node elected as a CH in a region, some nodes send join message to another CH and form a cluster in another region

- A subset of nodes presents themselves as candidates in competition among them for the post of CH. These nodes select themselves to be candidates with a certain probability based on residual energy or they select themselves to be candidates directly based on their residual energies if they are uncovered nodes (because if the node doesn't hear from any CH it will compete with CHs for the post of CH after its iterations finishes and the number of iterations of each node depends on its residual energy), but the candidate which win in a competition and takes the head role will be the candidate with a large residual energy and lower cost (in this example the cost is the node degree) in a certain transmission range

- No overlap

- The tie condition solved by choosing the CH with minimum *cost*

- The final CH takes the head role because it has the lowest cost among other final CHs in its cluster range does not need to send any additional announcement message to advertise itself to its regular node neighbors because the regular nodes in its range already knew about its existence from the announcement messages sent for head role competition and what remaining is that they will select and inform it if it has the lowest cost among the final CHs they hear from.

- Member nodes informs the CH they selected it

- Regular cluster shape, each cluster represents a circle with the CH is its center and its radius is a specified transmission range, all cluster members fall within this circle

- Nodes need no location information
- All nodes are static after deployment
- The protocol use iterations for clustering the network
- CHs with no members exist with large percentage; its percentage in the network of our example is about 50%
- CHs are elected through local radio communication, also regular nodes join the cluster through local radio communication, thus HEED satisfies free space model in intra-cluster communication
- The CH distribution is better than that in LEACH and EECS; there are no two CHs in the same range and there is no big portion from the monitored area void from CHs
- Non-optimal CH selection, the initial set of CHs is selected probabilistically
- Synchronization is not critical for HEED operation; HEED authors show that unsynchronized nodes can still execute HEED independently, but cluster quality may be affected
- Although the CH election procedures and the clustering procedures are made in parallel, there are no deceived nodes. Each node selects a CH when it is sure that the head state is its final state
- Fault tolerant due to dynamic nature
- HEED introduces load-balance among nodes, but it didn't give a special concern for the far CHs as EECS did. If as assumed in the example the CHs send to the sink directly, the far nodes from the sink will die faster. This appears from Figure 15 which shows the residual energy distribution after 200 round
- The average number of CHs which have members is approximately equal to 9. The count of CHs which have members versus round number up to 500 round is drawn in Figure 16.

The average number and standard deviation of CHs count and members per cluster count for LEACH, HEED, and EECS for the previous example are illustrated in Figure 17.

From Figure 17 it could be noticed that, HEED has the maximum value of the average count of CHs, but the maximum variation in this count is achieved by LEACH. On the other hand, LEACH has the maximum value of the average count of members per cluster and also it results in the maximum variation in this count.

CONCLUSION AND FUTURE RESEARCH DIRECTIONS

This chapter illustrates the variation in the design of clustering protocols for WSN, giving some design options for the protocol designer and gives him some guidelines to choose among them to achieve the required properties of the resultant clustered network to aid in achieving the aimed goals from the cluster-based routing protocol or the network performance as a whole. Also the chapter studies the clustering protocols of LEACH, EECS, and HEED in terms of some of the mentioned features and requirements of WSN clustering protocols and sheds the light on the achievements and limitations of each of them to help in understanding the issues related to WSN clustering mentioned in this chapter, understanding these three clustering protocols themselves, facilitating and giving a method for studying other clustering protocols, modifying these protocols or others to improve the properties of the resultant clustered network and accordingly the impact of the clustering algorithm in the overall routing protocol performance, and designing new clustering algorithms for WSN.

Figure 15. Residual energy distribution in HEED network as rounds go (each bubble represents a node and the width of each bubble represents the residual energy of the node represented by this bubble

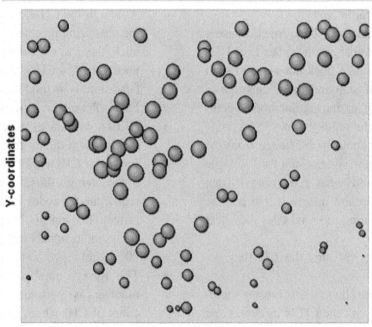

Figure 16. Count of formed CHs in HEED network

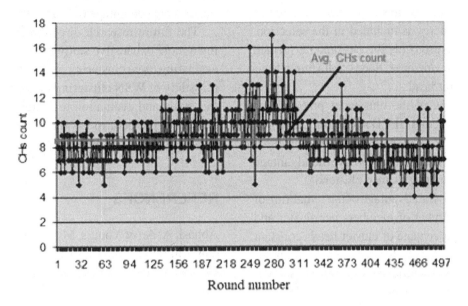

Figure 17. Average and standard deviation of CHs count and cluster size for LEACH, EECS, and HEED

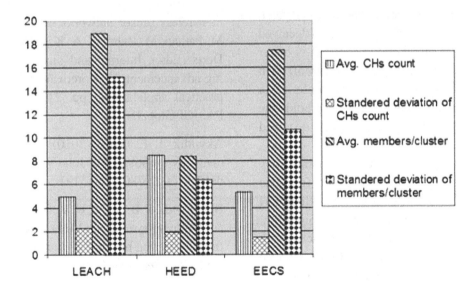

The three mentioned clustering protocols are dynamic, distributed, and adaptive. With respect to the order in which the leader election is performed, in both LEACH and EECS this process is performed before the clusters formation process, while in HEED the two processes are performed in parallel. The method of hiring the leader is also different among these protocols, in LEACH the cluster heads are hired by appointing themselves in this role if certain conditions are satisfied; in EECS and HEED, the cluster heads are hired upon a competition, the winner in a certain transmission range is selected based on some different determinants, but always a condition related to the residual energy is included in the selection. In all of these protocols the clusters are formed by having the member nodes informing their selected cluster head.

The features of the clustering protocols of LEACH, EECS, and HEED result in some strong and some weak characteristics in the clustered network. All of these protocols have the advantage of no overlap among formed clusters.

LEACH has the advantage of the remaining of the average number of members per cluster and the average percentage of cluster heads constant with the node density increase, and also an advantage that in the entire of the clustered network, there are no non-clustered nodes. The last just mentioned advantage of LEACH is shared with EECS; in addition, the CH distribution of EECS is better than that of LEACH and it introduces load-balance among nodes more than that is offered by LEACH; also in EECS, there are no deceived nodes. The CH distribution in HEED is better than that in LEACH and EECS; synchronization is not critical for HEED operation while it is a requirement in the others. HEED also holds the advantages of the EECS related to the deceived nodes and the load-balance.

As the protocols have strengths points, they have weakness points; LEACH suffers from bad locality, bad CHs distribution, non-optimal CH selection, load unbalance, there is a big variation in the generated number of CHs each round, rounds without formed CHs, and CHs with no members. EECS is susceptible to incur bad locality, and it suffers from CHs with no members and non-optimal CH selection. Regarding to the weakness of HEED, the first weakness point is represented by its usage of iterations for clustering the network, it doesn't give a special concern for the far CHs as EECS does; the non-optimal CH

selection disadvantage also exists in HEED and it incurs a big percentage of CHs with no members.

The future research directions may be making a more exhaustive study of WSN clustering, evaluating more clustering protocols by applying this study of WSN clustering on them, and using this study and evaluation to design a new WSN clustering protocol meets the requirements of a wide range of WSN applications.

REFERENCES

Abbasi, A. A., & Younis, M. (2007). A survey on clustering algorithms for wireless sensor networks. *Computer Communications*, *30*(14), 2826–2841. doi:10.1016/j.comcom.2007.05.024

Abdel Hady, A. Abd El-kader, S. M., Eissa, H. S., Salem, A., & Fahmy, H. M.A. (2012). A comparative analysis of hierarchical routing protocols in wireless sensor networks. In J. H. Abawajy, M. Pathan, M. Rahman, A. K. Pathan, & M. M. Deris (Eds.), Internet and distributed computing advancements: Theoretical frameworks and practical applications (pp. 212-246). Hershey, PA: IGI Global.

Akyildiz, I. F. (Ed.). (2010). *Wireless sensor networks*. Hoboken, NJ: John Wiley & Sons Ltd. doi:10.1002/9780470515181

Boregowda, S. B., Babu, N. V., Puttamadappa, C., & Mruthyunjaya, H. S. (2011). Energy balanced fixed clustering protocol for wireless sensor networks. *International Journal of Computer Science and Network Security*, *11*(8), 166–172.

Boukerche, A. (Ed.). (2009). *Algorithms and protocols for wireless sensor networks*. Hoboken, NJ: John Wiley & Sons, Inc.

Hasnaoui, M. L., Hssane, A. B., Ezzati, A., & Benalla, S. (2010). An oriented cluster formation for heterogeneous wireless sensor networks. *Journal of Computing*, 2(11), 34–39.

Heinzelman, W. R., Chandrakasan, A., & Balakrishnan, H. (2000). Energy-efficient communication protocol for wireless microsensor networks. In *Proceedings of the 33rd Annual Hawaii International Conference on System Sciences*. IEEE. doi: 10.1109/HICSS.2000.926982

Katiyar, V., Chand, N., & Soni, S. (2010). Clustering algorithms for heterogeneous wireless sensor network: A survey. *International Journal of Applied Engineering Research*, 1(2), 273–287.

Misra, S., Woungang, I., & Misra, S. C. (Eds.). (2009). *Guide to wireless sensor networks*. London: Springer-Verlag. doi:10.1007/978-1-84882-218-4

Mohammad El-Basioni, B. M. Abd El-kader, S. M., Eissa, H. S., & Zahra, M. M. (2012). Low loss energy-aware routing protocol for data gathering applications in wireless sensor network. In J. H. Abawajy, M. Pathan, M. Rahman, A. K. Pathan, & M. M. Deris (Eds.), Internet and distributed computing advancements: Theoretical frameworks and practical applications (pp. 272-302). Hershey, PA: IGI Global.

Varga, A. (2005). *Omnet++ discrete event simulation system (version 3.2)*. Retrieved from http://www.omnetpp.org/omnetpp/doc_details/2105-omnet-32-win32-binary-exe

Yang, Q., Zhuang, Y., & Li, H. (2011). A multi-hop cluster based routing protocol for wireless sensor networks. *Journal of Convergence Information Technology*, 6(3), 318–325. doi:10.4156/jcit.vol6.issue3.37

Ye, M., Li, C., Chen, G., & Wu, J. (2005). EECS: An energy efficient clustering scheme in wireless sensor networks. In *Proceedings of the 24th IEEE International Performance, Computing, and Communications Conference*, (pp. 535–540). IEEE. doi: 10.1109/PCCC.2005.1460630

Younis, O., & Fahmy, S. (2004). HEED: A hybrid, energy-efficient, distributed clustering approach for ad hoc sensor networks. *IEEE Transactions on Mobile Computing*, 3(4), 366–379. doi:10.1109/TMC.2004.41

Zhang, Y., Yang, L. T., & Chen, J. (Eds.). (2010). *RFID and sensor networks: Architectures, protocols, security and integrations*. New York: Taylor and Francis Group, LLC. doi:10.3837/tiis.2010.06.004

ADDITIONAL READING

Anker, T., Bickson, D., Dolev, D., & Hod, B. (2008). Efficient Clustering for Improving Network Performance in Wireless Sensor Networks. *Proceedings of the 5th European Conference on Wireless Sensor Networks (EWSN'08)*, 221-236.

Aslam, N., Phillips, W., Robertson, W., & Sivakumar, S. C. (2007). Balancing energy dissipation in clustered wireless sensor networks. In P. Thulasiraman, X. He, T. Xu, M. Denko, R. Thulasiram, & L. Yang (Eds.), *Frontiers of High Performance Computing and Networking ISPA 2007 Workshops* (pp. 465–474). Berlin, Heidelberg: Springer. doi:10.1007/978-3-540-74767-3_48

Bajaber, F., & Awan, I. (2010). Distributed Energy Balanced Clustering for Wireless Sensor Networks. *Proceedings of the 10th International Conference on Computer and Information Technology*, 385-392.

Bereketli, A., & Akan, O. B. (2009). Event-to-sink directed clustering in wireless sensor networks. *Proceedings of the IEEE conference on Wireless Communications & Networking Conference (WCNC'09)*, 2290-2295

Bhatti, S., Qureshi, I. A., & Memon, S. (2012). Node Clustering for Wireless Sensor Networks. *Mehran University Research Journal of Engineering & Technology, 31*(1), 163–176.

García-Hernando, A.-B., Martínez-Ortega, J.-F., López-Navarro, J.-M., Prayati, A., & Redondo-López, L. (2008). *Problem solving for wireless sensor networks*. Springer-Verlag London Limited. doi:10.1007/978-1-84800-203-6

Huang, Y.-F., Chen, C.-M., Chen, T.-R., Chen, J.-S., & Wang, N.-C. (2007). Performance of Relaying Cluster-Based Wireless Sensor Networks. *Communications of the IIMA, 7*(3), 89–100.

Jangra, M., Malik, S., & kumar, R. (2012). Energy Efficient Multi-hop Routing scheme with in Network Aggregation for WSN. *IJCEM International Journal of Computational Engineering & Management, 15*(5), 110-114.

Jurdak, R. (2007). *Wireless ad hoc and sensor networks: a cross-layer design perspective*. New York, NY 10013, USA: Springer Science+Business Media, LLC.

Khedo, K. K., & Subramanian, R. K. (2009). MiSense Hierarchical Cluster-Based Routing Algorithm (MiCRA) for Wireless Sensor Networks. *Proceedings of world academy of science, engineering and technology*, 190-196.

Kim, K. T., & Youn, H. Y. (2005). Energy-Driven Adaptive Clustering Hierarchy (EDACH) for Wireless Sensor Networks. *Proceedings of the International Federation for Information Processing*, 1098-1107.

Kumar, D., Aseri, T. C., & Patel, R. (2009). EEHC: Energy efficient heterogeneous clustered scheme for wireless sensor networks. *Computer Communications, 32*(4), 662–667. doi:10.1016/j.comcom.2008.11.025

Li, Y., Thai, M. T., & Wu, W. (2008). Wireless sensor networks and applications. Springer Science+Business Media, LLC.

Liu, M., Cao, J., Chen, G., & Wang, X. (2009). An energy-aware routing protocol in wireless sensor networks. *Sensors (Basel, Switzerland), 9*(1), 445–462. doi:10.3390/s90100445 PMID:22389610

Meenakshi, B., & Anandhakumar, P. (2012). Lifetime extension of wireless sensor network by selecting two cluster heads and hierarchical routing. Proceedings of the International Conference on Advances in Computing, Communications and Informatics (ICACCI '12), 1254-1260.

Mehr, M. A. (2011). Design and Implementation a New Energy Efficient Clustering Algorithm Using Genetic Algorithm for Wireless Sensor Networks. *Engineering and Technology, 53*(2), 430–433.

Mohammad El-Basioni, B. M., Abd El-kader, S. M., & Eissa, H. S. (2012). Designing a local path repair algorithm for directed diffusion protocol. *Egyptian Informatics Journal*, DOI information: 10.1016/j.eij.2012.07.001.

Mohammad El-Basioni, B. M., Abd El-kader, S. M., & Eissa, H. S. (2012). Improving LLEAP Performance by Determining Sufficient Number of Uniformly Distributed Sensor Nodes. *CiiT International Journal of Wireless Communication, 4*(13), 777-789.

Nawaz, K., & Buchmann, A. P. (2010). ACD-MCP: An Adaptive and Completely Distributed Multi-hop Clustering Protocol for Wireless Sensor Networks. *International Journal of Wireless & Mobile Networks, 2*(3), 18–37. doi:10.5121/ijwmn.2010.2302

Sehti, B. T., Pal, T., & Dasbit, S. (2010). An Energy-Efficient Clustering Scheme to Prolong Sensor Network Lifetime, *Proceedings of the International Conference and Workshop on Emerging Trends in Technology*, 838-842.

Singh, A. K., Purohit, N., Goutele, S., & Verma, S. (2012). An Energy Efficient Approach for Clustering in WSN using Fuzzy Logic. *International Journal of Computers and Applications, 44*(18), 8–12. doi:10.5120/6361-7575

Singh, B., & Lobiyal, D. K. (2012). An energy-efficient adaptive clustering algorithm with load balancing for wireless sensor network. *International Journal of Sensor Networks, 12*(1), 37–52. doi:10.1504/IJSNET.2012.047714

Singh, S. K., Singh, M. P., & Singh, D. K. (2010). Routing Protocols in Wireless Sensor Networks – A Survey. [IJCSES]. *International Journal of Computer Science & Engineering Survey, 1*(2), 63–83. doi:10.5121/ijcses.2010.1206

Wang, B. (2010). *Coverage control in sensor networks*. Great Britain/British Isles: Springer-Verlag New York Inc.

Zheng, J., & Jamalipour, A. (Eds.). (2009). *Wireless sensor networks: A networking perspective*. Hoboken, New Jersey: John Wiley & Sons. doi:10.1002/9780470443521

KEY TERMS AND DEFINITIONS

Clustering: An organizational unit of the wireless sensor network, and it is a mean to achieve hierarchical routing in wireless sensor network. It comprises grouping nodes into clusters each cluster has a head for coordinating its cluster members' work and receiving their data, aggregating it, and sending it outside the cluster.

Deceived Nodes: Nodes that may exist in a clustered wireless sensor network. They can be described by the adjective "deceived" because they are deceived by a message sent to them by another node to inform them that it will be its head while in subsequent clustering setup procedures it becomes a member node for another head without informing its deceived members. The result is that, these nodes remain deceived and unused.

Energy-Efficiency: To expend less energy in performing a certain task. While saving energy is an advisable design goal for most systems, it represents a necessity for wireless sensor network because sensor nodes are usually battery-powered and operate unattended.

Hierarchical Routing: The data routing method in which the network is hierarchical, in other words, the nodes constitute multiple levels because they are assigned different roles. In each level of the hierarchicy reside the nodes which perform the same jobs. Hierarchical routing assigns the role which consumes more energy to a small percent of nodes and to achieve load-balance this group of nodes is changed periodically.

Inter-Cluster Communication: The transmission of cluster data outside it by its head. This transmission may be through single-hop to the base station or through multi-hops among clusters' heads towards the base station.

Intra-Cluster Communication: The communication among one cluster members; usually the communication is from members to their head through single-hop or multi-hops.

Synchronization: To provide a common time frame to different nodes. It is mandatory for some important operations of wireless sensor network which aid in improving nodes' operations and saving their energy such as sleep scheduling and time-based channel sharing, but in the same time, clock synchronization itself contributes in energy consumption due to the highly energy consuming radio transmissions for delivering timing information.

Wireless Sensor Network: A distributed system consists of a large number of small electronic devices called sensor nodes, each node performs a particular job depends on the used application, which is based mainly on its work on the collection of information from the surrounding environment. Usually, sensor nodes are deployed randomly and communicate wirelessly.

Chapter 14
Performance Improvement of Clustered WSN by Using Multi–Tier Clustering

Yogesh Kumar Meena
Hindustan Institute of Technology and Management, India

Aditya Trivedi
ABV-Indian Institute of Information Technology and Management, India

ABSTRACT

In the last few decades, the Wireless Sensor Network (WSN) paradigm has received huge interest from the industry and academia. Wireless sensor networking is used in various fields like weather monitoring, wildfire detection/monitoring, battlefield surveillance, security systems, military applications, etc. Moreover, various networking and technical issues still need to be addressed for successful deployment of WSN, especially power management. In this chapter, the various methods of saving energy in sensor nodes and a method by which energy can be saved are discussed with emphasis on various energy saving protocols and techniques, and the improvement in the Performance of Clustered WSN by using Multi-tier Clustering. By using a two-tier architecture in the clustering and operation of sensor nodes, an increase in the network lifetime of the WSN is gained. Since this clustering approach has better results in term of energy savings and organizing the network, the main objective of this chapter is to describe power management techniques, two-tier architecture, clustering approaches, and network models to save the energy of a sensor network.

DOI: 10.4018/978-1-4666-5170-8.ch014

1. INTRODUCTION

Networking is an important phenomenon for communication. Since a long time, wired network is used for communication. Wired network is considered to be more reliable and suitable. The main reason behind it was that we were more familiar to wired network in comparison to wireless. But in recent years, use of wireless network has increased dramatically. Different types of wireless networks such ad hoc networks, mesh networks and Wi-max etc. have come into existence since past few years (Goldsmith, 2005).

Wireless networks had a significant presence in the world as far back as World War II. Through the use of wireless networks, information could be sent overseas or behind enemy lines easily, efficiently and more reliably. Since then, wireless networks have continued to develop and their uses have grown significantly. Cellular phones are part of huge wireless network systems. People use these phones daily to communicate with one another. Sending information overseas is possible through wireless network systems using satellites and other signals to communicate across the world. Emergency services such as the police department utilize wireless networks to communicate important information quickly. People in business use wireless networks to send and share data quickly whether it be in a small office building or across the world (Kahn, Katz, & Pister, 1999; Cerpa, Elson, Estrin, Girod, Hamilton, & Zhao, 2001).

In recent time, a new application of wireless network known as sensor network has developed and used for a number of purposes. It is an efficient way of monitoring an area and receiving information. So a lot of research work is going on this field.

1.1 Wireless Network

Wireless network refers to any type of computer network that is wireless, and is commonly associated with a telecommunications network whose interconnections between nodes is implemented without the use of wires. Wireless telecommunications networks are generally implemented with some type of remote information transmission system that uses electromagnetic waves, such as radio waves, for the carrier and this implementation usually takes place at the physical level or "layer" of the network.

Important use for wireless networks is as an inexpensive and rapid way to be connected to the Internet in countries and regions where the telecom infrastructure is poor or there is a lack of resources, as in most developing countries. Compatibility issues also arise when dealing with wireless networks. Different components not manufactured by the same company may not work together, or might require extra work to fix these issues. Wireless networks are typically slower than those that are directly connected through an Ethernet cable.

A wireless network is more vulnerable, because anyone can try to break into a network broadcasting a signal. Many networks offer WEP - Wired Equivalent Privacy - security systems which have been found to be vulnerable to intrusion. Though WEP does block some intruders, the security problems have caused some businesses to stick with wired networks until security can be improved. Another type of security for wireless networks is WPA - Wi-Fi Protected Access. WPA provides more security to wireless networks than a WEP security set up. The use of firewalls will help with security breaches which can help to fix security problems in some wireless networks that are more vulnerable (Goldsmith, 2005; Meena, Singh, & Chandel, 2012).

There are various hazards also related to wireless network. But we are going to mainly focus on wireless sensor network in our chapter. We are going to describe mainly features of wireless sensor network.

1.2 Wireless Sensor Network

A wireless sensor network typically consists of a large number of inexpensive, small, low-power communication devices called sensor nodes and one or more computing centres. Advances in energy-efficient design and wireless technologies have enabled the manufacture of the small devices to support several important wireless applications, including real-time multimedia communication (Nikmard & Taherizadeh, 2010; Robinson, 1998), medical application, surveillance using WSNs and home networking applications (Agre & Clare, 2000). In WSNs, the sensor nodes have the ability to sense, process data, and communicate with one another.

Figure 1 shows the main hardware components of a sensor node:

1. Memory that stores programs and intermediate data, a controller that processes all the data and controls the other components,
2. A limited power supply (e.g. battery) also includes power unit for managing power usages,
3. Location aware system for GPS sensor nodes,

4. A transceiver that performs the functions of both a transmitter and receiver with a limited transmission range, and
5. A sensor device that senses the ambient environment.

Basic characteristics of sensor node hardware are as follows:

*   Small
*   Low cost (dispensable)
*   Low bit rate
*   Low memory capacity
*   Limited computational power

Sensor nodes collaborate to detect events or phenomena depending on the application, to collect and process data, and to transmit the sensed information to the computing centre (base station) by hopping the data from node to node. Although the sensor nodes individually have limited capabilities, their collaboration to perform a specific task produces an enhanced view of the physical world. Examples of modern sensor nodes are rene mote, dot mote, mica node and weC mote.

As for military uses, we need continuous monitoring of the area so sensor networks are

Figure 1. Basic sensor node hardware architecture

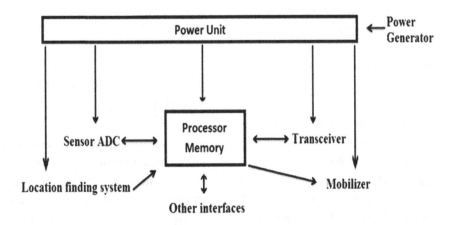

more suitable for it. Smart dust is the example of such type of needs (Kahn, Katz, & Pister 1999).

1.3 Applications of Wireless Sensor Network

Wireless Sensor Networks can be used in a wide range of exciting applications, such as target field imaging, intrusion detection, weather monitoring, security and tactical surveillance; distributed computing; the detection of ambient conditions such as temperature, movement, sound, and light or the presence of specific objects, inventory control, and disaster management.

We can divide WSN applications in mainly four types:

- Environmental data collection,
- Security monitoring,
- Node tracking, and
- Hybrid networks

For environmental data collection, we need continuous measurement of area for long time in order to get trends and dependencies (Cerpa, Elson, Estrin, Girod, Hamilton, & Zhao, 2001). For security monitoring sensor nodes play important role. We can examine any activities in monitoring area and using alarms we can get information about unusual activity. By this it is clear, here sensing and sending data is event driven. For node tracking, we should deployed nodes at certain location and data sending will be event driven. Hybrid networks contain all three properties.

1.4 Importance of Power Management

As we know that sensor nodes have limited source of energy. They are using battery for energy and in most of cases it is almost impossible to change the battery of sensor nodes. Therefore, an energy-efficient mechanism is required to save energy and prolong the network lifetime. Other more related

lifetime requirements have been used, including the requirement that as long as there is a sensor node alive, the network is considered to be alive.

The energy is consumed mainly for three purposes:

- Data transmission,
- Signal processing,
- Hardware operation.

These are interdependence or we can say that energy consumption is tightly coupled with other factor of sensor network like delay and throughput. So for longer lifetime of any sensor network we can manage the usage of power and can reduce the wastage of power. Network lifetime can be defined as the time elapsed until the first node (or the last node) in the network depletes its energy (dies) (Estrin, Girod, Pottie, & Srivastava, 2001).

Wasteful energy consumption can be due to

1. Idle listening to the media,
2. Retransmitting due to packet collisions,
3. Overhearing, and
4. Generating/handling control packets.

1.5. Organization of Chapter

Organization of this chapter as follows. In the second section, we discuss various power management techniques. Then in third section, we focused on clustering algorithm. We define our model and its constraints in the section four. Chapter is concluded in the last section.

2. POWER MANAGEMENT TECHNIQUES

Reducing the power consumption is one of the main challenges in wireless sensor networks. In fact, the network lifetime depends on how the energy is spent at each sensor node. Therefore, all aspects in sensor networks must be energy efficient, including communication protocols and the sensor node architecture. Several techniques to reduce power consumption can be applied at the design time, known as static approaches. In contrast, during run time, dynamic techniques can improve the reduction of power consumption by selectively shutting down hardware components.

For example Portable systems, such as laptop computers and personal digital assistants (PDAs) draw power from batteries; so reducing power consumption extends their operating times. For desktop computers or servers, high power consumption raises temperature and deteriorates performance and reliability.

So we will discuss some of the power management techniques proposed so far in this section.

2.1 Hardware Design Improvements

As proposed for other wireless sensor prototype designs, the hardware design of the wireless sensor is broken down into three major subsystems. First, a sensing interface is proposed for the collection of structural response data from multiple analog sensors (accelerometers, strain gages, displacement transducers). The next subsystem is the computational core which receives the sensor data in digital form from the sensing interface. The core is chiefly responsible for data management and execution of pre-programmed data interrogation tasks. Finally, when data is ready for communication (transmit or receive), the wireless communication subsystem provides for the connection to other wireless sensors in the network.

The new sensing unit presented features two notable improvements in its hardware design over a similar unit initially proposed (Swartz, Jung, Lynch, Wang, Shi, & Flynn, 2005). First a low-power, single-chip, IEEE 802.15.4 compliant wireless radio is adopted in the wireless sensor design. The newly created IEEE 802.15.4 standard is especially suitable for distributed computation, as it was developed for true ad-hoc peer-to-peer

networking among battery powered wireless devices. The second most significant improvement is the use of a four layer circuit board, replacing the more limited two layer boards in use today.

Two major advantages are realized by upgrading from a two layer circuit board to a four layer board. First, four layer boards allow for more compact designs as they provide more options for routing circuit connections. The second advantage is the ability to devote the internal layers of the board to power and grounding planes. Using an internal layer solely for power planes result in more efficient distribution of power than trace conductors in addition to also dissipating waste heat quickly. Due to the mixture of digital (*e.g.* microcontroller) and analog (*e.g.* analog-to-digital converter (ADC)) circuit elements on the same circuit board, it is desirous to separate the common ground for the analog components from the ground utilized by the digital components. Digital components on previous two layer board designs have shown a tendency to flood the ground resulting in a loss of ADC resolution.

In the Table 1, we have shown the various versions regarding their power supply and other hardware level changes (Zhang, Sadler, Lyon, & Martonosi, 2004).

Besides this, we can also enhance the battery performances by making suitable changes. We can use solar cells and other types of battery for increasing the node life time. This is also better option for improving the network lifetime.

2.2 OS Level Improvements

The second type of energy saving schemes can be based on the improvement at OS level of a sensor node. An OS is act as middleware for sensor node (e.g. Media Access Control, Task Scheduler etc.). Operating systems in embedded wireless communication increasingly must satisfy a tight set of constraints, such as power and real time performance, on heterogeneous software and hardware architectures. Due to these characteristics of OS

Table 1. Design summary for different versions of zebra net hardware nodes

	Version 0.1 Aug. 2002	Version 1 Feb. 2003	Version 2 July 2003	Version 3 Nov. 2003
Power Supply	Off- board	Buck-Boost and Boost Converters	Buck-Boost and Boost Converters	Two Buck-boost converters
Noise Reduction	Bypass Capacitor	Standard Low ERS capacitors	Os-con capacitors	LC post filters and common mode choke
Radio	2 radio system	2.4GHz	900 MHz	900 MHz
GPS	Cycle stealing from onboard CPU	Off-board antenna power	Ultra low noise linear regulator	Ultra low noise linear regulator
Data Flash	5Mbit	2Mbit	4Mbit	4Mbit
Battery Charger	None	Off- board	Off- board	Pulse- charging
System Weight	1,151 grams	145 grams	136 grams	138 grams
Battery	N/A	Li-Ion 2 A hr 45 grams	Li-Ion 2 A hr 45 grams	Li-Ion 2 A hr 45 grams

for WSN, we need some specification in an OS designed for WSN like it should be event driven in the case of wireless sensor network. More efficient solutions are obtained with OS that are developed to exploit the reactive event-driven nature of the domain and have built-in aggressive power management.

Facing a huge design space filled with uncertainty, sensor node operating systems such as Tiny OS (Hill,

Szewczyk, Woo, Hollar, Culler, & Pister, 2000) and SOS (Han, Rengaswamy, Shea, Kohler, & Srivastava, 2005) maximize system flexibility. The limited resources of long-lived, battery operated sensor nodes (especially RAM) lead these systems to adopt software components for efficient yet flexible composition. Tiny OS, for example, is little more than a non-preemptive scheduler applications are built from large libraries of components. SOS pushes flexibility even further, allowing applications to dynamically add and replace components at runtime. Their flexibility has given tremendous freedom to researchers and developers, placing few barriers to innovation and investigation.

Flexibility and newer available hardware always required some changes in OS. An OS has

many constraints regarding that. There are mainly three limitations:

- Application complexity,
- The high cost of porting to a new platform, and
- Reliability

These faults are created due to continuous changes in wireless sensor network and the maturation of sensor system. Currently the most popular OS for sensor network known as Tiny OS has three limitations - new platform support, application construction, and reliable operation.

An evaluation version of Tiny OS has also been proposed. It is proposed as T2 (Wolisz, 2005). It is component-based, is written in nesC, has a single thread of control, and uses non-blocking calls.

However, although similar at a high level, T2 differs in almost every detail. It has a more restrictive concurrency model, a different boot sequence, different interfaces, as well as many design patterns and architectures absent in its predecessor. When it comes to reliability, the proverbial devil is in the details, and we have designed T2 accordingly. According to Adam Wolisz, we can save energy and improve the output of WSN by making useful changes at OS level (Wolisz, 2005).

2.3 Routing Algorithm Improvements

For energy saving techniques, we can use better routing mechanism for messaging. So there are various protocols proposed for this purpose. Most of the communication protocols proposed for power aware wireless sensor networks often make one or more of the following assumptions which make them non-optimal for most real-life applications:

- **Homogeneous distribution of nodes in the network:** Nodes can be randomly initialized (for example thrown from an aircraft) and hence might be unevenly distributed in the coverage area.
- **Homogeneous distribution of energy resources:** Energy resources may be unevenly distributed in a sensor network for several reasons: unequal energy consumption in different nodes, sensor node battery replacement in multiple phases and unequal energy input from secondary sources (for e.g. solar or wind energy sources).
- **Single hop access to the sink:** In a large sensor network, all nodes cannot reach the sink even at the maximum transmission level and would have to find multi-hop routes to the sink.
- **Priori network information:** Network features like size, density and topology change with time hence should not be relied on initial information.
- **Reliable communication:** Two neighbouring nodes may not always be able to communicate with each other due to radio channel properties and other physical obstructions in between.

By using the properties of sensor nodes for communication in the organization of network we can improve the network life time. We can also improve network robustness using routing level improvement. Most routing algorithms for sensor networks focus on finding energy efficient paths to prolong the lifetime of sensor networks. As a result, the power of sensors on efficient paths depletes quickly, and consequently sensor networks become incapable of monitoring events from some parts of their target areas. In many sensor network applications, the events that must be tracked occur at random locations and have non-deterministic generation patterns. Therefore, ideally, routing algorithms should consider not only energy efficiency, but also the amount of energy remaining in each sensor, thus avoiding non-functioning sensors due to early power depletion. In Figure 2 we show simple cluster architecture for sensor network: (Meena, Singh, & Chandel, 2012).

In our study we are going to mainly emphasize on clustering process of the network and by using this we will try to improve the energy efficiency of network. Although we have gone through various routing algorithms such as direct communication approach, hierarchical routing methods, self-organized routing algorithm (David & Jennings, (2005), and other routing algorithms (K & M, 2005), little evidence exists for the effectiveness and efficiency of these algorithms with respect to the considerations mentioned earlier.

Sink node performs routing optimization for the whole network and as a result the system is vulnerable to failure of the sink node. Furthermore, as the number of nodes in the network increases, the optimal route computation becomes more and more difficult for the sink node.

2.4 Location Aware Techniques

Wireless sensor networks have recently received a lot of attention due to a wide range of applications such as object tracking, environmental monitoring, warehouse inventory, and health care. In these applications, physical data is continuously collected

Figure 2. Cluster based wireless sensor network

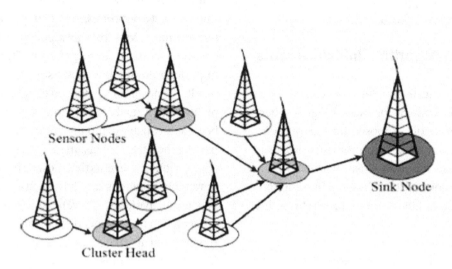

by the sensor nodes in order to facilitate application specific processing and analysis. We have studied about the location aware techniques. Sensor network applications typically are concerned more about physical phenomena or events associated with a geographical location or region than the raw data on a specific sensor node. In the location aware techniques the mainly concerns are about

- Location tracking of moving objects;
- Location-based routing and data collection;
- In-network query processing and query optimization.

Location-based routing protocols have been widely adopted in the design of wireless sensor networks. Most of the existing location-based routing protocols (Hussain, Farooq, Zia, & Akhlaq, 2004; Hwang, Pedram, & Soltan, 2008) are stateful, i.e., makes routing decisions based on cached geographical information of neighbouring sensor nodes. However, possible node movements, node failures, and energy conservation techniques in sensor networks result in dynamic networks with frequent topology transients, and thus pose a major challenge to stateful packet routing algorithms.

We studied a novel stateless location-based routing protocol, called PSGR, for location-aware sensor networks (Xu, Lee, & Mitchell, 2005). Based on PSGR, sensor nodes can locally determine their priority to serve as the next relay node using dynamically estimated network density and effectively suppress potential communication collisions without prolonging routing delays. PSGR also overcomes the communication void problem using two alternative stateless schemes, rebroadcast and bypass. Research result shows that PSGR exhibits superior performance in terms of energy consumption, routing latency and delivery rate.

Also based on location-based routing, we have studied an infrastructure-free window query processing technique for wireless sensor networks, called itinerary based window query execution (IWQE) (Wu, Xu, Tang, & Lee, 2006). In contrast to the conventional in-network query processing techniques proposed for wireless sensor networks which split a query execution in two stages, query propagation and data aggregation, IWQE combines them into one single stage for execution along a well-designed itinerary inside a query window.

IWQE, to the best of our knowledge, is the first infrastructure-free window query processing technique for wireless sensor networks. Many unique and challenging research issues which arise in IWQE (e.g., itinerary settings, query window coverage, in-network data processing, continuous data collection and handling of packet losses) have been studied thoroughly.

Hence we can say for energy saving parameter location aware techniques also play a vital role.

3. CLUSTERING ALGORITHM

In the various types of routing techniques we mainly focus on the clustering process. The clustering process has various features. In Figure 3, the simple architecture of clustered network has shown:

Looking at Figure 3, we can see the architecture of a generic Wireless Sensor Network (Akyildiz, Su, Sankarasubramaniam, & Cayirci, 2002), and examine how the clustering phenomenon is an essential part of the organizational structure. The network architecture has these important terms related to clustering.

- **Sensor Node:** A sensor node is the core component of a WSN. Sensor nodes can take on multiple roles in a network, such as

simple sensing; data storage; routing; and data processing.

- **Clusters:** Clusters are the organizational unit for WSNs. The dense nature of these networks requires the need for them to be broken down into clusters to simplify tasks such a communication.
- **Cluster Heads:** Cluster heads are the organization leader of a cluster. They often are required to organize activities in the cluster. Their tasks are not limited to data-aggregation and organization of the communication schedule of a cluster.
- **Base Station:** The base station is at the upper level of the hierarchical WSN. It provides the communication link between the sensor network and the end-user.
- **End User:** The data in a sensor network can be used for a wide-range of applications (Akyildiz, Su, Sankarasubramaniam, & Cayirci, 2002). Therefore, a particular application may make use of the network data over the Internet, using a PDA, or even a desktop computer. In a queried sensor network (where the required data is gathered from a query sent through the network), this query is generated by the end user.

Figure 3. Simple clustered sensor network architecture

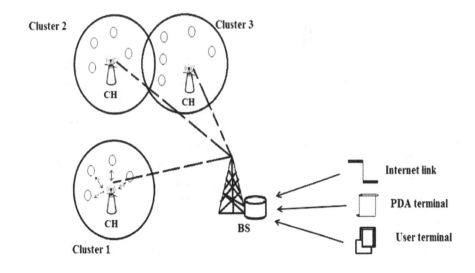

A clustering process has to take care of these factors related to wireless sensor network:

- Limited Energy
- Network Lifetime
- Limited Abilities
- Application Dependability

There are various problems in clustering process. Clustering algorithms play a vital role in achieving the targeted design goals for a given implementation. There are several key attributes that designers must carefully consider, which are of particular importance in wireless sensor networks. The important features for designing the network architecture (Dechene, Jardali, Luccini, & Sauer, 2005; Le, A, & M, 2010; Abbasia & Younis, 2007) are as follows:

- Cost of Clustering
- Selection of Cluster heads and Clusters
- Real-Time Operation
- Synchronization
- Data Aggregation
- Repair Mechanisms

- Quality of Service (QoS)

In the Figure 4, the simple distribution of various clustering schemes is shown

This figure has been taken from the paper presented on survey of clustering process. We have used this image as it is. In this chapter, we discuss some ideas about different types of clustering algorithms.

3.1 Heuristic Schemes

Heuristic algorithm has mainly focused on these two types of goals:

- Finding an algorithm with reasonable run-time (time needed to set up clusters is affordable);and/or
- With finding the optimal solution.

There are many types of heuristic algorithms that exist in choosing cluster heads. We will see that these algorithms deal only with a subset of parameters which impose constraints on the system. From this point of view, each one of these

Figure 4. Classification of proposed clustering schemes

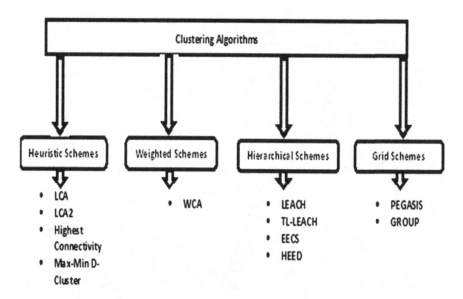

algorithms is only suitable for a specific application, rather than any arbitrary wireless mobile network. These algorithms are already displayed in the Figure 4 So we here discuss only silent features of these algorithms for getting a simple preview of these.

3.1.1 Link Cluster Algorithm (LCA)

This is one of the oldest algorithms for clustering schemes. It has some features for wired sensor network because it was initially developed for wired network.

In LCA (Dechene, Jardali, Luccini, & Sauer, 2005; Le, A & M, 2010; Abbasia & Younis, 2007), each node is assigned a unique ID number and has two ways of becoming a cluster head. The first way is if the node has the highest ID number in the set including all neighbour nodes and the node itself. The second way, assuming none of its neighbours are cluster heads, then it becomes a cluster head.

3.1.2 Link Cluster Algorithm 2 (LCA 2)

LCA2 was proposed to eliminate the election of an unnecessary number of cluster heads, as in LCA. In LCA2 (Dechene, Jardali, Luccini, & Sauer, 2005; Le, A, & M, 2010; Abbasia & Younis, 2007) they introduce the concept of a node being covered and non-covered. A node is considered covered if one of its neighbours is a cluster head. Cluster heads are elected starting with the node having the lowest ID among non-covered neighbours.

3.1.3 Highest-Connectivity Cluster Algorithm

This algorithm is similar to LCA. In Dechene, Jardali, Luccini, and Sauer (2005), Le, A, and M (2010), and Abbasia and Younis (2007), this scheme the number of node neighbours is broadcast to the surrounding nodes. The result is that instead of looking at the ID number, the connectivity of a node is considered. The node with the highest connectivity (connected to the most number of nodes) is elected cluster head, but in the case of a tie, the node with the lowest ID prevails. So this is also an important scheme for clustering process.

3.1.4 Max-Min D-Cluster Algorithm

In this algorithm a new distributed Cluster head election procedure, where no node is more than d (d is a value selected for the heuristic) hops away from the cluster head. This algorithm provides load balancing among cluster heads (Dechene, Jardali, Luccini, & Sauer, 2005; Le, A, & M, 2010; Abbasia & Younis, 2007).

The cluster head selection criterion is developed by having each node initiate 2 d rounds of flooding, from which the Results are logged. Then each node follows a simple set of rules to determine their respective cluster head.

Rule 1: Each node checks to see if it has received its sown id in the 2nd d round of flooding. If it has, then it can declare itself the cluster head and skip the other rules. Otherwise it proceeds to Rule 2.

Rule 2: Each node looks for node pairs. Once this is complete, it selects the minimum node pair to be the cluster head. If a node pair does not exist, they proceed to Rule 3.

Rule 3: Elects the maximum node id in the 1st d rounds of flooding as the cluster head for this node.

This algorithm is valid only if the following two assumptions are made:

Assumption 1: During the flooding, no node id will propagate further than d-hops from the originating node.

Assumption 2: All nodes that survive the flood max elect themselves cluster heads.

All these assumptions are clearly discussed in the paper given on this clustering technique for sensor network power management.

3.2 Weighted Schemes

In case of weighted schemes for sensor network the algorithm proposed in this section is a non-periodic procedure to the cluster head election, invoked on demand every time a reconfiguration of the networks topology is unavoidable. The mainly proposed algorithm under this category is WCA (Dechene, Jardali, Luccini, & Sauer, 2005; Le, A, & M, 2010; Abbasia & Younis, 2007).

3.2.1 Weighted Clustering Algorithm

The Weighted Clustering Algorithm (WCA) was firstly proposed (Chatterjee, Das, & Turgut, 2002). A node is selected to be the cluster head when it has the minimum weighted sum of four indices: the number of potential members; the sum of the distances to other nodes in its radio distance; the nodes average moving speed (where less movement is desired); and time of it being a cluster head (this takes battery life into account).

When a node moved out of its cluster, it will firstly check whether it can be a member of other clusters. If such a cluster exists, it will detach from current cluster and attach itself to that one. The process of joining a new cluster is known as reaffiliation. If the reaffiliation fails, the whole network will recall the cluster head election routine. This clustering algorithm tries to find a long-lasting architecture during the first cluster head election. When a sensor loses the connection with any cluster head, the election procedure is invoked to find a new clustering topology.

This is an important feature in power saving, as the re-election procedure, which consumes energy, occurs less frequently. This algorithm is based on a combination of metrics that takes into account several system parameters such as: the ideal node degree; transmission power; mobility; and the remaining energy of the nodes. Depend-

ing on the specific application, any or all of these parameters can be used as a metric to elect cluster heads. Another important aspect of the algorithm is that it is fully distributed; meaning that all the nodes in the mobile network share the same responsibility acting as cluster heads.

These features are considered in this algorithm:

- The cluster head election procedure is not periodic and is invoked as rarely as possible. This reduces system updates and hence computation and communication costs. The clustering algorithm is not invoked if the relative distances between the nodes and their cluster heads do not change.

- Each cluster head can ideally support only δ (a pre-defined threshold) nodes to ensure efficient medium access control (MAC) functioning. If the cluster head tries to serve more nodes than it is capable of, the system efficiency suffers in the sense that the nodes will incur more delay because they have to wait longer for their turn (as in TDMA) to get their share of the resource. A high system throughput can be achieved by limiting or optimizing the degree of each cluster head.

- The battery power can be efficiently used within certain transmission range, i.e., it will take less power for a node to communicate with other nodes if they are within close distance to each other. A cluster head consumes more battery power than an ordinary node since a cluster head has extra responsibilities to carry out for its members.

- Mobility is an important factor in deciding the cluster heads. In order to avoid frequent cluster head changes, it is desirable to elect a cluster head that does not move very quickly. When the cluster head moves fast, the nodes may be detached from the cluster head and as a result, a reaffiliation occurs. Reaffiliation takes place when one of the ordinary nodes moves out of a cluster and joins another existing cluster.

3.3 Hierarchical Schemes

3.3.1 LEACH

Low Energy Adoptive clustering hierarchical or LEACH (Heinzelman, Chandrakasan, & Balakrishnan, 2002) was first improvement on conventional clustering method in wireless sensor networks. In this algorithm non-cluster head nodes transmit data to cluster head of the same cluster. Then cluster head perform aggregation of all data and transmits to the sink or base station. Therefore, being a cluster-head node is much more energy intensive than being a non-cluster head node. So the selection of the cluster head becomes very significant.

The cluster heads are selected based on the suggested percentage of them for the network and the number of times the node has been a cluster-head so far. This decision is made by each node n choosing a random number between 0 and 1. If the number is less than a threshold T (n), the node becomes a cluster-head for the current round. The threshold is set as follows:

$$T(n) = \begin{cases} \dfrac{p}{1 - p\left(r \bmod \dfrac{1}{p}\right)} & if\, n\, |\in G \\ 0 & otherwise \end{cases}$$

where P is the desired cluster-head probability, r is the number of the current round and G is the set of nodes that have not been cluster-heads in the last 1/P rounds.

Once the nodes have elected themselves to be cluster heads they broadcast an advertisement message (ADV). Each non cluster-head node decides its cluster for this round by choosing the cluster head that requires minimum communication energy, based on the received signal strength of the advertisement from each cluster head belongs; it informs the cluster head by transmitting a join request message (Join-REQ) back to the cluster head. The cluster head node sets up a TDMA schedule and transmits this schedule to all the nodes in its cluster, completing the setup phase, which is then followed by a steady-state operation. This steady state operation is broken into frames, where nodes send their data to the cluster head at most once per frame during their allocated slot.

3.3.2 TL-LEACH

Two Levels hierarchical LEACH or TL-LEACH (Loscri, Marano, & Morabito, 2005) is extension of LEACH in which two levels clustering Primary and secondary are used. In this algorithm, the primary cluster head in each cluster communicates with the secondary, and the corresponding secondary communicate with the nodes in their sub-cluster. Then data communication is performed as in LEACH.

In addition, communication within a cluster is still scheduled using TDMA time-slots. The organization of a round will consist of first selecting the primary and secondary cluster heads using the same mechanism as LEACH, with the a priori probability of being elevated to a primary cluster head less than that of a secondary node. Communication of data from source node to sink is achieved in two steps.

a. Secondary nodes collect data from nodes in their respective clusters. Data transmission can be performed at this level.
b. Primary nodes collect data from their respective secondary clusters. Data aggregation can also be implemented at the primary cluster head level.

3.3.3 EECS

An Energy Efficient Clustering Scheme (or EECS) (Li, Chen, & Wu, 2005) is a clustering algorithm in which cluster head candidates compete for the ability to elevate to cluster head for a given round. This competition involves candidates broadcasting their residual energy to neighbouring candidates. If a given node does not find a node with more residual energy, it becomes a cluster head. Cluster formation is different than that of LEACH.

LEACH forms clusters based on the minimum distance of nodes to their corresponding cluster head. EECS extends this algorithm by dynamic sizing of clusters based on cluster distance from the base station (Li, Chen, & Wu, 2005).The result is an algorithm that addresses the problem that clusters at a greater range from the base station requires more energy for transmission than those that are closer. Ultimately, this improves the distribution of energy throughout of the network, resulting in better resource usage and extended network lifetime.

Hybrid Energy-Efficient Distributed Clustering or HEED (Younis & Fahmy, 2004) is a multi-hop clustering algorithm for wireless sensor networks, with a focus on efficient clustering by proper selection of Cluster heads based on the physical distance between nodes. The main objectives of HEED are:

- Distribute energy consumption to prolong network lifetime;
- Minimize energy during the cluster head selection phase;
- Minimize the control overhead of the network.

The most important aspect of HEED is the method of cluster head selection. Cluster heads are determined based on two important parameters given as below:

1. The residual energy of each node is used to probabilistically choose the initial set Cluster heads. This parameter is commonly used in many other clustering schemes.
2. Intra-Cluster Communication Cost is used by nodes to determine the cluster to join. This is especially useful if a given node falls within the range of more than one cluster head.

In HEED it is important to identify what the range of a node is in terms of its power levels as a given node will have multiple discrete transmission power levels. The power level used by a node for intra-cluster announcements and during clustering is referred to as cluster power level.

Low cluster power levels promote an increase in spatial reuse while high cluster power levels are required for inter cluster communication as they span two or more cluster areas. Therefore, when choosing a cluster, a node will communicate with the cluster head that yields the lowest intra-cluster communication cost. The intra-cluster communication cost is measured using the Average Minimum Reach ability Power (AMRP) measurement.

The AMRP is the average of all minimum power levels required for each node within a cluster range R to communicate effectively with the cluster head i. The AMRP of a node i then becomes a measure of the expected intra-cluster communication energy if this node is elevated to cluster head. Utilizing AMRP as a second parameter in cluster head selection is more efficient then a node selecting the nearest cluster head.

3.4 Grid Schemes

3.4.1 PEGASIS

Power-Efficient Gathering in Sensor Information Systems (or PEGASIS) (S & C-S, 2002) is a data-gathering algorithm that establishes the concept that energy savings can result from nodes not directly forming clusters. The algorithm presents the idea that if nodes form a chain from source to sink, only 1 node in any given transmission time-frame will be transmitting to the base station. Data-fusion occurs at every node in the sensor network allowing for all relevant information to permeate across the network. In addition, the average transmission range required by a node to relay information can be much less than in LEACH (Heinzelman, Chandrakasan, & Balakrishnan, 2002), resulting in an energy improvement versus the hierarchal clustering approach.

3.4.2 GROUP

The Group algorithm (Yu, Wang, Zhang, & Zheng, 2006) is a grid-based clustering algorithm. In this algorithm one of the sinks (called the primary sink), dynamically, and randomly builds the cluster grid. Data transmission in GROUP is dependent on the type of data being collected. In the case of a location unaware data query (data that is not dependant on the location of the sensing node), the query is passed from the central most sink in the network to its nearest cluster-head. That cluster-head will then broadcast the message to neighbouring cluster heads. If the data is location aware, then the requests are sent down the chain of cluster-heads towards the specified region using unicast packets.

4. MULTI-TIER CLUSTERING FOR CLUSTERD WSN

In the last sections, we study about various clustering schemes like LEACH, TL-LEACH, EECS, and HEED etc. There are other types of clustering schemes also, but we are focusing on the hierarchical schemes. We have gone through HEED and LEACH which are one-tiered and by studying this we found that they can be extended to multi-tiered for improvement in the performance of wireless sensor networks.

We know that clustering process reduces energy consumption in radio communication in comparison to other type of energy saving schemes. But it creates various hazards also in spite of significant gain in the network life time by clustering process. In HEED, author's simulation result shows its improvement over LEACH in different scenario. So we are considering HEED for multi-tier approach and improving its energy saving approach. In the case of single-tier architecture of sensor network, there is only one level of cluster heads.

In Figure 5, single-tier clustered network v/s two tier clustered network has been shown. In single –tier clustered network it is clear that there are various cluster heads in the network. Cluster heads are representative of their cluster for sending or receiving information. If we continue this idea for these cluster heads also and make another layer of cluster heads taking these cluster heads as participating nodes in that process, it will be the multi- tier approach for clustering process.

For multi-tier clustered network, we take first level of cluster heads as set of nodes and then forming cluster heads same way as in 1st level.

In Table 2, we have shown the comparison between 1-tier and multi-tier:

4.1 Network Model

We take a rectangular field of M×M. In this rectangular field a set of sensor nodes are deployed. This sensor network has following property:

- Nodes taken in network are location unaware i.e. they are not equipped with GPS.
- Nodes are stationary.
- All nodes have similar capabilities (processing/communication), and equal significance.
- The power level of each node is same at the time of deployment.
- Each node has fixed no. of discrete transmission power levels.

Figure 5. Single tier clustered sensor network v/s two tier clustered sensor network

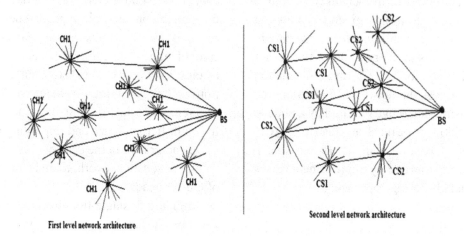

First level network architecture

Second level network architecture

Table 2. Comparison between 1-tier and multi-tier

Single Tier	Multi- Tier
• It's used when the distance between base station and nodes is small.	• It's used when the distance between base station and nodes is very large.
• The number of cluster heads are less	• The numbers of cluster heads are more.
• The data is sending to directly from the cluster head to the base station.	• The data is sent from lower level cluster head to higher level cluster head and then to base station.
• Energy consumption is more.	• Energy consumption is less.
• Implementation is easy.	• Implementation is complex.
• Low reliable.	• High reliable.

- The network serves multiple mobile/stationary observers, which implies that energy consumption is not similar for all sensor nodes.
- The base station is located inside the square area such that communication between base station and sensor node is subject to multi-path fading.
- All nodes sense the environment and transmit message of equal length.
- A subset of sensor nodes is chosen to be level one Cluster Heads and a subset of level one CH (cluster head) is chosen to be level two cluster head.

- The cluster heads receive and fuse data from the non-cluster head or lower level cluster head node, in addition to sensing the environment. They also sense the environment.
- The second level cluster heads transmit their data to base station for two-tier sensor network.
- The properties discussed above are used in the forming of sensor network model for our case. We try to take common properties of sensor network models described in other papers presented for energy saving schemes.
- In our model, we don't make any assumption about the following parameters:
  ◦ Homogeneity of node dispersion in the area where nodes are deployed,
  ◦ The density of the network,
  ◦ Distribution of energy consumption in sensor network

In Figure 6, the working methodology of model has been shown.

Figure 6. The working methodology of model

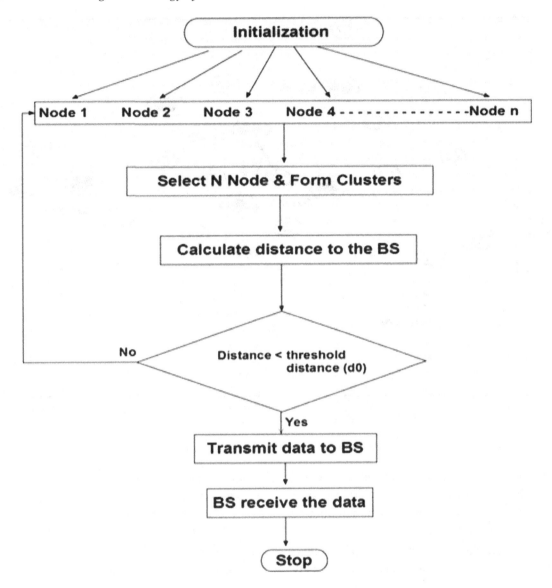

In Figure 7, we have shown the two level architecture of our model where basic features of network are defined (Meena, Singh & Chandel, 2012)

4.2 Clustering Process

We know the energy constraints of sensor nodes. So for maximizing the network lifetime we use clustering approach. The objective of the clustering algorithm is to partition the network into several clusters. The advantages of clustering algorithm are:

- Reducing routing table size,
- Reducing the redundancy of exchanged messages,
- Reducing the energy consumption, and
- Extending the networks lifetime.

Many clustering algorithms for ad hoc and sensor networks were proposed in the last few years. The traditional clustering algorithm (Goldsmith,

Figure 7. Network Model showing dimensions

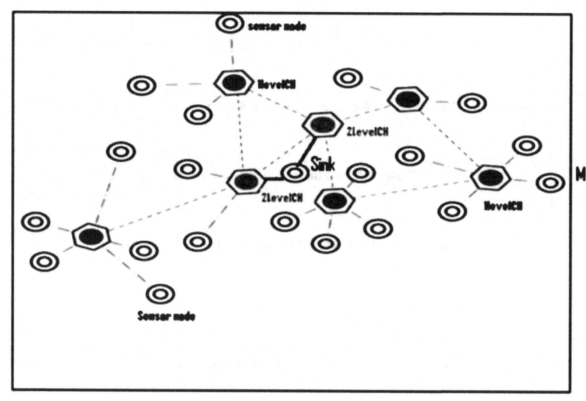

2005) (also called the general clustering algorithm) is as follows: each node sends an election message including node ID or cost to each of its neighbours, and receives the information from its neighbours. Each node checks if there are some CHs around. If the CHs exist, it will join one of the clusters and select the CH as its cluster head otherwise it will elect itself as CH. The time complexity of traditional clustering is O (n) in the worst case and O (log n) on average. The message complexity is O (1) for one node and O (n) for the networks.

In Figure 8, simple cluster network has been shown. In our process, we are going to follow clustering process of HEED with some changes. In case of HEED, clustering process takes consideration different aspects. Assume that N nodes are dispersed in a field and the above assumptions

hold. Our goal is to identify a set of cluster heads which cover the entire field. Each node V_i, where $1 \leq i \leq N$, is then mapped to exactly one cluster C_j, where $1 \leq j \leq N_c$, and N_c is the number of clusters ($N_c \leq N$). The node can directly communicate with its cluster head (via a single hop).

The following requirements must be met: (Meena, Singh & Chandel, 2012)

1. Clustering is completely distributed. Each node independently makes its decisions based on local information.

2. Clustering terminates within a fixed number of iterations (regardless of network diameter).

3. At the end of each clustering process time, each node is either a cluster head, or a non-

Figure 8. Simple cluster network

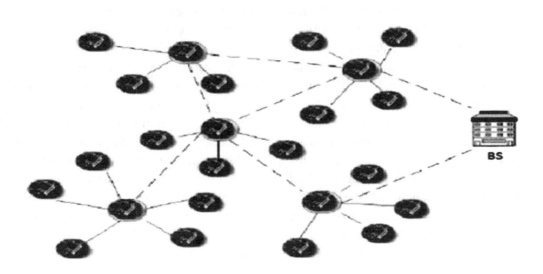

cluster head node (which we refer to as regular node) that belongs to exactly one cluster.

4. Clustering should be efficient in terms of processing complexity and message exchange.
5. Cluster heads are well-distributed over the sensor field.

These are important parameters regarding clustering of sensor network. In our model, we take care of these situations. The overarching goal of our approach is to prolong network lifetime. For this reason, cluster head selection is primarily based on the residual energy of each node. But residual energy measurement is not necessary, since the energy consumed per bit for sensing, processing, and communication is typically known. So we can easily get the information regarding residual energy of any sensor node. To increase energy efficiency and further prolong network lifetime, we also consider intra-cluster "communication cost" as a secondary clustering parameter. We get the maximum no. of possible cluster heads in each network based on the no. of nodes employed at that particular network.

4.3 Two-Tier Architecture

Wireless sensor networks are typically considered to be energy constrained because applications demand that they left unattended or recharging sensor batteries is not feasible. Clustering is a technique which can reduce energy consumption in a network as well as provides scalability. In clustering techniques, a subset of nodes are cluster head nodes which function as processing centres, receive data from a group of nodes and transmit them to the base station. The power spent by a sensor node in receiving the data is constant i.e. $E_{Rx} = lE_{elect}$ while the energy expended by a node in transmitting an l-bit message at distance d less than the threshold distance d_0 is given by $E_{Tx} = lE_{elect} + l\varepsilon_{fs}d^2$ and for distance greater than d_0 it is $E_{Tx} = lE_{elect} + l\varepsilon_{fs}d^4$. When the nodes aggregate data then the energy required fusing (aggregate) l bits is lE_f.

As the most power consuming activity of a sensor node is typically radio transmission. Hence radio communication must be kept to an absolute minimum; this means that the network traffic should be minimized. In order to reduce the amount of traffic in the network, we build clusters of sensor nodes. Some sensor nodes are selected to become cluster heads. Each cluster head aggregates the collected data and then sends it to the base station. This minimizes the total energy consumption by each node in comparison to a scheme in which all the nodes are required to directly send their information to the base station either in one hop or using a point to point route. However such a collection, aggregation and transmission of the information increase the workload of the cluster head in comparison to the members of the cluster i.e. non-cluster head nodes. As a result the energy consumed by a cluster head is much higher than the other cluster nodes. In order to equalize the energy consumption among nodes, it is imperative that the cluster head formation and the cluster heads are changed several times during the lifetime of the sensor network. This allows for the distribution of the workload and the total energy consumption among nodes.

In our proposed two-tier architecture, at the lower level cluster heads are selected which are termed as the first level cluster heads. These cluster heads receive the data from the other regular nodes in the cluster i.e. non cluster head nodes. Instead of transmitting the data to the Base Station these first level cluster heads form a second level cluster among themselves (in the second level clustering only the cluster heads of the first level takes part).At the second level cluster, all the first level cluster heads transmit the data to their respective second level cluster heads. These second level cluster heads will eventually transmit the data to the base station. As the aim of the formation of cluster and our proposed two-tier architecture is to increase the network lifetime of the whole network, hence the prime parameter for the selection of the cluster head is the residual energy of the node. When any node is selected as the cluster head in the clustering process, its energy is consumed more than the energy of the other simple nodes as cluster head has to aggregate the data from the simple nodes as well as it has to

transmit that data to the higher level cluster head or to the Base Station depending on whether it is a first level cluster head or second level cluster head respectively. Due to greater depreciation in the energy of the cluster head than the simple nodes, the possibility of the same node becoming the cluster head in the next clustering process is decreased. Hence it is also ensured by the main parameter (residual energy) for the elongation of network lifetime, that a node will not die much earlier than the other node as there is a very less probability of it becoming a cluster head in the next clustering process after the network operation.

The secondary parameter used in our approach for management of energy, i.e., to increase the energy-efficiency of the sensor nodes and to increase the network lifetime of the network is node degree. Node degree of a sensor node is defined as the number of nodes within the range of the sensor node. The node having greater node degree will serve better as the cluster head as more number of nodes will transmit the data to it instead of transmitting the data to the base station. The secondary parameter is used for the selection of the cluster head when there is a tie for the selection of cluster head. A "tie" means that two or more nodes are equally likely to be selected as the cluster head. A "tie" occurs when two or more nodes have same residual energy (e.g. initially the energy of all nodes is assumed to be equal who means every node has same probability of becoming the cluster head). To break the tie, secondary parameter is used.

5. CONCLUSION

Using two-layer architecture or new protocol, we get efficient gain over latest and well-established protocol in clustered sensor network type. Energy consumption per node and network life time (when first node dies and when last node dies), both show sufficient improvement in comparison to earlier algorithm HEED.

There are various scopes for future work:

- Anyone can use our model for multi hoped network.
- This protocol can be used in network having GPS enabled nodes.
- This protocol can be modified for different field according to need.
- Instead of taking fix nodes, one can do for movable nodes also.
- Our approach can be applied to the design of several types of sensor network protocols that require energy efficiency, scalability, prolonged network lifetime, and load balancing.
- Further work can be done for cross layer optimization approach for energy management by using MAC and DLL to work compatibly with our implemented routing protocol

SUMMARY

In this chapter, we have discussed, how the performance of a wireless sensor network is improved by clustering. In the first section brief introduction about the wireless network and wireless sensor network are discussed. Moreover, Basic Sensor Node Hardware Architecture (see, Figure 1), characteristics of sensor node hardware and examples of modern sensor nodes are also illustrated on the basis of WSN. In section one; various applications of WSN and importance of power management are also discussed.

In addition, in section second we discuss area for energy saving in wireless network. There can be various other ways also. But during our study we find these are the main classifications of the methods for saving energy in wireless sensor networks. After that, we focused on the routing algorithms and then continue with the clustering approach for this. The reason behind it was easily employable techniques for the routing algorithms.

This area not only concern about energy but also robustness of the network.

In third section, by taking view of all those type of clustering algorithms, we decide to go for the hierarchal clustering schemes. We studied various clustering algorithms and mainly focused on the latest and updated schemes for the hierarchical cluster process. We are mainly concern for multi-layer implementation of clustering schemes.

In fourth section, we define our network model, working methodology and the parameters of our model. After defining the model, the clustering process in our model has been discussed in which we have shown the clustering process for the two-tier architecture proposed and the requirements that must be met the clustering process. At the end of the chapter the whole architecture of our two-tier approach is discussed giving information about the consumption of energy in various functions of a sensor node such as in transmission, reception and in aggregation of the data received by the cluster heads from the regular nodes.

In the last section, the conclusion is summarized, which is evidently demanding of specific clustering requirement and new framework architecture for successful deployment of wireless sensor network in the future.

REFERENCES

Abbasia, A. A., & Younis, M. (2007). A survey on clustering algorithms for wireless sensor networks. *Computer Communications*, *30*(14-15), 2826–2841. doi:10.1016/j.comcom.2007.05.024

Agre, J., & Clare, L. (2000). An integrated architecture for cooperative sensing networks. *Proceedings of Computer*, *33*(5), 106–108. doi: doi:10.1109/2.841788

Akyildiz, I. F., Su, W., Sankarasubramaniam, Y., & Cayirci, E. (2002). A survey on sensor networks. *Proceedings of IEEE Communications Magazine*, *40*(8), 102–114. doi:10.1109/MCOM.2002.1024422

Boulis, A. (2011). *Castalia: A simulator for wireless sensor networks and body area networks.* Retrieved from http://castalia.npc.nicta.com.au/pdfs/Castalia%20-%20User%20Manual.pdf

Cerpa, A., Elson, J., Estrin, D., Girod, L., Hamilton, M., & Zhao, J. (2001). Habitat monitoring: Application driver for wireless communications technology. In *Proceedings of ACM SIGCOMM Workshop on Data Communications* (pp. 20-41). ACM Press. doi: 10.1145/844193.844196

Chatterjee, M., Das, S. K., & Turgut, D. (2002). A weighted clustering algorithm for mobile ad hoc networks. *Journal of Cluster Computing, 5*(2), 193–204. doi:10.1023/A:1013941929408

David, A., & Jennings, E. (2005). Self-organized routing for wireless micro sensor networks. *IEEE Transactions on Systems, Man, and Cybernetics, 35*(3), 349–359. doi:10.1109/TSMCA.2005.846382

Dechene, D. J., El Jardali, A., Luccini, M., & Sauer, A. (2005). *A survey of clustering algorithms for wireless sensor network.* Retrieved from http://www.dechene.ca/papers/report_635a.pdf

Estrin, D., Girod, L., Pottie, G., & Srivastava, M. (2001). Instrumenting the world with wireless sensor networks. In *Proceedings of International Conference on Acoustics, Speech, and Signal Processing* (ICASSP 2001) (pp. 2033 – 2036). Salt Lake City, UT: IEEE Press.

Goldsmith, A. (2005). *Overview of wireless communications.* Retrieved from http://www.cambridge.org/

Gou, S., He, T., Mokbel, M., Stankovic, J., & Abdelzaher, T. (2008). On accurate and efficient statistical counting in sensor based surveillance systems. In *Proceedings of the 5th IEEE International Conference on Mobile Ad Hoc and Sensor Systems* (MASS 2008) (pp. 24-35). Atlanta, GA: IEEE Press. doi: 10.1109/MAHSS.2008.4660038

Han, C. C., Rengaswamy, R. K., Shea, R., Kohler, E., & Srivastava, M. (2005). SOS: A dynamic operating system for sensor networks. In *Proceedings of the Third Internaltional Conference on Mobile Systems, Appliactions, and Services* (Mobisys). Retrieved from http://nesl.ee.ucla.edu/fw/documents/conference/2005/sos05SpotsDemo.pdf

Heinzelman, W. R., Chandrakasan, A., & Balakrishnan, H. (2002). An application-specific protocol architecture for wireless microsensor networks. *IEEE Transactions on Wireless Communications, 1*(4), 660–670. doi:10.1109/TWC.2002.804190

Hill, J., Szewczyk, R., Woo, A., Hollar, S., Culler, E. D., & Pister, K. S. J. (2000). System architecture directions for networked sensors. In *Architectural support for programming languages and operating systems* (pp. 93-104). Retrieved from http://webs.cs.berkeley.edu

Hussain, S. A., Farooq, U., Zia, K., & Akhlaq, M. (2004). An extended topology for zone-based location aware dynamic sensor networks. In *Proceedings of National Conference on Emerging Technologies.* Retrieved from http://citeseerx.ist.psu.edu/viewdoc/summary?doi=10.1.1.117.1634

Hwang, I., Pedram, M., & Soltan, M. (2008). Modulation-aware energy balancing in hierarchical wireless sensor networks. In *Proceedings of the 3rd International Symposium on Wireless Pervasive Computing* (pp. 355-359). Los Angeles, CA: IEEE Press.

K, A., & M, Y. (2005). A survey on routing protocols for wireless sensor networks. *Ad Hoc Networks, 3*(3), 325-349. http://dx.doi.org/10.1016/j.adhoc.2003.09.010

Kahn, J. M., Katz, R. H., & Pister, K. (1999). Next century challenges mobile networking for smart dust. In *Proceedings of the 5th Annual ACM/IEEE International Conference on Mobile Computing and Networking* (pp. 271- 278). Washington, DC: ACM Press.

Le, H. A, M., & M, T. (2010). A survey on clustering algorithms for wireless sensor networks. In *Proceedings of the 13th International Conference on Network-Based Information Systems* (NBiS-2010) (pp. 358 – 364). IEEE Press.

Li, C., Chen, G., & Wu, G. (2005). An energy efficient clustering scheme in wireless sensor networks. In *Proceedings of the 24th IEEE International Conference on Performance, Computing and Communications* (IPCCC 2005) (pp. 535-540). IEEE Press. doi: 10.1109/PCCC.2005.1460630

Loscri, V., Marano, S., & Morabito, G. (2005). A two-levels hierarchy for low-energy adaptive clustering hierarchy (TL-LEACH). In *Proceedings of Vehicular Technology Conference* (VTC2005) (pp. 1809-1813). Dallas, TX: IEEE Press.

Meena, Y. K., Singh, A., & Chandel, A. S. (2012). Distributed multi-tier energy-efficient clustering. *Journal of Computer Theory and Engineering*, *4*(1), 1–6. doi:10.7763/IJCTE.2012.V4.418

Nikmard, B., & Taherizadeh, S. (2010). Using mobile agent in clustering method for energy consumption in wireless sensor network. In *Proceedings of the International Conference on Computer and Communication Technology* (ICCCT-2010), (pp. 153-158). Allahabad, India: IEEE Press.

Robinson, L. (1998). Japan's new mobile broadcast company: Multimedia for cars, trains, and handhelds. In *Proceedings of the Advances Imaging*, (pp. 18-22). Academic Press.

S, L., & C-S, R. (2002). PEGASIS: Power-efficient gathering in sensor information networks. In *Proceeding of IEEE Aerospace Conference* (vol. 3, pp. 1125-1130). IEEE Press. doi: 10.1109/AERO.2002.1035242

Swartz, R. A., Jung, D., Lynch, J. P., Wang, Y., Shi, D., & Flynn, M. P. (2005). Flynn: Design of a wireless sensor for scalable distributed in-network computation in a structural health monitoring system. In *Proceedings of the 5th International Workshop on Structural Health Monitoring*. Stanford, CA: Academic Press.

Varga, A. (2001). The OMNeT++ discrete event simulation system. In *Proceedings of the European Simulation Multiconference* (ESM'2001). Prague, Czech Republic: ESM.

Varga, A. (2010). The OMNeT. In *Modeling and tools for network simulation*. Berlin: Springer Verlag. doi:10.1007/978-3-642-12331-3_3

Wolisz, A. (2005). T2: A second generation OS for embedded sensor networks. *TKN Technical Reports*, *35*(3), 349–359.

Wu, M., Xu, J., Tang, X., & Lee, W. C. (2006). Monitoring top-k query in wireless sensor networks. In *Proceedings of the IEEE International Conference on Data Engineering* (ICDE''06) (pp. 962 – 976). Atlanta, GA: IEEE Press.

Xu, Y., Lee, W. C., Xu, J., & Mitchell, G. (2005). PSGR: Priority-based stateless geo-routing in wireless sensor networks. In *Proceedings of the Second IEEE International Conference on Mobile Ad-hoc and Sensor Systems* (MASS'05). Washington, DC: IEEE Press.

Yang, C. L., Tarng, W., Hsieh, K. R., & Chen, M. (2010). A security mechanism for clustered wireless sensor networks based on elliptic curve cryptography. *Proceedings of IEEE SMC – eNewsletter*, (33). Retrieved from http://www.my-smc.org/news/back/2010_12/main_article3.html

Younis, O., & Fahmy, S. (2004). HEED: A hybrid energy efficient, distributed clustering approach for ad-hoc sensor networks. *IEEE Transactions on Mobile Computing*, *3*(4), 366–379. doi:10.1109/TMC.2004.41

Yu, L., Wang, N., Zhang, W., & Zheng, C. (2006). GROUP: A grid-clustering routing protocol for wireless sensor networks. In *Proceedings of the International Conference on Wireless Communications, Networking and Mobile Computing* (WiCOM 2006). IEEE Press. doi: 10.1109/WiCOM.2006.287

Zhang, P., Sadler, C., Lyon, S., & Martonosi, M. (2004). Hardware design experiences in ZebraNet. In *Proceedings of the 2nd International Conference on Embedded Networked Sensor Systems* (pp. 227-238). ACM Press. doi: 10.1145/1031495.1031522

KEY TERMS AND DEFINITIONS

Acknowledgement: An acknowledgement is a signal passed between communicating processes or computers to signify acknowledgement, or receipt of response, as part of a communications protocol. It gives proper information about the signal that data has been received successfully or not.

Base Station (BS): The base station is at the upper level of the hierarchical WSN. It provides the communication link between the sensor network and the end-user. It contains the antennas and other equipment needed to allow wireless communications devices to connect with the network.

Medium Access Control: In the seven layer OSI model of computer networking, Media access control (MAC) data communication protocol is a sublayer of the data link layer. The MAC sub layer provides addressing and channel access control mechanisms that make it possible for several terminals or network nodes to communicate within a multiple access network that incorporates a shared medium, e.g. Ethernet. The hardware that implements the MAC is referred to as a medium access controller.

Quality of Service (QoS): The quality of service refers to several related aspects of telephony and computer networks that allow the transport of traffic with special requirements. The goal of QoS is to provide guarantees on the ability of a network to deliver predictable results. It was defined by International Telecommunication Union.

Sensor Node: Sensor nodes have the ability to gather sensory information, sense, process data and communicating with one another. It is a large number of inexpensive, small, low-power communication devices.

Wireless Sensor Network (WSN): A WSN wireless sensor network typically consists of a large number of communicating devices which are used to monitor the physical and environment aspects such as gas pressure, temperature. sound etc. cooperatively pass their data through the network to a main location. Currently, wireless sensor network are used in many chemical industry for monitoring the gas pressure, on other side for the military applications such as battlefield surveillance field it plays a vital role and so on.

Chapter 15
Transport Protocol Performance for Multi–Hop Transmission in Wireless Sensor Network (WSN)

Farizah Yunus
Universiti Teknologi Malaysia, Malaysia

S. K. Syed-Yusof
Universiti Teknologi Malaysia, Malaysia

Sharifah H. S. Ariffin
Universiti Teknologi Malaysia, Malaysia

Nor-Syahidatul N. Ismail
Universiti Teknologi Malaysia, Malaysia

Norsheila Fisal
Universiti Teknologi Malaysia, Malaysia

ABSTRACT

The need for reliable data delivery at the transport layer for video transmission over IEEE 802.15.4 Wireless Sensor Networks (WSNs) has attracted great attention from the research community due to the applicability of multimedia transmission for many applications. The IEEE 802.15.4 standard is designed to transmit data within a network at a low rate and a short distance. However, the characteristics of WSNs such as dense deployment, limited processing ability, memory, and power supply provide unique challenges to transport protocol designers. Additionally, multimedia applications add further challenges such as requiring large bandwidth, large memory, and high data rate. This chapter discusses the challenges and evaluates the feasibility of transmitting data over an IEEE 802.15.4 network for different transport protocols. The analysis result highlights the comparison of standard transport protocols, namely User Datagram Protocol (UDP), Transport Control Protocol (TCP), and Stream Control Transmission Protocol (SCTP). The performance metrics are analyzed in terms of the packet delivery ratio, energy consumption, and end-to-end delay. Based on the study and analysis that has been done, the standard transport protocol can be modified and improved for multimedia data transmission in WSN. As a conclusion, SCTP shows significant improvement up to 18.635% and 40.19% for delivery ratio compared to TCP and UDP, respectively.

DOI: 10.4018/978-1-4666-5170-8.ch015

INTRODUCTION

In the last few years, *Wireless Sensor Networks (WSNs)* have raised a lot of attention of the research community such as the industry and the academic communities. *WSNs are multi-hop* networks that deploy hundreds or thousands of small sensor nodes with limited capabilities in term of power resources, processing and memory. Due to these limitations, *IEEE 802.15.4* standard represents a milestone in wireless sensor network. *IEEE 802.15.4* is a data communication protocol standard uniquely designed for low rate wireless personal area network (WPAN) (Adams, 2006). However, the issues of *multi-hop* communication and energy efficiency in WSN have imposed many challenges and need further specific research effort.

Typically, sensor nodes in *WSNs* have severe constraints in energy resources, processing ability and memory resources. The reason of these constraints is because of the effort to reduce the cost of the sensor node (Ai & Abouzeid, 2006). Furthermore, one of the characteristics of *WSNs* is the dense deployment, which is to increase the sensing coverage, connectivity and network lifetime (Costa & Guedes, 2010). But this characteristic limits the range of wireless communication because of energy constraint and thus, the transmission and processing function also needs to be controlled. Then, the *multi-hop* communication model is required to solve the problem of limited communication ranges by transmitting the data to the sink node using intermediate transmission nodes. Moreover, the WSNs communication protocol cannot fully apply buffering techniques due to the constraints on memory and processing. All of these constraints are crucial and need to be considered during designing a new protocol especially for multimedia applications, which involves the transmission of high traffic volume such as video.

There is an additional challenge imposed on multimedia transmission over WSNs, which is requiring a large bandwidth. This is due to the fact that *IEEE 802.15.4* standard is concerned with low data rate applications. The maximum allowable data rate in the *IEEE 802.15.4* standard is 250 Kbps with the *Maximum Transmission Unit* (MTU) for data transmission up to 128 bytes (Garcia-Sanchez, Garcia-Sanchez, & Garcia-Haro, 2008) compared to the IEEE802.11 standard that offers data rates of several Mbps with *Maximum Transmission Unit* (MTU) up to 1500 bytes. For that reason, video transmission over IEEE 802.15.4 standard is more challenging, and should be given special attention due to its nature. The goal is hence to achieve video transfer over low complexity and cost networks at a satisfactory level of quality in terms of Peak Signal to Noise Ratio (PSNR) at the receiver side.

In **WSNs**, a communication protocol stack provides similar functionalities to those of the conventional computer network protocol stack. However, the earlier is designed to operate in a more dynamic environment (Zahariadis & Voliotis, 2007). This communication protocol stack consists of 5 layers which are application, transport, network, Medium Access Control (MAC) and physical layers as shown in Figure 1.

Nevertheless, this chapter is mainly concerned with the *transport protocol* provided by the transport layer and responsible to ensure end-to-end reliable data transfer from source to destination. The WSNs *transport protocol* is mainly designed to support congestion control, guaranteed end-to-end reliability or both services either for scalar data or multimedia applications. The reliable data delivery mechanism for multimedia applications will ensure reliability for packet level or application level in order to achieve significant energy consumption reduction with low end-to-end delay in congested scenario (Bonivento & Fischione, 2007). On the other hand, the congestion control mechanism will act on congested nodes within the network to avoid or alleviate congestion in case it occurs.

Figure 1. Different research challenges relevant to layers of the wsn protocol stack

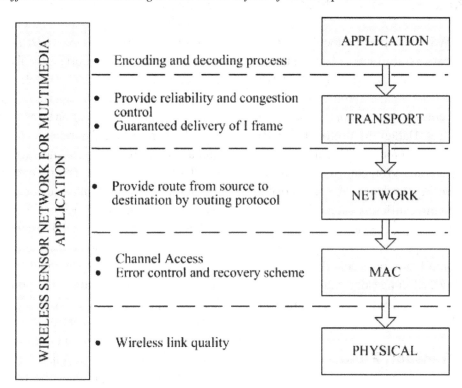

However, the *transport protocol* design for multimedia applications is a more challenging task due to strong limitations imposed by channel capacity and processing capability. The research community is very active in finding an efficient way to provide reliable delivery of video content on multi-hop WSNs with acceptable Quality of Service (QoS), which is defined in terms of several parameters such as PSNR and end-to-end delay.

Since the most important part of this work is based on simulations, the main objective of this work is to implement and evaluate multi-hop data transmission in WSNs with standard transport protocols which are UDP, TCP and SCTP. This evaluation is crucial to set a benchmark before undergoing any appropriate modification to those protocols in order to ensure the protocol can be successfully adapted over WSNs with multimedia applications.

BACKGROUND

Regarding to the application requirements of multimedia data, standard transport protocols cannot be directly implemented over WSNs. However, standard transport protocols can be improved with appropriate modification and take into consideration the characteristics of multimedia data for better performance in the network. Therefore, the performance of standard transport protocols should be analyzed to provide a possible consideration for necessary modifications and improvement for multimedia application over WSN environment. The modifications and improvement should not only consider reliability and congestion control, but also should consider achieving lower delay and energy consumption.

The Existing Transport Protocols

The existing WSN transport protocols for multimedia applications are depicted in Figure 2. The transport protocols can be divided into two categories which are standard protocols and non-standard protocols. *Standard transport protocol* consists of User Datagram Protocol (UDP), Transport Control Protocol (TCP) and Stream Control Transmission Protocol (SCTP). On the other hand, *non-standard transport protocol* can be divided into three different categories, which are protocols that support reliability for upstream and downstream, protocols that support only congestion control, and protocols that supports both reliability and congestion control. All of these protocols are proposed for WSN to transmit multimedia data.

Standard Transport Protocols

The related works are concentrating more on the three standard transport protocols. As a result, this chapter has focused on evaluation of the performance of those three transport protocols in WSNs. It is crucial to understand more and estimate the performance of these protocols before doing any appropriate and proper modifications.

User Datagram Protocol (UDP) is a minimal message-oriented transport protocol that is documented in IETF RFC 768 (Postel, 1980). *UDP* provides a very simple interface between the network layer and application layer, which results in making the protocol unreliable. *UDP* does not guarantee the reliability of message delivery to the desired destination. The packets may arrive out of order, appear duplicated at receiver or lost without any notification message. Small overhead makes the performance of UDP faster and efficient for applications that do not need any guaranteed data delivery. *UDP* also provides timeliness for real-time streaming multimedia applications, but it does not provide any reliable data delivery, flow control and congestion control.

However, *UDP* is not suitable for multimedia applications that require reliable data delivery because there is a significant effect of potentially dropped packets if they contain important data content like the Region of Interest (ROI) features used in JPEG2000 or the I-frame used in the MPEG standard video coding (Akyildiz, Melodia, & Chowdhury, 2007). Besides, UDP is not suitable to support heterogeneous traffic that consists of

Figure 2. WSN transport protocol classification for multimedia applications

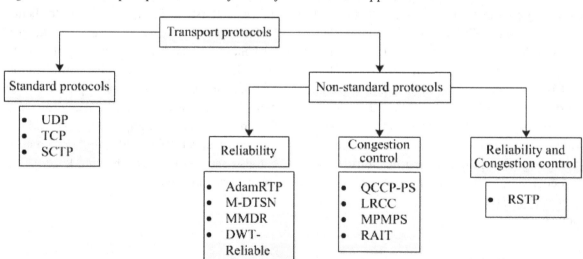

multimedia and scalar data. This is due to the fact that UDP header does not provide any description for many traffic classes that may influence the congestion control mechanism.

On the other hand, *Transmission Control Protocol (TCP)* is a connection-oriented transport protocol (Postel, 1981) that supports reliable and ordered data delivery from source to receiver. *TCP* provides three-way handshake mechanism for connection establishment to assure the receiver is ready to receive data from source. Even though TCP provides congestion control mechanism, but there is an effect of jitter introduced by the congestion control mechanism (Akyildiz et al., 2007) where the fast retransmission will react after 4 duplicates of packet are lost. Besides that, TCP also introduces the problem of Head of Line (HOL) blocking where the next packet transmission needs to wait if there is a packet lost and need to be retransmitted. As a result, it will increase the delay and at the same time will decrease the performance of the protocol.

Consequently, the standard protocols of UDP and TCP are not suitable for multimedia transmission over WSN because of several reasons. UDP does not provide any reliable congestion control mechanism. As a result, important frames in the video stream can be easily lost when there is an error during transmission because there is no mechanism for loss recovery. Even though TCP provides a reliable congestion control mechanism, but *TCP* only supports single flow or stream and does not provide priority protection mechanism for the important data. Thus, the transmission delay will increase and will lead to failure when the deadline is expired even if the data is successfully received.

However, there is another protocol that improved some drawbacks of TCP and UDP and also supports the functionalities of multi-streaming and partial reliability which is *Stream Control Transmission Protocol (SCTP)* (Stewart, 2007). Multi-stream can be defined as a multi-flow within a single association or connection between sender and receiver, and each stream is independent. Multi-stream can be a solution of the Head of Line (HOL) blocking introduced by TCP because of single stream data transmission. The effects of HOL blocking will increase the end-to-end delay and energy consumption in the network. Partial reliability is an extension of SCTP (PR-SCTP) (Stewart, Ramalho, Xie, Tuexen, & Conrad, 2004) that has a function of setting a reliability level according to specific messages. Partial reliability can differentiate its retransmission service for each message and allows packets to assign the priority. As results, it is expected to avoid undesired energy consumption for lower priority messages.

In summary, UDP is suitable and perform better for applications that considered time constraints rather than reliable data delivery. While TCP is suitable when the time is not constrained and reliable data delivery is important. But, in multimedia transmission over wireless network, both of time constraint and reliable data delivery are important to ensure high video quality reception at the receiver side. Thus, the most suitable standard transport protocol that can be applied over WSNs for multimedia applications is SCTP. This is due to both features of multi-stream and partial reliability, which are more appropriate to be adapted in multimedia data transmission which consists of three different types of frames and different levels of priority where I-frame needs more preservation compared to P and B-frames.

Non-Standard Transport Protocols

Mingorance-puga, Maciá-fernández, Grilo, and Tiglao (2010) proposed Multimedia Distributed Transport for Sensor Network (M-DTSN) for efficient multimedia transmission in WSNs. M-DTSN will estimate the sending of a certain amount of information and also the channel condition to decide either to complete the transmission or not. This protocol considers the factors of multimedia flow and sensor node to achieve efficiency of the transmission. The factor of multimedia flows are considered because of time constraint, while sensor node is considered due to the limitation on the amount of energy available for data transmission. However, this protocol does not provide any mechanism to mitigate or alleviate congestions, which is important to avoid energy waste due to an increasing number of retransmissions and packet loss.

The transport protocol of Adaptive Multi-flows Real-time (AdamRTP) Multimedia Delivery over WSNs is proposed in Zakaria and El-Marakby (2009). AdamRTP looks similar to MRTP (Mao & Bushmitch, 2006). However, MRTP is designed for ad-hoc networks that has session-oriented protocol, while AdamRTP is designed for WSNs. AdamRTP splits the multimedia source stream into smaller independent flows using MDC encoder and send data for each flow to the destination using joint/disjoint paths. This protocol is not session-oriented because the sender simply sends multimedia data to the sink as soon as it detects an event. AdamRTP protocol provides the adaptation mechanisms for changing number of flows and rate adaptation to enhance quality of service (QoS) and to extend the lifetime of WSN. The drawback of AdamRTP is that it does not have any mechanism for packet retransmission.

Qaisar and Radha (2009) proposed Multipath Multi-stream Distributed Reliability (MMDR) in WSN. MMDR protocol has three strategies to enhance the received video quality. First, splitting source-coded video into prioritized stream depending on its level of resolution of the original content. Second is providing reliable video delivery in WSN using multipath routes. Third is decoding the data packets partially and progressively. The idea is to use partial decoding at intermediate node by employing progressive error recovery algorithms. This protocol also uses low density parity check (LDPC) codes for channel coding of video sensor data. Although MMDR provides reliability scheme in WSN for multimedia data, but it does not provide any congestion control mechanism to mitigate any congestion in the network.

Discrete Wavelet Transform (DWT) is a reliable technique defined in Lecuire, Duran-Faundez, and Krommenacker (2008) to create a packet with different levels of priority and differentiated reliable data transmission over WSNs. Packets at intermediate nodes may be forwarded or dropped according to their level of priority and the remaining energy at these node. Packets that have high priority are ensured to be forward by every node in order to complete the decoding process at the sink node. Meanwhile, packets that have lower priority are forwarded based on available energy in the nodes and threshold defined by the application. However, the dropped packets are not retransmitted because of it is low importance. This protocol also does not provide congestion control to alleviate congestion when there is congestion occurring either at a node or in the network.

A congestion control mechanism is proposed in Yaghmaee and Adjeroh (2008) and the protocol is introduced as Queue based Congestion Control Protocol with Priority Support (QCCP-PS). The idea of QCCP-PS is to treat the priority index especially for random packet service. QCCP-PS assigns a transmission rate to each node depending on two criteria which are its priority and the current congestion degree of the intermediate node. The indication of congestion degree is based on the queue length, where the level of congestion is presented by a congestion index which is a number is between 0 and 1. The congestion detection is done based on two thresholds which are max_{th} and min_{th}. After detecting the congestion, the child nodes are notified by an implicit mechanism based on piggybacking congestion information on the normal data packet. The advantage of the proposed congestion protocol in Yaghmaee and Adjeroh (2008) is a fair rate adjustment for a heterogeneous network due to adaptation of priority for each node. However, QCCP-PS has a drawback which is introducing poor performance when all nodes have the same level of priority.

Another protocol that provides congestion control for multimedia application is presented in Maimour, Pham, and Amelot (2008), which is Load Repartition Congestion Control (LRCC). LRCC implements multipath routing to mitigate the congestion based on load repartition using multiple paths without reducing the transmission rate of the source. The LRCC also uses queue length as an indicator for congestion detection and uses special control messages for explicit congestion notification. After receiving the notification message, the current congested path will decrease the amount of data sent and the source node will balance the amount of traffic on the available path. This results in achieving faster and fairer load balancing after congestion. The LRCC achieves the objective to mitigate the congestion but does not include a reliable mechanism for packet retransmission.

Multi-priority Multi-path Selection (MPMPS) is proposed in Zhang, Hauswirth, and Shu (2008) where the maximum number of available node disjoint paths is chosen for maximizing the throughput of the multimedia data transmission. Additionally, the MPMPS use two kinds of information which is video and audio and assigns different levels of priority to data. Data that have higher priority use the best path (lower end-to-end delay) and the remaining paths are used by the data have lower priority. However, this congestion control protocol also has some drawbacks, such as the assumption that the transmission capacity of all nodes will be the same, which is not appropriate. This is due to the fact that transmission capacity of the nodes is higher for the route that has a frequent transmission.

The congestion control in Lee and Jung (2010) defines the Reliable Asynchronous Image Transfer (RAIT) protocol. RAIT proposes double sliding window method to avoid packet dropping because of congestion. For that reason, each intermediate node implements a receiving and sending queue. The token-bucket technique is used to control the packet flow due to the cooperation work between queuing layer and the network layer introduced by RAIT. Furthermore, the transmitting node has to know the current status of the next hop to avoid packet discarding because of congestion. The node that has a packet to transmit but does not have a token must send requests to the next hop. The disadvantage of this protocol is the application of token-based technique that can increase end-to-end delay, and thus decrease the performance of real-time sensing applications.

In literature, only one transport protocol that provides both functions of reliability and congestion control mechanism is called Reliable Synchronous Transport Protocol (RSTP) (Boukerche & Du, 2008). The RSTP uses synchronization for image transmission from multiple sources in order to reduce transmission delay and uses TCP-ELN (Buchholcz, Ziegler, & Do, 2005) for congestion control. For connection establishment, RSTP employs three-way handshake which is a similar technique to the standard protocol of TCP (Postel, 1981), but it is initiated by the receiver. The loss recovery in RSTP is handled by the receiver. The receiver checks the sequence number of the received packet and sends a retransmission request when there is any packet loss or when a timeout occurs without receiving the last packet. The advantage of handshake has provided a proper initiation between source and receiver in order to properly initiate the sequence number. However, it will result in an undesired initial delay, and thus increases the overall end-to-end delay for real-time multimedia applications.

Even though all of the above non-standard transport protocol support transmission of multimedia data within WSNs, but not all of the works use IEEE 802.15.4 as a medium. This is due to the fact that *IEEE 802.15.4* confronts with other characteristics that give more challenging task when designing transport protocol. Since the consideration of WSN characteristic is important in transport protocol design, the selection of video encoder is also crucial to assure its suitability to adapt with the challenging environment. Thus, the existing transport protocols still have some drawbacks that can be improved to achieve better performance transport protocols for multimedia data transmission over WSNs.

MAIN FOCUS OF THE CHAPTER

Issues of Transport Protocol for Multimedia Applications

When designing a transport protocol either for scalar data or multimedia data application, some design issues should be properly considered. This is due to the *WSNs* is an application based where the protocol design must be provided the requirement of the application needed. As discussed before, the transport protocol for multimedia application is more challenging because of additional characteristics of multimedia that required high bandwidth due to high traffic volume of the video frame and need special requirement.

Application Requirement

There are eight different traffic classes of multimedia and data transmission requirements that can be determined (Costa, 2012; Akyildiz, 2007) as presented in Table 1. Even though some of the traffic classes have limited practical applications, but they represent a reasonable indication of the acceptable level of packet loss and congestion for each type of data communication (Costa, 2012).

There are several factors and features that may influence the transport protocol design over WSNs for multimedia application. This section discusses the main or common factors and features that give a great influence in designing a protocol for multimedia.

- **Application-specific Quality of Service (QoS):** As mentioned before, WSNs are application based, where different applications need different QoS requirements. For example, Table 1 shows that different types of traffic classes are coupled by different types of data, bandwidth and application.

Table 1. Traffic classes for wsn application

Type of Class	Type of Data	Bandwidth	Description and Example of Application
Real-time, Loss intolerant	Multimedia stream	High	Monitoring, tracking and surveillance applications that require high quality multimedia data and need to reach the receiver in real time.
Real-time, Loss tolerant	Multimedia stream	High	Multimedia data transmitted in real-time to a human or device, tolerate with low or moderate loss.
Delay tolerant, Loss intolerant	Multimedia stream	High	Multimedia streaming intended for storage and offline processing with no loss.
Delay tolerant, Loss tolerant	Multimedia stream	High	Multimedia streaming intended for storage and offline processing with moderate losses.
Real-time, Loss intolerant	Data	Moderate	Critical control applications retrieving multimedia data.
Real-time, Loss tolerant	Data	Moderate	Monitoring data from densely deployed scalar data or snapshot multimedia data (e.g: images) that have to be received on time.
Delay tolerant, Loss intolerant	Data	Moderate	Data from monitoring process that requires offline post processing with loss constraint.
Delay tolerant, Loss tolerant	Data	Low	Snapshot application associated with scalar data with moderate losses.

Streaming multimedia data are generated over a prolonged period of time and require continuous information delivery. For that factor, they may tolerate a different level of loss and delay in transmission. Therefore, different applications have different requirements in terms of reliability level, congestion control, energy consumption, delay, data rate and fairness.

- **High bandwidth demand and power consumption:** As discussed before, the multimedia data require higher transmission bandwidth due the large size of data, while the maximum allowable data rate for IEEE 802.15.4 standard is 250Kbps. Hence, an efficient technique of compression is crucial to reduce the data rate of multimedia data in term of size of files in order to assure it is comparable with power consumption required in sensor nodes. Thus, the balance between bandwidth requirements and power consumption need to be considered when designing any protocol for multimedia application.

- **Resource constraints:** The sensor nodes in the network are usually resources constrained in term of energy consumption, the bandwidth requirement and memory capability. At the same time, sensor nodes are required to provide efficient reliability and congestion control mechanism, all at a low cost. Therefore, there is a tradeoff between energy and reliability where higher reliability consumes more energy (Quang & Won-Joo, 2006). However, providing an efficient congestion control mechanism can reduce the energy consumption.

MPEG-4 Encoder and Frame Format

An efficient encoding technique for multimedia data is important because transmission of uncompressed raw video frame requires high data rate and thus consumes more bandwidth and energy (Akyildiz et al., 2007). High ratio of compression for encoding technique is necessary to achieve a low bandwidth requirement and low energy consumption to prolong the lifetime of sensor nodes. Therefore, the *Motion Picture Experts Groups (MPEG-4)* video codec was chosen among the available video codecs because of its low bit rate feature capable of achieving as low as 64 Kbit/s (Ebrahimi & Horne, 2000) and improved coding efficiency compared to other related standards. The *MPEG-4* is one of the standards for the coding of moving pictures and audio for video streaming. Additionally, *MPEG-4* is able to encode mixed media data that consist of video, audio and speech.

Figure 3 shows the *MPEG-4* encoding structure and components that consists of slices, macroblocks and blocks. Each video picture or frame consists of several numbers of slices and every slice consists of macroblocks. A macroblock contains a section of the luminance component, and the spatially sub-sampled chrominance components that carry the shape, motion and texture information (Smart & Keepence, 2008). Sampling format for a macroblock is 4:2:0 where each macroblock contains 4 luminance blocks and 2 chrominance blocks. Meanwhile, each block contains 8×8 pixels and is encoded using Discrete Cosine Transform (DCT) coefficients (Moradi & Wong, 2007).

Then, the transform coefficients are quantized using an 8×8 quantization matrix. Quantization is the parameter responsible for the "lossiness" in the MPEG-4 encoding scheme. It basically determines the output of the DCT in video compression. By having a lower value of quantization scale, the compression ratio would be low and the video quality would remain close to the original video. However, a low value of quantization scale would result in a video size close to the original video. Increasing the quantization scale on the other hand would decrease the compressed video size as well as degrading the video quality. A highly compressed video would produce an artifact because of the missing information during the encoding process. The tradeoff between video quality and

Figure 3. MPEG-4 encoding structure and component (slices, macroblock and block)

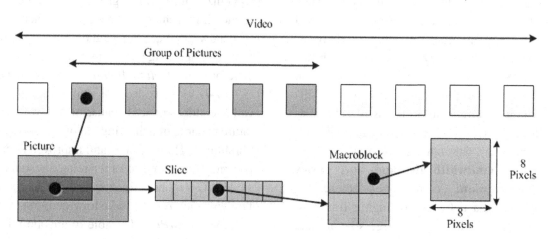

compressed video size must be balanced to achieve an acceptable video quality with an acceptable size for video transmission over WSN.

Video frame formats are used to determine the individual video frame in terms of pixels. The Common Intermediate Format (CIF) video frame format has a resolution of 352×288 pixels, and Quarter CIF (QCIF) video frame format has a resolution of 176×144 pixels. The pixels will represent different color spaces. The common color space is red, green and blue (RGB) representation.

The format used for both CIF and QCIF are in the YUV pixel format. The Y component is defined as a luminance (brightness level) while U and V components are defined as hue and intensity. Since human eyes system is more sensitive to luminance component, chrominance component can be a sub-sampled without reducing the video quality. The ratio of luminance to chrominance bytes is reduced with a sub-sampling process. The available YUV formats are YUV 4:4:4 (without sub-sampling), YUV 4:2:2, YUV 4:2:0 and YUV 4:1:1. However, YUV 4:2:0 is being used for video conferencing, digital television and modern video coding standards more extensively. Due to this popularity, most of the literature focuses on YUV 4:2:0 (Huszák & Imre, 2010).

The *MPEG-4* video can be divided into several numbers of video sequences that are called group of picture (GOP) (Huszák & Imre, 2010) as shown in Figure 4. The arrow indicates that frames depend on each other with forward and backward prediction. MPEG-4 video generates three different types of frames that consist of Intra-coded frame (I-frame), predicted frame (P-frame) and bidirectional frame (B-frame). I-frames are encoded independently and its frame size is larger compared to other types of frames. This is because I-frame contains most of the important video information, and the GOP is useless if I-frames are lost. P-frames are encoded using predictions from the previous I or P-frames. B-frames are encoded using predictions from previous and next I and P-frames. The choice of GOP structure is important because it will give effect on the frame size and file size. Additionally, it also will give impact to the MPEG video streaming in terms of network bit rate and video quality .

The frame rate is important because it shows the smoothness of the image transitions in the video and it is used to determine the quality of a video. The common values of frame rate for MPEG-4 video are between 25 to 30 fps. Frame rate also directly determines the size of the compressed video and thus is important to make sure

Figure 4. The prediction dependencies between frames in MPEG4 for group of picture (GOP)

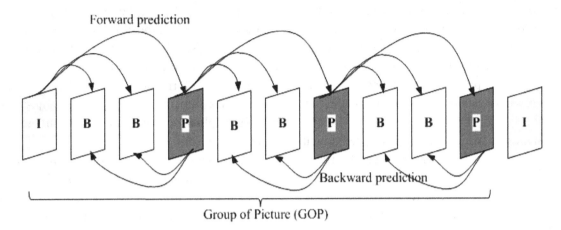

the data rate required meets the WSN specification.

The information of *MPEG-4* encoder is very important in order to understand how the video frames are encoded during the encoding process, about the frame format, intra-frame, inter-frame and their frame dependencies in GOP. All of this information is required to ensure that the requirement of this video coding is suitable to use over WSN. Reliability and congestion control mechanism provided by transport protocol is important in order to give more preservation to the important video frame which is I-frame. This is due to the loss of I-frame will result in an invaluable video at the receiver even if the P-frame is received successfully.

Reliability and Congestion Control

The limitations of sensor nodes and the characteristics of multimedia data require a new paradigm to be found in order to transmit high traffic volume over WSN. When designing a transport protocol for multimedia applications, the concept of end-to-end communication provided by transport layer should be emphasized in order to achieve higher efficiency of multimedia data transmission (Costa & Guedes, 2011). However, the main aims of

transport protocol is to transmit multimedia data over WSN in a promising way to achieve high and acceptable reliability data with energy efficiency in order to extend the lifetime of the network. It is also required to provide congestion control mechanism to avoid any congestion in order to ensure the good quality of multimedia data. Depending on the application, both reliability and congestion control mechanisms are important functionalities for the transport protocol to achieve high data delivery from source to the sink node.

Reliability is one of the functions or services provided by the transport protocol and its main goal is to ensure reliable data delivery by retransmitting lost packets. Generally, the packet is lost or dropped when there is congestion in the network that occurs during data transmission over wireless link. Besides that, the other reason for packet loss is when intermediate nodes fail due energy outage.

Reliability module operation can be divided into three steps which are loss detection, loss notification and loss recovery (Almalkawi, Zapata, Al-Karaki, & Morillo-Pozo, 2010). The loss detection component is performed by the sender or the receiver. At the sender part, the timer or overhearing technique will be used to check the packet dropping, while at the receiver, the sequence number in the received packet will be employed to discovery packet loss. After detecting packet loss, the component of loss notification will notify the source node or proper intermediate node about the packet loss. There are two ways of loss notification either using explicit or implicit methods. For explicit method, acknowledgment message will be sent to source to confirm the correct reception of transmitting packet or directly request the retransmission. While for the implicit method, node just overhear the successful transmission from the next hop. Then, the packet loss will be recovered by data recovery component that consists of two ways. One of the ways is by using a retransmission scheme either hop-by-hop approach or end-to-end approach. The second way is reconstructed corrupted packet using recovery methods such as Forward Error Correction (FEC) by transmitting redundancy packets. This method can be performed by intermediate nodes or at the sink node (Costa, 2012).

Mitigating the congestion in the network is crucial to avoid the degradation of link utilization that may increase the delay and energy consumption. Energy consumption and end-to-end delay will be increasing the congestion frequently occurs. Congestion may occur at intermediate node or sink node when there is the high transmission rate of multimedia data and result in packet dropping. This scenario happens when the rate of packet arrival exceeds the rate of packet service and will cause a buffer overflow.

Typically, the *congestion control* module consists of congestion detection process, congestion notification and congestion mitigation. Congestion can be detected using two ways, which are active method that uses a timer or require acknowledgements, and proactive method that observes network indicators such as queue length, packet service time and packet arrival time. After detecting congestion, the transport protocol needs to propagate congestion information from congested node to upstream sensor node. The approaches to notify the congestion can be categorized into two approaches, which are explicit congestion

notification and implicit congestion notification (Costa, 2012). The explicit congestion notification uses special control message to notify the involved sensor nodes of congestion. In contrast, the implicit congestion notification piggybacks congestion information in normal data packet. For the last step, the transmission rate adjustment will be applied for congestion mitigation to reduce current transmission rate.

In summary, the new transport protocol design should simultaneously address the unique challenges by the WSNs and multimedia data requirement with a suitable performance metric for both reliable data delivery and congestion control mechanisms. Thus, the modification of *standard transport protocol* should not only support an efficient reliability and congestion control mechanism, but also should consider low end-to-end delay and less energy consumption. Low delay is important to ensure the multimedia data received at the receiver will not be discarded because of timeout. In addition, energy efficiency is crucial for the transport protocol to extend the lifetime of the network.

Solution and Recommendations

The main focus of this chapter is the implementation and the evaluation of multi-hop data transmission in WSN for standard transport protocols which are UDP, TCP and SCTP. This evaluation is essential as a benchmark before undergoing any appropriate modification to ensure the standard transport protocol can be successfully adapted over WSNs with multimedia applications.

As mentioned before, there are three existing standard transport protocols which are UDP, TCP and SCTP. However, these transport protocols are usually used to transmit data over an IP-based network, i.e. Internet. *UDP* is an unreliable transport protocol that does not provide any reliability or congestion control mechanism. Conversely, *TCP* provides an end-to-end reliable connection and congestion control mechanism, while *SCTP* is a new transport protocol that offers higher reliability and performance than TCP. Moreover, *SCTP* provides multi-stream service for single connections. Even though all of these protocols are designed for wired networks, but they also can be applied to wireless network with appropriate modifications.

Even though there are several non-standard transport protocols that have been designed for multimedia applications as discussed in the previous section, but they also have some drawbacks that need to be improved. Therefore, we believe that the standard protocol with appropriate and proper modifications is preferable over non-standard transport protocol for multimedia applications. For example, the Sensor Transmission Control Protocol (Iyer, Gandham, & Venkatesan, 2005) has already implemented some feature from the SCTP transport protocol for scalar data and the performance of the protocol is efficient.

The TCP and SCTP transport protocols are suitable for efficient data transmission over an end-to-end reliable Internet. However, they need to be improved with some modification in order to achieve better performance over WSNs. This is due to the instability of WSNs in highly error-prone environments because of various reasons such as interference of radio signal, radio channel contention and survival rate of nodes (Vedantham, Sivakumar, & Park, 2007). Consequently, directly applying traditional TCP and SCTP protocol over WSNs will potentially suffer from severe performance degradation when the error rate increased significantly for *multi-hop* networks due to channel contention.

Thus, evaluation of those three standard transport protocols over WSNs will provide some opportunity to better understand and estimate the performance of these protocols. Based on this study, it will provide ideas for necessary modifications and improvement to ensure the chosen protocol may adapt to the requirement of WSNs. Besides that, the evaluation will help to understand the behavior of protocols under congested scenarios with high network density, which may lead to

improve the congestion control algorithm. Since the energy consumption is crucial for transmission in WSNs, the congestion control mechanism is important to help reduce energy consumed during video frame transmission. Reducing the number of retransmissions also can contribute to reducing consumed energy in the network.

SIMULATION RESULTS

Results for all three protocols will be evaluated in terms of the packet delivery ratio, energy consumption and end-to-end delay. The packet delivery ratio is the ratio of packets received at the sink node to the total number of packets sent from source node in a network, while energy consumption is the energy consumed in each sensor node during the simulation task. End-to-end delay is the total delay for data transmission over the sensor field from source to destination including time for propagation, delay for buffering, delay for processing and delay for retransmission of lost packets. The end-to-end delay can be calculated as shown in Equation (1).

$$EED = \sum_{n=1}^{N}\left(r_n - s_n\right) \qquad (1)$$

where,

r_n = the time that data packet *n* was sent;
s_n = the time that data packet *n* was received;
N = total number of data packet received;

Network Simulator-2 (NS-2) was used to simulate the three standard transport protocols over WSN using Real-Time Load Distribution (RTLD) (Ahmed & Fisal, 2011) routing protocol. The simulation parameters for the performance of standard transport protocol is shown in Table 2. The traffic load used in this simulation is high traffic which is 10 packets per second.

Table 2. Simulation parameters

Parameter	IEEE 802.15.4
Propagation model	Shadowing
Path loss exponent	2.45
Physical type	Phy/WirelessPhy/802.15.4
MAC type	Mac/802.15.4
CS threshold	1.10765e-11
RX threshold	1.10765e-11
Frequency	2.4Ghz
Initial energy	3.6 Joule
Power transmission	1mW
Simulation time	100s
Traffic	CBR and FTP
Transport protocol	UDP, TCP and SCTP
Number of nodes	4, 9, 16, 25, 36, 49

The delivery ratio decreases when the number of nodes is increased from the low density network (4 nodes) to high density network (49 nodes) as shown in Figure 5. The SCTP transport protocol achieves a high delivery ratio which is up to 18.635% and 40.19% compared to UDP and TCP

transport protocol respectively. This is due to SCTP has a feature of multi-streaming that avoids the Head of Line (HOL) blocking where each stream is independent of the others and transmits data continuously. While TCP and UDP transport protocols only provide one stream for data transmission. However, the delivery ratio for both SCTP and TCP are higher than UDP because there is a retransmission scheme for lost packets and guaranteed end-to-end reliability between source and destination. On the other hand, UDP does not have any retransmission technique for lost packets.

Figure 6 shows the network energy consumption for all standard transport protocols. The UDP transport protocol consumed less energy up to 76.01% and 73.303% compared to TCP and SCTP respectively for low network density. However, for high network density, TCP and SCTP are achieving 60.502% and 28.429% more energy efficient compared to UDP. Since WSNs are deployed with a lot number of sensor nodes, thus, energy saving can be achieved in wireless networks by transmitting data over TCP and SCTP rather than UDP. This is because TCP and SCTP read

Figure 5. Number of nodes vs. delivery ratio

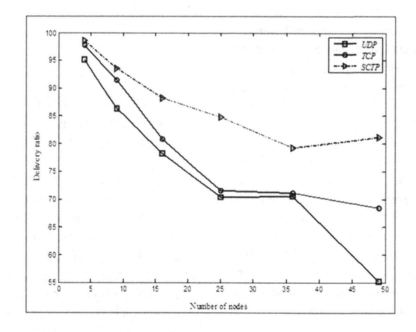

Figure 6. Number of nodes vs. network energy consumption

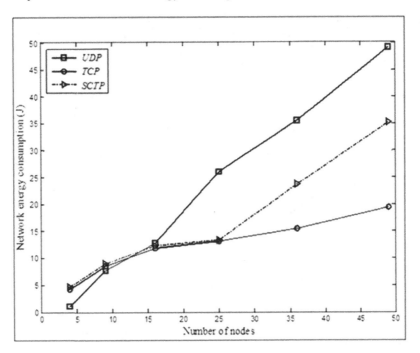

data as a stream of bytes and message is transmitted in the form of segments which have a larger packet size distribution. For that reason, less data packets to be transmitted and result in less energy consumption. Whereas UDP messages are sent in the form of datagrams into the network with smaller packet size distribution. It means that more packets to be transmitted with more energy consumption. Thus, the overall observation reports that the transport protocols which provide a retransmission scheme are suitable for high network density as they result in high delivery ratio and consume less energy.

Figure 7 demonstrates the performance of end-to-end delay for all transport protocol. We observe that the end-to-end delay increases when the number of nodes in the network is increased. End-to-end delay for UDP transport protocol is smaller than other transport protocols because there is no retransmission scheme for lost packets. Additionally, both of TCP and SCTP transport protocols need a connection establishment for session initiation to ensure that receiver is ready

to receive any data transmitted. As a consequence, this connection will increase end-to-end delay for TCP and SCTP transport protocols. However, end-to-end delay for SCTP transport protocol is higher up to 55. 35% than TCP protocol due to four way handshake of connection establishment compared to three way handshake for TCP transport protocol.

FUTURE RESEARCH DIRECTIONS

In the future works, based on the evaluation and observation from the results obtained, the most suitable standard transport protocol for multimedia data transmission is SCTP transport protocol because of multi-streaming feature. However, the performance of the protocol will be more efficient with the feature of partial reliability where data that can be transmitted depends on the level of reliability. Besides that, data that have higher priority will be given a higher level of reliability compared to those with lower priority. With that

Figure 7. Number of nodes versus end-to-end delay

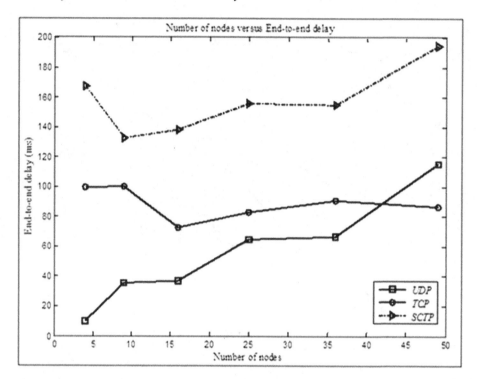

condition, less energy will be consumed because of the minimized number of retransmissions.

Even though the SCTP protocol is better than TCP and UDP in term of delivery ratio, but SCTP protocol presents high end-to-end delay compared to other protocols. This is caused by the connection establishment process of SCTP requiring four-way handshake, which results in increasing the end-to-end delay. Since SCTP has a large end-to-end delay, thus the SCTP protocol will be modified at connection establishment to decrease the transmission delay while maintaining high delivery ratio. Remaining in high delivery is crucial for multimedia data transmission in order to give more preservation on I-frame that contains more information compared to others frames.

Both of the above recommendations will be implemented for the modification of SCTP transport protocol to achieve better performance in term of high delivery ratio, less energy consumption and achieve low end-to-end delay.

CONCLUSION

In this chapter, a comprehensive study of the implementation of standard transport protocol in WSNs has been done. The performances of the transport protocol simulated using the NS-2 simulation tool has been presented. The observation of the results shows that the performance of the UDP transport protocol can be useful in low network density with a good delivery ratio and less energy consumption. For high network density, the retransmission algorithm and its congestion control mechanism make TCP and SCTP much better than UDP. However, SCTP is outperforming for delivery ratio compared to TCP because of some advantage provided by SCTP. SCTP provides the features of multi-streaming such that each stream is independent from the other. As a result, SCTP avoids HOL blocking because it will give an effect to the packet loss. The evaluation of standard transport protocol has been focused where each protocol has its own advantages and disadvantages. Therefore, the modification of the SCTP protocol should be based on the results obtained. The algorithm of the transport protocol for reliability mechanism is crucial to ensure a high packet delivery ratio with reasonable packet loss to increase the video quality. The proposed transport protocol also needs to take energy consumption into consideration due to WSN low profile requirement. Hence, our proposed transport protocol algorithm also needs to allow maximum network lifetime with multimedia transmission, which requires low power consumption and low end-to-end delay.

REFERENCES

Adams, J. (2006). An introduction to IEEE STD 802.15. 4. In *Proceedings of Aerospace Conference, 2006 IEEE*. IEEE. Retrieved from http://ieeexplore.ieee.org/xpls/abs_all.jsp?arnumber=1655947

Ahmed, A., & Fisal, N. (2011). Secure real-time routing protocol with load distribution in wireless sensor networks. *Security and Communication Networks*, 839–859. doi:10.1002/sec

Ai, J., & Abouzeid, A. (2006). Coverage by directional sensors in randomly deployed wireless sensor networks. *Journal of Combinatorial Optimization*, 1–11. Retrieved from http://link.springer.com/article/10.1007/s10878-006-5975-x

Akyildiz, I. (2007). Wireless multimedia sensor networks: A survey. *IEEE Wireless Communications*, 32–39. Retrieved from http://ieeexplore.ieee.org/xpls/abs_all.jsp?arnumber=4407225

Akyildiz, I. F., Melodia, T., & Chowdhury, K. R. (2007). A survey on wireless multimedia sensor networks. *Computer Networks*, *51*(4), 921–960. doi:10.1016/j.comnet.2006.10.002

Almalkawi, I. T., Zapata, M. G., Al-Karaki, J. N., & Morillo-Pozo, J. (2010). Wireless multimedia sensor network: Current trends and future directions. *Sensors (Basel, Switzerland)*, *10*(7), 6662–6717. doi:10.3390/s100706662 PMID:22163571

Bonivento, A., & Fischione, C. (2007). System level design for clustered wireless sensor networks. *Industrial Informatics*, *3*(3), 202–214. doi:10.1109/TII.2007.904130

Boukerche, A., & Du, Y. (2008). A reliable synchronous transport protocol for wireless image sensor networks. In *Proceedings of IEEE Symposium on Computers and Communications*, (pp. 1083–1089). IEEE. Retrieved from http://ieeexplore.ieee.org/xpls/abs_all.jsp?arnumber=4625679

Buchholcz, G., Ziegler, T., & van Do, T. (2005). TCP-ELN: On the protocol aspects and performance of explicit loss notification for TCP over wireless networks. In *Proceedings of 1st International Conference on Wireless Internet* (pp. 172–179). IEEE. Retrieved from http://ieeexplore.ieee.org/xpls/abs_all.jsp?arnumber=1509652

Costa, D. G. (2012). A survey on transport protocols for wireless multimedia sensor networks. *Transactions on Internet and Information Systems (Seoul)*, *6*(1), 241–269. doi: doi:10.3837/tiis.2012.01.014

Costa, D. G., & Guedes, L. A. (2010). The coverage problem in video-based wireless sensor networks: A survey. *Sensors (Basel, Switzerland)*, *10*(9), 8215–8247. doi:10.3390/s100908215 PMID:22163651

Costa, D. G., & Guedes, L. A. (2011). A survey on multimedia-based cross-layer optimization in visual sensor networks. *Sensors (Basel, Switzerland)*, *11*(5), 5439–5468. doi:10.3390/s110505439 PMID:22163908

Ebrahimi, T., & Horne, C. (2000). MPEG-4 natural video coding – An overview. *Signal Processing Image Communication*, *15*(4-5), 365–385. doi:10.1016/S0923-5965(99)00054-5

Garcia-Sanchez, A.-J., Garcia-Sanchez, F., & Garcia-Haro, J. (2008). Feasibility study of MPEG-4 transmission on IEEE 802.15.4 networks. In *Proceedings of 2008 IEEE International Conference on Wireless and Mobile Computing, Networking and Communications*, (pp. 397–403). IEEE. doi:10.1109/WiMob.2008.12

Huszák, Á., & Imre, S. (2010). Analysing GOP structure and packet loss effects on error propagation in MPEG–4 video streams. In *Proceeding of the 4th International Symposium on Communications, Control and Signal Processing*. IEEE. Retrieved from http://ieeexplore.ieee.org/xpls/abs_all.jsp?arnumber=5463469

Iyer, Y., Gandham, S., & Venkatesan, S. (2005). STCP: A generic transport layer protocol for wireless sensor networks. In *Proceedings of International Conference on Computer Communications and Networks*. IEEE. Retrieved from http://ieeexplore.ieee.org/xpls/abs_all.jsp?arnumber=1523908

Lecuire, V., Duran-Faundez, C., & Krommenacker, N. (2008). Energy-efficient image transmission in sensor networks. *International Journal of Sensor Networks*, *4*(1), 37–47. doi:10.1504/IJSNET.2008.019250

Lee, J.-H., & Jung, I.-B. (2010). Reliable asynchronous image transfer protocol in wireless multimedia sensor networks. *Sensors (Basel, Switzerland)*, *10*(3), 1486–1510. doi:10.3390/s100301487 PMID:22294883

Maimour, M., Pham, C., & Amelot, J. (2008). Load repartition for congestion control in multimedia wireless sensor networks with multipath routing. In *Proceedings of 2008 3rd International Symposium on Wireless Pervasive Computing*, (pp. 11–15). IEEE. doi:10.1109/ISWPC.2008.4556156

Mao, S., & Bushmitch, D. (2006). MRTP: A multiflow real-time transport protocol for ad hoc networks. *IEEE Transactions on Multimedia*, *8*(2), 356–369. doi:10.1109/TMM.2005.864347

Mingorance-Puga, J. F., Maciá-Fernández, G., Grilo, A., & Tiglao, N. M. C. (2010). Efficient multimedia transmission in wireless sensor networks. *Next Generation Internet (NGI)*, *2010*. Retrieved from http://ieeexplore.ieee.org/xpls/abs_all.jsp?arnumber=5534472

Moradi, S., & Wong, V. W. S. (2007). Technique to improve MPEG-4 traffic schedulers. In *Proceedings of IEEE Conference on Communications*. IEEE.

Postel, J. (1980). User Datagram Protocol. *Isi*, (August), 1–4. Retrieved from http://xml2rfc.tools.ietf.org/html/rfc0768

Postel, J. (1981). Transmission control protocol. *RFC: 793*. Retrieved from http://tools.ietf.org/html/rfc793

Qaisar, S., & Radha, H. (2009). Multipath multistream distributed reliable video delivery in wireless sensor networks. In *Proceeding of Conference on Information Sciences and Systems*, (pp. 207–212). IEEE. Retrieved from http://ieeexplore.ieee.org/xpls/abs_all.jsp?arnumber=5054718

Quang, B. D., & Won-Joo, H. (2006). Trade-off between reliability and energy consumption in transport protocols for wireless sensor networks. *International Journal of Computer Science and Network Security*, *6*(8), 47–53.

Smart, M., & Keepence, B. (2008). *Understanding MPEG-4 video*. IndigoVision Ltd.

Stewart, R. (2007). Stream control transmission protocol. *Network Working Group RFC 2960*. Retrieved from http://tools.ietf.org/html/rfc4960

Stewart, R., Ramalho, M., Xie, Q., Tuexen, M., & Conrad, P. (2004). Stream control transmission protocol (SCTP) partial reliability extension. *Network Working Group RFC 3758*. Retrieved from http://www.hjp.at/doc/rfc/rfc3758.html

Vedantham, R., Sivakumar, R., & Park, S.-J. (2007). Sink-to-sensors congestion control. *Ad Hoc Networks*, *5*(4), 462–485. doi:10.1016/j.adhoc.2006.02.002

Yaghmaee, M., & Adjeroh, D. (2008). A new priority based congestion control protocol for wireless multimedia sensor networks. In *Proceedings of IEEE International Symposium on a World of Wireless, Mobile and Multimedia Networks*. IEEE. Retrieved from http://ieeexplore.ieee.org/xpls/abs_all.jsp?arnumber=4594816

Zahariadis, T., & Voliotis, S. (2007). Open issues in wireless visual sensor networking. In *Proceedings of 2007 14th International Workshop on Systems, Signals and Image Processing and 6th EURASIP Conference focused on Speech and Image Processing, Multimedia Communications and Services*, (pp. 335–338). EURASIP. doi:10.1109/IWSSIP.2007.4381110

Zakaria, A., & El-Marakby, R. (2009). AdamRTP: Adaptive multi-flows real-time multimedia delivery over WSNs. In *Proceedings of IEEE International Symposium on Signal Processing and Information Technology, ISSPIT 2009*. IEEE. Retrieved from http://ieeexplore.ieee.org/xpls/abs_all.jsp?arnumber=5407580

Zhang, L., Hauswirth, M., & Shu, L. (2008). Multi-priority multi-path selection for video streaming in wireless multimedia sensor networks. In *Proceedings of Ubiquitous Intelligence and Computing* (pp. 23–25). Springer. Retrieved from http://link.springer.com/chapter/10.1007/978-3-540-69293-5_35

ADDITIONAL READING

Aghdasi, H. S., Abbaspour, M., Moghadam, M. E., & Samei, Y. (2008). An Energy-Efficient and High-Quality Video Transmission Architecture in Wireless Video-Based Sensor Networks. *Sensors (Basel, Switzerland)*, *8*(8), 4529–4559. doi:10.3390/s8084529

Ahmad, J., Khan, H., & Khayam, S. (2009). Energy Efficient Video Compression for Wireless Sensor Networks. *43rd Annual Conference on Information Sciences and Systems, 2009. CISS 2009*, 629–634. Retrieved from http://ieeexplore.ieee.org/xpls/abs_all.jsp?arnumber=5054795

Akan, O. (2007). Performance of Transport Protocols for Multimedia Communications in Wireless Sensor Networks. *Communications Letters, IEEE*, *11*(10), 826–828. Retrieved from http://ieeexplore.ieee.org/xpls/abs_all.jsp?arnumber=4389800

Almeida, J., Grilo, A., & Pereira, P. (2009). Multimedia Data Transport for Wireless Sensor Networks. *Next Generation Internet Networks, 2009. NGI'09* (pp. 1–8). Retrieved from http://ieeexplore.ieee.org/xpls/abs_all.jsp?arnumber=5175768

Baronti, P., Pillai, P., Chook, V. W. C., Chessa, S., Gotta, A., & Hu, Y. F. (2007). Wireless Sensor Networks: A Survey on the State of the Art and the 802.15.4 and ZigBee Standards. *Computer Communications*, *30*(7), 1655–1695. doi:10.1016/j.comcom.2006.12.020

Braun, T. Voigt, T., & Dunkels, A. (2007). TCP Support for Sensor Networks. *2007 Fourth Annual Conference on Wireless on Demand Network Systems and Services* (pp. 162–169). Ieee. doi:10.1109/WONS.2007.340494

Braun, T., Voigt, T., & Dunkels, a. (2005). Energy-Efficient TCP Operation in Wireless Sensor Networks. *PIK - Praxis der Informationsverarbeitung und Kommunikation, 28*(2), 93–100. doi:10.1515/PIKO.2005.93

Brennan, R., & Curran, T. (2001). SCTP congestion control: Initial simulation studies. *Proc. 17th Int'l Teletraffic Congress* (pp. 1–18). Retrieved from http://www.eeng.dcu.ie/~opnet/papers/ITC01_SCTPCongestionControl3.pdf

Cano, M. (2011). On the Use of SCTP in Wireless Networks. *Recent Advances in Wireless Communications and Networks, ISBN* (pp. 245–267). Retrieved from http://cdn.intechWeb.org/pdfs/18322.pdf

Cheng, R.-S., & Lin, H.-T. (2008). A Cross-layer Design for TCP End-to-end Performance Improvement in Multi-hop Wireless Networks. *Computer Communications, 31*(14), 3145–3152. doi:10.1016/j.comcom.2008.04.017

Chonggang, W., Sohraby, K., Lawrence, V., Li, B., & Hu, Y. (2006). Priority-based Congestion Control in Wireless Sensor Networks. *IEEE International Conference on Sensor Networks, Ubiquitous, and Trustworthy Computing (SUTC'06)* (Vol. 1, pp. 22–31). Ieee. doi:10.1109/SUTC.2006.1636155

Ciubotaru, B. (2007). Performances Analysis on Video Transmission in a Wireless Sensor Network. *4th International Symposium on Applied Computational Intelligence and Informatics, 2007. SACI'07* (pp. 183–186). Retrieved from http://ieeexplore.ieee.org/xpls/abs_all.jsp?arnumber=4262508

Ee, C. T., & Bajcsy, R. (2004). Congestion Control and Fairness for Many-to-One Routing in Sensor Networks. *Proceedings of the 2nd International Conference on Embedded Networked Sensor Systems - SenSys '04* (pp. 148). New York, New York, USA: ACM Press. doi:10.1145/1031495.1031513

Garcia-Sanchez, A.-J., Garcia-Sanchez, F., Garcia-Haro, J., & Losilla, F. (2010). A Cross-layer Solution for Enabling Real-time Video Transmission over IEEE 802.15.4 Networks. *Multimedia Tools and Applications, 51*(3), 1069–1104. doi:10.1007/s11042-010-0460-z

Gürses, E., & Akan, Ö. (2005). Multimedia Communication in Wireless Sensor Networks. *Annales des Télécommunications, 60*(7-8), 872–900. Retrieved from http://link.springer.com/article/10.1007/BF03219952

Liang, L., Gao, D., Zhang, H., & Leung, V. C. M. (2010). A Novel Reliable Transmission Protocol for Urgent Information in Wireless Sensor Networks. *Global Telecommunications Conference (GLOBECOM 2010), 2010 IEEE* (pp. 1–6). Ieee. doi:10.1109/GLOCOM.2010.5684267

Lie, A., & Klaue, J. (2008). Evalvid-RA : Trace Driven Simulation of Rate Adaptive MPEG-4 VBR Video. *Multimedia Systems, 14*(1), 33–50. Retrieved from http://link.springer.com/article/10.1007/s00530-007-0110-0 doi:10.1007/s00530-007-0110-0

Lien, Y. (2009). Hop-by-Hop TCP for Sensor Networks. *International Journal of Computer Networks & Communications, 1*(1), 1–16. Retrieved from http://citeseerx.ist.psu.edu/viewdoc/download?doi=10.1.1.192.1833&rep=rep1&type=pdf

Lin, C.-H., Ke, C.-H., Shieh, C.-K., & Chilamkurti, N. K. (2006). The Packet Loss Effect on MPEG Video Transmission in Wireless Networks. *20th International Conference on Advanced Information Networking and Applications - Volume 1 (AINA'06)* (pp. 565–572). Ieee. doi:10.1109/AINA.2006.325

Lin, M., Leu, J., Yu, W., Yu, M., & Wu, J. C. (2011). On Transmission Efficiency of the Multimedia Service over IEEE 802. 15. 4 Wireless Sensor Networks. *International Conference on Advanced Communication Technology, ICACT 2011* (pp. 184–189).

Maimour, M., Pham, C., & Hoang, D. (2009). A Congestion Control Framework for Handling Video Surveillance Traffics on WSN. *International Conference on Computational Science and Engineering, CSE'09, 2,* 943–948. doi:10.1109/CSE.2009.200

Meneses, D., Grilo, A., & Pereira, P. (2011). A transport protocol for real-time streaming in wireless multimedia sensor networks. *7th EURO-NGI Conference on Next Generation Internet (NGI)* (pp. 1–8). Retrieved from http://ieeexplore.ieee.org/xpls/abs_all.jsp?arnumber=5985941

Paniga, S., Borsani, L., Redondi, A., Tagliasacchi, M., & Cesana, M. (2011). Experimental Evaluation of a Video Streaming System for Wireless Multimedia Sensor Networks. *The 10th IFIP Annual Mediterranean Ad Hoc Networking Workshop (Med-Hoc-Net), 2011* (pp. 165–170). Retrieved from http://ieeexplore.ieee.org/xpls/abs_all.jsp?arnumber=5970484

Rahman, M., El Saddik, A., & Gueaieb, W. (2008). Wireless Sensor Network Transport Layer: State of the Art. *Sensors (Basel, Switzerland),* 221–245. Retrieved from http://link.springer.com/chapter/10.1007/978-3-540-69033-7_11

Sharif, A., Potdar, V., & Rathnayaka, A. (2009). Performance Evaluation of Different Transport Layer Protocols on the IEEE 802. 11 and IEEE 802. 15. 4 MAC / PHY Layers for WSN. *Proceedings of the 7th International Conference on Advances in Mobile Computing and Multimedia* (pp. 300–306). Retrieved from http://dl.acm.org/citation.cfm?id=1821805

Sharma, B., & Aseri, T. C. (2012). A Comparative Analysis of Reliable and Congestion-Aware Transport Layer Protocols for Wireless Sensor Networks. *International Scholarly Research Network. ISRN Sensor Networks, 2012,* 1–14. doi:10.5402/2012/104057

Yun, L., Rui, Z., Zhanjun, L., & Qilie, L. (2009). An Improved TCP Congestion Control Algorithm over Mixed Wired/Wireless Networks. *2nd IEEE International Conference on Broadband Network & Multimedia Technology* (pp. 786–790). Ieee. doi:10.1109/ICBNMT.2009.5347786

KEY TERMS AND DEFINITIONS

IEEE 802.15.4 Standard: Data communication protocol standard for low rate wireless personal area networks (WPAN).

Motion Picture Experts Groups (MPEG-4): Encoding technique for multimedia data.

Stream Control Transmission Protocol (SCTP): Protocol that provides reliability and congestion control mechanism with the functionalities of multi-streaming and partial reliability.

Transmission Control Protocol (TCP): Protocol that supports reliable data delivery and congestion control.

Transport Protocol: Protocol at transport layer in Open System Interconnection (OSI) model.

User Datagram Protocol (UDP): Simple interface between the network layer and application layer.

Wireless Sensor Network (WSN): The network that consists of hundreds or thousands of small sensor nodes.

Chapter 16
Hybrid MAC Layer Design for MPEG–4 Video Transmission in WSN

Nor-Syahidatul N. Ismail
Universiti Teknologi Malaysia, Malaysia

N. M. Abdul Latiff
Universiti Teknologi Malaysia, Malaysia

Sharifah H. S. Ariffin
Universiti Teknologi Malaysia, Malaysia

Farizah Yunus
Universiti Teknologi Malaysia, Malaysia

Norshiela Fisal
Universiti Teknologi Malaysia, Malaysia

ABSTRACT

Wireless Sensor Networks (WSNs) have been attracting increasing interest lately from the research community and industry. The main reason for such interest is the fact that WSNs are considered a promising means of low power and low cost communication that can be easily deployed. Nowadays, the advanced protocol design in WSNs has enhanced their capability to transfer video in the wireless medium. In this chapter, a comprehensive study of Medium Access Control (MAC) and MPEG-4 video transmission is presented. Various classifications of MAC protocols are explained such as random access, schedule access, and hybrid access. In addition, a hybrid MAC layer protocol design is proposed, which combines Carrier Sense Multiple Access (CSMA) and unsynchronized Time Division Multiple Access (TDMA) protocols using a token approach protocol. The main objective of this chapters is to present the design of a MAC layer that can support video transfer between nodes at low power consumption and achieve the level of quality of service (QoS) required by video applications.

DOI: 10.4018/978-1-4666-5170-8.ch016

1. INTRODUCTION

The demand for connecting devices without cable with low power consumption especially at inaccessible areas and hazardous environments fosters the rapid advance in protocol design for Wireless sensor networks (WSNs). Moreover, *WSN* can be considered as a promising means to establish green telecommunication networks. It is a worldwide goal to minimise energy consumption of any design in the telecommunication sector and the importance of the energy efficiency in telecommunication industry are discussed in Koutitas and Demestichas (2010). To achieve this goal, both academia and industry should cooperate in order to design low energy consuming protocols and produce low power devices for the wireless sector.

Since 2003, the Institute of Electrical and Electronics Engineers Standards Association (IEEE-SA) released the new IEEE 802.15.4 standard for low rate wireless personal area network (LR-WPAN) (Man, 2006). The *IEEE802.15.4* standard has been designed specifically for short-range communications with low-power sensor networks and can be considered with the resulting effect on low power consumption. This attribute makes the IEEE 802.15.4 wireless standard used widely in most WSN applications such as environmental monitoring, target detection, industrial process monitoring and emergency measures (Ismail, Yunus, & Ariffin, 2011). Recently, the availability of *CMOS* cameras that can be attached with wireless motes has enhanced the transmission capability in order to send the video applications in the wireless medium. Video applications based on WSN have significant importance due to their potential to collect visual information needed for security, military and scientific purposes.

In general, earlier wireless standards such as IEEE 802.11, IEEE 802.15.3 and Bluetooth are more suitable for video applications than IEEE 802.15.4 since they offer much higher data rates (Zainaldin, Lambadaris, & Nandy, 2008). However, this task can be achieved at lower cost and lower power consumption if IEEE 802.15.4-based devices were employed. The high energy overhead of IEEE 802.11-based networks makes this standard unsuitable for low power sensor networks (Dargie & Poellabauer, 2010). *MPEG-4* video is one of the international standards for video encoding which offers high level of compression (Sikora, 1997). Therefore, it is more suitable for low-rate communications, specifically; it can be used in implementation of video transmission over IEEE 802.15.4.

This chapter focuses on MAC layer protocol design. *MAC* layer is responsible to control channel access that allows multiple nodes within a network to communicate. Moreover, the MAC layer has significant importance in wireless networks: it organizes how the channel is shared across the users, which directly impacts the system throughput, fairness, reliability and quality of service (QoS). The high QoS requirement for video transmission brings the necessity to design efficient MAC protocols. Recently, hybrid MAC protocols have appeared to be the most suitable for supporting real-time communication in WSNs. This is because they provide a mechanism to guarantee support for real-time traffic, while promoting energy efficiency and scalability (Misra, Reisslein, & Xue, 2008). To achieve reliable communication, the token passing mechanism is used in this hybrid MAC design to minimize collision and solve strict synchronization problem in schedule access method.

Most of the previous works on *multimedia* applications over IEEE 802.15.4 focus on image transmission, while the research on the video applications is still in earlier stages. The objective of this chapter is to provide the reader with an understanding of how the MPEG-4 video transmission can be applied in low rate IEEE 802.15.4 wireless standard medium, and how the token approach protocol is used in MAC layer design.

This chapter is organized as follows. In section 2, an overview of MAC layer protocol that covers three access methods areas, namely, random access, schedule access and hybrid access are explained. In the same section, literature review on hybrid access and token approach protocol are presented. Then, by explicitly focusing on video transmission, we will highlight more recent works in this area. Section 3 discusses the issues and challenges in implementing video transmission in WSNs and issues of MAC layer protocol in wireless networks. The discussion is extended in the proposed solution and recommendations for aforementioned issues above in Section 4. Performance analysis of the proposed work is presented in this section. The final part of this chapter mentions about future research direction and conclusions in Sections 5 and 6.

2. BACKGROUND

2.1. Medium Access Control (MAC) Layer

Medium Access Control (MAC) layer is a part of data link layer in the standard OSI model for WSN. The other part is Logical Link Control layer (LLC) for error and flow Control (Roy & Sarma, 2010). MAC layer is often viewed as the brain of the network and has a significant effect on the performance of the sensor network. MAC layer resolves contentions in a multi-access wireless environment. The fundamental task of MAC protocol is that it has to regulate the channel access for each node in the shared medium to satisfy performance requirements of various applications run by sensor nodes (Song, Huang, Shirazi, & LaHusen, 2009). For *multimedia* applications such as video transmission, a large number of frames need to arrive at the destination node within an acceptable delay time. The acceptable delay time for real time applications such as voice is less than 250 ms and for streaming video is in the order of one second (Ganz, Ganz, & Wongthavar-awat, 2003). Inefficient design of the MAC layer protocol for video applications will increase the delay and subsequently decrease the throughput of the network.

In the early state, two access methods were implemented in the MAC layer. The first is random access, which is known as Carrier Sense Multiple Access (CSMA). The second method is scheduled access, which is known as Time Division Multiple Access (TDMA). *CSMA* is based on contention where the nodes strive to access the medium. In the CSMA protocol, there is no need for synchronization or topology knowledge because nodes compete to access the channel only with their neighbouring node and only the winner will succeed to get the channel access. A node that wishes to transmit first performs a clear channel assessment (CCA) by sensing the medium for a random duration. If the medium is idle then the node assumes that it may take ownership of the medium and begin to transmit. If the medium is busy, the node waits for a further random back-off period and sense the channel again until the medium is idle.

On the other hand, in TDMA, nodes gain access to communication channels by scheduling and reserving time slots for each node that make this protocol collision-free medium. *TDMA* protocol can be grouped into two categories, which are centralized and distributed approach. In centralized TDMA approach, a central node is responsible for deciding which node can access the channel, and the time duration of selected node that controls the channel. The allocation of time slots for each node within the cluster depends on the central node decision. In distributed TDMA approach, each sensor node is allowed to assign itself a time slot by collecting its neighbourhood information (Rashid, W.Embong, Zaharim, & Fisal, 2009). The nodes will choose a time slot that was not selected by its neighbours to avoid collision. Consequently, the central node is irrelevant. Each node wakes up at every beginning of the time slot to receive the beacon transmitted by the owner of the slot. If there is no beacon received, it can choose that slot in the next frame. Both of these approaches (centralized and distributed) need efficient time synchronization protocol.

2.2. Hybrid MAC Layer

The CSMA and TDMA protocols have advantages and weaknesses. The suitability in implementation of each protocol will depend on its application because WSN is design based on application-specific. For monitoring periodic data, CSMA outperforms TDMA protocol. However, for high traffic load, TDMA shows a good performance compared to CSMA. The combination of both protocols which is called hybrid MAC seems to be the best solution in MAC layer design for multimedia or video applications. *Hybrid MAC* can increase the network performance in term of scalability, delay, throughput, and energy efficiency. Nowadays, there are many researches done on hybrid MAC.

For instance, one of the popular hybrid MAC protocols that achieves high channel utilization is Zebra MAC (ZMAC) (Warrier, Aia, & Sichitiu, 2008). ZMAC uses contention level information on the channel to switch the communication either to CSMA (at low contention) or TDMA (at high contention). However, ZMAC lacks of scalability when there is any new node join the network and when it has data to send. If the channel condition were in low contention, it would not suffer any problem because CSMA protocol is used to send the data. However, if the channel condition is in high contention level, data cannot be transmitted because there is no time slot allocated for the new node. This scenario occurs because slot allocation is done offline only once at the beginning of network setup using DRAND (Distributed Randomized Time Slot Assignment Algorithm) protocol (Rhee & Warrier, 2009).

Emergency response MAC (ER-MAC) improves ZMAC protocol in the allocation of time slots for new nodes or dead nodes (Sitanayah, Sreenan, & Brown, 2010). ER-MAC works in tree topology and time slot starts from the leaf node towards base station. If there is a new node that joins the network and has data to send, the parent allocates one slot to forward the new node's data.

ER-MAC protocol also supports hybrid MAC. If there is no emergency response, each node uses its own time slot for sending data, but when there is an emergency response, each node wakes up in every slot for possible contention. Through this concept, ER-MAC achieves a high delivery ratio. However, ER-MAC consumes more energy in emergency situations where nodes wake up at every slot for possible contention.

The above two protocols are designed in IEEE 802.11 standard. In IEEE 802.15.4 MAC standard, hybrid MAC is already implemented in *beacon enable mode* if guarantee time slot (GTS) is used. There are 16 slots in super frame of beacon enable mode that consists of 9 slots for a contention access period (CAP) slot and 7 slots allocated for the contention free period (CFP) slot. CAP slot uses CSMA protocol and CFP or GTS slot uses TDMA protocol. The topology for the *hybrid MAC* in IEEE 802.15.4 standard is star topology. Any node that wants to send using GTS slot should request from the coordinator.

Even though hybrid MAC improves network performance, but the problem with synchronization still occurs. To avoid strict synchronization problems in TDMA protocol, asynchronous TDMA, which is a token approach protocol is proposed as a better solution.

2.3. Token Approach Protocol

Even though the *token approach protocol* was used long time ago in wired networks, it still relevant to be used in wireless networks. This protocol has been adopted in many research works to address different problems in different systems. Usually, the communication in token approach protocol is based on token availability. Each node can transmit the data if it holds the token. Recently, there have been a lot of researchers looking forward to use token mechanism in WSN to achieve high energy efficiency and provide reliable communication. For energy efficiency, most of the researchers applied sleep scheduling.

For example, an energy saving token ring (ESTR) protocol (Bagci, 2008) implement sleep interval to achieve high energy efficiency. ESTR reverses the communication scheme by allowing the token holding node to receive the message rather than transmit the data. However, in the sleep phase, the communication latency is increased and throughput is decreased. Besides, the number of collisions also increases in the initialization phase. Another token protocol that uses sleep scheduling is proposed in Ray, Dash, and Tarasia (2011). Queue request token is introduced in this protocol at the sink node. Since the token is held by sink node, therefore only one node will be able to communicate with the sink at one time. Unfortunately, the delay for this protocol is increased because the source node uses its parent node to request token from the sink node.

In Dash, Swain, and Ajay (2012), the author proposed dynamic sleep scheduling, where some percentage of nodes are randomly selected and put in sleep mode for a fixed period. This protocol introduces the multi token concept with tree topology, where each of the intermediate nodes possesses a single token. Through this approach, data will be transferred faster compared to the single token case. However, this protocol lacks of scalability because it works on tree topology, and the time taken to find new neighbours affects the data delivery. Token bus based MAC proposed by Udayakumar, Vyas, and Vyas (2012) also has the same limitation. It uses token bus algorithm and only one node can communicate with base station at one time.

A modified wireless token ring proposed in Wei, Zhang, Xiao, and Men (2012) enhances the throughput with dynamic token holding time. A node hands over the token after it finishes data transmission or if it does not have data to send. Each node in this protocol is connected in a closed loop ring. The node joining mechanism is used to expand the closed loop into a node cluster, which

enables better coverage. Any node outside the ring can get access to the channel by connecting with other nodes in ring connection. Unfortunately, joint mechanisms need four way handshake and increase energy consumption. Moreover, in closed loop topology, when nodes leave the network or power off, they may cause collapse of the ring.

All the token protocols aforementioned are only used to transmit scalar data. The token protocol proposed in Wang and Zhuang (2008) is used for multimedia applications and in a distributed manner without control by a central controller. This protocol has two tokens, one for priority data which is voice traffic, and the other one is for data traffic. The waiting time to sense the channel for being idle is shorter for voice than for data traffic to ensure higher access priority to voice traffic. After time token holding is expired, the current token holder decides the next token holder. Each node is always updating the information on the probability transition to avoid collision in voice/data transmission. Maintaining such update in this protocol is not a difficult task because it is implemented in WLAN networks, but to use it in IEEE 802.15.4 which has a small coverage area is not easy. Even though IEEE 802.15.4 standard is targeted for low data rate applications, there are also research efforts to support multimedia content delivery over the network.

2.4. Wireless Multimedia Sensor Network

Multimedia data can be defined as image, voice or video. By focusing on MPEG-4 video transmission, the recent works in this area are discussed. However, most of the works are implemented in IEEE 802.11 which has a high data rate compared to IEEE 802.15.4. But the method proposed in IEEE802.11 can be used to transmit video over IEEE 802.15.4 with some modifications. A frame dropping module was introduced in Shin and

Chung (n.d.) at the transport layer to drop less important (low priority) frames if the estimated bandwidth is not enough to support the required transmission rate. In Xiao, Du, Hu, and Zhang (2008), low priority frames are discarded at MAC layer if high priority frames meet the deadline or were lost in the medium. The entire dependency frame at queue will also be dropped. This approach is important to make sure other frames are received within acceptable delay time. On the other hand, retransmission of the lost packets implemented in Ma, Zhou, and Wu (2010) is unsuitable to be used IEEE 802.15.4 medium because of the delay and energy consumption induced by P-ARQ (priority automatic request queue).

Up to the time of writing this chapter, only one research work implements *MPEG-4* video transmission in IEEE 802.15.4, which was proposed in Garcia-Sanchez, Garcia-Sanchez, and Garcia-Haro (2008). It used guarantee time slot (GTS) in the CFP period to transmit the video at three different rates which are 46 kbps, 57 kbps and 64 kbps. However, the results reveal that, increasing traffic rate above 57 kbps provokes high delay. This work was enhanced in Garcia-Sanchez, Garcia-Sanchez, Garcia-Haro, and Losilla (2010) and Garcia-Sanchez, Garcia-Sanchez, and Garcia-Haro (2011) for agricultural application. The beacon enable mode is used with cross layer multimedia guarantee time slot (CL-MGTS) and is implemented on real device in IEEE 802.15.4. CAP slot is used to transmit control message, detection trigger and measurements data. Meanwhile, CFP slot is used to transmit the image. When the video application is triggered, 12 slots are used for CFP rather than 7 that are set by the standard and only 4 slots for data monitoring. Through this approach, access priority is given to video applications. The research on video transmission in IEEE 802.15.4 that has a low data rate opened more research issues.

3. MAIN FOCUS OF THE CHAPTER

3.1. Issues, Controversies and Problem

The implementation of video transmission over WSN brings new research issues and challenges especially in *IEEE802.15.4* wireless medium. There are several reasons that enable the possibility to transmit video over low data rate network. One of the reasons is the availability of inexpensive hardware such as CMUcam3 that used CMOS camera and has the compatible connector with wireless motes of IEEE 802.15.4 ("CMUcam3 Datasheet," 2007). A *CMOS* camera is able to ubiquitously capture the multimedia content from the environment. CMOS camera also can be attached with wireless mote MICAZ and MICA2 (Akyildiz, Melodia, & Chowdhury, 2007). The authors in Abdul Hadi et al. (2009) developed new wireless hardware platforms for IEEE 802.15.4 medium called TelG. Another reason that fostered multimedia WSN is compression capability inherited in the MPEG-4 codec, which is able to remove redundant information found in the individual frames that make up the video, and able to predict changes from one frame to the next. Besides, MPEG-4 supports low bit rates from 5kbps to 4Mbps (Ganz et al., 2003).

The development of Evalvid (Klaue & Rathke, 2003) framework in ns-2 simulation tools has allow the researcher to analyse video compression. Moreover, using this framework, it can assist researchers in evaluating their network designs or setups in terms of the perceived video quality of the end user before being implemented in real hardware. The hardware and software available nowadays fostered the development of multimedia WSN especially for video transmission in IEEE 802.15.4. In the context of multimedia transmission, the key design challenge come from the nature of multimedia application, MAC layer design and the WSN itself.

3.1.1 Challenges from Wireless Environment

A wireless network is relatively unreliable medium by its nature. Wired networks have very few transmission errors compared to wireless networks that have high errors rate. Many challenges limit the design of efficient video communications in *WSN*. Some of the challenges are caused by resource limitations such as limited power processing capability and the error resilience capability of video compression techniques. In addition, the bandwidth provided by wireless channel is low because of path loss, fading and interferences. The main challenges can be summarised as:

1. Instable wireless channel condition cause frequent transmission delay and packet losses.
2. Wireless channel characteristics such as shadowing, multi-path fading, and interferences cause degradation in the quality of the video streaming service .
3. In a wireless medium, not all sensor nodes can listen to all other nodes' signalling because of short range communication limitations. Some nodes may hear signals from the nodes on near end of an exchange but not the nodes at the far end (the hidden node problem).
4. The coverage range of each node is varied depending on the remaining battery power.
5. The data rate that a channel can support is affected greatly by distance and other environmental changes. Nodes need to continually adjust the data rate at which they exchange information to optimize throughput.

3.1.2 Challenges in MPEG-4 Video Transmission

The challenges in the *MPEG-4* video applications can be described as a high bandwidth requirement, constrained delay and sensitivity to packet loss. The high bandwidth requirement came from a huge amount of data generated by an image sensor which are much higher compared to many other types of sensors. This is much more than the available RAM memory on typical wireless sensor devices; for example, the RAM in Telos B is 16 Kbyte and TeLG mode is around 4 Kbyte. The MPEG-4 video consists of three different frames which are I, P and B frames that have different levels of importance. The video application is sensitive to packet loss because if the frame that has a high level of importance, which is I frame, is lost into the medium, all the P and B frame in the same group of picture (GOP) become useless and the scene cannot be constructed.

The sensor node devices have limited computation resources, which forces the compression algorithm to have low complexity and high compression efficiency to reduce the bandwidth requirement. Lastly, the big challenge came from IEEE 802.15.4 wireless medium that has low bit rate which is 250 kbps. Table 1 shows the bandwidth requirement for six video samples in QCIF format (176×144) that consists of low, medium and high motion. As the motion in the scene increases, the compression process gets more complex and the compression efficiency becomes lower as well.

Based on bandwidth requirement in Table 1, only Akiyo sample video can be transmitted in WSN environment. To enable the transmission of others video types in WSN environment, some parameters should be adjusted in encoder size to reduce the bit rate and at the same time maintain the *peak to signal noise ratio (PSNR)* value.

3.1.3 Issues in MAC Layer Protocol

In wireless networks, the common problem for the MAC protocol in WSN is the packet collision, where two nodes send data at the same time over the same channel. Too many collisions could reduce throughput, increase energy consumption and increase delay in data delivery. The responsibility of *MAC* protocol is to efficiently manage the transmission among nodes by reducing the packet

Table 1. Bandwidth requirement and total size of low, medium, and high motion MPEG-4 video

Video Sample (qcif)	Total Length(byte)	Bandwidth Requirement (kbps)	Category (Motion & Complexity)
Akiyo	261371	209	Low
Hall	514452	412	Low
Foreman	1115226	892	Medium
News	680189	544	Medium
Mobile	3479256	2783	High
Coastguard	1416359	1133	High

collisions. Packet that collides before it reaches the destination will result in errors or incomplete messages to the receiver. Another challenge is to manage the usage of limited energy. As energy efficiency is one of the most important attributes in sensor networks, retransmission of collided packets should be avoided as it consumes a lot of energy (Dash et al., 2012).

The popular protocol in MAC layer is the CSMA protocol. Even though *CSMA* protocol achieves high performance in scalability, but it suffers from the collision problems as mentioned in the previous paragraph. CSMA protocols show lower performance for high load traffic (Ismail, Ariffin, Latiff, & Yunus, 2013) and suffer from energy inefficiency due to idle listening, collision and overhearing (Aghdasi & Abbaspour, 2008).

To overcome collision problem in CSMA protocol, non-collision protocol, which is *TDMA* is widely used in energy efficient design protocols. However, it has difficulties to maintain the time synchronization and cope with dynamic topology changes. Tight synchronization incurs high energy overhead because it requires frequent message exchanges. Moreover, during low contention, TDMA gives much lower channel utilization and higher delays than CSMA because in TDMA a node can transmit only during its scheduled time slots, whereas in CSMA nodes can transmit at any time as long as there is no contention.

3.1.4 Problem in MAC IEEE 802.15.4

IEEE 802.15.4 *MAC* standard has two modes, which are beacon enable mode and non beacon enable mode. Both modes give different challenges in transmitting multimedia applications. In *beacon enable mode*, the network performance depends on the beacon order (BO) and superframe order (SO). The small value of BO and SO can provide well real-time services but bring lots overhead of beacon frames. Moreover, small SO and BO (high synchronization), wastes a lot of energy because of synchronization (Lin, Leu, & Yu, 2011). Guarantee time slot (GTS) slot that works like TDMA is provided in beacon enable mode. Nevertheless, there are some limitations in implementing GTS in the network. Firstly, the allocation of the GTS slots are only up to seven slots in one superframe, which is not enough to support the transmission of the real time data (Chen, 2009; Burda & Wietfeld, 2007).

Secondly, when a node wants to use the GTS slot, it will have to make a request from personal area network (PAN) coordinator during the beacon in the beginning of super frame. If a node misses that beacon, it will not use GTS slot until it receives the subsequent beacon correctly (Choi & Lee, 2011). This will waste the slot and data will be transmitted in CSMA mode and hence, probability of collision is high when the channel is busy. Thirdly, the topology is limited only to the star topology since in GTS; the communications are only between the PAN coordinator with the node. Relaying data is not enabled for data transmission when GTS is used.

The other modes in MAC IEEE 802.15.4 is *non beacon enable (NBE)* mode. In NBE mode that supports dense network scenarios, there is high probability of more than two nodes sensing the channel to be idle at the same time. Due to limited randomness provided by the Back-off Exponent (BE) being an integer between 0 and 3, there is a high probability that two nodes could be transmitting their data packets at the same time, hence leading to an increased probability of collisions. These problems lead to an ineffective usage of the available bandwidth, decreased throughput and increased latency of the network.

For scalar data, conserving energy is critical because they generate data traffic at lower rates. The IEEE 802.15.4 MAC achieves low-duty cycle operation in beacon enable mode but it is limited to the lower data throughput. On the other hand, image sensors demand higher data throughput in order to transmit the bulk of multimedia traffic. Furthermore, implementation of *beacon enable mode* in real test bed causes imperfect time synchronization among local or one hop neighbours. Beacon enable mode is not likely suitable to carry high data throughputs (Suh, Mir, & Ko, 2008). In this case, the NBE mode would be the preferred choice, since it operates without any sleep mechanism. NBE mode can offer higher data throughput at the expense of significant energy consumption, mainly due to the idle listening and collision problem. Unfortunately, IEEE 802.15.4 supports only one mode at a time. To achieve both goals (lower energy consumption and high data throughput) modification of non-beacon enable mode is needed (Suh et al., 2008).

4. SOLUTIONS AND RECOMMENDATIONS

4.1 Proposed Work

4.1.1 MPEG-4 Video Transmission in IEEE 802.15.4

The nature of *MPEG-4* video that generates a huge number of frames per second mandates higher bandwidth than what IEEE 802.15.4 medium can support. The allowable data rate in IEEE 802.15.4 is 250 kbps and maximum transmission unit (MTU) size is 127 bytes. It is significant to ensure the bandwidth or data rate required for video transmission is sufficient for low data rate devices. From Table 1, bandwidth requirement shows that only MPEG-4 low motion video which is Akiyo.qcif sample can be transmitted in WSN environment without any modification. To ensure that all the MPEG-4 video can be transmitted in low data rate network, the parameters in encoding process should be optimized.

EvalVid-RA tool is used in NS2 simulation environment to simulate multimedia transmission application through WSN. Based on EvalVid architecture (Klaue & Rathke, 2003), there are three steps in the simulation phase which are pre-process phase, network simulation phase, and simulation evaluation phase.

Video encoding is performed during pre-process phase to avoid high demand on CPU resources during simulation time. The output from pre-process simulation is used as an input trace file in NS2 simulation. In this phase, optimum encoder parameter is selected to reduce the bit rate and at the same time maintain the PSNR value before transmitting in WSN medium. The pre-process phase is very important to ensure bandwidth or data rate required for video transmission can be supported by the wireless medium. Most of works in literature for video application in Shin and Chung (n.d.), Xiao et al. (2008), and Ma et al. (2010), do not consider optimising encoded

Table 2. Performance metric for video quality in PSNR and MOS

Peak Signal to Noise Ratio (PSNR)	Mean Opinion Score (MOS)
> 37	5 (Excellent)
31-37	4 (Good)
25-31	3 (Fair)
20-25	2 (Poor)
< 20	1 (Bad)

parameters because the protocol implemented in IEEE 802.11 provide high bandwidth.

The pre-process phase is done offline at Cgywin window. Xvid_encraw.exe, ffmpeg.exe, MP4box.exe and mp4trace.exe encoders are used in pre-process configuration. Xvid_encraw codec is used to encode a yuv sequence into MPEG4 data format and create compressed raw video without B frame. Meanwhile, ffmpeg codec is used to convert MPEG-4 video into yuv format. MP4Box codec is used to packetize the frame for transport layer. Then, MP4trace codec is used to generate video trace file as an input in NS2 simulator. Video trace file consists of frame number, frame type, frame size, number of packets after packetization and sender time (Ke, Shieh, Hwang, & Ziviani, 2008). The encoded parameters selection in pre-process, constrain and threshold values are list below:

1. **Encoded parameters:** Quantization Scale (QS), frame per second (fps) and pattern of Group of picture (GOP).
2. **Constrains:** Bandwidth in WSN, Peak Signal Noise Ratio (PSNR)
3. **Threshold:** Bandwidth < 250 kbps, PSNR above 35

Quantization scale is used to adjust the value of video compression (QS = between 2-30) at the encoder side. Frame per second (fps) indicate how many frames that can be transmitted in one second. Frames can be grouped into sequences called a

group of pictures (GOP). GOP in MPEG-4 video consists of I, P and B frames which have different levels of important. To reduce the bandwidth requirement, the GOP is constructed without the less important frame B. All these parameters (QS, fps and GOP) are varied in order to meet the requirement for the video in WSN environment while preserving the original video. The encoded video will be compared with original video before transmitting into WSN channel. The *PSNR* value after encoding (PSNRi) must be above 35 that are categorized as a good picture based on performance metric shown in Table 2.

The values for encoded parameter selection for low, medium and high motion video are different because, as illustrate in Table 1, the bandwidth requirement for high motion is larger than low motion. The details of this work and results are explained in Yunus, Ariffin, Ismail, and Hamid (in press).

4.1.2 Proposed Hybrid MAC Layer Design in IEEE 802.15.4

The high requirements in wireless technology applications such as real time and delay constraints bring the necessity to achieve an efficient MAC layer design in WSN environment. For video transmission in WSN, the selection of the free collision medium is very important especially in low bit rate network such IEEE 802.15.4 that requires reliable communication. Hybrid MAC protocol is selected as a suitable MAC protocol for video applications in IEEE 802.15.4. *Hybrid MAC* is combination of CSMA and TDMA protocol. Nevertheless in this work, CSMA protocol will be combined with unsynchronized TDMA protocol which is token protocol. The main motivation of using *token approach* is because there is no need for strict time synchronization which can increase the energy consumption. In addition, token approach provides dynamic frame length adjustment to enhance the data throughput, and works with partial connectivity that can avoid hidden terminal problem and can guarantee fair node access.

Even though, hybrid MAC is already available in beacon enable mode in IEEE 802.15.4 standard, the poor network performance for high traffic data and difficulties to achieve synchronization for real test bed as mentioned in the previous section makes this mode not preferable for channel access to transmit video application. The hybrid MAC layer design proposed in this chapter is the modified version of NBE mode in MAC IEEE802.15.4 standard. NBE mode works fully in CSMA protocol; however, token protocol is added in this mode to provide collision free medium access. The combination of token protocol and CSMA protocol in NBE mode can increase the throughput of the network and decrease energy consumption.

The proposed MAC layer is divided into two phases, which are setup phase and steady state phase. The setup phase only runs once at the beginning of the network setup. Once the event is triggered, all the nodes follow the protocol in steady state phase. Different MAC protocols are implemented for both phases as stated below:

1. Setup phase = Full CSMA protocol.
2. Steady state phase = hybrid MAC (Token protocol and CSMA protocol)

The hybrid MAC can only be implemented in steady state phase where the video starts to be transmitted. For the development of the proposed protocol, the following assumptions about the network model are taken into consideration. First, there is only one source node that generates video data and at the same time responsible to generate a new token. The other nodes in the network act as an intermediate node that relays data until it reaches the destination node (sink node). Second, the intermediate node cannot generate a new token in the network, but it is able to receive and forward the token to the next token holder (NTH). Intermediate nodes only allow regener-

ating the same token if NTH fails to receive the token. Third, each node in the networks shares the same frequency to transmit and receive data. Besides, the antenna works in half duplex mode, where the node cannot receive and transmit at the same time. Lastly, Real Time Load Distribution (RTLD) routing protocol is used to find the route to destination (Ahmed & Fisal, 2010).

4.1.2.1 Setup Phase

In the setup phase, only CSMA protocol is used. Level implementation is done in this phase where each node is assigned with level id that indicates the hop distance from the source node. For example, node A is two hops away from the source node and its level id will be equal to 2. Each node will set level id itself after receiving level message from its neighbour. Source node is set with level 0 and other nodes are set to default level at the beginning of setup phase. Source node will be responsible to start the transmission of the level message at the beginning of the network setup. Then, other nodes only can broadcast one level message after receiving level message from their neighbour. The following are the steps in setup phase for level implementation:

Step 1: Source node starts the setup phase by broadcasting level message with level id equal to 0.

Step 2: The nodes within source node coverage range that received the message update their level by 1 and store the sender id as their one hop neighbour.

Step 3: Nodes in step 2 that have already updated their level will broadcast new level message that contains its new level id to their neighbour.

Step 4: Neighbour nodes that received the level message store one hop neighbour and update their level id based on algorithm below:

If (The level id is default level id);

$$level\,ID = RECV\,level_{id} + 1$$

Else if (Received level message contains level lower than its level id);

$$level\,ID = RECV\,level_{id} + 1$$

Else if (Received level message that contains level higher than its level id);

$$level\,ID = level\,ID$$

Else;

$$level\,ID = level\,ID$$

Step 5: Node in step 4 will send the level message if it has not broadcast any level message yet.

Step 6: Step 5 and 6 will be repeated until all the nodes in the network are assigned with level id.

There can be high probability of collisions among the level messages at the setup phase because some nodes are within communication range of each other and utilize the same frequency band. This may lead to some level messages failed to be received by neighbour nodes. As a result, some nodes cannot set their level value. In order to reduce the probability of collisions at setup phase, each node will take another random delay time before transmitting the level message into the medium. In the setup phase, when the node starts receiving the first level message, it sets a timer to control when this phase for their node would end and waits for steady state to begin. While waiting for steady state to begin, nodes will be in idle mode to save the energy consumption.

4.1.2.2 Steady State Phase

The steady state phase is a phase where the data begins to be transmitted into the network. Token

protocol and CSMA protocols are used in this phase. As mentioned in the introduction section, there are 5 layers that cooperate in WSN architecture and perform different tasks. Before any transmission can be done, routing protocol in network layer will broadcast control packet to find the next hop node in the network. Another control packet is added in this proposed protocol which is token packet. To increase the flexibility in this protocol, control packet (routing control packet and token control packet) can be sent in the medium using CSMA protocol without waiting for the token. However, for video/multimedia data, where the reliable transmission is important, a node should have the token first, before using CSMA protocol. The CSMA protocol is also used in video/multimedia data transmission to sense the channel condition and ensure the channel is idle when video/multimedia is transmitted. Through this method, frame collision can be minimized.

At the beginning of the steady state phase, source node generates the token. Source node holds the token until an event (video application) is triggered. Then source node transmits the data until time holding token (THT) expires. After that, the token is passed to the next token holder in the network. The selection of the NTH is based on next hop information in routing protocol. After the token reaches the intermediate node that is two hops away from source node, the source node will generate a new token. The detail of operation in token generation is explained in Ismail et al. (2013).

The *token protocol* proposed for a wireless network is different from traditional token protocols in a wired network. Below is the list of token protocols in wired networks that are not suitable to be implemented in wireless networks and its proposed solution:

1. Nodes are placed in one closed loop ring, but in WSN where nodes consists of distributed nodes deployed in a large area, one closed loop ring is not suitable.

○ **Solution in WSN:** Multiple closed loops and only allow a node that has the data to join the closed loops. The nodes that join the closed loops are not fixed.

2. When the station grabs the token, it will send all the frames in the buffer before passing the token to the next station. In video application in wireless network, this method will increase the delay and frames will be dropped because it reaches the deadline.

○ **Solution in WSN:** Proposing a dynamic token holding time. Each node can transmit the data in THT and if it finishes before THT expires, it hands over immediately to the NTH.

3. All the nodes are connected to each others in closed loop ring. If the node has data to send, it grabs the token and send the data, if not, the token will be passed to other nodes. In WSN that has a large number of nodes, passing the token from one node to another node that has no data to send results in high delay.

○ **Solution in WSN:** Pass the token to the node that has data to transmit. Propose cross layer design with network layer. In WSN, the node will forward the data to next hop in the network based on routing information.

4.2. Performance Analysis

In this section, we evaluate the basic performance of using token protocol in WSN environment. The performance of one hop transmission in WSN that used combination of token approach with CSMA protocol is compared with protocol that used CSMA only. In token approach protocol, access mechanism is fully depending on token permit. A node can transmit the data if it holds the token at that time. Meanwhile, for a random access

mechanism like CSMA protocol, each node can send data any time if it senses the channel is idle/free while running CCA. If the channel is busy, the node backoff and senses the channel again.

Figure 1 show the ring network topology used in this experiment. Node 0 to 5 acts as source nodes and node 6 in the middle is a sink node (destination). The network performance is analyzed when the number of source nodes that generate the data packet is varied. Based on Equation (1), N represents the total number of source nodes in the network, n is the number of source nodes that generate the data and m indicates the number of source nodes that do not generate any data or has no data to send. In this analysis, n will increase one by one until $n = N$ and $m = 0$, which means that all source node generate the data in the networks.

$$n = N - M \qquad (1)$$

In this experiment, the number of source nodes that generates the data is varied from 1 to 6 to see the performance in term of delivery ratio, fairness, energy consumption and delay. Node 0 will hold the token first and transmit the data. After THT is expired, it will pass the token to other nodes (in this case it pass to its neighbour node). After the node receives the token, whether it has data to transmit or not, it will hold the token in THT, after THT expired, it passes it to the next token holder.

The following five figures present the results using network simulation tools NS2 version 2.34. The performance is evaluated according to the following metrics, which are the performance deciding entities of any WSNs:

1. **Packet delivery ratio:** Represents the ratio between the number of packets generated by the application layer at source node and the packets received at the destination.

2. **Energy consumption:** It denotes the total energy consumed by all nodes in the network.

Figure 1. Network topology

3. **Average end to end delay:** The average end to end delay which calculates the average time required to receive the packet.

4. **Fairness index**: We use Jain's fairness index to evaluate the fairness among the flows. For a given set of flows of throughput (x_1, x_2, x_3,...x_n), the fairness index is defined in Equation (2) for $f(x)$:

$$f\left(x_1, x_2, \ldots x_n\right) = \frac{\left(\sum_{i=1}^{n} x_i\right)^2}{n \times \sum_{i=1}^{n} x_i^2} \qquad (2)$$

where n stands for the number of nodes that participate in sending the data packets in the network and x_i for the throughput of the ith node. *Fairness index* is always between 0 and 1. A lower

value implies poorer fairness. If the throughputs achieved by all the senders are the same, then the fairness is 1.

Figure 2 illustrates the result of packet delivery ratio for both protocols. The packet delivery is maintained for a token approach protocol which is above 90% reception, but CSMA protocol shows 14.54% average decrement when n increases. This is because, in CSMA protocol, the increasing number of source nodes in the network will increase the number of nodes that content to get channel access in the network. However, when n is small, packet delivery ratio shows good performance as token approach protocol.

Meanwhile, in token approach protocol, only nodes that have a token will transmit the data and the others will be in receiving or in idle mode. After time holding token (THT) expires, the token

Figure 2. Packet delivery ratio

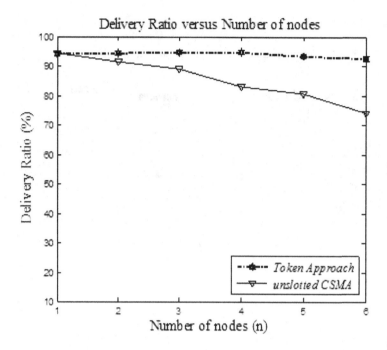

will hand over the token to other nodes. Each node is granted equal channel access time which is THT time. Through this method, fairness index for token approach protocol achieves better performance as shown in Figure 3. A fairness measure is used to determine whether a node in the network is receiving a fair share of system resources. The fairness index for token approach close to 1 indicates that the bandwidth sharing is fair. As shown in Figure 3, the bandwidth sharing for CSMA protocol is unfair because each node does not get equal transmission opportunity. Any nodes that select low backoff period will have high opportunity to access the channel.

There is a possibility that more than one node sense the medium at the same time. This is called the hidden terminal problem. When this problem happens, nodes transmit data at the same time and cause collision or link quality problem. The link quality problem happens when two packets are received at the same time. In CSMA protocol, packet will retransmit again and this increases the energy consumption in the network as depicted

in Figure 4. The energy consumption for CSMA protocol is higher than token approach protocol when the number of contended nodes increases.

Figure 5 shows end to end delay for both protocols. The CSMA protocol achieves good performance for end to end delay compared to token approach. This is because, node in token approach should wait for token before it transmits. Besides, in this scenario, whether the node has data or not, it keeps the token until THT is expired, which contributes to longer delays. To reduce end to end delay for token approach protocol, the other scenario is created where if the node receives the token but does not have data to sent, it will immediately pass the token to the NTH in the network.

Figure 6 shows the comparison of end to end delays for token approach protocol with two scenarios. In scenario 1, after received the token, whether node has a data or not, it will hold the token within THT. After THT expires, it is passed to next token holder. In scenario 2, if the node has data to transmit, it will hold the token, but if node does not have data, it will pass the token to next

Figure 3. Fairness index

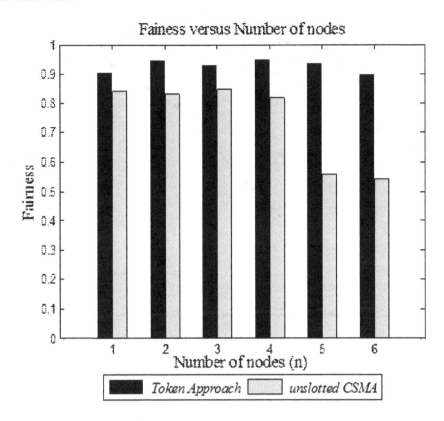

Figure 4. Energy consumption

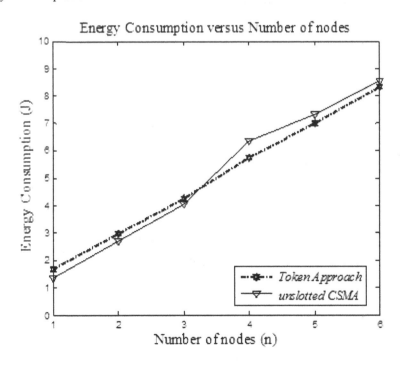

Figure 5. End to end delay

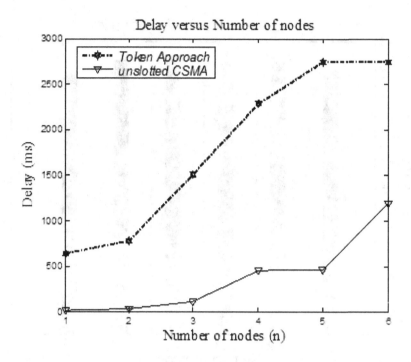

Figure 6. End to end delay

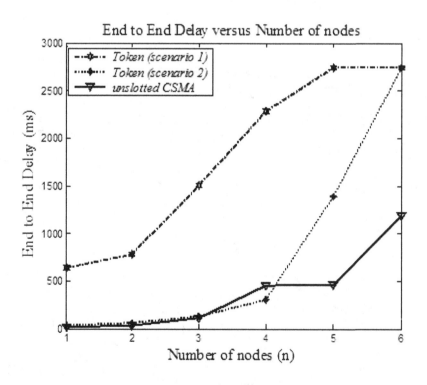

token holder without holding until the THT is expired. As shown in Figure 6, end to end delay for token approach protocol in scenario 2 is reduced compared to scenario 1. Meanwhile for *n* lower than 4, the end to end delay for scenario 2 and CSMA protocol is almost the same. From this figure, it can be concluded that token approach protocol can achieve good performance in terms of delay similar to that of CSMA protocol if nodes that join the closed loop ring at one time are less than 4 nodes. The performance analysis of one hop transmission can be a benchmarked for multi-hop transmission in WSN.

5. FUTURE RESEARCH DIRECTIONS

The recommendation for future research direction for MPEG-4 video application is in hardware development. It is suggested that the industry should cooperate with academic researchers to develop the video compression method that can be supported by IEEE 802.15.4 wireless medium since the analysis of encoded parameters can only be done offline in CGYWIN window. Most of the research on video applications in IEEE 802.11 or IEEE 802.15.4 undergoes simulations only without attempting the implementation of a real test bed. If there is an implementation with real test bed, the frame reception at the destination is very low around 4 or 5 images per second. The development of the MPEG-4 codec that can reduce the bit rate to a level that can be supported by IEEE 802.15.4 is indispensable.

6. CONCLUSION

The rapid advance in wireless protocol design fosters the development of video transmission in low data rate networks based on IEEE 802.15.4. The transmission of video into WSN environment gives a good promise for green technology in the future development. Even though there are much more challenges that should be considered, the solution and recommendation suggested in this chapter can overcome certain challenges. A hybrid MAC layer that combines CSMA protocol and unsynchronized TDMA which is token approach protocol is a preferable MAC protocol that can be used in video transmission. The hybrid MAC design can increase the network performance in terms of scalability, delay, throughput, and energy efficiency. In the near future it is expected that video-based wireless sensor networks will make these demands become reality with low power and low cost devices.

REFERENCES

Abad, M. F. K., & Jamali, M. A. J. (2011). Modify LEACH algorithm for wireless sensor network. *International Journal of Computer Science*, 8(5), 219–224.

Abdul Hadi, F, A. H., Rozahe, A. R., Norsheila, F., Kamilah, S. Y., Sharifah, H.S.A., & Liza, L. (2009). Development of IEEE802. 15.4 based wireless sensor network platform for image transmission. *International Journal of Engineering & Technology*, 9(10), 112–118.

Aghdasi, H. S., & Abbaspour, M. (2008). ET-MAC: An energy-efficient and high throughput MAC protocol for wireless sensor networks. In *Proceedings of Communication Networks and Services Research Conference,* (pp. 526-532). Academic Press.

Ahmed, A., & Fisal, N. (2010). Secure real-time routing protocol with load distribution in wireless sensor networks. *Security and Communication Networks*, 4(8), 839–869. doi:10.1002/sec.214

Akyildiz, I. F., Melodia, T., & Chowdhury, K. R. (2007). A survey on wireless multimedia sensor networks. *Computer Networks*, 51(4), 921–960. doi:10.1016/j.comnet.2006.10.002

Bagci, F. (2008). ESTR-Energy saving token ring protocol for wireless sensor networks. In *Proceedings of International Conference on Wireless Networks,* (pp. 3-9). IEEE.

Burda, R., & Wietfeld, C. (2007). Multimedia over 802.15. 4 and ZigBee networks for ambient environment control. In *Proceedings of Vehicular Technology Conference.* IEEE.

Chen, F. (2009). Improving IEEE 802.15. 4 for low-latency energy-efficient industrial applications. *Aktuelle Anwendungen in Technik und Wirtschaft*, 61-70.

Choi, W., & Lee, S. (2011). Implementation of the IEEE 802.15.4 module with CFP in NS-2. *Telecommunication Systems*, *52*(4), 2347–2356. doi:10.1007/s11235-011-9548-7

Dargie, W., & Poellabauer, C. (2010). *Fundamentals of wireless sensor networks: Theory and practice.* Hoboken, NJ: John Wiley & Sons, Ltd. doi:10.1002/9780470666388

Dash, S., Swain, A. R., & Ajay, A. (2012). Reliable energy aware multi-token based MAC protocol for WSN. In *Proceedings of International Conference on Advanced Information Networking and Applications,* (pp. 144-151). IEEE.

Ganz, A., Ganz, Z., & Wongthavarawat, K. (2003). *Multimedia wireless networks: Technologies, standards and QoS.* Englewood Cliffs, NJ: Pearson Education.

Garcia-Sanchez, A.-J., Garcia-Sanchez, F., & Garcia-Haro, J. (2008). Feasibility study of MPEG-4 transmission on IEEE 802.15.4 networks. In *Proceedings of International Conference on Wireless and Mobile Computing, Networking and Communications,* (pp. 397-403). IEEE.

Garcia-Sanchez, A.-J., Garcia-Sanchez, F., & Garcia-Haro, J. (2011). Wireless sensor network deployment for integrating video-surveillance and data-monitoring in precision agriculture over distributed crops. *Computers and Electronics in Agriculture*, *75*(2), 288–303. doi:10.1016/j. compag.2010.12.005

Garcia-Sanchez, A.-J., Garcia-Sanchez, F., Garcia-Haro, J., & Losilla, F. (2010). A cross-layer solution for enabling real-time video transmission over IEEE 802.15.4 networks. *Multimedia Tools and Applications*, *51*(3), 1069–1104. doi:10.1007/s11042-010-0460-z

Ke, C., Shieh, C., Hwang, W., & Ziviani, A. (2008). An evaluation framework for more realistic simulations of MPEG video transmission. *Journal of Information Science and Engineering*, *440*, 425–440.

Klaue, & Rathke. (2003). EvalVid – A framework for video transmission and quality evaluation. In *Proceedings of International Conference on Modelling Techniques and Tools for Computer Performance Evaluation* (pp. 225-272). IEEE.

Koutitas, G., & Demestichas, P. (2010). A review of energy efficiency in telecommunication networks. *Telfor Journal*, *2*(1), 2–7.

Lin, M., Leu, J., & Yu, W. (2011). On transmission efficiency of the multimedia service over IEEE 802.15. 4 wireless sensor networks. In *Proceedings of International Conference on Advanced Communication Technology (ICACT),* (pp. 184-189). ICACT.

Ma, Z., Zhou, J., & Wu, T. (2010). A cross-layer QoS scheme for MPEG-4 streams. In Proceedings of Wireless Communications, Networking and Information Security (WCNIS), (pp. 392-396). IEEE.

Man, L. (2006). *Part 15.4: Wireless medium access control (MAC) and physical layer (PHY) specifications for low-rate wireless personal area networks (WPANs)*. Retrieved from http://scholar.google.com/scholar?hl=en&btnG=Search&q=intitle:IEEE+Standard+for+Information+technology-+Telecommunications+and+information+exchange+between+systems-+Local+and+metropolitan+area+networks-+Specific+requirements--Part+15.4:+Wireless+MAC+and+PHY+Specifications+for+Low-Rate+WPANs#0

Misra, S., Reisslein, M., & Xue, G. (2008). A survey of multimedia streaming in wireless sensor networks. *IEEE Communications Surveys & Tutorials, 10*(4), 18–39. doi:10.1109/SURV.2008.080404

Nor-Syahidatul. Ismail, Yunus, & Ariffin. (2011). *MPEG-4 video transmission using distributed TDMA MAC protocol over IEEE 802.15.4 wireless technology*. Paper presented at International Conference on Modelling, Simulation And Applied Optimization. Kuala Lumpur, Malaysia.

Nor-Syahidatul, Ismail, Ariffin, Latiff, & Yunus. (2013). Medium access control with token approach in wireless sensor network. *International Journal of Computers and Applications, 65*(17), 51–57.

Rashid, R., Embong, W. M. A. E., Zaharim, A., & Fisal, N. (2009). Development of energy aware TDMA-based MAC protocol for wireless sensor network system. *European Journal of Scientific Research, 30*(4), 571–578.

Ray, S., Dash, S., & Tarasia, N. (2011). Congestion-less energy aware token based MAC protocol integrated with sleep scheduling for wireless sensor networks. In *Proceedings of World Congress on Engineering 2011* (Vol. 2, pp. 4-9). Academic Press.

Rhee, I., & Warrier, A. (2009). DRAND: Distributed randomized TDMA scheduling for wireless ad hoc networks. *IEEE Transactions on Mobile Computing, 8*(10), 1384–1396. doi:10.1109/TMC.2009.59

Roy, A., & Sarma, N. (2010). Energy saving in MAC layer of wireless sensor networks: A survey. In *Proceedings of National Workshop in Design and Analysis of Algorithm (NWDAA)*. Tezpur University.

Shin, P., & Chung, K. (2008). *A cross-layer based rate control scheme for MPEG-4 video transmission by using efficient bandwidth estimation in IEEE 802.11e*. Paper presented at International Conference Information Networking. Busan.

Sikora, T. (1997). The MPEG-4 video standard verification model. *IEEE Transactions on Circuits and Systems for Video Technology, 7*, 19–31. doi:10.1109/76.554415

Sitanayah, L., Sreenan, C. J., & Brown, K. N. (2010). ER-MAC: A hybrid MAC protocol for emergency response wireless sensor networks. In *Proceedings of 2010 Fourth International Conference on Sensor Technologies and Applications*, (pp. 244-249). IEEE.

Song, W.-Z., Huang, R., Shirazi, B., & LaHusen, R. (2009). TreeMAC: Localized TDMA MAC protocol for real-time high-data-rate sensor networks. *Pervasive and Mobile Computing, 5*(6), 750–765. doi:10.1016/j.pmcj.2009.07.004

Suh, C., Mir, Z. H., & Ko, Y.-B. (2008). Design and implementation of enhanced IEEE 802.15.4 for supporting multimedia service in wireless sensor networks. *Computer Networks, 52*(13), 2568–2581. doi:10.1016/j.comnet.2008.03.011

Udayakumar, P., Vyas, R., & Vyas, O. (2012). Token bus based MAC protocol for wireless sensor networks. *International Journal of Computers and Applications*, *43*(10), 6–10. doi:10.5120/6137-8373

Wang, P., & Zhuang, W. (2008). A token-based scheduling scheme for WLANS supporting voice/data traffic and its performance analysis. *Wireless Communications*, *7*(4), 1708–1718. doi:10.1109/TWC.2007.060889

Warrier, A., Aia, M., & Sichitiu, M. L. (2008). Z-MAC: A hybrid MAC for wireless sensor networks. *IEEE/ACM Transactions on Networking*, *16*(3), 511–524. doi:10.1109/TNET.2007.900704

Wei, F., Zhang, X., Xiao, H., & Men, A. (2012). A modified wireless token ring protocol for wireless sensor network. In *Proceedings of 2012 2nd International Conference on Consumer Electronics, Communications and Networks (CECNet)*, (pp. 795-799). CECNet.

Xiao, Y., Du, X., Hu, F., & Zhang, J. (2008). A cross-layer approach for frame transmissions of MPEG-4 over the IEEE 802.11e wireless local area networks. In *Proceedings of Wireless Communication and Networking Conference*, (pp. 1728-1733). IEEE.

Yunus, F., Ariffin, S. H. S., Ismail, N., Syahidatul, N., & Hamid, A. H. F. A. (2013). Optimum parameters for MPEG-4 data over wireless sensor network. *International Journal of Engineering and Technology*.

Zainaldin, A., Lambadaris, I., & Nandy, B. (2008). Adaptive rate control low bit-rate video transmission over wireless Zigbee networks. In *Proceedings of Communications, 2008, ICC'08. IEEE International Conference*, (pp. 52-58). IEEE.

ADDITIONAL READING

Aghdasi, H., Abbaspour, M., Moghadam, M., & Samei, Y. (2008). An Energy-Efficient and High-Quality Video Transmission Architecture in Wireless Video-Based Sensor Networks. *Sensors (Basel, Switzerland)*, 4529–4559. doi:10.3390/s8084529

Bishnoi, L. C., Singh, D., & Mishra, S. (2011). Simulation of Video Transmission over Wireless IP Network in Fedora Environment. *IP Multimedia Communications*, 9-14.

Cheng, S. -tzong, & Lin, J.-liang. (2006). Reliable Video Transmission Techniques for Wireless MPEG-4 Streaming Systems. *International Symposium on Dependable Computing (PRDC'06)*, 248-258.

Clematis, A., & Corana, A. (2006). Ring Algorithms on Heterogeneous Windows-based Clusters with Various Message Passing Environments. *Current & Future Issues of High-End Computing*, *33*, 195–202.

Ditze, M., Klobedanz, K., & Guido, K. (2005). Scheduling MPEG-4 Video Streams Through the 802. 11e Enhanced Distributed Channel Access. *Networking-ICN*, 1071-1079.

Gao, J., Hu, J., & Min, G. (2009). Performance Modelling of IEEE 802.15.4 MAC in LR-WPAN with Bursty ON-OFF Traffic. *International Conference on Computer and Information Technology.*, 58-62.

Garcia-Sanchez, F., Garcia-Sanchez, A.-J., Losilla, F., & Garcia-Haro, J. (2010). A nomadic access mechanism for enabling dynamic video surveillance over IEEE 802.15.4 networks. *Measurement Science & Technology*, *21*(12), 124006. doi:10.1088/0957-0233/21/12/124006

Jafari, M., & Kasaei, S. (2005). Prioritisation of Data Partitioned MPEG-4 for Streaming Video in GPRS Mobile Networks. *2005 1st IEEE and IFIP International Conference in Central Asia on Internet*, 1-7.

Kim, J., & Lee, W. (2007). Power saving medium access for beacon-enabled IEEE 802.15. 4 LR-WPANs. *Proceedings of the 2007 conference on Human interface: Part II*, 555--562.

Koubâa, Anis, Cunha, A., Alves, M., & Tovar, E. (2008). TDBS: a time division beacon scheduling mechanism for ZigBee cluster-tree wireless sensor networks. *Real-Time Systems*, *40*(3), 321–354. doi:10.1007/s11241-008-9063-4

Koubâa, A., Alves, M., & Tovar, E. (2005). *IEEE 802.15. 4 for Wireless Sensor Networks: A Technical Overview. Technical Report (TR-050702)*. Retrieved from https://www.cister.isep.ipp.pt/docs/ieee%2B802%252E15%252E4%2Bfor%2Bwireless%2Bsensor%2Bnetworks%253A%2Ba%2Btechnical%2Boverview/222/view.pdf

Krishnamurty, V., & Sazonov, E. (2008). Reservation-based Protocol for Monitoring Applications Using IEEE 802.15.4 Sensor Networks. *International Journal of Sensor Networks*, Retrieved from http://inderscience.metapress.com/index/47231J14646RJ027.pdf

Latré, Benoît, Mil, P., & Moerman, I. (2006). Throughput and Delay Analysis of Unslotted IEEE 802.15. 4. *Journal of Network*, *1*(1), 20–28.

Latré, B், & Mil, P. D. (2005). Maximum throughput and minimum delay in IEEE 802.15. 4. *Mobile Ad-hoc and Sensor Networks*, 866-876.

Lindeberg, M., Kristiansen, S., Plagemann, T., & Goebel, V. (2010). Challenges and Techniques for Video Streaming over Mobile Ad Hoc Networks. *Multimedia Systems*, *17*, 51–82. doi:10.1007/s00530-010-0187-8

Miller, D. a., Tilak, S., & Fountain, T. (2005). Token Equilibria in Sensor Networks with Multiple Sponsors. *2005 International Conference on Collaborative Computing: Networking, Applications and Worksharing* (pp. 1-5).

Mishra, a., & Papadimitratos, P. (2006). WSN11-4: A Cross Layer Design of IEEE 802.15.4 MAC Protocol. *IEEE Globecom 2006*, 1-6.

Misra, S., Reisslein, M., & Xue, G. (2008). A Survey of Multimedia Streaming in Wireless Sensor Networks. *IEEE Communications survey & tutorial* (Vol. 10, pp. 18-39).

Muñoz-Jaramillo, A., Nandy, D., Martens, P. C. H., & Yeates, A. R. (2010). a Double-Ring Algorithm for Modeling Solar Active Regions: Unifying Kinematic Dynamo Models and Surface Flux-Transport Simulations. *The Astrophysical Journal*, *720*(1). doi:10.1088/2041-8205/720/1/L20

Ramachandran, I., Das, A. K., & Roy, S. (2006). Analysis of the contention access period of IEEE 802.15.4 MAC. *ACM Journal, V*, 1-29. Retrieved from https://www.prism.uvsq.fr/~mogue/SENSORS/MAC/Sensor_Heykel/zig_802_15_4_Analysis.pdf

Razzaque, M. A., & Hong, C. S. (2008). Multi-Token Distributed Mutual Exclusion Algorithm. *22nd International Conference on Advanced Information Networking and Applications (aina 2008)*, 963-970.

Shih, B.-Y., Chang, C.-J., Chen, A.-W., & Chen, C.-Y. (2010). Enhanced MAC channel selection to improve performance of IEEE 802.15. 4. *international Journal of Innovative Computing. Information and Control*, *6*(12), 5511–5526.

Shin, Y.-S., Lee, K.-W., & Ahn, J.-S. (2010). Analytical Performance Evaluation of IEEE 802.15. 4 with Multiple Transmission Queues for Providing Qos Under Non-Saturated Conditions. *Communications (APCC), 2010 16th Asia-Pacific Conference*, 334-339.

Sokullu, R., Dagdeviren, O., & Korkmaz, I. (2008). On the IEEE 802.15.4 MAC Layer Attacks: GTS Attack. *2008 Second International Conference on Sensor Technologies and Applications (sensorcomm 2008)*, 673-678.

Wan, Z.-B., & Xiao, Y. WAng, Q., & Gao, Z.-G. (2010). WSN nodes for traffic data-acquisition system. *2010 2nd International Conference on Advanced Computer Control*, 142-146.

Wang, Q., Hempstead, M., & Yang, W. (2006). A Realistic Power Consumption Model for Wireless Sensor Network Devices. *2006 3rd Annual IEEE Communications Society on Sensor and Ad Hoc Communications and Networks*, 286-295.

Wang, Wei, Hempel, M., Peng, D., Wang, H., Sharif, H., Member, S., & Chen, H.-hwa. (2010). On Energy Efficient Encryption for Video Streaming in Wireless Sensor Networks. *IEEE Transactions on Multimedia*, *12*, 417–426. doi:10.1109/TMM.2010.2050653

Wang, W., & Wang, H. (2006). An energy efficient pre-schedule scheme for hybrid CSMA/TDMA MAC in wireless sensor networks. *Systems*, Retrieved from http://ieeexplore.ieee.org/xpls/abs_all.jsp?arnumber=4085822

Zacharias, S., & Newe, T. (2010). Technologies and Architectures for Multimedia-Support in Wireless Sensor Networks. *Smart Wireless Sensor Networks*, 373-381.

Zand, P., & Shiva, M. (2008). Defining a new frame based on IEEE 802.15. 4 for having the synchronized mesh networks with channel hopping capability. *Communication Technology, 2008. ICCT, 2008*, 54–57.

KEY TERMS AND DEFINITIONS

Energy Consumption: Total energy used by all the nodes in the network.

Hybrid MAC: A combination of more than one MAC protocol, which is CSMA and TDMA protocol.

IEEE 802.15.4 Standard: IEEE standard protocol for low data wireless personal area networks (LR-WPAN) and design for low cost communication.

MPEG-4: Moving Picture Expert Group (MPEG) is a standard for encoding techniques (audio and video coding) and aimed primarily at low bit rate video communication.

Quality of Service (QoS): The special requirement that should be achieved by certain application such as to guarantee a certain level of performance of data flow.

Token Protocol: The communication based on token availability which is the node that owns the token get the channel access into the medium.

Wireless Sensor Network (WSN): The network that connecting connects without cable and consists of distributed sensor network to monitor physical or environment condition.

Chapter 17
Graph Intersection–Based Benchmarking Algorithm for Maximum Stability Data Gathering Trees in Wireless Mobile Sensor Networks

Natarajan Meghanathan
Jackson State University, USA

Philip Mumford
Air Force Research Lab (RYWC), USA

ABSTRACT

The authors propose a graph intersection-based benchmarking algorithm to determine the sequence of longest-living stable data gathering trees for wireless mobile sensor networks whose topology changes dynamically with time due to the random movement of the sensor nodes. Referred to as the Maximum Stability-based Data Gathering (Max.Stable-DG) algorithm, the algorithm assumes the availability of complete knowledge of future topology changes and is based on the following greedy principle coupled with the idea of graph intersections: Whenever a new data gathering tree is required at time instant t corresponding to a round of data aggregation, choose the longest-living data gathering tree from time t. The above strategy is repeated for subsequent rounds over the lifetime of the sensor network to obtain the sequence of longest-living stable data gathering trees spanning all the live sensor nodes in the network such that the number of tree discoveries is the global minimum. In addition to theoretically proving the correctness of the Max.Stable-DG algorithm (that it yields the lower bound for the number of discoveries for any network-wide communication topology like spanning trees), the authors also conduct exhaustive simulations to evaluate the performance of the Max.Stable-DG trees and compare to that of the minimum distance spanning tree-based data gathering trees with respect to metrics such as tree lifetime, delay per round, node lifetime and network lifetime, under both sufficient-energy and energy-constrained scenarios.

DOI: 10.4018/978-1-4666-5170-8.ch017

INTRODUCTION

A wireless sensor network is a network of sensor nodes that gather data about the deployed environment, intelligently process them and forward to a central control center called the sink that is typically located far away from the monitored network field. Traditionally, wireless sensor networks are considered to gather data for static environments, where mobility in any form - sensor nodes, users and the monitored phenomenon are totally ignored. Recently, with the prolific growth of embedded systems and ubiquitous computing technologies, wireless mobile sensor networks (WMSN) have been envisioned and successfully deployed for varied applications such as traffic monitoring, route planning, civil infrastructure monitoring, geo-imaging, etc. (Hull et al., 2006). A WMSN could be either a homogeneous or heterogeneous network of nodes (a general term used to refer sensor-equipped computers, mobile phones and vehicles) equipped with sensing capabilities like a camera sensor, microphone, GPS sensor, etc (Kansal et al., 2007). By leveraging the mobility of the sensor nodes, a WMSN could be used to monitor and collect data over a relatively larger region as well as use fewer sensor nodes for a given region, compared to static sensor networks.

A WMSN has constraints (limited battery charge, memory, processing capability, bandwidth and transmission range) similar to that of the static sensor networks. Two nodes that are outside the transmission range of each other cannot communicate directly. Though energy constraints are attributed for the limited transmission range of the sensor nodes, transmissions over a longer distance are likely to suffer from interference and collisions. Due to all of the above resource and operating constraints, it would not be a viable solution to require every sensor node to directly transmit their data to the sink over a longer distance. This necessitates the need for developing energy-efficient data gathering algorithms to determine communication topologies that can be used to effectively combine data collected at these sensor nodes and send only the aggregated data (that is a representative of the entire network) to the sink.

A majority of the data gathering algorithms available in the literature are meant for static sensor networks (networks with sensor nodes staying fixed at a particular location) with either a static sink (Lindsey et al., 2002; Heinzelman et al., 2000) or mobile sink (Srinivasan & Wu, 2008; Vlajic & Stevanovic, 2009). Tree-based communication topologies for data gathering have been observed (Meghanathan, 2012) to be the most energy-efficient (in terms of the number of link transmissions) for static sensor networks. However, in the presence of dynamic node mobility (like in a WMSN), tree-based communication topologies could require frequent reconfigurations. Hence, mobility brings in an additional dimension of constraint to a WMSN: we need algorithms to determine stable data gathering trees that can exist for a longer time and do not require frequent reconfigurations.

In this research, we address the problem of developing a benchmarking algorithm to determine the sequence of stable data gathering trees for mobile sensor networks such that the number of tree reconfigurations is the global minimum. In this pursuit, we present the Max.Stable-DG algorithm that operates according to the following greedy principle: Given the complete knowledge of future topology changes, whenever a data gathering tree is required at time instant t, the algorithm chooses the longest-living data gathering tree from t. This strategy is repeated over the duration of the data gathering session, resulting in the generation of a sequence of longest-living data gathering trees (DG-trees) called the *Stable-Mobile-DG-Tree*. If n is the number of nodes in the network and T is the number of rounds of data gathering, the worst-case run-time complexity of the Max.Stable-DG algorithm is $O(n^2 T \log n)$ and $O(n^3 T \log n)$ when operated under sufficient-energy and energy-constrained scenarios respectively. The $n^2 \log n$ factor in the above run-time complexity is the worst-case run-time complexity of the minimum-weight spanning tree algorithm, Prim's algorithm (Cormen et al., 2009) used to determine the underlying spanning trees from which the data gathering trees are derived.

The contents presented in this chapter are an extension of our earlier work (Meghanathan & Mumford, 2012) on stable data gathering trees for WMSNs. To the best of our knowledge, we have not come across any other work on stable data gathering trees for WMSNs. For the rest of the chapter, the terms 'node' and 'vertex', 'edge' and 'link', 'data aggregation' and 'data gathering' will be used interchangeably. They mean the same. Note that our research focuses on the analysis of the use of tree-based communication topologies for data gathering and not on the physical layer aspects such as filtering/sorting of information gathered from the sensor nodes.

SYSTEM MODEL, TERMINOLOGY, AND OVERVIEW

System Model

The system model adopted in this research is as follows:

1. Each sensor node is assumed to operate with an identical and fixed *transmission range*.
2. The *sensing range* of the node is half the transmission range. A sensor node is assumed to be able to sense and collect data at locations within its sensing range and disseminate to nodes within transmission range. Zhang and Hou (2005) showed that the transmission range has to be at least twice the sensing range for coverage to imply connectivity.
3. We run the Breadth First Search (BFS; Cormen et al., 2009) algorithm, starting from a root node (leader node), on a spanning tree on to derive a rooted data gathering tree.

4. The leader node of a data gathering tree remains the same until the tree exists and is randomly chosen each time a new tree needs to be determined.
5. Data gathering proceeds in rounds, starting with each leaf node forwarding its data to the immediate upstream node in the DG-tree. An intermediate node at a particular level aggregates data received from its immediate child nodes (i.e., downstream nodes at one level below) with its own data and forwards the aggregated data to its immediate parent node (i.e., upstream node at one level above). Like this, the aggregated data gets forwarded all the way to the leader node of the DG-tree.

Terminology

To determine the sequence of stable data gathering trees that span over several time instants, we use the notion of static graphs and mobile graphs (Farago & Syrotiuk, 2003) to capture the sequence of topological changes in the network. We model the *static graph* as a unit disk graph (Kuhn et al., 2004) wherein there exists an edge between two nodes if and only if the physical Euclidean distance between the two nodes is less than or equal to the transmission range. The Euclidean distance for an edge $i - j$ between two nodes i and j, currently at (X_i, Y_i) and (X_j, Y_j) is given by: $\sqrt{(X_i - X_j)^2 + (Y_i - Y_j)^2}$ and is assigned the weight of the edge.

A mobile graph $G(i, j)$, where $1 \leq i \leq j \leq T$ (T is the total number of rounds of the data gathering session corresponding to the network lifetime) is defined as $G_i \cap G_{i+1} \cap \ldots G_j$. A mobile graph is thus a logical graph, capturing the presence or absence of edges in the constituent static graphs that are obtained by periodically sampling the network topology at time instants for the corresponding rounds of data gathering. The weight of an edge in the mobile graph $G(i, j)$ is the geometric mean of the weights of the particular edge in the individual static graphs spanning $G_i, ..., G_j$. Since the geometric mean of a set of values is less than or equal to any of the values in the set, the geometric mean of the edge weights will be less than or equal to the transmission range of the two end nodes for the entire duration spanned by the mobile graph. Note that at any time, a mobile graph includes only *live sensor nodes*, nodes that have positive available energy.

A *static spanning tree* is the minimum-weight spanning tree determined on a static graph. Since the edge weights correspond to the physical Euclidean distance between the end nodes of the edge, the static spanning tree is the minimum-distance spanning tree (whose sum of edge weights is the minimum) on the static graph. A *static data gathering tree* is the rooted form of its corresponding static spanning tree where the root node is the leader node chosen for the particular round of data gathering. A *mobile spanning tree* is a minimum-weight spanning tree determined on a mobile graph whose edge weights are the geometric mean of the corresponding edge weights in the constituent static graphs. A *mobile data gathering tree* is the rooted form of the mobile spanning tree whose root node is the leader node chosen for the rounds spanning over the duration of the mobile graph (i.e., until the mobile graph gets disconnected due to node mobility or a node failure occurs, whichever happens first).

Overview

A high-level overview of the working of the Max. Stable-DG algorithm is as follows: To determine a data gathering tree at a particular time instant t_i corresponding to round i ($1 \leq i \leq T$, the total number of rounds of the data gathering session), we determine a mobile graph $G(i, j)$, where $i \leq j$ such that there exists a spanning tree in $G(i, j)$ and not in

$G(i, j+1)$. We run the Breadth First Search (BFS) algorithm on such a longest-living spanning tree (starting from an arbitrarily chosen leader node) and transform it to a data gathering tree (rooted at the leader node) that is used for all the rounds from time instants t_i to t_j, which is considered as the lifetime of the spanning tree. The above procedure is repeated for the entire duration of the data gathering session or the network lifetime, as appropriate. Any spanning tree algorithm can be used to determine the spanning tree on a mobile graph. We use the $O(n^2*\log n)$ Prim's algorithm to determine the minimum-weight spanning tree on the mobile graph (of n nodes) in which the weight of an edge is the geometric mean of the weights of the corresponding edge in the constituent static graphs.

A *key assumption* behind the Max.Stable-DG algorithm is that the entire sequence of topological changes is known beforehand at the time of determining the mobile graph with a spanning tree (further transformed to a stable long-living data gathering tree) spread over the longest sequence of time instants corresponding to successive rounds of data gathering. The above assumption may not be practical for distributed systems of sensor networks where one cannot accurately know the future topological changes in its entirety. However, note that the goal in this research is to develop a *benchmarking algorithm* that can be used to determine the sequence of longest-living data gathering trees such that the number of tree discoveries (reconfigurations) is the global minimum under particular conditions. The optimal (minimum) number of tree discoveries incurred with the centralized Max.Stable-DG benchmarking algorithm *will form the lower bound for the number of tree discoveries* incurred with any distributed algorithm for data gathering in mobile sensor networks.

The greedy principle behind the proposed algorithm is very generic in nature and can be used to determine the sequence of any longest-living communication topology, such as: a connected dominating set (Meghanathan, 2010), a chain (Lindsey et al., 2002), a cluster (Heinzelman, et al., 2000), as long as there is an algorithm or heuristic to determine the communication topology on a network graph. Moreover, since the Max.Stable-DG algorithm determines the stable sequence of spanning trees (covering all the nodes in the network), the benchmarks (lifetime and the number of spanning tree discoveries) obtained with this algorithm can also serve as the benchmarks for any network-wide communication topology (like a connected dominating set that spans all the nodes) determined by running any relevant algorithm under identical operating conditions. Henceforth, the Max.Stable-DG algorithm can be used to evaluate the relative stability of network-wide communication topologies determined using any centralized or distributed algorithm (very few of which is currently available in the literature) proposed for mobile sensor networks.

RELATED WORK ON DATA GATHERING IN WIRELESS MOBILE SENSOR NETWORKS

Research on mobile sensor networks started with the deployment of mobile sink nodes on a network of static sensor nodes. A common approach (e.g., Zhao & Yang, 2009; Xing et al., 2008; Wu et al., 2010) of data gathering in such environments is to let the sink move in the network on the shortest possible path towards the location from which the desired data is perceived to originate. Zhong & Cassandras (2011) propose a distributed algorithm to optimize both coverage control and mobile collection using a Bayesian occupancy grid mapping technique that recursively updates the locations of potential data sources. Meghanathan and Skelton (2007) propose a two-layer network

of resource-rich mobile sinks (in the upper layer) and resource-constrained static sensor nodes (in the bottom layer): each sink is assigned a particular region to monitor and collect data. A sink node moves to the vicinity of the sensor nodes (within a few hops) to collect data. The collected data is exchanged with peer mobile sinks. Meghanathan et al. (2010) developed a prototype implementation of the same.

Very few topology-based data gathering algorithms have been proposed for WMSNs. Most of these are variants of the cluster-based LEACH (Low Energy Adaptive Clustering Hierarchy; Heinzelman et al., 2000) algorithm, adapted for WMSNs. These include proposals that take into consideration the available energy level () and the mobility level () of the nodes to decide on the choice of cluster heads; stability level of the links between a regular node and its cluster head (); as well as set up a panel of cluster heads to facilitate cluster reconfiguration in the presence of node mobility (). Liu et al (2007) propose a distributed algorithm in which cluster heads are selected based on node IDs (0 to C-1, C to 2C-1 ..., to operate with C clusters at a time) or node locations (nodes that are closest to certain landmarks within a WMSN serve as cluster heads). Macuha et al. (2011) investigate the use of a directed acyclic graph as the underlying communication topology for a sensor network field, modeled according to the theory of thermal fields, to form propagation paths such that the temperature of the nodes on the path increases as data progresses towards the sink, which is considered to be the warmest.

The only tree-based data gathering algorithm proposed for WMSNs is that of Singh et al. (2010) who suggested the use of a shortest path-based spanning tree wherein each sensor node is constrained to have at most a certain number of child nodes. Results from the literature for mobile ad hoc networks (e.g., Meghanathan, 2011; Meghanathan & Farago, 2008) indicate that minimum hop shortest paths and trees in mobile network topologies are quite unstable and need to be frequently reconfigured. We could not find any other related work on tree-based data gathering for wireless mobile sensor networks.

DATA GATHERING ALGORITHMS BASED ON MAXIMUM STABILITY AND MINIMUM-DISTANCE SPANNING TREES

Maximum Stability Spanning Tree-Based Data Gathering (Max.Stable-DG) Algorithm

The Max.Stable-DG algorithm is based on a greedy look-ahead principle and the intersection of static graphs. When a data gathering tree (DG-tree) is required at round i corresponding to time instant t_i, the idea is to determine a mobile graph $G(i, j) = G_i \cap G_{i+1} \cap \ldots G_j$ ($i \leq j$) such that there exists a spanning tree in $G(i, j)$ and not in $G(i, j+1)$. We find such an epoch $t_i, \ldots, t_j$ iteratively as follows: Let t_k and t_{k+1} be two time instants such that $t_i \leq t_k < t_{k+1} \leq t_j$. Given that $G(i, k)$ is connected (and has a spanning tree), we construct $G(i, k+1)$ by including only those edges from $G(i, k)$ that also exist in G_{k+1}; we then check whether $G(i, k+1)$ is connected or not. Note that a spanning tree exists for a graph if and only if the graph is connected. If $G(i, k+1)$ is connected, we continue the above procedure until we find t_j (t_k) such that $G(i, j)$ is connected and $G(i, j+1)$ is not connected. Note that the connectivity of a mobile graph is determined by running the Breadth First Search (BFS) algorithm.

Once we find a mobile graph $G(i, j)$ that will not be connected if further expanded, we run the Prim's minimum-weight spanning tree algorithm on $G(i, j)$, wherein the weight of an edge is the geometric mean of the weights of the edge in the constituent static graphs $G_i, \ldots, G_j$, to obtain a mobile spanning tree for time instants $t_i, \ldots, t_j$. To obtain the corresponding mobile data gathering tree for $G(i,j)$, we choose an arbitrary leader node (as the root node) for this mobile spanning tree

and run the Breadth First Search (BFS) algorithm starting from the root node. The direction of the edges and the intermediate nodes/leaf nodes are automatically decided as we run BFS on the mobile spanning tree. We then set $i = j+1$ and repeat the above procedure to find a mobile spanning tree and its corresponding mobile data gathering tree that exists for the maximum amount of time since t_{j+1}. A sequence of such maximum lifetime (i.e., longest-living) mobile data gathering trees over the timescale T corresponding to the number of rounds of a data gathering session is referred to as the *Stable Mobile Data Gathering Tree*. Figure 1 presents the pseudo code of the Max.Stable-DG algorithm that takes as input the sequence of static graphs spanning the entire duration of the data gathering session.

While running the Max.Stable-DG algorithm under energy-constrained scenarios, one or more nodes in a mobile graph $G(i, j)$ may die due to exhaustion of battery charge at some intermediate round k ($i \leq k \leq j$) corresponding to time instant t_k ($t_i \leq t_k \leq t_j$) even though $G(i, j)$ would be topologically connected if all nodes in G_i had sufficient battery charge at time instants $t_i, \ldots, t_j$. In case of pre-mature node failures at time instant t_k before t_j, we restart the Max.Stable-DG algorithm starting from time instant t_k considering only the live sensor nodes (i.e., the sensor nodes that have positive available energy) and determine the longest-living data gathering tree that spans all the live sensor nodes since t_k. The pseudo code of the Max.Stable-DG algorithm in Figure 1 handles node failures, when run under energy-constrained scenarios, through the *if* block segment in statement 8. If all nodes have sufficient-energy and there are no node failures, the algorithm does not execute statement 8.

Run-Time Complexity Analysis of the Max.Stable-DG Algorithm

To expand a mobile graph $G(i, j) = G_i \cap G_{i+1} \cap \ldots G_j$ to $G(i, j+1)$, we need to check whether an edge in $G(i, j)$ exists in G_{j+1} (the static graph at

Figure 1. Pseudo code for the maximum stability-based data gathering tree algorithm

Input: Sequence of static graphs G_1, G_2, ... G_T; Total number of rounds of the data gathering session – T
Output: *Stable-Mobile-DG-Tree*
Auxiliary Variables: i, j
Initialization: $i = 1$; $j = 1$; *Stable-Mobile-DG-Tree* = Φ

Begin *Max.Stable-DG Algorithm*

1   while ($i \leq T$) do

2   Find a mobile graph $G(i,j) = G_i \cap G_{i+1} \cap ... \cap G_j$ such that there exists at least one spanning
         tree in $G(i,j)$ and {no spanning tree exists in $G(i,j+1)$ or $j = T$}

3   *Mobile-Spanning-Tree(i,j)* = **Prim's Algorithm** ($G(i,j)$)

4   *Root(i,j)* = Choose a node randomly in $G(i,j)$

5   *Mobile-DG-Tree(i,j)* = **Breadth First Search** (*Mobile-Spanning-Tree(i,j)*, *Root(i,j)*)

6   *Stable-Mobile-DG-Tree* = *Stable-Mobile-DG-Tree* U { *Mobile-DG-Tree(i,j)* }

7   for each time instant $t_k \in \{t_i, t_{i+1}, ..., t_j\}$ do
             Use the *Mobile-DG-Tree(i,j)* in t_k

8   if node failure occurs at t_k then
                 $j = k - 1$
                 break
             end if
         end for

9       $i = j + 1$

10   end while

11   return *Stable-Mobile-DG-Tree*

End *Max.Stable-DG Algorithm*

time instant t_{j+1}) and this can be done in $O(n^2)$ time. There can be at most $O(n^2)$ edges on a graph of n vertices. The overall complexity of the Max. Stable-DG algorithm is the sum of the time complexity to construct the mobile graphs, the time complexity to run the spanning tree algorithm on these mobile graphs and the time complexity to transform these spanning trees to data gathering trees using BFS.

- **Sufficient-Energy Scenarios:** When the network is operated under sufficient-energy scenarios (i.e., no node failures) for a data gathering session comprising of T rounds, we will have to construct T mobile graphs, resulting in a time complexity of $O(n^2T)$. On each of these T mobile graphs, we need to run the $O(n^2\log n)$ Prim's minimum spanning tree algorithm, incurring a time complexity of $O(n^2T\log n)$. A spanning tree on n vertices has n-1 edges. The time-complexity of running the BFS algorithm on a n-vertex spanning tree of n-1 edges is $O(n)$. To transform the $O(T)$ spanning trees to data gathering trees, we will

incur a time complexity of $O(nT)$. Thus, the overall complexity of the Max.Stable-DG algorithm under sufficient-energy scenarios is $O(n^2T) + O(n^2T\log n) + O(nT) = O(n^2T\log n)$.

- **Energy-Constrained Scenarios:** There can be at most n-1 node failures on an n-node network to trigger the execution of statement 8 in the pseudo code of Figure 1 for the Max.Stable-DG algorithm. A premature node failure at time instant t_k ($i \leq k \leq j$), while using a mobile data gathering tree that has been determined on a mobile graph for time instants t_i, ..., t_j, would require us to construct a mobile graph starting from t_k and the number of mobile graphs we have to construct and run increases by $j-k+1$. At the worst case, if there are n-1 node failures, the number of additional mobile graphs that we need to construct and run the spanning tree algorithm increases by $(T) + (T-1) + (T-(n-2)) = (n-1)T - [1 + 2 + ... + (n-2)] = O(nT) + O(n^2)$. Under sufficient-energy scenarios, we need to construct T mobile graphs and run the spanning tree algorithm on each of them. Under energy-constrained scenarios, we need to construct $T + O(nT) + O(n^2)$ mobile graphs. Since $n \ll T$, $n^2 \ll nT$. Hence, $T + O(nT) + O(n^2) = T + O(nT) = O(nT)$ mobile graphs are constructed. As mentioned earlier, it takes $O(n^2)$ time complexity to construct a mobile graph on n vertices. Hence, the time complexity to construct the $O(nT)$ mobile graphs is $O(n^3T)$. If we run the Prim's $O(n^2\log n)$ minimum spanning tree algorithm on a mobile graph, then the time-complexity to run the spanning tree algorithm on each of the $O(nT)$ mobile graphs would be $O(nT * n^2\log n) = O(n^3T\log n)$. Running the $O(n)$ BFS algorithm on each of the $O(nT)$ spanning trees to obtain the data gathering trees would incur a time-complexity of $O(n^2T)$. Thus, the overall time-complexity of the Max.Stable-DG algorithm under the energy-constrained scenarios is $O(n^3T) + O(n^3T\log n) + O(n^2T) = O(n^3T\log n)$.

Minimum-Distance Spanning Tree Based Data Gathering Algorithm

In a recent work, Meghanathan (2012) observed the minimum-distance spanning tree-based data gathering trees to be the most energy-efficient communication topology for data gathering in static sensor networks. In the Simulations section, we compare the performance of the Max. Stable-DG trees with that of the minimum-distance spanning tree based data gathering (MST-DG) trees. The MST-DG tree algorithm emulates the LORA (Least Overhead Routing Approach) strategy (Abolhasan et al., 2004) of choosing a data gathering tree that appears to be the best at the current time instant and continues to use it as long as it exists. The sequence of MST-DG trees over the duration of the data gathering session is determined as follows: If a MST-DG tree is required for a particular round, we run the Prim's

algorithm on the static graph generated for the topology snapshot captured at the time instant for the round. If a spanning tree (minimum-distance spanning tree) exists, we transform it to a DG-tree by running BFS (starting from an arbitrarily chosen root-leader node) and continue to use it in the subsequent time instants as long as all the edges of the spanning tree exist. The leader node of the MST-DG tree remains the same until the tree breaks due to node mobility or node failures. When the MST-DG tree ceases to exist for a round, we repeat the above procedure. This way, we generate a sequence of MST-DG trees, referred to as the *MST Mobile Data Gathering Tree*.

To be fair to the Max.Stable-DG algorithm, we run the MST-DG algorithm in a centralized fashion with the assumption that the entire static graph information is available at the beginning of each round. The time-complexity of generating the sequence of MST-DG trees on a network of n nodes for a total of T rounds for the data gathering session is $O(n^2T\log n)$ for both the sufficient-energy and energy-constrained scenarios. In the presence of node failures (energy-constrained scenarios), we would require at most n-1 more rounds (in addition to the T rounds for sufficient-energy scenarios). Since $n \ll T$, the total number of static graphs on which the Prim's $O(n^2\log n)$ algorithm run would be $T + O(n) = O(T)$, leading to an overall time complexity of $O(n^2T\log n)$ for energy-constrained scenarios.

Example Run of the Globally Optimal Maximum Stability and Locally Optimal Minimum-Distance Spanning Tree Based Data Gathering Algorithms

The first part of Figures 2 and 3 illustrate the (identical) sequence of static graphs $G_1G_2G_3G_4G_5$ that represent contiguous snapshots of the network topology that will be used to run the Max. Stable-DG and MST-DG algorithms respectively. For simplicity and clarity in the representation, we do not use weights for the edges. In both the figures, the reader could assume that the spanning trees determined on the static graphs and mobile graphs at different instances of execution of the algorithms are the minimum-weight spanning trees on the corresponding graphs.

In Figure 2, we could find a connected mobile graph G(1, 3) and not G(1, 4). We determine a spanning tree on the mobile graph G(1, 3) = G_1 ∩ G_2 ∩ G_3 and derive a data gathering tree rooted at an arbitrarily selected leader node (node 3). This stable data gathering tree is to be used for the rounds corresponding to the time instants of the static graphs G_1, G_2 and G_3. Similarly, we determine a mobile graph G(4, 5) on which we find a data gathering tree (with an arbitrary root node - node 6) that exists in both G_4 and G_5. Thus, we require a total of two data gathering tree discoveries for the sequence of static graphs $G_1G_2G_3G_4G_5$.

When we apply the MST-DG algorithm on the sequence of static graphs $G_1G_2G_3G_4G_5$ (Figure 3), we observe to need a total of four data gathering trees (one for G_1 and G_2; and one each for G_3, G_4

and G_5) and that many reconfigurations/discoveries. We observe similar trends for the MST-DG tree lifetime in the simulations. In general, frequent topology reconfigurations is the bane of non-stability based algorithms (typically for dynamically changing distributed systems like mobile sensor networks) that choose a communication topology appearing to be locally optimal with respect to one or more metrics (like energy consumption, delay, etc) and do not take into consideration the mobility of the nodes.

Proof of Correctness of the Maximum Stability-Based Data Gathering Algorithm

We use the approach of "Proof by Contradiction" to show that the Max.Stable-DG algorithm returns the sequence of longest-living stable data gathering trees such that the number of spanning tree discoveries (reconfigurations) is the global minimum (optimal solution). Let m be the number of spanning tree discoveries incurred using the Max.Stable-DG algorithm on a sequence of

Figure 2. Example to illustrate the execution of the maximum stability spanning tree-based data gathering tree algorithm that uses the globally optimal approach

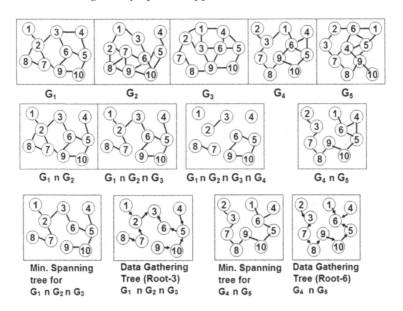

Figure 3. Example to illustrate the execution of the minimum-distance spanning tree based data gathering algorithm that uses the locally optimal approach

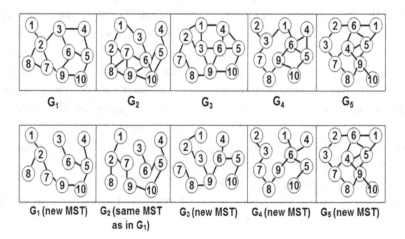

static graphs $G_1 G_2 \ldots G_T$. Let there be a hypothetical algorithm that incurs n spanning tree discoveries ($n < m$) on the same sequence of static graphs. For this to be feasible, there should be a mobile graph $G(p, s)$ in which the hypothetical algorithm should not have gone through any tree transition and the Stable.Max-DG algorithm should have gone through at least one tree transition. Note that the Stable.Max.DG algorithm takes intersection of the static graphs and checks for the connectivity (and hence a spanning tree) on the intersection mobile graph. If the Stable.Max-DG algorithm could not find a spanning tree in $G(p, s)$, it implies $G(p, s) = G_p \cap G_{p+1} \cap \ldots G_s$ is not connected. Hence, it is not possible for any algorithm, including our hypothetical algorithm, to find a spanning tree spread over the static graphs of $G(p, s)$. In other words, the hypothetical algorithm should have also had at least one transition in $G(p, s)$. The above proof holds good for any value of static graph indices p and s, where $1 \le p \le s \le T$, and T is the total number of rounds corresponding to the duration of the data gathering session. Thus, the number of spanning tree/data gathering tree discoveries incurred using the Max.Stable-DG algorithm is the global minimum.

Note that the above proof implicitly assumed that all the sensor nodes are alive for the entire duration of the mobile graph $G(p, s) = G_p \cap G_{p+1} \cap \ldots G_s$ for $1 \le p \le s \le T$. In other words, we have theoretically proved the correctness of the Max.Stable-DG algorithm (that it incurs the optimal number of tree discoveries) under sufficient-energy scenarios. Due to the stochastic nature of the mobile sensor networks, it is not possible to theoretically prove the correctness of the Max.Stable-DG algorithm under energy-constrained scenarios. One can only validate the optimality of the lifetime of the Max.Stable-DG trees under energy-constrained scenarios through simulations. We have done so in the Simulation section, wherein we observe the Max.Stable-DG trees to sustain a significantly longer lifetime compared to the MST-DG trees under energy-constrained scenarios.

SIMULATIONS

In this section, we illustrate the performance of the Max.Stable-DG trees via exhaustive simulations and in comparison with that of the MST-DG trees under diverse conditions of network density and mobility. Simulations are conducted in a discrete-event simulator developed (in Java) by us exclusively for data gathering in mobile sensor networks. The medium access control (MAC) layer is assumed to be an ideal channel (collision-free without any interference). Sensor nodes are assumed to be both TDMA (Time Division Multiple Access) and CDMA (Code Division Multiple Access)-enabled (Viterbi, 1995). Each upstream node comes up with a transmission schedule for its immediate downstream nodes (to forward the beacon data, if leaf nodes, or aggregated data, if intermediate nodes) to which the latter strictly adhere to. Upstream nodes that are in the same level of the data gathering tree could communicate simultaneously with their downstream nodes on a TDMA-based communication schedule, with each upstream node using a unique CDMA code.

The network dimension is 100m x 100m. A total of 100 nodes are uniform-randomly distributed throughout the network, with nodes operating on an identical and fixed transmission range. Network density is altered by varying the transmission range per sensor node from 20m to 50m, in increments of 5m. For brevity, we only present simulation results obtained for 25m an 30m (representative of moderate density, with connectivity of 97% and above), and for 40m (representative of high density, with 100% connectivity).

Simulations are conducted under both *sufficient-energy scenarios* (each node provided with an abundant supply of energy-50J/node and there are no node failures due to exhaustion of battery charge; simulations are run for 1000 seconds) and *energy-constrained scenarios* (each node provided with a limited energy supply-2J/node and simulations are conducted until the network of live sensor nodes gets disconnected due to one or more node failures). The energy-constrained scenarios are appropriately prefixed as '*EC*' next to the names of the data gathering trees.

The energy consumption model used is a first order radio model (Rappaport, 2002) that has been also used in several of the well-known previous work (e.g., Lindsey et al., 2002; Heinzelman et al., 2000) in the literature. According to this model, the energy expended by a radio to run the transmitter or receiver circuitry is $E_{elec} = 50$ nJ/bit and $\in_{amp} = 100$ pJ/bit/m^2 for the transmitter amplifier. The radios are turned off when a node wants to avoid receiving unintended transmissions. The energy lost in transmitting a k-bit message over a distance d is given by: $E_{TX}(k, d) = E_{elec} * k + \in_{amp} *k* d^2$. The energy lost to receive a k-bit message is: $E_{RX}(k) = E_{elec} * k$.

Data gathering is constant-bit rate based: proceeds at the rate of 4 rounds per second (0.25 seconds time gap between successive rounds). The data packet size is 2000 bits and the size of the control message (assumed to be broadcast once by every node in its neighborhood during flooding-based discovery of the data gathering trees) is 400 bits. Flooding is an energy-expensive operation in ad hoc and sensor networks, especially in high density networks, because each node has to spend energy to receive multiple copies of the message, once from each of its neighbor nodes.

The node mobility model used in the simulations is the Random Waypoint model (Bettstetter et al., 2004) that assumes nodes to move independent of each other and the velocity is uniform-randomly selected from the range [0... v_{max}]. To start with, each node is uniform-randomly distributed throughout the network field; a node chooses a random destination location (within the dimensions of the network) and a random velocity from the range [0...v_{max}] to move to the chosen location. After reaching the targeted location, the node continues to move by choosing another random destination location and another random velocity from the range [0...v_{max}]. Each node continues to move like this for 6000 seconds (anticipated to be the maximum simulation time for energy-constrained scenarios considered in this research) to generate a mobility profile. All of these mobility profiles (generated offline for each node) are fed to the code for the data gathering algorithms when the latter are run. For a given v_{max} value, we also vary the dynamicity of the network by varying the number of static nodes (out of the 100 nodes) in the network. The values for the number of static nodes used are: 0 (all nodes are mobile), 20, 50 and 80.

Performance Metrics

We generate 200 mobility profiles of the network (for a total duration of 6000 seconds) for each combination of v_{max} and # static nodes. The data points presented in Figures 5 through 16 are average values obtained over these 200 mobility profiles. We measure the tree lifetime and delay per round (for both sufficient-energy and energy-constrained scenarios) and energy consumption metrics (for sufficient-energy scenarios): energy lost per round, energy lost per node and fairness of node usage. In addition, for energy-constrained scenarios only, we also measure node lifetime and network lifetime as well as the distribution of node lifetime and probability of node failures. The above metrics accurately capture the impact of the topological structure, network dynamicity and the stability of the data gathering trees on energy consumption.

The performance metrics measured in the simulations are:

1. **Tree Lifetime:** The duration for which a data gathering tree existed, averaged over the entire simulation time period.
2. **Delay per Round:** Measured in terms of the number of time slots needed per round of data aggregation at the intermediate nodes, all the way to the leader node of the data gathering tree, averaged across all the rounds of the simulation. A brief description of the algorithm used to compute the delay per round is given in a following section along with an illustration in Figure 4.

3. **Node Lifetime:** Measured as the time of first node failure due to exhaustion of battery charge.
4. **Network Lifetime:** Measured as the time of disconnection of the network of live sensor nodes (nodes with positive available battery charge), while the network would have stayed connected if all the sensor nodes were alive (tested, using Breadth First Search, before confirming whether an observed time instant is the network lifetime).

We obtain the distribution of node failures as follows: For every combination of transmission range per node, v_{max} and # static nodes, we keep track of: (a) the fraction of mobility profiles (among the 200 mobility profiles generated for a given v_{max} and # static nodes) that report x number of node failures (x ranging from 1 to 100, the total number of nodes); (b) the time at which x number of node failures (x ranging from 1 to 100) have occurred in each of the 200 mobility profiles and average these values for the time of node failures (as reported in Figures 14, 15 and 16). We discuss the results for the distribution of the time and probability of node failures along with the discussion on average node lifetime and network lifetime in a following section.

Algorithm to Compute the Delay per Round of Data Gathering

The delay incurred at an intermediate node is the number of timeslots it takes to gather data from all of its immediate child nodes. In a TDMA-based system, per timeslot, a parent node can gather data from only one child node. An intermediate node has to gather data from all of its child nodes before forwarding the aggregated data further up the DG-tree. The delay for the data gathering tree is one plus the delay incurred at the leader node (root node). The delay calculations start from levels corresponding to the Height of the tree (at the bottommost leaf nodes) and proceed all the way to level zero (i.e., the root node). The delay incurred at a leaf node is 0. To compute the delay at an intermediate node u, we first sort the set of delay incurred at the immediate child nodes of u, represented as *Child-Nodes*(u) and then iterate through the sorted list of the delays at the child nodes, in increasing order. The *Delay*(u) associated with an intermediate node u is the final value of the temporary variable, *Temp-Delay*(u), updated as we iterate through the sorted list of delay in *Child-Nodes*(u): for every child node v $\in$ Child-Nodes(u), *Temp-Delay*(u) = Maximum [*Temp-Delay*(u) $+ 1$, *Delay*(v) $+ 1$)], as we assume it takes one time slot for a child node to transfer its aggregated data to its immediate predecessor node in the tree. We illustrate the working of the above explained procedure for delay computation on a data gathering tree through an example presented in Figure 4. The integer inside a circle indicates the node ID and the integer outside a circle indicates the delay for data aggregation at the node.

Tree Lifetime

We observe the stability of the DG-trees to be highly influenced by the maximum velocity (v_{max}) of the nodes. Under sufficient-energy scenarios, for a fixed number of static nodes and transmis-

sion range per node, the lifetime incurred for both the Max.Stable-DG and MST-DG trees decreases proportionately with increase in the v_{max} values from 3 m/s to 10 m/s and further to 20 m/s. The lifetime observed for the DG-trees in energy-constrained scenarios is always less than or equal to those observed for sufficient-energy scenarios. This could be attributed to the possible failure of one or more nodes in the tree due to exhaustion of battery charge and the tree needs to be reconfigured, even though the tree may topologically exist (i.e., location and distance wise).

We observe the largest difference in the DG-tree lifetimes between the sufficient-energy and energy-constrained scenarios when the network is operated under low node mobility conditions (v_{max} = 3 m/s). This could be attributed to the significantly longer lifetime observed for the DG-trees at low node mobility conditions when operated with sufficient-energy for the nodes. Similarly, when operated at higher transmission range, the nodes have more freedom to move around (compared to operating at low and moderate transmission ranges) and we observe larger differences between the lifetime of the two DG-trees, especially in sufficient-energy scenarios. When operated at low node mobility and higher transmission ranges per node, the lifetime of the Max. Stable-DG trees is larger than that of the MST-DG trees by a factor of about 3-4.5 (sufficient-energy

Figure 4. Example to illustrate the calculation of delay per round of data gathering

Nodes at Level	Nodes
0	12
1	7, 9
2	15, 2, 5, 14
3	3, 0, 6, 11, 13
4	10, 1, 4
5	8

Delay for the DG Tree
= 1 + Delay for the Root
= 1 + 6 = 7 time units

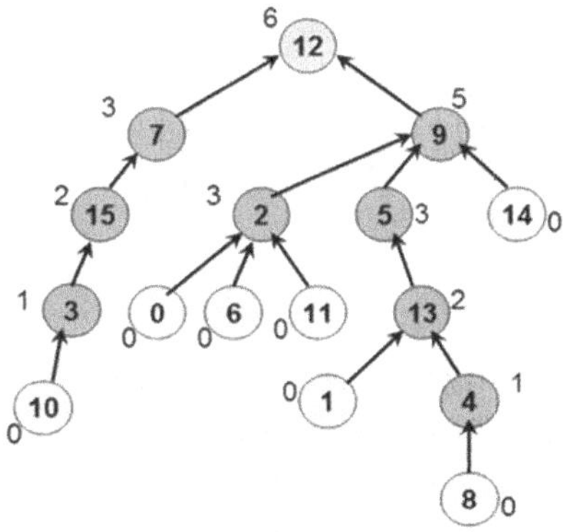

Figure 5. Average tree lifetime (low node mobility: vmax = 3 m/s)

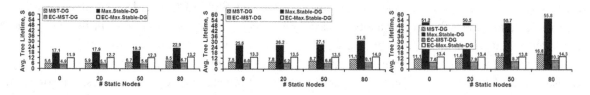

Figure 6. Average tree lifetime (moderate node mobility: vmax = 10 m/s)

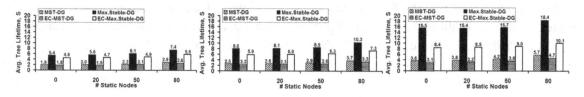

Figure 7. Average tree lifetime (high node mobility: vmax = 20 m/s)

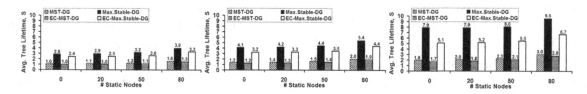

Figure 8. Average delay per round (Low Node Mobility: vmax = 3 m/s)

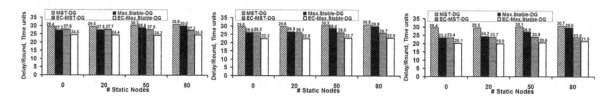

Figure 9. Average delay per round (Moderate Node Mobility: vmax = 10 m/s)

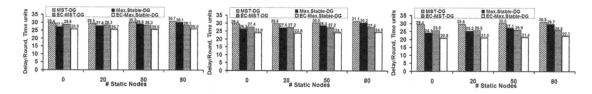

scenarios) and 50-75% (energy-constrained scenarios). Under energy-constrained scenarios, the energy spent by the MST-DG trees for tree reconfigurations in the moderate and high node mobil-

ity conditions is relatively very high compared to the energy spent for flooding-based broadcast tree discoveries under low node mobility conditions; this leads to the Max.Stable-DG trees having a

Figure 10. Average delay per round (High Node Mobility: vmax = 20 m/s)

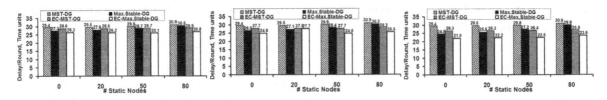

Figure 11. Average node and network lifetime (low node mobility: v_{max} = 3 m/s)

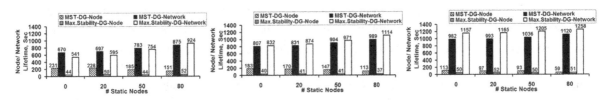

Figure 12. Average node and network lifetime (moderate node mobility: v_{max} = 10 m/s)

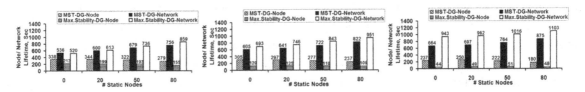

Figure 13. Average node and network lifetime (High Node Mobility: v_{max} = 20 m/s)

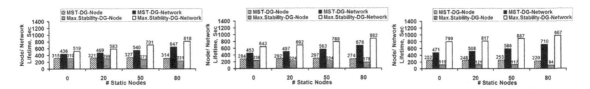

relatively larger difference in lifetime compared to the MST-DG trees, under moderate-high mobility conditions (vis-a-vis low node mobility conditions).

For a fixed v_{max} value, the lifetime of the Max. Stable-DG and MST-DG trees increase respectively by a factor of 3 and 2, as we increase the transmission range per node from 25m to 40m. At larger transmission ranges per node, Max. Stable-DG trees choose stable links (attributed to the algorithm's look-ahead approach). On the other hand, the local optimum approach results in

the MST-DG trees forming relatively less stable links even when operated at higher transmission ranges per node.

When we operate the network at 80 static nodes (out of a total of 100 nodes), we observe the lifetime of both the DG-trees to increases by about 50%; on the other hand, we observe only a 10-15% increase in the lifetime of the two DG-trees at 20 and 50 static nodes. This vindicates the impact of node mobility on the stability of the data gathering trees. Even if half of the nodes in the network are operated static, we observe the data gathering trees to have about the same vulnerability for a link failure vis-à-vis operating the network with all mobile nodes.

Delay per Round

A minimum-distance based spanning tree (like the MST-DG tree) tends to have relatively fewer leaf nodes and more intermediate nodes (that are however not uniformly distributed at different levels and not all leaf nodes are located at the bottommost level of the tree) - leading to a much larger depth. All these structural complexities contribute to a much larger delay per round of data gathering. On the other hand, the Max.Stable-DG trees are structurally more balanced (with the distribution of the nodes at different levels) and have relatively fewer intermediate nodes and more leaf nodes, all of which contribute to a much lower delay per round of data gathering.

The Max.Stable-DG trees are observed to incur a relatively lower delay per round of data gathering compared to the MST-DG trees under all operating conditions (the difference is as large as 25%). For both the DG-trees, the delay per round is not much affected by the dynamicity of the network (the maximum node velocity has the least impact among all the operating parameters) and is more influenced by the topological structure (height and distribution of the nodes) of the two spanning trees. The MST-DG trees exhibit a relatively lower delay per round under energy-constrained scenarios (compared to the sufficient-energy scenarios), attributed to the reduction in the number of live sensor nodes (nodes fail prematurely due to exhaustion of energy) and slightly better distribution of nodes when fewer in number. The delay per round for the Max.Stable-DG trees appear to be more influenced by the transmission range per node (energy-constrained scenarios) and the number of static nodes (sufficient-energy scenarios).

Under the energy-constrained scenarios, node failures contribute to the decrease in delay per round (ranging from 10-40% compared to the

sufficient-energy scenarios) for both the DG-trees: the largest decrease is when the DG algorithms are operated with transmission range per node of 40m (for all levels of node mobility). Consequently, for a given level of node mobility, the difference in the delay per round of data gathering incurred with the Max.Stable-DG and MST-DG trees decrease with increase in the transmission range per node.

For a given level of node mobility and transmission range per node, the delay per round incurred for both the DG trees increases with increase in the number of static nodes. When all nodes are mobile, the Max.Stable-DG trees incur 10-25% lower delay per round, compared to the MST-DG trees. However, when operated with 80% static nodes, the delay per round incurred for the Max.Stable-DG trees converges to that of the MST-DG trees, especially under sufficient-energy scenarios. Under energy-constrained scenarios, the Max.Stable-DG trees still incur a relatively lower delay per round (10-15% lower compared to the MST-DG trees), attributed to the decrease in the number of live sensor nodes coupled with the shallow topological structure of the data gathering tree.

Node Lifetime and Network Lifetime

We observe a node lifetime-network lifetime tradeoff between the maximum stability-based and minimum-distance based spanning tree driven data gathering trees for mobile sensor networks. The MST-DG trees incur larger node lifetime (time of first node failure) for all the 48 combinations of operating conditions (vmax, # static nodes and transmission ranges per node); the Max.Stable-DG trees incur the larger network lifetime (time of disconnection of the network of live sensor nodes due to the failure of one or more nodes) for most of the operating conditions.

Note that, for both the DG-trees, we stay with a (randomly) chosen leader node as the root node of the data gathering tree until the tree is reconfigured due to the failure of one or more links. Likewise, as long as a DG tree exists, the intermediate nodes of the DG tree continue to aggregate data from all of their immediate child nodes and forward the aggregated data to their upstream node. Since the Max.Stable-DG trees exhibit a longer lifetime (especially under operating conditions like low-moderate node mobility coupled with moderate-larger transmission ranges per node that facilitate greater stability), the leader node and the intermediate nodes tend to be overused (unfairness in node usage), leading to a lower node lifetime.

For both the DG trees, increase in the maximum node velocity (v_{max}) contributes to an increase in the node lifetime, especially when operated under moderate-high transmission ranges per node. Comparatively, we observe a much reduced percentage increase in node lifetime (with increase in v_{max}) when operated under lower transmission ranges per node (due to the relatively more tree reconfigurations needed). Increase in v_{max} triggers regular tree reconfigurations, contributing to the fairness of node usage, especially when the energy consumption overhead due to tree discoveries is under control. The Max.Stable-DG trees benefit the most with increase in v_{max}: the node lifetime increases by as large as 200-400% as we increase v_{max} from 3 m/s to 10 m/s, and further increases by 50-100% as v_{max} is increased from 10 m/s to 20 m/s. The MST-DG trees do exhibit an increase in node lifetime, with increase in v_{max}, albeit at a reduced rate - due to the inherent unstable nature of the links. With increase in v_{max} from 3 m/s to 10 m/s and further to 20 m/s, the node lifetime incurred with the MST-DG trees increases by about 50-100% and 10-20% respectively.

The MST-DG trees have been observed to incur node lifetime that is as large as 400% and 135%, compared to the Max.Stable-DG trees, under low-moderate and high node mobility conditions respectively. For both the DG trees, the node lifetime increases with increase in transmission range per node (for a fixed number of static nodes) and almost remains the same with increase in the number of static nodes (for a fixed transmission range per node). For a given level of node mobility, the Max.Stable-DG trees exhibit a 30-40% decrease in node lifetime with increase in the transmission range per node from 25m to 30m, and another 50-60% decrease with a further increase in the transmission range per node from 30m to 40m. For fixed network dynamicity (v_{max} and # static nodes), the stability of the Max.Stable-DG trees increases significantly with increase in the transmission range per node, negatively contributing to the node lifetime (due to overuse of certain nodes at the cost of others). On the other hand, for a fixed transmission range per node, the use of static nodes and a simultaneous increase in v_{max}, do not negatively impact the node lifetime. At v_{max} = 20 m/s, we observe relatively fewer premature node failures even when operated with 80% static nodes. The node lifetime incurred with MST-DG trees is more impacted with the use of static nodes at low node mobility scenarios (Figure 11) and the node lifetime incurred with the Max.Stable-DG trees is more impacted with the use of static nodes at moderate and higher node mobility scenarios (Figures 12 and 13).

The Max.Stable-DG trees yield a significantly longer network lifetime compared to that of the MST-DG trees. Though the Max.Stable-DG trees incur premature node failures (lower node lifetime) in the early stages of the DG session, the relatively lower energy loss per round/and per node as well as a shorter tree height (with more even distribution of the number of child nodes per intermediate node) helps to rotate the roles of the intermediate nodes and leader node among the nodes to increase the fairness of node usage. All of these save significantly more energy at the remaining nodes that withstand the initial set of failures. There are only four combinations of operating conditions under which the MST-DG trees incur larger network lifetime – these correspond to transmission range per node of 25m and v_{max} = 3 m/s (0, 20 and 50 static nodes), 20 m/s (0 static nodes).

The difference in the network lifetime incurred for the Max.Stable-DG trees and the MST-DG trees increases with increase in v_{max} and/or transmission range per node. The difference in network lifetime increases by about 5-20%, 15-40% and 20-60% at levels of low, moderate and high levels of node mobility respectively. Similar range of differences in the network lifetime can be observed for the two DG trees at transmission ranges per node of 25m, 30m and 40m, with the difference increasing as v_{max} increases. For a given v_{max} and transmission range per node, the number of static nodes does not make a significant impact on the difference in the network lifetime between the two DG trees for transmission ranges per node of 25m and 30m; for transmission range per node 40m, the difference in the network lifetime decreases by about 15-35%. This could be due to the relatively high stability of the Max.Stable-DG trees when operated at larger transmission ranges per node in the presence of more static nodes.

For a fixed v_{max} and transmission range per node, the absolute values of the network lifetime for the two DG trees increase with increase in the number of static nodes. For a given level of node mobility, the absolute network lifetime incurred with both the DG trees increase with increase in the transmission range per node: the rate of increase in the absolute network lifetime sustained with the MST-DG trees decreases with increase in v_{max}; whereas the Max.Stable-DG trees maintain a steady increase in the network lifetime for all levels of node mobility. For a given transmission range per node and number of static nodes, the network lifetime incurred with the Max.Stable-DG trees and MST-DG trees decreases by about 30-50% and 50-100% respectively as we increase v_{max} from 3 m/s to 20 m/s. The relative unstable nature of the MST-DG trees contribute to the frequent tree discoveries, leading to higher energy losses.

For a given transmission range per node, the node lifetime increases with increase in v_{max} and the network lifetime decreases with increase in v_{max}. Hence, the maximum increase in the absolute time of node failures occurs at low node mobility. This vindicates the impact of network-wide flooding based tree discoveries on energy consumption: the energy lost by nodes across the network is about the same. Hence, more frequent network-wide flooding based tree discoveries result in a cascade of node failures, as is observed in the case of the MST-DG trees (a flat curve for the distribution of node failure times is seen in Figures 14, 15 and 16). The Max.Stable-DG trees are observed to sustain a relatively steeper curve for the distribution of node failure lifetimes: the unfairness of node usage (contributing to premature node failures) in the initial stages do help the trees to prolong the network lifetime. Aided by node mobility, during the later rounds of data gathering, it is possible for the energy-rich nodes to swap roles (and serve now as intermediate nodes) with the energy-constrained nodes (that now serve as leaf nodes).

When there are 0 static nodes, the Max.Stable-DG trees sustain a lower probability of node failure (the plots to the left in Figures 14, 15 and 16) in comparison to the MST-DG trees. When operated with 80% static nodes, the probability of node failures for both the DG trees is about the same and is higher than that observed when all nodes are mobile. In more stable conditions, certain nodes (like the leader node and intermediate nodes) are overused. Thus, with the use of static nodes, though the absolute network lifetime can be marginally increased (by about 10-70%; the increase is larger at moderate transmission range per node and larger values of v_{max}), the probability of node failures to occur also increases.

The Max.Stable-DG trees incur a significant difference between the node lifetime and network lifetime (the range could be from factor of 1.7 to 23). For a given level of node mobility and transmission range pre-node, the difference between the node lifetime and network lifetime increases with increase in the number of static nodes. Similarly, for a given level of node mobility, the difference between the node lifetime and network lifetime incurred for the Max.Stable-DG trees increases with increase in the transmission range per node. Both these observations could be attributed to the relatively fewer number of flooding-based tree discoveries (at larger transmission ranges per node as well as in the presence of more static nodes). The MST-DG trees incur a very minimal increase in the network lifetime compared to the node lifetime (the range could be from factor of 1.4 to 5.7), especially when operated at higher levels of node mobility.

The results from Figures 14, 15 and 16 also indicate that the number of node failures to be incurred with the Max.Stable-DG trees to exceed that of the node failure time incurred with the MST-DG trees decreases with increase in maximum node mobility. At low v_{max} values, the time of first node failure incurred with the MST-DG trees could be as large as 400% more than the time of first node failure incurred with the Max.Stable-DG trees.

However, at higher levels of node mobility, the time of first node failure incurred with the MST-DG trees could be at most 100% larger than that of the Max.Stable-DG trees; consequently, the node failure times incurred with the Max.Stable-DG trees could quickly exceed that of the MST-DG trees at higher v_{max} values. The probability for node failures to occur with the Max.Stable-DG trees is relatively lower at moderate transmission ranges per node and low/moderate levels of node mobility; but, converges to that of the MST-DG trees when operated at higher levels of node mobility and larger transmission ranges per node. For a given v_{max} value and transmission range per node, the number of node failures required for the failure times incurred with the Max.Stable-DG trees to exceed to that of the MST-DG trees increases with increase in the number of static nodes.

Figure 14. Distribution of node failure times and probability of node failures [v_{max} = 3 m/s]

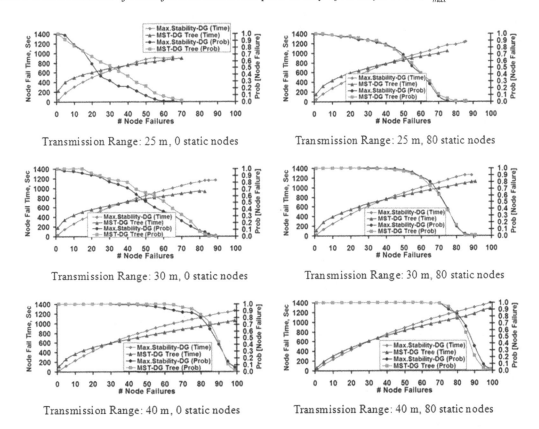

Figure 15. Distribution of node failure times and probability of node failures [v$_{max}$ = 10 m/s]

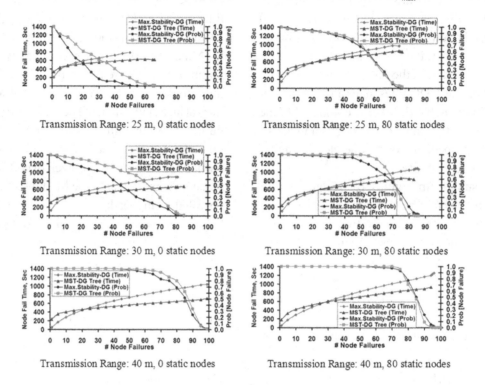

Figure 16. Distribution of node failure times and probability of node failures [v$_{max}$ = 20 m/s]

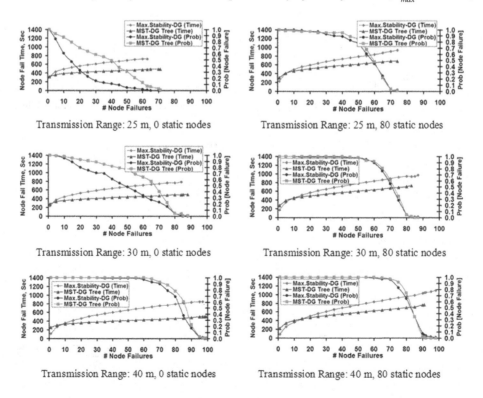

CONCLUSION

Given the entire sequence of topology changes over the duration of the data gathering session as input, we have developed a benchmarking algorithm (Max.Stability-DG algorithm) to determine a sequence of longest-living stable data gathering trees such that the number of tree discoveries is the global minimum. We have theoretically analyzed the correctness of the algorithm for sufficient-energy scenarios and verified through exhaustive simulations for both sufficient-energy and energy-constrained scenarios. The run-time complexity of the algorithm is analyzed to be $O(n^2 T \log n)$ and $O(n^3 T \log n)$ when operated under sufficient-energy and energy-constrained scenarios respectively, where n is the number of nodes in the network and T is the duration of the data gathering session. Since the Max.Stable-DG trees are based on network-wide spanning trees, the benchmarks incurred for the lifetime of these trees (and hence the number of tree discoveries) can be used to evaluate the relative stability of a spanning tree or any network-wide communication topology (like a connected dominating set) for mobile sensor networks. With a polynomial-time complexity and a much broader scope of application, as described above, the Max.Stability-DG algorithm has all the characteristics to become a global standard for evaluating the stability of communication topologies for data gathering in mobile sensor networks.

The extensive simulations (for varied levels of network dynamicity - by changing the maximum node velocity and number of static nodes as well as for varied levels of network density - by changing the transmission range per node) under both sufficient-energy and energy-constrained scenarios demonstrate that the Max.Stable-DG trees are significantly more stable than that of the minimum-distance spanning tree-based data gathering (MST-DG) trees. Due to the inherent nature of the Max.Stable-DG algorithm to determine long-living spanning trees and use them as data gathering trees as long as they exist, we observe the Max.Stable-DG trees to incur a lower node lifetime (time of first node failure) compared that of the MST-DG trees. However, the tradeoff between stability and fairness of node usage ceases to exist beyond the first few node failures; the reduced number of flooding-based tree discoveries coupled with the shallow structure and even distribution of nodes as intermediate nodes and leaf nodes contribute to a significantly longer network lifetime for the Max.Stable-DG trees. On the contrary, the MST-DG trees, though incur a larger time of first node failure, suffer a significantly lower network lifetime due to a cascade of node failures after the first node failure (expedited to frequent network-wide flooding-based tree discoveries). Table 1 summarizes the overall performance gains obtained with the Max.Stability-DG tree vis-à-vis the MST-DG trees under both the sufficient-energy and energy-constrained scenarios, as applicable.

Table 1. Overall performance gains for the maximum stability spanning tree based data gathering (max. stability-dg) tree and minimum-distance spanning tree based data gathering (MST-DG) Tree

Performance Metric	Better Data Gathering Tree	Range of Performance Gain Compared to the other Data Gathering Tree	
		Sufficient-Energy Scenario	Energy-Constrained Scenario
Tree Lifetime	Max.Stability-DG tree	150% to 360% larger	40% to 200% larger
Delay per Round	Max.Stability-DG tree	4% to 25% lower	7% to 18% lower
Node Lifetime	MST-DG tree	Not applicable	10% to 420% larger
Network Lifetime	Max.Stability-DG tree	Not applicable	5% to 60% larger

ACKNOWLEDGMENT

This research was sponsored by the U. S. Air Force Office of Scientific Research (AFOSR) through the Summer Faculty Fellowship Program for the lead author (Natarajan Meghanathan) in June-July 2012. The research was conducted under the supervision of the co-author (Philip Mumford) at the U. S. Air Force Research Lab (AFRL), Wright-Patterson Air Force Base (WPAFB) Dayton, OH. The AFRL public release number for this manuscript is 88ABW-2013-2114. The views and conclusions in this document are those of the authors and should not be interpreted as representing the official policies, either expressed or implied, of the funding agency. The U. S. Government is authorized to reproduce and distribute reprints for Government purposes notwithstanding any copyright notation herein.

REFERENCES

Abolhasan, M., Wysocki, T., & Dutkiewicz, E. (2004). A review of routing protocols for mobile ad hoc networks. *Ad Hoc Networks*, *2*(1), 1–22. doi:10.1016/S1570-8705(03)00043-X

Banerjee, T., Xie, B., Jun, J. H., & Agarwal, D. P. (2007). LIMOC: Enhancing the Lifetime of a Sensor Network with Mobile Cluster heads. In *Proceedings of the Vehicular Technology Conference Fall* (pp. 133-137), Baltimore, MD, USA: IEEE.

Bettstetter, C., Hartenstein, H., & Perez-Costa, X. (2004). Stochastic properties of the random-way point mobility model. *Wireless Networks*, *10*(5), 555–567. doi:10.1023/B:WINE.0000036458.88990.e5

Cormen, T. H., Leiserson, C. E., Rivest, R. L., & Stein, C. (2009). *Introduction to algorithms* (3rd ed.). Cambridge, MA: MIT Press.

Deng, S., Li, J., & Shen, L. (2011). Mobility-based clustering protocol for wireless sensor networks with mobile nodes. *IET Wireless Sensor Systems*, *1*(1), 39–47. doi:10.1049/iet-wss.2010.0084

Farago, A., & Syrotiuk, V. R. (2003). MERIT: A scalable approach for protocol assessment. *Mobile Networks and Applications*, *8*(5), 567–577. doi:10.1023/A:1025193929081

Heinzelman, W., Chandrakasan, A., & Balakarishnan, H. (2000). Energy-efficient communication protocols for wireless microsensor networks. In *Proceedings of the Hawaiian International Conference on Systems Science*. Maui, HI: IEEE.

Hull, B., Bychkovsky, V., Zhang, Y., Chen, K., Goraczko, M., & Miu, A. … Madden, S. (2006). CarTel: A distributed mobile sensor computing system. In *Proceedings of the 4th International Conference on Embedded Networked Sensor Systems*. Boulder, CO: ACM.

Kansal, A., Goraczko, M., & Zhao, F. (2007). Building a sensor network of mobile phones. In *Proceedings of the International Symposium on Information Processing in Sensor Networks* (pp. 547-548). Cambridge, MA: IEEE.

Kuhn, F., Moscibroda, T., & Wattenhofer, R. (2004). Unit disk graph approximation. In *Proceedings of the Workshop on Foundations of Mobile Computing* (pp. 17-23). Philadelphia, PA: ACM.

Lindsey, S., Raghavendra, C., & Sivalingam, K. M. (2002). Data gathering algorithms in sensor networks using energy metrics. *IEEE Transactions on Parallel and Distributed Systems*, *13*(9), 924–935. doi:10.1109/TPDS.2002.1036066

Liu, C.-M., Lee, C.-H., & Wang, L.-C. (2007). Distributed clustering algorithms for data gathering in wireless mobile sensor networks. *Journal of Parallel and Distributed Computing*, *67*(11), 1187–1200. doi:10.1016/j.jpdc.2007.06.010

Macuha, M., Tariq, M., & Sato, T. (2011). Data collection method for mobile sensor networks based on the theory of thermal fields. *Sensors (Basel, Switzerland), 11*(7), 7188–7203. doi:10.3390/s110707188 PMID:22164011

Meghanathan, N. (2010). A data gathering algorithm based on energy-aware connected dominating sets to minimize energy consumption and maximize node lifetime in wireless sensor networks. *International Journal of Interdisciplinary Telecommunications and Networking, 2*(3), 1–17. doi:10.4018/jitn.2010070101

Meghanathan, N. (2011). Performance comparison of minimum hop and minimum edge based multicast routing under different mobility models for mobile ad hoc networks. *International Journal of Wireless and Mobile Networks, 3*(3), 1–14. doi:10.5121/ijwmn.2011.3301

Meghanathan, N. (2012). A comprehensive review and performance analysis of data gathering algorithms for wireless sensor networks. *International Journal of Interdisciplinary Telecommunications and Networking, 4*(2), 1–29. doi:10.4018/jitn.2012040101

Meghanathan, N., & Farago, A. (2008). On the stability of paths, steiner trees and connected dominating sets in mobile ad hoc networks. *Ad Hoc Networks, 6*(5), 744–769. doi:10.1016/j.adhoc.2007.06.005

Meghanathan, N., & Mumford, P. D. (2012). Maximum stability data gathering trees for mobile sensor networks. *International Journal on Mobile Network Design and Innovation, 4*(3), 164–178. doi:10.1504/IJMNDI.2012.051972

Meghanathan, N., Sharma, S., & Skelton, G. W. (2010). On energy efficient dissemination in wireless sensor networks using mobile sinks. *Journal of Theoretical and Applied Information Technology, 19*(2), 79–91.

Meghanathan, N., & Skelton, G. W. (2007). A two layer architecture of mobile sinks and static sensors. In *Proceedings of the 15th International Conference on Advanced Computing and Communication* (pp. 249-254). Guwahati, India: IEEE.

Rappaport, T. S. (2002). *Wireless communications: Principles and practice* (2nd ed.). Upper Saddle River, NJ: Prentice Hall.

Santhosh Kumar, G., Vinu Paul, M. V., & Jacob Poulose, K. (2008). Mobility metric based LEACH-mobile protocol. In *Proceedings of the 16th International Conference on Advanced Computing and Communications* (pp. 248-253). Chennai, India: IEEE.

Sarma, H. K. D., Kar, A., & Mall, R. (2010). Energy efficient and reliable routing for mobile wireless sensor networks. In *Proceedings of the 6th International Conference on Distributed Computing in Sensor Systems Workshops*. Cambridge, MA: IEEE.

Singh, M., Sethi, M., Lal, N., & Poonia, S. (2010). A tree based routing protocol for mobile sensor networks (MSNs). *International Journal on Computer Science and Engineering, 2*(1S), 55–60.

Srinivasan, A., & Wu, J. (2008). TRACK: A novel connected dominating set based sink mobility model for WSNs. In *Proceedings of the 17th International Conference on Computer Communications and Networks*. St. Thomas, U. S. Virgin Islands: IEEE.

Viterbi, A. J. (1995). *CDMA: Principles of spread spectrum communication*. Upper Saddle River, NJ: Prentice Hall.

Vlajic, N., & Stevanovic, D. (2009). Sink mobility in wireless sensor networks: When theory meets reality. In *Proceedings of the Sarnoff Symposium*. Princeton, NJ: IEEE.

Wu, W., Beng Lim, H., & Tan, K.-L. (2010). Query-driven data collection and data forwarding in intermittently connected mobile sensor networks. In *Proceedings of the 7th International Workshop on Data Management for Sensor Networks* (pp. 20-25). Singapore: IEEE.

Xing, G., Wang, T., Jia, W., & Li, M. (2008). Rendezvous design algorithms for wireless sensor networks with a mobile base station. In *Proceedings of the 9th ACM International Symposium on Mobile Ad hoc Networking and Computing* (pp. 231-240). Hong Kong: ACM.

Zhang, H., & Hou, J. C. (2005). Maintaining sensing coverage and connectivity in large sensor networks. *Wireless Ad hoc and Sensor Networks. International Journal (Toronto, Ont.), 2*(1-2), 89–123.

Zhao, M., & Yang, Y. (2009). Bounded relay hop mobile data gathering in wireless sensor networks. In *Proceedings of the 6th International Conference on Mobile Ad hoc and Sensor Systems* (pp. 373-382). Macau, China: IEEE.

Zhong, M., & Cassandras, C. G. (2011). Distributed coverage control and data collection with mobile sensor networks. *IEEE Transactions on Automatic Control, 56*(10), 2445–2455. doi:10.1109/TAC.2011.2163860

ADDITIONAL READING

Ahmed, K., & Gregory, M. (2011). Integrating Wireless Sensor Networks with Cloud Computing. In *Proceedings of the Seventh International Conference on Mobile Ad hoc and Sensor Networks* (pp. 364-366), Beijing, China.

Awwad, S. A. B., Ng, C. K., Noordin, N. K., & Rasid, M. F. A. (2009). Cluster Based Routing protocol for Mobile Nodes in Wireless Sensor Network. In *Proceedings of the International Symposium on Collaborative Technologies and Systems* (pp. 233-241), Baltimore, MD, USA.

Ayaz, M., Abdullah, A., & Low Tang, J. (2010). Temporary cluster based routing for Underwater Wireless Sensor Networks. In *Proceedings of the International Symposium in Information Technology* (pp. 1009-1014), Kuala Lumpur, Malaysia.

Boukerche, A., Werner, R., & Pazzi, N. (2007). Lightweight Mobile Data Gathering Strategy for Wireless Sensor Networks. In *Proceedings of the 9th International Conference on Mobile and Wireless Communications Network* (pp. 151-155), Cork, Ireland.

Caro, Y., West, D. B., & Yuester, R. (2000). Connected Domination and Spanning Trees with many Leaves. *SIAM Journal on Discrete Mathematics, 13*(2), 202–211. doi:10.1137/S0895480199353780

Chen, Y.-S., & Liao, Y.-J. (2006). HVE-Mobicast: A Hierarchical-Variant-Egg-based Mobicast Routing Protocol for Wireless Sensornets. In *Proceedings of the Wireless Communications and Networking Conference* (pp. 697-702), Las Vegas, NV, USA: IEEE.

Deng, S., Li, J., & Shen, L. (2011). Mobility-based Clustering Protocol for Wireless Sensor Networks with Mobile Nodes. *IET Wireless Sensor Systems, 1*(1), 39–47. doi:10.1049/iet-wss.2010.0084

Hu, Y., Xue, Y., Li, Q., Liu, F., Keung, G. Y., & Li, B. (2009). The Sink Node Placement and Performance Implication in Mobile Sensor Networks. *Journal of Mobile Networks and Applications, 14*(2), 230–240. doi:10.1007/s11036-009-0158-5

Iwanari, Y., Asaka, T., & Takahashi, T. (2011). Power Saving Mobile Sensor Networks by Relay Communications. In *Proceedings of Consumer Communications and Networking Conference* (pp. 1150-1154), Las Vegas, NV, USA: IEEE.

Keung, G. Y., Li, B., & Zhang, Q. (2012). The Intrusion Detection in Mobile Sensor Network. *IEEE/ACM Transactions on Networking, 20*(4), 1152–1161. doi:10.1109/TNET.2012.2186151

Keung, G. Y., Zhang, Q., & Li, B. (2010). The Delay-Constrained Information Coverage Problem in Mobile Sensor Networks: Single Hop Case. *Journal of Wireless Networks, 16*(7), 1961–1973. doi:10.1007/s11276-010-0238-2

Khodashahi, M. H., Tashtarian, F., Moghaddam, M. H. Y., & Honary, M. T. (2010). Optimal Location for Mobile Sink in Wireless Sensor Networks. In *Proceedings of the Wireless Communications and Networking Conference* (pp. 1-6), Sydney, Australia: IEEE.

Li, J., Huang, L., & Xiao, M. (2008). Energy Efficient Topology Control Algorithms for Variant Rate Mobile Sensor Networks. In *Proceedings of the 4th International Conference on Mobile Ad hoc and Sensor Networks* (pp. 23-30), Wuhan, China.

Meghanathan, N. (2009). Use of Tree Traversal Algorithms for Chain Formation in the PEGASIS Data Gathering Protocol for Wireless Sensor Networks. *Transactions on Internet and Information Systems (Seoul), 3*(6), 612–627.

Meghanathan, N. (2010a). Grid Block Energy based Data Gathering Algorithms for Wireless Sensor Networks. *International Journal of Communication Networks and Information Security, 2*(3), 151–161.

Meghanathan, N. (2010b). A Data Gathering Algorithm based on Energy-aware Connected Dominating Sets to Minimize Energy Consumption and Maximize Node Lifetime in Wireless Sensor Networks. *International Journal of Interdisciplinary Telecommunications and Networking, 2*(3), 1–17. doi:10.4018/jitn.2010070101

Meghanathan, N. (2010c). An Algorithm to Determine Energy-aware Maximal Leaf Nodes Data Gathering Tree for Wireless Sensor Networks. *Journal of Theoretical and Applied Information Technology, 15*(2), 96–107.

Moro, G., Monti, G., & Ouksel, A. M. (2006). Routing and Localization Services in Self-Organizing Wireless Ad-Hoc and Sensor Networks Using Virtual Coordinates. In *Proceedings of the International Conference on Pervasive Services* (pp. 243-246), Lyon, France: ACS/IEEE.

Poduri, S., & Sukhatme, G. S. (2004). Constrained Coverage for Mobile Sensor Networks. In *Proceedings of International Conference on Robotics and Automation* (pp. 165-172), New Orleans, LA, USA: IEEE.

Rezazadeh, J., Moradi, M., & Ismail, A. S. (2012). Mobile Wireless Sensor Networks Overview. *International Journal of Computer Communications and Networks, 2*(1), 17–22.

Wang, C., Ramanathan, P., & Saluja, K. K. (2010). Modeling Latency-Lifetime Tradeoff for Target Detection in Mobile Sensor Networks. *ACM Transactions on Sensor Networks, 7*(1), 1–24. doi:10.1145/1806895.1806903

Wu, W., Lim, H. B., & Tan, K.-L. (2010). Query-driven Data Collection and Data Forwarding in Intermittently Connected Mobile Sensor Networks. In *Proceedings of the Seventh International Workshop on Data Management for Sensor Networks* (pp. 20-25), Singapore.

Zheng, T., Radhakrishnan, S., & Sarangan, V. (2012). A Routing Layer Sleep Scheme for Data Gathering in Wireless Sensor Networks. In *Proceedings of the International Conference on Communications* (pp. 735-739), Ottawa, ON, Canada: IEEE.

KEY TERMS AND DEFINITIONS

Algorithm: A sequence of steps to perform a particular task.

Communication Topology: An arrangement of the nodes in the network that can be used to propagate data to the sink.

Data Aggregation: The process of combining data from multiple sensor nodes and generating a representative data of the same size as the data from the individual sensor nodes.

Data Gathering Tree: A communication topology that is used to aggregate data from all the sensor nodes towards a leader node that can further propagate the aggregated data to the sink.

Delay per Round: The number of timeslots incurred for the leader node to generate an ag-gregated data, collected from all the sensor nodes, propagated through the data gathering tree.

Leader Node: The root node of the data gathering tree.

Mobile Sensor Network: A network of sensor nodes that move arbitrarily.

Network Lifetime: The time (number of rounds) of network disconnection due to the failure of one or more sensor nodes.

Node Lifetime: The time (number of rounds) of first node failure due to the exhaustion of battery charge at the sensor nodes.

Round: During a round of data gathering, data gets aggregated starting from the leaf nodes of the tree and propagates all the way to the leader node. An intermediate node in the tree collects the aggregated data from its immediate child nodes and further aggregates with its own data before forwarding to its immediate parent node in the tree.

Stability: A characteristic of data gathering trees related to the duration of existence of the tree; the longer the duration of existence, the more stable is the data gathering tree.

Tree Lifetime: The duration of existence of a data gathering tree.

Chapter 18

Comparative Study of Adaptive Multiuser Detections in Hybrid Direct–Sequence Time–Hopping Ultrawide Bandwidth Systems

Qasim Zeeshan Ahmed
King Abdullah University of Science and Technology (KAUST), Saudi Arabia

Lie-Liang Yang
University of Southampton, UK

ABSTRACT

This chapter considers low-complexity detection in hybrid Direct-Sequence Time-Hopping (DS-TH) Ultrawide Bandwidth (UWB) systems. A range of Minimum Mean-Square Error (MMSE) assisted Multiuser Detection (MUD) schemes are comparatively investigated with emphasis on the low-complexity adaptive MMSE-MUDs, which are free from channel estimation. In this contribution, three types of adaptive MUDs are considered, which are derived based on the principles of Least Mean-Square (LMS), Normalized Least Mean-Square (NLMS), and Recursive Least-Square (RLS), respectively. The authors study comparatively the achievable Bit Error-Rate (BER) performance of these adaptive MUDs and of the ideal MMSE-MUD, which requires ideal knowledge about the UWB channels and the signature sequences of all active users. Both the advantages and disadvantages of the various adaptive MUDs are analyzed when communicating over indoor UWB channels modeled by the Saleh-Valenzuela (S-V) channel model. Furthermore, the complexity of the adaptive MUDs is analyzed and compared with that of the single-user RAKE receiver and also with that of the ideal MMSE-MUD. The study and simulation results show that the considered adaptive MUDs constitute feasible detection techniques for deployment in practical UWB systems. It can be shown that, with the aid of a training sequence of reasonable length, an adaptive MUD is capable of achieving a similar BER performance as the ideal MMSE-MUD while requiring a complexity that is even lower than that of a corresponding RAKE receiver.

DOI: 10.4018/978-1-4666-5170-8.ch018

1. INTRODUCTION

In recent years ultrawide bandwidth (UWB) communications have drawn wide interest in both research and industry communities (Molisch, 2005; Scholtz, 1993). Initially, UWB signaling has been implemented using solely carrier-less time-hopping pulse-position modulation (TH-PPM) (Scholtz, 1993), which, for convenience, is referred to as the pure TH-UWB scheme. Then, the conventional direct-sequence spread-spectrum (DS-SS) scheme has been introduced for implementation of UWB communications (Foerster, 2002), which is referred to as the pure DS-UWB scheme. Explicitly, both these pulse-based UWB schemes employ their advantages and disadvantages, when communicating over the UWB channels, which are highly dispersive (Molisch, 2005; Fishler, 2005). Therefore, it might be desirable to adopt in one UWB system both the DS-SS and TH, forming a so-called hybrid DS-TH UWB communications scheme (Ahmed et al., 2007), in order to take the advantages of both the pure DS-UWB and pure TH-UWB, while, simultaneously, to avoid their shortcomings. Note that, the hybrid DS-TH UWB scheme may be viewed as a generalized pulse-based UWB scheme, which includes both the pure DS-UWB and pure TH-UWB schemes as its special examples (Ahmed et al., 2007). Additionally, the hybrid DS-TH UWB scheme employs a higher number of degrees-of-freedom than either the pure DS-UWB or pure TH-UWB scheme, which may be beneficial to the design and reconfiguration in order to achieve high-flexibility UWB communications.

Pulse-based UWB schemes constitute a range of promising alternatives to be employed for home, personal-area, sensor network, etc. applications, where devices are required to consume minimum power while maintaining low-complexity (Reed et al., 2005; Breining, 1999). Due to its low-complexity in the context of conventional wideband communications (Proakis, 2000), RAKE receiver has naturally been considered for detection in UWB systems (Win & Scholtz, 1998; Mireles, 2002). However, UWB channels are highly dispersive, resulting in that a UWB receiver typically receives a high number of resolvable paths with each resolvable path conveying only a small portion of the transmitted energy. Hence, in order to capture efficiently the transmitted energy, a large number of RAKE fingers are required, which substantially increases the detection complexity (Li & Rusch, 2002; Honig & Tsatsanis, 2000). Furthermore, for achieving coherent combining (Simon & Alouini, 2005), the RAKE receiver requires to estimate a huge number of multipath component channels, which is highly-complex and impractical. Additionally, when communicating over multiuser communications scenarios, multiuser interference (MUI) and inter-symbol interference (ISI) may severely degrade the achievable BER performance of the UWB systems (Yang & Giannakis, 2004).

Due to the above-mentioned considerations, in this contribution we propose and investigate a range of *minimum mean-square error (MMSE)* assisted *multiuser detection (MUD)* schemes for the hybrid DS-TH UWB systems. We focus our attention on the three types of low-complexity adaptive MUDs, which are implemented based on the principles of *least mean-square (LMS), normalized least mean-square (NLMS)* and *recursive least-square (RLS),* respectively (Haykin et al., 2002). As our forthcoming discourse shown, these adaptive MUDs are free from channel estimation and are capable of achieving the approximate MMSE solutions with the aid of training sequences of certain length. In this contribution we investigate comparatively the achievable BER performance of the adaptive MUDs as well as of the ideal MMSE-MUD (Moshavi, 1996), which demands ideal knowledge about the UWB channels and the signature sequences of all active users (Woodward & Vucetic, 1998; Moshavi, 1996). The BER performance of the hybrid DS-TH UWB, pure DS-UWB and pure TH-UWB systems, which employ the above-mentioned adaptive/ideal MMSE-MUDs, is investigated, when communicating over indoor UWB channels modeled by the Saleh-Valenzuela (S-V) channel model (Molisch, 2005; Karedal et al., 2004). The advantages and disadvantages of the considered adaptive MUDs are analyzed in the context of UWB communications. Furthermore, the complexity of the adaptive MUDs is analyzed and compared with that of the single-user RAKE receiver and also with that of the ideal MMSE-MUD.

Our study shows that the three types of adaptive MUDs considered in this contribution are highly efficient detection schemes for pulse-based UWB systems. They are free from channel estimation and can effectively capture the transmitted energy dispersed over UWB channels. They are capable of achieving a BER performance close to that achieved by the ideal MMSE-MUD. Furthermore, as our forthcoming complexity analysis shown, the detection complexity of an adaptive MUD may be significantly lower than that of the RAKE receiver, even without taking into account the complexity required by the RAKE receiver for channel estimation.

The rest of this chapter is organized as follows. Section-2 describes the system model of the hybrid DS-TH UWB system. Various detection schemes for the hybrid DS-TH UWB system are provided in Section-3. In this section the complexity of various detection schemes is also analyzed. Section-4 demonstrates our simulation results and, in Section-5 our conclusions are presented. Finally, we have added Section-6 to address the future trends in the field of UWB.

2. DESCRIPTION OF HYBRID DS-TH UWB SYSTEM

2.1 Transmitted Signal

The transmitter schematic block diagram for the considered hybrid DS-TH UWB system is shown in Figure 1. In our hybrid DS-TH UWB system binary phase-shift keying (BPSK) baseband

modulation is assumed for simplicity. As shown in Figure 1, a data bit of the kth user is first modulated by the N_c-length DS spreading sequence, yielding N_c chips. Then, the N_c chips are transmitted by N_c time-domain pulses within one symbol-duration, where the positions of the N_c time-domain pulses are determined by the TH pattern assigned to the kth user. According to Figure 1, it can be shown that the hybrid DS-TH UWB signal transmitted by the kth user can be written as:

$$s^{(k)}(t) = \sqrt{\frac{E_b}{N_c T_\psi}} \sum_{j=0}^{\infty} b^{(k)}_{\left\lfloor \frac{j}{N_c} \right\rfloor} d^{(k)}_j \psi\left(t - jT_c - c^{(k)}_j T_\psi\right) \qquad (1)$$

where x represents the floor function, which returns the largest integer not exceeding x, $\psi(t)$ is the basic time-domain pulse of width T_ψ, which satisfies $\int_{0}^{T_\psi} \psi^2(t)\, dt = T_\psi$. The bandwidth of the hybrid DS-TH UWB signals is determined by the basic time-domain pulse. For brevity, the parameters invoked in (1) as well as those used in our forthcoming discourse are listed as follows:

- E_b: Energy per bit;
- N_c: Number of chips per bit, which is defined as the DS spreading factor;
- N_ψ: Number of time-slots per chip, which is defined as the TH spreading factor;
- T_b and T_c: Bit-duration and chip-duration, where $T_b = N_c T_c$;
- T_ψ: Width of the basic time-domain pulse or width of a time-slot, it satisfies $T_c = N_\psi T_\psi$;

Figure 1. Transmitter schematic block diagram of hybrid direct-sequence time-hopping ultrawide bandwidth (DS-TH UWB) systems

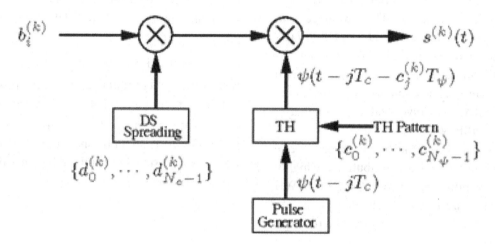

- $b_i^{(k)} \in \{+1, -1\}$: The i th data bit transmitted by user k, $b_i^{(k)}$ takes a value of +1 or -1 with equal probability;

- $d_j^{(k)}$: Binary DS spreading sequence assigned to the k th user;

- $c_j^{(k)}$: TH pattern assigned to the k th user, $c_j^{(k)} \in \{= 0, 1, \ldots, N_\psi - 1\}$ and takes any value with equal probability;

- $N_c N_\psi$: Total spreading factor of the hybrid DS-TH UWB system.

From the above description, it can be observed that, when $N_\psi = 1$, T_ψ and T_c are equal and, in this case, the hybrid DS-TH UWB scheme is reduced to the pure DS-UWB scheme associated with a spreading factor of N_c. By contrast, when $N_c = 1$, the hybrid DS-TH UWB is reduced to the pure TH-UWB scheme associated with a spreading factor of N_ψ.

2.2 Channel Model

In this contribution the IEEE 802.15.4a channel model is considered (Karedal et al., 2004). This channel model is suitable for industrial environ-

ments, where the abundance of metallic scatters may cause dense multipath scattering, resulting in Rayleigh distributed small-scale fading (Molisch, 2005; Karedal et al., 2004). Specifically, in this contribution the Saleh-Valenzuela (S-V) channel model is invoked in our investigation, which can be represented as:

$$h(t) = \sum_{v=0}^{V-1} \sum_{u=0}^{U-1} h_{u,v} \delta(t - T_v - T_{u,v}) \qquad (2)$$

where V represents the number of clusters and U denotes the number of resolvable multipaths in a cluster. Hence, the total number of resolvable multipath components can be as high as $L = UV$. In (2), $h_{u,v} = |h_{u,v}| e^{j\theta_{u,v}}$ represents the fading gain of the u th multipath in the v th cluster, T_v denotes the arrival time of the v th cluster, while $T_{u,v}$ denotes the arrival time of the u th multipath in the v th cluster. In the considered UWB channel, we assume that the average power of a multipath component at a given delay, say at $T_v + T_{u,v}$, is related to the power of the first resolvable multipath in the first cluster through the relationship of:

$$u, v = o, o^{exp} \left(-\frac{T_v}{\Gamma} \right) exp \left(-\frac{T_{u,v}}{\gamma} \right) \qquad (3)$$

where $u, v = E\left[|h_{u,v}|^2 \right]$ represents the power of the u th resolvable multipath in the v th cluster, Γ and γ denote the power decay constants in the context of the clusters and multipath, respectively. Note that, in order to make the channel model sufficiently general, in this contribution we assume that the delay-spread of the UWB channels spans $g \geq 1$ data bits. In this case, the total number of resolvable multipath of L satisfies:

$$(g-1)N_c N_\psi \leq (L-1) \leq gN_c N_\psi$$

2.3 Receiver Structure

Let us assume that the hybrid DS-TH UWB system supports K users. Then, when the DS-TH UWB signals in the form of (1) are transmitted over the UWB channels with the channel impulse response (CIR) as shown in (2), the received signal can be expressed as:

$$r(t) = \sqrt{\frac{E_b}{N_c T_\psi}} \sum_{k=1}^{K} \sum_{v=0}^{V-1} \sum_{u=0}^{U-1} \sum_{j=0}^{MN_c-1} h_{u,v}^{(k)} b_{\frac{j}{N_c}}^{(k)} d_j^{(k)}$$

$$\times \psi_{rec} \left(t - jT_c - c_j^{(k)}T_\psi - T_v^{(k)} - T_{u,v}^{(k)} - \tau_k \right) + n(t) \qquad (4)$$

where $n(t)$ represents the additive white Gaussian noise (AWGN), which is Gaussian distributed with zero-mean and single-sided power spectral density (PSD) of N_0 per dimension, $\{\tau_k\}$ take into account the lack of synchronization among the K user signals as well as their corresponding transmission delays, while $\psi_{rec}(t)$ denotes the received time-domain pulse, which, as analyzed

in (Zhang, 2005), is usually the second derivative of the transmitted pulse $\psi(t)$ as seen in (1).

The receiver schematic block diagram for adaptive detection of hybrid DS-TH UWB signals is shown in Figure 2. The received signal is first input to a matched-filter (MF) having an impulse response of $\psi_{rec}^*(-t)$, which is matched to the received pulse $\psi_{rec}(t)$. The output of the MF is then sampled at a rate of $1/T_\psi$ in order to generate the observation samples. Finally, the observation samples are input to an adaptive filter, which generates the estimates of the transmitted data bits. Let us assume that a block of M data bits are transmitted. Then, the receiver can collect a total of $(MN_c N_\psi + L - 1)$ number of samples for detection of the M bits, where $(L-1)$ is due to the delay-spread of the UWB channel, which results in L number of resolvable multipath. To be more specific, the λth observation sample can be obtained by sampling the MF's output at the time instant $t = T_0 + (\lambda + 1)T_\psi$, yielding:

$$y_\lambda = \left(\sqrt{\frac{E_b T_\psi}{N_c}} \right)^{-1} \int_{T_0 + \lambda T_\psi}^{T_0 + (\lambda+1)T_\psi}$$

$$\times r(t)\psi_{rec}^*(t)dt, \quad \lambda = 0, 1, \cdots, MN_c N_\psi \qquad (5)$$

$$+ L - 2$$

where T_0 denotes the time-of-arrival (ToA) of the first path in the first cluster. Let us define:

$$y = \left[y_0, y_1, \cdots, y_{MN_c N_\psi + L - 2} \right]^T \qquad (6)$$

$$n = \left[n_0, n_1, \cdots, n_{MN_c N_\psi + L - 2} \right]^T \qquad (7)$$

Then, according to (7), it can be shown that the element n_λ in n can be represented as:

Figure 2. Receiver schematic block diagram for the hybrid direct-sequence time-hopping ultra-wide bandwidth (DS-TH UWB) systems

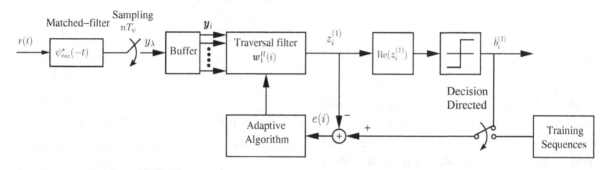

$$n_\lambda = \left(\sqrt{\frac{E_b T_\psi}{N_c}} \right)^{-1} \int_{T_0 + \lambda T_\psi}^{T_0 + (\lambda+1)T_\psi} n\left(t\right) \psi_{rec}^*\left(t\right) dt \qquad (8)$$

which is a Gaussian random variable with zero mean and a variance of $\sigma^2 = N_0 / 2E_b$ per dimension. Furthermore, upon substituting the received signal in the form of (4) into (5), we can show that, after some simplifications, y can be expressed as:

$$y = \sum_{k=1}^{K} C_k H_k b_k + n \qquad (9)$$

where $b_k = \left[b_0^{(k)}, b_1^{(k)}, \cdots, b_{M-1}^{(k)} \right]^T$ contains the M number of data bits transmitted by the kth user, the channel matrix H_k of the kth user is given by:

$$H_k = diag\left\{ h_k, h_k, \cdots, h_k \right\} \qquad (10)$$

which is a $(ML \times M)$-dimensional matrix with h_k formed by the CIR of user k as:

$$h_k = \left[h_{(0,0)}^{(k)}, h_{(1,0)}^{(k)}, \cdots, h_{(U-1,V-1)}^{(k)} \right]^T \qquad (11)$$

It can be shown that the detail of the matrices and vectors in (9) can be found in (Ahmed et al., 2010). Let us now consider the bit-by-bit detection of hybrid DS-TH UWB signals in the next section.

3. DETECTION IN HYBRID DS-TH UWB SYSTEMS

As mentioned in Section 2.2, for the considered hybrid DS-TH UWB system, there exists strong ISI due to the delay-spread of UWB channels, which, as shown in Section 2.2, is assumed to span $g \geq 1$ data bits. Under this assumption, it can be implied that the ith data bit experiences interference from $\min\left(i, g-1\right)$ data bits transmitted before the ith data bit and also from $\min\left(M-1-i, g-1\right)$ data bits transmitted after the ith data bit. Hence, according to the principles of MUD and equalization (Haykin, 2002; Verdu, 1998), the optimum MUD for the hybrid DS-TH UWB system should consider simultaneously all the users supported by the system and also all the data blocks transmitted by all the users. However, in our hybrid DS-TH UWB system the spreading factor of $N_c N_\psi$ and the number of resolvable multipaths L might be very high, resulting in extreme complexity of the block-based MUDs, even when linear MUDs are considered. Hence, in this contribution we consider only the bit-based

MUDs for the sake of achieving low-complexity detection for the hybrid DS-TH UWB system. Specifically, in this contribution the bit-by-bit based MUDs are considered with the emphasis on the adaptive MUDs.

Let the observation vector y_i and the noise vector n_i in the context of the ith data bit of the first user - which is assumed to be the desired user - be represented as:

$$y_i = \left[y_{iN_cN_\psi}, y_{iN_cN_\psi+1}, \cdots, y_{(i+1)N_cN_\psi+L-2} \right]^T \quad (12)$$

$$n_i = \left[n_{iN_cN_\psi}, n_{iN_cN_\psi+1}, \cdots, n_{(i+1)N_cN_\psi+L-2} \right]^T \quad (13)$$

Let us first review the ideal MMSE-MUD before considering the adaptive bit by bit based MUDs.

3.1 Minimum Mean-Square Error Multiuser Detection

For the MMSE-MUD, the decision variable for the desired bit $b_i^{(1)}$ of the first user can be formed as:

$$z_i^{(1)} = w_1^H y_i, \quad i = 0, 1, \cdots, M-1 \quad (14)$$

where the optimum weight vector w_1 in MMSE sense is given by (Haykin, 2002):

$$w_1 = R_{y_i}^{-1} r_{y_i b_i^{(1)}} \quad (15)$$

where R_{y_i} and $r_{y_i b_i^{(1)}}$ are the auto-correlation matrix of y_i and the cross-correlation matrix between y_i and $b_i^{(1)}$, respectively.

Based on the decision variable $z_i^{(1)}$, the estimate to the transmitted data bit $b_i^{(1)}$ can be formed as

$$\hat{b}_i^{(1)} = +1, \text{ if } \Re\left\{ \hat{z}_i^{(1)} \geq 0 \right\} \text{ and } \hat{b}_i^{(1)} = -1, \text{ otherwise.}$$

As shown in (15), in order to determine the weight vector w_1 in MMSE sense, the detector requires the knowledge about the signature codes and channel state information (CSI) associated with all the active users, which is generally impractical to obtain in UWB communications. This is because, as mentioned previously in Section 2.2, the UWB channels are highly dispersive, resulting in a huge number of resolvable multipaths with each multipath conveying very low power. Furthermore, as shown in (20), the MMSE-MUD needs to invert R_{y_i}, which is a $\left(N_c N_\psi + L - 1 \right) \times \left(N_c N_\psi + L - 1 \right)$ – dimensional matrix. Hence, the detection complexity of the MMSE-MUD of (15) might be extreme, when the spreading factor $N_c N_\psi$ and/or the value of L are high. Therefore, in this contribution the adaptive MUDs for the hybrid DS-TH UWB system are proposed, which can substantially reduce the detection complexity, as shown in our forthcoming discourse.

3.2 Adaptive Multiuser Detection

The adaptive MUD schemes considered in this contribution do not require the knowledge about the spreading codes and CSI of the active users (Woodward et al., 1998; Miller, 1995). The adaptive MUDs also do not compute the inverse of the auto-correlation matrix R_{y_i}. Hence, we can argue that the adaptive MUD schemes may have a significantly lower complexity than the ideal MMSE-MUD requiring ideal knowledge about the spreading codes and CSI of the active users. Furthermore, the adaptive MUDs are feasible for implementation in practice, since in adaptive MUDs training sequences of reasonable length can be applied in order to assist the adaptive receiver to converge.

The schematic block diagram of the adaptive receiver for the hybrid DS-TH UWB systems is shown in Figure 2. Specifically, the adaptive receiver is operated in two modes, namely the training mode and the detection mode. During the training mode, the $\left(N_c N_\psi + L - 1\right)$ observation samples corresponding to each of the training data bits are first stored in a buffer as shown in Figure 2. Let y_i contain the observation samples corresponding to the ith training bit $b_i^{(1)}$. As shown in Figure 2, y_i is input to a transversal filter with a weight vector w_1, yielding an estimate $z_i^{(1)}$ to the training bit $b_i^{(1)}$. The estimate $z_i^{(1)}$ can be expressed as:

$$z_i^{(1)} = w_1^H y_i, \ i = 0, 1, \cdots, M - 1 \qquad (16)$$

and, correspondingly, the estimation error is computed as:

$$e(i) = b_i^{(1)} - z_i^{(1)} = b_i^{(1)} - w_1^H y_i \qquad (17)$$

Finally, based on the estimation error $e(i)$, the weight vector $w_1(i)$ of the transversal filter is then updated to $w_1(i + 1)$ with the aid of certain adaptive algorithms, as will be detailed in the forthcoming discourse.

After the adaptive receiver is sufficiently trained, it is then switched from the training mode to the detection mode. During the detection mode, signal detection is carried out in the same way as that in the ideal MMSE-MUD, as shown in Section 3.1. Furthermore, during the detection mode, the adaptive receiver is operated under the decision-directed principles (Kumar, 1983), and the weight vector of the transversal filter is updated using the detected data bits, as shown in Figure 2.

In this contribution three types of adaptive algorithms are invoked for achieving the adaptive MUD for the hybrid DS-TH UWB system. Below we consider these adaptive algorithms in detail.

3.3 Least Mean-Square (LMS) Adaptive Algorithm

The LMS adaptive algorithm is a well-known classic adaptive algorithm, which belongs to the category of stochastic gradient algorithms (Haykin, 2002). The principle behind the LMS is to find a sub-optimal weight vector solution w_1 of (15) through stochastic gradient techniques, so that the mean-square error (MSE) achieved is close to the minimum MSE of the ideal MMSE-MUD. To be more specific, the LMS adaptive algorithm can be described as shown in Algorithm 1.

The LMS adaptive algorithm is a low-complexity adaptive algorithm. However, it suffers from the problem of gradient-noise amplification (Haykin, 2002). Furthermore, according to (Haykin, 2002) and (Montazeri et al., 1995), the convergence speed and robustness of the LMS algorithm are depended on the step-size μ as well as on the statistics of the input data vector y_i. For the hybrid DS-TH UWB system communicating over indoor UWB channels, the input data vector y_i may be viewed as a stable Gaussian random process, since the received signals are constituted by many independent multipath component signals. However, the step-size used may have a strong impact on the adaptive receiver. For example, using a larger step-size value usually results in a higher convergence speed and hence requiring relatively shorter training sequences. From this point of view, the spectrum-efficiency of the communication system may be improved. However, a larger step-size value generally leads to a higher MSE value, yielding a higher error rate of detection. By contrast, when a smaller step-size value is employed, a lower error rate of

Algorithm 1.

```
Parameters:

μ= a suitable step-size,
```
$$0 < \mu < \frac{2}{E\left[y_i^2\right]}$$

```
Initialization
```
$w_1\left(0\right) = 0$, `when without a-prior`
```
knowledge.
Weight vector update
For
```
$i = 0, 1, \cdots$, `compute`
`estimation error:` $e\left(i\right) = b_i^{(1)} - w_1^H y_i$,
```
and
weight vector:
```
$$w_1\left(i+1\right) = w_1\left(i\right) + \mu y_i e^*\left(i\right).$$

the adaptive receiver may be achieved, in this case both the convergence speed of the adaptive receiver and the spectral-efficiency of the communication system decrease. Due to the above-mentioned reason, in our LMS adaptive receiver for the hybrid DS-TH UWB system, two step-size values are employed, a large one for the training mode and a small one for the detection mode. Specifically, during the training mode, a relatively large step-size value is applied, so that the LMS adaptive receiver converges as quickly as possible. By contrast, during the detection mode, a relatively small step-size value is applied, so that the adaptive LMS receiver is capable of converging to the MMSE solution and achieving the best possible BER performance.

3.4 Normalized Least Mean-Square (NLMS) Adaptive Algorithm

The NLMS adaptive algorithm also belongs to the family of stochastic gradient algorithms (Haykin, 2002). The NLMS adaptive algorithm can mitigate the problem of gradient noise amplification, making it more robust to the gradient noise than the LMS adaptive algorithm (Haykin, 2002). In comparison with the LMS adaptive algorithm,

the NLMS adaptive algorithm has a higher convergence speed and, more promisingly, its convergence speed is independent of the variation of communications environments (Haykin, 2002). Furthermore, in a given communications environment, the NLMS adaptive receiver is capable of achieving a lower MSE than the corresponding LMS adaptive receiver (Haykin, 2002). Consequently, the NLMS adaptive algorithm may use a relatively short training sequence, which hence increases the spectral efficiency of the communication systems.

The NLMS algorithm can be described as shown in Algorithm 2.

The main problem with stochastic gradient algorithms is their strong dependence on the eigenvalue distribution of the autocorrelation matrix R_{y_i} of the input signal vector (Glentis et al., 1999). It has been shown that the larger the ratio between the maximum and minimum eigenvalues, the slower is the convergence speed (Breining et al., 1999). Therefore, the stochastic gradient based adaptive algorithms might not be suitable for the communication environments experiencing colored noise, which results in that the ratio between the maximum and minimum eigenvalues is usually high (Glentis et al., 1999). Furthermore, both the LMS and NLMS adaptive algorithms have only single parameter, which may be adjusted for controlling the convergence speed and stability of the adaptive algorithms. Let us below consider the RLS adaptive algorithm, which has more than one parameters that can be controlled for adjusting the convergence speed (Proakis et al., 2003).

3.5 Recursive Least-Square (RLS) Adaptive Algorithm

The RLS adaptive algorithm belongs to the category of least square (LS) adaptive algorithms. It makes use of the information of the detected data, in order to enhance the convergence speed

Algorithm 2.

```
Parameters:
μ= a suitable step-size,  0 < μ < 2.
Initialization
```
$w_1(0) = 0$, when without a-prior
```
knowledge.
Weight vector update
For  i = 0,1,⋯ ,  compute
```
estimation error: $e(i) = b_i^{(1)} - w_1^H y_i$,
```
and
weight vector:
```

$$w_1(i+1) = w_1(i) + \frac{\mu}{y_i^2} y_i e^*(i).$$

and detection performance. The RLS adaptive algorithm is capable of achieving a significantly faster convergence speed than the LMS-assisted adaptive algorithm, but at the cost of a substantially increased computational complexity (Haykin, 2002; Proakis et al., 2003). Let λ be the forgetting factor and $P(i)$ be the inverse of the autocorrelation matrix upon receiving y_i. Then, the RLS adaptive algorithm can be summarized as shown in Algorithm 3 (Haykin, 2002; Proakis et al., 2003).

In comparison with the LMS algorithm in Section 3.2 and the NLMS algorithm in Section 3.2, the RLS algorithm employs more parameters, which may be adjusted in order to control the convergence speed. The stability of the RLS adaptive algorithm depends on the initialization of $P(0)$ as well as on the values of the parameters δ and λ (Breining et al., 1999). It has been shown (Breining et al., 1999; Moustakides, 1997) that the RLS adaptive algorithm is highly sensitive to the value of δ. The convergence speed and stability of the RLS algorithm may be significantly different for different settings of the δ value. The value of the forgetting factor λ also has effect on

the convergence speed and stability of the RLS algorithm (Breining et al., 1999).

3.6 Complexity Analysis

The complexity of the single-user correlation detector, the ideal MMSE-MUD using full knowledge about CSI and signature sequences, as well as the LMS-, NLMS- and RLS-aided adaptive MUDs is summarized in Table 1. The complexity stated in Table 1 denotes the number of operations including both additions and multiplications, which are required for detecting a bit of a user. In this table we defined a parameter $T = N_c N_\psi + L - 1$.

From Table 1 it can be observed that the complexity of the single-user correlation detector is much less than that of the ideal MMSE-MUD. Note that, both these detectors require the knowledge about the channels and signature sequences in the context of all the active users, when all the active users are detected, such as at the base-station (BS). Therefore, in these two detection schemes, complexity is required for obtaining the channel knowledge, in addition to the complexity for detection. As shown in Table 1, even without requiring channel estimation, the LMS- and NLMS-assisted adaptive MUDs generally have a significantly lower complexity than the single-user correlation detector and also than the ideal MMSE-MUD, when communicating over the UWB channels, which generate a high number of resolvable multipaths, implying $L > 1$. The complexity of the RLS-assisted adaptive MUD is lower than that of the ideal MMSE-MUD, but higher than that of the single-user correlation detector, when ignoring the complexity for channel estimation. Among the three adaptive MUD schemes considered, both the LMS- and NLMS-assisted adaptive MUDs have a similar detection complexity, while the RLS-assisted adaptive MUD has the highest detection complexity.

Algorithm 3.

```
Parameters:
λ= a suitable forgetting factor in the range
```

$$\left(1 - \frac{1}{\left(N_c N_\psi + L - 1\right)}\right) \leq \lambda \leq \left(1 - \frac{1}{10\left(N_c N_\psi + L - 1\right)}\right)$$

```
Initialization
```

$w_1(0) = 0$, when without a-prior knowledge.

$P(0) = \delta^{-1}I$, where δ is a set as a relatively small positive constant for

```
higher SNR,
and as a relatively large positive constant for lower SNR.
Computation
For  i = 1,2,⋯,  compute
```

gain vector: $k(i) = \dfrac{\lambda^{-1}P(i-1)y_i}{1 + \lambda^{-1}y_i^H P(i-1)y_i}$,

a-priori estimation error: $e(i) = b_i^{(1)} - w_1^H y_i$,

weight vector: $w_1(i) = w_1(i-1) + k(i)e^*(i)$.

inverse of autocorrelation matrix: $P(i) = \lambda^{-1}P(i-1) - \lambda^{-1}k(i)y_i^H P(i-1)$

Table 1. Complexity comparison of different schemes in UWB systems

Algorithms	Number of Additions	Number of Multiplications
Correlation detector	(L+1)T	(L+1)T
Ideal MMSE-MUD	T³/6+2T²+(L+1)T	T³/6+2T²+(L+1)T
LMS-assisted adaptive MMSE-MUD	2T+1	3T
NLMS-assisted adaptive MMSE-MUD	3T+1	4T+1
RLS-assisted adaptive MMSE-MUD	5T²+3T+2	6T²+5T+2

4. SIMULATION RESULTS AND DISCUSSION

In this section a range of performance results are presented, in order to characterize the achievable BER performance of the hybrid DS-TH UWB systems employing the LMS-, NLMS-, or the RLS-aided adaptive MUDs. Firstly, the learning curves of the adaptive algorithms are depicted, when they are applied to our hybrid DS-TH UWB system. Then, the BER performance of the hybrid DS-TH UWB systems using various adaptive MUDs is presented. Furthermore, the BER performance of the pure DS-UWB system and pure TH-UWB system is provided, since both these UWB systems constitute special examples of the hybrid DS-TH UWB system. Note that, our simulations were carried out based on the S-V channel model considered in Karedal et al.

(2004), which has the parameters $1/\Lambda = 14.11$ ns, $\Gamma = 2.63$ns and $\gamma = 4.58$ns.

For convenience, some of the variables and parameters used in the figures, which have not been specified previously, are listed as follows:

- $f_D T_b$: Normalized Doppler frequency-shift;

- μ_{LMS} and μ_{NLMS}: Step-size in the LMS and NLMS adaptive algorithms during the training mode;

- μ_{DDLMS} and μ_{DDNLMS}: Step-size in the LMS and NLMS adaptive algorithms during the detection mode. During the detection mode, the LMS and NLMS adaptive algorithms are operated under the decision-directed principles;

- δ_{RLS}, λ_{RLS}: Parameters for the RLS adaptive algorithm during the training mode;

- λ_{DDRLS}: Forgetting factor of the RLS adaptive algorithm during the detection mode.

Figure 3 shows the ensemble-average squared error learning curves of the LMS-, NLMS- and RLS-adaptive MUDs for the hybrid DS-TH UWB systems supporting $K=15$ users at the SNR per bit of $\frac{E_b}{N_0} = 10$dB. In our simulations, the step-size employed by the LMS- and NLMS- adaptive algorithms were fixed to $\mu_{LMS} = 0.05$ and $\mu_{NLMS} = 0.5$, respectively, while in the RLS-adaptive algorithm the parameters were set to $\lambda_{RLS} = 0.998$ and $\delta_{RLS} = 5.0$. In Figure 3 the ensemble-average squared error was obtained from the average over 2000 independent realizations. From the results of Figure 3 it can be observed that the RLS-aided adaptive MUD converges to the lowest MSE among the three types of adaptive MUDs considered. Furthermore, the RLS-aided adaptive MUD is capable of converging faster than any of the other two adaptive MUDs.

Figure 4 compares the BER versus average SNR per bit performance of the hybrid DS-TH

UWB systems supporting single user, when the ideal MMSE-MUD or various adaptive MUDs are employed. In our simulations two communications scenarios were considered, which result in $L=15$ and $L=150$ resolvable multipaths, respectively. Since the total spreading factor is $N_c N_\psi = 64$, it can be shown that $g=1$ for $L=15$ and $g=3$ for $L=150$, implying that the second communications scenario experiences higher ISI than the first communications scenario. From the results of Figure 4, we can observe that the BER performance of the hybrid DS-TH UWB system corresponding to $L=150$ is generally better than that corresponding to $L=15$. Therefore, all the MMSE-MUDs considered are capable of making efficient use of the energy dispersed over the multipaths for improving the BER performance. As shown in Figure 4, all the adaptive MUDs are highly efficient for detection in the hybrid DS-TH UWB systems. The BER performance achieved by any of the adaptive MUDs is very close to that achieved by the ideal MMSE-MUD, which in the worst case is about 1dB away from the ideal MMSE-MUD. Furthermore, from Figure 4 we can observe that, among the three types of adaptive MUDs considered, the RLS-adaptive MUD outperforms both the LMS-and NLMS-adaptive MUDs, while the LMS-adaptive MUD is the worst, in terms of the achievable BER performance. Finally, it can be observed that the LMS-adaptive MUD does not perform well, when the SNR value is too low. This is because the step-size μ_{LMS} is too high, resulting in that the LMS-adaptive MUD is unable to converge. In principle, smaller step-size might be employed to enhance the convergence. However, smaller step-size implies longer training sequences resulting in lower spectral efficiency.

Figure 5 compares the BER versus average SNR per bit performance of the hybrid DS-TH UWB systems supporting $K=5$ users, when communicating over indoor S-V modeled Rayleigh fading channels yielding $L=150$ resolvable paths. The other parameters used in our simulations were detailed associated with the figure. From the

Figure 3. Learning curves of the adaptive MUDs based on LMS, NLMS and RLS adaptive algorithms, when the hybrid DS-TH UWB system supports K=15 users. The other parameters used were

$$\frac{E_b}{N_0} = 10dB, \ f_d T_b = 0.0001, \ N_c = 16, \ N_\psi = 4, \ L = 15, \ \mu_{LMS} = 0.05, \ \mu_{NLMS} = 0.5, \ \lambda_{RLS} = 0.998$$

and $\delta_{RLS} = 5.0$.

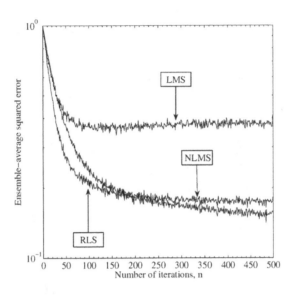

waterfall BER curves as shown in Figure 5, it can be implied that all the detectors considered are capable of suppressing both the MUI and ISI efficiently. Furthermore, as the single-user case shown in Figure 4, the RLS- adaptive MUD outperforms both the LMS-adaptive MUD and NLMS-adaptive MUD, in terms of the achievable BER performance. As Table 1 shown, among the three adaptive MUDs, the LMS-adaptive MUD has the lowest complexity. However, it achieves the worst BER performance, as shown in Figure 4 and Figure 5.

Finally, in Figure 6 we compare the BER performance of the pure DS-UWB, pure TH-UWB and the hybrid DS-TH UWB systems, when they employ the NLMS-adaptive MUD. In Specifically, in Figure 6 μ_{DDNLMS} corresponds to the case that the NLMS-adaptive MUD is only updated during the training mode. During the detection mode, the NLMS-adaptive MUD is fixed without

updating. By contrast, when $\mu_{DDNLMS} = 0.05$, the NLMS-adaptive MUD updates its weights during both training mode and detection mode. However, the step-size used during the detection mode is $\mu_{DDNLMS} = 0.05$, which is significantly lower than the step-size $\mu_{NLMS} = 0.5$ used during the training mode. In other words, after the training, the weights in the NLMS- adaptive MUD are only slowly updated with the aid of the detected date, in order to minimize the impact of erroneously detected data on the NLMS-adaptive MUD. From the results of Figure 6, it can be observed that the NLMS-adaptive MUD is suitable for all the above-mentioned three UWB systems. As shown in Figure 6, for both cases, the hybrid DS-TH UWB system slightly outperforms the pure DS-UWB system and the pure TH-UWB system, while the BER performance of the pure DS-UWB system is slightly better than that of the pure TH-UWB system, but worse than that of the hybrid DS-TH

Figure 4. BER versus average SNR per bit performance of the hybrid DS-TH UWB systems supporting single user, when communicating over indoor UWB channel. The other parameters used in our simulations were $f_d T_b = 0.0001$, $N_c = 16$, $N_\psi = 4$, $L = 15$, $\mu_{LMS} = 0.05$, $\mu_{DDLMS} = 0.01$, $\mu_{NLMS} = 0.5$, $\mu_{DDNLMS} = 0.5$, $\lambda_{RLS} = 0.987$, $\lambda_{DDRLS} = 0.998$ and $\delta_{RLS} = 5.0$. Furthermore, in our simulations each frame consisted of 1000 bits, out of which 160 were training bits and 840 were data bits.

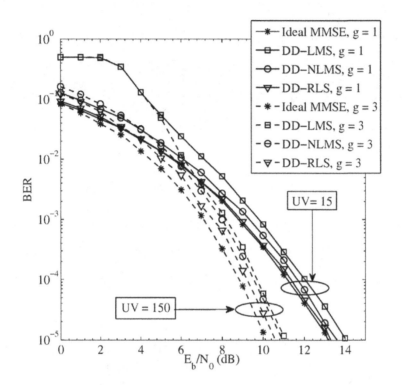

UWB system. Furthermore, as the results shown in Figure 7, the BER performance can be improved, when the adaptive MUD is updated during the detection mode with the aid of the detected data.

Finally, in Figure 7, the number of multiplications is plotted with respect to the number of resolvable multipaths *L*. As multiplications is more complex as compared to addition. It can be observed that, as *L* increases, more operations are required to detect a bit. The complexity of the ideal MMSE detector is more than all the detectors. The complexity of LMS- and NLMS-adaptive MUD is similar to one another. While, the complexity of RLS-based MUD is more than LMS- and NLMS- adaptive MUD.

5. CONCLUSION

In this contribution we have provided a comparative study of the hybrid DS-TH UWB systems, when they employ various detection schemes. Specifically, the ideal MMSE-MUD and three types of adaptive MUDs, namely the LMS-, NLMS- and RLS-adaptive MUDs, are investigated, when communicating over indoor UWB channels modeled by the S-V fading channel model. Furthermore, the BER performance of the pure DS-UWB and pure TH-UWB systems is investigated as the special examples of the hybrid DS-TH UWB system. From our analysis and simulation results, we can conclude that the considered adaptive MUDs

Figure 5. BER versus average SNR per bit performance of the hybrid DS-TH UWB systems sup-porting K=5 users, when communicating over indoor UWB channel. The other parameters used in our simulations were $f_dT_b = 0.0001$, $N_c = 16$, $N_\psi = 4$, $L = 15$, $\mu_{LMS} = 0.05$, $\mu_{DDLMS} = 0.01$, $\mu_{NLMS} = 0.8$, $\mu_{DDNLMS} = 0.1$, $\lambda_{RLS} = 0.998$, $\lambda_{DDRLS} = 0.998$, and $\delta_{RLS} = 5.0$. Furthermore, in our simulations each frame consisted of 1000 bits, out of which 160 were training bits and 840 were data bits.

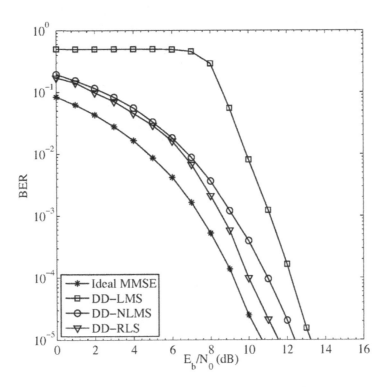

constitute a range of highly promising detection schemes for the pulse-based UWB systems, which include the above-mentioned pure DS-UWB, pure TH-UWB and hybrid DS-TH UWB systems. The adaptive detection schemes are free from channel estimation, which is extremely hard to achieve in the pulse-based UWB systems. However, as our simulation results shown, all the three types of adaptive detection schemes are capable of achieving the BER performance, which is close to that achieved by the ideal MMSE-MUD requiring ideal channel knowledge. Furthermore, the adaptive MUDs themselves can also provide us a design trade-off between the affordable complexity and the achievable BER performance. Specifically, among the three types of adaptive MUDs, the RLS-adaptive MUD is capable of achieving the best BER performance, but has the highest detection complexity. By contrast, the LMS-adaptive MUD achieves the worst BER performance at the lowest detection complexity. Our future research will concentrate on further decreasing the complexity of the adaptive MUDs by employing efficient rank-reduction techniques.

Figure 6. BER versus average SNR per bit performance of the hybrid DS-TH UWB systems supporting K=15 users, when communicating over indoor UWB channel. The other parameters used in our simulations were $f_d T_b = 0.0001$, $L = 15$, $\mu_{NLMS} = 0.5$ and $\mu_{DDNLMS} = 0.05$. In our simulation each frame consisted of 1000 bits, out of which 160 were training bits and 840 were data bits.

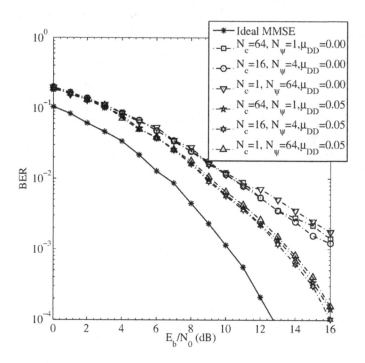

Figure 7. Number of multiplications required versus the number of resolvable multipaths, when communicating over indoor UWB channel

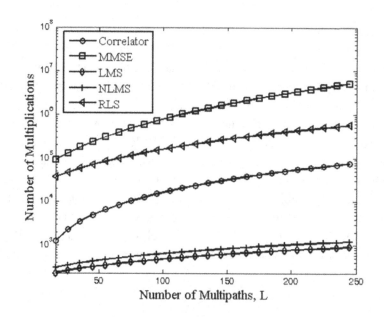

6. FUTURE TRENDS

We consider three future avenues which are based on synthesizing results from different fields:

- A promising avenue is reducing the filter length of UWB signals. Reduced filter length will increase the convergence; improve BER and robustness of the system. An effort using channel shortening techniques to reduce the filter length is introduced in Hussain (2005). Recently, Ahmed et al. (2013) has applied reduced rank technique to reduce the filter length. However, compressive sensing and sparse representation could be combined in order to design novel techniques for UWB system;

- In this thesis, the adaptive detectors considered are operated in the principles of MMSE. However, the MMSE detector is optimum only when the conditional probability density function (CDF) of the detector's output when given the transmitted symbol is Gaussian (Ahmed et al, 2011). In pulse based UWB systems, the CDF of the detector's output given the transmitted symbol is not Gaussian whenever there are multiple users present (Beaulieu et.al 2009). Therefore, it may be advisable to determine the exact expression for the conditional pdf;

- Another major application of UWB is indoor localization. Indoor localization can be improved by employing ultra-wide bandwidth transmission as compared to wide-band signals due to their fine delay resolution and obstacle-penetration capabilities. Cheap and robust UWB localization algorithms are required in order to achieve this gain.

REFERENCES

Ahmed, Q. Z., & Yang, L.-L. (2007). Performance of hybrid direct-sequence time-hopping ultrawide bandwidth systems in Nakagami-m fading channels. In *Proceedings IEEE 18th International Symposium on Personal, Indoor and Mobile Radio Communications*. IEEE.

Ahmed, Q. Z., & Yang, L.-L. (2008). Normalised least mean-square aided decision-directed adaptive detection in hybrid direct-sequence time-hopping UWB systems. In *Proceedings IEEE 69th Vehicular Technology Conference*. IEEE.

Ahmed, Q. Z., & Yang, L.-L. (2010). Reduced-rank adaptive multiuser detection in hybrid direct-sequence time-hopping ultrawide bandwidth systems. *IEEE Transactions on Wireless Communications*, *9*(1), 156–167. doi:10.1109/TWC.2010.01.081172

Breining, C., Dreiscitel, P., Hansler, E., Mader, A., Nitsch, B., & Puder, H. et al. (1999). Acoustic echo control: An application of very-high-order adaptive filters. *IEEE Signal Processing Magazine*, *16*, 42–69. doi:10.1109/79.774933

Fishler, E., & Poor, H. V. (2005). On the tradeoff between two types of processing gains. *IEEE Transactions on Communications*, *53*, 1744–1753. doi:10.1109/TCOMM.2005.855001

Foerster, J. R. (2002). The performance of a direct-sequence spread ultra-wideband system in the presence of multipath, narrowband interference, and multiuser interference. In *Proceedings IEEE Conference on Ultra Wideband Systems and Technologies*. IEEE.

Glentis, G. O., Berberidis, K., & Theodoridis, S. (1999). Efficient least squares adaptive algorithms for FIR transversal filtering. *IEEE Signal Processing Magazine*, *16*, 13–41. doi:10.1109/79.774932

Haykin, S. (2002). *Adaptive filter theory* (4th ed.). Upper Saddle River, NJ: Prentice Hall.

Honig, M., & Tsatsanis, M. K. (2000). Adaptive techniques for multiuser CDMA receivers. *IEEE Signal Processing Magazine, 17*, 49–61. doi:10.1109/79.841725

Karedal, J., Wyne, S., Almers, P., Tufvesson, F., & Molisch, A. F. (2004). Statistical analysis of the UWB channel in an industrial environment. In *Proceedings IEEE 60th Vehicular Technology Conference*. IEEE.

Kumar, R. (1983). Convergence of a decision-directed adaptive equalizer. In *Proceeding IEEE 22nd Conference on Decision and Control*. IEEE.

Li, Q., & Rusch, L. A. (2002). Multiuser detection for DS-CDMA UWB in the home environment. *IEEE Journal on Selected Areas in Communications, 20*, 1701–1711. doi:10.1109/JSAC.2002.805241

Miller, S. L. (1995). An adaptive direct-sequence code-division multiple-access receiver for multiuser interference rejection. *IEEE Transactions on Communications, 43*, 1746–1755. doi:10.1109/26.380225

Mireles, F. R. (2002). Signal design for ultra-wide band communications in dense multipath. *IEEE Transactions on Vehicular Technology, 51*, 1517–1521. doi:10.1109/TVT.2002.804838

Molisch, A. F. (2005). Ultrawide-band propagation channels-theory, measurement, and modeling. *IEEE Transactions on Vehicular Technology, 54*, 1528–1545. doi:10.1109/TVT.2005.856194

Montazeri, M., & Duhamel, P. (1995). A set of algorithms linking NLMS and block RLS algorithms. *IEEE Transactions on Signal Processing, 43*, 444–453. doi:10.1109/78.348127

Moshavi, S. (1996). Multi-user detection for DS-CDMA communications. *IEEE Communications Magazine, 34*, 124–136. doi:10.1109/35.544334

Moustakides, G. V. (1997). Study of the transient phase of the forgetting factor RLS. *IEEE Transactions on Signal Processing, 45*, 2468–2476. doi:10.1109/78.640712

Proakis, J. G. (2000). *Digital communications* (4th ed.). New York: McGraw Hill.

Proakis, J. G., Rader, C., Ling, F., Nikias, C., Moonen, M., & Proudler, I. (2003). *Algorithms for statistical signal processing*. Englewood Cliffs, NJ: Pearson Education.

Reed, J. H. (2005). *An introduction to ultra wideband communication systems*. Upper Saddle River, NJ: Prentice Hall.

Scholtz, R. A. (1993). Multiple access with time hopping impulse modulation. In *Proceedings IEEE Military Communications Conference*. IEEE.

Simon, M. K., & Alouini, M.-S. (2005). *Digital communication over fading channels* (2nd ed.). John Wiley and IEEE Press.

Verdu, S. (1998). *Multiuser detection*. Cambridge, UK: Cambridge University Press.

Win, M. Z., & Scholtz, R. A. (1998). On the robustness of ultra-wide bandwidth signals in dense multipath environments. *IEEE Communications Letters, 2*, 51–53. doi:10.1109/4234.660801

Woodward, G., & Vucetic, B. S. (1998). Adaptive detection for DS-CDMA. *Proceedings of the IEEE, 86*, 1413–1434. doi:10.1109/5.681371

Yang, L., & Giannakis, G. B. (2004). Ultra-wideband communications: An idea whose time has come. *IEEE Signal Processing Magazine, 21*, 26–54. doi:10.1109/MSP.2004.1359140

Zhang, J., Abhayapala, T., & Kennedy, R. (2005). Role of pulses in ultra wideband systems. In *Proceedings IEEE International Conference on Ultra-Wideband*. IEEE.

ADDITIONAL READING

Ahmed, Q. Z., Alouini, M. S., & Aissa, S. (2013). Bit Error-Rate Minimizing Detector for Amplify and Forward Relaying Systems Using Generalized Gaussian Kernel. *IEEE Signal Processing Letters*, *20*(1), 55–58. doi:10.1109/LSP.2012.2229271

Ahmed, Q. Z., Park, K.H., Alouini, M. S., and Aissa, S. (2013). Compression and Combining Based on Channel Shortening and Rank Reduction Technique for Cooperative Wireless Sensor Networks. To appear in *IEEE Transactions on Vehicular Technology*.

Ahmed, Q. Z., Yang, L.-L., & Cheng, S. (2011). Reduced-Rank Adaptive Least Bit-Error-Rate Detection in Hybrid Direct-Sequence Time-Hopping Ultrawide Bandwidth Systems. *IEEE Transactions on Vehicular Technology*, *60*(3), 849–857. doi:10.1109/TVT.2011.2109974

Beaulieu, N. C., & Young, D. J. (2009). Designing Time-Hopping Ultrawide Bandwidth Receivers for Multiuser Interference Environments. *Proceedings of the IEEE*, *97*, 255–283. doi:10.1109/JPROC.2008.2008782

Chen, W., Gao, Q., Xiong, H. G., Fei, L., & Li, Q. (2013). *TH-UWB Receiver Based on two Pdfs Approximation in Multiuser Systems*. Progress In Electromagnetics.

Cheng, C. (2011). *A subspace-based blind detector with rapid channel tracking structure for TH-UWB systems over time-varying multi-path channels*. Springer Telecommunication Systems. doi:10.1007/s11235-011-9446-z

Chu, X., & Murch, R. D. (2005). Multidimensional modulation for ultra-wideband multiple-access impulse radio in wireless multipath channels. *IEEE Transactions on Wireless Communications*, *4*, 2373–2386. doi:10.1109/TWC.2005.853873

D'Amico, A. A., Mengali, U., & Taponecco, L. (2005). Impact of MAI and channel estimation errors on the performance of rake receivers in UWB communications. *IEEE Transactions on Wireless Communications*, *4*, 2435–2440. doi:10.1109/TWC.2005.853918

Hu, B., & Beaulieu, N. C. (2005). Pulse shapes for ultrawideband communication systems. *IEEE Transactions on Wireless Communications*, *4*, 1789–1797. doi:10.1109/TWC.2005.850311

Hussain, S. I., & Choi, J. (2005). Single Correlator Based UWB Receiver Implemetation Through Channel Shortening Equalizer. *Proceedings IEEE Asia-Pacific Conference on Communications*.

Kong, Z., Zhong, L., Zhu, G., Ding, L. (2011). Differential multiuser detection using a novel genetic algorithm for ultra-wideband systems in lognormal fading channel. *Journal of Zhejiang University Science Computer and Electronics*.

Molisch, A. F., Foerster, J. R., & Pendergrass, M. (2003). Channel models for ultrawideband personal area networks. *IEEE Wireless Communications*, *10*, 14–21. doi:10.1109/MWC.2003.1265848

Piazzo, L. (2004). Performance analysis and optimization for impulse radio and direct-sequence impulse radio in multiuser interference. *IEEE Transactions on Communications*, *52*, 801–810. doi:10.1109/TCOMM.2004.826246

Öztürk, E., Chen, H. H., & Guizani, M. (2011). Multi-scale direct sequence ultra-wideband communications over time-dispersive channels. *IET Communications*, *51*(11), 1597–1606.

Win, M. Z., Dardari, D., Molisch, A. F., Wiesbeck, W., & Zhang, J. (2009). History and applications of UWB. *Proceedings of the IEEE*, *97*, 198–204. doi:10.1109/JPROC.2008.2008762

Zhang, J., Orlik, P. V., Sahinoglu, Z., Molisch, A. F., & Kinney, P. (2009). UWB systems for wireless sensor networks. *Proceedings of the IEEE*, *97*, 313–331. doi:10.1109/JPROC.2008.2008786

KEY TERMS AND DEFINITIONS

Ultrawide Bandwidth System: A system with bandwidth greater than 500 MHz.

Chapter 19
A Survey of MAC Layer Protocols to Avoid Deafness in Wireless Networks Using Directional Antenna

Rinki Sharma
M. S. Ramaiah School of Advanced Studies, India

Govind Kadambi
M. S. Ramaiah School of Advanced Studies, India

Yuri A. Vershinin
Coventry University, UK

K. N. Mukundan
Broadcom Communication Technologies, India

ABSTRACT

Directional antennas have gained immense popularity among researchers working in the area of wireless networks. These antennas help in enhancing the performance of wireless networks through increased spatial reuse, extended communication range, energy efficiency, reduced latency, and communication reliability. Traditional Medium Access Control (MAC) protocols such as IEEE 802.11 are designed based on use of omnidirectional antennas. Therefore, suitable design changes are required to exploit the benefits of directional antennas in wireless networks. Though directional antennas provide many benefits to enhance network performance, their inclusion in the network also results in certain challenges in network operation. Deafness is one such problem that occurs among nodes using directional antennas. This chapter concentrates on the problem of deafness, which is introduced due to the use of directional antennas in wireless ad-hoc, sensor, and mesh networks. Many researchers have provided numerous solutions to deal with the problem of deafness in these networks. In this chapter, the authors first explain the problem of deafness and then present an extensive survey of solutions available in the literature to deal with the problem of deafness in wireless ad-hoc, sensor, and mesh networks. The survey is accompanied by a critical analysis and comparison of available solutions. Drawbacks of available solutions are discussed and future research directions are presented.

DOI: 10.4018/978-1-4666-5170-8.ch019

INTRODUCTION

Numerous technological advances and easy access to wireless devices have seen increased research interest in wireless networks. Traditional wireless networks are mainly infrastructure based where nodes communicate through base station or access point. In such networks, nodes are required to be within the range of infrastructure. Such networks are called as single-hop networks. With introduction of cheaper hardware, faster processors and smaller transceivers; focus turned towards ad-hoc networks. In ad-hoc networks, nodes communicate without an access point or base station. In such networks, a node is capable of acting as a source, sink or router of information. This concept gave rise to multi-hop communication. Further research and technological advances introduced the concept of wireless mesh networks (WMNs) and wireless sensor networks (WSNs), mobile ad-hoc networks (MANETs) and vehicular ad-hoc networks (VANETs) to name a few. IEEE provided standardized protocols and techniques for physical and MAC layer operations in these networks. These protocols assumed use of omnidirectional antennas in the network. With growing research interest in wireless networks, researchers started to study challenges faced by wireless networks and ways to overcome these challenges. Some of the challenges that researchers tried address were interference, spatial reuse, energy efficiency, latency in communication and information security. This is when researchers proposed to exploit the benefits of directional antenna/antenna beamforming. Many researchers proposed the use of directional antennas in wireless networks and proved their benefits through simulations and analysis. However, it was noted that traditional MAC protocols failed to work with directional antennas, and despite numerous benefits that directional antennas provided, they could not be used directly in wireless networks. Use of directional antennas introduced new problems in network operation. Deafness, new

hidden terminal, asymmetry in gain, and need for location information of destination node for antenna beamforming were some of such problems that researchers needed to work upon. The aim of this chapter is to provide an extensive survey of solutions to mitigate deafness available in the literature. Though there are surveys available for protocols based on directional MAC, not many surveys are available which concentrate only on the problem of deafness occurred due to the use of directional antennas. Some of the researchers have presented surveys of deafness solutions, i.e., Choudhury and Vaidya (2004) and Gossain, Cordeiro, Cavalcanti, and Agrawal (2004). These surveys are outdated and focus only on ad-hoc networks. In this chapter, the authors concentrate on the problem of deafness introduced due to the use of directional antennas in the network. This chapter is targeted towards wireless networks researchers who want to pursue their research in area of wireless networks using directional antennas. The main objective of this chapter is to make the readers aware of the problem of deafness and present them with ongoing research and future research directions to combat this problem. For this, the authors provide an extensive literature survey of solutions available in literature to deal with the problem of deafness in wireless ad-hoc, sensor and mesh networks. This is accompanied with critical analysis and comparison of surveyed solutions highlighting their benefits and drawbacks. Based on this the authors provide scope and directions for future research.

We first provide the readers with brief introduction to directional antennas. The authors first discuss the use of Carrier Sense Multiple Access/Collision Avoidance (CSMA/CA) protocol for medium access control in wireless networks followed by introduction to the problem of deafness. It is important to analyze the benefits and drawbacks of using directional antennas in wireless networks in order to understand and appreciate the methods used to deal with deafness. Therefore, the authors first discuss the advantages and disadvantages of using directional antennas in wireless networks. Since the characteristics of wireless ad-hoc, mesh and sensor networks are different, analysis of use of directional antennas and effect of deafness is carried out separately for these networks. The authors first provide a brief introduction to wireless ad-hoc, mesh and sensor networks and discuss the use of directional antennas in these networks. This is followed by a survey of solutions available in literature to deal with deafness. The surveyed solutions are then compared and analyzed. Based on this analysis the authors present future research directions.

INTRODUCTION TO DIRECTIONAL ANTENNAS

In this section, the authors present a brief introduction to directional antennas. They intend to give the readers just enough information which is required for them to recognize the difference between omnidirectional and directional antenna, so that they can understand the problem of deafness. The primary function of antenna is to radiate and gather electromagnetic energy to and from the space respectively. Simple dipole antennas can radiate and gather energy to and from the space equally well in all the directions. Such antennas are known as omnidirectional antennas. Gain of an antenna is measured by the product of its efficiency and directivity. It is measured in dBi. Gain of an ideal omnidirectional antenna pattern is same in all the directions. Gain of an ideal omnidirectional antenna is 0 dBi. As mentioned in Balanis (2012) and Stutzman and Thiele (2012), directional antennas are designed to radiate/gather energy to/from the space in a particular direction more than other directions. These antennas have higher gain in a particular direction when compared to other directions. The direction with higher gain forms main lobe of directional antenna pattern, while other directions with lesser gains form side lobe and back lobe.

Directional antennas are generally realized through antenna arrays. Single antenna elements are arranged in the form of an array, separated with each other in terms of a fraction of the wavelength. Specific antenna radiation patterns of an antenna array can be achieved by determining the number of antenna elements, spacing between them, geometrical configuration, phase and amplitude of signal applied to each antenna element. Smart Antenna technology is a combination of antenna array and Digital Signal Processing (DSP). This allows the antenna elements to transmit and receive adaptively. Beamforming antennas are a type of smart antennas. Beamforming antennas use sophisticated algorithms to estimate Directional of Arrival (DoA) of the signal and change the radiation pattern accordingly in order to direct the beam in required direction (Winters, 2006; Tsoulus,1999).

Beamforming antennas can be mainly classified into switched beam antenna and steered beam antenna. Features and characteristics of these antennas are as explained below:

1. **Switched Beam Antennas:** Based on the DoA estimation, the antenna adaptively switches to one of the beams in a predefined set of beams. Therefore, in a wireless network, a node can adaptively switch its beam according to the direction of communication. These antennas are less complex and expensive to implement when compared to steered beam antennas;

2. **Steered Beam Antennas:** These antennas are also known as Adaptive Antenna Arrays (AAA) and use sophisticated signal processing algorithms to direct their beams in required direction for communication. With the help of complex DSP algorithms antennas can place their nulls towards interfering nodes in order to suppress interference from these nodes. These antennas perform better than switched beam antennas in multipath environments, but they are expensive and

costly to implement. Also, use of complex signal processing algorithms and need to constantly track and locate other nodes lead to increased power consumption at the nodes.

CARRIER SENSE MULTIPLE ACCESS/COLLISION AVOIDANCE (CSMA/CA)

As the wireless medium is shared among many nodes for communication, there has to be some controlling mechanism for access to the shared medium in order to avoid collision between transmissions of multiple nodes. MAC protocols in wireless networks primarily belong to two categories: Contention-free and contention-based MAC. In contention-free MAC, nodes do not have to contend with each other for access to the medium. In contention-free MAC communication between nodes takes place in dedicated time slots allocated to particular nodes. To successfully use dedicated time-slots, nodes need to be time-synchronized with each other. However, time-synchronization is difficult to achieve in wireless networks due to characteristics of wireless medium such as limited channel bandwidth, asymmetric channel, interference and dropped packets. Apart from the

characteristics of wireless channel, clock skews (as all oscillators on different nodes may not be synchronized), and varying node distance due to node mobility thus varying travel time, also make clock synchronization difficult in wireless networks. Therefore, most common MAC protocol used in wireless networks is the contention-based MAC protocol called Carrier Sense Multiple Access/ Collision Avoidance (CSMA/CA) mechanism as shown in Figure 1. CSMA/CA is modified form of Carrier Sense Multiple Access/Collision Detection (CSMA/CD), which is used as a medium access mechanism in Ethernet, commonly used to establish wired networks. Unlike in wired networks, where collision needs to be detected, in wireless networks collisions must be avoided due to limited bandwidth and difficulty of detecting collisions in radio environment. IEEE 802.11's Distributed Coordination Function (DCF) is the most common contention-based protocol used in wireless networks (Crow, Widjaja, Kim, & Sakai, 1997). This protocol assumes that all the participating nodes use omnidirectional antenna. First, all the nodes sense the medium to be idle. If the medium is found to be idle, an intended transmitter waits for DCF Inter Frame Space (DIFS) duration of 50 µs plus random backoff time and then transmits Request-To-Send (RTS). All the nodes receive

Figure 1. Carrier Sense Multiple Access/Collision Avoidance (CSMA/CA)

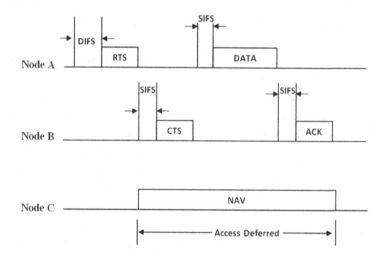

RTS but only intended source nodes replies back with Clear-To-Send (CTS) after waiting for 10 μs of Short Inter Frame Space (SIFS).

Since RTS and CTS frames carry information about duration of communication in the duration field, all other nodes transit to waiting state by enabling Network Allocation Vector (NAV) timer and wait for communication to complete. During this period no other node in the vicinity is allowed to transmit any frames in the medium to avoid collision. After receiving CTS frame and waiting for SIFS duration, transmitter transmits DATA frame. Receiver node receives the DATA frame, waits for SIFS duration and transmits ACK (acknowledgement). This is explained through Figure 1 where Node A is the transmitter, Node B is the receiver and Node C is other node in the vicinity which waits for communication between Nodes A and B to get over by putting on its NAV timer.

DEAFNESS

The problem of deafness occurs due to the use of directional antenna. Deafness is caused when a transmitting node repeatedly tries to send RTS to a destination node, but the destination node cannot hear the RTS packet because it has formed a directional beam away from the transmitting node (Gossain, Cordeiro, Cavalcanti, & Agrawal, 2004).

In the scenario shown in Figure 2, Node B is communicating with Node C using directional antenna. Node A being unaware of ongoing communication between Nodes B and C tries to

establish communication link with Node B and sends RTS to it. Since Node B is pointing its beam towards Node C, Node B does not respond to RTS. Node A waits for CTS, increases backoff time and retransmits RTS to Node B. However, Node B is deaf towards RTS sent by Node A. In this process, Node A wastes battery power and network capacity by retransmitting RTS. In this scenario Node A transmits to Node B while Node B transmits to Node C.

BENEFITS OF USING DIRECTIONAL ANTENNAS IN WIRELESS NETWORKS

Wireless ad-hoc, mesh and sensor networks share certain characteristics which are common to a wireless medium. Some of the major benefits of directional antennas are common to these networks. Therefore, before analyzing the benefits of directional antennas on these networks separately, we discuss and analyze various benefits of using directional antennas in wireless networks in general. These benefits are discussed below (Yi, Pei, & Kalayanaraman, 2003):

- **Reduction in latency:** Wireless networks are formed among the nodes which are within the communication range of each other. Therefore, to establish communication between nodes which are not in direct communication range of each other, multi-hop communication is necessary. Intermediate nodes need to act as routers to

Figure 2. Deafness scenario

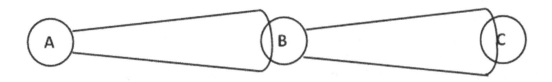

route the information from source node to destination node. Each intermediate node introduces certain processing delay while routing information. Higher the number of intermediate nodes in a route, longer will be the time taken for the data to traverse from source node to destination. This will lead to increase in end-to-end delay or latency. Directional antennas have longer communication range due to higher signal-to-noise ratio. Therefore, use of directional antennas can reduce or eliminate the intermediate nodes in a particular route, which further leads to reduced end-to-end delay or latency. Lesser involvement and participation of intermediate nodes allows these nodes to enter sleep mode and save energy when they are not sending, receiving or routing any information in the network (Choudhury & Vaidya, 2005);

- **Better security:** Directional antennas provide better security due to their resistance against jamming. Increased communication range due to use of directional antennas reduces or eliminates the need for intermediate nodes. This reduces the risk of security attacks due to the presence of malicious nodes in wireless networks (Hu & Evans, 2004);

- **Reduced interference:** Directional antennas acquire reduced interference due to narrow beamwidth. This is important for all wireless networks because wireless networks are more prone to interference among neighboring nodes when compared to their wired counterparts (ElBatt, Anderson, & Ryu, 2003);

- **Energy efficiency:** Directional antennas provide higher gain leading to increased communication range for given transmit power, when compared to omnidirectional antennas. This leads to increased battery lifetime for nodes which use directional antennas for communication. This prop-

erty of directional antennas can be very useful for nodes which depend on limited battery power for communication. In sensor networks and in mobile ad-hoc networks such as military or emergency and rescue operations energy efficiency due to the use of directional antennas can be evident. Increased battery lifetime increases overall network lifetime (Spyropoulos & Raghavendra, 2002);

- **Better spatial reuse:** Directional antennas provide better spatial reuse due to antenna beamforming when compared to omnidirectional antennas. This not only reduces the number of exposed nodes in the network but also reduces interference among neighboring nodes. This results in increased network capacity and efficiency. Antenna beamforming reduces the number of exposed nodes (nodes waiting for communication between two nodes to complete so that medium becomes free and available for them to access). This is an important feature of DAs and is useful for all wireless networks (Wang & Garcia-Luna-Aceves, 2002).

CHALLENGES INTRODUCED DUE TO USE OF DIRECTIONAL ANTENNAS IN WIRELESS NETWORKS

Basic method of medium access control uses the CSMA/CA which is based on omnidirectional transmissions. Therefore, use of directional antennas in wireless networks gives rise to certain challenges. These challenges are briefly explained below:

- **Deafness:** In CSMA/CA, exchange of RTS, CTS, DATA and ACK happens omnidirectionally. When omnidirectional antenna is replaced with directional antenna, all these messages are exchanged directionally. Due to directional transmission of RTS and CTS, nodes which are away from the direction of RTS and CTS transmissions stay unaware of ongoing communication between two nodes. During this time of communication between two nodes whose beams are pointed towards each other, if a third node sends RTS to one of these communicating nodes, then these nodes will not receive this RTS (because their beams are pointed away from the direction of third node's beam), and will not send CTS. This way, two communicating nodes whose beams are pointed towards each other become deaf towards RTS coming from other nodes. This is termed as 'Deafness' (Gossain, Cordeiro, Cavalcanti, & Agrawal, 2004). A survey of solutions to this problem is the focus of this chapter;

- **Directional exposed node:** This problem is similar to the exposed terminal problem caused due to omnidirectional antenna, but it is constrained to directional range of transmitting and receiving nodes. When two nodes communicate using directional antenna, then the Directional Network Allocation Vector (DNAV) timer is applied in the region covered by antenna beams of communicating nodes. DNAV is a variant of NAV timer (present in CSMA/CA mechanism for omnidirectional antennas), implemented for directional antennas. DNAV reserves the channel only in specific directions, unlike NAV, which reserves the channel in all directions. If any other node falls in the region or direction reserved by DNAV, it can neither transmit nor receive any packets with other nodes till DNAV timer expires. This is known as directional exposed node problem;

- **Directional hidden node:** Directional hidden node problem occurs when a node

fails to detect an ongoing communication between two nodes communication using their directional antenna, and transmits RTS to the receiver node of communicating pair of nodes. This leads to collision at the receiver. Also, the transmitter of RTS does not receive CTS in return, due to which it retransmits RTS to receiver node of communicating pair of nodes, again leading to collision. This can bring down network capacity drastically (Sekido, Takata, Bandai, & Watanabe, 2005);

- **Node location determination:** For a transmitter using directional antenna, it is important to know the location of intended receiver, so that it can direct its beam in the correct direction. This becomes even more important when the participating nodes are mobile as their location keeps changing constantly. Numerous algorithms are available in literature to determine node location. Some of the most important algorithms are Angle of Arrival (AoA), Direction of Arrival (DoA), Time of Arrival (ToA) and Time Difference of Arrival (TDoA). Nowadays, most of the nodes are equipped with Global Positioning System (GPS) through which nodes can get to their location. They can broadcast their location information to other nodes (Vasudevan, Kurose, & Towsley, 2005).

If the listed problems are not handled correctly they can bring down network performance and efficiency drastically. Therefore it is very important to solve these problems in order to exploit the benefits of directional antennas. In this chapter our focus is on the problem of Deafness. We survey and compare the solutions available in literature to solve the problem of Deafness arising due to the use of directional antenna in wireless ad-hoc, sensor and mesh networks.

WIRELESS AD-HOC NETWORKS

Wireless networks are of two types: a. Infrastructure based and b. Infrastructure less or ad-hoc networks. Infrastructure networks consist of an access point/base station to enable communication between nodes in the network. In ad-hoc networks, nodes communicate with each other without the help of any access point or base station. Ad-hoc networks have gained popularity because of their ease of establishment and cost effectiveness. During calamities such as tsunami or earthquake, emergency and rescue operation teams can easily form ad-hoc networks without the need of any base station. These networks support multi-hop communication in order to establish communication links among nodes which are not within the communication range of each other. Participating nodes can be either stationary or mobile. Users can carry their laptops, phones, PDAs etc., and exchange information with each other as long as they are in communication range of each other. Such networks are also known as Mobile Ad-hoc Networks (MANETs). Similar networks can be formed among moving vehicles, where vehicles act as mobile nodes and can exchange accident, traffic jam or other required information with vehicles within their vicinity. Such networks are used in Intelligent Transportation Systems (ITS), (Haas, Deng, Liang, Papadimitratos, & Sajama, 2002).

Use of Directional Antennas in Wireless Ad-Hoc Networks

Many features of the underlying physical layer in ad-hoc networks can be controlled to enhance the performance of ad-hoc networks. Some of the main features include modulation techniques, encoding schemes, spread spectrum techniques, data rate, transmission power, receiver sensitivity and type of antenna used. Adaptive and intelligent control of these features can enhance network capacity and performance tremendously. In this chapter we will concentrate on the type of antenna used

at the nodes in ad-hoc network which can help in enhancing network capacity and performance.

As studied earlier, use of directional antenna can enhance network capacity and performance in wireless networks. As the nodes participating in mobile networks include devices such as laptops and PDAs, when the use of directional or beam-forming antennas is considered, the main concern that comes to mind is that whether the size and cost of these antennas make them worthy to be used in ad-hoc networks? However, as pointed out in (Ramanathan, 2001), at higher operating frequencies such as 5.8 GHz ISM band and 2.4 GHz ISM band, the size of directional antennas reduces considerably to a few centimeters. Also, with advances in Micro-Electro-Mechanical Systems (MEMS) technology the size of these antenna components has reduced considerably making these antennas suitable to be used in ad-hoc networks.

Since nodes participating in MANETs rely on limited battery power, use of directional antennas help in achieving increased energy efficiency. MANETs support multi-hop communication, therefore increased communication range of directional antennas help in reduced latency and enhanced security against the presence of malicious nodes. In Vehicular Ad-hoc NETworks (VANETs), where participating mobile nodes are vehicles, there is no dearth of battery power for communication. Therefore, it is not required to use directional antennas in VANETs for the sole reason of energy efficiency. However, other benefits of directional antennas can still be exploited by VANETs.

Effect of Deafness on the Performance of Wireless Ad-Hoc Networks

Deafness leads to increase in the number of retransmission attempts. With every failure nodes need to double their contention window and retry, which leads to increased latency. Also, as the number of retransmission attempts reach maximum limit, transmission is deferred and packet is dropped. Upper layers may consider it as route failure and may restart route discovery. This further leads to waste of network resources, thus deteriorating network performance.

Deafness leads to waste of bandwidth and brings down network capacity due to the presence of unproductive control packets in the network. This also wastes energy of the node transmitting RTS in the network without knowing that other node is already engaged in communication. In MANETs where participating nodes depend on limited battery power, this waste of energy can be detrimental. As wireless networks have limited bandwidth, presence of unwanted control packets in the network generated due to deafness can lead to network congestion and bring down network capacity. Performance of MANETs and VANETs can deteriorate drastically due to this.

WIRELESS SENSOR NETWORKS

Wireless Sensor Networks (WSNs) are used for monitoring of unknown and untested environments. In these networks, hundreds of sensor nodes can be deployed in the concerned area for sensing and monitoring applications. Deployed sensor devices sense required parameters and transmit them to a sink node for further processing over wireless medium. WSNs support multi-hop and ad-hoc communication. Sensor networks usually consists of static nodes, however there are some applications that may use mobile sensor nodes as well. For example, in remote patient health monitoring systems patients can wear sensors which measure their physiological parameters and send them to doctors. In such networks patients can move around in a given area while doctors can monitor their health remotely. Unlike ad-hoc networks, sensor networks are 'data centric' in the sense that these networks are concerned with sensing a particular parameter or characteristic of the concerned network. Some of the common

applications of WSNs are habitat monitoring, environmental monitoring, soil moisture monitoring, structural health monitoring, smart home/office applications and patient monitoring systems.

A wireless sensor node is very small in size and consists of a transducer to sense required physical quantity, embedded processor to process the data, small memory to save data, small battery and a wireless transceiver module. Because sensor nodes are deployed in large numbers over areas which are generally difficult to access, these nodes need to rely on available battery power. Sensor nodes are expected to have longer lifetime (of few months to years) because it is cumbersome for users to remove already deployed sensors and change their battery. WSNs often experience high network load scenarios while working on applications that require data collection. To handle this load, it becomes important to look for provisions that can reduce end-to-end delay and support simultaneous traffic without interference. This can be achieved through directional antennas (Lewis, 2004; Pottie, 1998).

Use of Directional Antennas in WSNs

Since wireless sensor nodes have limited resources which must be used optimally, researchers have shown interest in using directional antennas for communication among nodes in wireless sensor network (WSNs). Two main aspects of a WSN are its reliability and lifetime. To achieve energy efficient and reliable communication over WSNs, researchers have proposed to use directional antenna. Numerous researchers have supported the use of directional antennas in wireless sensor networks (Nilsson, 2009; Wu, Zhang, Wu, & Niu, 2006). The authors in Giorgetti, Cidronali, Gupta, and Manes (2007) have proposed low cost directional antenna for WSNs to be used in 2.4 GHz band. The authors in Boudour, Lecointre, Berthou, Dragomirescu, and Gayraud (2007) have observed that advances in MEMS have made it possible to realize small and efficient directional antennas for sensor nodes. The authors in Collett, Loh, Liu, and Qin (n.d.) have observed that wireless sensor nodes using omnidirectional antennas experience significant loss of communication link, while nodes equipped with directional antenna could establish efficient communication links. Sensor nodes participating in WSNs have limited battery power. Unlike wireless ad-hoc networks and mesh networks, sensor networks are used for applications where sensor nodes are placed in remote and untested environments. Therefore it is expected that battery of these sensor nodes have longer lifetime. With increased communication range and reduced dependence on intermediate nodes, participating nodes can enter sleep mode thus saving battery. This also increases overall network lifetime. WSNs support multihop communication. Increased communication range achieved through the use of directional antennas minimizes the possibility of abuse of data by malicious nodes. WSNs are popularly used in data aggregation applications. In such applications WSNs may encounter high data load which may lead to network congestion. Use of directional antennas can enhance channel use by simultaneous data exchange (by virtue of spatial reuse) and reduced latency, thus enhancing data propagation capacity of the network. A WSN consists of large number of sensor nodes deployed within a given area. Therefore, there are more chances of interference in sensor networks when compared to ad-hoc or mesh networks. Therefore, exploiting the capability directional antenna to reduce interference and provide better spatial reuse becomes essential in WSNs.

Effect of Deafness on the Performance of WSNs

Like ad-hoc networks, in sensor networks also, deafness leads to wastage of bandwidth the presence of unwanted and futile RTS frames in the network may lead to network congestion, thus drastically deteriorating network capacity. In WSNs where energy efficiency is critical, the problem of deafness tends to deteriorate network performance drastically. This overshadows all the benefits of using directional antennas in wireless sensor networks.

WIRELESS MESH NETWORKS

Wireless Mesh Networks (WMNs) were proposed to overcome inability of ad-hoc networks to cover large areas over wireless medium in robust and reliable manner. There are three main components of a WMN; mesh clients, mesh routers and gateways. Unlike ad-hoc networks, in WMNs, all the mesh routers are connected to each other in the form of mesh topology. This makes the networks robust because if one mesh router fails, data can always be routed through other mesh routers. Since WMNs offer redundant paths, they eliminate possibilities of single point failures and bottlenecks. Therefore, mesh routers form a robust wireless backbone for WMNs. WMNs also support multihop communication. Gateways in mesh networks enable their integration with other existing wireless networks. We can have IEEE 802.11 Wireless Local Area Network (WLAN), IEEE 802.15 Wireless Personal Area Network (WPAN) and IEEE 802.16 Wireless Metropolitan Area Network (WMAN) based mesh networks (Hossain, & Leung, 2007; Akyildiz, Wang, & Wang, 2005).

Use of Directional Antennas in Mesh Networks

Unlike ad-hoc networks, in WMNs, mesh routers are generally stationary devices such as access-points or base-stations. These devices are connected to source of constant power supply. Therefore, unlike in ad-hoc and sensor networks, mesh routers do not lack battery power. In WMNs, use of directional antenna is supported mainly for the purpose of reducing end-to-end delay, reduced interference, better spatial reuse and simultaneous communication between neighboring nodes due to antenna beamforming (Kumar, Gupta, & Das, 2006). This helps in establishing an efficient wireless backbone and increase network capacity. Higher gain of directional antenna provides longer communication range. This reduces or eliminates the number of intermediate nodes in multihop transmission of data, leading to reduced end-to-end delay. In WMNs this property of directional antennas enhances network efficiency. With directional antennas, signals can be directed only towards intended receiver. This helps to achieve better spatial reuse and reduced interference. Simultaneous data transmission over neighboring paths is possible without interference. This enhances network capacity and efficiency. Other benefits of directional antennas such as enhanced security and energy efficiency may not prove very significant in WMNs. Mesh routers in WMN mainly consist of access points and base stations which are authenticated stationary nodes. Therefore, there are fewer chances of malicious routers in the network. Also, access points and base stations have constant power supply; therefore energy is not a constraint. Mesh networks based on WLAN and WPAN technologies use CSMA/CA as medium access mechanism in contention based networks (Li, Yang, Conner, & Sadeghi, 2005; Stine, 2006). Therefore, in order to exploit benefits of directional antennas WMNs also face the same challenges as ad-hoc and sensor networks

Effect of Deafness on the Performance of WMNs

As discussed earlier, presence of futile RTS frames generated due to the problem of deafness in the network can lead to network congestion and bring down network capacity. Wireless mesh networks, though equally affected by these problems as ad-hoc and sensor networks, are more robust in tackling these problems when compared to ad-hoc and sensor networks. Mesh networks provide redundant paths so that if one of the communication paths suffers congestion, data can be routed

through other paths. Waste of energy due to RTS retransmissions is not a problem in mesh networks because mesh routers are generally connected to a constant source of power supply. However it is favorable for mesh networks to overcome the problem of deafness in order to exploit the benefits of directional antenna as deafness results in negative effects on network performance.

CLASSIFICATION OF METHODS ADOPTED TO DEAL WITH DEAFNESS

It is observed from the literature survey that most of the protocols proposed to deal with the problem of deafness can be broadly classified into two categories namely a) random access protocols and b) synchronized access protocols. In random access protocols nodes contend for access to the medium through RTS/CTS mechanism. Nodes first sense the medium to be idle and then transmit RTS frame to reserve the medium. On successful reception of RTS frame the concerned node sends CTS frame. After that, corresponding data and acknowledgement frames are exchanged. In case of synchronized access protocols, nodes access the medium according to a predetermined schedule. It is required that nodes are synchronized in order to follow the schedule. However, it must be noted that in wireless networks it is difficult to achieve synchronization among the nodes primarily due to wireless channel characteristics. Due to this, a significant number of protocols belong to the category of random access protocols.

Random Access Protocols

Random access protocol can be classified into mainly three categories based on the method used for RTS transmission. Three common methods employed by random access based proposed protocols for deafness avoidance are: a) Omnidirectional RTS transmission; b) Unidirectional

RTS transmission; and c) Multidirectional RTS transmission.

In omnidirectional RTS transmission method, the RTS frame is transmitted in all the directions (omnidirectionally) so that all the neighboring nodes of RTS transmitter become aware of ongoing communication. This is one of the most common ways to avoid deafness.

In unidirectional RTS transmission method, the transmitting node uses directional antenna to transmit the RTS frame. RTS frame is transmitted only in the direction of destination. In this case other nodes which do not fall in the direction of RTS transmission will not be aware of the ongoing transmission. This can lead to the problem of deafness. Such protocols use other methods such as tones to solve deafness.

In case of multidirectional RTS transmission, the RTS frame is transmitted directionally in multiple or all the available directions, either sequentially or concurrently. Concurrent transmissions emulate omnidirectional transmission with increased coverage due to directional antenna. Effect of sequential transmissions is also same as that of omnidirectional antenna, but it induces sweeping delay. This way all the neighboring nodes of RTS transmitter become aware of the ongoing communication, which helps in solving the problem of deafness.

In all the surveyed protocols, data and acknowledgement frames are transmitted directionally.

Synchronized Access Protocols

In synchronized access protocols, nodes use time-slots based on a predetermined schedule to access the medium. This requires time synchronization among the participating nodes. As the communication among nodes happens according to time-slots, deafness can be avoided because until a pair of nodes complete communication, no other node tries to start the communication with already engaged nodes. However, it is required to have time synchronization among the nodes for synchronized access. Since it is difficult to achieve global synchronization among nodes in multihop wireless ad-hoc networks, most of the proposed protocols use local synchronization.

Hybrid Access Protocols

Some of the protocols use both random access and synchronized access protocols. These are known as hybrid access protocols.

Apart from the classification of random access, synchronized access or hybrid access, protocols can have other characteristics also which are useful for operation. For example, if a protocol supports the use of directional antenna, it can be switched-beam or steered-beam antenna. A protocol may use single channel or multiple channels for communication. Some protocols may support dynamic power control while other may not. It must be noted that a protocol may support more than one of the above mentioned features. These classifications of surveyed protocols are summarized in Figure 3.

SURVEY OF PROTOCOLS PROPOSED TO SOLVE THE PROBLEM OF DEAFNESS

In this section we provide a literature survey of around 50 protocols proposed to solve the problem of deafness while using directional antennas in wireless networks. We have broadly classified these protocols in three categories of random access protocols, synchronized access protocols and hybrid access protocols (use both random access and synchronized access mechanisms). Among these categories, the protocols can be further classified into two categories based on whether these protocols use tones or not. This classification is presented in Figure 3. Table 1 provides comparison of all the surveyed protocols.

Figure 3. Classification of surveyed protocols

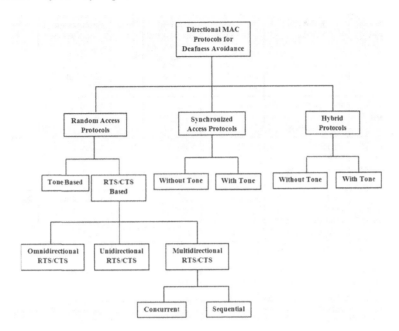

Random Access Protocols without Tones

A power controlled MAC protocol, Directional MAC with Power Control (D-MAP) is proposed in Arora, Krunz, and Muqattash (2004). This protocol uses switched beam directional antenna having constant gain in the main lobe. It uses two separate channels, one for exchanging control packets and other for exchanging data packets. According to the authors, this helps in reducing the number of collisions in the network. When the nodes are idle, they listen omnidirectionally to these two channels. RTS frames are sent omnidirectionally over the control channel with a fixed power common to all the nodes in the network. Upon reception, the intended receiver node is required to estimate the Angle-of-Arrival (AoA) and computes a power-control factor which is encapsulated in the CTS packet and sent directionally to the sender of RTS. It must be noted that though RTS is transmitted omnidirectionally, CTS is transmitted directionally after gauging the AoA of RTS. The source node (sender of RTS),

uses the computed power-control factor received along with the CTS packet for dynamic power control for transmission of data packet such that the transmission power is just sufficient to overcome interference at the receiver. The power of Directional CTS (DCTS) is scaled according to a power scaling factor which ensures that all the potential interferers can listen to DCTS and set their DNAV, thus deferring RTS transmission in the direction of the receiver. This way the problem of Deafness is avoided while exploiting the benefits of directional antenna and power control.

Power Controlled Directional-Medium Access Control (PCD-MAC) for wireless mesh networks is proposed in Capone, Martignon, and Fratta (2008). Authors have noted that adaptive antenna arrays have proved to be highly efficient in WMNs. PCD-MAC is a protocol for WMNs where nodes use adaptive array antennas with power control in order to increase network efficiency and reduce network interference. In this protocol, the RTS and CTS frames are sent in each antenna sector with maximum possible power that does not cause interference with other nodes in the network. Mo-

Table 1. Summary of surveyed protocols

Protocol Name	Directional Antenna Type	RTS/CTS Exchange	Idle Listening	Channels for Control and Data Packets	Tone	Beamforming Information	Channel Access Mechanism	Power Control
D-MAP	Switched Beam	ORTS DCTS	Omni	Separate	No	AoA of ORTS	Random Access	Yes
PCD-MAC	Adaptive Antenna Array	RTS and CTS transmitted on all directional beams	Omni	Same	No	Assumed Available	Random Access	Yes
D-MAC Scheme1	Multiple Directional Antennas per node	DRTS OCTS (Multidirectional Concurrent CTS)	Per-DA	Same	No	Assumed Available	Random Access	No
D-MAC Scheme2	Multiple Directional Antennas per node	DRTS ORTS (Multidirectional Concurrent RTS)	Per-DA	Same	No	Assumed Available	Random Access	No
CRDMAC	Switched Beam	DRTS DCTS	Omni	Separate	No	--	Random Access	No
S-MAC	Sectorized-DA	DRTS DCTS	Sector Monitoring	Same	No	--	Random Access	No
RI-DMAC	Switched Beam	DRTS DCTS	Omni	Same	No	Assumed Available	Random Access	No
OPDMAC	Switched Beam	DRTS DCTS	Omni	Same	No	Maintained by Upper Layer	Random Access	No
CMAC	Adaptive Array	Circular RTS Circular CTS	Multiple Receptions Over Angular Sectors	Same	No	AoA Estimation	Random Access	No
DtD-MAC	Switched Beam	DRTS DCTS	Directional	Same	No	AoA Estimation	Random Access	No
BMAC	Adaptive Array	DRTS OCTS	Omni	Same	No	Periodic Training Sequence	Random Access	No
DMAC-NT	Switched Beam	ORTS OCTS	Omni	Same	No	Predicted	Random Access	No
CDRTS	Switched Beam	Circular RTS DCTS	Omni	Same	No	DoA Cache	Random Access	No
CRCM	Switched Beam	Circular RTS Circular CTS	Omni	Same	No	DoA Cache	Random Access	No
MDA	Switched Beam	Circular RTS Circular CTS (Multidirectional Sequential)	Omni	Same	No	Maintained by Upper Layer	Random Access	No
DMAC-DACA	Switched Beam	Circular RTS Circular CTS (Multidirectional Sequential)	Omni	Same	No	GPS	Random Access	No
DMAC-DA	Switched Beam	Sweeping RTS Sweeping CTS	Omni	Same	No	GPS	Random Access	No
DMAC/DA	Switched Beam	Circular RTS Circular CTS (Multidirectional Sequential)	Omni	Same	No	Assumed Available	Random Access	No
DMAC/DA with NPN	Switched Beam	Circular RTS Circular CTS (Multidirectional Sequential)	Omni	Same	No	Assumed Available	Random Access	No

continued on following page

Table 1. Continued

Protocol Name	Directional Antenna Type	RTS/CTS Exchange	Idle Listening	Channels for Control and Data Packets	Tone	Beamforming Information	Channel Access Mechanism	Power Control
SpotMAC	Adaptive Array	DRTS Multidirectional Sequential CTS	Omni	Same	No	AoA Cache	Random Access	No
CDR-MAC	Switched Beam	Multidirectional Sequential RTS DCTS	Omni	Same	No	AoA and DoA Cache	Random Access	No
Directional MAC Protocol	Switched Beam	ORTS OCTS	Omni	Same	Yes	AoA Cache	Random Access	No
NCDMAC	Switched Beam	DRTS,DCTS, ORTS,OCTS, Multidirectional Sequential RTS	Directional Circum rotation	Same	No	AoA Cache	Random Access	No
PH-DMAC	Switched Beam	DRTS, DCTS	Omni	Same	No	DoA Cache	Random Access	No
Baseline MAC Protocol	Sectored Antenna	Sectored RTS Sectored CTS	Omni (All Sectors)	Separate	No	Sector Monitoring	Random Access	No
SDMAC	Switched Beam	DRTS, DCTS	Omni	Same	No	Assumed Available	Random Access	No
QSDMAC	Switched Beam	DRTS, DCTS	Omni	Same	No	Assumed Available	Random Access	No
Budhwani et. al.	Switched Beam	DRTS, CTS on all Unblocked Beams	Omni	Same	No	Assumed Available	Random Access	No
DUDMAC	Switched Beam	ORTS, OCTS	Omni	Separate	No	Not Required	Random Access	No
MPCD-MAC	Sectored Antenna	ORTS, OCTS using Single Sectored Antenna	Omni (All Directional Beams)	Separate	No	Assumed Available	Random Access	No
ToneDMAC	Switched Beam	DRTS DCTS	Omni	Separate	Yes	Assumed Available	Random Access	No
FFT-DMAC	Adaptive Array	DRTS	Omni	Separate	Yes	Assumed Available	Random Access	No
DBTMA/DA	Switched Beam	DRTS DCTS	Omni	Separate	Yes	Predicted	Random Access	No
BT-DMAC	Switched Beam	DRTS DCTS	Omni	Separate	Yes	AoA Cache	Random Access	No
DSDMAC	Sectored Antenna	DRTS DCTS	Sectored	Separate	Yes	Assumed Available	Random Access	No
ToneDUDMAC	Switched Beam	ORTS, DCTS	Omni	Separate	Yes	Not Required	Random Access	No
Omar et.al.	Switched Beam	Omni Broadcast, DRTS, DCTS. ORTS (No Direction Information)	Omni (All Beams)	Same	Yes	Cache Table	Random Access	Yes
SYN-DMAC	Switched Beam	DRTS DCTS	Omni	Same	No	Assumed Available	Synchronized Access	No
Di-ATC	Steerable Beam	---	Directional	Same	No	AoA, Neighbor Tracking	Synchronized Access	No

continued on following page

Table 1. Continued

Protocol Name	Directional Antenna Type	RTS/CTS Exchange	Idle Listening	Channels for Control and Data Packets	Tone	Beamforming Information	Channel Access Mechanism	Power Control
RDMAC	Switched Beam	ORTS OCTS	Omni	Same	No	DoA Cache	Synchronized Access	No
CDMAC	Switched Beam	ORTS, OCTS	Omni	Same	No	AoA Cache	Synchronized Access	Yes
BIBD Based Directional MAC	Switched Beam	DRTS, DCTS	Omni (All Directional Beams)	Same	No	Signal Strength of RTS	Synchronized Access	No
MDMAC	Steerable Antenna Array	---	Omni (All Sectors)	Same	No	Assumed Available	Synchronized Access	No
DiS-MAC	Single Beam	---	Directional	Same	No	TDoA	Synchronized Access	No
CW-DMAC	Switched Beam	ORTS OCTS	Omni	Same	No	Assumed Available	Synchronized Access	No
P-MAC	Adaptive Array	DRTS DCTS	Directional	Same	Yes	AoA and DoA Cache	Synchronized Access	No
HMAC	Switched Beam	DRTS, DCTS	Omni (All Directional Beams)	Same	No	DoA Cache	Random Access and Synchronized Access	No
D-STAR	Switched Beam	---	Omni (All Directional Beams)	Same	No	Phase and Angle	Space-Time Synchronization and Random Access	Yes
MD-STAR	Switched Beam	---	Omni (All Directional Beams)	Same	No	Phase and Angle	Space-Time Synchronization and Random Access	Yes

tive behind this is to inform maximum possible neighbor nodes about the ongoing communication. This helps in alleviating the problem of deafness in the network. After this, exchange of data and ACK packets takes place directionally with minimum required power that is just enough for two concerned nodes to communicate. This is done to avoid interference with ongoing transmissions in the network.

One of the initial protocols proposed to deal with deafness in ad-hoc networks using directional antennas was proposed in Ko, Shankarkumar, and Vaidya (2000), known as Directional Medium Access Control (DMAC) protocol. DMAC assumes that all the participating nodes know each other's location. All the nodes listen to ongoing transmissions and set their Network Allocation Vector (NAV) for the directions where an ongo-

ing communications are detected. There are two schemes proposed in DMAC, one in which RTS packets are sent directionally while CTS packets are sent omnidirectionally. Through omnidirectional exchange of CTS packets a node makes its neighbors aware of ongoing communication to avoid deafness. In second scheme, in case all the directions are free for transmission, RTS packets can be transmitted in all the directions. Though this reduces the occurrence of deafness in the network, it is observed that deafness may still exist. To further reduce the effects of deafness, DMAC uses Directional Wait- to-Send (DWTS) packets.

Circular Ready To Receive Directional Medium Access Control (CRDMAC) protocol is proposed in Lu, Li, Dong, and Ji (2011) to mitigate deafness in wireless networks. The protocol is based on the use of sub channels and circular directional transmission of Ready To Receive (RTR) packets. The communication channel is divided into two sub-channels one is data channel and other is narrow control channel. Unlike other protocols, in CRDMAC RTS, CTS, DATA and ACK are transmitted in data channel while RTR packet is transmitted in control channel. RTR packet only carries node ID of the sender. Since it is transmitted circularly and directionally covering all the neighbors around the sender, neighbors become aware of start of communication. After completion of communication, both sender and receiver nodes again transmit RTR packet circularly and directionally to inform neighbors about end of ongoing communication. It is assumed that an idle node is capable of tuning to both control and data channels.

Sectorized-Medium Access Control (S-MAC) protocol is proposed in Zhu, Nadeem, and Agre (2006). In this protocol, nodes only use directional transmission and reception. Every node consists of multiple directional antennas which are capable of self-interference cancellation. S-MAC uses multiple directional antennas and receivers to provide 360 degree coverage around a node. These antennas constantly monitor the medium around the node in all sectors. S-MAC nodes constantly monitor RTS packets in the vicinity and set their NAVs accordingly. For example, if a node does not hear data packets in a sector after RTS transmission then it assumes the channel to be clear and resets the NAV. If a channel is found to be busy through constant monitoring, then node does not transmit any packet in that particular direction or sector. S-MAC protocol is compatible with IEEE 802.11 standard and can interoperate with nodes using omnidirectional antennas.

Receiver-Initiated Directional MAC (RI-DMAC) protocol is proposed in Takata, Bandai, and Watanabe (2006) for handling the problem of Deafness while using directional antenna in ad-hoc networks. All the nodes use switched-beam antenna. RI-DMAC protocol handles deafness reactively using the method of polling. In this

protocol, the authors propose two operations namely sender-initiated and receiver-initiated operation. While the sender-initiated mode is a default mode, receiver-initiated mode gets triggered when the transmitter node experience deafness. In the sender-initiated mode, all the packets (RTS/CTS/DATA/ACK) are transmitted directionally to exploit the benefits of directionality. Every node in the network is required to maintain a polling-table to check the presence of potential deafness nodes. After exchange of DATA and ACK, the sender and receiver nodes check their polling table to find the existence of potential deafness nodes. If more than one node is registered in the polling table then their reception time is compared and the longest delayed node is selected to be polled. A directional Ready-to-Receive (RTR) frame is sent to the selected node. At the reception of RTR, the selected node (which was possibly suffering from deafness), transmits DATA frame. RTR frames are not retransmitted. This way authors propose to solve deafness while exploiting the benefits of directional antenna.

Opportunistic Directional MAC (OPDMAC) protocol is proposed in Bazan and Jaseemuddin (2011) for multihop wireless networks using directional antenna. The authors have noted that binary exponential backoff algorithm is not appropriate while using directional antennas. OPDMAC uses switched beam antenna for directional communication. This protocol uses a backoff mechanism wherein a node does not have to undergo idle backoff if there is a transmission failure. Instead it transmits other outstanding packets in other directions. This backoff mechanism enhances network performance and capacity through increased channel utilization and reduced idle waiting time. A node is forced to transit to Listening Period (LP) after each successful transmission. LP is a random time period during which the node listens in an omnidirectional mode. This listening phase reduces the transmission failures due to deafness. This way OPDMAC reduces the impact of deafness while exploiting the benefits of directional antenna.

Cooperative Medium Access Control (CMAC) protocol is proposed in Munari, Rossetto, and Zorzi (2008). CMAC protocol uses three methods to inform nodes about status of their neighbors. First, RTS and CTS is transmitted circularly using directional antenna covering all the nodes in the vicinity. Second, nodes carry out multiple receptions and decode given number of provided received from different directions. This way a node can update itself with other ongoing communications even when it is involved in a communication itself. Third method used is about cooperation between the nodes. If nodes involved in communication go back to idle state after finishing communication, they may not be completely aware of other ongoing communication and may become victim of deafness. In such cases other nodes can inform these nodes about ongoing communication. With these three methods used together, CMAC protocol solves deafness and enhance network performance.

Directional-to-Directional MAC (DtD-MAC) protocol is proposed in Shihab, Cai, and Pan (2009). Here, both sender and receiver nodes communicate in directional only mode. Nodes make use of a single directional antenna which can be switched beam or steerable directional antenna. In this protocol, all the nodes cache location information of their neighbor nodes so that they can decide the direction towards which the antenna beam should point to transmit Directional RTS (DRTS). Idle nodes continuously scan the medium clockwise or anticlockwise, in all the directions using directional antenna, thus emulating an omnidirectional antenna. If these idle nodes which are scanning the medium detect any DRTS frame which is direct towards them, they lock their antenna beam in that direction and transmit a Directional CTS (DCTS). The transmitting node transmits multiple DRTS packets towards the receiver in order to capture the continuously scanning idle receiver node. Ideally, 2M DRTS packets are sent to the receiver where M

is the number of antenna beams or sectors. Other nodes that overhear the communication set their Directional Network Allocation Vector (DNAV) accordingly. This way the overhearing nodes refrain themselves from interfering other ongoing communications. After setting DNAV the idle node continues scanning in all other directions. This approach helps in solving the problem of deafness because this way, a node will be aware of ongoing communication in its neighborhood and will not send RTS to a node which is already engaged in directional communication with another node in other direction.

802.11b based Beamformed MAC (BMAC) protocol which uses adaptive array smart antenna for communication is proposed in Fakih, Diouris, and Andrieux (2006). BMAC exploits these antennas for their ability to form nulls in the direction of interference, apart from increase in spectrum efficiency and extended communication range. BMAC uses a proactive channel tracking algorithm along with proposed MAC protocol to exploit the benefits of beamforming. Through an antenna array of 'N' elements, a node computes transmit beamforming weights in order to direct nulls toward 'N' noisy neighboring nodes. The node then transmits Beamformed RTS (BRTS) using transmit beamforming weights. The nodes which are in the vicinity of the destination of BRTS update Specified NAV (SNAV), which is NAV for specified node. Based on SNAV, the nodes defer themselves from sending RTS to communicating nodes (both source and destination). After completion of DATA and ACK exchange between the communicating nodes, a Channel Acquisition (CA) session is enabled, which informs about the availability of previously communicating nodes to other nodes in the network. This way, BMAC protocol handles the problem of Deafness in the network.

Directional Medium Access Control using NAV Table (DMAC-NT) protocol is proposed in (Jung, Lee, & Han, 2009) to solve the problem of deafness in Vehicular Ad-hoc Networks

(VANETs) using directional antenna. Here, RTS and CTS packets are transmitted omnidirectionally. Based on omnidirectionally transmitted RTS and CTS packets, nodes update their NAV tables. All the nodes maintain multi-NAV tables which include node ID and NAV of every node that is busy in communication. Based on this information nodes defer any transmissions which may interfere with other ongoing communication.

Circular Directional RTS (CDRTS) protocol is proposed in Korakis, Jakllari, and Tassiulas (2003) to deal with Deafness, while exploiting the benefits of directional antenna. In this protocol, nodes use switched beam directional antenna and cover complete area around the transmitting nodes with successive and sequential RTS transmissions. This way, transmitting node does not need to know the location of the receiving node. Nodes use only directional transmission in order to exploit the feature of increased coverage area with directional antennas. The transmitting node informs the neighbors about upcoming communication so that the neighboring nodes defer their transmission using DNAV to avoid interference and collisions. After multidirectional transmission of RTS packet is complete, receiver node replies with directional CTS. Another important feature of this protocol is that circular transmission of RTS respects ongoing transmissions in a particular beam direction. If during circular transmission of RTS, transmitter's DNAV does not allow it to transmit in a particular direction, then it does not transmit RTS in that direction.

Circular RTS and CTS MAC (CRCM) protocol is proposed in Jakllari, Broustis, Korakis, Krishnamurthy, and Tassiulas (2005). In this protocol, both RTS and CTS are transmitted circularly in all the directions to inform the neighbors of both transmitter and receiver about ongoing communication. Though transmission of RTS occurs in all the directions, CTS is transmitted only to 'unaware' neighbor. Unaware neighbors are the nodes which fall in the coverage of receiver but not in transmitter. As it is possible that both

transmitter and receiver may have some common neighbors, the significance of transmitting CTS only to 'unaware' neighbors falling in coverage of receiver is that the nodes falling in range of transmitter would have already been made aware through circular RTS. Nodes respect DNAV such that if during circular transmission of RTS and CTS, transmitter's or receiver's DNAV does not allow it to transmit in a particular direction, then it does not transmit RTS or CTS in that direction. In this protocol, nodes maintain location information of their neighbors in a location table.

Gossain, Cordeiro, and Agarwal (2005) have proposed directional MAC scheme for wireless ad-hoc networks called MAC protocol for Directional Antennas (MDA). In this scheme, nodes use switched beam directional antenna to circularly transmit RTS and CTS packet in order to make their neighbors aware of ongoing communication. However, this protocol is different from CDRTS (Korakis, Jakllari, & Tassiulas, 2003) and CRCM (Jakllari, Broustis, Korakis, Krishnamurthy, & Tassiulas, 2005) because in MDA circular RTS and CTS are transmitted simultaneously after successful exchange of unidirectional RTS and CTS. For unidirectional exchange of RTS/CTS, nodes require prior information about location of other nodes. This scheme introduces Enhanced DNAV (EDNAV) which enhances spatial reuse when compared to DNAV. Diametrically Opposite Directional (DOD) procedure is used to avoid overlap in coverage of circular RTS/CTS packets. Also, nodes maintain a Deafness Table (DT) consisting node id and duration for which it will busy at a particular beam number. This table is used to deal with deafness scenarios.

Directional MAC with Deafness Avoidance and Collision Avoidance (DMAC-DACA) protocol is proposed in Li and Safwat (2006). Nodes make use of switched beam antenna. In this protocol, initially nodes exchange RTS and CTS packets directionally. This is called basic RTS/CTS. After this, nodes transmit RTS/CTS packets in all the directions by sweeping the beam counterclockwise. This mechanism is called sweeping RTS/CTS and it is performed to inform all the neighbors of transmitter and receiver nodes about ongoing transmission. Sweeping RTS/CTS respects DNAV, therefore, if it is not allowed to transmit in a beam the node switches to the next beam. As multidirectional RTS/CTS transmission leads to sweeping delay, the duration information in RTS and CTS frames carries additional time required for sweeping transmissions. DMAC-DACA uses Deaf Neighbor Table (DNT) and Deafness Vector (DV) for deafness avoidance. Every node maintains DNT and DV. DNT consists of a set of deafness nodes and duration for which they will be busy with ongoing communication. DV is a counter which shows that intended destination will not be available for particular duration and thus any transmission for intended destination should be deferred till that time. This way, nodes handle the problem of deafness while using directional antenna. Authors have not mentioned the behavior/performance of the protocol in the presence of mobility. Also, this protocol is computationally extensive and has higher control overhead.

Same authors have proposed another protocol named Directional MAC with Deafness Avoidance (DMAC-DA) for deafness avoidance in ad-hoc and sensor networks in Li and Safwat (2005). Note that though the name of this protocol is same as that of one proposed in Takata, Bandai, and Watanabe (2007), these two protocols are different. In Li and Safwat (2005), the authors have identified two types of deafness problems, one due to a node being a transmitter or receiver (DF1), and another because of node being in a deaf zone (DF2). According to the authors, though DF1 has been extensively studied, they are first ones to discover DF2. All nodes are equipped with switched-beam antenna having 'M' non-overlapping beams. An idle node hears omnidirectionally, while RTS/CTS packets are transmitted directionally sweeping counterclockwise to inform neighboring nodes about upcoming communication (Sweeping RTS/CTS). DMAC-DA uses DNAV, therefore

though there is backoff mechanism for individual beams, if a node is not permitted to transmit on a particular beam due to DNAV, it remains silent for appropriate duration and then switches on to the next beam. Unlike standard CTS frames, sweeping CTS frames carry both transmitter and receiver addresses. Two deafness avoidance methods DA1 and DA2 are proposed for DF1 and DF2 respectively. In DA1 node uses Deaf Neighbor Table (DNT) carrying information about deaf neighbors and duration after which they will be available, and Deafness Vector (DV) is a counter related to a particular destination for which sender has to defer transmission till the destination is available. In DA2, node location information is necessary and it is retrieved through GPS. Here, every node maintains a location table which has latest location information about the neighbors. Before transmission every node checks if any of its neighbors are located in the deaf zone, and then carries out communication according to DA1. Simulation studies prove that the proposed protocol significantly reduces affect of deafness and therefore, enhances network throughput.

Directional MAC with Deafness Avoidance (DMAC/DA) is proposed in Takata, Bandai, and Watanabe (2007). In this protocol also nodes make use of switched beam antennas. Authors introduce Wait-to-Send (WTS) apart from RTS and CTS. WTS is transmitted by two communicating nodes after successful exchange of RTS and CTS packets to inform the certain nodes about ongoing communication. Based on this information concerned nodes can defer transmission of RTS to any of the busy nodes for certain time, thus avoiding Deafness. All the nodes maintain a neighbor table which has four fields namely node ID, beam number, deafness duration and link activity. Node ID field maintains concerned node's ID, beam number field maintain the directional beam number over which the node maintaining neighbor table heard the WTS frame, deafness duration field maintains the duration for which the concerned node will remain deaf and link activity field maintains the reception time of earlier transmission between two communicating nodes. The protocol handles mobility scenarios based on the entries present in this field.

Same authors have proposed a modification of DMAC/DA protocol in the same paper (Takata, Bandai, & Watanabe, 2007). The modified protocol is called DMAC/DA with Next Packet Notification (NPN). This protocol is based on the principle that if all the nodes can acquire next packet information of their neighbor nodes then WTS frames can be transmitted more effectively to alleviate deafness and reduce control overhead due to WTS transmissions. Here, the transmitter sets more data bit in frame control header of data frame if there is more data to be sent to the intended receiver. Based on this information, the link activity field of the neighbor table is updated. This way, every node can distinguish active transmitter nodes among neighbor nodes, and WTS frames can be transmitted only over the beams which are directed towards active transmitters. This method reduces the control overhead associated with transmission of WTS frames in the network and is more efficient in mitigating deafness when compared to basic DMAC/DA.

SpotMAC protocol for wireless mesh networks is proposed in Chin (2007). This protocol uses directional antenna with pencil beams (very narrow beams). It is observed in this paper that pencil beams provide higher spatial reuse, reduces exposed terminal problem and constrains hidden node problem to linear topology. However, pencil beams significantly increase the probability of Deafness, because due to very narrow beams there will be more nodes in vicinity which will be unaware of ongoing communication between other nodes. To deal with this problem, whenever a transmitter encounters failure in data transmission, SpotMAC allows the transmitter to contend for the medium immediately after backing off for a random time based on constant contention window. When the number of failures exceed a certain threshold, then the contention window

is also increased exponentially. It is observed in the paper that this helps in mitigating Deafness. It is also observed that SpotMAC provides high spectral reuse and network efficiency in non-deafness scenarios.

Circular Directional RTS MAC (CDR-MAC) protocol is proposed in Korakis, Jakllari, and Tassiulas (2008). The nodes are equipped with switched beam directional antenna. RTS frame is transmitted circularly till it scans all the area in the vicinity using directional antenna. This emulates omnidirectional antenna while achieving benefits of directional antenna such as longer communication range and better gain. CTS frame is transmitted directionally just after CRTS in concluded. CRTS makes the nodes in the vicinity of the sender aware of ongoing transmission. Other nodes that hear CRTS defer transmission till given duration thus avoiding deafness. Nodes constantly track the location of other neighbor nodes and maintain this information in a node location table. This helps when participating nodes are mobile. Another important feature of CDR-MAC is that CRTS respects any ongoing transmission through DNAV. If a particular beam is used for communication, CRTS transmission skips that beam and continues with the next beam. This method may fail to solve the problem of deafness among the nodes neighboring the transmitter because CTS is transmitted directionally and neighbors of the transmitter may not be aware of the ongoing communication.

Another directional MAC protocol to handle deafness in wireless networks using directional antenna is proposed in Subramanium and Das (2005). This protocol is very similar to that mentioned in Prabhu and Das (2007). Here, RTS and CTS packets are sent in omnidirectional mode while DATA and ACK are sent in directional mode. The RTS and CTS packets carry information about angle of directional communication and control window. This way neighboring nodes become aware of direction in which the communication will take place and accordingly defer any transmission in that direction. Apart from that, a Negative CTS (NCTS) packet is transmitted omnidirectionally by receiver of RTS if it cannot establish communication link because its beam is blocked. If the sender of RTS receives a NCTS, it transmits the Transmission Cancel (TC) packet omnidirectionally to inform all the neighbors about deferring the communication so that neighboring nodes cancel their NAVs which were set based of previously sent RTS packet.

Nested Circular Directional Medium Access Control (NCDMAC) protocol is proposed in Ding, Wang, Shen, and Guo (2010). This protocol uses circular DRTS and multihop RTS/CTS for exchanging handshake information among nodes using directional and omnidirectional antennas. Directional nodes use switched beam antennas. NCDMAC is designed for the networks where nodes use both directional and omnidirectional antennas, and the network performance degrades due to asymmetry in gains of directional and omnidirectional antennas. Nodes using directional antenna scan their vicinity periodically in order to detect the control frames correctly. A nested circumrotation mechanism is implemented based on which the directional nodes transmit DRTS frames in every direction and form a neighbor table. Later node switches its antenna to receive CTS from other directions. Authors have proved through simulation results that NCDMAC protocol efficiently supports hybrid antennas and performs better than MMAC and CMAC protocols in terms of throughput.

Physical layer aware Directional Medium Access Control (PH-DMAC) protocol which uses physical layer information such as packet arrival and Directional of Arrival (DoA) to mitigate deafness is proposed in Hadjadj-Aoul and Nait-Abdesselam (2008). Here, nodes are equipped with switched beam antenna use Directional Virtual Carrier Sensing (DVCS) for packet transmission. DNAV is set to either mitigate deafness or at the reception of corrupted packets. If a corrupted packet is received with Signal to Interference plus Noise Ratio (SINR) greater than the threshold SINR, then it is considered to be RTS collision. This protocol also proposes to send Directional CTS to identified deaf nodes so that they can send the RTS packet again. This protocol is compatible with IEEE 802.11 standard.

The use of sectored antenna is proposed in Swaminathan, Noneakar, and Russell (2012) to mitigate the problem of deafness in a wireless network using directional antenna. The proposed MAC protocol is called Baseline MAC Protocol. Authors have observed that directional antenna can drastically enhance network performance of wireless networks and have gained enormous popularity among researchers. However, most of the protocols proposed for directional communication also need some transmission (either control packets or tones) in omnidirectional mode to alleviate deafness. It is proved in this paper that the use of sectored antenna instead of steered beam or switched beam antenna can eliminate the need of using omnidirectional antenna to handle the problem of deafness. Here, multiple directional antennas also known as sectors are used on a node and each sector's pattern is fixed relative to node's orientation. Every antenna is employed with a separate radio transceiver. Any two antennas have non-overlapping regions of coverage and all the antennas together provide complete coverage of region around the node as in an omnidirectional antenna. Because all the sectors constantly monitor neighboring area, it removes the need for omnidirectional antenna and at the same time alleviates the problem of deafness, because now, node is aware of ongoing communication among neighboring nodes.

Selectively Directional MAC (SDMAC) protocol is proposed to mitigate the problem of deafness in mobile ad-hoc networks in Li, Zhai, and Fang (2009). In this protocol, every node maintains a deafness table that consists of node IDs of deaf nodes along with the duration for which they will be deaf. Exchange of Type I DRTS and DCTS between sender and receiver decides about outgoing beam and beam status. Using Type II DRTS and DCTS the communicating nodes notify other nodes about ongoing communication. Based on this information other nodes include these nodes in their deafness table. Nodes defer any transmission that may interfere with ongoing transmissions based on information in the deafness table, thus handling the problem of deafness while exploiting the benefits of directional communication. Authors have also proposed Queue-SDMAC (Q-SDMAC) and Q-SDMAC with cache which improve the performance of SDMAC further while handling deafness.

A MAC protocol for ad-hoc networks using directional antennas is proposed in Budhwani, Sarkar, and Nagaraj (2010). In this protocol nodes make use of switched beam antennas and maintain a DNAV table that consists of information about received CTS packets, respective duration of communication and directions blocked due to ongoing communication. RTS packets are transmitted directionally, while CTS packets are transmitted in all unblocked directions. DATA and ACK packets are transmitted directionally. Authors have compared the performance of their protocol with 802.11 and D-MAC protocols and found it to be more efficient in terms of throughput.

Dual-channel Directional MAC (DUDMAC) protocol is proposed in Lee, Han, and Jwa (n.d.). This protocol uses separate control and data channels. ORTS, OCTS, NDATA and NCTS are transmitted over control channel, while DDATA and DACK are transmitted over data channel. ORTS and OCTS are used to mitigate deafness, while NDATA and NCTS are used to avoid collisions in data channels. Node location information is not required for DUDMAC to function as it transmits RTS and CTS frames omnidirectionally.

Multi-channel Power Controlled Directional Medium Access Control (MPCD-MAC) protocol for WMNs is proposed in Martignon (2011). The author has observed that though WMNs have emerged as next generation networks providing high-speed last mile Internet access, their per-

formance and efficiency suffer due to the use of omnidirectional antennas. Author has proposed MPCD-MAC protocol for WMNs using directional antenna. Here, nodes are required to be equipped with multiple interfaces and directional antennas. Communication channel is divided into two sub channels, one for exchange of control packets (RTS and CTS), and another for data and ACK packets. The RTS and CTS packets are transmitted in all the directions over control channel. This way all the neighboring nodes in the vicinity are made aware of any ongoing communication, which helps in alleviating the problem of deafness. MPCD-MAC is based on PCD-MAC (Capone, Martignon, & Fratta, 2008), but uses only single-sectored antenna patterns for easy realization. MPCD-MAC also uses multiple orthogonal channels which further enhances network performance when compared to previously existing PCD-MAC.

MAC-layer Anycasting is proposed in Choudhury and Vaidya (2004) to mitigate deafness in wireless ad-hoc networks using directional antenna. This framework involves interaction between network and MAC layer. When a source node needs to send data to a destination node, a routing protocol first discovers and maintains multiple routes between a source and destination. Out of these multiple routes, *K* best routes are selected and named as *anycast group*. The *anycast group* and data packet to be transmitted are then sent to the MAC layer. MAC layer now selects a suitable next hop node and tries to transmit data to it. If this transmission fails then MAC layer looks for next best option given in the anycast group. Authors observe that in ad-hoc networks using directional antenna, transmission failure may occur due the deafness. In such case of transmission failure, the transmitter can search for next option from the anycast group instead to suffering from deafness.

Directional Ultrawideband Medium Access Control (DU-MAC) protocol is proposed in Karapistoli, Gragopoulos, Tsetsinas, and Pavlidou (2009) to mitigate deafness in Ultrawideband

(UWB)-based sensor networks. In this protocol, idle nodes rotate their directional receiving beam continuously in 360 degrees. Nodes use preamble trailer messages to inform neighboring nodes about possible transmissions. After reception of preamble trailer messages, nodes which are 'ready to receive', send an acknowledgement packet. Based on this information, neighboring nodes can lock their beams for data transmission and reception and unlock the beams after end of communication. Though this protocol may solve the problem of deafness in UWB based sensor networks, rotating the directional receive beam continuously may increase energy consumption. This could be a disadvantage because energy conservation is crucial in wireless sensor networks where nodes depend on limited battery power but are expected to have longer lifetime.

A directional antenna-based architecture for IEEE 802.11s based WMNs is proposed in Ben-Othman, Mokdad, and Cheikh (2011). In this paper, the authors have pointed out that installing IEEE 802.11 based WMN requires high number of access points due to limitation of coverage. Use of directional antenna will increase the coverage and therefore less number of access points will be required to establish WMNs. Authors have noted that use of directional antenna in IEEE 802.11s based WMN backbone will enhance network performance because higher spatial reuse and reduced interference will allow parallel non-interfering communication, increased communication range will allow link establishment between far off nodes with fewer hops. Therefore, it is concluded that use of directional antenna in WMN backbone will lead to robust communication and enhance network performance and efficiency. In this paper, the authors have used sectorial antenna composed of multiple fixed beam antennas. Each fixed beam antenna aims in a different direction such that all the fixed beam antennas together give 2π coverage. These antennas are pre-fixed in particular direction. Authors have pointed out that this configuration solves the problem of Deafness

because beams and directions of communication are predefined. In this paper, each mesh point (MP) is considered to have three fixed beam antennas aiming in different directions. Each antenna covers an angle of 120 degrees. Using sectorial antenna authors have proposed Sector-Hybrid Wireless Mesh Protocol (Sector-HWMP).

Random Access Protocols with Tones

ToneDMAC protocol is proposed in Choudhary and Vaidya (2004) to address the problem of Deafness. Nodes make use of switched beam directional antenna for communication while backoff phase is performed in omnidirectional mode. ToneDMAC assumes that congestion would be the prime reason for communication failures. However, if the cause is Deafness, then corrective measures are taken. In this protocol, communicating nodes omnidirectionally transmit an out-of-band tone after completion of every dialog. These tones assist the neighbors of communicating nodes in distinguishing between deafness and congestion. After receiving the tone, neighbor nodes try to identify the tone-originator based on the frequency and duration of the tone. After identifying the tone-originator a backlogged neighbor node checks if the tone-originator node is same as the node with which it intends to establish communication. If that is the case, then neighbor deduces Deafness to be the cause of previous transmission failures. Now, the neighbor node reduces backoff interval and restarts transmission of pending packets. It must be noted that unlike other proposals of 'busy tones', ToneDMAC uses single transceiver.

Flip-flop tone – directional medium access control (FFT-DMAC) is proposed in Li, Li, Shu, and Wu (2007) to solve Deafness in wireless ad-hoc networks. In this method two pairs of flip-flop tones used, one pair is sent omnidirectionally while other pair is sent directionally. A tone is an unmodulated sine wave that does not contain any information. It can only be detected through energy estimation on a narrow frequency band. To identify the nodes which send the tones, node id is coded in tone's frequency. To solve the problem of deafness, first pair of tones is sent omnidirectionally in order to inform all the nodes in the vicinity about start and end of ongoing communication between a pair of nodes. Every node maintains two lists, one that of 'deafness nodes' and other that of 'ongoing transmission nodes' based on information obtained through the tones. This list is constantly updated upon reception of the tones. Another pair of tones which is sent directionally is used to exposed and hidden node problem. The obtained results show that FFT-DMAC is successfully able to handle deafness, exposed node and hidden node problems. One of major disadvantages of tone based protocols is that it requires additional hardware.

Dual Busy Tone Multiple Access using Directional Antenna (DBTMA/DA) protocol is proposed in Huang, Shen, Srisathapornphat, and Jaikeo (2002) to solve deafness. This protocol uses a separate control channel to exchange information regarding communication status between nodes. A receive busy tone (BTr) is broadcasted before CTS transmission and another transmit busy tone (BTt) is transmitted to inform nodes around transmitter and receiver about ongoing communication. This protocol requires two channel or two antennas for implementing control channels. This can make implementation of this protocol complicated.

Another tone based protocol called Busy Tone-Directional Medium Access Control (BT-DMAC) in Dai, Ng, and Wu (2007). In this protocol sending and receiving nodes turn on busy tones to inform other nodes about the ongoing communication. Every node is equipped with two interfaces, one for switched beam antenna and another for omnidirectional antenna. In idle state, all the nodes listen to the medium omnidirectionally. On arrival of any signal, node records the beam where signal strength is highest. It records the beam number and neighbor ID in a Neighbor Location Table (NLT). After determination of direction, node switches

to directional mode for exchange of RTS, CTS, data and ACK. Two busy tones are implemented in BT-DMAC protocol, transmitting busy tone and receiving busy tone. These busy tones are further divided into two subtones, ID tone and beam number tone. Though the tones share single channel, they have sufficient spectral separation so that the tones do not interfere with each other. Tones are pulse modulated and carry information about node IDs and beam numbers. Because tones are transmitted omnidirectionally, neighboring nodes stay informed about ongoing communication happening among particular nodes and their beam numbers. Therefore, any transmission that may lead to interference is deferred, thus solving the problem of Deafness. Authors have observed that BT-DMAC can successfully solve deafness and hidden node problem.

Abdullah, Cai, and Gebali (2012) have proposed Dual-Sensing Directional Medium Access Control (DSDMAC) protocol which uses busy-tones to solve the problem of deafness in wireless ad-hoc networks. The protocol uses two types of busy tone signals BT_1 which uses a continuous pattern and BT_2 which uses ON/OFF pattern. Communication channel is divided into two sub-channels, the data channel which carries directional RTS, CTS, DATA and ACK packets, and a busy-tone channel which carries the busy-tone signal. BT_1 is used to solve hidden-terminal problem, while BT_2 is used to solve deafness. When a node transmitting DRTS experiences failure, it checks for BT_2 from the direction of the receiver. If BT_2 exists then it is known that intended receiver is busy with ongoing communication, otherwise it is assumed that there has been a collision of DRTS. Busy-tones are transmitted by source and destination pairs and the directions of all the possible transmitters and receivers are assumed to be known.

Tone Dual Channel Directional Medium Access Control (ToneDUDMAC) is proposed in Lee, Han, and Jwa (n.d.) to mitigate deafness in ad-hoc networks where nodes do not have knowledge about each other's location. In this protocol, RTS is transmitted omnidirectionally (ORTS) while CTS, data and ACK (DCTS, DDATA and DACK) are transmitted directionally. ORTS and DCTS are transmitted over control channels while data and ACK are transmitted over data channel. A CTS_tone is transmitted omnidirectionally to mitigate deafness, while directional DATA_tone is transmitted to minimize the blocking area for directional antenna. Negative CTS (NCTS) and negative DATA (NDATA) packets are used to avoid packet collisions. Simulation results show that ToneDUDMAC protocol provides better throughput when compared to ToneDMAC, DMAC, IEEE 802.11 MAC and dual channel MAC protocol (DUCHA).

Another tone based protocol is presented in Omar and Elsayed (2010). This protocol uses busy tone signal accompanied with directional data transmission. Both transmitting and receiving nodes transmit busy tone signals, thus informing other nodes present in their zones of ongoing communication. This helps in deferring deafness. All the nodes are equipped with switched beam antenna. When a node is idle state, or when it is broadcasting, all the beams are active, therefore, emulating omnidirectional antenna with extended range. RTS, CTS and DATA frames are transmitted directionally, while broadcasts are transmitted over all the beams. All the nodes maintain a cache table containing location information of other nodes. If a node does not have location information of some other node with which it wants to communicate, then it transmits the RTS frame omnidirectionally.

Synchronized Access Protocols without Tones

Synchronized access protocol called SYN-DMAC is proposed in Wang, Fang, and Wu (2005) for ad-hoc networks using directional antenna. This protocol works on the assumption that there is system-wide synchronization present in the network. Each node is equipped with single

transceiver with switched beam antenna. Each node can run in omnidirectional and directional mode. In idle state, nodes work in omnidirectional mode. A node in random access phase can be in any of three modes, sending, receiving or pending mode. Transmitter node sends directional RTS to the intended receiver node. Upon receiving RTS, the intended receiver node checks if it is pending mode and intend-to-receive beam has not been reserved by other nodes. Based on this information, it either transmits directional CTS or negative-CTS. If directional CTS is transmitted, then intended transmitter node sends a directional confirmed RTS (CRTS). Any other nodes in the vicinity hearing directional CTS or CRTS will block any transmission in the intended beam to mitigate deafness. Authors argue that SYN-DMAC handles the problem of deafness through a novel timing structure. It is observed that duration for which deafness may occur is T_{cr}, which is the time taken to exchange RTS/CTS control packets.

Angular Topology Control with Directional Antenna (Di-ATC) scheme is proposed in Gelal, Jakllari, Krishnamurthy, and Young (2006). This scheme requires all the nodes in the network to use only directional antennas to eliminate asymmetry in communication range which occurs due to use of both directional and omnidirectional antennas. It also aims to combat deafness which occurs due to the use of directional antennas. It supports directional neighbor discovery and topology control on chosen neighbor nodes. Direct communication links are maintained with certain chosen neighbors while other nodes are reached through multi-hop paths. Nodes with direct links negotiate for common time-slots for communication through polling each other at regular intervals of time. Nodes maintain updated information about direction of every other chosen node so that they can direct their beams appropriately. Since nodes communicate according to pre-decided time slots, deafness can be avoided.

Reservation based Directional Medium Access Control protocol (RDMAC) for multi-hop wireless networks using directional antenna is proposed in Jakllari, Luo, and Krishnamurthy (2007). Here, nodes use switched beam antenna and assume prior information about location of other nodes. There is no centralized synchronization required among nodes in this protocol. RDMAC protocol operates in sessions, where each session is divided into reservation and transmission period. Reservation period is further divided into probing phase and beam indication phase. In probing phase, omni-directional transmission of RTS and CTS frames takes place, while in beam indication phase RTS and CTS frames are transmitted directionally. In transmission period data and ACK are exchanged directionally. Each node maintains a DNAV table to avoid transmission in particular directions to avoid collision with other ongoing communica-tion. This way RDMAC alleviates the problem of deafness in the network. Authors have also explored interference due to minor lobes while using directional antenna. It is observed through simulations that RDMAC performs better in term of aggregate throughput and average delay when compared to CDR-MAC and IEEE 802.11 DCF.

Coordinated Directional Medium Access Control (CDMAC) protocol for ad-hoc networks is proposed in Wang, Zhai, Li, Fang, and Wu (2009). In this protocol, nodes are required to have local synchronization. CDMAC presents a timing structure wherein a contention period is followed by two contention free periods. In the contention free period exchange of control packets takes place while two contention periods is used for parallel exchange of data and ACK packets. There are three control packets in CDMAC namely RTS, CTS and Confirmed-RTS (CRTS). After omnidirec-tional exchange of RTS and CTS frames between sender and receiver, the sender transmits CRTS omnidirectionally. CRTS is a small message which confirms reservation of communication channel. As these control messages are transmitted omni-directionally, all the nodes in the vicinity become aware of ongoing communication and defer any

packet transmission which may either interfere with ongoing transmission or lead to deafness.

Balanced Incomplete Block Design theory Based Directional Medium Access Control (BIBD based Directional MAC) protocol is proposed to mitigate deafness in wireless ad-hoc networks in Boggia, Camarda, Cormio, and Grieco (2009). Authors have observed that protocols to mitigate deafness which have been proposed in the past are not very efficient in terms of energy because they use tones and sweeping of RTS/CTS pack-ets. BIBD based Directional MAC protocol is a slotted directional MAC protocol which mitigates deafness and is energy efficient. In this protocol nodes use time slots for communication. At the beginning of a slot, the transmitting node sends RTS packet in 'k' beams which are active in that particular slot using CSMA/CA. When the intended receiver node receives RTS it switches off all the beams other than the one over which it received RTS with highest signal strength. Over this particular beam it directionally transmits CTS packet. Transmitter selects particular beam over which it receives CTS with highest signal strength. Now DATA and ACK packet are exchanged on selected beams from both the nodes. Other nodes which overhear this communication defer any transmissions which may interfere with ongoing transmissions in that particular time slot.

Memory guided Directional Medium Access Control (MDMAC) protocol for 60GHz outdoor WMNs is proposed in Singh, Mudumbai, and Madhow (2010). Authors have observed that multi-gigabit outdoor WMNs which operate in 60GHz millimeter wave (mm wave) unlicensed band have great scope in providing extended In-ternet access over wireless. However, signals at mm-wave frequencies experience high path loss. Therefore, use of directional antenna can extend communication range. Use of electronically steer-able directional antenna arrays on mesh nodes is being explored by researchers for enhancing the performance of WMNs. Though directionality reduces interference, increases communication range and enhances spatial reuse, it introduces deafness. This requires a protocol that can deal with the problem of deafness while exploiting the benefits of directional antenna in outdoor WMNs operating in 60GHz unlicensed frequency band. Many previously proposed protocols solve deafness using omnidirectional exchange of RTS and CTS frames or tones. Authors have observed that such protocols are not applicable to WMNs operating in 60GHz frequency band because 60GHz communication needs to be highly directional due to small carrier wavelength. With no scope of using omnidirectional mode, authors have proposed Memory guided Directional Medium Access Control (MDMAC) protocol which works on Time Division Multiplexing (TDM) schedules. Nodes maintain slot information in the form of slot allocation tables and blocked slots. Based on previous performance in allocated schedules, nodes make memory based decision about which slots can be used and which cannot be used. These decisions take place in distributed manner and help in dealing deafness that arises due to the use of highly directional antennas.

Directional Scheduled Medium Access Control (DiS-MAC) protocol is proposed in Karveli, Voulgaris, Ghavami, and Aghvami (2008) for WSNs used for roadside and highway monitoring. Authors have noticed that due to their long communication range and reduced interference directional antennas can prove highly beneficial for roadside and highway monitoring applications. Here, each node is equipped with a directional antenna which can point its beam in a particular direction and a transceiver. Nodes in the network are required to be synchronized and traffic flow is unidirectional. Therefore, this protocol is appropriate for networks having linear topology. The communication channel is divided into two phases where each phase is 'T' duration long. In first phase only 2n-1 number of nodes are allowed to communicate while second phase rest of the nodes communicate. Authors claim that this method of communication drastically brings

down the occurrence of collisions and deafness. This protocol may be difficult to implement as it requires system wide synchronization which can be difficult to achieve among nodes in WSN.

Contention Window based Directional Medium Access Control (CW-DMAC) protocol is proposed in Prabhu and Das (2007) to solve the problem of deafness in multi-hop wireless networks. This protocol assumes that directional gain is same as omnidirectional gain. It is achieved by reduction in transmission power when transmitting directionally. RTS and CTS packets are transmitted omnidirectionally, while data and acknowledgement packets are transmitted directionally. In order to avoid the possibility of omnidirectional RTS and CTS colliding with ongoing data transmission in the network, exchange on control and data packets is separated in time. After exchange of control packets, nodes wait for duration of time known as *control window* before data transmission. RTS and CTS packets carry beam index information based on which data and ACK exchange would take place directionally. This way neighboring nodes are informed about ongoing communication to alleviate deafness. This way CW-DMAC protocol alleviates deafness while using directional antenna in wireless network. Authors have proved through simulations that CW-DMAC achieves better throughput when compared to basic DMAC.

Synchronized Access Protocols with Tones

Polling based Medium Access Control (PMAC) protocol for MANETs based on synchronized access is proposed in Jakllari, Luo, and Krishnamurthy (2007). This protocol exploits benefits of directional antennas as well as constantly tracks location of mobile nodes in the vicinity. Here, time is divided into frames, and each frame is further divided into states. Each frame constitutes of three states namely search state, polling state and data transfer state. In search state nodes search for neighbors, in polling state nodes poll the neighbors, and in data transfer state exchange of data takes place. In search state, nodes discover each other through directional exchange of pilot tones. After discovery, nodes exchange a list of polling slots and agree upon a slot for periodic communication. Before initiating communication, nodes exchange control packets to avoid collisions. It is important to note that since PMAC is based on scheduled access, Deafness does not occur. After establishing a connection in search state, nodes communicate with their neighbors in poll and data transfer state. During these states, nodes continuously track location of their mobile neighbors using Angle of Arrival (AoA) and Direction of Arrival (DoA).

Hybrid Access Protocols

HybridMAC (HMAC) protocol is proposed for use of multiple beam smart antennas in multihop wireless networks in Jain, Gupta, and Agarwal (2008). This protocol proposes concurrent packet transmission and reception for enhanced performance of wireless networks. The authors have made two important observations regarding deafness mitigation, one that a node must transmit control packets at least in the beams having prospective transmitters for it, and a node should transmit the control packets in all the beams if there is a possibility of having a node in the vicinity that may interfere with ongoing transmissions. HMAC maintains a Hybrid Network Allocation Vector (HNAV) table, where it maintains two Boolean variables namely, *isValidRTSReceived* and *isInvalidCTSReceived*. Based on whether these Boolean variables are true or not, control messages are transmitted to inform other nodes of ongoing communication and thus mitigate deafness. The authors have also presented wireless mesh network architecture and proposed the use of HMAC protocol to enhance performance of such networks. Support for mobility is not considered.

Directive Synchronous Transmission Asynchronous Reception (D-STAR) for WSNs is proposed in Manes et al. (2008). It requires the nodes in the sensor network to be time-space synchronized. Protocol operation is divided into two phases: discovery and regime phase. In discovery phase all the nodes broadcast a HELLO message carrying its ID and phase information in each angular sector. In regime phase, a reference node unicasts HELLO packets to neighbors present in different angular sectors according to phase information exchanged during discovery phase. D-STAR mitigates deafness based on the hypotheses that antenna switching is performed correctly, local clocks on the nodes to drift and phase information exchanged among the nodes is correct. Meaning that all the sensor nodes are space-time synchronized.

Mobile Directive Synchronous Transmission Asynchronous Receiver (MD-STAR) protocol is proposed in Bencini, Collodi, Di Palma, and Manes (2010) for WSNs using directional antennas. Operation of this protocol is similar to that proposed in Manes et al. (2008) with inclusion of support to node mobility. Authors have identified the benefits of using directional antennas in WSN for energy saving and target tracking. This protocol realizes space-time synchronization among nodes in the network. The protocol is composed of discovery phase and regime phase. In discovery phase, nodes transmit HELLO packets in every angular sector. In regime phase nodes can be mobile and broadcast HELLO packets in every sector looking for connections and notifying other nodes about their presence. Authors have pointed out that this way; nodes can mitigate the problem of deafness arising due to the use of directional antennas.

DISCUSSION ON DIFFERENT METHODS PROPOSED TO AVOID DEAFNESS

In this section, we discuss and analyze various design choices of proposed protocols. Performance of a protocol to be used in wireless networks highly

Table 2. Comparison of different design choices for protocols proposed for deafness avoidance

Design Parameter	Method	Merits	Demerits
Type of Antenna	Switched Beam	- Ease of implementation and use	- Does not work well in multipath environments
	Adaptive Array	- Higher gain and nulling capability	- Complex implementation and use - Work well in multipath environments
RTS / CTS Transmission	Omnidirectional RTS / CTS	- All neighbors of transmitter / receiver informed about ongoing communication	- Increased interference and chances of exposed node problem - Reduced spatial reuse - Reduced communication range
	Unidirectional RTS / CTS	- Increased communication range - Increased spatial reuse - Reduced interference and chances of exposed node problem	- All neighbors of communicating nodes not informed about ongoing communication, leading to deafness and directional hidden node problem -Destination node location information required for beamforming
	Multidirectional Sequential RTS / CTS	- All neighboring nodes within the range of transmitter/ receiver informed about ongoing communication - Mitigates deafness for nodes within range of transmitter - Eliminates asymmetry in gain	- Sweeping delay -Higher control overhead
	Multidirectional Concurrent RTS / CTS	- No sweeping delay - Emulates omnidirectional RTS / CTS transmission but with increased range	- Complex antenna systems required
Tones	Use of Busy Tones	- Mitigate deafness by informing neighbor nodes about ongoing communication	- Separate control channel required - Increased cost and complex implementation
Idle Listening	Omnidirectional	- Helps in deafness mitigation	- Reduced sensing and communication range
	Directional	- Increased sensing and communication range	- Increased chances of deafness
Channel Access Mechanism	Synchronized Access	- Enhanced network performance due to conflict-free scheduling - Mitigates deafness as nodes communicate according to designated schedule	- Difficult to achieve local and global synchronization in wireless networks - Complex implementation
	Random Access	- Can give rise to deafness	- Simpler implementation
Multichannel Communication	Separate channels for control and data packets	- Reduced collisions	- Complex implementation

depends on its design, method to transmit frames, type of antenna, number of channels used for communication, method of idle listening and method of channel access. Detailed discussion on these factors is carried out and summarized in Table 2:

- **Method to transmit frames:** All the surveyed protocols exchange data and acknowledgement packets directionally to exploit features of directional antennas. The method of RTS and CTS transmissions vary among protocols. RTS and CTS frames are transmitted mainly in four ways, omnidirectional RTS/CTS, unidirectional RTS/CTS, multidirectional concurrent RTS/CTS and multidirectional sequential RTS/CTS. Omnidirectional RTS/CTS, when used at both transmitter and receiver nodes, inform neighbor nodes of ongoing communication thus mitigating deafness. Such solutions suffer from asymmetry in gain because while RTS and CTS are transmitted omnidirectionally, data and acknowledgement are exchanged directionally. Unidirectional RTS/CTS exchange can cause deafness. Therefore, such protocols use other methods such as tones or time-slots to mitigate deafness. Multidirectional concurrent RTS/CTS exchange emulates omnidirectional antenna but with directional coverage which mitigates deafness by informing neighbor nodes over longer range about ongoing communication. Multidirectional sequential RTS/CTS exchange does inform neighbor nodes about ongoing communication, but it introduces sweeping delay. This method may not be efficient in networks having highly mobile nodes because a mobile node may miss receiving RTS/CTS frame by moving from a location where RTS/CTS has not been transmitted to a location where it has already been transmitted. It is observed that control overhead of such protocols is significantly high leading to higher power consumption;

- **Type of antenna:** Mainly four types of antennas have been used in the surveyed protocols. These are switched beam, steerable, sectored and adaptive array antennas. However, most of the protocols have used switched beam antennas due to their simplicity. Though switched beam antennas are simple to implement, they do not work well in multipath environments. Some of the protocols make use of nulling capabilities of directional antennas to overcome interference due to neighboring nodes. Though adaptive array antennas provide higher gain and nulling capability, their cost and complexity are the limiting factors due to which many researchers have avoided using them;

- **Number of channels used for communication:** As observed from Table 1, many protocols use same channel for exchange of control and data packets. However, some protocol use one channel for RTS/CTS exchange and another for exchange of data and acknowledgement. Some protocols use a separate channel to transmit tones. Protocols that use separate control or tone channels require complex transceivers. Also, it is difficult to ensure that control or tone channels do not interfere with data channels in practical;

- **Method of idle listening:** Most of the surveyed protocols support omnidirectional listening for idle nodes. However, as observed from Table 1, some protocols also support directional circumrotation, per-DA, sector monitoring and multiple receptions over angular sector. Per-DA and sector monitoring methods can be used to emulate omnidirectional idle listening by using directional antennas. Directional circumrotation method is complex to implement and consumes higher energy.

Multiple receptions over angular sectors can increase control overhead;

- **Method of channel access:** Channel access mechanisms can be classified into two broad categories of random access mechanism and synchronized access mechanism. Synchronized access mechanisms communicate based on time slots and thus ensure deafness mitigation. However, this mechanism requires local or global time synchronization among participating nodes which is difficult to achieve in wireless networks. Due to this, most of the protocols prefer to use random access mechanism for channel access.

FUTURE RESEARCH DIRECTIONS

From available literature it is observed that though numerous protocols have been proposed to deal with the problem of deafness in wireless network, many important issues still need to be addressed. In this section, we have discussed and investigated issues which lay foundation for possible future work.

From the literature survey it is observed that though there are many analytical models available for IEEE 802.11 DCF MAC, which uses omnidirectional antenna, but only few are available for directional MAC. The available ones have made simplistic assumptions about antenna radiation pattern and channel parameters. Most of the channel models have ignored challenges such as deafness which are specific to directional antennas. Therefore, it is required to carry out further research to develop analytical models for MAC protocols designed for wireless networks that support multi-hop communication using directional antennas.

Most of the protocols have carried out performance evaluations through simulations. Though simulations are an integral part of networking research, it is important that these protocols are

implemented in hardware and tested in real-time scenarios.

Many protocols available in the literature have proposed to use tones to defer deafness due to the use of directional antenna. Apart from needing separate channel and extra bandwidth, nodes require additional hardware, which increases cost and complexity. Many proposed protocols have assumed that there is no interference between the tone and data signal. However, in practical that may not be the case. Also, transmitting tones can lead to increased power consumption that negatively affects energy efficiency in wireless networks. This requires efficient hardware design which apart from minimizing possible interference between tone and data signal, ensures energy efficient communication. There is tremendous scope of research in this area. Another issue with tone based protocols is difficulty in tone detection in presence of fading and multipath environments. Because tone durations are short, and they are transmitted over narrow bandwidth channels, fading and multipath can lead to difficulty in detection of tones in wireless networks. Researchers also need to work towards proposing ways to ensure tone detection and overcome the effect of fading and multipath environments.

Many authors have proposed synchronized access protocols to handle deafness while using directional antennas. However, it is difficult to achieve synchronization among nodes in a wireless network due to wireless channel characteristics. Apart from that, in sensor networks where energy efficiency is given high priority, nodes may support transition between sleep and wakeup modes. This requires design of efficient synchronization and scheduling algorithms in wireless networks. This field requires further research.

Most of the proposed protocols have provided solutions to deal with MAC layer challenges while using directional antennas, but it is unclear how proposed MAC layer solutions support routing mechanism in multi-hop wireless networks. This leads to disconnect between proposed MAC layer solutions and routing mechanisms. Further work is required towards joint design of physical layer characteristics, scheduling, MAC and routing to provide an efficient and complete solution. Researchers should also explore the possibility of cross-layer designs in this domain.

Multi Input Multi Output (MIMO) has gained tremendous popularity in wireless networks (Hu, & Zhang, 2004; Sundaresan, Sivakumar, Ingram, & Chang, 2004). However, in literature survey, we did not come across protocols that have explored design of MAC layer based on MIMO with directional transmissions. There is tremendous scope of research in design and development of MAC protocols that support the use of MIMO and antenna beamforming for efficient network performance. While designing MIMO and directional antenna based researcher must ensure that proposed designs efficiently handle deafness, hidden nodes and other MAC related problems.

Many authors have proposed multichannel communication while using directional antenna. These solutions can either use separate channels for exchanging control and data information, or can have simultaneous transmissions over multiple channels. Such designs lead to another type of deafness wherein a node tries to communicate with another node which is tuned to some other channel while communicating with a third node. Solving deafness due to multichannel communication is a different research field and many researchers are working to ensure fair and efficient network access in multichannel wireless networks. Energy efficiency is another concern which such protocols need to work upon.

Another important point that we noticed during literature survey is that though many protocols have been suggested to use directional antennas and handle deafness, support for mobility has been ignored in many of the proposed designs. Other than some static mesh and sensor networks, most of the wireless network designs support

mobility. One of the principal features of wireless networks is support for node mobility due to which wireless devices have gained tremendous popularity among users. Therefore, solutions that are not designed considering node mobility can be termed incomplete. It is required to carry out further research in this area.

CONCLUSION

In this chapter, we have presented a survey of solutions for deafness avoidance in wireless ad-hoc, mesh and sensor networks that use directional antenna for communication. We analyzed benefits and challenges of using directional antennas in wireless ad-hoc, mesh and sensor networks individually. Deafness being the focus of this chapter, it is discussed how deafness affects the performance of wireless ad-hoc, mesh and sensor networks. In this chapter, around 50 solutions or protocols to combat deafness are surveyed from available literature, analyzed and compared based on their characteristics. It is observed that in all the protocols, data and acknowledgement frames are transmitted directionally to exploit the benefits of directional antennas. However, RTS and CTS frames can be transmitted omnidirectionally, unidirectionally or multidirectionally. It is observed that most of the surveyed protocols use switched beam antennas due to their simplicity. Also, random access mechanism is the most common method of medium access. The features of surveyed protocols are analyzed and their merits and demerits are examined. Future research directions are provided to help the readers who are interested in carrying out their research in the field of using directional antennas in wireless ad-hoc, mesh and sensor networks. We firmly believe that this chapter will benefit the readers through extensive and in-depth survey of available solutions to mitigate deafness and their analysis.

REFERENCES

Abdullah, A. A., Cai, L., & Gebali, F. (2012). DSD-MAC: Dual sensing directional MAC protocol for ad hoc networks with directional antennas. *IEEE Transactions on Vehicular Technology, 61*(3), 1266–1275. doi:10.1109/TVT.2012.2187082

Akyildiz, I. F., Wang, X., & Wang, W. (2005). Wireless mesh networks: A survey. *Computer Networks, 47*(4), 445–487. doi:10.1016/j.comnet.2004.12.001

Arora, A., Krunz, M., & Muqattash, A. (2004). Directional medium access protocol (DMAP) with power control for wireless ad hoc networks. In *Proceedings of Global Telecommunications Conference, 2004*. IEEE.

Balanis, C. A. (2012). *Antenna theory: analysis and design*. Wiley-Interscience.

Bazan, O., & Jaseemuddin, M. (2011). On the design of opportunistic MAC protocols for multihop wireless, networks with beamforming antennas. *IEEE Transactions on Mobile Computing, 10*(3), 305–319. doi:10.1109/TMC.2010.68

Ben-Othman, J., Mokdad, L., & Cheikh, M. O. (2011). A new architecture of wireless mesh networks based IEEE 802.11 s directional antennas. In *Proceedings of Communications (ICC)*. IEEE.

Bencini, L., Collodi, G., Di Palma, D., Manes, G., & Manes, A. (2010). An energy efficient cross layer solution based on smart antennas for wireless sensor network applications. In *Proceedings of Sensor Technologies and Applications (SENSORCOMM)*. IEEE. doi:10.1109/SENSORCOMM.2010.43

Boudour, G., Lecointre, A., Berthou, P., Dragomirescu, D., & Gayraud, T. (2007). On designing sensor networks with smart antennas. In *Proceedings of 7th IFAC International Conference on Fieldbuses and Networks in Industrial and Embedded Systems*. IFAC.

Budhwani, S., Sarkar, M., & Nagaraj, S. (2010). A MAC layer protocol for sensor networks using directional antennas. In *Proceedings of Sensor Networks, Ubiquitous, and Trustworthy Computing (SUTC)*. IEEE. doi:10.1109/SUTC.2010.11

Capone, A., Martignon, F., & Fratta, L. (2008). Directional MAC and routing schemes for power controlled wireless mesh networks with adaptive antennas. *Ad Hoc Networks, 6*(6), 936–952. doi:10.1016/j.adhoc.2007.08.002

Chang, J. J., Liao, W., & Hou, T. C. (2009). Reservation-based directional medium access control (RDMAC) protocol for multi-hop wireless networks with directional antennas. In *Proceedings of Communications, 2009*. IEEE. doi:10.1109/ICC.2009.5199410

Chin, K. W. (2007). SpotMAC: A pencil-beam MAC for wireless mesh networks. In *Proceedings of Computer Communications and Networks, 2007*. IEEE. doi:10.1109/ICCCN.2007.4317801

Choudhury, R. R., & Vaidya, N. H. (2004). Deafness: A MAC problem in ad hoc networks when using directional antennas. In *Proceedings of Network Protocols, 2004*. IEEE. doi:10.1109/ICNP.2004.1348118

Choudhury, R. R., & Vaidya, N. H. (2004). MAC-layer anycasting in ad hoc networks. *ACM SIGCOMM Computer Communication Review, 34*(1), 75–80. doi:10.1145/972374.972388

Choudhury, R. R., & Vaidya, N. H. (2005). Performance of ad hoc routing using directional antennas. *Ad Hoc Networks, 3*(2), 157–173. doi:10.1016/j.adhoc.2004.07.004

Collett, M., Loh, T., Liu, H., & Qin, F. (n.d.). Smart antennas and wireless sensor networks - Achieving efficient, reliable communication. *National Physical Laboratory*.

Crow, B. P., Widjaja, I., Kim, L. G., & Sakai, P. T. (1997). IEEE 802.11 wireless local area networks. *IEEE Communications Magazine, 35*(9), 116–126. doi:10.1109/35.620533

Dai, H. N., Ng, K. W., & Wu, M. Y. (2007). A busy-tone based MAC scheme for wireless ad hoc networks using directional antennas. In *Proceedings of Global Telecommunications Conference, 2007*. IEEE.

Ding, Y., Wang, X., Shen, H., & Guo, P. (2010). Media access control protocol based on hybrid antennas in ad hoc networks. In *Proceedings of Ubiquitous Intelligence & Computing and 7th International Conference on Autonomic & Trusted Computing (UIC/ATC)*. IEEE.

ElBatt, T., Anderson, T., & Ryu, B. (2003). Performance evaluation of multiple access protocols for ad hoc networks using directional antennas. In *Proceedings of Wireless Communications and Networking, 2003*. IEEE. doi:10.1109/WCNC.2003.1200505

Fakih, K., Diouris, J. F., & Andrieux, G. (2006). BMAC: Beamformed MAC protocol with channel tracker in MANET using smart antennas. In *Proceedings of Wireless Technology, 2006*. IEEE. doi:10.1109/ECWT.2006.280466

Gelal, E., Jakllari, G., Krishnamurthy, S. V., & Young, N. E. (2006). An integrated scheme for fully-directional neighbor discovery and topology management in mobile ad hoc networks. In *Proceedings of Mobile Adhoc and Sensor Systems (MASS)*. IEEE. doi:10.1109/MOBHOC.2006.278551

Giorgetti, G., Cidronali, A., Gupta, S. K., & Manes, G. (2007). Exploiting low-cost directional antennas in 2.4 GHz IEEE 802.15 4 wireless sensor networks. In *Proceedings of Wireless Technologies, 2007 European Conference* (pp. 217-220). IEEE.

Gossain, H., Cordeiro, C., & Agrawal, D. P. (2005). MDA: An efficient directional MAC scheme for wireless ad hoc networks. In *Proceedings of Global Telecommunications Conference, 2005*. IEEE.

Gossain, H., Cordeiro, C., Cavalcanti, D., & Agrawal, D. P. (2004). The deafness problems and solutions in wireless ad hoc networks using directional antennas. In *Proceedings of Global Telecommunications Conference Workshops, 2004*. IEEE.

Haas, Z. J., Deng, J., Liang, B., Papadimitratos, P., & Sajama, S. (2002). *Wireless ad hoc networks*. Hoboken, NJ: John Wiley & Sons, Inc.

Hadjadj-Aoul, Y., & Naït-Abdesselam, F. (2008). On physical-aware directional MAC protocol for indoor wireless networks. In *Proceedings of Global Telecommunications Conference, 2008*. IEEE.

Hossain, E., & Leung, K. (Eds.). (2007). Wireless mesh networks: Architectures and protocols. Springer Science+ Business Media, LLC.

Hu, L., & Evans, D. (2004). Using directional antennas to prevent wormhole attacks. In *Proceedings of Network and Distributed System Security Symposium (NDSS)*. NDSS.

Hu, M., & Zhang, J. (2004). MIMO ad hoc networks: Medium access control, saturation throughput, and optimal hop distance. *Journal of Communications and Networks, 6*(4), 317–330.

Huang, Z., Shen, C. C., Srisathapornphat, C., & Jaikaeo, C. (2002). A busy-tone based directional MAC protocol for ad hoc networks. In *Proceedings of MILCOM 2002* (Vol. 2, pp. 1233-1238). IEEE.

Jain, V., Gupta, A., & Agrawal, D. P. (2008). On-demand medium access in multihop wireless networks with multiple beam smart antennas. *IEEE Transactions on Parallel and Distributed Systems, 19*(4), 489–502. doi:10.1109/TPDS.2007.70739

Jakllari, G., Broustis, I., Korakis, T., Krishnamurthy, S. V., & Tassiulas, L. (2005). Handling asymmetry in gain in directional antenna equipped ad hoc networks. In Proceedings of Personal, Indoor and Mobile Radio Communications, 2005 (Vol. 2, pp. 1284-1288). IEEE.

Jakllari, G., Luo, W., & Krishnamurthy, S. V. (2007). An integrated neighbor discovery and MAC protocol for ad hoc networks using directional antennas. *IEEE Transactions on Wireless Communications*, 6(3), 1114–1024. doi:10.1109/TWC.2007.05471

Jung, S., Lee, S., & Han, K. (2009). A DMAC protocol to improve spatial reuse by managing the NAV table of the nodes in VANET. In *Proceedings of Computer and Electrical Engineering, 2009* (Vol. 1, pp. 387–390). IEEE. doi:10.1109/ICCEE.2009.167

Karapistoli, E., Gragopoulos, I., Tsetsinas, I., & Pavlidou, F. N. (2009). A directional MAC protocol with deafness avoidance for UWB wireless sensor networks. [IEEE.]. *Proceedings of Communications, 2009*, 1–5.

Karveli, T., Voulgaris, K., Ghavami, M., & Aghvami, A. H. (2008). A collision-free scheduling scheme for sensor networks arranged in linear topologies and using directional antennas. [IEEE.]. *Proceedings of Sensor Technologies and Applications, 2008*, 18–22.

Ko, Y. B., Shankarkumar, V., & Vaidya, N. H. (2000). Medium access control protocols using directional antennas in ad hoc networks. In Proceedings of *Nineteenth Annual Joint Conference of the IEEE Computer and Communications Societies* (Vol. 1, pp. 13-21). IEEE.

Korakis, T., Jakllari, G., & Tassiulas, L. (2003). A MAC protocol for full exploitation of directional antennas in ad-hoc wireless networks. In *Proceedings of the 4th ACM International Symposium on Mobile Ad Hoc Networking & Computing* (pp. 98-107). ACM.

Korakis, T., Jakllari, G., & Tassiulas, L. (2008). CDR-MAC: A protocol for full exploitation of directional antennas in ad hoc wireless networks. *IEEE Transactions on Mobile Computing*, 7(2), 145–155. doi:10.1109/TMC.2007.70705

Kumar, U., Gupta, H., & Das, S. R. (2006). A topology control approach to using directional antennas in wireless mesh networks. In *Proceedings of Communications, 2006* (Vol. 9, pp. 4083–4088). IEEE. doi:10.1109/ICC.2006.255720

Lee, E. J., Han, C. N. K. D. H., & Jwa, D. Y. Y. J. W. (n.d.). *A tone dual-channel DMAC protocol in location unaware ad hoc networks*. Academic Press.

Lewis, F. L. (2004). Wireless sensor networks. In *Smart environments: Technologies, protocols, and applications*. Academic Press.

Li, G., Yang, L. L., Conner, W. S., & Sadeghi, B. (2005). Opportunities and challenges for mesh networks using directional antennas. In *Proceedings of WiMESH'05*. WiMESH.

Li, P., Zhai, H., & Fang, Y. (2009). SDMAC: Selectively directional MAC protocol for wireless mobile ad hoc networks. *Wireless Networks*, 15(6), 805–820. doi:10.1007/s11276-007-0076-z

Li, Y., Li, M., Shu, W., & Wu, M. Y. (2007). FFT-DMAC: A tone based MAC protocol with directional antennas. In *Proceedings of Global Telecommunications Conference, 2007* (pp. 3661-3665). IEEE.

Li, Y., & Safwat, A. M. (2005). Efficient deafness avoidance in wireless ad hoc and sensor networks with directional antennas. In *Proceedings of the 2nd ACM International Workshop on Performance Evaluation of Wireless Ad Hoc, Sensor, and Ubiquitous Networks* (pp. 175-180). ACM.

Li, Y., & Safwat, A. M. (2006). DMAC-DACA: Enabling efficient medium access for wireless ad hoc networks with directional antennas. In *Proceedings of Wireless Pervasive Computing*. IEEE.

Lu, H., Li, J., Dong, Z., & Ji, Y. (2011). CRDMAC: An effective circular RTR directional MAC protocol for wireless ad hoc networks. In Proceedings of Mobile Ad-hoc and Sensor Networks (MSN), (pp. 231-237). IEEE.

Martignon, F. (2011). Multi-channel power-controlled directional MAC for wireless mesh networks. *Wireless Communications and Mobile Computing, 11*(1), 90–107. doi:10.1002/wcm.917

Munari, A., Rossetto, F., & Zorzi, M. (2008). Cooperative cross layer MAC protocols for directional antenna ad hoc networks. *ACM SIGMOBILE Mobile Computing and Communications Review, 12*(2), 12–30. doi:10.1145/1394555.1394558

Nilsson, M. (2009). Directional antennas for wireless sensor networks. In *Proceedings of the 9th Scandinavian Workshop on Wireless Adhoc Networks (Adhoc'09)*. Adhoc.

Omar, E. A., & Elsayed, K. M. F. (2010). Directional antenna with busy tone for capacity boosting and energy savings in wireless ad-hoc networks. In Proceedings of High-Capacity Optical Networks and Enabling Technologies (HONET), (pp. 91-95). IEEE.

Pottie, G. J. (1998). Wireless sensor networks. In *Proceedings of Information Theory Workshop, 1998* (pp. 139-140). IEEE.

Prabhu, A., & Das, S. R. (2007). Addressing deafness and hidden terminal problem in directional antenna based wireless multi-hop networks. In *Proceedings of Communication Systems Software and Middleware, 2007*. IEEE.

Ramanathan, R. (2001). On the performance of ad hoc networks with beamforming antennas. In *Proceedings of the 2nd ACM International Symposium on Mobile Ad Hoc Networking & Computing* (pp. 95-105). ACM.

Sekido, M., Takata, M., Bandai, M., & Watanabe, T. (2005). A directional hidden terminal problem in ad hoc network MAC protocols with smart antennas and its solutions. In *Proceedings of Global Telecommunications Conference*. IEEE.

Shihab, E., Cai, L., & Pan, J. (2009). A distributed asynchronous directional-to-directional MAC protocol for wireless ad hoc networks. *IEEE Transactions on Vehicular Technology, 58*(9), 5124–5134. doi:10.1109/TVT.2009.2024085

Singh, S., Mudumbai, R., & Madhow, U. (2010). Distributed coordination with deaf neighbors: Efficient medium access for 60 GHz mesh networks. In *Proceedings of INFOCOM, 2010*. IEEE.

Spyropoulos, A., & Raghavendra, C. S. (2002). Energy efficient communications in ad hoc networks using directional antennas. In *Proceedings of INFOCOM 2002: Twenty-First Annual Joint Conference of the IEEE Computer and Communications Societies* (Vol. 1, pp. 220-228). IEEE.

Stine, J. A. (2006). Exploiting smart antennas in wireless mesh networks using contention access. *IEEE Wireless Communications, 13*(2), 38–49. doi:10.1109/MWC.2006.1632479

Stutzman, W. L., & Thiele, G. A. (2012). *Antenna theory and design*. Hoboken, NJ: Wiley.

Subramanian, A. P., & Das, S. R. (2005). Using directional antennas in multi-hop wireless networks: Deafness and directional hidden terminal problems. In *Proceedings of the 13th International Conference on Network Protocols (ICNP'05)*. ICNP.

Subramanian, A. P., & Das, S. R. (2005). Using directional antennas in multi-hop wireless networks: Deafness and directional hidden terminal problems. In *Proceedings of the 13th International Conference on Network Protocols (ICNP'05)*. ICNP.

Sundaresan, K., Sivakumar, R., Ingram, M. A., & Chang, T. Y. (2004). Medium access control in ad hoc networks with MIMO links: Optimization considerations and algorithms. *IEEE Transactions on Mobile Computing*, *3*(4), 350–365. doi:10.1109/TMC.2004.42

Swaminathan, A., Noneaker, D. L., & Russell, H. B. (2012). The design of a channel-access protocol for a wireless ad hoc network with sectored directional antennas. *Ad Hoc Networks*, *10*(3), 284–298. doi:10.1016/j.adhoc.2011.06.011

Takata, M., Bandai, M., & Watanabe, T. (2006). A receiver-initiated directional MAC protocol for handling deafness in ad hoc networks. In *Proceedings of Communications, 2006* (Vol. 9, pp. 4089–4095). IEEE. doi:10.1109/ICC.2006.255721

Takata, M., Bandai, M., & Watanabe, T. (2007). A MAC protocol with directional antennas for deafness avoidance in ad hoc networks. In *Proceedings of Global Telecommunications Conference, 2007* (pp. 620-625). IEEE.

Tsoulos, G. V. (1999). Smart antennas for mobile communication systems: Benefits and challenges. *Electronics & Communication Engineering Journal*, *11*(2), 84–94. doi:10.1049/ecej:19990204

Vasudevan, S., Kurose, J., & Towsley, D. (2005). On neighbor discovery in wireless networks with directional antennas. In *Proceedings of INFOCOM 2005: 24th Annual Joint Conference of the IEEE Computer and Communications Societies* (Vol. 4, pp. 2502-2512). IEEE.

Wang, J., Fang, Y., & Wu, D. (2005). SYN-DMAC: A directional MAC protocol for ad hoc networks with synchronization. In *Proceedings of Military Communications Conference, 2005* (pp. 2258-2263). IEEE.

Wang, J., Zhai, H., Li, P., Fang, Y., & Wu, D. (2009). Directional medium access control for ad hoc networks. *Wireless Networks*, *15*(8), 1059–1073. doi:10.1007/s11276-008-0102-9

Wang, Y., & Garcia-Luna-Aceves, J. J. (2002). Spatial reuse and collision avoidance in ad hoc networks with directional antennas. In *Proceedings of Global Telecommunications Conference, 2002* (Vol. 1, pp. 112-116). IEEE.

Winters, J. H. (2006). Smart antenna techniques and their application to wireless ad hoc networks. *IEEE Wireless Communications*, *13*(4), 77–83. doi:10.1109/MWC.2006.1678168

Wu, Y., Zhang, L., Wu, Y., & Niu, Z. (2006). Interest dissemination with directional antennas for wireless sensor networks with mobile sinks. In *Proceedings of the 4th International Conference on Embedded Networked Sensor Systems* (pp. 99-111). ACM.

Yi, S., Pei, Y., & Kalyanaraman, S. (2003). On the capacity improvement of ad hoc wireless networks using directional antennas. In *Proceedings of the 4th ACM International Symposium on Mobile Ad Hoc Networking & Computing* (pp. 108-116). ACM.

Zhu, C., Nadeem, T., & Agre, J. R. (2006). Enhancing 802.11 wireless networks with directional antenna and multiple receivers. In *Proceedings of Military Communications Conference, 2006*. IEEE.

ADDITIONAL READING

Ash, J., & Potter, L. (2004, September). Sensor network localization via received signal strength measurements with directional antennas. In *Proceedings of the 2004 Allerton Conference on Communication, Control, and Computing* (pp. 1861-1870).

Cho, J., Lee, J., Kwon, T., & Choi, Y. (2006, April). Directional antenna at sink (DAaS) to prolong network lifetime in wireless sensor networks. In *Wireless Conference 2006-Enabling Technologies for Wireless Multimedia Communications (European Wireless), 12th European* (pp. 1-5). VDE.

Das, S. M., Pucha, H., Koutsonikolas, D., Hu, Y. C., & Peroulis, D. (2006). DMesh: incorporating practical directional antennas in multichannel wireless mesh networks. *Selected Areas in Communications. IEEE Journal on*, *24*(11), 2028–2039.

Hossain, E., & Leung, K. K. (Eds.). (2008). *Wireless mesh networks: architectures and protocols*. Springer.

Lee, K. F. (1984). *Principles of antenna theory* (p. 1). *Chichester, Sussex, England and New York*: John Wiley and Sons.

Mohapatra, P., & Krishnamurthy, S. (Eds.). (2005). *AD HOC NETWORKS: technologies and protocols*. Springer. doi:10.1007/b99485

Murthy, C. S. R., & Manoj, B. S. (2012). *Ad hoc wireless networks architectures and protocols*. *Dorling Kindersley*. India: Pvt. Limited.

Pahlavan, K. (2011). *Principles of wireless networks: A unified approach*. John Wiley & Sons, Inc.

Shorey, R., Ananda, A., Chan, M. C., & Ooi, W. T. (2006). Mobile, wireless, and sensor networks: technology, applications, and future directions. John Wiley & Sons.

Viani, F., Lizzi, L., Donelli, M., Pregnolato, D., Oliveri, G., & Massa, A. (2010). Exploitation of parasitic smart antennas in wireless sensor networks. *Journal of Electromagnetic Waves and Applications*, *24*(7), 993–1003. doi:10.1163/156939310791285227

Zander, J., Kim, S. L., Almgren, M., & Queseth, O. (2001). *Radio resource management for wireless networks*. Artech House, Inc.

KEY TERMS AND DEFINITIONS

Directional Antenna: A directional antenna radiates its RF energy in a particular direction. This allows for increased communication range and reduced interference from other nodes in the vicinity.

Deafness: A phenomenon wherein a Node A fails to initiate communication with another Node B, because Node B is beamformed in a different direction.

Carrier Sense Multiple Access/Collision Avoidance (CSMA/CA): A medium access mechanism used in wireless networks. In this mechanism, the nodes try to avoid collision before they happen. For this, nodes do not transmit until the transmission channel is found to be idle.

Wireless Ad-Hoc Network: A network of communicating devices which does not depend on any pre-existing infrastructure such as access-point or base-station.

Wireless Sensor Network (WSN): WSNs are composed of spatially distributed sensor nodes which are used to sense/monitor physical or environmental conditions such as temperature, pressure, humidity etc.

Wireless Mesh Network: A network of wireless access points organized in mesh topology. These are self form and self heal networks, and provide reliable and robust communication.

Chapter 20
Metamaterial–Based Wearable Microstrip Patch Antennas

J. G. Joshi
Government Polytechnic, India

Shyam S. Pattnaik
National Institute of Technical Teachers Training and Research, India

ABSTRACT

This chapter presents metamaterial-based wearable, that is, textile-based, antennas for Wi-Fi, WLAN, ISM, BAN and public safety band applications, which have been designed, fabricated, and tested. Textile substrates like polyester and polypropylene are used to design and fabricate these antennas. The metamaterial inclusions are directly used to load the different microstrip patch antennas on the same substrate, which significantly enhances the gain and bandwidth with considerable size reduction. The microstrip patch antenna generates sub-wavelength resonances under loading condition due to the modifications of its resonant modes. The DNG and SNG metamaterials are used to load the microstrip patch antennas for size reduction by generating the sub-wavelength resonances. The simulated and measured results are found to be in good agreement for all the presented wearable antennas. The bending effect on antenna performance due to human body movements is also presented in this chapter.

1. INTRODUCTION

The handheld communication devices and *body centric communication* (on body and off body) systems are widely used in the day-to-day activities of human being. Some of the devices are used by the military, fire fighters and police personnel for the safety of human life and property, surveillance of sensitive locations like airport, seaport, shopping malls etc. These devices and systems are wearable computers, flexible mobile phones, personal digital assistant (PDA), smart phones, tablet computers, *public safety band* systems, bio-telemetry sensors, military and police equipments etc. These devices may be worn under, over, or in clothing, or may be the clothes of the user or wearer (Choi, 2010; Hall, 2006; Hertleer, 2009; Joshi, 2013, 2012, 2012, 2013; Sankaralingam 2010).

DOI: 10.4018/978-1-4666-5170-8.ch020

These systems need compact, high gain, light weight, and hidden antennas which should be an integral part of the wearer clothing. These systems uses different frequency bands under 3G, 4G and 5G technologies such as *Body Area Networks* (BAN), *Industrial, Scientific, and Medical* (ISM) band, WLAN, Wi-Fi, Wi-max, Bluetooth, HYPER LAN, public safety band etc. (Choi, 2010; Hall, 2006; Hertleer, 2009; Joshi, 2013, 2012, 2012, 2013; Sankaralingam, 2010). It is difficult to achieve a better tradeoff between the gain, bandwidth, and more prominently the size of microstrip patch antennas in spite of numerous advantages, hence these antennas are not suitable for wearable applications (Bahl, 1980; Garg, 2001; Joshi, 2011; Joshi, 2010; Joshi, 2011, Joshi 2012; Pozar, 1995). The cloth or textile based antennas which are the integral part of wearer clothing and used for the mentioned applications hence they are called as *wearable antennas*. These antennas can be applied for youngsters, the aged persons, physically challenged persons, sportsmen, monitoring the activities of animals, police and military personnel etc. (Choi, 2010; Hall, 2006; Hertleer, 2009; Joshi, 2013, 2012, 2012, 2013; Sankaralingam, 2010). The textile or cloth based wearable antenna communicates voice, data, or biotelemetry signals at high data rates.

The wearable antenna should have features like light weight, conformal, need to be hidden, and it should not affect the health of user. In practice, synthetic or natural materials are used as substrate to manufacture the textile or cloth based wearable antennas. These materials are cotton, liquid crystal polymer (LCP), fleece fabric, foam, Nomex, nylon, polyester, conducting ribbon, insulated wire, conducting paint, copper coated fabric, geo-textile etc. (Choi, 2010; Hall, 2006; Hertleer, 2009; Joshi, 2013, 2012, 2012, 2013; Sankaralingam, 2010).

Different wearable antennas have been reported in the literature. A study on necessity of wearable antennas for PAN, BAN and ISM band has been presented. (Hall, 2006). A flexible composite right left-handed transmission line flexible antenna fabricated on LCP that resonates at 2.3 GHz (Choi, 2012). This antenna is useful for flexible mounting surface applications because of its insensitiveness to bending. The microstrip patch antenna on conductive fabrics and insulating polyester is experimentally demonstrated by (Sankaralingam, 2010). This antenna is designed for wireless Broadband (WiBro) communication systems operating in the frequency range 2.3 GHz to 2.4 GHz. But, due to size of the patch there is possibility of deposition of electromagnetic signals in the body. A wearable microstrip patch antenna on fabric substrate with a U-slot having reconfigurable beam steering capability at 6 GHz has been reported (Sang-Jun Ha, 2011). An embroidered wearable multi-resonant folded dipole (MRFD) antenna for FM signal reception has been presented (Jung-SimRoh, 2010). A textile endfire antenna for wireless body area networks (BANs) operating at 60 GHzis fabricated on 0.2 mm thick fabric from a cotton shirt (Nacer Chahat, 2012).

The objective of this chapter is to present the metamaterial based wearable microstrip patch antennas for 802.11a WLAN, Wi-Fi, Wi-max and public safety band applications. The polyester cloth and geo-textile that is polypropylene substrates are used to fabricate these antennas. The authors have used different configurations of metamaterial split ring resonators to design and develop the wearable antennas. In this chapter, different metamaterial based wearable microstrip patch antennas designed, fabricated and tested by the authors are presented. The antennas presented in this chapter are simulated by using a method of moment based IE3D electromagnetic simulator.

2. METAMATERIAL AND ITS ROLE PERTAINING TO WEARABLE MICROSTRIP PATCH ANTENNAS

The microstrip patch antennas are widely used in different RF communication systems, devices, equipments and they have applications over wide frequency range. But these antennas have certain limitations as; narrow bandwidth, lower gain, lesser flexibility in the design, limited scope to optimize the design, large size and extraneous radiation from the feeds and junctions (Bahl, 1980; Garg, 2001; Joshi, 2011; Joshi, 2010; Joshi, 2011, Joshi, 2012; Pozar, 1995). Metamaterial is used to overcome these limitations and to improve the performance of microstrip patch antennas (Chen, 2010; Joshi, 2010, 2011, 2012, 2013).

The basic properties of conventional materials available in the nature are *positive permeability* and *permittivity* hence they are called as *Double Positive* (DPS) materials. In 1968, Russian physicist Viktor Veselago theoretically predicted that metamaterial possesses negative values of magnetic permeability (μ) and/or electric permittivity (ε) (Veselago, 1968). Metamaterial possesses both negative permeability (μ) and/or negative permittivity (ε). Hence, they are termed as *Double Negative* (DNG) or *Single Negative* (SNG) metamaterials. In SNG metamaterials either single of the property i.e. permeability or permittivity is negative. The permeability (μ) negative materials are called *Mu Negative* (MNG) and permittivity (ε) negative materials are termed as *Epsilon Negative* (ENG) materials. These materials have *negative refractive index* that is a reversal of Snell's law hence, called as NIM. Due to negative refractive index, the group and phase velocities of electromagnetic wave appear in opposite direction such that the direction of propagation is reversed with respect to the energy flow direction (Veselago, 1968; Pendry, 1999; Smith, 2005; Ziolkowski, 2003). It also exhibits the reversal of Doppler's shift. Metamaterial structure consists of *Split Ring Resonators* (SRRs) to produce negative permeability and thin wire elements to generate negative permittivity. Sometimes, metamaterials are also called as *backward-wave media* or Veselago media. These interesting properties of metamaterial plays prominent role in designing antenna structure's to enhance gain, directivity, bandwidth while satisfying the condition of miniaturization. Metamaterial characteristics of different SRR structures such as circular, square, rectangular, omega (Ω), split

squared ring resonators (SSRR), multiple split ring resonator (MSRR), spiral resonator (SR), labyrinth resonator (LR) etc. have been studied and verified by the different research groups. The pioneer work of design, analysis and synthesis of Edge-Coupled Split Ring Resonator (EC-SRR), Broad-side Coupled Split Ring Resonator (BC-SRR) etc. is also thoroughly studied (Veselago, 1968; Pendry, 1999; Smith, 2005; Ziolkowski, 2003; Bilotti, 2007; Marques, 2003). Metamaterials are used in different microwave applications like antennas, antenna arrays, waveguides, filters, directional couplers, phase shifters, transducers, etc.

2.1 Role of Metamaterial in Wearable Microstrip Patch Antennas

Metamaterial inclusions (SRRs) are used to load the microstrip patch antennas either by partially filling it beneath the substrate of patch or placing it as superstrate on the top of the patch (Chen, 2010; Joshi, 2010, 2011, 2012, 2013). These inclusions may be used directly to load the patch antenna on the same *substrate*. These techniques significantly enhance the gain, bandwidth, directivity of the microstrip patch antennas with considerable size reduction. Under loading condition, the microstrip patch antenna generates *sub-wavelength resonances* due to the modifications of the *resonant modes*. The DNG and SNG metamaterial is used to load the microstrip patch antennas for size reduction.

Looking into the scope and requirements of the wearable antennas it is essential to reduce their size for easy integration in the wearer clothing. The size reduction is helpful to (a) minimize the deposition of *electromagnetic field* in the human (wearer) body; (b) reduce the *bending effect* on antenna performance due to human body movements. Authors have designed and fabricated the metamaterial based wearable microstrip patch antennas to reduce the size and to achieve the mentioned features necessary for wearable microstrip patch antennas. Most recently, antenna researchers have verified and evidenced an innovative approach to overcome the limitations of microstrip patch antennas by using metamaterial. The metamaterial not only helps to reduce the size of the microstrip patch antennas but also to enhance the gain and bandwidth of these antennas. Different wearable antennas have been developed by the authors for Wi-Fi, Wi-Max, WLAN and public safety band applications are presented in this chapter.

3. METAMATERIAL EMBEDDED WEARABLE RECTANGULAR MICROSTRIP PATCH ANTENNA

3.1 Preamble of this Antenna

This section presents an indigenous low cost metamaterial embedded wearable rectangular microstrip patch antenna using *polyester* substrate for IEEE 802.11a *WLAN* applications. The proposed antenna resonates at 5.10 GHz with a bandwidth and gain of 97 MHz and 4.92 dBi respectively. The electrical size of this antenna is 0.254 λ × 0.5 λ. The slots are cut in a rectangular patch to reduce the bending effect. But it leads to mismatch the impedance at WLAN frequency band hence, a metamaterial square SRR is embedded inside the slot. A prototype antenna has been fabricated, tested and the measured results are presented in this section. The simulated and measured results of the proposed antenna are found to be in good agreement. The bending effect on the performance of this antenna is experimentally verified.

3.2 Antenna Design

This is a polyester substrate based wearable antenna designed for IEEE 802.11a WLAN applications. Figure 1(a) depicts the step-by-step design procedure of the proposed wearable antenna. The antenna design is divided into three steps: (a) design and simulations of rectangular microstrip patch antenna, (b) making rectangular and square

slots in the rectangular patch to excite the desired lower resonant frequency for size reduction and (c) embedding the designed metamaterial square SRR inside the square slot for better impedance matching at WLAN frequency band. In simulations, when the SRR is embedded inside the square slot of the patch, better matching is noticed at the resonance frequency 5.10 GHz. In simulations, it is observed that a small difference in the placement of square SRR shifts the resonance frequency with considerable changes in the matching conditions. Finally, the SRR is placed inside the square slot at the distance of d = 1 mm as shown in Figure 1(b). The square SRR is magnetically coupled with the slotted rectangular patch to form a *LC resonator* that resonates at 5.10 GHz by making the antenna compact. Figure 1(b) depicts the sketch and geometrical structure of metamaterial square SRR embedded wearable rectangular microstrip patch antenna.

Figure 2(a) and Figure 2(b) respectively depict the photographs of radiating patch and ground plane with SMA connector of the fabricated metamaterial SRR embedded wearable rectangular microstrip patch antenna. Initially, the rectangular microstrip patch antenna of length (L_r) is designed for resonant frequency 8.65 GHz using Equation 1 (Bahl, 1980; Garg, 2001; Joshi, 2012; Pozar, 1995):

$$L_r = \frac{c}{f_r \sqrt{\varepsilon_e}} \tag{1}$$

where, f_r is resonant frequency, c is velocity of light (3 × 10⁸ m/sec), ε_r is relative permittivity (8.854 × 10⁻¹² F/m), ε_e is effective dielectric constant of the substrate which is calculated using Equation 2 (Bahl, 1980; Garg, 2001; Joshi, 2012; Pozar, 1995):

$$\varepsilon_e = \frac{(\varepsilon_r + 1)}{2} + \frac{(\varepsilon_r - 1)}{2} \left(1 + \frac{12h}{W}\right)^{-1/2} \tag{2}$$

Figure 1. (a) Step-by-step design procedure; (b) Sketch and geometrical structure of metamaterial square SRR embedded wearable rectangular microstrip patch antenna

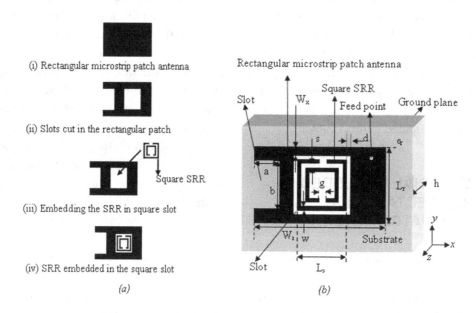

Figure 2. Photographs of the fabricated metamaterial square SRR embedded wearable rectangular microstrip patch antenna: (a) Radiating patch; (b) Ground plane

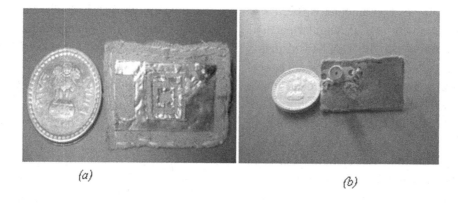

The dimensions of rectangular microstrip patch are; length $L_r = 15$ mm and width $W_r = 30$ mm. The slots are cut in the rectangular microstrip patch to reduce the resonant frequency to WLAN applications as well as to reduce the metal area of patch. Initially, a rectangular slot of dimensions a = 6 mm and b = 8 mm is cut inside the radiating edge of the rectangular microstrip patch but no better impedance matching is obtained. Hence, at the centre of rectangular patch a square slot of

dimensions 10 mm × 10 mm is created. Again the better matching could not be obtained in the WLAN frequency band. Further, to obtain the better impedance matching and to achieve a sub-wavelength resonance a metamaterial square SRR is embedded inside the square slot to load this antenna. The geometrical dimensions of square SRR as shown in Figure 1(b) are; width of split rings (w), separation between inner and outer split rings (s), and gap at the splits of rings

(g) are set to w = s = g = 1 mm respectively. The length of outer square split ring (L$_s$) is 9 mm. The distance between outer square SRR and the edge of cut on the rectangular microstrip patch is d = 1 mm. Equations 1 and 2 are used to design the rectangular microstrip patch antenna at resonance frequency 8.65 GHz without (i) making slots in the radiating patch and (ii) embedding the SRR inside the square slot. The designed antenna resonates at 8.48 GHz which is reduced to the working frequency 5.10 GHz by making the slots and embedding the SRR inside the slot. Due to slots the resonant length of the designed rectangular patch is changed and poor matching is observed during the simulations at the working frequency. The inductance and capacitance of SRR with the mutual induction between the antenna and SRR provides better matching at the working frequency 5.10 GHz.

The *aspect ratio* of rectangular microstrip patch that is length (L$_r$) to width (W$_r$) ratio is 0.5. Similarly, the aspect ratio of square slot is 1. The aspect ratio of rectangular slot that is length 'a' to width 'b' is set to 0.75 which is to one half of the sum of aspect ratios of rectangular patch and the slot. The antenna is co-axially fed by a 50 Ω SMA connector at $x = 11$ mm and $y = 7$ mm. The polyester cloth substrate of thickness h = 1 mm, relative permittivity $\varepsilon_r = 1.39$, and loss tangent tan $\delta = 0.01$ is used to design and fabricate the proposed antenna. The substrate of desired thickness is prepared by cutting and sewing the polyester cloth. According to the designed dimensions and shapes the radiating patch, square SRR, and ground plane of the antenna are cut from the self adhesive copper tape of thickness 0.1 mm and tightly adhered on the prepared substrate. The size of this antenna at resonance frequency 5.10 GHz is 0.254 $\lambda \times$ 0.5 λ.

Advantages of this antenna are; (a) in this antenna design the slots are cut in the rectangular microstrip patch to make the antenna compact due to which the *metal portion* of the radiating patch has been removed. Thus, as compared to the conventional rectangular microstrip patch antenna (without slots),. small portion of the proposed antenna (with slots) gets bent due to the body movements. Hence, the unwanted bending effects on the resonance frequency and impedance matching (S$_{11}$) of the antenna have been reduced. (b) This type of geometry is useful to reduce the deposition of electromagnetic field that is SAR due to the fringing field entering in the body tissues. (c) Lower resonance frequency has been achieved.

3.3 Results and Discussion

Initially, the metamaterial characteristics of the square SRR are verified and presented before analyzing its loading effect on the proposed rectangular microstrip patch antenna. Figure 3 shows the reflection (S$_{11}$) and transmission (S$_{21}$) coefficient characteristics of the square SRR that resonates at 8.48 GHz. The *effective medium theory* is used to verify the permeability (μ_r) and permittivity (ε_r) from the reflection and transmission coefficients (S-parameters). The *Nicolson-Ross-Weir* (NRW) approach is used to obtain these effective medium parameters (Joshi, 2013, 2012, 2012, 2013; Pendry, 1999; Smith, 2005; Ziolkowski, 2003). The expressions of Equations 3 and 4 are used to determine the effective parameters. The metamaterial characteristics of the SRR are verified using the *S*-parameters obtained from IE3D electromagnetic simulator and MATLAB code with mathematical Equations 3 and 4:

$$\mu_r = \frac{2}{jk_0 h} \frac{1 - V_2}{1 + V_2} \tag{3}$$

$$\varepsilon_r = \frac{2}{jk_0 h} \frac{1 - V_1}{1 + V_1} \tag{4}$$

where k_0 is wave number, h is substrate thickness, V_1 and V_2 are composite terms to represent addition and subtraction of *S*-parameters. Values of

Figure 3. Reflection (S_{11}) and Transmission (S_{21}) coefficient characteristics of square SRR

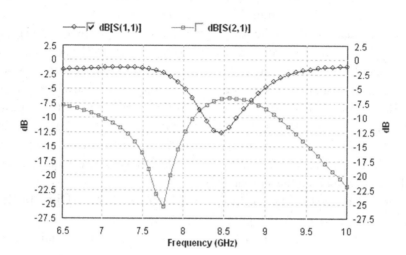

V_1 and V_2 are calculated as $V_1 = S_{21} + S_{11}$ and $V_2 = S_{21} - S_{11}$. The factor $k_0h = 0.336$ which is $<<1$ (Pendry, 1999; Smith, 2005).

Figure 4 depicts the *relative permeability* (μ_r) characteristics of the square SRR which indicates that the SRR is a single negative that is mu negative (MNG) metamaterial. The value of permeability (μ_r) is negative in the frequency range of 8.35 GHz to 8.7 GHz. This SRR is embedded inside the slot to load the rectangular microstrip patch. The negative magnetic permeability of the SRR in the frequency range 8.35 GHz to 8.7 GHz significantly improves the impedance matching to the source at desired resonance frequency 5.10 GHz which is much lower than the isolated antenna. Thus, MNG SRR provides the miniaturization of the rectangular microstrip patch antenna to obtain the desired sub-wavelength performance. In this frequency range, the equivalent capacitance of MNG SRR and the patch inductance forms a *LC* resonator that resonates at 5.10 GHz. The value of relative magnetic permeability at working frequency 5.10 GHz is real. Basically, SRR is a *LC* resonant circuit. The resonance frequency of the square SRR is calculated by using equivalent circuit theory to validate it with the simulated

frequency. The inductance (L) of the square SRR is calculated using Equation 5 (Bilotti, 2007; Joshi, 2010, 2011, 2012, 2013):

$$L = \frac{\mu_0}{2} \frac{L_{savg}}{4} 4.86 \left[\ln \frac{0.98}{\rho} + 1.84\rho \right] \qquad (5)$$

where μ_0 is free space permeability ($4\pi \times 10^{-7}$ H/m), ρ is filling ratio expressed as $\rho = \dfrac{(N-1)(w+s)}{\left[L_s - (N-1)(w+s) \right]}$, the average length of square SRR (L_{savg}) is calculated as $L_{savg} = 4 \left[L_s - (N-1)(w+s) \right]$ and N is number of split rings.

The equivalent capacitance (C) that is capacitance per unit length of the square SRR is calculated using Equation 6 (Joshi, 2010, 2011, 2012, 2013; Bilotti, 2007):

$$C = \varepsilon_0 \frac{N-1}{2}$$

$$\times \left[2L_s - (2N-1)(w+s) \right] \frac{K\sqrt{1-k_1^2}}{K(k_1)} \qquad (6)$$

Figure 4. Relative permeability (μ_r) characteristics of the square SRR

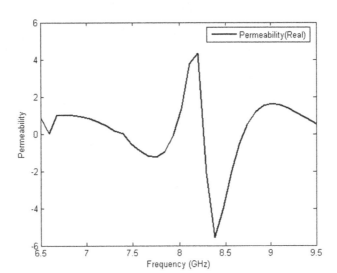

where ε_0 is free space permittivity (8.854×10^{-12} F/m), K is the complete elliptic integral of first kind, k_1 is the argument of integral expressed as

$$k_1 = \frac{s/2}{w + s/2} .$$

Thus, by using equivalent circuit theory and mathematical equations, the calculated values of equivalent circuit elements are: inductance $L = 30\ nH$ and capacitance $C = 0.0119\ pF$. Theoretically, using the values of L and C the resonant frequency of SRR is calculated to 8.43 GHz. The simulated resonant frequency of SRR is 8.48 GHz (Figure 3) which is in good agreement with the theoretical results. Figure 5 depicts the simulated return loss (S_{11}) characteristics of the rectangular microstrip patch antenna without slots and SRR. In this configuration, the antenna resonates at 8.97 GHz which is in good agreement with the designed frequency. Further, the slots are cut in the radiating patch to decrease the resonant frequency of this antenna to WLAN frequency band applications. A square SRR is placed in the square slot to obtain the better impedance matching. Figure 6 depicts the simulated reflection coefficient (S_{11}) characteristics of proposed wearable antenna with the slots

and embedded square SRR. In this condition, the antenna resonates at 5.10 GHz with a bandwidth and gain of 97 MHz and 4.95 dBi respectively. The simulated and measured results are validated by fabricating and testing the antenna.

Figure 7 shows photograph of the experimental set up of testing and measurement of the fabricated antenna. The return loss (S_{11}) characteristics of the fabricated wearable antenna has been measured by Bird site analyzer® interfaced with a personal computer (Model no. SA-6000EX, Frequency range 25 MHz to 6 GHz). Figure 8 shows the measured reflection coefficient (S_{11}) characteristics of the fabricated metamaterial square SRR loaded wearable rectangular microstrip patch antenna which resonates at 5.34 GHz with the better matching at -27.96 dB. Figure 9 shows the measured VSWR of 1.07 at the resonance frequency 5.34 GHz. The weight of fabricated antenna is measured by a digital weighing machine Essae (DS-852) and found to 2.8 gm with a SMA connector (1.2 gm without SMA connector).

Figure 10(a) and Figure 10(b) respectively illustrate the azimuth and elevation radiation patterns of the proposed antenna. The gain and directivity of this antenna is 4.95 dBi and 8.60 dBi respectively. Figure 11(a) and Figure 11(b) re-

Figure 5. Simulated reflection coefficient characteristics of rectangular microstrip patch antenna without cut and square SRR

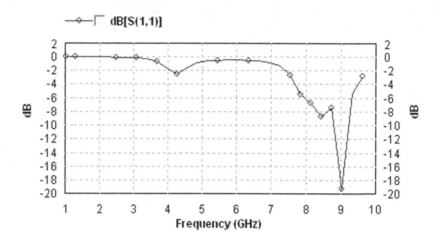

Figure 6. Simulated reflection coefficient characteristics of metamaterial embedded wearable rectangular microstrip patch antenna

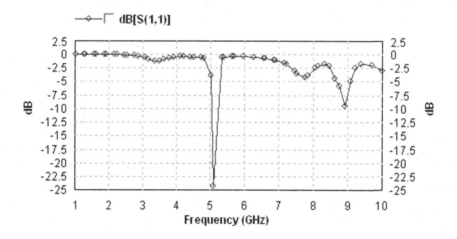

spectively depicts the simulated surface and *vector current distribution* along the proposed wearable rectangular microstrip antenna without and with the square SRR embedded inside the slot. Figure 11(a) shows that the current is not uniformly distributed without the SRR embedded inside the slot.

When the square SRR is embedded inside the slot, current flows along slotted portion and due to the *electromagnetic induction* the *time varying flux* induces current on outer and inner split rings of square SRR (Figure 11(b)). The arrow shows current flow along the microstrip patch and the square SRR. The current is uniformly distributed along the slot of the antenna. Thus, the SRR

Figure 7. Photograph of experimental set up of fabricated metamaterial square SRR embedded wearable rectangular microstrip patch antenna

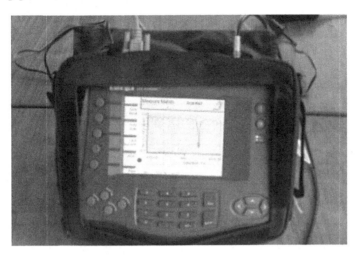

Figure 8. Measured reflection coefficient (S_{11}) characteristics of metamaterial square SRR embedded wearable rectangular microstrip patch antenna

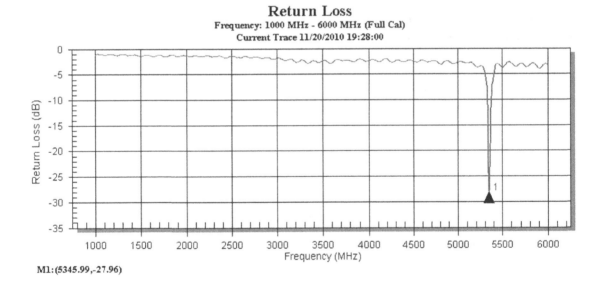

M1:(5345.99,-27.96)

embedding makes the uniform current distribution along the antenna. It results to induce the large electric field across the gap capacitance at the splits and mutual capacitance between the split rings. Under loading condition, the mutual inductance between the square SRR and the edge of rectangular patch is calculated to $M = 0.873\ nH$ using Equation 7 (Bilotti, 2007; Joshi, 2010, 2011, 2012, 2013):

$$M = \frac{\mu_0 L_s}{2\pi}\left[0.467 + \frac{0.059\left(W_x + w\right)^2}{L_s^{\,2}}\right] \quad (7)$$

where W_x is edge width of the slotted rectangular patch as shown in Figure 1(b). The inductance of slotted antenna and the equivalent capacitance of the square SRR forms the LC resonator circuit of

Figure 9. Measured VSWR characteristics of metamaterial square SRR embedded wearable rectangular microstrip patch antenna

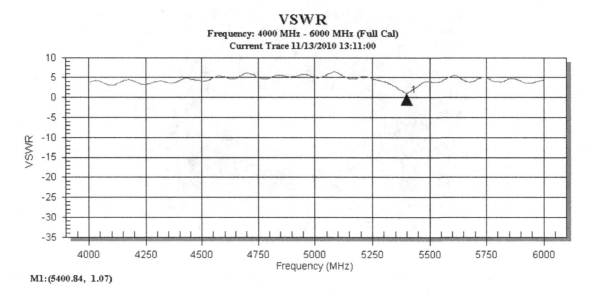

Figure 10. Radiation patterns of metamaterial square SRR embedded wearable rectangular microstrip patch antenna: (a) Azimuth; (b) Elevation

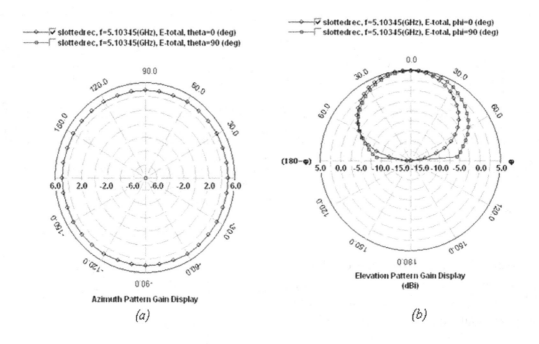

Figure 11. Simulated surface and vector current distribution along wearable rectangular microstrip patch antenna: (a) without embedding the square SRR; (b) with embedded square SRR inside the slot

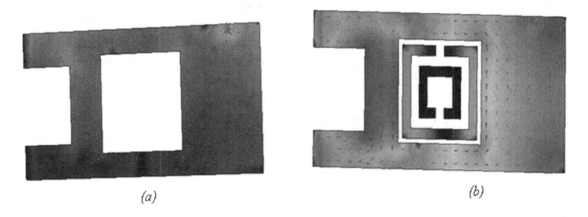

(a)   *(b)*

Figure 12. Photographs of proposed antenna bent on PVC pipes of 54.5mm and 44.5mm radii

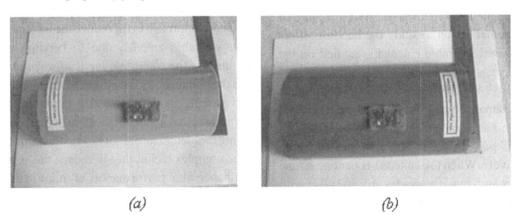

(a)   *(b)*

the SRR embedded rectangular microstrip patch antenna which in turn provides the better impedance matching at resonance frequency 5.10 GHz.

3.4 Experimental Study of Bending Effects on the Wearable Antenna Performance

In section 3.3, the performance of the proposed antenna is theoretically and experimentally verified under the flat surface condition. In practice, the wearable antenna is installed as an integral part of the clothing on different parts of the human body like shoulder, forearm, wrist, waist and thigh. The bending of wearable antenna

takes place according to the frequent movements of the human body. Therefore, an experimental study is executed to examine the bending effect on impedance matching and resonance frequency of proposed wearable antenna under different bending conditions. In this experiment, different shapes of human body like shoulder, wrist, knee and thigh are realized by using the curved surfaces of two cylindrical polyvinyl chloride (PVC) pipes of internal radius 54.5 mm and 44.5 mm respectively. The proposed antenna has been tested by properly bending and swaddling it on surface of both of the PVC pipes. Figure 12(a) and Figure 12(b) represents the photographs of PVC pipes used in this experimentation along

Figure 13. Photograph of experimental set up to test the bending effect on fabricated SRR embedded wearable rectangular microstrip patch antenna

with the antenna. Figure 13 shows a snapshot of experimental set up to study the bending effect on proposed wearable antenna swaddled on the PVC pipes.

Figure 14 and Figure 15 respectively depicts the measured return loss (S_{11}) characteristics of fabricated antenna under bending conditions on the *PVC pipes* of radii 54.5 mm and 44.5 mm respectively. When this antenna is bent on a pipe of radius 54.5 mm it resonates at 5.367 GHz with return loss of -17.97 dB as shown in Figure 14. Similarly, when the antenna is bent on a pipe ra-

dius of 44.5 mm the resonance frequency of the antenna is shifted to 5.388 GHz with the return loss of -20.22 dB as shown in Figure 15. From the experimental results it is observed that in bending condition the resonance frequency of the proposed antenna is shifted to higher side when the antenna is more bent because the *resonant length* of the antenna is reduced. In this experiment, when the reflection coefficient (S_{11}) and *impedance bandwidth* of measured results in bending conditions are studied then no extensive changes in the performance of the proposed antenna are observed.

Figure 16(a) and Figure 16(b) shows photographs of on body positioning of the fabricated wearable antenna on the helmet and shoulder respectively. Thus, it is observed that the slotting means the metal removing technique is an advantageous approach to (a) reduce the adverse effects on wearable antenna due to bending and (b) minimize the *electromagnetic absorption* (SAR) in the human body. The impedance mismatch due to slotting in the microstrip patch at the sub-wavelength resonance is well matched by embedding the metamaterial SRR. This technique avoids the complex techniques to reduce the size and to enhance the performance of microstrip patch antennas like meandering, shorting pin etc.

Figure 14. Measured reflection coefficient (S_{11}) characteristics of the proposed wearable antenna bent on 54.5mm radius PVC pipe

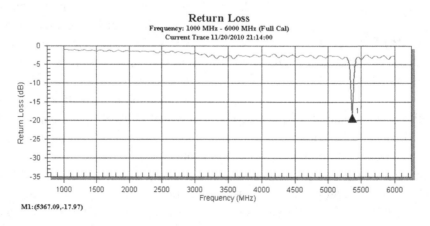

Figure 15. Measured reflection coefficient (S_{11}) characteristics of the proposed wearable antenna bent on 44.5mm radius PVC pipe

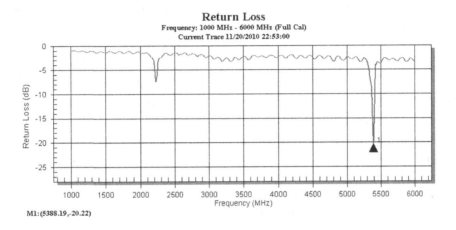

Figure 16. Photographs of on body positioning of the fabricated metamaterial SRR embedded wearable rectangular microstrip patch antenna

(a)   (b)

3.5 Concluding Remarks

In this section, a metamaterial square SRR embedded wearable rectangular microstrip patch antenna for IEEE 802.11a WLAN applications is presented. The bending effect on the performance of wearable antenna can be reduced making slots in the radiating patch but it leads to mismatch the impedance at the sub-wavelength resonance that is at desired lower resonance frequency. It is found that the embedding a metamaterial SRR is an advantageous approach to obtain the better impedance matching at the desired resonance frequency. This SRR introduces additional inductance, capacitance and mutual inductance to match the impedance at the required frequency. The simulated and measured frequency of the proposed wearable antenna is found to be in good agreement. The important features of this antenna are light weight, simple fabrication, and low cost.

4. GEO-TEXTILE BASED METAMATERIAL LOADED WEARABLE MICROSTRIP PATCH ANTENNA

4.1 Preamble of this Antenna

In this section, a *geo-textile* material, that is, *polypropylene-based*, metamaterial loaded wearable *T-shaped microstrip patch antenna* for public safety band applications has been presented. Under unloaded condition of T-shaped microstrip patch antenna, poor matching is observed in the public safety band. This antenna is loaded with metamaterial SRRs that provides better matched condition in the public safety band. In loading condition, the antenna resonates at 4.97 GHz with the bandwidth and gain of 50 MHz and 6.40 dBi respectively. An equivalent circuit of the designed antenna under loading condition is also prepared and analyzed. The *Federal Communication Commission* (FCC) has allotted a separate frequency band of 4.94 GHz to 4.99 GHz that is 50 MHz band for public safety applications dedicated to the protection of human life, health, and property where point-to-point or point- to-multipoint connectivity is necessary (Joshi, 2010, 2013). It covers the applications such as fire fighter, police vehicles, offsite workers, rescue teams, private ambulance services, military services, airport and seaport surveillances so that the interior and sensitive locations can be monitored round the clock for the protection of human life and property. The compact, light weight, efficient, and easily installable antennas are essential in the public safety band. Looking into the safety of human life it is essential to develop the wearable antenna that can be easily integrated as a part of military uniforms, fire fighting and police garments, military tent clothing, seat belts and covers of military and police vehicles. In this work, the authors proposed a geo-textile material that is polypropylene substrate based wearable antenna for public safety band applications. The polypropylene is a *non-woven*

type of geo-textile which is used as a substrate because of its features such as light weight, the polypropylene sheets are available in different thickness which avoids the processes like sewing to obtain the substrate of desired thickness. This antenna consists of a T-shaped microstrip patch is loaded with four metamaterial square SRRs of equal dimensions by placing them around the patch. The proposed antenna has been fabricated, tested, and the measured results are presented in this section. An equivalent circuit model of the T-shaped microstrip patch antenna under loading condition is also prepared and analyzed.

4.2 Antenna Design

Figure 17 depicts the geometrical structure of metamaterial loaded wearable T-shaped microstrip patch antenna on the geo-textile polypropylene substrate. Figure 18(a) and Figure 18(b) respectively depicts the photographs of radiating patch and ground plane of the fabricated antenna. The antenna consists of two rectangular microstrips which are overlapped on each other to form a T-shaped microstrip patch. The purpose to select the T- shape is to increase the resonant length of the antenna so as to reduce the size of antenna and to accommodate the metamaterial SRRs to achieve the further size reduction.

The dimensions of T-shaped microstrip patch are; length of horizontal and vertical microstrips is $L_h = 22.5$ mm and $L_v = 10$ mm respectively. The width of both microstrips is $W_r = 3$ mm. The geometrical dimensions of a square SSR are; length of outer split ring is $L_s = 9$ mm, gap at the split of both rings (g), width of the rings (w) are set to; $g = w = 1$ mm, and separation between the inner and outer split rings (s) is 0.5 mm. The electrical size of square SRR is $0.253 \lambda \times 0.253 \lambda$ (λ is the free space wavelength at resonance frequency of square SRR 8.45 GHz). The T-shaped microstrip patch antenna is loaded with such a four square SRRs that are placed at the distances $g_1 = 1$ mm, $g_2 = 0.75$ mm, $g_3 = 4.5$ mm, and $g_4 = 0.6$ mm to

Figure 17. Geometrical sketch of metamaterial loaded wearable T-shaped microstrip patch antenna on geo-textile polypropylene substrate

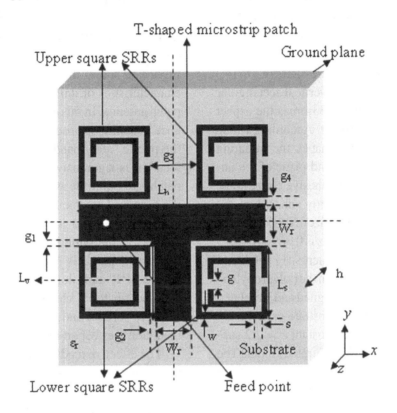

Figure 18. Photographs of fabricated metamaterial loaded wearable T-shaped microstrip patch antenna on polypropylene substrate: (a) Radiating patch; (b) Ground plane

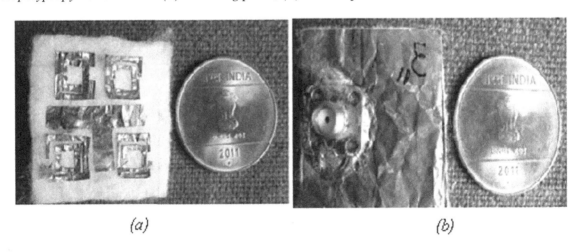

obtain the resonance frequency of public safety band. The aspect ratio of horizontal rectangular microstrip that is length (L_h) to width (W_r) is fixed to 7.5 similarly, the ratio of the gaps between upper square SRRs (g_3) to (g_4) is also set to 7.5. The ratio of the gaps of lower square SRRs (g_1) to (g_2) is set to 1.33. The length of vertical rectangular microstrip (L_v) is fixed to 1.33 times the aspect ratio of horizontal microstrip. According to the designed dimensions and shapes the radiating patch, square SRR, and ground plane of the antenna are cut from the self adhesive copper tape of thickness 0.1 mm and tightly adhered on the polypropylene substrate. The size of this antenna at resonance frequency 4.97 GHz is $0.369 \lambda \times 0.369 \lambda$. The finely cut T-shaped microstrip patch and the SRRs are tightly adhered on the polypropylene substrate. This antenna is designed and simulated on polypropylene (PR 30) substrate of thickness h = 1.9 mm and dielectric constant ε_r = 2.2 supplied by TECHFAB India, Mumbai, India. The antenna is co-axially fed at x = -7.2 mm and y = 0 mm. The proposed antenna is entirely handmade

and high degree of accuracy is maintained in the entire fabrication processes.

4.3 Results and Discussion

Figure 19 shows the simulated return loss (S_{11}) characteristics of unloaded T-shaped microstrip patch antenna. In this configuration, poor impedance matching is observed at 4.95 GHz that is in the public safety band. However, the feed point location is rigorously determined on the entire patch to obtain the good matched condition in the proposed resonance frequency band. Hence, to obtain the good impedance matching of the antenna at the public safety band the T-shaped microstrip patch is loaded with metamaterial square SRR inclusions as shown in Figure 17. The metamaterial characteristics of the square SRR are verified and presented before analyzing the loading effect on microstrip patch.

The effective medium theory is used to verify the permeability (μ_r) and permittivity (ε_r) characteristics of the square SRRs from the reflection and transmission coefficients (S-parameters). The

Figure 19. Simulated return loss (S_{11}) characteristics of unloaded wearable T-shaped microstrip patch antenna on polypropylene substrate

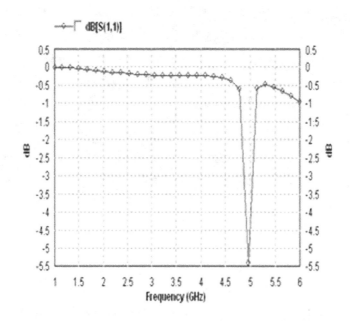

Figure 20. Relative permeability (μ_r) characteristics of the square SRR

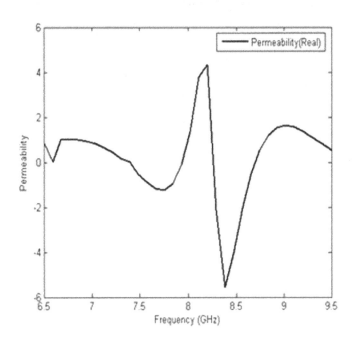

Nicolson-Ross-Weir (NRW) approach is used to obtain the effective medium parameters (Joshi, 2013, 2012, 2012, 2013; Pendry, 1999; Smith, 2005; Ziolkowski, 2003). The expressions of Equations 3 and 4 are used to determine these effective parameters. The metamaterial characteristics of the SRR are verified using the *S*-parameters obtained from IE3D electromagnetic simulator and MATLAB code with mathematical Equations 3 and 4.

Figure 20 depicts the relative permeability (μ_r) characteristics of the SRR which indicates that the SRR structure is single negative that is mu negative (MNG) metamaterial. The value of permeability (μ_r) is negative in the frequency range of 8.35 GHz to 8.7 GHz. Such four square SRRs are used to load the T-shaped microstrip patch as shown in Figure17. In loading condition, a good matching is obtained in the public safety band.

Figure 21 depicts the simulated return loss (S_{11}) characteristics of the metamaterial loaded polypropylene based wearable microstrip patch antenna. This antenna resonates at 4.97 GHz (in

the frequency band of 4.94 GHz to 4.99 GHz) with the bandwidth and gain of 50 MHz and 6.38 dBi respectively. The directivity of the proposed antenna is 7.56 dBi. Figure 22 shows the photograph of the experimental set up of return loss measurement of the fabricated antenna using Bird site analyzer® (Model No. SA-6000 EX, Frequency range 25 MHz to 6 GHz). Figure 23 shows the measured return loss (S_{11}) characteristics of the T-shaped microstrip patch antenna loaded with SRRs. This antenna will find its applications in different public safety band applications.

4.4 Equivalent Circuit Analysis and Theoretical Discussion

Figure 24 shows the equivalent circuit diagram of metamaterial square SRR loaded polypropylene based wearable T-shaped microstrip patch antenna. Basically, SRR is a *LC* resonant circuit where *L* and *C* are the equivalent inductance and capacitance respectively. L_{coax} represents the probe inductance. The inductance (*L*) of the square SRR

Figure 21. Simulated return loss (S$_{11}$) characteristics of metamaterial loaded wearable T-shaped microstrip patch antenna on polypropylene substrate

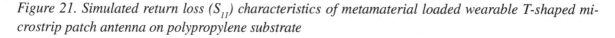
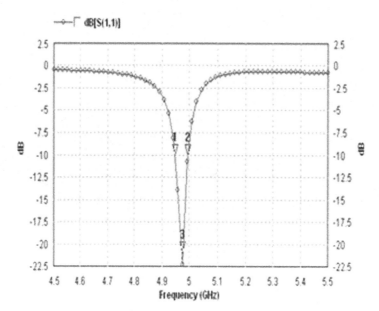

Figure 22. Photograph of experimental set up to test SRR loaded polypropylene based wearable T-shaped microstrip patch antenna

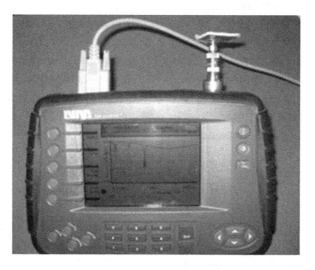

is calculated using Equation 8 (Bilotti, 2007; Joshi, 2010, 2011, 2012, 2013):

$$L = \frac{\mu_0}{2}\frac{L_{savg}}{4}4.86\left[\ln\frac{0.98}{\rho}+1.84\rho\right] \qquad (8)$$

where; μ_0 is permeability of free space ($4\pi \times 10^{-7}$ H/m), ρ is the filling ratio expressed as

$$\rho = \frac{(N-1)(w+s)}{\left[L_s-(N-1)(w+s)\right]},$$ L_{savg} is the average length of square SRR and calculated as

Figure 23. Measured return loss (S_{11}) characteristics of wearable T-shaped microstrip patch antenna on polypropylene substrate

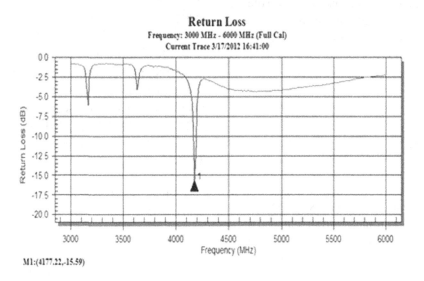

Figure 24. Equivalent circuit diagram of SRR loaded wearable T-shaped microstrip patch antenna on polypropylene substrate

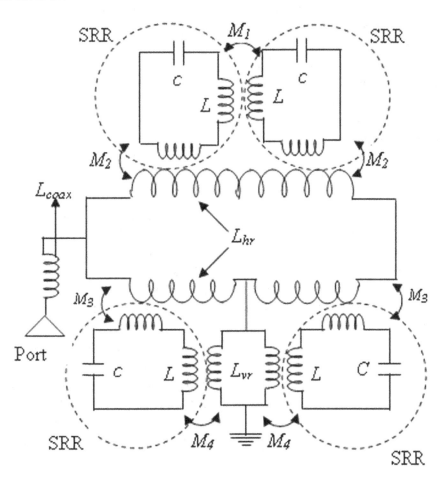

$$L_{savg} = 4\left[L_s - (N-1)(w+s)\right]$$ and N is number of split rings. The equivalent capacitance (C) that is capacitance per unit length of the square SRR is calculated using Equation 9 (Bilotti, 2007; Joshi, 2010, 2011, 2012, 2013):

$$C = \varepsilon_0 \frac{N-1}{2}$$
$$\times \left[2L_s - (2N-1)(w+s)\right] \frac{K\sqrt{1-k_1^2}}{K(k_1)} \tag{9}$$

where; ε_0 is the permittivity of free space (8.854×10^{-12} F/m), K is the complete elliptic integral of first kind, k_1 is the argument of integral expressed as $k_1 = \dfrac{s/2}{w + s/2}$.

By using the *principles of equivalent circuit theory* and mathematical equations, the calculated values of equivalent circuit elements are; inductance $L = 30$ nH and capacitance $C = 0.0119$ pF. Theoretically, using the values of L and C the resonance frequency of square SRR is calculated to 8.43 GHz. The simulated resonant frequency of square SRR is 8.48 GHz (Figure 20) which is in good agreement with the theoretical results. In loading condition, the four SRR inclusions are inductively coupled with the T-shaped microstrip patch. The *LC* resonant circuit of the corresponding square SRR gets mutually coupled with the T-shaped microstrip patch through the mutual inductance 'M' which is modelled as the *magnetic coupling* between these two elements. Let, M_1 is mutual inductance between the upper neighbouring SRRs and calculated to 0.819 nH by using Equation 10 (Bilotti, 2007; Joshi, 2010, 2011, 2012, 2013):

$$M_1 = \frac{\mu_0 L_s}{2\pi}\left[0.467 + \frac{0.059w^2}{L_s^2}\right] \tag{10}$$

Consider, M_2 is mutual inductance between the upper SRRs and the horizontal microstrip of the T-shaped patch. Let M_3 is mutual inductance between the two lower SRRs and horizontal microstrip of the patch. M_4 is mutual inductance between the two lower SRRs and vertical microstrip of the patch. The mutual inductance M_2 to M_4 are respectively calculated to 0.861 nH using Equation 11 (Bilotti, 2007, Joshi, 2010, 2011, 2012, 2013):

$$M_2 = M_3 = M_4$$
$$= \frac{\mu_0 L_s}{2\pi}\left[0.467 + \frac{0.059\left(W_r + w\right)^2}{L_s^2}\right] \tag{11}$$

The inductance of horizontal (L_{hr}) and vertical (L_{vr}) rectangular microstrips of the T-shaped patch is calculated to 14.6 nH and 5 nH respectively using Equation 12 (Bilotti, 2007; Joshi, 2010, 2011, 2012, 2013):

$$L_{hr} = L_{vr} = \frac{\mu_0 L_r}{2\pi}$$
$$\times \left[\ln\left(\frac{2L_r}{W_r}\right) + 0.5 + \left(\frac{W_r}{3L_r}\right) - \left(\frac{W_r^2}{24L_r^2}\right)\right] \tag{12}$$

The values of L_{hr} and L_{vr} are calculated at the lengths $L_r = L_h = 22.5$ mm and $L_r = L_v = 10$ mm respectively. In loading condition, the SRRs are positioned proximity to the horizontal and vertical microstrips of the T-shaped patch. The microstrip patch is coaxially excited hence, due to electromagnetic induction the time varying flux induces the current on each of the square SRR used for loading the patch. Thus, the electric field is induced across the gap capacitance of the splits and mutual capacitance (capacitance per unit length) between the inner and outer splits rings of the SRRs. The inductance of rectangular microstrip patch antenna with the capacitance of SRRs and the mutual inductances forms the *LC* resonant circuit of the loaded antenna. This capacitance compensates inductance of T-shaped microstrip patch and the good matching is obtained at the lower resonant frequency 4.97 GHz. Thus, the capacitance of SRRs and the mutual inductance are sufficiently large to match with inductance of the T-shaped microstrip patch. Therefore, the negative permeability SRRs acts as matching elements at the lower resonant frequency 4.97 GHz. In loading condition, the gain of antenna is enhanced because the SRRs acts as matching elements and accepts maximum power from the source for the radiation. Thus, the SRR loading reduces the resonant frequency of the proposed antenna with a better impedance matching at 4.97 GHz by reducing the antenna size.

Figure 25(a) and Figure 25(b) respectively depicts the simulated surface and vector current distribution along the designed metamaterial loaded wearable T-shaped microstrip patch antenna. The current is uniformly distributed along the antenna structure as shown in Figure 25(a). The arrow shows current flow along the T-shaped microstrip patch and the square SRRs as depicted in Figure 25(b). The current is induced in the SRRs because of mutual coupling. Figure 26(a) and 26(b) respectively depicts the azimuth and elevation radiation patterns of the polypropylene

based metamaterial loaded T-shaped microstrip patch antenna indicating the gain of 6.38 dBi.

Figure 27(a) shows the photograph of on body positioning of the fabricated wearable antenna on the clothing of security personnel. Figure 27(b) depicts the integration of fabricated antenna in the seat cover of a vehicle.

4.5 Concluding Remarks

Geo-textile material that is polypropylene based metamaterial SRR loaded T-shaped microstrip patch wearable antenna for public safety band applications is presented. Under loading condition, the T-shaped microstrip patch resonates in the public safety band. Thus, metamaterial loading is a simple technique for size reduction with considerable gain and bandwidth. The SRR loading introduces the inductance, capacitance and mutual inductance to match the impedance at desired resonance frequency. The advantages of proposed antenna are small size, inexpensive, light weight, and easy integration within the clothing. This antenna will find its application in the clothing and helmets of rescue teams, military, fire fighters, security and police personnel, as well as garments of the military tents.

Figure 25. Simulated current distribution along wearable T-shaped microstrip patch antenna on polypropylene substrate

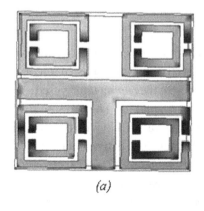

(a)

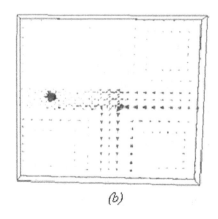

(b)

Figure 26. Radiation patterns of metamaterial loaded wearable T-shaped microstrip patch antenna on polypropylene substrate: (a) Azimuth; (b) Elevation

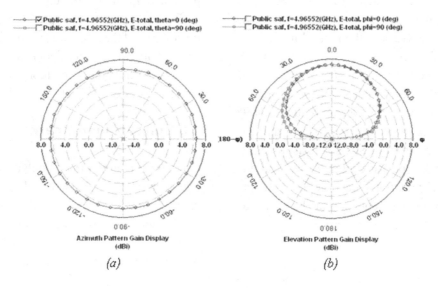

Figure 27. Positioning of the metamaterial loaded polypropylene based wearable T-shaped microstrip patch antenna

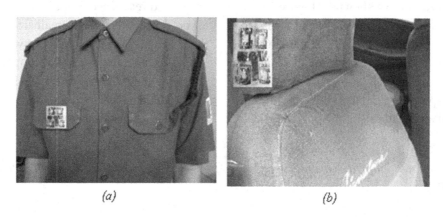

5. POLYPROPYLENE BASED METAMATERIAL INTEGRATED WEARABLE MICROSTRIP PATCH ANTENNA

5.1 Preamble of this Antenna

In this section, a metamaterial integrated T-shaped wearable microstrip patch antenna for IEEE 802.11a Wi-Fi 5 GHz applications is presented.

This antenna is made from non-woven geo-textile that is polypropylene as a non-conducting substrate whereas the *copper tape* is used as conducting parts of this antenna. The proposed wearable antenna operates at 5 GHz with bandwidth and gain of 45 MHz and 6.15 dBi respectively. The electrical size of this antenna is $0.367\lambda \times 0.217\lambda$. The simulated and measured results of this antenna are found to be in good agreement. The objective of this work is to design and fabricate a polypropylene based metamaterial integrated wearable microstrip patch antenna for IEEE 802.11a Wi-Fi applications.

5.2 Antenna Design

Figure 28 shows the geometrical structure of the designed polypropylene based metamaterial integrated T-shaped wearable microstrip patch antenna. Figure 29(a) and Figure 29(b) respectively depicts the photographs of the radiating patch and ground plane of the fabricated wearable antenna. The antenna consists of two rectangular microstrips which are overlapped on each other to form a T-shaped microstrip patch. The dimensions of the T-shaped microstrip patch are; length of horizontal and vertical microstrips $L_h = 22.5$ mm and $L_v = 10$ mm respectively. The width of both microstrips is $W_r = 3$ mm (The same as used in section 4.2).

The geometrical dimensions of square SRRs are; length of outer split ring is $L_s = 9$ mm, gap at the split of both rings (g), width of the rings (w) are set to; $g = w = 1$ mm, and separation between the inner and outer split rings (s) is 0.5 mm. The electrical size of the SRR is $0.253\,\lambda \times 0.253\,\lambda$ (λ is the free space wavelength at resonance frequency of an isolated SRR). The two square SRRs are integrated in the space available at the right and left side of the T-shaped microstrips at the distance of $g_1 = 1$ mm and $g_2 = 0.75$ mm respectively to obtain the desired resonance frequency band. The *self adhesive copper tape* of thickness 0.1 mm is used to fabricate the radiating patch and the ground plane of the antenna. This antenna is designed and simulated on polypropylene (PR 30) substrate of thickness h = 1.9 mm and dielectric constant $\varepsilon_r = 2.2$ supplied by TECHFAB India, Mumbai, India. The antenna is co-axially fed at $x = -7.2$ mm and $y = 0$ mm.

5.3 Results and Discussion

Initially, the antenna is designed without integrating the SRRs. In this configuration, the poor impedance matching is observed at 5 GHz (as shown in Figure 19). Hence, to obtain good impedance matching at 5 GHz band, the SRR inclusions are integrated to load the T-shaped microstrip as

Figure 28. Geometrical structure of T-shaped wearable microstrip patch antenna

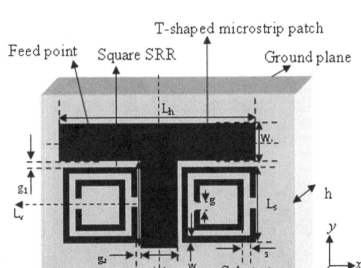

Figure 29. Photographs of fabricated wearable antenna

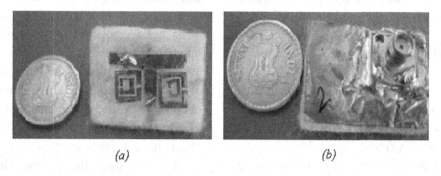

(a)   (b)

shown in Figure 28. The simulated return loss (S_{11}) characteristics of the proposed antenna integrated with SRRs is illustrated in Figure 30. This antenna resonates at 5 GHz with bandwidth and gain of 45 MHz and 6.15 dBi respectively. The directivity of this antenna is 7.5 dBi. The fabricated antenna is tested with the Bird site analyzer® Model No. SA-6000 EX, Frequency range 25 MHz to 6 GHz. Figure 31 depicts the measured return loss (S_{11}) characteristics of fabricated wearable antenna with the inset experimental set up to test the antenna. Good agreement has been obtained between the simulated and measured results.

Figure 32 shows the equivalent circuit diagram of metamaterial SRR loaded polypropylene based wearable T-shaped microstrip patch antenna. L and C are the equivalent inductance and capacitance of SRR. L_{hr} and L_{vr} are inductance of horizontal and vertical rectangular microstrips respectively. M_1 and M_2 is mutual inductance between the T-shaped microstrip and the neighbouring SRR.

Figure 33(a) and Figure 33(b) respectively depicts the azimuth and elevation radiation patterns of the proposed wearable antenna indicating the gain of 6.15 dBi.

Figure 30. Simulated return loss (S_{11}) characteristics

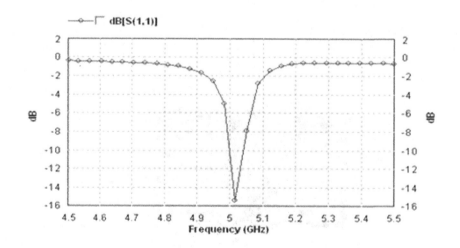

Figure 31. Measured return loss (S_{11}) characteristics

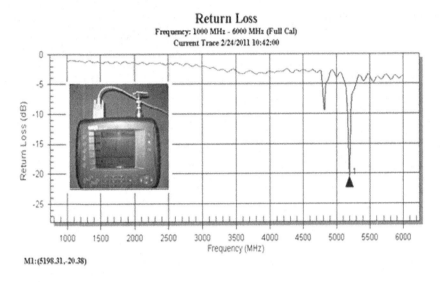

M1:(5198.31,-20.38)

Figure 32. Equivalent circuit diagram

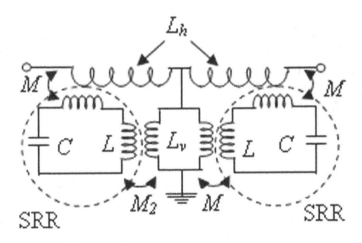

5.4 Concluding Remarks

The metamaterial SRR integration includes the inductance, capacitance and mutual inductance which provide better matching at desired lower resonance frequency of microstrip patch antennas in a small size with considerable gain and bandwidth. Features of this antenna are inexpensive, light weight, and easy integration within the clothing. This antenna will find its application in clothing and helmets of rescue teams, sports activities, security and police personnel.

Figure 33. Radiation patterns: (a) Azimuth; (b) Elevation

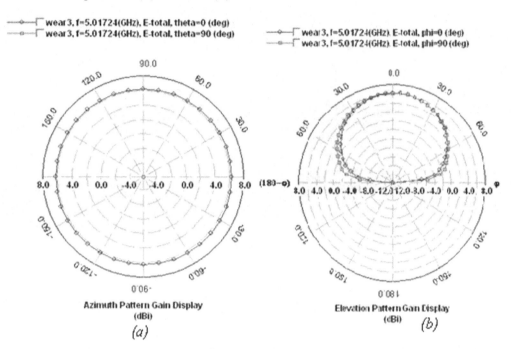

Azimuth Pattern Gain Display
(dBi)
(a)

Elevation Pattern Gain Display
(dBi)
(b)

6. POLYESTER BASED WEARABLE MICROSTRIP PATCH ANTENNA AND ANALYSIS OF EFFECT OF SUBSTRATE CORRUGATIONS ON ITS PERFORMANCE

6.1 Preamble of this Antenna

In this section, a wearable microstrip patch antenna using *Inter-Digital Capacitor* (IDC) and *rectangular stub* for 802.11 b WLAN band applications has been presented. The antenna is fabricated on an inexpensive polyester cloth substrate. The antenna resonates at 2.45 GHz with the bandwidth and gain of 41 MHz and 5 dBi respectively. In simulated return loss characteristics, better matching is obtained (-25 dB) at the resonance frequency 2.45 GHz whereas no matching is observed at the other frequency band. In measured results, in addition to WLAN frequency band a good matching is observed at 2.71 GHz (-18 dB). This matching is obtained due to the corrugations in the polyester

substrate. The surface impedance became more capacitive because of the air inside the corrugations hence; good matching is obtained at 2.71 GHz. The effect of these corrugations on the antenna performance is analyzed in this section. The electrical dimension of the proposed wearable antenna at 2.45 GHz is 0.213 λ × 0.246 λ. The antenna has been fabricated, tested, and the measured results are found to be in good agreement with the simulated results. An equivalent circuit of the designed antenna with and without corrugations is also prepared and analyzed.

6.2 Antenna Design

Figure 34 depicts the sketch and geometrical structure of wearable microstrip patch antenna using inter-digital capacitor (IDC) and rectangular stub fabricated on the polyester substrate. This antenna is fabricated using copper tape and polyester cloth substrate.

Figure 34. Sketch and geometrical structure of wearable microstrip patch antenna using inter-digital capacitor and rectangular stub

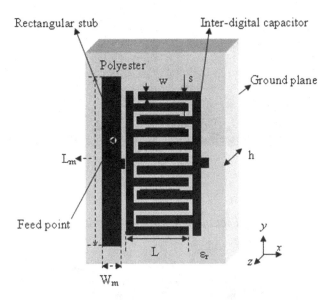

Figure 35(a) and Figure 35(b) respectively depicts the photographs of radiating patch and ground plane of the fabricated wearable microstrip patch antenna on polyester substrate. The dimensions of rectangular stub are; length L_m = 30 mm and width W_m = 4 mm. This stub is connected to the IDC through a square microstrip. The geometrical dimensions of IDC are; number of finger pairs N = 6, finger width w = 2 mm, spacing between the teeth's is s = 1 mm and finger length L = 15 mm. The dimensions of polyester cloth substrate are; thickness h = 3.14 mm, relative permittivity (ε_r) 1.39 and the loss tangent tan δ = 0.01 is used to fabricate this antenna. The polyester cloth layers are stitched in vertical and horizontal directions with the 3 mm periodicity of stitches to fabricate the substrate of thickness 3.14 mm. The antenna is coaxially fed using SMA connector at x = -13 mm and y = 13 mm. The electrical dimension of the antenna is 0.213 λ × 0.246 λ. The radiating patch comprising IDC with rectangular stub and the ground plane of the antenna are finely cut from the self adhesive copper tape of thickness 0.1 mm according to the designed dimensions and shapes.

The finely cut rectangular microstrip stub, IDC and the ground plane are tightly adhered on the polyester substrate at appropriate locations. The proposed antenna is entirely handmade and high degree of accuracy is maintained in the entire fabrication processes.

6.3 Results and Discussion

Figure 36 shows the photograph of the experimental set up of return loss measurement of fabricated antenna using Bird site analyzer® (Model No. SA-6000 EX, Frequency range 25 MHz to 6 GHz). Figure 37 depicts the simulated return loss (S_{11}) characteristics of proposed antenna. It is observed that proposed antenna resonates at 2.45 GHz with the bandwidth of 41 MHz. In addition to this, a poor matching is observed at second frequency that is approximately at 2.71 GHz. The encircled portion indicates this frequency band.

Figure 38 shows the measured return loss (S_{11}) characteristics of fabricated wearable antenna. The antenna resonates at 2.48 GHz in addition to this; a significant matched condition is noticed at

Figure 35. Photographs of fabricated wearable microstrip patch antenna on polyester substrate using IDC and rectangular stub: (a) Radiating patch; (b) Ground plane

(a)   *(b)*

Figure 36. Photograph of experimental set up to test wearable microstrip patch antenna using IDC and rectangular stub on polyester substrate

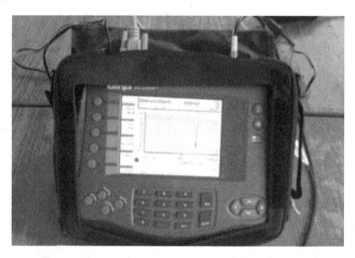

Figure 37. Simulated return loss (S$_{11}$) characteristics of wearable microstrip patch antenna using IDC and rectangular stub on polyester substrate

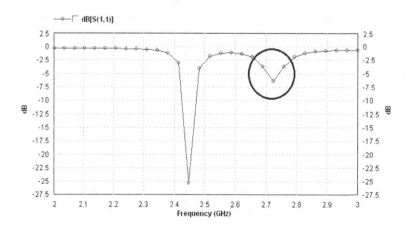

Figure 38. Measured return loss (S_{11}) characteristics of wearable microstrip patch antenna using IDC and rectangular stub on polyester substrate

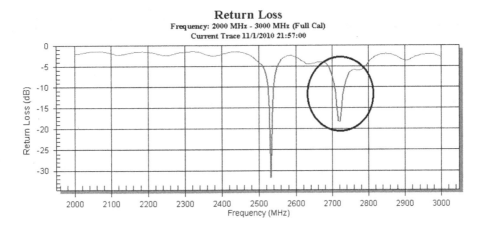

second resonance frequency 2.71 GHz which is not obtained in the simulated return loss characteristics (Figure 37). The second resonance frequency band is encircled as shown in Figure 38. The impedance matching is obtained at the second frequency is due to the corrugations produced in the polyester substrate during the stitching process. The effect of corrugations on impedance matching is analyzed and presented in Section 6.4.

Figure 39(a) and Figure 39(b) respectively depicts the azimuth and elevation radiation patterns of the wearable microstrip patch antenna at 2.45 GHz indicating the gain of 5 dBi. Figure 40 depicts the simulated surface and vector current distribution along the designed wearable microstrip patch antenna on polyester substrate. The current is uniformly distributed along the antenna structure. The arrow shows the uniform current flow along the fingers of IDC.

Figure 41(a) and Figure 41(b) respectively depicts the photographs of the on body positioning of fabricated antenna on the clothing of sport personnel and on the shirt collar of a wearer.

6.4. Equivalent Circuit Analysis and Theoretical Discussion

Figure 42 depicts the equivalent circuit diagram of the proposed wearable microstrip patch antenna. Basically, antenna is a *LC* resonant circuit where; *L* and *C* are the equivalent inductance and capacitance of the antenna structure respectively. The details of the equivalent circuit elements are as follows. C_{idc} is the capacitance of IDC and it is calculated using Equation 13 (Caloz, 2006; Joshi, 2011):

$$C_{idc} = (\varepsilon_r + 1)L[(N_I - 1)A_1 + A_2](pF) \quad (13)$$

where N_I is the number of fingers of inter-digital capacitor.

A_I and A_2 are calculated using expressions of Equation 14:

$$A_1 = 4.409 \tanh\left[0.55\left(\frac{h}{w}\right)^{0.45}\right].10^{-6}(pF / \mu m)$$

Figure 39. Radiation patterns of wearable microstrip patch antenna on polyester substrate: (a) Azimuth; (b) Elevation

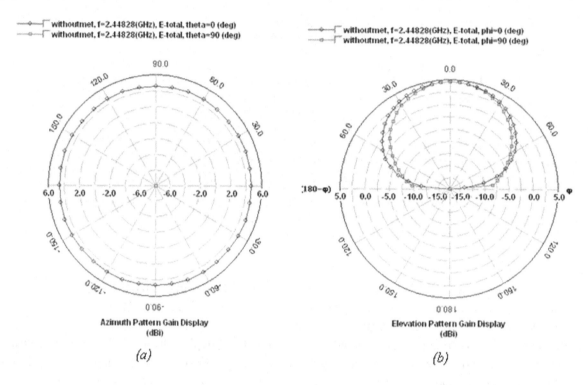

(a)

(b)

Figure 40. Simulated current distribution along wearable microstrip patch antenna on polyester substrate

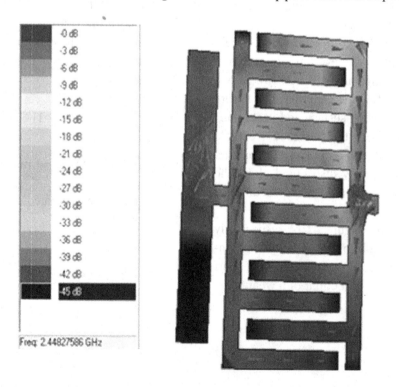

Figure 41. Photograph of on body positioning of the fabricated wearable microstrip patch antenna

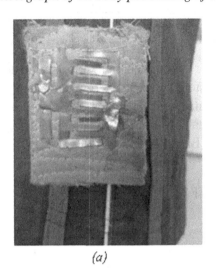
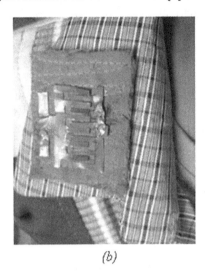

(a)　　　　　　　　　　　*(b)*

Figure 42. Equivalent circuit diagram of a wearable microstrip patch antenna using IDC and rectangular stub on polyester substrate

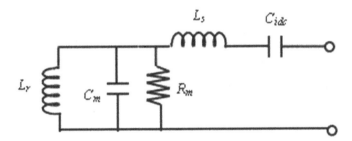

Figure 43. Stitched substrate of desired thickness using polyester cloth layers

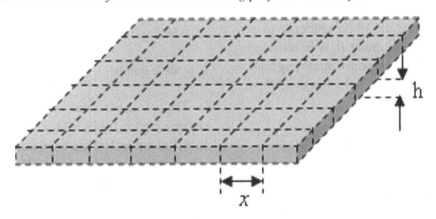

$$A_2 = 9.92 \tanh\left[0.52\left(\frac{h}{w}\right)^{0.5}\right].10^{-6}(pF/\mu m) \tag{14}$$

$$\delta e_{ff} = \delta + \frac{\Delta}{h} + \frac{P_r}{2\omega_0 W_e} \tag{18}$$

In this equation; Δ is the skin depth; W_e is the stored electric energy and P_r is the power radiated by the antenna. The values of W_e and P_r are calculated using Equations 19 and 20 respectively (Bahl,1980; Garg, 2001; Pozar,1995):

For calculations of A_1 and A_2 the dimensions are considered in μm. The calculated value of capacitance of IDC (C_{idc}) is 0.470 pF.

L_r is the inductance of a rectangular stub and calculated to 19.50 nH using Equation 15 (Bilotti, 2007; Joshi, 2010, 2011, 2012, 2013):

$$W_e = \frac{\varepsilon_0 \varepsilon_r L_m W_m V_o^2}{8h} \tag{19}$$

$$L_r = \frac{\mu_0 L_m}{2\pi} \times \left[\ln\left(\frac{2L_m}{W_m}\right) + 0.5 + \left(\frac{W_m}{3L_m}\right) - \left(\frac{W_m^2}{24L_m^2}\right)\right] \tag{15}$$

$$P_r = \frac{V_0^2}{R_0}\frac{A\pi^5}{192}\left[\frac{(1-B)\left(1-\frac{A}{15}+\frac{A^2}{420}\right)}{+\left(2-\frac{A}{7}+\frac{A^2}{189}\right)}\right] \tag{20}$$

Similarly, the inductance (L_s) of a small square microstrip connecting to the rectangular stub and IDC is calculated to 52 nH by using Equation 15. The capacitance of rectangular stub with respect to ground plane of the antenna is calculated to 0.453 pF using Equation 16 (Bilotti, 2007; Joshi, 2010, 2011, 2012, 2013):

where:

$$A = \pi\left[\left(a + 2\Delta a\right)/\lambda_0\right]^2$$
$$B = \pi\left[\left(b + 2\Delta b\right)/\lambda_0\right]^2$$

$$C_m = \frac{\varepsilon_0 \varepsilon_r A}{h} \tag{16}$$

Δa and Δb are obtained using Hammerstad formula:

where ε_0 is the permittivity of free space (8.854 $\times 10^{-12}$ F/m), A is the area of capacitor due to the rectangular stub. In this circuit, R_m is the resistance of rectangular stub which is calculated using Equation 17 (Bahl,1980; Garg, 2001; Pozar, 1995):

V_o = Input voltage of antenna

R_0 = Intrinsic impedance of free space

k_0 = wave number

$$R_m = \frac{\omega\mu_0 h}{k_0^2 \varepsilon_r \delta e_{ff}} \frac{2}{L_m W_m} \cos^2\left(\frac{\pi x_0}{L_m}\right) \tag{17}$$

The calculated value of rectangular stub resistance (R_m) is 0.119 MΩ. Thus, the calculated values of equivalent circuit elements of the proposed antenna are; $L_a = 14.18$ nH and $C_a = 0.258$ pF. Theoretically, using the values of L and C the calculated resonance frequency of the proposed antenna is 2.64 GHz which is in good agreement of the simulated results.

where μ_0 is permeability of free space ($4\pi \times 10^{-7}$ H/m), x_0 is feed point location, and δ_{eff} is *effective loss tangent*. The δ_{eff} is calculated by Equation 18 (Bahl,1980; Garg,2001; Pozar, 1995):

6.5 Effect of Substrate Corrugations on Antenna Performance

In this antenna, the number of polyester cloth layers are stitched to obtain the substrate of desired thickness. During the stitching process the cloth gets crimped and the substrate becomes periodically corrugated. It is observed that the sinusoidal shaped (wavy natured) corrugations get developed along the substrate that results to vary the substrate thickness. Figure 43 illustrates the stitched substrate using polyester layers. The periodicity of the stitches is $x = 3$ mm.

Figure 44 shows the 3-dimensional view of corrugations developed in the polyester substrate of the proposed fabricated wearable antenna. The radiating patch and ground plane are adhered on the top and bottom of this stitched substrate respectively. It is also seen from the photograph (Figure 35(a)) of the fabricated antenna that the square shaped corrugations are developed inside the substrate that consist of air and polyester cloth with the periodicity of $x = 3$ mm and depth of 3.14 mm respectively.

Figure 45 depicts cross-sectional view of the fabricated antenna showing the sinusoidal nature of corrugations inside the substrate after stitching process. This shows the periodicity of corrugations that is air and polyester cloth in which 'l' and 'w' are respectively the length and width of ground plane of the fabricated antenna.

Figure 46 depicts the equivalent circuit diagram of inductance-capacitance developed because of corrugations in the substrate. The corrugations make the alterations in the height of microstrip patch substrate in the direction of propagation and it increases the inductance and capacitance per unit length of the radiating patch. The inductance L_{sub} and capacitance C_{sub} are developed due to the corrugations in the substrate of fabricated antenna constitutes the equivalent circuit (Figure 46).

The capacitance (C) between the top layer and bottom layer of the antenna structure due to corrugations is expressed as Equation 21 (Bahl,1980; Garg, 2001; Joshi, 2012; Pozar, 1995):

$$C_{sub} = \frac{\varepsilon_0 \varepsilon_{reff} hw}{x} \tag{21}$$

where ε_{reff} is the *effective permittivity* of polyester cloth and air medium in the corrugations of the substrate and calculated by Equation 22, x is the periodicity of the stitch which makes the corrugations:

Figure 44. 3-dimensional view of corrugations developed in polyester substrate of fabricated wearable antenna

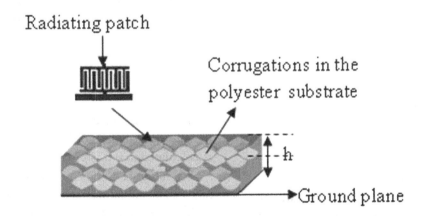

Figure 45. Cross-sectional view of the fabricated wearable antenna

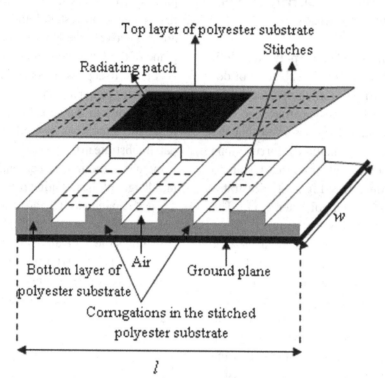

Figure 46. Equivalent circuit diagram of inductive and - capacitive loading due to corrugations in substrate

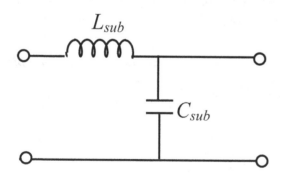

expression (23) (Bahl,1980; Garg, 2001; Joshi, 2012; Pozar, 1995):

$$C_{sub} = \varepsilon_0 \varepsilon_{reff} h \left(\frac{w}{l} \right) \qquad (23)$$

If the ratio *w/l* of Equation 23 is normalized then the total capacitance of the corrugated substrate is expressed by Equation 24 (Bahl,1980; Garg, 2001; Joshi, 2012; Pozar, 1995):

$$C_{sub} = \varepsilon_0 \varepsilon_{reff} h \qquad (24)$$

Thus, from Equation 23 the capacitance of the substrate under corrugated condition (C_{sub}) is calculated to 0.028 *pF*.

From Figure 45 it is observed that due the air medium the radiating patch and ground plane are mutually coupled. The mutual inductance (M_{rgair}) between radiating patch and ground plane is calculated using Equation 25 (Jan, 2000):

$$\varepsilon_{reff} = \frac{\varepsilon_r \varepsilon_{r1}(h + h_1)}{(\varepsilon_r h + \varepsilon_{r1} h_1)} \qquad (22)$$

ε_{r1} is the dielectric constant of air.

Now, the total capacitance (C_{total}) of the substrate under corrugated condition is calculated by

$$M_{rgair} = \mu_0 \sqrt{(l_1 + l)}$$
$$\times \left(\frac{2}{f}\right)\left[\left(1 - \frac{f^2}{2}\right)k(f) - E(f)\right] \qquad (25)$$

where $k(f)$ and $E(f)$ are the complete elliptic integral of first and second kind respectively, l_1 is the length of radiating patch and l is the length of ground plane:

$$f = \sqrt{\frac{4(l_1 l)}{h^2 + (l_1 + l)^2}}$$

The calculated value of mutual inductance is 0.297 μH.

In this antenna structure, the TM$_0$ *surface wave propagation* takes place through the substrate. Equation 26 validates the condition of substrate thickness for the propagation of surface waves (Bahl,1980; Garg, 2001; Joshi, 2012; Pozar, 1995):

$$\left(\frac{h}{\lambda_0}\right) < \left(\frac{1}{4\sqrt{(\varepsilon_r - 1)}}\right) \qquad (26)$$

The TM surface wave propagates when the *surface impedance* (Z_s) is inductive and these modes get suppressed when Zs is capacitive. In simulated condition, the corrugations in the substrate are not considered. Under practical condition, the surface impedance becomes more capacitive (Equation 24) due to the air present in corrugations. Hence, the good impedance matching is obtained at the second resonance frequency that is 2.71 GHz in the measured return loss (Figure38) which is not observed in the simulated reflection coefficient characteristics (Figure37).

6.6 Concluding Remarks

In this section, a polyester based microstrip patch wearable antenna for WLAN applications is presented. It is found that due to the corrugations in the stitched substrate the capacitance get introduced in the manufactured antenna structure. Thus, the surface impedance becomes more capacitive results in generating other resonance frequency band in the measured results. The advantages of this antenna are small size, inexpensive, light weight, and easy integration within the clothing. This antenna will find its application in the clothing of fire fighters, military, sportsmen and police personnel.

7. CONCLUSION

Human being is efficiently using the RF communication technologies and carries the handheld devices as an integral part of the body to perform the day-to-day activities with great comfort and safety. Different communication networks are used to establish the communication between the on and off body communication devices or systems. For such type of devices and systems wearable antennas are essential to establish the communication. In this chapter, the planar metamaterial based wearable microstrip patch antennas have been presented. The metamaterial is embedded in a polyester substrate rectangular microstrip patch antenna is fabricated and tested. This antenna is useful for WLAN applications and it is found that there is no appreciable effect on the performance of this antenna due to bending conditions. The designed and fabricated inter-digital capacitor based antennas are found suitable for ISM band (2.45 GHz) and WLAN applications. Apart from light weight, rugged and inexpensive, the polypropylene substrate based planar metamaterial loaded antennas for Wi-Fi and public safety band applications are consistent in performance. The metamaterial loading provides better impedance matching at lower resonance frequency of the microstrip patch antenna with enhanced gain and bandwidth. The simulated, theoretical, and measured results of designed wearable antennas are found to be in good agreement.

REFERENCES

Bahl, I. J., & Bhartia, P. (1980). *Microstrip antennas*. Dedham, MA: Artech House.

Bilotti, F., Toscano, A., Vegni, L., Aydin, K., Alici, K. B., & Ozbay, E. (2007). Equivalent-circuit models for the design of metamaterials based on artificial magnetic inclusions. *IEEE Transactions on Microwave Theory and Techniques*, 55(12), 2865–2873. doi:10.1109/TMTT.2007.909611

Caloz, C., & Itoh, T. (2006). *Electromagnetic metamaterials: Transmission line theory and microwave applications*. Hoboken, NJ: Wiley Interscience, John Wiley and Sons.

Chahat, Zhadobov, Le Coq, & Sauleau. (2012). Wearable endfire textile antenna for on-body communications at 60 GHz. *IEEE Antennas and Wireless Propagation Letters*, 11, 799–802. doi:10.1109/LAWP.2012.2207698

Chen, P. Y., & Alu, A. (2010). Dual-band miniaturized elliptical patch antenna with μ–negative metamaterials. *IEEE Antennas and Propagation Letters*, 9, 351–354. doi:10.1109/LAWP.2010.2048884

Choi, S. H., Jung, T. J., & Lim, S. (2010). Flexible antenna based on compact right/left-handed transmission line. *Electronics Letters*, 46(17), 1181–1182. doi:10.1049/el.2010.1464

Garg, B. Bahl, & Ittipiboon. (2001). Microstrip antenna design handbook. Norwood, MA: Artech House.

Ha & Jung. (2011). Reconfigurable beam steering using a microstrip patch antenna with a U-slot for wearable fabric applications. *IEEE Antennas and Wireless Propagation Letters*, 10, 1228–1231. doi:10.1109/LAWP.2011.2174022

Hall, P. S., & Hao, Y. (2006). Antennas and propagation for body centric communications. In *Proceedings of First European Conference on Antennas and Propagation* (EuCAP 2006). EuCAP.

Hertleer, C., Rogier, H., Vallozzi, L., & Van Langenhove, L. (2009). A textile antenna for off-body communication integrated into protective clothing for firefighters. *IEEE Transactions on Antennas and Propagation*, 57(4), 919–925. doi:10.1109/TAP.2009.2014574

Hesselbarth, J., & Vahldieck, R. (2000). Microstrip patch antennas on corrugated substrates. In *Proceedings of 30th European Microwave Conference*. Academic Press.

Hurley, W. G., & Duffy, M. C. (1995). Calculation of self and mutual impedances in planar magnetic structures. *IEEE Transactions on Magnetics*, 31(4), 2416–2422. doi:10.1109/20.390151

Joshi, J. G., Pattnaik, S. S., Devi, S., & Lohokare, M. R. (2011). Bandwidth enhancement and size reduction of microstrip patch antenna by magneto-inductive waveguide loading. *Journal of Wireless Engineering and Technology*, 2(2), 37–44. doi:10.4236/wet.2011.22006

Joshi, J. G., & Shyam, S. Pattnaik, Devi, S., & Lohokare, M.R. (2010). Microstrip patch antenna loaded with magnetoinductive waveguide. In *Proceedings of Twelfth National Symposium on Antennas and Propagation* (APSYM 2010). APSYM.

Joshi, J.G., & Shyam, S., Pattnaik, Devi, S., & Lohokare, M.R. (2011). Frequency switching of electrically small patch antenna using metamaterial loading. *Indian Journal of Radio & Space Physics*, 40(3), 159–165.

Joshi, J. G., & Shyam, S., Pattnaik, Devi, S., & Raghavan, S. (2012). Magneto-inductive waveguide loaded microstrip patch antenna. *International Journal of Microwave and Optical Technology*, 7(1), 11–20.

Joshi, J. G., Shyam, S., Pattnaik, & Devi, S. (2012a). Metamaterial embedded wearable rectangular microstrip patch antenna. *International Journal of Antennas and Propagation.* doi:10:1155/2012/974315

Joshi, J. G., & Shyam, S., Pattnaik, & Devi, S. (2012b). Partially metamaterial ground plane loaded rectangular slotted microstrip patch antenna. *International Journal of Microwave and Optical Technology, 7*(1), 1–10.

Joshi, J. G., Shyam, S., & Pattnaik. (2013). Polypropylene based metamaterial integrated wearable microstrip patch antenna. In *Proceedings of IEEE Indian Antenna Week* (IAW 2013). Aurangabad, India: IEEE.

Joshi, J. G., & Shyam, S., Pattnaik, & Devi, S. (2013). Geo-textile based metamaterial loaded wearable microstrip patch antenna. *International Journal of Microwave and Optical Technology, 8*(1), 25–33.

Marques, R., Mesa, F., Martel, J., & Medina, F. (2003). Comparative analysis of edge and broadside coupled split ring resonators for metamaterial design: Theory and experiment. *IEEE Transactions on Antennas and Propagation, 51*(10), 2572–2581. doi:10.1109/TAP.2003.817562

Pendry, J. B., Holden, A. J., Robbins, D. J., & Stewart, W. J. (1999). Magnetism from conductors and enhanced nonlinear phenomena. *IEEE Transactions on Microwave Theory and Techniques, 47*(11), 2075–2084. doi:10.1109/22.798002

Pozar, D. M., & Schaubert, D. H. (1995). *Microstrip anternnas: The analysis and design of microstrip antennas and arrays.* Hoboken, NJ: John Wiley & Sons. doi:10.1109/9780470545270

Roh, Chi, & Lee, Tak, Nam, & Kang. (2010). Embroidered wearable multi-resonant folded dipole antenna for FM reception. *IEEE Antennas and Wireless Propagation Letters, 9*, 803–806. doi:10.1109/LAWP.2010.2064281

Sankaralingam, S., & Gupta, B. (2010). Development of textile antennas for body wearable applications and investigations on their performance under bent conditions. *Progress in Electromagnetics Research B, 22*, 53–71. doi:10.2528/PIERB10032705

Smith, D.R., & Vier, D.C., Koschny, & Soukoulis, C.M. (2005). Electromagnetic parameter retrieval from inhomogeneous metamaterials. *Physical Review E: Statistical, Nonlinear, and Soft Matter Physics, 71*, 036617-1–10. doi:10.1103/PhysRevE.71.036617

Veselago, V. G. (1968). The electrodynamics of substances with simultaneously negative values of ε and μ. *Soviet Physics - Uspekhi, 10*, 509–514. doi:10.1070/PU1968v010n04ABEH003699

Ziolkowski, R. W. (2003). Design, fabrication, and testing of double negative metamaterials. *IEEE Transactions on Antennas and Propagation, 51*(16), 1516–1529. doi:10.1109/TAP.2003.813622

KEY TERMS AND DEFINITIONS

Aspect Ratio: The length to width ratio of a rectangular microstrip patch.

Bending Effect: Effect on a wearable antenna due to the bending of a human body or body parts.

Body Area Networks (BAN): Wireless network of wearable computing devices.

Body Centric Communication: Human body and human-to-human body communication with the help of wearable antennas or implantable wireless sensors.

Double Negative (DNG) Metamaterials: Metamaterials that possesess both negative permeability and permittivity.

DPS Materials: Materials that possess both positive permeability and permittivity are called as double positive (DPS) materials.

Epsilon Negative (ENG): Metamaterials that possesess negative permittivity (epsilon).

LC resonator: A parallel connection of inductor (*L*) and capacitor (*C*).

Mu Negative (MNG) Metamaterials: Metamaterials that possesess negative permeability (mu).

NRI Materials: Materials that possesess a negative refractive index.

Polyester Substrate: Polyetser material used as a substrate to manufacture or fabricate the antenna.

Polypropylene: A geo-textile material used as a substrate to manufacture the wearable antennas.

Public Safety Band: A band of frequencies in the 4.9 GHz range dedicated to protection of human life, health, and property where point-to-point or point-to-multipoint connectivity is necessary.

Single Negative (SNG) Metamaterials: Metamaterials that possesess either negative permeability or negative permittivity.

Wearable Antennas: Cloth or textile based antennas that are integral part of wearer clothing and used for applications such as WLAN, Wi-Fi, Wi-Max, BAN, public safety, etc.

Compilation of References

3GPP. (2006). *Physical layer aspects for evolved universal terrestrial radio access (UTRA)* (Tech. Rep. No. TR 25.814 V7.1.0). 3rd Generation Partnership Project.

3GPP. (2009). *Physical channels and modulation* (Tech. Rep. No. TS 36.211 V8.8.0 Release 8). 3rd Generation Partnership Project.

3GPP. (2010). Further advancements for E-UTRA physical layer aspects (Rel.9) (Technical Report). 3GPP. TR36.814v9.

3GPP. (2011a). *TR25.367 v 10.0.0: Mobility procedures for home node B (HNB), overall description.* 3GPP.

3GPP. (2011b). TR36.902 v 9.3.1: Self-configuring and self-optimizing networks (SON) use cases and solutions. 3GPP.

3GPP. (2013). TS36.300 v 11.5.0: Technical specification group radio access network, evolved universal terrestrial radio access (E-UTRA) and evolved universal terrestrial radio access network (E-UTRAN), overall description, stage 2. 3GPP.

3GPP. TS 23.401. (n.d.). *General packet radio service (GPRS) enhancements for evolved universal terrestrial radio access network (E-UTRAN) access (release 8).* 3GPP.

3rd Generation Project Partnership (3GPP) Official Website. (2013). Retrieved June 07, 2013, from http://www.3gpp.org/

3rd Generation Project Partnership (3GPP). (1998). *The 3rd generation partnership project agreement.* Retrieved June 07, 2013, from http://www.3gpp.org/ftp/Inbox/2008_Web_files/3gppagre.pdf

3rd Generation Project Partnership (3GPP). (2005). *Requirements for support of radio resource management (FDD).* TS 25.133—V3.22.0. Retrieved June 07, 2013, from http://www.arib.or.jp/english/html/overview/doc/STD-T63v9_50/5_Appendix/R99/25/25133-3m0.pdf

3rd Generation Project Partnership (3GPP). (2008a). *Evolved universal terrestrial radio access (E-UTRA), LTE physical layer—General description (release 8).* TS 36.211—V8.3.0. Retrieved June 07, 2013, from http://www.etsi.org/deliver/etsi_ts/136200_136299/136211/08.07.00_60/ts_136211v080700p.pdf

3rd Generation Project Partnership (3GPP). (2008b). *Automatic physical cell ID assignment.* 3GPP. TSG-SA5 (Telecom Management). S5-081185.

3rd Generation Project Partnership (3GPP). (2009a). *Automatic neighbour relation (ANR) management, concepts and requirements.* 3GPP. TS 32.511—Version 9.0.0, release 9. Retrieved June 07, 2013, from http://www.3gpp.org/ftp/Specs/archive/32_series/32.511/32511-900.zip

3rd Generation Project Partnership (3GPP). (2009b). Technical specification group radio access network, requirements for evolved UTRA (E-UTRA) and evolved UTRAN (E-UTRAN). TR 25.913, Release 7.

3rd Generation Project Partnership (3GPP). (2009c). *Evolved universal terrestrial radio access, physical channels and modulation (release 8).* TS 36.211—V8.6.0. Retrieved June 07, 2013, from http://www.3gpp.org/ftp/specs/archive/36_series/36.211/36211-860.zip

3rd Generation Project Partnership (3GPP). (2009d). *Self-organizing networks (SON), concepts and requirements.* 3GPP. TS 32.500—Version 9.0.0, Release 9. Retrieved June 03, 2013, from http://www.3gpp.org/ftp/Specs/archive/32_series/32.500/32500-900.zip

3rd Generation Project Partnership (3GPP). (2009e). *Study on self-healing*. 3GPP. TR 32.823—Version 9.0.0, Release 9. Retrieved June 07, 2013, from http://www.3gpp.org/ftp/Specs/archive/32_series/32.823/32823-900.zip

3rd Generation Project Partnership (3GPP). (2009f). *SID on study on solutions for energy saving within UTRA NodeB*. RP-091439, Release 9. Retrieved June 03, 2013, from http://www.3gpp.org/ftp/tsg_ran/TSG_RAN/TSGR_46/Docs/RP-091439.zip

3rd Generation Project Partnership (3GPP). (2010a). *Self-organizing networks (SON) policy network resource model (NRM) integration reference point (IRP), requirements*. 3GPP. TS 32.521—Version 9.0.0, Release 9. Retrieved June 07, 2013, from http://www.3gpp.org/ftp/Specs/archive/32_series/32.521/32521-900.zip

3rd Generation Project Partnership (3GPP). (2010b). *Self-configuring and self-optimizing network (SON) use cases and solutions*. 3GPP. TS 36.902—Version 9.2.0, Release 9, 15. Retrieved June 07, 2013, from http://www.3gpp.org/ftp/Specs/archive/36_series/36.902/36902-920.zip

3rd Generation Project Partnership (3GPP). (2010c). *Self-organizing networks (SON) policy network resource model (NRM) integration reference point (IRP), information service (IS)*. 3GPP. TS 32.522—Version 9.1.0, Release 9, 8. Retrieved June 07, 2013, from http://www.3gpp.org/ftp/Specs/archive/32_series/32.522/32522-910.zip

3rd Generation Project Partnership (3GPP). (2010d). *Radio resource control (RRC), protocol specification*. 3GPP. TS 36.331—Version 9.3.0, Release 9. Retrieved June 07, 2013, from http://www.3gpp.org/ftp/Specs/archive/36_series/36.331/36331-930.zip

3rd Generation Project Partnership (3GPP). (2010e). *Overview of 3GPP*. Release 8. Version 0.1.0. Retrieved June 07, 2013, from http://www.3gpp.org/ftp/Information/WORK_PLAN/Description_Releases/Previous_versions/Rel-08_description_20100421.zip

3rd Generation Project Partnership (3GPP). (2010f). *Overview of 3GPP*. Release 9. Version 0.1.0. Retrieved June 07, 2013, from http://www.3gpp.org/ftp/Information/WORK_PLAN/Description_Releases/Previous_versions/Rel-09_description_20100621.zip

3rd Generation Project Partnership (3GPP). (2010g). *Self-configuration of network elements, concepts and requirements*. 3GPP. TS 32.501—Version 9.1.0, Release 9. Retrieved June 07, 2013, from http://www.3gpp.org/ftp/Specs/archive/32_series/32.501/32501-910.zip

3rd Generation Project Partnership (3GPP). (2010h). *Self-configuration of network elements integration reference point (IRP), information service (IS)*. 3GPP. TS 32.502—Version 9.2.0, Release 9. Retrieved June 07, 2013, from http://www.3gpp.org/ftp/Specs/archive/32_series/32.502/32502-920.zip

3rd Generation Project Partnership (3GPP). (2010i). Self-configuration of network elements integration reference point (IRP), common object request broker architecture (CORBA) solution set (SS). 3GPP. TS 32.503—Version 9.1.0, Release 9. Retrieved June 07, 2013, from http://www.3gpp.org/ftp/Specs/archive/32_series/32.503/32503-910.zip

3rd Generation Project Partnership (3GPP). (2010j). Evolved universal terrestrial radio access network (E-UTRAN) network resource model (NRM) integration reference point (IRP), information service (IS). 3GPP. TS 32.762—Version 9.4.0, Release 9. Retrieved June 07, 2013, from http://www.3gpp.org/ftp/Specs/archive/32_series/32.762/32762-940.zip

3rd Generation Project Partnership (3GPP). (2010k). *S1 application protocol (S1AP)*. 3GPP. TS 36.413—Version 9.3.0, Release 9. Retrieved June 07, 2013, from http://www.3gpp.org/ftp/Specs/archive/36_series/36.413/36413-930.zip

3rd Generation Project Partnership (3GPP). (2010l). Self-organizing networks (SON), policy network resource model (NRM) integration reference point (IRP), common object request broker architecture (CORBA) solution set (SS). 3GPP. TS 32.523—Version 9.0.0, Release 9. Retrieved June 07, 2013, from http://www.3gpp.org/ftp/Specs/archive/32_series/32.523/32523-900.zip

3rd Generation Project Partnership (3GPP). (2010m). *Study on energy savings management (ESM)*. 3GPP. TR 32.826—V 10.0.0. Retrieved June 03, 2013, from http://www.quintillion.co.jp/3GPP/Specs/32826-a00.pdf

3rd Generation Project Partnership (3GPP). (2010n). *Overview of 3GPP.* Release 10. Version 0.0.7. Retrieved June 03, 2013, from http://www.3gpp.org/ftp/Information/WORK_PLAN/Description_Releases/Previous_versions/Rel-10_description_20100621.zip

3rd Generation Project Partnership (3GPP). (2010o). *Self-healing concepts and requirements.* 3GPP. TS 32.541—Version 1.4.0, Release 10. Retrieved June 03, 2013, from http://www.3gpp.org/ftp/Specs/archive/32_series/32.541/32541-140.zip

3rd Generation Project Partnership (3GPP). (2010p). *Integration of device management information with Itf-N.* 3GPP. TR 32.827—Version 10.1.0, Release 10. Retrieved June 03, 2013, from http://www.3gpp.org/ftp/Specs/archive/32_series/32.827/32827-a10.zip

3rd Generation Project Partnership (3GPP). (2010q). *X2 application protocol (X2AP).* 3GPP. TS 36.423—Version 9.3.0, Release 9. Retrieved June 07, 2013, from http://www.3gpp.org/ftp/Specs/archive/36_series/36.423/36423-930.zip

3rd Generation Project Partnership (3GPP). (2011a). *About 3GPP.* Retrieved June 07, 2013, from http://www.3gpp.org/About-3GPP

3rd Generation Project Partnership (3GPP). (2011b). *Energy saving management (ESM), concepts and requirements.* 3GPP. TS 32.551—V 10.1.0. Retrieved June 03, 2013, from http://www.3gpp.org/ftp/Specs/archive/32_series/32.551/

3rd Generation Project Partnership (3GPP). (2011c). Evolved universal terrestrial radio access network (E-UTRAN), self-configuring and self-optimizing network (SON) use cases and solutions. TR 36.902. Retrieved June 03, 2013, from http://www.3gpp.org/ftp/Specs/archive/36_series/36.902/

3rd Generation Project Partnership (3GPP). (2012a). *3GPP work items on self-organizing networks*—Version 0.0.9. Retrieved June 02, 2013, from http://www.3gpp.org/ftp/Information/WORK_PLAN/Description_Releases/SON_20120924.zip

3rd Generation Project Partnership (3GPP). (2012b). *Study on operations, administration and maintenance (OAM) aspects of inter-radio-access-technology (RAT) energy saving.* TR 32.834. Retrieved June 03, 2013, from http://www.3gpp.org/ftp/Specs/archive/32_series/32.834/

3rd Generation Project Partnership (3GPP). (2013a). Universal terrestrial radio access (UTRA) and evolved universal terrestrial radio access (E-UTRA), radio measurement collection for minimization of drive tests (MDT), overall description, stage 2. TS 37.320. Retrieved June 03, 2013, from http://www.3gpp.org/ftp/Specs/archive/37_series/37.320/

3rd Generation Project Partnership (3GPP). (2013b). *Release 11.* Retrieved June 03, 2013, from http://www.3gpp.org/ftp/Information/WORK_PLAN/Description_Releases/

3rd Generation Project Partnership (3GPP). (2013c). *Release 12.* Retrieved June 03, 2013, from http://www.3gpp.org/ftp/Information/WORK_PLAN/Description_Releases/

3rd Generation Project Partnership (3GPP). (2013d). Evolved universal terrestrial radio access (E-UTRA) and evolved universal terrestrial radio access network (E-UTRAN), overall description, stage 2 (release 11). Technical Specification. TS 36.300—V11.6.0. Retrieved June 07, 2013, from http://www.3gpp.org/ftp/Specs/archive/36_series/36.300/36300-b60.zip

Abad, M. F. K., & Jamali, M. A. J. (2011). Modify LEACH algorithm for wireless sensor network. *International Journal of Computer Science*, 8(5), 219–224.

Abbasi, A. Z. (2013). *Design of workflows for context-aware applications.* (Unpublished doctoral dissertation). National University of Computer and Emerging Sciences, Karachi, Pakistan.

Abbasi, A. A., & Younis, M. (2007). A survey on clustering algorithms for wireless sensor networks. *Computer Communications*, 30(14), 2826–2841. doi:10.1016/j.comcom.2007.05.024

Abdel Hady, A. Abd El-kader, S. M., Eissa, H. S., Salem, A., & Fahmy, H. M.A. (2012). A comparative analysis of hierarchical routing protocols in wireless sensor networks. In J. H. Abawajy, M. Pathan, M. Rahman, A. K. Pathan, & M. M. Deris (Eds.), Internet and distributed computing advancements: Theoretical frameworks and practical applications (pp. 212-246). Hershey, PA: IGI Global.

Abdul Hadi, F, A. H., Rozahe, A. R., Norsheila, F., Kamilah, S. Y., Sharifah, H.S.A., & Liza, L. (2009). Development of IEEE802. 15.4 based wireless sensor network platform for image transmission. *International Journal of Engineering & Technology*, 9(10), 112–118.

Abdullah, A. A., Cai, L., & Gebali, F. (2012). DSDMAC: Dual sensing directional MAC protocol for ad hoc networks with directional antennas. *IEEE Transactions on Vehicular Technology*, 61(3), 1266–1275. doi:10.1109/TVT.2012.2187082

Abolhasan, M., Wysocki, T., & Dutkiewicz, E. (2004). A review of routing protocols for mobile ad hoc networks. *Ad Hoc Networks*, 2(1), 1–22. doi:10.1016/S1570-8705(03)00043-X

Adams, J. (2006). An introduction to IEEE STD 802.15. 4. In *Proceedings of Aerospace Conference, 2006 IEEE*. IEEE. Retrieved from http://ieeexplore.ieee.org/xpls/abs_all.jsp?arnumber=1655947

Aghdasi, H. S., & Abbaspour, M. (2008). ET-MAC: An energy-efficient and high throughput MAC protocol for wireless sensor networks. In *Proceedings of Communication Networks and Services Research Conference*, (pp. 526-532). Academic Press.

Agre, J., & Clare, L. (2000). An integrated architecture for cooperative sensing networks. *Proceedings of Computer*, 33(5), 106–108. doi: doi:10.1109/2.841788

Ahmed, A., & Fisal, N. (2011). Secure real-time routing protocol with load distribution in wireless sensor networks. *Security and Communication Networks*, 839–859. doi:10.1002/sec

Ahmed, Q. Z., & Yang, L.-L. (2007). Performance of hybrid direct-sequence time-hopping ultrawide bandwidth systems in Nakagami-m fading channels. In *Proceedings IEEE 18th International Symposium on Personal, Indoor and Mobile Radio Communications*. IEEE.

Ahmed, Q. Z., & Yang, L.-L. (2008). Normalised least mean-square aided decision-directed adaptive detection in hybrid direct-sequence time-hopping UWB systems. In *Proceedings IEEE 69th Vehicular Technology Conference*. IEEE.

Ahmed, A., & Fisal, N. (2010). Secure real-time routing protocol with load distribution in wireless sensor networks. *Security and Communication Networks*, 4(8), 839–869. doi:10.1002/sec.214

Ahmed, F., Tirkkonen, O., Peltomäki, M., Koljonen, J. M., Yu, C. H., & Alava, M. (2010). Distributed graph coloring for self-organization in LTE networks. *Journal of Electrical and Computer Engineering*. doi:10.1155/2010/402831

Ahmed, Q. Z., & Yang, L.-L. (2010). Reduced-rank adaptive multiuser detection in hybrid direct-sequence time-hopping ultrawide bandwidth systems. *IEEE Transactions on Wireless Communications*, 9(1), 156–167. doi:10.1109/TWC.2010.01.081172

Ai, J., & Abouzeid, A. (2006). Coverage by directional sensors in randomly deployed wireless sensor networks. *Journal of Combinatorial Optimization*, 1–11. Retrieved from http://link.springer.com/article/10.1007/s10878-006-5975-x

Akan, O. B., Karli, O. B., & Ergul, O. (2009). Cognitive radio sensor networks. *IEEE Network*, 23(4), 34–40. doi:10.1109/MNET.2009.5191144

Akyildiz, I. (2007). Wireless multimedia sensor networks: A survey. *IEEE Wireless Communications*, 32–39. Retrieved from http://ieeexplore.ieee.org/xpls/abs_all.jsp?arnumber=4407225

Akyildiz, I. F. (Ed.). (2010). *Wireless sensor networks*. Hoboken, NJ: John Wiley & Sons Ltd. doi:10.1002/9780470515181

Akyildiz, I. F., Lee, W., & Chowdhury, K. R. (2009). CRAHNS: Cognitive radio ad hoc networks. *Ad Hoc Networks*, 7(5), 810–836. doi:10.1016/j.adhoc.2009.01.001

Akyildiz, I. F., Melodia, T., & Chowdhury, K. (2007). A survey on wireless multimedia sensor networks. *Computer Networks*, 51(4), 921–960. doi:10.1016/j.comnet.2006.10.002

Akyildiz, I. F., & Wang, X. D. (2005). A survey on wireless mesh networks. *IEEE Radio Communications, 43*(9), 23–30. doi:10.1109/MCOM.2005.1509968

Akyildiz, I. F., Wang, X., & Wang, W. (2005). Wireless mesh networks: A survey. *Computer Networks, 47*, 445–487. doi:10.1016/j.comnet.2004.12.001

Akyildiz, I., Su, W., Sankarasubramaniam, Y., & Cayirci, E. (2002a). A survey on sensor networks. *IEEE Communications Magazine, 40*(8), 102–114. doi:10.1109/MCOM.2002.1024422

Akyildiz, I., Su, W., Sankarasubramaniam, Y., & Cayirci, E. (2002b). Wireless sensor networks: A survey. *Computer Networks, 38*(4), 393–422. doi:10.1016/S1389-1286(01)00302-4

Alabri, H. M., Mukhopadhyay, S. C., Punchihewa, G. A., Suryadevara, N. K., & Huang, Y. M. (2012). Comparison of applying sleep mode function to the smart wireless environmental sensing stations for extending the life time. In *Proceedings of 2012 IEEE International Instrumentation and Measurement Technology Conference* (pp. 2634–2639). Graz, Austria: IEEE.

Alazawi, Z., Altowaijri, S., & Mehmood, R. (2011). Intelligent disaster management system based on cloud-enabled vehicular networks. In *Proceedings of 11th International Conference on ITS Telecommunications (ITST)* (pp. 361-368). St. Petersburg, Russia: IEEE.

Albesa, J., Casas, R., Penella, M. T., & Gasulla, M. (2007). REALnet: An environmental WSN testbed. In *Proceedings of International Conference on Sensor Technologies and Applications, SensorComm 2007* (pp. 502-507). Valencia, Spain: SensorComm.

Alemdar, H., & Ersoy, C. (2010). Wireless sensor networks for healthcare: A survey. *Computer Networks, 54*(15), 2688–2710. doi:10.1016/j.comnet.2010.05.003

Ali-Yahiya, T. (2011). *Understanding LTE and its performance*. Academic Press. doi:10.1007/978-1-4419-6457-1

Al-Jaroodi, J., & Mohamed, N. (2012). Service-oriented middleware: A survey. *Journal of Network and Computer Applications, 35*(1), 211–220. doi:10.1016/j.jnca.2011.07.013

Allman, M. (1998). On the generation and use of tcp acknowledgements. *ACM Computer Communication Review, 28*, 1114–1118.

Almalkawi, I. T., Zapata, M. G., Al-Karaki, J. N., & Morillo-Pozo, J. (2010). Wireless multimedia sensor network: Current trends and future directions. *Sensors (Basel, Switzerland), 10*(7), 6662–6717. doi:10.3390/s100706662 PMID:22163571

Al-Omari, S. A. K., & Sumari, P. (2010). An overview of mobile ad hoc networks for the existing protocols and applications. *International Journal on Applications of Graph Theory in Wireless Ad hoc Networks and Sensor Networks, 2*(1), 87-110.

Altman, E., & Jimenez, T. (2003). Novel delayed ACK techniques for improving TCP performance in multihop wireless networks. In *Proceeding of IFIP International Federation for Information Processing* (pp. 237-250). Venice, Italy: IFIP.

Amin, K., Laszewski, G. V., & Mikler, A. R. (2004). Toward an architecture for ad hoc grids. In *Proceedings of 12th International Conference on Advanced Computing and Communications* (pp. 1-5). Ahmedabad, India: IEEE.

Amirijoo, M., Frenger, P., Gunnarsson, F., Kallin, H., Moe, J., & Zetterberg, K. (2008). Neighbor cell relation list and physical cell identity self-organization in LTE. In *Proceeding of IEEE International Conference on Communications Workshops (ICC'08)*. IEEE.

Amirijoo, M., Frenger, P., Gunnarsson, F., Moe, J., & Zetterberg, K. (2013). On self-optimization of the random access procedure in 3G long term evolution. In *Proceedings of IFIP/IEEE International Symposium on Integrated Network Management (IM)-Workshops*. New York: IEEE.

Amirrudin, N. A., Ariffin, S. H. S., Malik, N. N. N. A., & Ghazali, N. E. (2013). Mobility prediction via Markov model in LTE femtocell. *International Journal of Computers and Applications, 65*(18), 40–44.

Amuthan, A., & Baradwaj, B. A. (2011). Secure routing scheme in MANETs using secret key sharing. *International Journal of Computers and Applications, 22*(1).

Ancillotti, E., Bruno, R., Conti, M., & Pinizzott, A. (2009). Dynamic address autoconfiguration in hybrid ad hoc networks. *Pervasive and Mobile Computing*, 5(4), 300–317. doi:10.1016/j.pmcj.2008.09.008

Andrews, J. G., Claussen, H., Dohler, M., Rangan, S., & Reed, M. C. (2012). Femtocells: Past, present, and future. *IEEE Journal on Selected Areas in Communications*, 30(3), 497–508. doi:10.1109/JSAC.2012.120401

Andrews, J. G., Ganti, R. K., Haenggi, M., Jindal, N., & Weber, S. (2010). A primer on spatial modelling and analysis in wireless networks. *IEEE Communications Magazine*, 48(11), 156–163. doi:10.1109/MCOM.2010.5621983

Andrews, J., Baccelli, F., & Ganti, R. (2010). A tractable approach to coverage and rate in cellular networks. *IEEE Transactions on Communications*, (99): 1–13.

Aqeel-ur-Rehman & Shaikh. Z. A. (2008). Towards design of context-aware sensor grid framework for agriculture. In *Proceedings of Fifth International Conference on Information Technology (ICIT)* (pp. 244-247). Rome, Italy: WASET.

Ardagna, C. A., Cremonini, M., Damiani, E., De Capitani di Vimercati, S., & Samarati, P. (2006). Supporting location-based conditions in access control policies. In *Proceedings of the 2006 ACM Symposium on Information, Computer and Communications Security*, (pp. 212-222). ACM.

Arora, A., Krunz, M., & Muqattash, A. (2004). Directional medium access protocol (DMAP) with power control for wireless ad hoc networks. In *Proceedings of Global Telecommunications Conference, 2004*. IEEE.

Ateniese, G., Santis, A. D., Ferrara, A. L., & Masucci, B. (2006). Provably-secure time-bound hierarchical key assignment schemes. In *Proceedings of CCS'06*. Alexandria, VA: ACM.

Atiquzzaman, M., & Reaz, A. (2005). Survey and classification of transport layer mobility management schemes. In Proceedings of Personal, Indoor and Mobile Radio Communications, 2005 (Vol. 6151, pp. 2109–2115). IEEE.

Attar, A., Tang, H., Vasilakos, A. V., Yu, F. R., & Leung, V. C. M. (2012). A survey of security challenges in cognitive radio networks: Solutions and future research directions. *Proceedings of the IEEE*, 100(12), 3172–3186. doi:10.1109/JPROC.2012.2208211

Auer, G., Giannini, V., Godor, I., Skillermark, P., Olsson, M., Imran, M., & Blume, O. (2011). Cellular energy efficiency evaluation framework. In *Proceedings of IEEE Vehicular Technology Conference* (VTC Spring). IEEE.

Awada, A., Wegmann, B., Rose, D., Viering, I., & Klein, A. (2011). Towards self-organizing mobility robustness optimization in inter-RAT scenario. In *Proceeding of the 73rd IEEE Vehicular Technology Conference*. IEEE.

Azevedo, T. S., Bezerra, R. L., Campos, C. A. V., & De Moraes, L. F. M. (2009). An analysis of human mobility using real traces. In *Proceedings of 2009 IEEE Wireless Communications and Networking Conference*. IEEE. doi:10.1109/WCNC.2009.4917569

Babadi, B., & Tarokh, V. (2008). A distributed asynchronous algorithm for spectrum sharing in wireless ad hoc networks. In *Proceedings of the 42nd Annual Conference on Information Sciences and Systems (CISS '08)*. CISS.

Baccelli, F., & Blaszczyszyn, B. (2010a). Stochastic geometry and wireless networks: Applications. *Foundations and Trends in Networking*, 4(1-2), 1–312. doi:10.1561/1300000026

Baccelli, F., & Blaszczyszyn, B. (2010b). Stochastic geometry and wireless networks: Theory. *Foundations and Trends in Networking*, 3(3-4), 249–449. doi:10.1561/1300000006

Baccelli, F., Klein, M., Lebourges, M., & Zuyev, S. (1997). Stochastic geometry and architecture of communication networks. *Telecommunication Systems*, 7(1-3), 209–227. doi:10.1023/A:1019172312328

Baccelli, F., Miihlethaler, P., & Blaszczyszyn, B. (2009). Stochastic analysis of spatial and opportunistic Aloha. *IEEE Journal on Selected Areas in Communications*, 27(7), 1105–1119. doi:10.1109/JSAC.2009.090908

Baccelli, F., & Zuyev, S. (1996). Stochastic geometry models of mobile communication networks. In *Frontiers in queueing: Models and applications in science and engineering*. Academic Press.

Badarneh, O. S., & Kadoch, M. (2009). Multicast routing protocols in mobile ad hoc networks: A comparative survey and taxonomy. *EURASIP Journal on Wireless Communications and Networking*, (1): 1–52.

Badic, O'Farrell, Loskot, & He. (n.d.). Energy efficient radio access architectures for green radio: Large versus small cell size deployment. In *Proceedings of IEEE Vehicular Technology Conference*. IEEE.

Bagci, F. (2008). ESTR-Energy saving token ring protocol for wireless sensor networks. In *Proceedings of International Conference on Wireless Networks*, (pp. 3-9). IEEE.

Bahl, I. J., & Bhartia, P. (1980). *Microstrip antennas*. Dedham, MA: Artech House.

Bakre, B., & Badrinath, R. (1995). I-TCP: Indirect TCP for mobile hosts. In *Proceeding of the 15th Int. Conf. Distributed Computing Systems* (pp. 136 – 143). Vancouver, Canada: IEEE.

Balachandan, R. K., Zou, X., Ramamurthy, B., & Thukral, A. (2007). An efficient and attack resistant agreement scheme for secure group communications in mobile ad-hoc networks. In *Wireless communication and mobile computing*. Hoboken, NJ: Wiley.

Balakrishnan, H., Seshan, S., Amir, E., & Katz, R. (1995). Improving tcp/ip performance over wireless networks. In *Proceeding of 1st ACM Mobicom* (pp. 2-11). ACM.

Balanis, C. A. (2012). *Antenna theory: analysis and design*. Wiley-Interscience.

Baldini, G., Sturman, T., & Biswas, A. R. (2012). Security aspects in software defined radio and cognitive radio networks: A survey and a way ahead. *IEEE Communications. Surveys & Tutorials*, 14(2), 355–379. doi:10.1109/SURV.2011.032511.00097

Baliosian, J., & Stadler, R. (2007). Decentralized configuration of neighboring cells for radio access networks. In *Proceeding of the 1st IEEE Workshop on Autonomic Wireless Access (in Conjunction with IEEE WoWMoM)*. Helsinki, Finland: IEEE.

Balon, M., & Liau, B. (2012). Mobile virtual network operator. In *Proceedings of XVth International Telecommunications Network Strategy and Planning Symposium (NETWORKS)* (pp. 1-6). IEEE.

Bandh, T. (2011). *The SOCRATES SON-function coordination concept* (Technical report). Retrieved June 07, 2013, from http://www.net.in.tum.de/fileadmin/bibtex/publications/papers/socrates.pdf

Bandh, T., Carle, G., & Sanneck, H. (2009). Graph coloring based physical-cell-ID assignment for LTE networks. In *Proceeding of the ACM International Wireless Communications and Mobile Computing Conference. (IWCMC '09)*. Leipzig, Germany: ACM.

Bandh, T., Romeikat, R., & Sanneck, H. (2011). Policy-based coordination and management of SON functions. In *Proceeding of IM 2011*. IM.

Bandh, T., Sanneck, H., & Romeikat, R. (2011). An experimental system for SON function coordination. In *Proceeding of the 73rd Vehicular Technology Conference (VTC Spring)*. IEEE.

Banerjee, T., Xie, B., Jun, J. H., & Agarwal, D. P. (2007). LIMOC: Enhancing the Lifetime of a Sensor Network with Mobile Cluster heads. In *Proceedings of the Vehicular Technology Conference Fall* (pp. 133-137), Baltimore, MD, USA: IEEE.

Barakah, D. M., & Ammad-Uddin, M. (2012). A survey of challenges and applications of wireless body area network (WBAN) and role of a virtual doctor server in existing architecture. In *Proceedings of 2012 Third International Conference on Intelligent Systems, Modelling and Simulation (ISMS)*. Sabah, Malaysia: ISMS.

Basagni, S., Conti, M., Giordano, S., & Stojmenovic, I. (2004). *Mobile ad hoc networking*. Hoboken, NJ: Wiley-IEEE Press. doi:10.1002/0471656895

Bazan, O., & Jaseemuddin, M. (2011). On the design of opportunistic MAC protocols for multihop wireless, networks with beamforming antennas. *IEEE Transactions on Mobile Computing*, 10(3), 305–319. doi:10.1109/TMC.2010.68

Bazan, O., & Jaseemuddin, M. (2012). A survey on MAC protocols for wireless ad hoc networks with beamforming antennas. *IEEE Communications Surveys & Tutorials*, 14(2), 216–239. doi:10.1109/SURV.2011.041311.00099

Bencini, L., Collodi, G., Di Palma, D., Manes, G., & Manes, A. (2010). An energy efficient cross layer solution based on smart antennas for wireless sensor network applications. In *Proceedings of Sensor Technologies and Applications (SENSORCOMM)*. IEEE. doi:10.1109/SENSORCOMM.2010.43

Benenati, D., Feder, P. M., Lee, N. Y., Martin-Leon, S., & Shapira, R. (2002). A seamless mobile VPN data solution for CDMA2000,* UMTS, and WLAN users. *Bell Labs Technical Journal*, *7*(2), 143–165. doi:10.1002/bltj.10010

Ben-Othman, J., Mokdad, L., & Cheikh, M. O. (2011). A new architecture of wireless mesh networks based IEEE 802.11 s directional antennas. In *Proceedings of Communications (ICC)*. IEEE.

Bensaou, B., Wang, Y., & Ko, C. C. (2000). Fair media access in 802.11 based wireless ad-hoc networks. In *Proceeding of Mobihoc* (pp. 99–106). Boston, MA: ACM.

Bertino, E., & Kirkpatrick, M. S. (2011). Location-based access control systems for mobile users: concepts and research directions. In *Proceedings of the 4th ACM SIGSPATIAL International Workshop on Security and Privacy in GIS and LBS*, (pp. 49-52). ACM.

Bertino, E., Samarati, P., & Jajodia, S. (1993). Authorizations in relational database management systems. In *Proceedings of 1st ACM Conf. on Computer and Commun. Security*, (pp. 130 -139). ACM.

Bettstetter, C., Hartenstein, H., & Perez-Costa, X. (2004). Stochastic properties of the random-way point mobility model. *Wireless Networks*, *10*(5), 555–567. doi:10.1023/B:WINE.0000036458.88990.e5

Bhargavan, V., Demers, A., Shenker, S., & Zhang, L. (1994). MACAW, a media access protocol for wireless LANs. In *Proceeding of ACM SIGCOMM* (pp. 212-225). London, UK: ACM.

Bhaskaran, R., & Madheswaran, R. (2010). Performance analysis of congestion control in mobile adhoc grid layer. *International Journal of Computers and Applications*, *1*(20), 102–110.

Biaz, S., & Vaidya, N. H. (1998). Distinguishing congestion losses from wireless transmission losses: A negative result. In *Proceeding of IEEE 7th Int. Conf. on Computer Communications and Networks* (pp. 722-731). IEEE.

Bichler, M., & Lin, K. (2006). Service-oriented computing. *IEEE Computer*, *39*(3), 88–90.

Bilotti, F., Toscano, A., Vegni, L., Aydin, K., Alici, K. B., & Ozbay, E. (2007). Equivalent-circuit models for the design of metamaterials based on artificial magnetic inclusions. *IEEE Transactions on Microwave Theory and Techniques*, *55*(12), 2865–2873. doi:10.1109/TMTT.2007.909611

Biswas, J., Barai, M., & Nandy, S. K. (2004). Efficient hybrid multicast routing protocol for ad-hoc wireless networks. In *Proceedings of 29th Annual IEEE International Conference on Local Computer Networks* (pp. 180-187). Tampa, FL: IEEE.

Biswas, P. K., Qi, H., & Xu, Y. (2008). Mobile-agent-based collaborative sensor fusion. *Information Fusion*, *9*, 399–411. doi:10.1016/j.inffus.2007.09.001

Blanche, S. (2013). *Linux for amateur radio applications*. Retrieved July 28, 2013, from http://www.qsl.net/vk2kfj/linux.html

Bolotnyy, L., & Robins, G. (2007). Physically unclonable function-based security and privacy in rfid systems. In *Proceedings of Fifth Annual IEEE International Conference on Pervasive Computing and Communications, 2007*, (pp. 211–220). IEEE.

Boneh, D. (1998). The decision diffe hellman problem. In *Proceedings of Third Algorithomic Number Theory Symposium*, (pp. 48-63). Portland, OR: Springer.

Bonivento, A., & Fischione, C. (2007). System level design for clustered wireless sensor networks. *Industrial Informatics*, *3*(3), 202–214. doi:10.1109/TII.2007.904130

Boregowda, S. B., Babu, N. V., Puttamadappa, C., & Mruthyunjaya, H. S. (2011). Energy balanced fixed clustering protocol for wireless sensor networks. *International Journal of Computer Science and Network Security*, *11*(8), 166–172.

Boudour, G., Lecointre, A., Berthou, P., Dragomirescu, D., & Gayraud, T. (2007). On designing sensor networks with smart antennas. In *Proceedings of 7th IFAC International Conference on Fieldbuses and Networks in Industrial and Embedded Systems*. IFAC.

Boukerche, A., & Du, Y. (2008). A reliable synchronous transport protocol for wireless image sensor networks. In *Proceedings of IEEE Symposium on Computers and Communications,* (pp. 1083–1089). IEEE. Retrieved from http://ieeexplore.ieee.org/xpls/abs_all.jsp?arnumber=4625679

Boukerche, A. (Ed.). (2009). *Algorithms and protocols for wireless sensor networks.* Hoboken, NJ: John Wiley & Sons, Inc.

Boukerche, A., Turgut, B., Aydin, N., & Mohammad, Z., Ahmad, Bölöni, L., & Turgut, D. (2011). Routing protocols in ad hoc networks: A survey. *Computer Networks, 55*(13), 3032–3080. doi:10.1016/j.comnet.2011.05.010

Boulis, A. (2011). *Castalia: A simulator for wireless sensor networks and body area networks.* Retrieved from http://castalia.npc.nicta.com.au/pdfs/Castalia%20-%20User%20Manual.pdf

Boyd, S., El Ghaoui, L., Feron, E., & Balakrishnan, V. (1994). *Linear matrix inequalities in system and control theory.* Philadelphia, PA: SIAM. doi:10.1137/1.9781611970777

Bragaa, R. B., Chavesb, I. A., Oliveiraa, C. T. D., Andradeb, R. M. C., Souzab, J. N. D., Martina, H., & Schulzec, B. (2013). RETENTION: A reactive trust-based mechanism to detect and punish malicious nodes in ad hoc grid environments. *Journal of Network and Computer Applications, 36*(1), 274–283. doi:10.1016/j.jnca.2012.06.002

Breining, C., Dreiscitel, P., Hansler, E., Mader, A., Nitsch, B., & Puder, H. et al. (1999). Acoustic echo control: An application of very-high-order adaptive filters. *IEEE Signal Processing Magazine, 16,* 42–69. doi:10.1109/79.774933

Brown, K., & Singh, S. (1997). M-TCP: TCP for mobile cellular networks. *ACM Computer Communications Review, 27,* 19–43. doi:10.1145/269790.269794

Brown, T. X. (2000). Cellular performance bounds via shotgun cellular systems. *IEEE Journal on Selected Areas in Communications, 18*(11), 2443–2455. doi:10.1109/49.895048

Buchholcz, G., Ziegler, T., & van Do, T. (2005). TCP-ELN: On the protocol aspects and performance of explicit loss notification for TCP over wireless networks. In *Proceedings of 1st International Conference on Wireless Internet* (pp. 172–179). IEEE. Retrieved from http://ieeexplore.ieee.org/xpls/abs_all.jsp?arnumber=1509652

Budhwani, S., Sarkar, M., & Nagaraj, S. (2010). A MAC layer protocol for sensor networks using directional antennas. In *Proceedings of Sensor Networks, Ubiquitous, and Trustworthy Computing (SUTC).* IEEE. doi:10.1109/SUTC.2010.11

Bullington, K. (1957). Radio propagation fundamentals. *The Bell System Technical Journal, 36*(3), 593–625. doi:10.1002/j.1538-7305.1957.tb03855.x

Burda, R., & Wietfeld, C. (2007). Multimedia over 802.15.4 and ZigBee networks for ambient environment control. In *Proceedings of Vehicular Technology Conference.* IEEE.

Butler, M. (2011). *Dynamic risk assessment access control.* (Master's thesis). The University of Tulsa, Tulsa, OK.

Butler, M., Hawrylak, P., & Hale, J. (2011). Graceful privilege reduction in RFID security. In *Proceedings of the Seventh Annual Workshop on Cyber Security and Information Intelligence Research.* IEEE.

Butler, M., Reed, S., Hawrylak, P. J., & Hale, J. (2013). Implementing graceful RFID privilege reduction. In *Proceedings of the Eighth Annual Cyber Security and Information Intelligence Research Workshop.* IEEE.

Buyya, A., Yeo, C. S., & Venugopal, S. (2009). Market-oriented cloud computing: Vision, hype, and reality for delivering it service s as computing utilities. In *Proceedings of 9th IEEE/ACM International Symposium Cluster Computing and the Grid.* Shanghai, China: IEEE.

Cabric, D., Mishra, S. M., & Brodersen. (2004). Implementation issues in spectrum sensing for cognitive radios. In *Proceedings of Asilomar Conference on Signals, Systems & Computers* (pp. 772 - 776). Pacific Grove, CA: IEEE.

Cacciapuoti, A. S., Akyildiz, I. F., & Paura, L. (2012). Correlation-aware user selection for cooperative spectrum sensing in cognitive radio ad hoc networks. *IEEE Journal on Selected Areas in Communications, 30*(2), 297–306. doi:10.1109/JSAC.2012.120208

Calder, D. (2000). *FPAC sysop manual*. Retrieved May 24, 2013, from http://www.n4zkf.com/SYS_FPAC.htm

Callon, R., & Suzuki, M. (2005). *A framework for layer 3 provider-provisioned virtual private networks (ppvpns). Request for Comments (RFC) 4110*. IETF.

Caloz, C., & Itoh, T. (2006). *Electromagnetic metamaterials: Transmission line theory and microwave applications*. Hoboken, NJ: Wiley Interscience, John Wiley and Sons.

Campista, M. E. M., Esposito, P. M., Moraes, I. M., Costa, L. H. M. K., Duarte, O. C. M. B., & Passos, D. G. et al. (2008). Routing metrics and protocols for wireless mesh networks. *IEEE Network, 22*(1), 6–12. doi:10.1109/MNET.2008.4435897

Camp, T., Boleng, J., & Davies, V. (2002). A survey of mobility models for ad hoc network research. *Wireless Communications and Mobile Computing, 2*(5), 483–502. doi:10.1002/wcm.72

Canberk, B., & Oktug, S. (2012). A dynamic and weighted spectrum decision mechanism based on SNR tracking in CRAHNS. *Ad Hoc Networks, 10*(6), 752–759. doi:10.1016/j.adhoc.2011.02.006

Canourgues, L., Lephay, J., Soyer, L., & Beylot, A. L. (2006). Stamp: Shared-tree ad-hoc multicast protocol. In *Proceedings of IEEE Military Communications Conference* (pp. 1-7). Washington, DC: IEEE.

Cao, Y., & Xie, S. (2005). A position based beaconless routing algorithm for mobile ad hoc networks. In *Proceedings of International Conference on Communications, Circuits and Systems* (pp. 303-307). Hong Kong, China: IEEE.

Capone, A., Martignon, F., & Fratta, L. (2008). Directional MAC and routing schemes for power controlled wireless mesh networks with adaptive antennas. *Ad Hoc Networks, 6*(6), 936–952. doi:10.1016/j.adhoc.2007.08.002

Capozzi, F., Piro, G., Grieco, L. A., Boggia, G., & Camarda, P. (2012). On accurate simulations of LTE femtocells using an open source simulator. *EURASIP Journal on Wireless Communications and Networking*, (1): 1–13. doi: doi:10.1186/1687-1499-2012-328

Carugi, M., & De Clercq, J. (2004). Virtual private network services: Scenarios, requirements and architectural constructs from a standardization perspective. *IEEE Communications Magazine, 42*(6), 116–122. doi:10.1109/MCOM.2004.1304246

Cearley, D. (2012). Gartner identifies the top 10 strategic technology trends for 2013. *Gartner Press Release*. Retrieved August 11, 2013, http://www.gartner.com/newsroom/id/2209615

Center for Future Health. (2010, June 1). Retrieved November 20, 2012, from http://www.futurehealth.rochester.edu/smart_home/

Cerpa, A., Elson, J., Estrin, D., Girod, L., Hamilton, M., & Zhao, J. (2001). Habitat monitoring: Application driver for wireless communications technology. In *Proceedings of ACM SIGCOMM Workshop on Data Communications* (pp. 20-41). ACM Press. doi: 10.1145/844193.844196

Cesana, M., Cuomo, F., & Ekici, E. (2010). Routing in cognitive radio networks: Challenges and solutions. *Ad Hoc Networks, 9*(3), 228–248. doi:10.1016/j.adhoc.2010.06.009

Chahat, Zhadobov, Le Coq, & Sauleau. (2012). Wearable endfire textile antenna for on-body communications at 60 GHz. *IEEE Antennas and Wireless Propagation Letters, 11*, 799–802. doi:10.1109/LAWP.2012.2207698

Chakraborty, D., Joshi, A., Yesha, Y., & Finin, T. (2002). GSD: A novel group-based service discovery protocol for MANETs. In *Proceedings of 4th IEEE Conference on Mobile and Wireless Communications Networks (MWCN)* (pp. 140-144). Baltimore, MD: IEEE.

Chakraborty, D., Perich, F., Avancha, S., & Joshi, A. (2001). DReggie: Semantic service discovery for m-commerce applications. In *Proceedings of Workshop on Reliable and Secure Applications in Mobile Environment. In Conjunction with 20th Symposium on Reliable Distributed Systems (SRDS)* (pp. 1-6). New Orleans, LA: STDS.

Chandran, K. (1998). A feedback based scheme for improving TCP performance in ad-hoc wireless networks. In *Proceeding International Conference on Distributed Computing Systems* (pp. 472-479). Amsterdam: IEEE.

Chandran, K., Raghunathan, S., Venkatesan, S., & Prakash, R. (1998). A feedback based scheme for improving TCP performance in ad-hoc wireless networks. In *Proceedings of International Conference on Distributed Computing Systems (ICDCS'98)* (pp. 34-39). Amsterdam: IEEE.

Chandrasekhar, V. (2008). Femtocell networks: A survey. *IEEE Communications Magazine*, 1–23.

Changa, H., Kanb, H., & Hob, M. (2012). Adaptive TCP congestion control and routing schemes using cross-layer information for mobile ad hoc networks. *Computer Communications*, *35*(4). PMID:22267882

Chang, J. J., Liao, W., & Hou, T. C. (2009). Reservation-based directional medium access control (RDMAC) protocol for multi-hop wireless networks with directional antennas. In *Proceedings of Communications, 2009*. IEEE. doi:10.1109/ICC.2009.5199410

Chang, R. W. (1996). Synthesis band-limited orthogonal signals for multichannel data transmission. *The Bell System Technical Journal*, *45*(10), 1775–1796. doi:10.1002/j.1538-7305.1966.tb02435.x

Chatterjee, M., Das, S. K., & Turgut, D. (2002). A weighted clustering algorithm for mobile ad hoc networks. *Journal of Cluster Computing*, *5*(2), 193–204. doi:10.1023/A:1013941929408

Chen, F. (2009). Improving IEEE 802.15. 4 for low-latency energy-efficient industrial applications. *Aktuelle Anwendungen in Technik und Wirtschaft*, 61-70.

Chen, Y., Fleury, E., & Razafindralambo, T. (2009). Scalable address allocation protocol for mobile ad hoc networks. In *Proceedings of 5th International Conference on Mobile Ad-Hoc and Sensor Networks* (pp. 41-48). WuYi Mountain, China: IEEE.

Chen, K. C. (1994). Medium access protocols of wireless LANs for mobile computing. *IEEE Network*, *8*(5), 50–63. doi:10.1109/65.313014

Chen, L., Jiang, C., & Li, J. (2008). VGITS: ITS based on intervehicle communication networks and grid technology. *Journal of Network and Computer Applications*, *31*, 285–302. doi:10.1016/j.jnca.2006.11.002

Chen, M., Gonzalez, S., Leung, V., Zhang, Q., & Li, M. (2010). A 2G-RFID-based e-healthcare system. *IEEE Wireless Communications*, *17*(1), 37–43. doi:10.1109/MWC.2010.5416348

Chen, P. Y., & Alu, A. (2010). Dual-band miniaturized elliptical patch antenna with μ–negative metamaterials. *IEEE Antennas and Propagation Letters*, *9*, 351–354. doi:10.1109/LAWP.2010.2048884

Chen, Y., Zhang, S., Xu, S., & Li, G. (2011). Fundamental trade-offs on green wireless networks. *IEEE Communications Magazine*, 49.

Chetan, S., Kumar, G., Dinesh, K., & Mathew, K. (n.d.). *Cloud computing for mobile world*. National Institute of Technology. Retrieved from http://chetan.ueuo.com/projects/CCMW.pdf

Chin, K. W. (2007). SpotMAC: A pencil-beam MAC for wireless mesh networks. In *Proceedings of Computer Communications and Networks, 2007*. IEEE. doi:10.1109/ICCCN.2007.4317801

Chlamtac, I., Conti, M., & Liu, J. J. (2003). Mobile ad hoc networking: Imperatives and challenges. *Ad Hoc Networks*, *1*(1), 13–64. doi:10.1016/S1570-8705(03)00013-1

Choi, S. H., Jung, T. J., & Lim, S. (2010). Flexible antenna based on compact right/left-handed transmission line. *Electronics Letters*, *46*(17), 1181–1182. doi:10.1049/el.2010.1464

Choi, W., & Lee, S. (2011). Implementation of the IEEE 802.15.4 module with CFP in NS-2. *Telecommunication Systems*, *52*(4), 2347–2356. doi:10.1007/s11235-011-9548-7

Chong, C.-Y., & Kumar, S. (2003). Sensor networks: Evolution, opportunities, and challenges. *Proceedings of the IEEE*, *91*(8), 1247–1256. doi:10.1109/JPROC.2003.814918

Choudhury, R. R., & Vaidya, N. H. (2004). Deafness: A MAC problem in ad hoc networks when using directional antennas. In *Proceedings of Network Protocols, 2004*. IEEE. doi:10.1109/ICNP.2004.1348118

Choudhury, R. R., & Vaidya, N. H. (2004). MAC-layer anycasting in ad hoc networks. *ACM SIGCOMM Computer Communication Review*, *34*(1), 75–80. doi:10.1145/972374.972388

Choudhury, R. R., & Vaidya, N. H. (2005). Performance of ad hoc routing using directional antennas. *Ad Hoc Networks*, *3*(2), 157–173. doi:10.1016/j.adhoc.2004.07.004

Chowdhury, K. R., Felice, M. D., & Akyildiz. (2009). TP-CRAHN: A transport protocol for cognitive radio ad-hoc networks. In *Proceedings of IEEE Infocom* (pp. 2482-2490). IEEE.

Chowdhury, K. R., & Melodia, T. (2010). Platforms and testbeds for experimental evaluation of cognitive ad hoc networks. *IEEE Communications Magazine*, *48*(9), 96–104. doi:10.1109/MCOM.2010.5560593

Chowdhury, K., & Felice, M. (2009). Search: A routing protocol for mobile cognitive radio ad-hoc networks. *Computer Communications*, *32*(18), 1–6. doi:10.1016/j.comcom.2009.06.011

Cisco. (2004). *Comparing MPLS-based VPNs, IPSec-based VPNs, and a combined approach from Cisco Systems* (Technical Report). Cisco Cooperation. Alvarez, M. A., Jounay, F., Major, T., & Volpato, P. (2011). *LTE backhauling deployment scenarios* (Technical Report). Next Generation Mobile Networks Alliance. Alvarez, M. A., Jounay, F., & Volpato, P. (2013). *Security in LTE backhauling* (Technical Report). Next Generation Mobile Networks Alliance.

Cisco. (2010). *Architectural considerations for backhaul of 2G/3G and long term evolution networks* (Technical Report). Cisco Cooperation.

Collett, M., Loh, T., Liu, H., & Qin, F. (n.d.). Smart antennas and wireless sensor networks - Achieving efficient, reliable communication. *National Physical Laboratory*.

Combes, R. (2013). *Mécanismes auto-organisants dans les réseaux sans fil*. (Dissertation thesis). Universite Pierre et Marie Curi, Paris, France.

Combes, R., Altman, Z., & Altman, E. (2010). On the use of packet scheduling in self-optimization processes: Application to coverage-capacity optimization. In *Proceeding of the 8th International Symposium on Modeling and Optimization in Mobile, Ad Hoc and Wireless Networks (WiOpt`10)*. IEEE.

Combes, R., Altman, Z., & Altman, E. (2013). Coordination of autonomic functionalities in communications networks. In *Proceeding of the 11th International Symposium on Modeling and Optimization in Mobile, Ad Hoc, and Wireless Networks (WiOpt'13)*. IEEE.

Cooper, N., & Meghanathan, N. (2010). Impact of mobility models on multi-path routing in mobile ad hoc networks. *International Journal of Computer Networks and Communications*, *2*(1), 185–194.

Corke, P. (n.d.). *CSIRO*. Retrieved November 20, 2012, from http://www.ict.csiro.au/files/AutonomousSystems/Flecks.pdf

Corke, P., Wark, T., Jurdak, R., Hu, W., Valencia, P., & Moore, D. (2010). Environmental wireless sensor networks. *Proceedings of the IEEE*, *98*(11), 1903–1917. doi:10.1109/JPROC.2010.2068530

Cormen, T. H., Leiserson, C. E., Rivest, R. L., & Stein, C. (2009). *Introduction to algorithms* (3rd ed.). Cambridge, MA: MIT Press.

Cormio, C., & Chowdhury, K. R. (2009). A survey on MAC protocols for cognitive radio networks. *Ad Hoc Networks*, *7*(7), 1315–1329. doi:10.1016/j.adhoc.2009.01.002

Costa, D. G. (2012). A survey on transport protocols for wireless multimedia sensor networks. *Transactions on Internet and Information Systems (Seoul)*, *6*(1), 241–269. doi: doi:10.3837/tiis.2012.01.014

Costa, D. G., & Guedes, L. A. (2010). The coverage problem in video-based wireless sensor networks: A survey. *Sensors (Basel, Switzerland)*, *10*(9), 8215–8247. doi:10.3390/s100908215 PMID:22163651

Costa, D. G., & Guedes, L. A. (2011). A survey on multimedia-based cross-layer optimization in visual sensor networks. *Sensors (Basel, Switzerland)*, *11*(5), 5439–5468. doi:10.3390/s110505439 PMID:22163908

Crow, B. P., Widjaja, I., Kim, L. G., & Sakai, P. T. (1997). IEEE 802.11 wireless local area networks. *IEEE Communications Magazine*, *35*(9), 116–126. doi:10.1109/35.620533

Croy, P. (2011). *LTE backhual requirements, reality check (Technical Report)*. Aviat Networks Inc.

Culler, D., Estrin, D., & Srivastava, M. (2004). Guest editors' introduction: Overview of sensor networks. *Computer*, *37*(8), 41–49. doi:10.1109/MC.2004.93

Dababneh, D. (2013). *LTE network planning and traffic generation*. (Master Thesis). Ottawa-Carleton Institute for Electrical and Computer Engineering (OCIECE), Carleton University, Ottawa, Canada.

Dacier, M., Deswarte, Y., & Kaâniche, M. (1996). *Quantitative assessment of operational security: Models and tools*. LAAS Research Report 96493.

Dahlman, E., Parkvall, S., Skoeld, J., & Beming. (2007). *3G evolution—HSPA and LTE for mobile broadband* (2nd ed.). Academic Press.

Dahlman, E., Parkvall, S., & Skoeld, J. (2011). *4G—LTE/LTE-advanced for mobile broadband*. Academic Press.

Dahlman, E., Parkvall, S., & Skold, J. (2011). *4G: LTE/LTE-advanced for mobile broadband: LTE/LTE-advanced for mobile broadband*. Academic Press.

Dai, H. N., Ng, K. W., & Wu, M. Y. (2007). A busy-tone based MAC scheme for wireless ad hoc networks using directional antennas. In *Proceedings of Global Telecommunications Conference, 2007*. IEEE.

Dang, L., Xu, J., Li, H., & Dang, N. (2010). DASR: Distributed anonymous secure routing with good scalability for mobile ad hoc networks. In *Proceedings of 5th IEEE Asia-Pacific Services Computing Conference*, (pp. 454-461). Hangzhou, China: IEEE Computer Society.

Daniel, A. (2004). IP virtual private networks–A service provider perspective. *IEEE Communications*, *151*(1), 62–70. doi:10.1049/ip-com:20040133

Daoui, M., M'zoughi, A., & Lalam, M. (2008). Mobility prediction based on an ant system. *Computer Communications*, *31*, 3090–3097. doi:10.1016/j.comcom.2008.04.009

Dargie, W., & Poellabauer, C. (2010). *Fundamentals of wireless sensor networks: Theory and practice*. Hoboken, NJ: John Wiley & Sons, Ltd. doi:10.1002/9780470666388

Dash, S., Swain, A. R., & Ajay, A. (2012). Reliable energy aware multi-token based MAC protocol for WSN. In *Proceedings of International Conference on Advanced Information Networking and Applications*, (pp. 144-151). IEEE.

David, A., & Jennings, E. (2005). Self-organized routing for wireless micro sensor networks. *IEEE Transactions on Systems, Man, and Cybernetics*, *35*(3), 349–359. doi:10.1109/TSMCA.2005.846382

Dawson, T. (1996). *Linux AX25-HOWTO, amateur radio*. Retrieved May 24, 2013, from http://www.linuxdocs.org/HOWTOs/AX25-HOWTO.html

Dechene, D. J., El Jardali, A., Luccini, M., & Sauer, A. (2005). *A survey of clustering algorithms for wireless sensor network*. Retrieved from http://www.dechene.ca/papers/report_635a.pdf

del Apio, M., Mino, E., Cucala, L., Moreno, O., Berberana, I., & Torrecilla, E. (2011). Energy efficiency and performance in mobile networks deployments with femtocells. In *Proceeding of the 22nd IEEE International Symposium on Personal Indoor and Mobile Radio Communications (PIMRC'11)*. Toronto, Canada: IEEE.

Dely, P., Castro, M., Soukhakian, S., Moldsvor, A., & Kassler, A. (2010). Practical considerations for channel assignment in wireless mesh networks. In *Proceedings of IEEE GLOBECOM Workshops* (pp. 763-767). Miami, FL: IEEE.

Deng, S., Li, J., & Shen, L. (2011). Mobility-based clustering protocol for wireless sensor networks with mobile nodes. *IET Wireless Sensor Systems*, *1*(1), 39–47. doi:10.1049/iet-wss.2010.0084

Devadas, S., Suh, E., Paral, S., Sowell, R., Ziola, T., & Khandelwal, V. (2008). Design and implementation of PUF-based unclonable RFID ICS for anti-counterfeiting and security applications. In *Proceedings of 2008 IEEE International Conference on RFID*, (pp. 58–64). IEEE.

Dhillon, H. S., Ganti, R. K., Baccelli, F., & Andrews, J. G. (2011). Coverage and ergodic rate in K-tier downlink heterogeneous cellular networks. In *Proceedings of 49th Annual Allerton Conference on Communication, Control, and Computing* (Allerton), (pp. 1627–1632). Allerton.

Diab, A., & Mitschele-Thiel, A. (2012). Comparative evaluation of distributed physical cell identity assignment schemes for LTE-advanced systems. In *Proceeding of the 7th Performance Monitoring, Measurement and Evaluation of Heterogeneous Wireless and Wired Networks Workshop*. Paphos, Cyprus: IEEE.

Diab, A., & Mitschele-Thiel, A. (2013). Development of distributed and self-organized physical cell identity assignment schemes for LTE-advanced systems. In *Proceedings of the 16th ACM International Conference on Modeling, Analysis and Simulation of Wireless and Mobile Systems*. Barcelona, Spain: ACM.

Ding, Y., Wang, X., Shen, H., & Guo, P. (2010). Media access control protocol based on hybrid antennas in ad hoc networks. In *Proceedings of Ubiquitous Intelligence & Computing and 7th International Conference on Autonomic & Trusted Computing (UIC/ATC)*. IEEE.

Dinh, H. T., Lee, C., Niyato, D., & Wang, P. (2011). *A survey of mobile cloud computing: Architecture, applications, and approaches*. Wireless Communications and Mobile Computing. doi:10.1002/wcm.1203

Djenouril, D., Mahmoudil, O., Bouamama, M., Jones, D. L., & Merabti, M. (2007). *On securing MANET routing protocol against control packet*. Retrieved from www.researchgate.net/.On_Securing_MANET_Routing_Protocol_AgainstControlPacket

Doherty, L., Pister, K. S. J., & El Ghaoui, L. (2001). Convex position estimation in wireless sensor networks. In *Proceeding of the IEEE INFOCOM* (pp.1655-1663). IEEE.

Doss, R., Jennings, A., & Shenoy, N. (2004). *Mobility prediction for seamless mobility in wireless networks*. Melbourne, Australia.

Dow, C. R., Lin, P. J., Chen, S. C., Lin, J. H., & Hwang, S. F. (2005). A study of recent research trends and experimental guidelines in mobile ad-hoc network. In *Proceedings of 19th International Conference on Advanced Information Networking and Applications* (pp. 72-77). Tamkang University, Taiwan: IEEE.

Drira, K., Seba, H., & Kheddouci, H. (2010). ECGK: An efficient clustering scheme for group key management in MANETs. *International Journal of Computer Communications*, *33*, 1094–1107. doi:10.1016/j.comcom.2010.02.007

Duong, T., & Tran, D. (2012). An effective approach for mobility prediction in wireless network based on temporal weighted mobility rule. *International Journal of Computer Science and Telecommunications*, *3*(2).

Ebrahimi, T., & Horne, C. (2000). MPEG-4 natural video coding – An overview. *Signal Processing Image Communication*, *15*(4-5), 365–385. doi:10.1016/S0923-5965(99)00054-5

Egbogah, E. E., & Fapojuwo, A. O. (2011). A survey of system architecture requirements for health care-based wireless sensor networks. *Sensors (Basel, Switzerland)*, *11*(5), 4875–4898. doi:10.3390/s110504875 PMID:22163881

Eidenbenz, S., Resta, G., & Santi, P. (2008). The commit protocol for truthful and cost-efficient routing in ad hoc networks with selfish nodes. *IEEE Transactions on Mobile Computing*, *7*(1), 19–33. doi:10.1109/TMC.2007.1069

ElBatt, T., Anderson, T., & Ryu, B. (2003). Performance evaluation of multiple access protocols for ad hoc networks using directional antennas. In *Proceedings of Wireless Communications and Networking, 2003*. IEEE. doi:10.1109/WCNC.2003.1200505

Elleithy, K., & Rao, V. (2011). Femto cells: Current status and future directions. *International Journal of Next-Generation Networks*, *3*(1), 1–9. doi:10.5121/ijngn.2011.3101

Elliott, D. K. (2005). *Understanding GPS: Principles and applications* (2nd ed.). Artech House Publishers.

EmberZnet Application Developers Guide. (n.d.). Retrieved November 20, 2012, from http://www.ember.com/

Enoki, K. (1999). Concept of i-mode service: New communication infrastructure in the 21st century. *NTT DoCoMo Technical Journal*, *1*(1), 4–9.

EPCglobal. (2008). *EPC™ radio-frequency identity protocols class-1 generation-2 UHF RFID protocol for communications at 860 MHz – 960 MHz version 1.2.0*. EPCglobal Inc.

EPCglobal. (2010). *EPC tag data standard version 1.5.* EPCglobal.

Ericsson. (2007). *Summary of downlink performance evaluation* (Technical Report). 3GPP. TSG RAN R1-072444.

Ericsson. (2009). *LTE - An introduction.* Ericsson.

Erol-Kantarci, M., & Mouftah, H. T. (2011). Wireless sensor networks for cost-efficient residential energy management in the smart grid. *IEEE Transactions on Smart Grid, 2*(2), 314–325. doi:10.1109/TSG.2011.2114678

Estrin, D., Girod, L., Pottie, G., & Srivastava, M. (2001). Instrumenting the world with wireless sensor networks. In *Proceedings of International Conference on Acoustics, Speech, and Signal Processing* (ICASSP 2001) (pp. 2033 – 2036). Salt Lake City, UT: IEEE Press.

Estrin, D., Govindan, R., Heidemann, J., & Kumar, S. (1999). Next century challenges: Scalable coordination in. In *Proceedings of ACM/IEEE International Conference on Mobile Computing and Networks* (pp. 263-270). Seattle, WA: ACM/IEEE.

ETRX-2 Zigbee Module Product Manual. (n.d.). Retrieved November 20, 2012, from http://www.telegesis.com/

ETSI Technical Report. (2010). *LTE, evolved universal terrestrial radio access network (E-UTRAN), self-configuring and self-optimizing network use cases and solutions.* ETSI TR 136 902—V9.2.0. Retrieved June 07, 2013, from http://www.etsi.org/deliver/etsi_tr/136900_136999/136 902/09.02.00_60/tr_136902v090200p.pdf

Fadlullah, Z. M., Nishiyama, H., Nei Kato, T., & Fouda, M. M. (2013). Intrusion detection system (ids) for combating attacks against cognitive radio networks. *IEEE Network, 27*(3), 51–56. doi:10.1109/MNET.2013.6523809

Fakih, K., Diouris, J. F., & Andrieux, G. (2006). BMAC: Beamformed MAC protocol with channel tracker in MANET using smart antennas. In *Proceedings of Wireless Technology, 2006.* IEEE. doi:10.1109/ECWT.2006.280466

Fall, K., & Varadhan, K. (1998). *Notes and documentation.* LBNL. Retrieved from http://www.mash.cs.berkeley.edu/ns

Farago, A., & Syrotiuk, V. R. (2003). MERIT: A scalable approach for protocol assessment. *Mobile Networks and Applications, 8*(5), 567–577. doi:10.1023/A:1025193929081

Fazio, P., & Marano, S. (2012). A new Markov-based mobility prediction scheme for wireless networks with mobile hosts. In Proceedings of Performance Evaluation of Computer and Telecommunication Systems (SPECTS). SPECTS.

Feder, P. M., Lee, N. Y., & Martin-Leon, S. (2003). A seamless mobile VPN data solution for UMTS and WLAN users. In *Proceedings of 4th International Conference on 3G Mobile Communication Technologies,* (pp. 210-216). IET.

Fehske, A. J., Fettweis, G., Malmodin, J., & Biczok, G. (2011). The global footprint of mobile communications: The ecological and economic perspective. *IEEE Communications Magazine, 49.*

Feh, T., & Dg, G., L, X., Mn, N., & Kl, Y. (2009). MEMSWear - Biomonitoring system for remote vital signs monitoring. *Journal of the Franklin Institute, 6,* 531–542.

Felice, M. D., Doost-Mohammady, R., Chowdhury, K. R., & Bononi, L. (2012). Cognitive vehicular networks: Smart radios for smart vehicles. *IEEE Vehicular Technology Magazine, 7*(2), 26–33. doi:10.1109/MVT.2012.2190177

FemtoForum. (2011). *HeNB (LTE-Femto) network architecture.* Retrieved June 07, 2013, from http://www.smallcellforum.org/

Feng, S., & Seidel, E. (2008). *Self-organizing networks (SON) in 3GPP long term evolution: Novel mobile radio (NOMOR) research center.* Retrieved June 07, 2013, from http://www.nomor.de/uploads/gc/TQ/gcTQfDWApo9o-sPfQwQoBzw/SelfOrganisingNetworksInLTE_2008-05.pdf

Feng, W., Caoa, J., Zhang, C., Zhang, J., & Xin, Q. (2012). Coordination of multi-link spectrum handoff in multi-radio multi-hop cognitive networks. *Journal of Parallel and Distributed Computing, 72,* 613–625. doi:10.1016/j.jpdc.2011.11.004

Fernando, N., Loke, S. W., & Rahayu, W. (2013). Mobile cloud computing: A survey. *Future Generation Computer Systems, 29,* 84–106. doi:10.1016/j.future.2012.05.023

Fettweis, G., & Zimmermann, E. (2008). ICT energy consumption - Trends and challenges. In *Proceedings of the IEEE Wireless Personal Multimedia Communications*. IEEE.

Fishler, E., & Poor, H. V. (2005). On the tradeoff between two types of processing gains. *IEEE Transactions on Communications*, *53*, 1744–1753. doi:10.1109/TCOMM.2005.855001

Floyd, S., & Henderson, T. (1999). New reno modification to TCP's fast recovery. *RFC 2582*.

Floyd, S., & Jacobson, V. (1993). Random early detection gateways for congestion avoidance. *IEEE/ACM Transactions on Networking*, *1*, 397–413. doi:10.1109/90.251892

Foerster, J. R. (2002). The performance of a direct-sequence spread ultra-wideband system in the presence of multipath, narrowband interference, and multiuser interference. In *Proceedings IEEE Conference on Ultra Wideband Systems and Technologies*. IEEE.

Forum, F. (2011, December). Femtocell market status. *Informa Telecoms & Media*.

Foschini, G. J., & Gans, M. J. (1998). On limits of wireless communications in a fading environment when using multiple antennas. *Wireless Personal Communications*, *6*(3), 311–335. doi:10.1023/A:1008889222784

Foschini, L., Taleb, T., Corradi, A., & Bottazzi, D. (2011). M2M-based metropolitan platform for IMS-enabled road traffic management in IoT. *IEEE Communications Magazine*, *49*(11), 50–57. doi:10.1109/MCOM.2011.6069709

Fotino, M., & De Rango, F. (2011). Energy issues and energy aware routing in wireless ad-hoc networks. In X. Wang (Ed.), *Mobile ad-hoc networks: Protocol design* (pp. 156–167). University of Calabria. doi:10.5772/13309

Francis, B., Narasimhan, V., & Nayak, A. (2012). Enhancing TCP congestion control for improved performance in wireless networks. In *Ad-hoc, mobile, and wireless networks* (pp. 472–483). Springer. doi:10.1007/978-3-642-31638-8_36

Fu, Z., Greenstein, B., Meng, X., & Lu, S. (2002). Design and implementation of a tcp-friendly transport protocol for ad hoc wireless networks. In *Proceeding of 10th IEEE International Conference on Network Protocosls (ICNP'02)* (pp. 216-225). Paris, France: IEEE.

Gambs, S., Killijian, M.-O., & Del Prado Cortez, M. N. (2012). Next place prediction using mobility Markov chains. In *Proceedings of the First Workshop on Measurement, Privacy, and Mobility*. doi:10.1145/2181196.2181199

Gangwar, S., Pal, S., & Kumar, K. (2012). Mobile ad hoc networks: A comparative study of QoS routing protocols. *International Journal of Computer Science Engineering and Technology*, *2*(1), 771–775.

Ganz, A., Ganz, Z., & Wongthavarawat, K. (2003). *Multimedia wireless networks: Technologies, standards and QoS*. Englewood Cliffs, NJ: Pearson Education.

Garcia-Sanchez, A.-J., Garcia-Sanchez, F., & Garcia-Haro, J. (2008). Feasibility study of MPEG-4 transmission on IEEE 802.15.4 networks. In *Proceedings of 2008 IEEE International Conference on Wireless and Mobile Computing, Networking and Communications*, (pp. 397–403). IEEE. doi:10.1109/WiMob.2008.12

Garcia-Sanchez, A.-J., Garcia-Sanchez, F., & Garcia-Haro, J. (2011). Wireless sensor network deployment for integrating video-surveillance and data-monitoring in precision agriculture over distributed crops. *Computers and Electronics in Agriculture*, *75*(2), 288–303. doi:10.1016/j.compag.2010.12.005

Garcia-Sanchez, A.-J., Garcia-Sanchez, F., Garcia-Haro, J., & Losilla, F. (2010). A cross-layer solution for enabling real-time video transmission over IEEE 802.15.4 networks. *Multimedia Tools and Applications*, *51*(3), 1069–1104. doi:10.1007/s11042-010-0460-z

Garg, B. Bahl, & Ittipiboon. (2001). Microstrip antenna design handbook. Norwood, MA: Artech House.

Garza, C., Ashai, B. H., Monturus, E., & Syputa, R. (2010). [*LTE operator commitments: Deployment scenarios and growth opportunities*. MARAVEDIS Wireless Market Research & Analysis.]. *Top (Madrid)*, *25*.

Gascueña, J. M., & Fernández-Caballero, A. (2011). On the use of agent technology in intelligent, multisensory and distributed surveillance. *The Knowledge Engineering Review*, *26*(2), 191–208. doi:10.1017/S0269888911000026

Ge, H., & Lu, Z. (2009). A history-based handover prediction for LTE systems. In *Proceeding of 4th International Conference on Ubi-Media Computing (U-Media)*. U-Media.

Gelal, E., Jakllari, G., Krishnamurthy, S. V., & Young, N. E. (2006). An integrated scheme for fully-directional neighbor discovery and topology management in mobile ad hoc networks. In *Proceedings of Mobile Adhoc and Sensor Systems (MASS)*. IEEE. doi:10.1109/MOBHOC.2006.278551

Gerla, M., Tang, K., & Bagrodia, R. (1999). TCP performance in wireless multihop networks. In *Proceeding of IEEE WMCSA* (pp. 41-50). New Orleans, LA: IEEE.

Gesbert, D., Hanly, S., Huang, H., Shamai Shitz, S., Simeone, O., & Yu, W. (2010). Multi-cell MIMO cooperative networks: A new look at interference. *IEEE Journal on Selected Areas in Communications*, 28(9), 1380–1408. doi:10.1109/JSAC.2010.101202

Ghosekar, P., Katkar, G., & Ghorpade, P. (2010). Mobile ad hoc networking: Imperatives and challenges. *International Journal of Computer Application. Special Issue on Mobile Ad-Hoc Networks*, 1(3), 153–158.

Giorgetti, G., Cidronali, A., Gupta, S. K., & Manes, G. (2007). Exploiting low-cost directional antennas in 2.4 GHz IEEE 802.15 4 wireless sensor networks. In *Proceedings of Wireless Technologies, 2007 European Conference* (pp. 217-220). IEEE.

Gjupta, W. A., & Williamson, C. (2004). Experimental evaluation of TCP performance in multi-hop wireless ad hoc networks. In *Proceeding of MASCOTS* (pp. 3-11). Volendam, The Netherlands: MASCOTS.

Glentis, G. O., Berberidis, K., & Theodoridis, S. (1999). Efficient least squares adaptive algorithms for FIR transversal filtering. *IEEE Signal Processing Magazine*, 16, 13–41. doi:10.1109/79.774932

Golaup, A., Mustapha, M., & Patanapongpibul, L. (2009). Femtocell access control strategy in UMTS and LTE. *IEEE Communications Magazine*, 47(9). doi:10.1109/MCOM.2009.5277464

Goldsmith, A. (2005). *Overview of wireless communications*. Retrieved from http://www.cambridge.org/

Goldsmith, A. (2005). *Wireless communications*. Cambridge, UK: Cambridge University Press. doi:10.1017/CBO9780511841224

Gomez, J., & Campbell, A. T. (2007). Variable-range transmission power control in wireless ad hoc networks. *IEEE Transactions on Mobile Computing*, 6(1), 87–99. doi:10.1109/TMC.2007.250673

Gonzalez, C. R. A. (2009). Open-source SCA-based core framework and rapid development tools enable software-defined radio education and research. *IEEE Communications Magazine*, 47(10), 48–55. doi:10.1109/MCOM.2009.5273808

Gossain, H., Cordeiro, C., & Agrawal, D. P. (2005). MDA: An efficient directional MAC scheme for wireless ad hoc networks. In *Proceedings of Global Telecommunications Conference, 2005*. IEEE.

Gossain, H., Cordeiro, C., Cavalcanti, D., & Agrawal, D. P. (2004). The deafness problems and solutions in wireless ad hoc networks using directional antennas. In *Proceedings of Global Telecommunications Conference Workshops, 2004*. IEEE.

Gou, S., He, T., Mokbel, M., Stankovic, J., & Abdelzaher, T. (2008). On accurate and efficient statistical counting in sensor based surveillance systems. In *Proceedings of the 5th IEEE International Conference on Mobile Ad Hoc and Sensor Systems* (MASS 2008) (pp. 24-35). Atlanta, GA: IEEE Press. doi: 10.1109/MAHSS.2008.4660038

Grgić, K., Zagar, D., & Krizanović, V. (2011). *Medical applications of wireless sensor networks-current status and future directions*. Official Publication of the Medical Association of Zenica-Doboj Canton Bosnia and Herzegovina.

Grobler, T. L., & Penzhorn, W. T. (2004). Fast decryption methods for the RSA cryptosystem. In *Proceedings of 7th AFRICON Conference*. Paris, France: IEEEXPLORE.

GSM. (1998). *Global system for mobile communications: Technical specifications*. Retrieved from http://www.etsi.org/deliver/etsi_gts/07/0705/05.05.00_60/gsmts_0705v050500p.pdf

Guan, Q., Yu, F., Jiang, S., & Wei, G. (2010). Prediction-based topology control and routing in cognitive radio mobile ad hoc networks. *IEEE Transactions on Vehicular Technology*, 59(9), 4443–4452. doi:10.1109/TVT.2010.2069105

Gungor, V., & Hancke, G. (2009). Industrial wireless sensor networks: Challenges, design principles, and technical approaches. *IEEE Transactions on Industrial Electronics*, *56*(10), 4258–4265. doi:10.1109/TIE.2009.2015754

Guo, W., & Wang, S. (2012, November). Interference-aware self-deploying femto-cell. *IEEE Wireless Communications Letters*.

Guo, W., Wang, S., O'Farrell, T., & Fletcher, S. (2013). Energy consumption of 4G cellular networks: A London case study. In *Proceedings of IEEE Vehicular Technology Conference* (VTC). IEEE.

Guo, W., & O'Farrell, T. (2012). Relay deployment in cellular networks: Planning and optimization. *IEEE Journal on Selected Areas in Communications*, 31.

Guo, W., & O'Farrell, T. (2013). Dynamic cell expansion with self-organizing cooperation. *IEEE Journal on Selected Areas in Communications*, 31.

Guo, W., Rigelsford, J., Ford, K., & O'Farrell, T. (2012). Dynamic basestation antenna design for low energy networks. *Progress in Electromagnetics Research, C*, 31.

Gustås, P., Magnusson, P., Oom, J., & Storm, N. (2002). Real-time performance monitoring and optimization of cellular systems. *Ericsson Review, 1*.

Gyselinckx, B., Penders, J., & Vullers, R. (2007). Potential and challenges of body area networks for cardiac monitoring. *Journal of Electrocardiology*, *40*, S165–S168. doi:10.1016/j.jelectrocard.2007.06.016 PMID:17993316

Ha & Jung. (2011). Reconfigurable beam steering using a microstrip patch antenna with a U-slot for wearable fabric applications. *IEEE Antennas and Wireless Propagation Letters*, *10*, 1228–1231. doi:10.1109/LAWP.2011.2174022

Haas, Z. J., Deng, J., Liang, B., Papadimitratos, P., & Sajama, S. (2002). *Wireless ad hoc networks*. Hoboken, NJ: John Wiley & Sons, Inc.

Habetha, J. (2006). The MyHeart project—Fighting cardiovascular diseases by prevention and early diagnosis. In *Proceedings of 28th Ann. Int. IEEE EMBS Conf.* (pp. 6746-6749). IEEE.

Hadjadj-Aoul, Y., & Naït-Abdesselam, F. (2008). On physical-aware directional MAC protocol for indoor wireless networks. In *Proceedings of Global Telecommunications Conference, 2008*. IEEE.

Haenggi, M., Andrews, J. G., Baccelli, F., Dousse, O., & Franceschetti, M. (2009a). Stochastic geometry and random graphs for the analysis and design of wireless networks. *IEEE Journal on Selected Areas in Communications*, *27*(7), 1029–1046. doi:10.1109/JSAC.2009.090902

Haenggi, M., & Ganti, R. K. (2009). *Interference in large wireless networks*. Now Publishers Inc.

Halamka, J., Juels, A., Stubblefield, A., & Westhues, J. (2006). The security implications of VeriChip cloning. *Journal of the American Medical Informatics Association*, *13*(6), 601–607. doi:10.1197/jamia.M2143 PMID:16929037

Hall, P. S., & Hao, Y. (2006). Antennas and propagation for body centric communications. In *Proceedings of First European Conference on Antennas and Propagation* (EuCAP 2006). EuCAP.

Halonen, T., Romero, J., & Melero. (2003). *GSM, GPRS and EDGE performance: Evolution towards 3G/UMTS*. Wiley.

Hamadani, E., & Rakocevic, V. (2008). A cross layer solution to address TCP intra-flow performance degradation in multihop ad hoc networks. *Journal of Internet Engineering*, *2*(1), 146–156.

Hamalainen, S., Sanneck, H., & Sartori, C. (Eds.). (2012). *LTE self-organising networks (SON), network management automation for operational efficiency*. Hoboken, NJ: John Wiley & Sons, Ltd.

Hamrioui, S., & Lalam, M. (2008). Incidence of the improvement of the transport – MAC protocols interactions on MANET performance. In *Proceeding of the 8th Annual International Conference on New Technologies of Distributed Systems (NOTERE 2008)* (pp. 634-648). Lyon, France: NOTERE.

Hamrioui, S., & Lalam, M. (2010). Incidences of the improvement of the MAC-transport and MAC–routing interactions on MANET performance. In *Proceeding of the International Conference on Next Generation Networks and Services*. IEEE.

Hamrioui, S., Bouamra, S., & Lalam, M. (2007). Interactions entre le protocole MAC et le protocole de transport TCP pour l'optimisation des MANET. In *Proceeding of the 1ˢᵗ International Workshop on Mobile Computing & Applications (NOTERE'2007)*. NOTERE.

Hamrioui, S., Bouamra, S., & Lalam, M. (2008). Les effets de la mobilité des nœuds sur la QoS dans un MANET. In *Proceeding of the Maghrebian Conference on Software Engineering and Artificial Intelligence (MCSEAI'08)*. Oran, Algeria: MCSEAI.

Hamrioui, S., Lalam, M., & Lorenz, P. (2011). Effets de la mobilité sur les protocoles de transport et de routage dans les MANET. In *Proceeding of Journées Doctorales en Informatique et Réseaux, (JDIR'11)*. Belfort, France: JDIR.

Han, C. C., Rengaswamy, R. K., Shea, R., Kohler, E., & Srivastava, M. (2005). SOS: A dynamic operating system for sensor networks. In *Proceedings of the Third Internaltional Conference on Mobile Systems, Appliactions, and Services* (Mobisys). Retrieved from http://nesl.ee.ucla.edu/fw/documents/conference/2005/sos-05SpotsDemo.pdf

Hanbali, A. A., Altman, E., & Nain, P. (2005). A survey of TCP over ad hoc networks. *IEEE Communications Surveys & Tutorials*, *7*(3), 22–36. doi:10.1109/COMST.2005.1610548

Harris, S. (2002). *IP VPNs: An overview for network executives* (Technical Report). Retrieved from http://www.onsiteaustin.com/whitepapers/VPN20justfication.pdf

Hartney, C. J. (2012). *Security risk metrics: An attack graph-centric approach*. (Unpublished Master's thesis). The University of Tulsa, Tulsa, OK.

Hashiro, R. (2012). *JNOS, amateur radio and mobile IP email/BBS*. Retrieved July 20, 2013, from http://www.qsl.net/ah6rh/am-radio/packet/jnos.html

Hasnaoui, M. L., Hssane, A. B., Ezzati, A., & Benalla, S. (2010). An oriented cluster formation for heterogeneous wireless sensor networks. *Journal of Computing*, *2*(11), 34–39.

Hassanein, H. L. J. (2006). Reliable energy aware routing in wireless sensor networks. In *Proceedings of Second IEEE Workshop on Dependability and Security in Sensor Networks and Systems* (pp. 54-56). Columbia, MD: IEEE.

Hawrylak, P. J., Hartney, C., Haney, M., Hamm, J., & Hale, J. (2013). Techniques to model and derive a cyber-attacker's intelligence. In B. Igelnik, & J. Zurada (Eds.), *Efficiency and scalability methods for computational intellect* (pp. 162–180). Hershey, PA: Information Science Reference. doi:10.4018/978-1-4666-3942-3.ch008

Haykin, S. (2002). *Adaptive filter theory* (4th ed.). Upper Saddle River, NJ: Prentice Hall.

He, W., Ge, Z., & Hu, Y. (2007). Optimizing UDP packet sizes in ad hoc networks. In *Proceedings of International Conference on Wireless Communications, Networking and Mobile Computing* (pp. 1617-1619). Shanghai, China: IEEE.

Heath, R. W., & Kountouris, M. (2012). Modeling heterogeneous network interference. In *Proceedings of Information Theory and Applications Workshop* (ITA), (pp. 17–22). ITA.

Heide, J., Fitzek, F. H. P., & Pedersen, M. V. (2012). Green mobile clouds: Network coding and user cooperation for improved energy efficiency. In *Proceedings of IEEE 1st International Conference on Cloud Networking (CLOUDNET)* (pp. 1-8). Paris, France: IEEE.

Heilman, K. J., & Porges, S. (2007). Accuracy of the lifeshirt (vivometrics) in the detection of cardiac rhythms. *Biological Psychology*, *3*, 300–305. doi:10.1016/j.biopsycho.2007.04.001 PMID:17540493

Heinzelman, W. R., Chandrakasan, A., & Balakrishnan, H. (2000). Energy-efficient communication protocol for wireless microsensor networks. In *Proceedings of the 33rd Annual Hawaii International Conference on System Sciences*. IEEE. doi: 10.1109/HICSS.2000.926982

Hertleer, C., Rogier, H., Vallozzi, L., & Van Langenhove, L. (2009). A textile antenna for off-body communication integrated into protective clothing for firefighters. *IEEE Transactions on Antennas and Propagation*, *57*(4), 919–925. doi:10.1109/TAP.2009.2014574

Hesselbarth, J., & Vahldieck, R. (2000). Microstrip patch antennas on corrugated substrates. In *Proceedings of 30th European Microwave Conference*. Academic Press.

Hill, J., Szewczyk, R., Woo, A., Hollar, S., Culler, E. D., & Pister, K. S. J. (2000). System architecture directions for networked sensors. In *Architectural support for programming languages and operating systems* (pp. 93-104). Retrieved from http://webs.cs.berkeley.edu

Hofmann-Wellenhof, B., Lichtenegger, H., & Collins, J. (2004). *Global positioning system: Theory and practice* (5th ed.). New York: Springer.

Holland, G., & Vaidya, N. H. (1999). Analysis of TCP performance over mobile ad hoc networks. In *Proceeding of Annual International Conference on Mobile Computing and Networking (Mobicom'99)* (pp. 207-218). Seattle, WA: ACM.

Holma, H., & Toskala, A. (2009). *LTE for UMTS: OFDMA and SC-FDMA based radio access*. Chichester, UK: Wiley. doi:10.1002/9780470745489

Holma, H., & Toskala, A. (Eds.). (2004). *WCDMA for UMTS: Radio access for third generation mobile communications*. Hoboken, NJ: John Wiley & Sons, Ltd.

Holma, H., & Toskala, A. (Eds.). (2011). *LTE for UMTS evolution to LTE-advanced* (2nd ed.). Hoboken, NJ: John Wiley & Sons Inc. doi:10.1002/9781119992943

Honig, M., & Tsatsanis, M. K. (2000). Adaptive techniques for multiuser CDMA receivers. *IEEE Signal Processing Magazine, 17*, 49–61. doi:10.1109/79.841725

Hossain, E., & Leung, K. (Eds.). (2007). Wireless mesh networks: Architectures and protocols. Springer Science+ Business Media, LLC.

Hu, L., & Evans, D. (2004). Using directional antennas to prevent wormhole attacks. In *Proceedings of Network and Distributed System Security Symposium (NDSS)*. NDSS.

Huang, Z., Shen, C. C., Srisathapornphat, C., & Jaikaeo, C. (2002). A busy-tone based directional MAC protocol for ad hoc networks. In *Proceedings of MILCOM 2002* (Vol. 2, pp. 1233-1238). IEEE.

Huerta-Canepa, G., & Lee, D. (2012). A virtual cloud computing provider for mobile devices. In *Proceedings of 1st ACM Workshop on Mobile Cloud Computing & Services: Social Networks and Beyond* (pp. 1-5). San Francisco, CA: ACM.

Hu, G., Tay, W. P., & Wen, Y. (2012). Cloud robotics: Architecture, challenges and applications. *IEEE Network, 26*(3), 21–28. doi:10.1109/MNET.2012.6201212

Hu, H., Zhang, J., Zheng, X., Yang, Y., & Wu, P. (2010). Self-configuration and self-optimization for LTE networks. *IEEE Communications Magazine, 48*(2). doi:10.1109/MCOM.2010.5402670

Hull, B., Bychkovsky, V., Zhang, Y., Chen, K., Goraczko, M., & Miu, A. … Madden, S. (2006). CarTel: A distributed mobile sensor computing system. In *Proceedings of the 4th International Conference on Embedded Networked Sensor Systems*. Boulder, CO: ACM.

Hu, M., & Zhang, J. (2004). MIMO ad hoc networks: Medium access control, saturation throughput, and optimal hop distance. *Journal of Communications and Networks, 6*(4), 317–330.

Hung, C.-C., & Huang, S.-Y. (2010). On the study of a ubiquitous healthcare network with security and QoS. In *Proceedings of 2010 IET International Conference on Frontier Computing – Theory, Technologies and Applications* (pp. 139-144). Taichung, Taiwan: IET.

Hu, P., & Ibnkahla, M. (2012). A consensus-based protocol for spectrum sharing fairness in cognitive radio ad hoc and sensor networks. *International Journal of Distributed Sensor Networks*, (1): 1–12. doi:10.1155/2012/370251

Huraj, L., & Sládi, V. (2009). Authorization through trust chains in ad hoc grids. In *Proceedings of Euro American Conference on Telematics and Information Systems: New Opportunities to increase Digital Citizenship Prague*. ACM.

Hurley, W. G., & Duffy, M. C. (1995). Calculation of self and mutual impedances in planar magnetic structures. *IEEE Transactions on Magnetics, 31*(4), 2416–2422. doi:10.1109/20.390151

Hussain, R., Son, J., Eun, H., & Oh, S. K. H. (2012). Rethinking vehicular communications: Merging vanet with cloud computing. In *Proceedings of IEEE 4th International Conference on Cloud Computing Technology and Science (CloudCom)* (pp. 606-609). Taipei, Taiwan: IEEE.

Hussain, S. A., Farooq, U., Zia, K., & Akhlaq, M. (2004). An extended topology for zone-based location aware dynamic sensor networks. In *Proceedings of National Conference on Emerging Technologies*. Retrieved from http://citeseerx.ist.psu.edu/viewdoc/summary?doi=10.1.1.117.1634

Hussain, S. R., Saha, S., & Rahman, A. (2010). SAAMAN: Scalable address autoconfiguration in mobile ad hoc networks. *Journal of Network and Systems Management, 19*(3), 394–426. doi:10.1007/s10922-010-9187-4

Hussein, Y., Ali, B., Varahram, P., & Sali, A. (2011). Enhanced handover mechanism in long term evolution (LTE) networks. *Scientific Research and Essays, 6*(24), 5138–5152. doi: doi:10.5897/SRE11.480

Huszák, Á., & Imre, S. (2010). Analysing GOP structure and packet loss effects on error propagation in MPEG–4 video streams. In *Proceeding of the 4th International Symposium on Communications, Control and Signal Processing*. IEEE. Retrieved from http://ieeexplore.ieee.org/xpls/abs_all.jsp?arnumber=5463469

Hu, Y. C., Johnson, D. B., & Perrig, A. (2003). SEAD: Secure efficient distance vector routing for mobile wireless adhoc networks. *Ad Hoc Networks Journal, 1*(1), 175–192. doi:10.1016/S1570-8705(03)00019-2

Hwang, I., Pedram, M., & Soltan, M. (2008). Modulation-aware energy balancing in hierarchical wireless sensor networks. In *Proceedings of the 3rd International Symposium on Wireless Pervasive Computing* (pp. 355-359). Los Angeles, CA: IEEE Press.

Hyytia, E., & Vittamo, J. (2005). Random waypoint model in n-dimensional space. *Operations Research Letters, 33*, 567–571. doi:10.1016/j.orl.2004.11.006

IBM. (2000). *Salutation service discovery in pervasive computing environments*. Retrieved from http://www-3.ibm.com/pvc/tech/salutation.shtml

IEEE 802.11. (2007). *Part 11: Wireless LAN medium access control (MAC) and physical layer (PHY) specifications*. IEEE Computer Society.

IEEE 802.15.3. (2003). *Part 15.3: Wireless medium access control (MAC) and physical layer (PHY) specifications for high rate wireless personal area networks (WPANs)*. IEEE Computer Society.

IEEE 802.15.4. (2006). *Part 15.4: Wireless medium access control (MAC) and physical layer (PHY) specifications for low-rate wireless personal area networks (WPANs)*. IEEE Computer Society.

IEEE 802.16. (2004). *Part 16: Air interface for fixed broadband wireless access systems—Standard for local and metropolitan area networks*. IEEE Computer Society.

IEEE. (2011). *IEEE standard for local and metropolitan area networks - Part 15.4: Low-rate wireless personal area networks (LR-WPANs)*. IEEE.

IETF. (2012). *IETF MANET working group*. Retrieved on August 2013, from http://www.ietf.org/html.charters/manet-charter.html

Inn, E. R., & Winston, K. G. S. (2006). Distributed steiner-like multicast path setup for mesh-based multicast routing in ad-hoc networks. In *Proceedings of IEEE International Conference on Sensor Networks, Ubiquitous and Trustworthy Computing* (pp. 192-197). Taichung, Taiwan: IEEE.

International Telecommunication Union (ITU) Official Website. (n.d.). Retrieved June 07, 2013, from http://www.itu.int/en/Pages/default.aspx

Ishizu, K. (2006). Adaptive wireless-network testbed for CR technology. In *Proceedings of 1st International Workshop on Wireless Network Testbeds, Experimental Evaluation & Characterization* (pp. 18-25). Los Angeles, CA: ACM.

Islam, N., & Aqeel-ur-Rehman. (2013). A comparative study of major service providers for cloud computing. In *Proceedings of 1st International Conference on Information and Communication Technology Trends (ICICTT)* (pp. 228-232). Federal Urdu University.

Islam, N., & Shaikh, Z. A. (2011). A survey of data management issues & frameworks for mobile ad hoc networks. In *Proceedings of 4th International Conference on Information and Communication Technologies* (pp. 1-5). Karachi: IEEE.

Islam, N., Shaikh, Z. A., & Talpur, S. (2008). Towards a grid based approach to traffic routing in VANET. In *Proceedings of E-INDUS*. Karachi, Pakistan: IIEE.

Islam, N., Shaikh, N. A., Ali, G., & Shaikh, Z. A., & Aqeel-ur-Rehman. (2010). A network layer service discovery approach for mobile ad hoc network using association rules mining. *Australian Journal of Basic and Applied Sciences, 4*(6), 1305–1315.

Islam, N., & Shaikh, Z. A. (2012). Towards a robust and scalable semantic service discovery scheme for mobile ad hoc network. *Pakistan Journal of Engineering and Applied Sciences, 10*, 68–88.

Islam, N., & Shaikh, Z. A. (2013). Security issues in mobile ad hoc network. In S. Khan, & A. K. Pathan (Eds.), *Wireless networks and security* (pp. 49–80). Springer. doi:10.1007/978-3-642-36169-2_2

Islam, N., & Shaikh, Z.A., Aqeel-ur-Rehman, Siddiqui, M.S. (2013). HANDY: A hybrid association rules mining approach for network layer discovery of services wireless networks. *Wireless Networks*. doi:10.1007/s11276-013-0571-3

ITU. (2009). *International telecommunication union, what really is a third generation (3G) mobile technology*. Retrieved from http://www.itu.int/ITU-D/imt-2000/DocumentsIMT2000/What_really_3G.pdf

Iyer, Y., Gandham, S., & Venkatesan, S. (2005). STCP: A generic transport layer protocol for wireless sensor networks. In *Proceedings of International Conference on Computer Communications and Networks*. IEEE. Retrieved from http://ieeexplore.ieee.org/xpls/abs_all.jsp?arnumber=1523908

Iyer, G. N., & Durga, S. (2013). State of the art security mechanisms for mobile cloud environments. *International Journal of Advanced Research in Computer Science and Software Engineering, 3*(4), 470–476.

Izquierdo, L., & Izquierdo, S. (2009). Techniques to understand computer simulations: Markov chain analysis. *Journal of Artificial Societies and Social Simulation, 12*(1).

J, H., & AC, W. (2010, June 15). *A dynamic, context-aware security infrastructure for distributed healthcare applications*. Retrieved from http://www.cs.virginia.edu/~acw/security/doc/Publications/A%20Dynamic,%20Context-Aware%20Security%20Infrastructure%20for%20Distri~1.pdf

Jain, A., Dubey, A. K., Upadhyay, R., & Charhate, S. V. (2008). Performance evaluation of wireless network in presence of hidden node: A queuing theory approach. In *Proceeding of Second Asia International Conference on Modelling and Simulation* (pp. 225-229). Kuala Lumpur: IEEE.

Jain, A., Hong, L., & Pankanti, S. (2000). Biometric identification. *Communications of the ACM, 43*(2), 90–98. doi:10.1145/328236.328110

Jain, V., Gupta, A., & Agrawal, D. P. (2008). On-demand medium access in multihop wireless networks with multiple beam smart antennas. *IEEE Transactions on Parallel and Distributed Systems, 19*(4), 489–502. doi:10.1109/TPDS.2007.70739

Jakllari, G., Broustis, I., Korakis, T., Krishnamurthy, S. V., & Tassiulas, L. (2005). Handling asymmetry in gain in directional antenna equipped ad hoc networks. In Proceedings of Personal, Indoor and Mobile Radio Communications, 2005 (Vol. 2, pp. 1284-1288). IEEE.

Jakllari, G., Luo, W., & Krishnamurthy, S. V. (2007). An integrated neighbor discovery and MAC protocol for ad hoc networks using directional antennas. *IEEE Transactions on Wireless Communications, 6*(3), 1114–1024. doi:10.1109/TWC.2007.05471

Jangra, A., & Goel, N., Priyanka, & Bhatia, K. K. (2010). IEEE WLANs standards for mobile ad-hoc networks (MANETs), performance analysis. *Global Journal of Computer Science and Technology, 10*(14), 42–47.

Jansen, T., Amirijoo, M., Tuerke, U., Jorguseski, L., Zetterberg, K., & Nascimento, R. … Balan, I. (2009). Embedding multiple self-organisation functionalities in future radio access networks. In *Proceedings of VTC Spring 2009*. IEEE.

Jansen, T., Balan, I., Turk, J., Moerman, I., & Kurner, T. (2010). Handover parameter optimization in LTE self-organizing networks. In *Proceedings of 2010 IEEE 72nd Vehicular Technology Conference - Fall*. IEEE. doi:10.1109/VETECF.2010.5594245

Jayapal, C., & Vembu, S. (2011). Adaptive service discovery protocol for mobile ad hoc networks. *European Journal of Scientific Research, 49*(1), 6–17.

Jayasuriya, A., Perreau, S., Dadej, A., & Gordon, S. (2004). Hidden vs. exposed terminal problem in ad hoc networks. In *Proceedings of the Australian Telecommunication Networks and Applications Conference* (pp. 52-59). IEEE.

Jetcheva, J. G., & Johnson, D. B. (2001). Adaptive demand-driven multicast routing in multi-hop wireless ad-hoc networks. In *Proceedings of ACM International Symposium on Mobile Ad-hoc Networking and Computing* (pp. 33-44). Long Beach, CA: ACM.

Jiang, R., Gupta, V., & Ravishankar, C. (2003). Interactions between TCP and the IEEE 802.11 MAC protocol. In *Proceeding of DARPA Information Survivability Conference and Exposition* (vol. 1, pp. 273- 282). Washington, DC: DARPA.

Jimenez, T., & Altman, E. (2003). Novel delayed ACK techniques for improving TCP performance in multihop wireless networks. In *Proceeding of Personal Wireless Communications (PWC'03)* (pp. 237-242). Venice, Italy: PWC.

Jin, Z., Oresko, J., Huang, S., & Cheng, A. C. (2009). HeartToGo: A personalized medicien technology for cardiovascular disease prevention and detection. In *Proceedings of IEEE/NIH LiSSA, 2009* (pp. 80-83). IEEE.

Johnson, D. B., & Maltz, D. A. (1997). Dynamic source routing in ad hoc wireless networks. In T. Imielinski, & H. Korth (Eds.), *Mobile computing* (pp. 153–181). Academic Press.

Joshi, J. G., & Shyam, S. Pattnaik, Devi, S., & Lohokare, M.R. (2010). Microstrip patch antenna loaded with magnetoinductive waveguide. In *Proceedings of Twelfth National Symposium on Antennas and Propagation* (APSYM 2010). APSYM.

Joshi, J. G., Shyam, S., & Pattnaik. (2013). Polypropylene based metamaterial integrated wearable microstrip patch antenna. In *Proceedings of IEEE Indian Antenna Week (IAW 2013)*. Aurangabad, India: IEEE.

Joshi, J. G., Shyam, S., Pattnaik, & Devi, S. (2012a). Metamaterial embedded wearable rectangular microstrip patch antenna. *International Journal of Antennas and Propagation.* doi:10:1155/2012/974315

Joshi, J. G., Pattnaik, S. S., Devi, S., & Lohokare, M. R. (2011). Bandwidth enhancement and size reduction of microstrip patch antenna by magneto-inductive waveguide loading. *Journal of Wireless Engineering and Technology, 2*(2), 37–44. doi:10.4236/wet.2011.22006

Joshi, J. G., & Shyam, S., Pattnaik, & Devi, S. (2012b). Partially metamaterial ground plane loaded rectangular slotted microstrip patch antenna. *International Journal of Microwave and Optical Technology, 7*(1), 1–10.

Joshi, J. G., & Shyam, S., Pattnaik, & Devi, S. (2013). Geotextile based metamaterial loaded wearable microstrip patch antenna. *International Journal of Microwave and Optical Technology, 8*(1), 25–33.

Joshi, J. G., & Shyam, S., Pattnaik, Devi, S., & Raghavan, S. (2012). Magneto-inductive waveguide loaded microstrip patch antenna. *International Journal of Microwave and Optical Technology, 7*(1), 11–20.

Joshi, J.G., & Shyam, S., Pattnaik, Devi, S., & Lohokare, M.R. (2011). Frequency switching of electrically small patch antenna using metamaterial loading. *Indian Journal of Radio & Space Physics, 40*(3), 159–165.

Juels, A. (2005). Strengthening EPC tags against cloning. In *Proceedings of the 4th ACM Workshop on Wireless Security*, (pp. 67-76). ACM.

Jung, S., Lee, S., & Han, K. (2009). A DMAC protocol to improve spatial reuse by managing the NAV table of the nodes in VANET. In *Proceedings of Computer and Electrical Engineering, 2009* (Vol. 1, pp. 387–390). IEEE. doi:10.1109/ICCEE.2009.167

Junhai, L., Liu, X., & Danxia, Y. (2008). Research on multicast routing protocols for mobile ad-hoc networks. *Computer Networks, 52*(5), 988–997. doi:10.1016/j.comnet.2007.11.016

Jun, J., & Sichitiu, M. L. (2003). The nominal capacity of wireless mesh networks. *IEEE Wireless Communications, 10*(5), 8–14. doi:10.1109/MWC.2003.1241089

K, A., & M, Y. (2005). A survey on routing protocols for wireless sensor networks. *Ad Hoc Networks, 3*(3), 325-349. http://dx.doi.org/10.1016/j.adhoc.2003.09.010

Kahn, J. M., Katz, R. H., & Pister, K. (1999). Next century challenges mobile networking for smart dust. In *Proceedings of the 5th Annual ACM/IEEE International Conference on Mobile Computing and Networking* (pp. 271- 278). Washington, DC: ACM Press.

Kam, P. (1990). MACA - A new channel access method for packet radio. In *Proceeding of 9th ARRL/CRRL Amateur Radio Computer Networking Conference* (pp. 134-140). Ontario, Canada: ARRL/CRRL.

Kamruzzaman, S. M., Hamdi, M. A., & Abdullah-Al-Wadud, M. (2010). An energy-efficient MAC protocol for QoS provisioning in cognitive radio ad hoc networks. *Radio Engineering, 19*(4), 112–119.

Kansal, A., Goraczko, M., & Zhao, F. (2007). Building a sensor network of mobile phones. In *Proceedings of the International Symposium on Information Processing in Sensor Networks* (pp. 547-548). Cambridge, MA: IEEE.

Karapistoli, E., Gragopoulos, I., Tsetsinas, I., & Pavlidou, F. N. (2009). A directional MAC protocol with deafness avoidance for UWB wireless sensor networks.[IEEE.]. *Proceedings of Communications, 2009*, 1–5.

Karedal, J., Wyne, S., Almers, P., Tufvesson, F., & Molisch, A. F. (2004). Statistical analysis of the UWB channel in an industrial environment. In *Proceedings IEEE 60th Vehicular Technology Conference*. IEEE.

Karlsson, B., Backstrom, O., Kulesza, W., & Axelsson, L. (2005). Intelligent sensor networks - An agent oriented approach. In *Proceedings of Workshop on Real-World Wireless Sensor Networks*. Stockholm, Sweden: IEEE.

Karnadi, F. K., Mo, Z. H., & Lan, K. (2007). Rapid generation of realistic mobility models for VANET. In *Proceedings of IEEE Wireless Communications and Networking Conference* (pp. 2506-2511). Kowloon: IEEE.

Karveli, T., Voulgaris, K., Ghavami, M., & Aghvami, A. H. (2008). A collision-free scheduling scheme for sensor networks arranged in linear topologies and using directional antennas.[IEEE.]. *Proceedings of Sensor Technologies and Applications, 2008*, 18–22.

Katiyar, V., Chand, N., & Soni, S. (2010). Clustering algorithms for heterogeneous wireless sensor network: A survey. *International Journal of Applied Engineering Research, 1*(2), 273–287.

Katsis, C. D., Gianatsas, G., & Fotiadis, D. I. (2006). An integrated telemedicine platform for the assessment of affective physiological states. *Diagnostic Pathology, 1*, 16. doi:10.1186/1746-1596-1-16 PMID:16879757

Kawadia, V., & Kumar, P. (2005). Experimental investigations into TCP performance over wireless multihop networks. In *Proceeding of SIGCOMM Workshop on Experimental Approaches to Wireless Network Design and Analysis* (pp. 29-34). New York, NY: ACM.

Ke, C., Shieh, C., Hwang, W., & Ziviani, A. (2008). An evaluation framework for more realistic simulations of MPEG video transmission. *Journal of Information Science and Engineering, 440*, 425–440.

Keiss, W., Fuessler, H., & Widmer, J. (2004). Hierarchical location service for mobile ad hoc networks. *ACM SIGMOBILE Mobile Computing and Communications Review, 8*(4), 47–58. doi:10.1145/1052871.1052875

KF. N., E, L., & B, L. (2009). Medical MoteCare: A distributed personal healthcare monitoring system. In *Proceedings of International Conference on eHealth Telemedicine, and Social Medicine*. Cancun, Mexico: IEEE.

Kherani, A., & Shorey, R. (2004). Throughput analysis of TCP in multi-hop wireless networks with IEEE 802.11 MAC. In *Proceeding of IEEE WCNC'04*, (vol. 1, pp. 237-242). IEEE.

Khorashadi, B. (2009). *Enabling traffic control and data dissemination applications with vGrid - A vehicular ad hoc distributed computing framework*. (Unpublished doctoral dissertation). University of California, Berkeley, CA.

Kiess, W., & Mauve, M. (2007). A survey on real-world implementations of mobile ad-hoc networks. *Ad Hoc Networks, 5*(3), 324–339. doi:10.1016/j.adhoc.2005.12.003

Kim, J., & Lee, T. (2010). Handover in UMTS networks with hybrid access femtocells. In Proceedings of Advanced Communication Technology (ICACT). ICACT.

Kim, J., Ryu, B., Cho, K., & Park, N. (2011). Interference control technology for heterogeneous networks. In *Proceeding of the 5th International Conference on Mobile Ubiquitous Computing, Systems, Services and Technologies*. Lisbon, Portugal: IEEE.

Kim, D., & Toh, C., & Choi. (2001). TCP-BuS: Improving TCP performance in wireless ad hoc networks. *Journal of Communications and Networks*, *3*(2), 175–186.

Kim, T., & Kim, J. (2013). Handover optimization with user mobility prediction for femtocell-based wireless networks. *International Journal of Engineering and Technology*, *5*(2), 1829–1837.

Kirby, G., Dearle, A., Macdonald, A., & Fernandes, A. (2010). An approach to ad hoc cloud computing. *CoRR*, *1*, 1–6.

Kirch, O. (1995). *Linux network administrator's guide*. Sebastopol, CA: O'Reilly & Associates.

Kirkpatrick, M. S., Damiani, M. L., & Bertino, E. (2011). Prox-RBAC: A proximity-based spatially aware RBAC. In *Proceedings of the 19th ACM SIGSPATIAL International Conference on Advances in Geographic Information Systems*, (pp. 339-348). ACM.

Kirmse, A., Udeshi, T., Bellver, P., & Shuma, J. (2011). Extracting patterns from location history. In *Proceedings of the 19th ACM SIGSPATIAL International Conference on Advances in Geographic Information Systems*, (pp. 397–400). ACM.

Kitagawa, K., Komine, T., Yamamoto, T., & Konishi, S. (2011). A handover optimization algorithm with mobility robustness for LTE systems. In *Proceeding of the 22nd IEEE International Symposium on Personal Indoor and Mobile Radio Communications (PIMRC`11)*. IEEE.

Klaue, & Rathke. (2003). EvalVid − A framework for video transmission and quality evaluation. In *Proceedings of International Conference on Modelling Techniques and Tools for Computer Performance Evaluation* (pp. 225-272). IEEE.

Knight, P., & Lewis, C. (2004). Layer 2 and 3 virtual private networks: Taxonomy, technology, and standardization efforts. *IEEE Communications Magazine*, *42*(6), 124–131. doi:10.1109/MCOM.2004.1304248

Ko, Y. B., Shankarkumar, V., & Vaidya, N. H. (2000). Medium access control protocols using directional antennas in ad hoc networks. In Proceedings of *Nineteenth Annual Joint Conference of the IEEE Computer and Communications Societies* (Vol. 1, pp. 13-21). IEEE.

Kojima, J., & Mizoe, K. (1984). *Radio Mobile communication system wherein probability of loss of calls is reduced without a surplus of base station equipment*. U.S Patent 4435840 (1984). Washington, DC: US Patent Office.

Komarinski, M., & Collett, C. (1998). *Linux system administration handbook*. Upper Saddle River, NJ: Prentice Hall.

Kong, L. (2011). *Spatial access control on multi-touch user interface*. (Master's thesis). The University of Tulsa, Tulsa, OK.

Korakis, T., Jakllari, G., & Tassiulas, L. (2003). A MAC protocol for full exploitation of directional antennas in ad-hoc wireless networks. In *Proceedings of the 4th ACM International Symposium on Mobile Ad Hoc Networking & Computing* (pp. 98-107). ACM.

Korakis, T., Jakllari, G., & Tassiulas, L. (2008). CDR-MAC: A protocol for full exploitation of directional antennas in ad hoc wireless networks. *IEEE Transactions on Mobile Computing*, *7*(2), 145–155. doi:10.1109/TMC.2007.70705

Korhonen, I., Parkka, J., & Gils, M. V. (2003). Health monitoring in the home of the future. *IEEE Engineering in Medicine and Biology Magazine*, *22*(3), 66–73. doi:10.1109/MEMB.2003.1213628 PMID:12845821

Kottkamp, M., Roessler, A., Schlienz, J., & Schuetz, J. (2011). *LTE release 9—Technology introduction*. White paper.

Koutitas, G., & Demestichas, P. (2010). A review of energy efficiency in telecommunication networks. *Telfor Journal*, *2*(1), 2–7.

Kovachev, D., Cao, Y., & Klamma, R. (2011). Mobile cloud computing: A comparison of application models. *CoRR*, *1*, 1–8.

Kozat, U. C., Koutsopoulos, I., & Tassiulas, L. (2004). A framework for cross-layer design of energy-efficient communication with QoS provisioning.[IEEE.]. *Proceedings - IEEE INFOCOM, 2004*, 1446–1456.

Krishnan, S., Wagstrom, P., & Laszewski, G. (2002). *GSFL: A workflow framework for grid services*. The Globus Alliance. Retrieved from www.globus.org/cog/papers/gsfl-paper.pdf

Kuang, T., Xiao, F., & Williamson, C. (2003). Diagnosing wireless TCP performance problems: A case study. In *Proceeding of SPECTS* (pp. 176-185). SPECTS.

Kuhn, F., Moscibroda, T., & Wattenhofer, R. (2004). Unit disk graph approximation. In *Proceedings of the Workshop on Foundations of Mobile Computing* (pp. 17-23). Philadelphia, PA: ACM.

Kumar, R. (1983). Convergence of a decision-directed adaptive equalizer. In *Proceeding IEEE 22nd Conference on Decision and Control*. IEEE.

Kumar, G. V., Reddyr, Y. V., & Nagendra, M. (2010). Current research work on routing protocols for MANET: A literature survey. *International Journal on Computer Science and Engineering*, 2, 706–713.

Kumar, S., Raghavan, V. S., & Deng, J. (2006). Medium access control protocols for ad hoc wireless networks: A survey. *Ad Hoc Networks*, 4(3), 326–358. doi:10.1016/j.adhoc.2004.10.001

Kumar, U., Gupta, H., & Das, S. R. (2006). A topology control approach to using directional antennas in wireless mesh networks. In *Proceedings of Communications, 2006* (Vol. 9, pp. 4083–4088). IEEE. doi:10.1109/ICC.2006.255720

Kurosawa, S., Nakayama, H., Kato, N., Jamalipour, A., & Nemoto, Y. (2007). Detecting blackhole attack on AODV-based mobile ad hoc networks by dynamic learning method. *International Journal of Network Security*, 5(3), 38–346.

Kurose, J., & Ross, K. (2005). *Computer networking: A top-down approach featuring the internet*. Reading, MA: Addison Wesley.

Kushalnagar, N., Montenegro, G., & Schumacher, C. (2007). *IPv6 over low-power wireless personal area networks (6LoWPANs), overview, assumptions, problem statement, and goals*. IETF RFC 4919.

Kushner, H. J., & Yin, G. G. (2003). *Stochastic approximation and recursive algorithms and applications* (2nd ed.). Springer Stochastic Modeling and Applied Probability.

Kwak, H., Lee, P., Kim, Y., Saxena, N., & Shin, J. (2008). Mobility management survey for home-eNB based 3GPP LTE systems. *Journal of Information Processing Systems*, 4(4). doi:10.3745/JIPS.2008.4.4.145

Kwan, R., Arnott, R., Paterson, R., Trivisonno, R., & Kubota, M. (2010). On mobility load balancing for LTE systems. In = *Proceeding of the 72nd IEEE Vehicular Technology Conference (VTC)*. Ottawa, Canada: IEEE.

Kwon, H., Lee, H., & Cioffi, J. M. (2010). Cooperative strategy by Stackelberg games under energy constraint in multi-hop relay networks. In *Proceedings of 28th IEEE Conference on Global Telecommunications* (pp. 3431-3436). Miami, FL: IEEE.

Kwon, S., & Shroff, N. B. (2006). Energy-efficient interference-based routing for multi-hop wireless networks.[Barcelona, Spain: IEEE.]. *Proceedings - IEEE INFOCOM, 2006*, 1–12.

Kyösti, P., Meinilä, J., Hentilä, L., Zhao, X., Jämsä, T., & Schneider, C. (n.d.). *WINNER II channel models* (d1. 1.2 v1. 1). no. IST-4-027756 WINNER II, D, 1.

Laboratori REAL de la UPC. (n.d.). Retrieved November 20, 2012, from http://www.upc.edu/mediambient/recerca/lreal1.html

Lagkas, T. D., Angelidis, P., Stratogiannis, D. G., & Tsiropoulos, G. I. (2010). Analysis of queue load effect on channel access prioritization in wireless sensor networks. In *Proceedings of 2010 6th IEEE International Conference on Distributed Computing in Sensor Systems Workshops (DCOSSW)* (pp. 1-6). Santa Barbara, CA: IEEE.

Lagkas, T., Stratogiannis, D. G., Tsiropoulos, G. I., Sarigiannidis, P., & Louta, M. (2011). Load dependent resource allocation in cooperative multiservice wireless networks: Throughput and delay analysis. In *Proceedings of 2011 IEEE Symposium on Computers and Communications (ISCC)* (pp. 185-190). Corfu, Greece: IEEE.

Lagkas, T. D., Stratogiannis, D. G., Tsiropoulos, G. I., & Angelidis, P. (2011). Bandwidth allocation in cooperative wireless networks: Buffer load analysis and fairness evaluation. *Physical Communication*, *4*(3), 227–236. doi:10.1016/j.phycom.2011.03.001

Lam, J. H., Busan, S. K., Lee, S.-G., & Tan, W. K. (2012). Multi-channel wireless mesh networks test-bed with embedded systems. In *Proceedings of 26th International Conference on Advanced Information Networking and Applications Workshops (WAINA)* (pp. 533-537). Fukuoka, Japan: IEEE.

Lang, E., Redana, S., & Raaf, B. (2009). Business impact of relay deployment for coverage extension in 3GPP LTE-advanced. In *Proceedings of IEEE International Conference on Communications: Communications Workshops*. IEEE.

Law, K. L. E. (2009). Transport protocols for wireless mesh networks. In S. Misra, S. C. Misra, & I. Woungang (Eds.), *Guide to wireless mesh networks* (pp. 255–275). Springer.

Laxmi, V., Jain, L., & Gaur, M. S. (2006). Ant colony optimization based routing on NS-2. In *Proceedings of International Conference on Wireless Communication and Sensor Networks*. Allahabad: IEEE

Le, H. A, M., & M, T. (2010). A survey on clustering algorithms for wireless sensor networks. In *Proceedings of the 13th International Conference on Network-Based Information Systems* (NBiS-2010) (pp. 358 – 364). IEEE Press.

Lecuire, V., Duran-Faundez, C., & Krommenacker, N. (2008). Energy-efficient image transmission in sensor networks. *International Journal of Sensor Networks*, *4*(1), 37–47. doi:10.1504/IJSNET.2008.019250

Lee, E. J., Han, C. N. K. D. H., & Jwa, D. Y. Y. J. W. (n.d.). *A tone dual-channel DMAC protocol in location unaware ad hoc networks*. Academic Press.

Lee, C.-H., Shih, C.-Y., & Chen, Y.-S. (2012). Stochastic geometry based models for modelling cellular networks in urban areas. *Wireless Networks*, 1–10.

Lee, D., Choi, S., Choi, J., & Jung, J. (2007). Location aided secure routing scheme in mobile ad hoc networks. In *Proceedings of Computational Science and Its Applications* (pp. 131–139). Kuala Lumpur, Malaysia: Springer-Verlag. doi:10.1007/978-3-540-74477-1_13

Lee, J.-H., & Jung, I.-B. (2010). Reliable asynchronous image transfer protocol in wireless multimedia sensor networks. *Sensors (Basel, Switzerland)*, *10*(3), 1486–1510. doi:10.3390/s100301487 PMID:22294883

Lee, P., Jeong, J., Saxena, N., & Shin, J. (2009). Dynamic reservation scheme of physical cell identity for 3GPP LTE femtocell systems. *Journal of Information Processing Systems*, *5*(4). doi:10.3745/JIPS.2009.5.4.207

Lee, W., & Akyildiz, I. F. (2011). A spectrum decision framework for cognitive radio networks. *IEEE Transactions on Mobile Computing*, *10*(2), 161–174. doi:10.1109/TMC.2010.147

Lehsaini, M. (2009). *Diffusion et couverture basées sur le clustering dans les réseaux de capteurs: Application à la domotique*. (Thèse de Doctorat Spécialité Informatique). Université de Franche-Comté.

Lehser, F. (Ed.). (2008). *Next generation mobile networks—Recommendation on SON and O&M requirements (Technical Report)*. NGMN Alliance.

Lei, H., Gao, C., Guo, Y., Ren, Z., & Huang, J. (2012). Wait-time-based multi-channel MAC protocol for wireless mesh networks. *Journal of Networks*, *7*(8), 1208–1213. doi:10.4304/jnw.7.8.1208-1213

Leijdekkers, P., & Gay, V. (2008). A self-test to detect a heart attack using a mobile phone and wearable sensors. In *Proceedings of 21st IEEE CBMS Int. Symp.* (pp. 93-98). IEEE.

Leonard, B. (2005). *KF8GR linux ham home page*. Retrieved July 28, 2013, from http://www.qsl.net/kf8gr/

Levis, P., Madden, S., Polastre, J., Szewczyk, R., Whitehouse, K., Woo, A., et al. (2005). TinyOS: An operating system for sensor networks. *Ambient Intell.*, 115-148.

Lewis, F. L. (2004). Wireless sensor networks. In *Smart environments: Technologies, protocols, and applications*. Academic Press.

Li Longjiang, C. Y., & Xiaoming, X. (2009). Cluster-based autoconfiguration for mobile ad hoc networks. *Wireless Personal Communications*, 49(4), 561–573. doi:10.1007/s11277-008-9577-z

Li, G., Yang, L. L., Conner, W. S., & Sadeghi, B. (2005). Opportunities and challenges for mesh networks using directional antennas. In *Proceedings of WiMESH'05*. WiMESH.

Li, H., & Singhal, M. (2006). A secure routing protocol for wireless ad hoc networks. In *Proceeding of 39th Hawaii International Conference on System Sciences*. Kauai, HI: IEEE.

Li, Y., & Safwat, A. M. (2005). Efficient deafness avoidance in wireless ad hoc and sensor networks with directional antennas. In *Proceedings of the 2nd ACM International Workshop on Performance Evaluation of Wireless Ad Hoc, Sensor, and Ubiquitous Networks* (pp. 175-180). ACM.

Li, Y., Li, M., Shu, W., & Wu, M. Y. (2007). FFT-DMAC: A tone based MAC protocol with directional antennas. In *Proceedings of Global Telecommunications Conference, 2007* (pp. 3661-3665). IEEE.

Li, Z., Sun, L., & Ifeachor, E. C. (2005). Challenges of mobile ad-hoc grids and their applications in e-healthcare. In *Proceedings of 2nd International Conference on Computational Intelligence in Medicine and Healthcare (CIMED2005)*. Lisbon, Portugal: CIMED.

Liang, H., Huang, D., & Peng, D. (2012). On economic mobile cloud computing model. In M. Gris, & G. Yang (Eds.), *Mobile computing, applications, and services* (pp. 329–341). Springer. doi:10.1007/978-3-642-29336-8_22

Li, C., & Li, L. (2012a). Design and implementation of economics-based resource management system in ad hoc grid. *Advances in Engineering Software*, 45(1), 281–291. doi:10.1016/j.advengsoft.2011.10.003

Li, C., & Li, L. (2012b). A resource selection scheme for QoS satisfaction and load balancing in ad hoc grid. *The Journal of Supercomputing*, 59(1), 499–525. doi:10.1007/s11227-010-0450-y

Li, C., Li, L., & Luo, Y. (2013). Agent based sensors resource allocation in sensor grid. *Applied Intelligence*, 39(1), 121–131. doi:10.1007/s10489-012-0397-1

Li, J. (2006). *Quality of service (QoS) provisioning in multihop ad hoc networks. (Doctorate of Philosophy)*. California: Computer Science in the Office of Graduate Studies.

Lim, A. (2001). Distributed services for information dissemination in self-organizing sensor networks. *Journal of the Franklin Institute*, 338(6), 707–727. doi:10.1016/S0016-0032(01)00020-5

Lim, D., Lee, J. W., Gassend, B., Suh, G. E., van Dijk, M., & Devadas, S. (2005). Extracting secret keys from integrated circuits. *IEEE Transactions on Very Large Scale Integration Systems*, 13(10), 1200–1205. doi:10.1109/TVLSI.2005.859470

Lim, S., Yu, C., & Das, C. R. (2009). RandomCast: An energy efficient communication scheme for mobile ad hoc networks. *IEEE Transactions on Mobile Computing*, 8(8), 1039–1051. doi:10.1109/TMC.2008.178

Lin, C. H., Lai, W. S., Huang, Y. L., & Chou, M. C. (2008). I-SEAD: A secure routing protocol for mobile ad hoc networks. In *Proceedings of International Conference on Multimedia and Ubiquitous Engineering* (pp. 102-107). Busan, Korea: IEEE.

Lin, M., Leu, J., & Yu, W. (2011). On transmission efficiency of the multimedia service over IEEE 802.15. 4 wireless sensor networks. In *Proceedings of International Conference on Advanced Communication Technology (ICACT)*, (pp. 184-189). ICACT.

Lin, P., Qiao, C., & Wang, X. (2004). Medium access control with a dynamic duty cycle for sensor networks. In *Proceedings of IEEE Wirel. Commun. Net. Conf.* IEEE.

Lin, P., Zhang, J., Chen, Y., & Zhang, Q. (2011, June). Macro-femto heterogeneous network deployment and management: From business models to technical solutions. *IEEE Wireless Communications*, 64–70.

Lindsey, S., Raghavendra, C., & Sivalingam, K. M. (2002). Data gathering algorithms in sensor networks using energy metrics. *IEEE Transactions on Parallel and Distributed Systems*, 13(9), 924–935. doi:10.1109/TPDS.2002.1036066

Li, P., Zhai, H., & Fang, Y. (2009). SDMAC: Selectively directional MAC protocol for wireless mobile ad hoc networks. *Wireless Networks*, *15*(6), 805–820. doi:10.1007/s11276-007-0076-z

Li, Q., Li, G., Lee, W., Il Lee, M., Mazzarese, D., Clerckx, B., & Li, Z. (2010). MIMO techniques in WiMAX and LTE: A feature overview. *IEEE Communications Magazine*, *48*(5), 86–92. doi:10.1109/MCOM.2010.5458368

Li, Q., & Rusch, L. A. (2002). Multiuser detection for DS-CDMA UWB in the home environment. *IEEE Journal on Selected Areas in Communications*, *20*, 1701–1711. doi:10.1109/JSAC.2002.805241

Liu, Y., Guo, L., Ma, H., & Jiang, T. (2009). Energy efficient on–demand multipath routing protocol for multi-hop ad hoc networks. In *Proceedings of 10th International Symposium on Spread Spectrum Techniques and Applications*, (pp. 592-597). Bologna, Italy: IEEE.

Liu, Y., Li, W., Zhang, H., & Lu, W. (2010). Graph based automatic centralized PCI assignment in LTE. In *Proceeding of the IEEE Symposium on Computers and Communications (ISCC'10)*. Riccione, Italy: IEEE.

Liu, C.-M., Lee, C.-H., & Wang, L.-C. (2007). Distributed clustering algorithms for data gathering in wireless mobile sensor networks. *Journal of Parallel and Distributed Computing*, *67*(11), 1187–1200. doi:10.1016/j.jpdc.2007.06.010

Liu, F., Guo, S., & Sun, Y. (2013). Primary user signal detection based on virtual multiple antennas for cognitive radio networks. *Progress in Electromagnetics Research*, *42*, 213–227.

Liu, J., Matta, I., & Crovella, M. (2003). End-to-end inference of loss nature in a hybrid wired/wireless environment. In *Proceeding of WiOpt'03: Modeling and Optimization in Mobile*. Sophia-Antipolis, France: INRIA.

Liu, J., & Singh, S. (2001). ATCP: TCP for mobile ad hoc networks. *IEEE JSAC*, *19*(7), 1300–1315.

Li, Y., & Safwat, A. M. (2006). DMAC-DACA: Enabling efficient medium access for wireless ad hoc networks with directional antennas. In *Proceedings of Wireless Pervasive Computing*. IEEE.

Liyanage, M., & Gurtov, A. (2012). Secured VPN models for LTE backhaul networks. In *Proceedings of the Vehicular Technology Conference* (pp. 1-5). IEEE.

Lobinger, A., Stefanski, S., Jansen, T., & Balan, I. (2010). Load balancing in downlink LTE self-optimizing networks. In *Proceeding of 71st IEEE Vehicular Technology Conference (VTC'10)*. Taipei, Taiwan: IEEE.

Logan, A. G., McIsaac, W. J., Tisler, A., Irvine, M. J., Saunders, A., & Dunai, A. et al. (2007). Mobile phone–based remote patient monitoring system for management of hypertension in diabetic patients. *American Journal of Hypertension*, *20*(9), 942–948. doi:10.1016/j.amjhyper.2007.03.020 PMID:17765133

Lohier, S., Doudane, Y. G., & Pujolle, G. (2006). MAC-layer adaptation to improve TCP flow performance in 802.11 wireless networks. In *Proceeding of WiMob'06* (pp. 427–433). ACM. doi:10.1109/WIMOB.2006.1696392

Lopez-Perez, D., Guvenc, I., de la Roche, G., Kountouris, M., Quek, T., & Zhang, J. (2011). Enhanced intercell interference coordination challenges in heterogeneous networks. *IEEE Transactions on Wireless Communications*, *18*, 22–30. doi:10.1109/MWC.2011.5876497

Loscri, V., Marano, S., & Morabito, G. (2005). A two-levels hierarchy for low-energy adaptive clustering hierarchy (TL-LEACH). In *Proceedings of Vehicular Technology Conference* (VTC2005) (pp. 1809-1813). Dallas, TX: IEEE Press.

Louthan, G., Hardwicke, P., Hawrylak, P., & Hale, J. (2011). Toward hybrid attack dependency graphs. In *Proceedings of the Seventh Annual Workshop on Cyber Security and Information Intelligence Research*. IEEE.

Lou, W., Liu, W., Zhang, Y., & Fang, Y. (2009). Spread: Improving network security by multipath routing in mobile ad hoc networks. *Wireless Networks*, *15*(3), 279–294. doi:10.1007/s11276-007-0039-4

Lu, H., Li, J., Dong, Z., & Ji, Y. (2011). CRDMAC: An effective circular RTR directional MAC protocol for wireless ad hoc networks. In Proceedings of Mobile Ad-hoc and Sensor Networks (MSN), (pp. 231-237). IEEE.

Lu, K., Qian, Y., Rodriguez, D., Rivera, W., & Rodriguez, M. (2007). Wireless sensor networks for environmental monitoring applications: A design framework. In *Proceedings of IEEE Global Telecommunications Conference, GLOBECOM '07* (pp. 1108–1112). Washington, DC: IEEE.

Luketic, I., Simunic, D., & Blajic, T. (2011). Optimization of coverage and capacity of self- organizing network in LTE. In *Proceeding of the 34th International Convention MIPRO*. MIPRO.

Luprano, J. (2006). European projects on smart fabrics, interactive textiles: Sharing opportunities and challenges. In *Proceedings of Workshop Wearable Technol. Intel. Textiles*. Helsinki, Finland: IEEE.

Lu, X., Towsley, D., Lio, P., & Xiong, Z. (2012). An adaptive directional MAC protocol for ad hoc networks using directional antennas. *Science China Information Sciences, 55*(6), 1360–1371. doi:10.1007/s11432-012-4550-6

Lv, W., Li, W., Zhang, H., & Liu, Y. (2010). Distributed mobility load balancing with RRM in LTE. In *Proceeding of the 3rd IEEE International Conference on Broadband Network and Multimedia Technology*. IEEE.

M, B., VM, B., AN, C., MR, E., WF, J., SM, M., et al. (2007). Remote health-care monitoring using Personal Care Connect. *IBM Systems Journal, 1*, 95-114.

Ma, Z., Zhou, J., & Wu, T. (2010). A cross-layer QoS scheme for MPEG-4 streams. In Proceedings of Wireless Communications, Networking and Information Security (WCNIS), (pp. 392-396). IEEE.

MacDonald, V. H. (1979). The cellular concept. *The Bell System Technical Journal, 58*(1), 15–41. doi:10.1002/j.1538-7305.1979.tb02209.x

Macuha, M., Tariq, M., & Sato, T. (2011). Data collection method for mobile sensor networks based on the theory of thermal fields. *Sensors (Basel, Switzerland), 11*(7), 7188–7203. doi:10.3390/s110707188 PMID:22164011

Magnusson, S., & Olofsson, H. (1997). Dynamic neighbor cell list planning in a micro cellular network. In *Proceeding of the IEEE International Conference on Universal Personal Communications*. San Diego, CA: IEEE.

Maimour, M., Pham, C., & Amelot, J. (2008). Load repartition for congestion control in multimedia wireless sensor networks with multipath routing. In *Proceedings of 2008 3rd International Symposium on Wireless Pervasive Computing*, (pp. 11–15). IEEE. doi:10.1109/ISWPC.2008.4556156

Mallah, R. A., & Quintero. (2009). A light-weight service discovery protocol for ad hoc networks. *Journal of Computer Science, 5*(4), 330–337. doi:10.3844/jcssp.2009.330.337

Man, L. (2006). *Part 15.4: Wireless medium access control (MAC) and physical layer (PHY) specifications for low-rate wireless personal area networks (WPANs)*. Retrieved from http://scholar.google.com/scholar?hl=en&btnG=Search&q=intitle:IEEE+Standard+for+Information+technology-+Telecommunications+and+information+exchange+between+systems-+Local+and+metropolitan+area+networks-+Specific+requirements--Part+15.4:+Wireless+MAC+and+PHY+Specifications+for+Low-Rate+WPANs#0

Manadhata, P. K., & Wing, J. M. (2011). An attack surface metric. *IEEE Transactions on Software Engineering, 37*(3), 371–386. doi:10.1109/TSE.2010.60

Mao, S., Bushmitch, D., Narayanan, S., & Panwar, S. S. (2006). MRTP: A multiflow real-time transport protocol for ad hoc networks. *IEEE Transactions on Multimedia, 8*(2), 356–369. doi:10.1109/TMM.2005.864347

Marina, M. K., & Das, S. R. (2004). Impact of caching and MAC overheads on routing performance in ad hoc networks. *Computer Communications Journal, 27*, 239–252. doi:10.1016/j.comcom.2003.08.014

Marinelli, E. E. (2009). *Hyrax: Cloud computing on mobile devices using MapReduce*. (Unpublished Master's thesis). School of Computer Science, Carnegie Mellon University, Pittsburgh, PA.

Marinescuand, D. C., Marinescuand, G. M., & Ji, Y. Boloni, & Siegel, H. J. (2003). Ad hoc grids: Communication and computing in a power constrained environment. In *Proceedings of Workshop on Energy-Efficient Wireless Communications and Networks (EWCN)* (pp. 113-122). Phoenix, AZ: IEEE.

Marinho, J., & Monteiro, E. (2012). Cognitive radio: Survey on communication protocols, spectrum decision issues, and future research directions. *Wireless Networks*, *18*(2), 147–164. doi:10.1007/s11276-011-0392-1

Marques, R., Mesa, F., Martel, J., & Medina, F. (2003). Comparative analysis of edge and broad-side coupled split ring resonators for metamaterial design: Theory and experiment. *IEEE Transactions on Antennas and Propagation*, *51*(10), 2572–2581. doi:10.1109/TAP.2003.817562

Martignon, F. (2011). Multi-channel power-controlled directional MAC for wireless mesh networks. *Wireless Communications and Mobile Computing*, *11*(1), 90–107. doi:10.1002/wcm.917

Martignon, F., Paris, S., & Capone, A. (2011). DSA-mesh: A distributed security architecture for wireless mesh networks. *Security and Communication Networks*, *4*(3), 242–256. doi:10.1002/sec.181

Martin, L., & Demeure, I. (2008). Using structured and segmented data for improving data sharing on MANETs. In *Proceedings of IEEE 19th International Symposium on Personal, Indoor and Mobile Radio Communications (PIMRC 2008)*. Cannes, France: IEEE.

Mathivaruni, R., & Vaidehi, V. (2008). An activity based mobility prediction strategy using markov modeling for wireless networks. In *Proceedings of the World Congress on Engineering and Computer Science 2008 WCECS 2008* (pp. 379–384). WCECS.

Matin, M. A. (2012). *Developments in wireless network prototyping, design, and deployment: Future generations*. Hershey, PA: IGI Global. doi:10.4018/978-1-4666-1797-1

MC, V., & IF, A. (2009). Error control in wireless sensor networks: A cross layer analysis. *IEEE/ACM Transactions on Networking*, *17*(4), 1186–1199. doi:10.1109/TNET.2008.2009971

McKeown, N. (2009). Software-defined networking. In *Proceedings of INFOCOM*. IEEE.

Meena, Y. K., Singh, A., & Chandel, A. S. (2012). Distributed multi-tier energy-efficient clustering. *Journal of Computer Theory and Engineering*, *4*(1), 1–6. doi:10.7763/IJCTE.2012.V4.418

Meghanathan, N. (2006). An algorithm to determine the sequence of stable connected dominating sets in mobile ad hoc networks. In *Proceedings of the 2nd Advanced International Conference on Telecommunications*. IARIA.

Meghanathan, N., & Skelton, G. W. (2007). A two layer architecture of mobile sinks and static sensors. In *Proceedings of the 15th International Conference on Advanced Computing and Communication* (pp. 249-254). Guwahati, India: IEEE.

Meghanathan, N. (2010). A data gathering algorithm based on energy-aware connected dominating sets to minimize energy consumption and maximize node lifetime in wireless sensor networks. *International Journal of Interdisciplinary Telecommunications and Networking*, *2*(3), 1–17. doi:10.4018/jitn.2010070101

Meghanathan, N. (2011). Performance comparison of minimum hop and minimum edge based multicast routing under different mobility models for mobile ad hoc networks. *International Journal of Wireless and Mobile Networks*, *3*(3), 1–14. doi:10.5121/ijwmn.2011.3301

Meghanathan, N. (2012). A comprehensive review and performance analysis of data gathering algorithms for wireless sensor networks. *International Journal of Interdisciplinary Telecommunications and Networking*, *4*(2), 1–29. doi:10.4018/jitn.2012040101

Meghanathan, N., & Farago, A. (2008). On the stability of paths, steiner trees and connected dominating sets in mobile ad hoc networks. *Ad Hoc Networks*, *6*(5), 744–769. doi:10.1016/j.adhoc.2007.06.005

Meghanathan, N., & Mumford, P. D. (2012). Maximum stability data gathering trees for mobile sensor networks. *International Journal on Mobile Network Design and Innovation*, *4*(3), 164–178. doi:10.1504/IJMNDI.2012.051972

Meghanathan, N., Sharma, S., & Skelton, G. W. (2010). On energy efficient dissemination in wireless sensor networks using mobile sinks. *Journal of Theoretical and Applied Information Technology*, *19*(2), 79–91.

Mehlfu¨hrer, C., Wrulich, M., Ikuno, J. C., Bosanska, D., & Rupp, M. (2009). Simulating the long term evolution physical layer. In *Proceedings of the 17th European Signal Processing Conference* (EUSIPCO 2009), (pp. 1471–1478). EUSIPCO.

Menaka, A., & Pushpa, M. E. (2009). Trust based secure routing in AODV routing protocol. In *Proceedings of the 3rd IEEE International Conference on Internet Multimedia Services Architecture and Applications* (pp. 268-273). Piscataway, NJ: IEEE Xplore.

Mendes, L. D., & Rodrigues, J. J. (2011). A survey on cross-layer solutions for wireless sensor networks. *Journal of Network and Computer Applications*, *34*(2), 523–534. doi:10.1016/j.jnca.2010.11.009

Metz, C. (2003). The latest in virtual private networks: Part I. *IEEE Internet Computing*, *7*(1), 87–91. doi:10.1109/MIC.2003.1167346

MEVICO. (2010). *Mobile networks evolution for individual communications experience (MEVICO)*. Retrieved from http://www.celtic-initiative.org/Projects/Celtic-projects/Call7/MEVICO/mevico-default.asp

Microsoft. (2000). *Understanding universal plug and play* (White Paper). Retrieved from www.upnp.org/download/UPNP_understandingUPNP.doc

Miettinen, A. P., & Nurminen, J. K. (2010). Energy efficiency of mobile clients in cloud computing. In *Proceedings of 2nd USENIX Conference on Hot Topics in Cloud Computing*. Boston, MA: USENIX Association.

Miller, S. L. (1995). An adaptive direct-sequence code-division multiple-access receiver for multiuser interference rejection. *IEEE Transactions on Communications*, *43*, 1746–1755. doi:10.1109/26.380225

Mingorance-Puga, J. F., Maciá-Fernández, G., Grilo, A., & Tiglao, N. M. C. (2010). Efficient multimedia transmission in wireless sensor networks. *Next Generation Internet (NGI), 2010*. Retrieved from http://ieeexplore.ieee.org/xpls/abs_all.jsp?arnumber=5534472

Mireles, F. R. (2002). Signal design for ultra-wide band communications in dense multipath. *IEEE Transactions on Vehicular Technology*, *51*, 1517–1521. doi:10.1109/TVT.2002.804838

Miridakis, N. I., Giotsas, V., Vergados, D. D., & Douligeris, C. (2009). A novel power-efficient middleware scheme for sensor grid applications. In N. Bartolini, S. Nikoletseas, P. Sinha, V. Cardellini, & A. Mahanti (Eds.), *Quality of service in heterogeneous networks* (pp. 476–492). Springer. doi:10.1007/978-3-642-10625-5_30

Misra, S., Reisslein, M., & Xue, G. (2008). A survey of multimedia streaming in wireless sensor networks. *IEEE Communications Surveys & Tutorials*, *10*(4), 18–39. doi:10.1109/SURV.2008.080404

Misra, S., Woungang, I., & Misra, S. C. (Eds.). (2009). *Guide to wireless sensor networks*. London: Springer-Verlag. doi:10.1007/978-1-84882-218-4

Mjeku, M., & Gomes, N. J. (2008). Analysis of the request to send/clear to send exchange in WLAN over fiber networks. *Journal of Light Ware Technology*, *26*(13-16), 2531–2539. doi:10.1109/JLT.2008.927202

Mobile Virtual Centre of Excellence (VCE). (n.d.). *Green radio programme*. Retrieved from http://mobilevce.com/

Mohammad El-Basioni, B. M. Abd El-kader, S. M., Eissa, H. S., & Zahra, M. M. (2012). Low loss energy-aware routing protocol for data gathering applications in wireless sensor network. In J. H. Abawajy, M. Pathan, M. Rahman, A. K. Pathan, & M. M. Deris (Eds.), Internet and distributed computing advancements: Theoretical frameworks and practical applications (pp. 272-302). Hershey, PA: IGI Global.

Molisch, A. F. (2005). Ultrawide-band propagation channels-theory, measurement, and modeling. *IEEE Transactions on Vehicular Technology*, *54*, 1528–1545. doi:10.1109/TVT.2005.856194

Montazeri, M., & Duhamel, P. (1995). A set of algorithms linking NLMS and block RLS algorithms. *IEEE Transactions on Signal Processing*, *43*, 444–453. doi:10.1109/78.348127

Montenegro, G., Kushalnagar, N., Hui, J., & Culler, D. (2007). *Transmission of IPv6 packets over IEEE 802.15.4 networks*. Internet Engineering Task Force RFC-4944.

Moradi, S., & Wong, V. W. S. (2007). Technique to improve MPEG-4 traffic schedulers. In *Proceedings of IEEE Conference on Communications*. IEEE.

Moshavi, S. (1996). Multi-user detection for DS-CDMA communications. *IEEE Communications Magazine*, *34*, 124–136. doi:10.1109/35.544334

Motorola. (2007). *Long term evolution (LTE): A Technical overview* (Technical White Paper). Motorola.

Motorola. (2009). *LTE operations and maintenance strategy—Using self-organizing networks to reduce OPEX*. White paper.

Moustakides, G. V. (1997). Study of the transient phase of the forgetting factor RLS. *IEEE Transactions on Signal Processing*, *45*, 2468–2476. doi:10.1109/78.640712

Mu͂noz, P., Barco, R., De la Bandera, I., Toril, M., & Luna-Ram͂ırez, S. (2011). Optimization of a fuzzy logic controller for handover-based load balancing. In *Proceeding of 73rd IEEE Vehicular Technology Conference (VTC)*. Budapest: IEEE.

Mulligan, G. (2007). The 6LoWPAN architecture. In *Proceedings of the 4th Workshop on Embedded Networked Sensors*, (pp. 78-82). IEEE.

Multihop Routing. (2003, September 3). Retrieved November 19, 2012, from http://www.tinyos.net/tinyos-1.x/doc/multihop/multihop_routing.html

Munari, A., Rossetto, F., & Zorzi, M. (2008). Cooperative cross layer MAC protocols for directional antenna ad hoc networks. *ACM SIGMOBILE Mobile Computing and Communications Review*, *12*(2), 12–30. doi:10.1145/1394555.1394558

MVNO. (2012). *The MVNO directory*. Retrieved from http://www.mvnodirectory.com/overview.html

Nahm, K., Helmy, A., & Kuo, C.-C. J. (2004). On interactions between MAC and transport layers in 802.11 ad-hoc networks. In *Proceeding of SPIE ITCOM'04*. Philadelphia: SPIE.

Nangalia, V., Prytherch, D. R., & Smith, G. B. (2010). Health technology assessment review: Remote monitoring of vital signs-current status and future challenges. *Critical Care (London, England)*, *24*(14), 233. doi:10.1186/cc9208

Naseer ul Islam, M., & Mitschele-Thiel, A. (2012). Reinforcement learning strategies for self-organized coverage and capacity optimization. In *Proceeding of the IEEE Conference on Wireless Communications and Networking*. IEEE.

Neel, J. (2006). *Analysis and design of cognitive radio networks and distributed radio resource management algorithms*. (Unpublished doctoral dissertation). Virginia Polytechnic Institute and State University, Blacksburg, VA.

Neel, J. O., & Reed, J. H. (2006). Performance of distributed dynamic frequency selection schemes for interference reducing networks. In *Proceedings of the IEEE Military Communications Conference (MILCOM '06)*. IEEE.

Nee, R., Jones, V. K., Awater, G., Zelst, A., Gardner, J., & Steele, G. (2006). The 802.11n MIMO-OFDM standard for wireless LAN and beyond. *Wireless Personal Communications*, *37*(3-4), 445–453. doi:10.1007/s11277-006-9073-2

Nesargi, S., & Prakash, R. (2002). MANETConf: Configuration of hosts in a mobile ad hoc network. In *Proceedings of IEEE INFOCOM* (pp. 1059–1068). New York: IEEE. doi:10.1109/INFCOM.2002.1019354

Network, N. S. (2011). *Improving 4G coverage and capacity indoors and at hotspots with LTE femtocells*. Author.

Next Generation Mobile Networks (NGMN) Alliance. (2006). *Next generation mobile networks beyond HSPA & EVDO*. White Paper, Version 3.0. Retrieved June 07, 2013, from http://www.ngmn.org/uploads/media/Next_Generation_Mobile_Networks_Beyond_HSPA_EVDO_Web.pdf

Next Generation Mobile Networks (NGMN) Alliance. (2008a). *NGMN recommendation on SON and O&M requirements*. Requirement Specification, Version 1.23. Retrieved June 07, 2013, from http://www.ngmn.org/uploads/media/NGMN_Recommendation_on_SON_and_O_M_Requirements.pdf

Next Generation Mobile Networks (NGMN) Alliance. (2008b). *NGMN use cases related to self-organizing network, overall description*. Deliverable, Version 2.02. Retrieved June 07, 2013, from http://www.ngmn.org/uploads/media/NGMN_Use_Cases_related_to_Self_Organising_Network__Overall_Description.pdf

Ng, P. C., Liew, S. C., Sha, K. C., & To, W. T. (2005). Experimental study of hidden-node problem in IEEE802.11 wireless networks. In *Proceeding of ACM SIGCOMM'05*. ACM.

Nguyen, K., & Hwang, W. (2012). An efficient power control scheme for spectrum mobility management in cognitive radio sensor networks. In J. J. J. H. Park, Y. Jeong, S. O. Park, & H. Chen (Eds.), *Embedded and multimedia computing technology and service* (pp. 667–676). Springer. doi:10.1007/978-94-007-5076-0_81

Nidd, M. (2001). Service discovery in DeapSpace. *IEEE Personal Communications, 8*(4), 39–45. doi:10.1109/98.944002

Nikmard, B., & Taherizadeh, S. (2010). Using mobile agent in clustering method for energy consumption in wireless sensor network. In *Proceedings of the International Conference on Computer and Communication Technology* (ICCCT-2010), (pp. 153-158). Allahabad, India: IEEE Press.

Nilsson, M. (2009). Directional antennas for wireless sensor networks. In *Proceedings of the 9th Scandinavian Workshop on Wireless Adhoc Networks (Adhoc'09)*. Adhoc.

Niu, Z., Wu, Y., Gong, J., & Yang, Z. (2010, November). Cell zooming for cost-efficient green cellular networks. *IEEE Communications Magazine*, 74–79. doi:10.1109/MCOM.2010.5621970

Nokia Siemens Networks and Nokia. (2008). *SON use case: Cell Phy ID automated configuration*. R3-080376. Technical Report.

Nor-Syahidatul, Ismail, Ariffin, Latiff, & Yunus. (2013). Medium access control with token approach in wireless sensor network. *International Journal of Computers and Applications, 65*(17), 51–57.

Nor-Syahidatul. Ismail, Yunus, & Ariffin. (2011). *MPEG-4 video transmission using distributed TDMA MAC protocol over IEEE 802.15.4 wireless technology*. Paper presented at International Conference on Modelling, Simulation And Applied Optimization. Kuala Lumpur, Malaysia.

Norton, K. (2009). *Using a soundmodem on packet radio*. Retrieved July 27, 2013, from http://mysite.verizon.net/ka1fsb/sndmodem.html

Noyes, K. (2013). Make 2013 the year you switch to linux. *Network World News*. Retrieved August 11, 2013, from http://www.networkworld.com/news/2013/010213-make-2013-the-year-you-265433.html

NS2. (n.d.). *Network simulator*. Retrieved from http://www.isi.edu/nsnam

Nugent, J. (2010). *The ARRL Michigan section digital radio group (DRG)*. Retrieved July 21, 2013, from http://www.mi-drg.org/

Olariu, S., & Yan, T. H. G. (2012). The next paradigm shift: From vehicular networks to vehicular clouds. In S. Basagni, M. Conti, S. Giordano, & I. Stojmenovic (Eds.), *Mobile ad hoc networking: The cutting edge directions* (pp. 645–700). Hoboken, NJ: Wiley.

Oliveira, R., & Braun, T. A. (2005). Dynamic adaptive acknowledgment strategy for TCP over multihop wireless networks. In *Proceeding of IEEE INFOCOM* (pp. 1863–1874). Miami, FL: IEEE.

Oliver, N., & Flores-Mangas, F. (2006). *HealthGear: A real-time wearable system for monitoring and analyzing physiological signals*. Microsoft Res. Tech. Rep. MSR-TR-2005-182.

Olofsson, H., Magnusson, S., & Almgren, M. (1996). A concept for dynamic neighbor cell list planning in a cellular system. In *Proceedings of IEEE Personal, Indoor and Mobile Radio Communications*. Taipei, Taiwan: IEEE.

Omar, E. A., & Elsayed, K. M. F. (2010). Directional antenna with busy tone for capacity boosting and energy savings in wireless ad-hoc networks. In Proceedings of High-Capacity Optical Networks and Enabling Technologies (HONET), (pp. 91-95). IEEE.

Open Geospatial Consortium. (2013). *Sensor model language (sensorml)*. Retrieved on August 2013, from http://vast.nsstc.uah.edu/SensorML

Oren, Y., & Wool, A. (2010). RFID-based electronic voting: What could possibly go wrong? In *Proceedings of 2010 IEEE International Conference on RFID*, (pp. 118-125). IEEE.

Ou, X., Boyer, W. F., & McQueen, M. A. (2006). A scalable approach to attack graph generation. In *Proceedings of the 13th ACM Conference on Computer and Communications Security*, (pp. 336-345). ACM.

Palace, B. (1996). *Data mining: What is data mining?* Retrieved from http://www.anderson.ucla.edu/faculty/jason.frand/teacher/technologies/palace/datamining.htm

Pantelopoulos, A., & Bourbakis, N. G. (2010). A survey on wearable sensor-based systems for health monitoring and prognosis. *A Survey on Wearable Sensor-Based Systems for Health Monitoring and Prognosis, 40*(1), 1–12.

Papanastasiou, S., Mackenzie, L. M., Ould-Khaoua, M., & Charissis, V. (2006). On the interaction of TCP and routing protocols in MANETs. In Proceeding of of AICT/ICIW. Guadalupe, French Caribbean: AICT/ICIW.

Papaoulakis, N., Nikitopoulos, D., & Kyriazakos, S. (2003). Practical radio resource management techniques for increased mobile network performance. In *Proceedings of the 12ᵗʰ IST Mobile and Wireless Communications Summit*. IST.

Paradiso, R., Loriga, G., & Taccini, N. (2005). A wearable health care system based on knitted integral sensors. *IEEE Transactions on Information Technology in Biomedicine, 9*(3), 337–344. doi:10.1109/TITB.2005.854512 PMID:16167687

Pariselvam, S., & Parvathi. (2012). Swarm intelligence based service discovery architecture for mobile ad hoc networks. *European Journal of Scientific Research, 74*(2), 205–216.

Parodi, F. Kylvaejae, Alford, M. G., Li, J., & Pradas, J. (2007). An automatic procedure for neighbor cell list definition in cellular networks. In *Proceeding of the 1ˢᵗ IEEE Workshop on Autonomic Wireless Access (in Conjunction with IEEE WoWMoM)*. Helsinki, Finland: IEEE.

Parr, A., Miesen, R., & Vossiek, M. (2013). Inverse SAR approach for localization of moving RFID tags. In *Proceedings of 2013 IEEE International Conference on RFID*, (pp. 104-109). IEEE.

Parsa, C., & Garcia-Luna-Aceves Tulip, J. J. (1999). A link-level protocol for improving TCP over wireless links. In *Proceeding of IEEE WCNC* (vol. 3, pp. 1253 - 1257). New Orleans, LA: IEEE.

Pathak, P. H., & Dutta, R. (2011). A survey of network design problems and joint design approaches in wireless mesh networks. *IEEE Communications Surveys & Tutorials, 13*(3), 396–428. doi:10.1109/SURV.2011.060710.00062

Pendry, J. B., Holden, A. J., Robbins, D. J., & Stewart, W. J. (1999). Magnetism from conductors and enhanced nonlinear phenomena. *IEEE Transactions on Microwave Theory and Techniques, 47*(11), 2075–2084. doi:10.1109/22.798002

Perich, F., Joshi, A., & Chirkova, R. (2006). Data management for mobile ad-hoc networks. In W. Kou, & Y. Yesha (Eds.), *Enabling technologies for wireless e-business* (pp. 132–176). Springer. doi:10.1007/978-3-540-30637-5_7

Perich, F., Joshi, A., Finin, T., & Yesha, Y. (2004). On data management in pervasive computing environments. *IEEE Transactions on Knowledge and Data Engineering*, 621–634. doi:10.1109/TKDE.2004.1277823

Perkins, C. E., Belding-Royer, E. M., & Das, S. R. (1999). Ad hoc on-demand distance vector (AODV) routing. In *Proceedings of 2ⁿᵈ Workshop on Mobile Computing and Applications (WMCSA '99)* (pp. 90-100). WMCSA.

Perkins, C. E., Royer, E. M., & Das, S. R. (2006). *Ad hoc on-demand distance-vector (AODV routing)*. IETF Internet draft (draft-ietf-manet-aodv-o6.txt).

Perkins, C. (2000). *Ad hoc networking*. Reading, MA: Addison-Wesley.

Pham, H., & Jha, S. (2004). An adaptive mobility-aware MAC protocol for sensor networks (MS-MAC). In *Proceedings of the IEEE International Conference on Mobile Ad-hoc and Sensor Systems* (pp. 558-560). Fort Lauderdale, FL: IEEE.

Philips, C., & Swiler, L. (1998). A graph-based system for network-vulnerability analysis. In *Proceedings of the 1998 Workshop on New Security Paradigms*, (pp. 71-79). New York, NY: ACM.

Pidoux, B. (2012). *Linux FPAC mini-HOWTO*. Retrieved April 20, 2013, from http://rose.fpac.free.fr/MINI-HOWTO/

Popa, D., Cremene, M., Borda, M., & Boudaoud, K. (2013). A security framework for mobile cloud applications. In *Proceedings of 11th Roedunet International Conference (RoEduNet)* (pp. 1-4). Sinaia: IEEE.

Postel, J. (1980). User Datagram Protocol. *Isi*, (August), 1–4. Retrieved from http://xml2rfc.tools.ietf.org/html/rfc0768

Postel, J. (1981). Transmission control protocol. *RFC: 793*. Retrieved from http://tools.ietf.org/html/rfc793

Pottie, G. J. (1998). Wireless sensor networks. In *Proceedings of Information Theory Workshop, 1998* (pp. 139-140). IEEE.

Pozar, D. M., & Schaubert, D. H. (1995). *Microstrip anternnas: The analysis and design of microstrip antennas and arrays*. Hoboken, NJ: John Wiley & Sons. doi:10.1109/9780470545270

Prabhu, A., & Das, S. R. (2007). Addressing deafness and hidden terminal problem in directional antenna based wireless multi-hop networks. In *Proceedings of Communication Systems Software and Middleware, 2007*. IEEE.

Proakis, J. G. (2000). *Digital communications* (4th ed.). New York: McGraw Hill.

Proakis, J. G., Rader, C., Ling, F., Nikias, C., Moonen, M., & Proudler, I. (2003). *Algorithms for statistical signal processing*. Englewood Cliffs, NJ: Pearson Education.

Pushpalatha, M., Venkataraman, R., Khemka, R., & Rao, T. R. (2009). Fault tolerant and dynamic file sharing ability in mobile ad hoc networks. In *Proceeding of International Conference on Advances in Computing, Communication and Control* (pp. 474-478). Curan, NY: ACM.

Qabajeh, L. K., Kiah, L. M., & Qabajeh, M. M. (2009). A qualitative comparison of position-based routing protocols for ad-hoc networks. *International Journal of Computer Science and Network Security, 9*(2), 131–140.

Qaisar, S., & Radha, H. (2009). Multipath multi-stream distributed reliable video delivery in wireless sensor networks. In *Proceeding of Conference on Information Sciences and Systems*, (pp. 207–212). IEEE. Retrieved from http://ieeexplore.ieee.org/xpls/abs_all.jsp?arnumber=5054718

Qi, H., Iyengar, S., & Chakrabarty, K. (2001). Distributed sensor networks—A review of recent research. *Journal of the Franklin Institute, 338*(6), 655–668. doi:10.1016/S0016-0032(01)00026-6

Qin, F., & Liu, Y. (2009). Multipath based QoS routing in MANET. *Journal of Networks, 4*(8), 771–778. doi:10.4304/jnw.4.8.771-778

Quang, B. D., & Won-Joo, H. (2006). Trade-off between reliability and energy consumption in transport protocols for wireless sensor networks. *International Journal of Computer Science and Network Security, 6*(8), 47–53.

Quang, B. V. (2012). A survey on handoffs—Lessons for 60 GHz based wireless systems. *IEEE Communications Surveys & Tutorials, 14*(1), 64–86. doi:10.1109/SURV.2011.101310.00005

Qutqut, M., & Hassanein, H. (2012). *Mobility management in wireless broadband femtocells*. Academic Press.

Raj, P. N., & Swadas, P. B. (2009). DPRAODV: A dynamic learning system against blackhole attack in AODV based MANET. *International Journal of Computer Science Issues, 2*(3), 1126–1131.

Ramanathan, R. (2001). On the performance of ad hoc networks with beamforming antennas. In *Proceedings of 2nd ACM International Symposium on Mobile Ad Hoc Networking & Computing* (pp. 95-105). Long Beach, CA: ACM.

Ramanathan, R. (2001). On the performance of ad hoc networks with beamforming antennas. In *Proceedings of the 2nd ACM International Symposium on Mobile Ad Hoc Networking & Computing* (pp. 95-105). ACM.

Ramanathan, R., Redi, J., Santivanez, C., Wiggins, D., & Polit, S. (2005). Ad hoc networking with directional antennas: A complete system solution. *IEEE Journal on Selected Areas in Communications, 23*(3), 496–506. doi:10.1109/JSAC.2004.842556

Ramiro, J., & Hamied, K. (2012). *Self-organizing networks (SON) self-planning—Self-optimization and self-healing for GSM, UMTS and LTE*. Hoboken, NJ: John Wiley & Sons.

Ranch, D. (2013). *Hampacketizing centos-6 and centos-5*. Retrieved July 21, 2013, from http://www.trinityos.com/HAM/CentosDigitalModes/hampacketizing-centos.html

Rangwala, S., Jindal, A., Jang, K., Psounis, K., & Govindan, R. (2008). Understanding congestion control in multi-hop wireless mesh networks. In *Proceedings of MobiCom* (pp. 291–302). San Francisco, CA: ACM. doi:10.1145/1409944.1409978

Rappaport, T. S. et al. (1996). *Wireless communications: Principles and practice* (Vol. 2). Upper Saddle River, NJ: Prentice Hall.

Rashid, R., Embong, W. M. A. E., Zaharim, A., & Fisal, N. (2009). Development of energy aware TDMA-based MAC protocol for wireless sensor network system. *European Journal of Scientific Research*, *30*(4), 571–578.

Raspberry Pi Foundation. (2011-2013). Why raspberry pi? *FAQs | Raspberry Pi*. Retrieved May 24, 2013, from http://www.raspberrypi.org/faqs

Ratsimor, O., Chakraborty, D., Joshi, A., & Finin, T. (2002). Allia: Alliance-based service discovery for ad-hoc environments. In *Proceedings of International Workshop on Mobile Commerce* (pp. 1-9). Atlanta, GA: ACM.

Rawat, D. B., Zhao, Y., Yan, G., & Song, M. (2013). Crave: Cognitive radio enabled vehicular communications in heterogeneous networks. In *Proceedings of Radio and Wireless Symposium (RWS)* (pp. 190-192). Austin, TX: IEEE.

Ray, S., Dash, S., & Tarasia, N. (2011). Congestion-less energy aware token based MAC protocol integrated with sleep scheduling for wireless sensor networks. In *Proceedings of World Congress on Engineering 2011* (Vol. 2, pp. 4-9). Academic Press.

Raychaudhuri, D., & Mandayam, N. B. (2012). Frontiers of wireless and mobile communications. *Proceedings of the IEEE*, *100*(4), 824–840. doi:10.1109/JPROC.2011.2182095

Reed, J. H. (2005). *An introduction to ultra wideband communication systems*. Upper Saddle River, NJ: Prentice Hall.

Revere, L., Black, K., & Zalila, F. (2010). RFIDs can improve the patient care supply chain. *Hospital Topics*, *88*(1), 26–31. doi:10.1080/00185860903534315 PMID:20194108

Rhee, I., & Warrier, A. (2009). DRAND: Distributed randomized TDMA scheduling for wireless ad hoc networks. *IEEE Transactions on Mobile Computing*, *8*(10), 1384–1396. doi:10.1109/TMC.2009.59

Rinaldi, R., & Veca, G. (2007). The hydrogen for base radio stations. In *Proceedings of 29th International Telecommunication Energy Conference* (INTELEC) (pp. 288-292). Rome, Italy: INTELEC.

Rivest, R. L., Shamir, A., & Adleman, L. (1978). A method for obtaining digital signatures and public-key cryptosystems. *ACM*, *21*(2), 120–126. doi:10.1145/359340.359342

Robinson, L. (1998). Japan's new mobile broadcast company: Multimedia for cars, trains, and handhelds. In *Proceedings of the Advances Imaging*, (pp. 18-22). Academic Press.

Roh, Chi, & Lee, Tak, Nam, & Kang. (2010). Embroidered wearable multi-resonant folded dipole antenna for FM reception. *IEEE Antennas and Wireless Propagation Letters*, *9*, 803–806. doi:10.1109/LAWP.2010.2064281

Rohner, C., Nordstr¨om, E., Gunningberg, P., & Tschudin, C. (2005). Interactions between TCP, UDP and routing protocols in wireless multi-hop ad hoc networks. In *Proceedings of 1st IEEE ICPS Workshop on Multi-Hop Ad Hoc Networks: From Theory to Reality* (pp. 1-8). Santorini, Greece: IEEE.

Romeikat, R., Bauer, B., Bandh, T., Carle, G., Sanneck, H., & Schmelz, L.-C. (2010). Policy-driven workflows for mobile network management automation. In *Proceeding of the ACM (IWCMC'10)*. ACM.

Ronai, A., & Officer, C. M. (2009). *LTE ready mobile backhaul (Technical Report)*. Ceragon Networks Ltd.

Rosenbaum, G., Lau, W., & Jha, S. (2003). An analysis of virtual private network solutions. In *Proceedings of the 28th Annual IEEE International Conference on Local Computer Networks* (pp. 395–404). IEEE.

Rosenbaum, G., Lau, W., & Jha, S. (2003). Recent directions in virtual private network solutions. In *Proceedings of the 11th IEEE International Conference on Networks* (pp. 217–223). IEEE.

Roy, A., & Sarma, N. (2010). Energy saving in MAC layer of wireless sensor networks: A survey. In *Proceedings of National Workshop in Design and Analysis of Algorithm (NWDAA)*. Tezpur University.

Royer, E. M., & Toh, C. K. (1999). A review of current routing protocols for ad hoc mobile wireless networks. *IEEE Personal Communications*, *6*(2), 46–55. doi:10.1109/98.760423

Ruiz-Avil'es, J. M., Luna-Ram'ırez, S., Toril, M., & Ruiz, F. (2012). Traffic steering by self-tuning controllers in enterprise LTE femtocells. *EURASIP Journal onWireless Communications and Networking*. Retrieved June 07, 2013, from http://jwcn.eurasipjournals.com/content/2012/1/337

S, L., & C-S, R. (2002). PEGASIS: Power-efficient gathering in sensor information networks. In *Proceeding of IEEE Aerospace Conference* (vol. 3, pp. 1125-1130). IEEE Press. doi: 10.1109/AERO.2002.1035242

Sadiq, A., & Bakar, K. (2011). Mobility and signal strength-aware handover decision in mobile IPv6 based wireless LAN. In *Proceedings of the International Multi-Conference of Engineers and Computer Scientists*. IEEE.

Sailer, T. (1997). *PacketBlaster 97 - Soundkarten-PR mit aktuellen betriebssystemen*. Paper presented at 13 Internationale Packet-Radio-Tagung. Darmstadt, Germany.

Sailer, T. (1997). Using a PC and a soundcard for popular amateur digital modes. In *Proceedings of 16th ARRL and TAPR Digital Communications Conference*. Newington, CT: American Radio Relay League.

Sakai, K., Sun, M.-T., & Ku, W.-S. (2008). Maintaining CDS in mobile ad hoc networks. *Lecture Notes in Computer Science*, *5258*, 141–153. doi:10.1007/978-3-540-88582-5_16

Sallabi, F. M., Odeh, A. O., & Shuaib, K. (2011). Performance evaluation of deploying wireless sensor networks in a highly dynamic WLAN and a highly populated indoor environment. In *Proceedings of 2011 7th International Wireless Communications and Mobile Computing Conference (IWCMC)* (pp. 1371 –1376). Istanbul, Turkey: IWCMC.

Sanaei, Z., Abolfazli, S., Gani, A., & Buyya, R. (2013). Heterogeneity in mobile cloud computing: Taxonomy and open challenges. *IEEE Communications Surveys & Tutorials*, (99), 1-24.

Sandhu, R. (1993). Lattice-based access control models. *Computer*, *26*(11), 9–19. doi:10.1109/2.241422

Sandhu, R., & Samarati, P. (1994). Access control: Principle and practice. *IEEE Communications Magazine*, *32*(9), 40–48. doi:10.1109/35.312842

Sankaralingam, S., & Gupta, B. (2010). Development of textile antennas for body wearable applications and investigations on their performance under bent conditions. *Progress in Electromagnetics Research B*, *22*, 53–71. doi:10.2528/PIERB10032705

Santhosh Kumar, G., Vinu Paul, M. V., & Jacob Poulose, K. (2008). Mobility metric based LEACH-mobile protocol. In *Proceedings of the 16th International Conference on Advanced Computing and Communications* (pp. 248-253). Chennai, India: IEEE.

Sanzgiri, K., Dahill, B., Levine, B. N., Shields, C., & Belding-Royer, E. M. (2002). A secure routing protocol for ad hoc networks. In *Proceedings of 10th IEEE International Conference on Network Protocols* (pp.78-79). Paris, France: IEEE.

Saraydar, C., & Yener, A. (2001). Adaptive cell sectorization for CDMA systems. *IEEE Journal on Selected Areas in Communications*, *19*(6), 1041–1051. doi:10.1109/49.926360

Sarkar, S., Kisku, B., Misra, S., & Obaidat, M. S. (2009). Chinese remainder theorem-based RSA-threshold cryptography in MANET using verifiable secret sharing scheme. In *Proceedings of IEEE International Conference on Wireless and Mobile Computing, Networking and Communications* (pp.258 – 262). Marrakech: IEEE.

Sarma, H. K. D., Kar, A., & Mall, R. (2010). Energy efficient and reliable routing for mobile wireless sensor networks. In *Proceedings of the 6th International Conference on Distributed Computing in Sensor Systems Workshops*. Cambridge, MA: IEEE.

Sasirekha, G., & Bapat, J. (2012). Evolutionary game theory-based collaborative sensing model in emergency CRAHNS. *Journal of Electrical and Computer Engineering*, (1): 1–12.

Schmelz, L. C., Amirijoo, M., Eisenblaetter, A., Litjensm, R., Neuland, M., & Turk, J. (2011). A coordination framework for self-organisation in LTE networks. In *Proceedings of IEEE International Symposium on Integrated Network Management*. Dublin, Ireland: IEEE.

Schofield, A. (2012). 2013 predictions: Africa. *ICT and the Global Community*. Retrieved August 11, 2013, from http://www.idgconnect.com/blog-abstract/704/adrian-schofield-africa-2013-predictions-africa

Scholtz, R. A. (1993). Multiple access with time hopping impulse modulation. In *Proceedings IEEE Military Communications Conference*. IEEE.

Sehgal, P. K., & Nath, R. (2009). An encryption based dynamic and secure routing protocol for mobile adhoc network. *International Journal of Computer Science and Security*, *3*(1).

Seidel, E., & Saad, E. (2010). LTE home node Bs and its enhancements in release 9. *Nomor Research*, 1–5.

Sekido, M., Takata, M., Bandai, M., & Watanabe, T. (2005). A directional hidden terminal problem in ad hoc network MAC protocols with smart antennas and its solutions. In *Proceedings of Global Telecommunications Conference*. IEEE.

Self-Optimization and self-ConfiguRATion in wirelEss networkS (SOCRATES) Project. (2008). Retrieved June 07, 2013, from http://www.fp7-socrates.eu, http://www.fp7-socrates.org

Shah, C. S., Bashir, A. L., Chauhdary, S. H., Jiehui, C., & Park, M.-S. (2009). Mobile ad hoc computational grid for low constraint devices. In *Proceedings of International Conference on Future Computer and Communication* (pp. 416-420). Kuala Lumpur, Malaysia: IEEE.

Shah, R. C., & Rabaey, J. M. (2002). Energy aware routing for low energy ad hoc sensor networks. In *Proceedings of IEEE Wireless Communications and Networking Conference* (pp. 350-355). IEEE.

Shahram, A. H. Krishnamachari, Bar, & Richmond. (2006). Data management techniques for continuous media in ad-hoc networks of wireless devices. In B. Furht (Ed.), Collection of encyclopedia of multimedia (pp. 144-149). Springer.

Shamir, A. (1979). How to share a secret? *Magazine of Communications of the ACM*, *22*(11). doi: doi:10.1145/359168.359176

Shand, & Vuillemin. (1993). Fast implementation of RSA cryptography. In *Proceedings of 11th IEEE Symphosium on Computer Arithmetic* (pp. 252-259). IEEE.

Sheu, P.-R., Tsai, H.-Y., Lee, Y.-P., & Cheng, J. Y. (2009). On calculating stable connected dominating sets based on link stability for mobile ad hoc networks. *Tamkang Journal of Science and Engineering*, *12*(4), 417–428.

Shihab, E., Cai, L., & Pan, J. (2009). A distributed asynchronous directional-to-directional MAC protocol for wireless ad hoc networks. *IEEE Transactions on Vehicular Technology*, *58*(9), 5124–5134. doi:10.1109/TVT.2009.2024085

Shin, K., Yun, S., & Cho, D. (2011). Multi-channel MAC protocol for QoS support in ad-hoc network In *Proceedings of IEEE Consumer Communications and Networking Conference* (pp. 975-976). Las Vegas, NV: IEEE.

Shin, P., & Chung, K. (2008). *A cross-layer based rate control scheme for MPEG-4 video transmission by using efficient bandwidth estimation in IEEE 802.11e*. Paper presented at International Conference Information Networking. Busan.

Shneyderman, A., Bagasrawala, A., & Casati, A. (2000). *Mobile VPNs for next generation GPRS and UMTS networks* (White Paper). Lucent Technologies.

Shneyderman, A., & Casati, A. (2003). *Mobile VPN: Delivering advanced services in next generation wireless systems*. Hoboken, NJ: John Wiley & Sons.

Sichitiu, M. L. (2005). Wireless mesh networks: Opportunities and challenges. In *Proceedings of Wireless World Congress* (pp. 1-6). Palo Alto, CA: IEEE.

Sieg, A. (2009). *JNOS—15 years later | WB5RMG: RadioActive blog*. Retrieved July 27, 2013, from http://wb5rmg.wordpress.com/2009/11/28/jnos-15-years-later/

SIGMONA. (2013). *SDN concept in generalized mobile network architectures (SIGMONA)*. Retrieved from http://www.celticinitiative.org/Projects/Celtic-Plus-Projects/2012/SIGMONA/sigmona-default.asp

Sikora, T. (1997). The MPEG-4 video standard verification model. *IEEE Transactions on Circuits and Systems for Video Technology*, *7*, 19–31. doi:10.1109/76.554415

Simon, M. K., & Alouini, M.-S. (2005). *Digital communication over fading channels* (2nd ed.). John Wiley and IEEE Press.

Singh, A., & Chakrabarti, P. (2013). Ant based resource discovery and mobility aware trust management for mobile grid systems. In *Proceedings of IEEE 3rd International Advance Computing Conference (IACC)* (pp. 637-644). Ghaziabad, India: IEEE.

Singh, S., Mudumbai, R., & Madhow, U. (2010). Distributed coordination with deaf neighbors: Efficient medium access for 60 GHz mesh networks. In *Proceedings of INFOCOM, 2010*. IEEE.

Singh, M., Sethi, M., Lal, N., & Poonia, S. (2010). A tree based routing protocol for mobile sensor networks (MSNs). *International Journal on Computer Science and Engineering, 2*(1S), 55–60.

Singh, S., Dutta, S. C., & Singh, D. K. (2012). A study on recent research trends in MANET. *International Journal of Research and Reviews in Computer Science, 3*(3), 1654.

Singh, V. P., & Kumar, K. (2010). Literature survey on power control algorithms for mobile ad-hoc network. *Wireless Personal Communications, 60*(4), 679–685. doi:10.1007/s11277-010-9967-x

Sinha, D., Bhattacharya, U., & Chaki, R. (2012a). A secure routing scheme in MANET with CRT based secret sharing. In *Proceeding of 15th International Conference of Computer and Information Technology* (pp.225-229). Bangladesh: IEEE XPlore.

Sinha, D., Bhattacharya, U., & Chaki, R. (2012b). A CRT based encryption methodology for secure communication in MANET. *International Journal of Computer Applications, 39*(16). DOI number is 10.5120/4904-7406

Sistla, A. P., Wolfson, O., & Xu, B. (2005). Opportunistic data dissemination in mobile peer-to-peer networks. In C. B. Medeiros, M. J. Egenhofer, & E. Bertino (Eds.), *Advances in spatial and temporal databases* (pp. 923–923). Springer. doi:10.1007/11535331_20

Sitanayah, L., Sreenan, C. J., & Brown, K. N. (2010). ER-MAC: A hybrid MAC protocol for emergency response wireless sensor networks. In *Proceedings of 2010 Fourth International Conference on Sensor Technologies and Applications*, (pp. 244-249). IEEE.

Sivakumar, R., Sinha, P., & Bharghavan, V. (1999). CEDAR: A core-extraction distributed ad hoc routing algorithm. *IEEE Journal on Selected Areas in Communications, 17*(8), 1454–1465. doi:10.1109/49.779926

Skoric, M. (2000-2010a). Linux+WindowsNT mini-HOWTO. *The Linux Documentation Project*. Retrieved March 24, 2013, from http://tldp.org/HOWTO/Linux+WinNT.html

Skoric, M. (2000-2010b). LILO mini-HOWTO. *The Linux Documentation Project*. Retrieved March 24, 2013, from http://tldp.org/HOWTO/LILO.html

Skoric, M. (2000-2010c). FBB packet radio BBS mini-HOWTO. *The Linux Documentation Project*. Retrieved March 24, 2013, from http://tldp.org/HOWTO/FBB.html

Skoric, M. (2009). Amateur radio in education. In H. Song & T. Kidd (Eds.), Handbook of research on human performance and instructional technology (pp. 223-245). Hershey, PA: Information Science Reference (IGI-Global).

Skoric, M. (2012). Simulation in amateur packet radio networks. In *Simulation in computer network design and modeling: Use and analysis* (pp. 216–256). Hershey, PA: IGI Global. doi:10.4018/978-1-4666-0191-8.ch011

Skoric, M. (2013). Security in amateur packet radio networks. In *Wireless networks and security: Issues, challenges and research trends* (pp. 1–47). Berlin, Germany: Springer. doi:10.1007/978-3-642-36169-2_1

Smart, M., & Keepence, B. (2008). *Understanding MPEG-4 video*. IndigoVision Ltd.

Smith, J. (2013). *Samsung wants to offer 5G in 2020, speeds of 1Gbps reached in tests*. Retrieved from http://www.pocket-lint.com/news/120967-samsung-wants-to-offer-5g-in-2020-speeds-of-1gbps-reached-in-tests

Smith, D.R., & Vier, D.C., Koschny, & Soukoulis, C.M. (2005). Electromagnetic parameter retrieval from inhomogeneous metamaterials. *Physical Review E: Statistical, Nonlinear, and Soft Matter Physics, 71*, 036617-1–10. doi:10.1103/PhysRevE.71.036617

Soldani, D., & Ore, I. (2007). Self-optimizing neighbor cell lists for UTRA FDD networks using detected set reporting. In *Proceeding of IEEE Vehicular Technology Conference*. IEEE.

Song, W.-Z., Huang, R., Shirazi, B., & LaHusen, R. (2009). TreeMAC: Localized TDMA MAC protocol for real-time high-data-rate sensor networks. *Pervasive and Mobile Computing, 5*(6), 750–765. doi:10.1016/j.pmcj.2009.07.004

Soon, Y. O., Park, J. S., & Gerla, M. (2008). E-ODMRP: Enhanced ODMRP with motion adaptive refresh. *Journal of Parallel and Distributed Computing, 68*(8), 130–134.

Sosinsky, B. (2010). *Cloud computing bible.* Hoboken, NJ: Wiley Publications.

Spyropoulos, A., & Raghavendra, C. S. (2002). Energy efficient communications in ad hoc networks using directional antennas. In *Proceedings of INFOCOM 2002: Twenty-First Annual Joint Conference of the IEEE Computer and Communications Societies* (Vol. 1, pp. 220-228). IEEE.

Srinivasan, A., & Wu, J. (2008). TRACK: A novel connected dominating set based sink mobility model for WSNs. In *Proceedings of the 17th International Conference on Computer Communications and Networks.* St. Thomas, U. S. Virgin Islands: IEEE.

Std, I. E. E. E. 802.11. (1999). Wireless LAN media access control (MAC) and physical layer (PHY) specifications. IEEE.

Stewart, R. (2007). Stream control transmission protocol. *Network Working Group RFC 2960.* Retrieved from http://tools.ietf.org/html/rfc4960

Stewart, R., Ramalho, M., Xie, Q., Tuexen, M., & Conrad, P. (2004). Stream control transmission protocol (SCTP) partial reliability extension. *Network Working Group RFC 3758.* Retrieved from http://www.hjp.at/doc/rfc/rfc3758.html

Stine, J. A. (2006). Exploiting smart antennas in wireless mesh networks using contention access. *IEEE Wireless Communications,* 13(2), 38–49. doi:10.1109/MWC.2006.1632479

Stordahl, R. (2012). BPQ32_5.2.9.1_20120925 including BPQAPRS is now available. *BPQ32 Yahoo Group.* Retrieved May 24, 2013, from http://groups.yahoo.com/group/BPQ32/message/9075

Stutzman, W. L., & Thiele, G. A. (2012). *Antenna theory and design.* Hoboken, NJ: Wiley.

Su, W., Lee, S.-J., & Gerla, M. (2000). Mobility prediction in wireless networks. In Proceedings 21st Century Military Communications Architectures and Technologies for Information Superiority (Cat. No.00CH37155), (pp. 491–495). doi: doi:10.1109/MILCOM.2000.905001

Subramanian, A. P., & Das, S. R. (2005). Using directional antennas in multi-hop wireless networks: Deafness and directional hidden terminal problems. In *Proceedings of the 13th International Conference on Network Protocols (ICNP'05).* ICNP.

Suh, C., Mir, Z. H., & Ko, Y.-B. (2008). Design and implementation of enhanced IEEE 802.15.4 for supporting multimedia service in wireless sensor networks. *Computer Networks,* 52(13), 2568–2581. doi:10.1016/j.comnet.2008.03.011

Sun Microsystems. (2001). *JINI technology core platform specification, version 1.2.* Retrieved from www-csag.ucsd.edu/teaching/cse291s03/Readings/core1_2.pdf

Sun, H. M., & Wu, M. E. (2005). *An approach towards rebalanced RSA-CRT with short public exponent.* Cryptology ePrint Archive: Report 2005/053. Retrieved from http://eprint.iacr.org/2005/053

Sundaresan, K., Anantharaman, V., & Hsieh, H. (2005). ATP: A reliable transport protocol for ad hoc networks. *IEEE Transactions on Mobile Computing,* 4(6), 588–603. doi:10.1109/TMC.2005.81

Sundaresan, K., Sivakumar, R., Ingram, M. A., & Chang, T. Y. (2004). Medium access control in ad hoc networks with MIMO links: Optimization considerations and algorithms. *IEEE Transactions on Mobile Computing,* 3(4), 350–365. doi:10.1109/TMC.2004.42

Sun, Y., Yu, W., Han, W., & Liu, R. (2006). Information theoretic framework of trust modeling and evaluation for ad hoc networks. *IEEE Journal on Selected Areas in Communications,* 24(2), 305–317. doi:10.1109/JSAC.2005.861389

Sustainability Report 2010-2011. (2011). Vodafone Group Plc.

Su, W., Lee, S.-J., & Gerla, M. (2001). Mobility prediction and routing in ad hoc wireless networks. *International Journal of Network Management,* 11(1), 3–30. doi:10.1002/nem.386

Svecs, I., Sarkar, T., Basu, S., & Wong, J. S. (2010). XIDR: A dynamic framework utilizing cross-layer intrusion detection for effective response deployment. In *Proceedings of 34th Annual IEEE Computer Software and Applications Conference Workshops* (pp. 287-292). Seoul, Korea: IEEE.

Swaminathan, A., Noneaker, D. L., & Russell, H. B. (2012). The design of a channel-access protocol for a wireless ad hoc network with sectored directional antennas. *Ad Hoc Networks*, *10*(3), 284–298. doi:10.1016/j.adhoc.2011.06.011

Swartz, R. A., Jung, D., Lynch, J. P., Wang, Y., Shi, D., & Flynn, M. P. (2005). Flynn: Design of a wireless sensor for scalable distributed in-network computation in a structural health monitoring system. In *Proceedings of the 5th International Workshop on Structural Health Monitoring*. Stanford, CA: Academic Press.

Takagi, H., & Kleinrock, L. (1983). Optimal transmission ranges for randomly distributed packet radio terminals. *IEEE Transactions on Communications*, *32*(3), 246–257. doi:10.1109/TCOM.1984.1096061

Takata, M., Bandai, M., & Watanabe, T. (2007). A MAC protocol with directional antennas for deafness avoidance in ad hoc networks. In *Proceedings of Global Telecommunications Conference, 2007* (pp. 620-625). IEEE.

Takata, M., Bandai, M., & Watanabe, T. (2006). A receiver-initiated directional MAC protocol for handling deafness in ad hoc networks. In *Proceedings of Communications, 2006* (Vol. 9, pp. 4089–4095). IEEE. doi:10.1109/ICC.2006.255721

Talebifard, P., & Leung, V. C. M. (2013). Towards a content-centric approach to crowd-sensing in vehicular cloud. *Journal of Systems Architecture*. Retrieved from http://www.sciencedirect.com/science/article/pii/S1383762113001501

Talucc, F. T., & Bari, D. P. D. (1997). MACA-BI (MACA by invitation), a wireless MAC protocol for high speed ad hoc networking. In *Proceedings of International Conference on Universal Personal Communications* (pp. 913-917). San Diego, CA: IEEE.

Tang, F., Guo, M., Dong, M., Li, M., & Guan, H. (2008). Towards context-aware workflow management for ubiquitous computing. In *Proceedings of International Conference on Embedded Software and Systems* (pp. 221-228). Las Vegas, NV: IEEE.

Tariquea, M., Vitae, A., & Tepe, K. E. (2009). Minimum energy hierarchical dynamic source routing for mobile adhocnetworks. *Ad Hoc Networks*, *7*(6), 1125–1135. doi:10.1016/j.adhoc.2008.10.002

Tarokh, V., Chae, C.-B., Hwang, I., & Heath, R. W. Jr. (2009). *Interference aware-coordinated beamforming system in a two-cell environment*. Academic Press.

Terracina, A., Beco, S., Kirkham, T., Gallop, J., Johnson, I., Randal, D. M., & Ritchie, B. (2006). Orchestration and workflow in a mobile grid environment. In *Proceedings of Fifth International Conference on Grid and Cooperative Computing Workshops* (pp. 251-258). Hunan, China: IEEE.

Tham, C., & Buyya, R. (2005). Sensorgrid: Integrating sensor networks and grid computing. *CSI Communications*, *29*(1), 24–29.

The WALSAIP Project. (n.d.). Retrieved November 8, 2012, from http://walsaip.uprm.edu/

Toh, C.-K. (1996). A novel distributed routing protocol to support ad-hoc mobile computing. In *Proceeding of IEEE 15th Annual Int'l Phoenix Conf. Comp., & Commun.*, (pp. 480–86). IEEE.

Toh, C. K., Mahonen, P., & Uusitalo, M. (2005). Standardization efforts & future research issues for wireless sensors & mobile ad hoc networks. *IEICE Transactions on Communications*, *88*(9), 3500. doi:10.1093/ietcom/e88-b.9.3500

Toril, M., & Wille, V. (2008). Optimization of handover parameters for traffic sharing in GERAN. *Wireless Personal Communications*, *47*(3), 315–336. doi:10.1007/s11277-008-9467-4

Torkestania, J. A. (2013). A new distributed job scheduling algorithm for grid systems. *Cybernetics and Systems: An International Journal*, *44*(1), 77–93. doi:10.1080/01969722.2012.744556

Tranter, J. (1997). Packet radio under linux. *Linux Journal*. Retrieved July 21, 2013, from http://www.linuxjournal.com/article/2218

Tsaoussidis, V., & Badr, H. (2000). TCP-probing: Towards an error control schema with energy and throughput performance gains. In *Proceeding of 8th IEEE Conference on Network Protocols*. IEEE.

Tsoulos, G. V. (1999). Smart antennas for mobile communication systems: Benefits and challenges. *Electronics & Communication Engineering Journal*, *11*(2), 84–94. doi:10.1049/ecej:19990204

U, A., JA, W., P, L., G, T., F, D., M, B., et al. (2004). AMON: A wearable multiparameter medical monitoring and alert system. *IEEE Trans Inf Technol Biomed*, *4*, 415-27.

Udayakumar, P., Vyas, R., & Vyas, O. (2012). Token bus based MAC protocol for wireless sensor networks. *International Journal of Computers and Applications*, *43*(10), 6–10. doi:10.5120/6137-8373

Ulvan, A., Bestak, R., & Ulvan, M. (2010). The study of handover procedure in LTE-based femtocell network. In *Proceeding of 3rd Joint IFIP on Wireless and Mobile Networking Conference (WMNC)*. IFIP.

US Department of Energy. (2004). *Assessment study on sensors and automation (report)*. Washington, DC: Office of Energy and Renewable Energy, U.S. Department of Energy.

UTStarcom. (2009). *3G/LTE mobile backhaul network MPLS-TP based solution* (Technical Report). UTStarcom Inc.

Varadharajan, V., Shankaran, R., & Hitchens, M. (2004). Security for cluster based ad hoc networks. *Journal of Computer Communications*, (27), 488–501.

Varaprasad, G., Dhanalakshmi, S., & Rajaram, M. (2008). *New security algorithm for mobile adhoc networks using zonal routing protocol*. Ubiquitous Computing and Communication Journal.

Varga, A. (2001). The OMNeT++ discrete event simulation system. In *Proceedings of the European Simulation Multiconference* (ESM'2001). Prague, Czech Republic: ESM.

Varga, A. (2005). *Omnet++ discrete event simulation system (version 3.2)*. Retrieved from http://www.omnetpp.org/omnetpp/doc_details/2105-omnet-32-win32-binary-exe

Varga, A. (2010). The OMNeT. In *Modeling and tools for network simulation*. Berlin: Springer Verlag. doi:10.1007/978-3-642-12331-3_3

Vasudevan, S., Kurose, J., & Towsley, D. (2005). On neighbor discovery in wireless networks with directional antennas. In *Proceedings of INFOCOM 2005: 24th Annual Joint Conference of the IEEE Computer and Communications Societies* (Vol. 4, pp. 2502-2512). IEEE.

Vazhkudai, S., & Laszewski, G. V. (2001). A greedy grid - The grid economic engine directive. In *Proceedings of First International Workshop on Internet Computing and E-Commerce* (pp. 1806-1815). San Francisco, CA: IEEE.

Vedantham, R., Sivakumar, R., & Park, S.-J. (2007). Sink-to-sensors congestion control. *Ad Hoc Networks*, *5*(4), 462–485. doi:10.1016/j.adhoc.2006.02.002

Velummylum, N., & Meghanathan, N. (2010). On the utilization of ID-based connected dominating sets for mobile ad hoc networks. *International Journal of Advanced Research in Computer Science*, *1*(3), 36–43.

Venkateswaran, R. (2001). Virtual private networks. *IEEE Potentials*, *20*(1), 11–15. doi:10.1109/45.913204

Verdu, S. (1998). *Multiuser detection*. Cambridge, UK: Cambridge University Press.

Veselago, V. G. (1968). The electrodynamics of substances with simultaneously negative values of ε and μ. *Soviet Physics - Uspekhi*, *10*, 509–514. doi:10.1070/PU1968v010n04ABEH003699

Viterbi, A. J. (1995). *CDMA: Principles of spread spectrum communication*. Upper Saddle River, NJ: Prentice Hall.

Vlajic, N., & Stevanovic, D. (2009). Sink mobility in wireless sensor networks: When theory meets reality. In *Proceedings of the Sarnoff Symposium*. Princeton, NJ: IEEE.

Wang, F., & Zhang, Y. (2002). Improving TCP performance over mobile ad hoc networks with out-of-order detection and response. In *Proceedings of MOBIHOC* (pp. 217-225). Lausanne, Switzerland: ACM.

Wang, F., Min, M., Li, Y., & Du, D. (n.d.). On the construction of stable virtual backbones in mobile ad hoc networks. In *Proceedings of the International Performance Computing and Communications Conference*. Phoenix, AZ: IEEE.

Wang, J. W., Chen, H. C., & Lin, Y. P. (2009).A secure DSDV routing protocol for ad hoc mobile networks. In *Proceedings of Fifth International Joint Conference on INC, IMS and IDC* (pp.2079-2084). IEEE.

Wang, J., & Prabhala, B. (2012). Periodicity based next place prediction. In *Proceedings of Nokia Mobile Data Challenge 2012 Workshop*. Nokia.

Wang, J., Cho, J., Lee, S., & Ma, T. (2011). Real time services for future cloud computing enabled vehicle networks. In *Proceedings of International Conference on Wireless Communications and Signal Processing (WCSP)* (pp. 1-5). Nanjing, China: IEEE.

Wang, J., Fang, Y., & Wu, D. (2005). SYN-DMAC: A directional MAC protocol for ad hoc networks with synchronization. In *Proceedings of Military Communications Conference, 2005* (pp. 2258-2263). IEEE.

Wang, Y., & Garcia-Luna-Aceves, J. J. (2002). Spatial reuse and collision avoidance in ad hoc networks with directional antennas. In *Proceedings of Global Telecommunications Conference, 2002* (Vol. 1, pp. 112-116). IEEE.

Wang, J., Zhai, H., Li, P., Fang, Y., & Wu, D. (2008). Directional medium access control for ad hoc networks. *Wireless Networks*, *15*(8), 1059–1073. doi:10.1007/s11276-008-0102-9

Wang, P., & Zhuang, W. (2008). A token-based scheduling scheme for WLANS supporting voice/data traffic and its performance analysis. *Wireless Communications*, *7*(4), 1708–1718. doi:10.1109/TWC.2007.060889

Wang, S., Guo, W., & O'Farrell, T. (2012, May). Low energy indoor network: Deployment optimisation. *EURASIP Journal on Wireless Communications and Networking*. doi:10.1186/1687-1499-2012-193

Warrier, A., Aia, M., & Sichitiu, M. L. (2008). Z-MAC: A hybrid MAC for wireless sensor networks. *IEEE/ACM Transactions on Networking*, *16*(3), 511–524. doi:10.1109/TNET.2007.900704

WB, H., AP, C., & H, B. (2002). Anapplication-specific protocol architecture forwirelessmicrosensornetworks. *IEEE Transactions on Wireless Communications*, *1*(4), 660–670. doi:10.1109/TWC.2002.804190

Wei, F., Zhang, X., Xiao, H., & Men, A. (2012). A modified wireless token ring protocol for wireless sensor network. In *Proceedings of 2012 2nd International Conference on Consumer Electronics, Communications and Networks (CECNet)*, (pp. 795-799). CECNet.

Wei, Y., & Peng, M. (2011). A mobility load balancing optimization method for hybrid architecture in self organizing network. In *Proceeding of the IET International Conference on Communication Technology and Application*. IET.

Wei, Z. (2010). Mobility robustness optimization based on UE mobility for LTE system. In *Proceeding of the International Conference on Wireless Communications and Signal Processing (WCSP'10)*. WCSP.

Weiser, M. (1995). The computer for the 21st century. *Scientific American*, *272*(3), 78–89.

Welsh, M., & Kaufman, L. (1995). *Running linux*. Sebastopol, CA: O'Reilly & Associates.

WH, W., AAT, B., MA, B., LK, A., JD, B., & WJ, K. (2008). MEDIC: Medical embedded device for individualized care. *Artificial Intelligence in Medicine*, *42*, 137–152. doi:10.1016/j.artmed.2007.11.006 PMID:18207716

White, T. (2012). *Hadoop: The definitive guide*. Sebastopol, CA: O'Reilly Media.

Wicks, A. M., Visich, J. K., & Li, S. (2006). Radio frequency identification applications in hospital environments. *Hospital Topics*, *84*(3), 3–8. doi:10.3200/HTPS.84.3.3-9 PMID:16913301

Wille, V., Pedraza, S., Toril, M., Ferrer, R., & Escobar, J. (2003). Trial results from adaptive handover boundary modification in GERAN. *Electronics Letters*, *39*(4), 405–407. doi:10.1049/el:20030244

Win, M. Z., & Scholtz, R. A. (1998). On the robustness of ultra-wide bandwidth signals in dense multipath environments. *IEEE Communications Letters*, 2, 51–53. doi:10.1109/4234.660801

Winters, J. H. (2006). Smart antenna techniques and their application to wireless ad hoc networks. *IEEE Wireless Communications*, 13(4), 77–83. doi:10.1109/MWC.2006.1678168

Wiseman, J. (2011). *BPQAXIP configuration*. Retrieved June 02, 2013, from http://www.cantab.net/users/john.wiseman/Documents/BPQAXIP%20Configuration.htm

Wiseman, J. (2011). *BPQETHER ethernet driver for BPQ32 switch*. Retrieved April 06, 2013, from http://www.cantab.net/users/john.wiseman/Documents/BPQ%20Ethernet.htm

Wolisz, A. (2005). T2: A second generation OS for embedded sensor networks. *TKN Technical Reports*, 35(3), 349–359.

Wong, C. K., Gouda, M., & Lam, S. S. (2000). Secure group communications using key graphs. *IEEE/ACM Transactions on Networking*, 1(8).

Woo, S., Kang, D., & Choi, S. (2010). Automatic neighbouring BS list generation scheme for femtocell network. In *Proceedings of 2010 Second International Conference on Ubiquitous and Future Networks (ICUFN)*, (pp. 251–255). ICUFN. doi:10.1109/ICUFN.2010.5547200

Woodward, G., & Vucetic, B. S. (1998). Adaptive detection for DS-CDMA. *Proceedings of the IEEE*, 86, 1413–1434. doi:10.1109/5.681371

Wu, C. H., Hong, J. H., & Wu, C. W. (2001). RSA cryptosystem design based on the Chinese remainder theorem. In *Proceedings of Asia and South Pacific Design Automation Conference* (pp. 391-395). Yokohama, Japan: ACM.

Wu, M., Xu, J., Tang, X., & Lee, W. C. (2006). Monitoring top-k query in wireless sensor networks. In *Proceedings of the IEEE International Conference on Data Engineering* (ICDE''06) (pp. 962 – 976). Atlanta, GA: IEEE Press.

Wu, S.-J. (2011). A new handover strategy between femtocell and macrocell for LTE-based network. In *Proceedings of 2011 Fourth International Conference on Ubi-Media Computing* (pp. 203–208). IEEE. doi:10.1109/U-MEDIA.2011.58

Wu, W., Beng Lim, H., & Tan, K.-L. (2010). Query-driven data collection and data forwarding in intermittently connected mobile sensor networks. In *Proceedings of the 7th International Workshop on Data Management for Sensor Networks* (pp. 20-25). Singapore: IEEE.

Wu, Y., Zhang, L., Wu, Y., & Niu, Z. (2006). Interest dissemination with directional antennas for wireless sensor networks with mobile sinks. In *Proceedings of the 4th International Conference on Embedded Networked Sensor Systems* (pp. 99-111). ACM.

Wu, B., Chen, J., Wu, J., & Cardei, M. (2006). A survey of attacks and countermeasures in mobile ad hoc networks. In Y. Xiao, X. Shen, & D.-Z. Du (Eds.), *Wireless mobile network security* (pp. 103–135). Springer.

Wunder, G., Kasparick, M., Stolyar, A., & Viswanathan, H. (2010). Self-organizing distributed inter-cell beam coordination in cellular networks with best effort traffic. In *Proceeding of the 8th International Symposium on Modeling and Optimization in Mobile, Ad Hoc and Wireless Networks (WiOpt'10)*. Avignon, France: WiOpt.

Wyner, A. D. (1994). Shannon-theoretic approach to a Gaussian cellular multiple-access channel. *IEEE Transactions on Information Theory*, 40(6), 1713–1727. doi:10.1109/18.340450

Xia, G., Huang, Z. G., Wang, Z., Cheng, X., Li, W., & Znati, T. (2006). Secure data transmission on multiple paths in mobile ad hoc networks. *Lecture Notes in Computer Science*, 4138, 424–434. doi:10.1007/11814856_41

Xia, H., Jia, Z., Li, X., Ju, L., & Sha, E. H. M. (2012). Trust prediction and trust-based source routing in mobile ad hoc networks. *International Journal of Ad hoc. Networks*, 7(11), 2096–2114.

Xiao, Y., Du, X., Hu, F., & Zhang, J. (2008). A cross-layer approach for frame transmissions of MPEG-4 over the IEEE 802.11e wireless local area networks. In *Proceedings of Wireless Communication and Networking Conference*, (pp. 1728-1733). IEEE.

Xiaodong, D., Jun, W., Ping, J., & Kaifeng, G. (2011). Design and implement of wireless measure and control system for greenhouse. In *Proceedings of the 30th Chinese Control Conference (CCC 2011)*. Yantai, China: CCC.

Xing, G., Wang, T., Jia, W., & Li, M. (2008). Rendezvous design algorithms for wireless sensor networks with a mobile base station. In *Proceedings of the 9th ACM International Symposium on Mobile Ad hoc Networking and Computing* (pp. 231-240). Hong Kong: ACM.

XO. (2010). *A business guide to MPLS IP VPN migration: Five critical factors (Technical Report)*. XO Communications Inc.

Xu, B., Ouksel, A., & Wolfson, O. (2004). Opportunistic resource exchange in inter-vehicle ad-hoc networks. In *Proceedings of IEEE International Conference on Mobile Data Management* (pp. 4-12). Brisbane, Australia: IEEE.

Xu, L., Sun, C., Li, X., Lim, C., & He, H. (2009). The methods to implement self optimization in LTE system. In *Proceeding of the International Conference on Communications Technology and Applications*. Alexandria, Egypt: IEEE.

Xu, Y., Lee, W. C., Xu, J., & Mitchell, G. (2005). PSGR: Priority-based stateless geo-routing in wireless sensor networks. In *Proceedings of the Second IEEE International Conference on Mobile Ad-hoc and Sensor Systems (MASS'05)*. Washington, DC: IEEE Press.

Xu, B., & Wolfson, O. (2005). Data management in mobile peer-to-peer networks. In W. S. Ng, B.-C. Ooi, A. M. Ouksel, & C. Sartori (Eds.), *Databases, information systems, and peer-to-peer computing* (pp. 1–15). Springer. doi:10.1007/978-3-540-31838-5_1

Xu, Y. Q., & Yin, M. (2013). A mobility-aware task scheduling model in mobile grid. *Applied Mechanics and Materials*, *336-338*, 1786–1791. doi:10.4028/www.scientific.net/AMM.336-338.1786

Yaghmaee, M., & Adjeroh, D. (2008). A new priority based congestion control protocol for wireless multimedia sensor networks. In *Proceedings of IEEE International Symposium on a World of Wireless, Mobile and Multimedia Networks*. IEEE. Retrieved from http://ieeexplore.ieee.org/xpls/abs_all.jsp?arnumber=4594816

Yan, G., Rawat, D. B., & Bista, B. B. (2012). Towards secure vehicular clouds. In *Proceedings of Sixth International Conference on Complex, Intelligent and Software Intensive Systems (CISIS)* (pp. 370-375). Palermo, CA: IEEE.

Yang, B., & Zhang, Z. (2011). An improved UDP-Lite protocol for 3D model transmission over wireless network. In *Proceedings of International Conference on Informatics, Cybernetics, and Computer Engineering (ICCE2011)* (pp. 351-357). Melbourne, Australia: Springer.

Yang, C. L., Tarng, W., Hsieh, K. R., & Chen, M. (2010). A security mechanism for clustered wireless sensor networks based on elliptic curve cryptography. *Proceedings of IEEE SMC – eNewsletter*, (33). Retrieved from http://www.mysmc.org/news/back/2010_12/main_article3.html

Yang, H., Cho, S., & Park, C. Y. (2012). Improving performance of remote TCP in cognitive radio networks. *Transactions on Internet and Information Systems (Seoul)*, *6*(9), 2323–2338.

Yang, L., & Giannakis, G. B. (2004). Ultra-wideband communications: An idea whose time has come. *IEEE Signal Processing Magazine*, *21*, 26–54. doi:10.1109/MSP.2004.1359140

Yang, Q., Zhuang, Y., & Li, H. (2011). A multi-hop cluster based routing protocol for wireless sensor networks. *Journal of Convergence Information Technology*, *6*(3), 318–325. doi:10.4156/jcit.vol6.issue3.37

Ye, M., Li, C., Chen, G., & Wu, J. (2005). EECS: An energy efficient clustering scheme in wireless sensor networks. In *Proceedings of the 24th IEEE International Performance, Computing, and Communications Conference*, (pp. 535–540). IEEE. doi: 10.1109/PCCC.2005.1460630

Ye, W., Heidemann, J., & Estrin, D. (2004). Medium access control with coordinated adaptive sleeping for wireless sensor networks. *IEEE/ACM Transactions on Networking*, *12*, 493–506. doi:10.1109/TNET.2004.828953

Yi, S., Pei, Y., & Kalyanaraman, S. (2003). On the capacity improvement of ad hoc wireless networks using directional antennas. In *Proceedings of the 4th ACM International Symposium on Mobile Ad Hoc Networking & Computing* (pp. 108-116). ACM.

Yonhap. (2012). *Samsung to offer 5G service by 2020*. Retrieved from http://english.yonhapnews.co.kr/news/2013/05/12/0200000000AEN20130512000900320.HTML

You, L., Liu, C., & Tong, S. (2011). Community medical network (CMN), architecture and implementation. *Global Mobile Congress*, 1-6.

Younis, O., & Fahmy, S. (2004). HEED: A hybrid, energy-efficient, distributed clustering approach for ad hoc sensor networks. *IEEE Transactions on Mobile Computing, 3*(4), 366–379. doi:10.1109/TMC.2004.41

Yu, L., Wang, N., Zhang, W., & Zheng, C. (2006). GROUP: A grid-clustering routing protocol for wireless sensor networks. In *Proceedings of the International Conference on Wireless Communications, Networking and Mobile Computing* (WiCOM 2006). IEEE Press. doi: 10.1109/WiCOM.2006.287

Yuan, X., Bagga, S., Shen, J., Balakrishnan, M., & Benhaddou, D. (2008). DS-MAC: Differential service medium access control design for wireless medical information systems. In *Proceedings of the 30th Annual IEEE Conference on Engineering in Medicine and Biology Society* (pp. 1801-1804). Vancouver, Canada: IEEE.

Yucek, T., & Arslan, H. (2009). A survey of spectrum sensing algorithms for cognitive radio applications. *IEEE Communications Surveys & Tutorials, 11*(1), 116–130. doi:10.1109/SURV.2009.090109

Yu, F., Sun, B., Krishnamurthy, V., & Ali, S. (2010). Application layer QoS optimization for multimedia transmission over cognitive radio networks. *Wireless Networks, 17*(2), 371–383. doi:10.1007/s11276-010-0285-8

Yuki, T., Yamamoto, T., Sugano, M., Murata, M., Miyahara, H., & Hatauchi, T. (2004). *Performance improvement of TCP over an ad hoc network by combining of data and ACK packets*. IEICE Transactions on Communications.

Yunus, F., Ariffin, S. H. S., Ismail, N., Syahidatul, N., & Hamid, A. H. F. A. (2013). Optimum parameters for MPEG-4 data over wireless sensor network. *International Journal of Engineering and Technology.*

Zahariadis, T., & Voliotis, S. (2007). Open issues in wireless visual sensor networking. In *Proceedings of 2007 14th International Workshop on Systems, Signals and Image Processing and 6th EURASIP Conference focused on Speech and Image Processing, Multimedia Communications and Services*, (pp. 335–338). EURASIP. doi:10.1109/IWSSIP.2007.4381110

Zainaldin, A., Lambadaris, I., & Nandy, B. (2008). Adaptive rate control low bit-rate video transmission over wireless Zigbee networks. In *Proceedings of Communications, 2008, ICC'08. IEEE International Conference*, (pp. 52-58). IEEE.

Zakaria, A., & El-Marakby, R. (2009). AdamRTP: Adaptive multi-flows real-time multimedia delivery over WSNs. In *Proceedings of IEEE International Symposium on Signal Processing and Information Technology, ISSPIT 2009*. IEEE. Retrieved from http://ieeexplore.ieee.org/xpls/abs_all.jsp?arnumber=5407580

Zayani, M., & Zeghlache, D. (2012). Cooperation enforcement for packet forwarding optimization in multi-hop ad-hoc networks. In *Proceedings of IEEE Wireless Communications and Networking Conference* (pp. 3150-3163). Paris, France: IEEE Press.

Zayani, M., & Zeghlache, D. (2009). FESCIM: Fair, efficient, and secure cooperation incentive mechanism for hybrid ad hoc networks. *Transactions on Mobile Computing, 11*(5), 753–766.

Zdarsky, F. A., Robitzsch, S., & Banchs, A. (2010). Security analysis of wireless mesh backhauls for mobile networks. *Journal of Network and Computer Applications, 34*(2), 432–442. doi:10.1016/j.jnca.2010.03.029

Zeng, Q., & Agrawal, D. (2002). Handoff in wireless mobile networks. In *Handbook of wireless networks and mobile computing* (pp. 1–26). Academic Press. doi:10.1002/0471224561.ch1

Zephyr Inc . (n.d.). Retrieved June 15, 2005, from http://www.zephyrtech.co.nz

Zetterberg, K., Ab, E., Scully, N., Turk, J., Jorguseski, L., & Pais, A. (2010). Controllability for of home eNodeBs. In *Proceedings of Joint Workshop COST 2100 SWG 3.1 & FP7-ICT-SOCRATES*. COST.

Zhai, H., Chen, X., & Fang, Y. (2007). Improving transport layer performance in multihop ad hoc networks by exploiting MAC layer information. *IEEE Transactions on Wireless Communications, 6*(5). doi:10.1109/TWC.2007.360371

Zhang, C., & Tsaoussidis, V. (2001). TCP-probing: Towards an error control schema with energy and throughput performance gains. In *Proceeding of 11th IEEE/ACM NOSSDAV*. New York: IEEE.

Zhang, H., Wen, X., Wang, B., Zheng, W., & Sun, Y. (2010). A novel handover mechanism between femtocell and macrocell for LTE Based networks. In *Proceedings of 2010 Second International Conference on Communication Software and Networks*, (pp. 228–231). IEEE. doi:10.1109/ICCSN.2010.91

Zhang, J., Abhayapala, T., & Kennedy, R. (2005). Role of pulses in ultra wideband systems. In *Proceedings IEEE International Conference on Ultra-Wideband*. IEEE.

Zhang, L., Hauswirth, M., & Shu, L. (2008). Multi-priority multi-path selection for video streaming in wireless multimedia sensor networks. In *Proceedings of Ubiquitous Intelligence and Computing* (pp. 23–25). Springer. Retrieved from http://link.springer.com/chapter/10.1007/978-3-540-69293-5_35

Zhang, P., & Yan, Z. (2011). A QoS-aware system for mobile cloud computing. In *Proceedings of IEEE International Conference on Cloud Computing and Intelligence Systems (CCIS)* (pp. 518-522). Beijing, China: IEEE.

Zhang, P., Sadler, C., Lyon, S., & Martonosi, M. (2004). Hardware design experiences in ZebraNet. In *Proceedings of the 2nd International Conference on Embedded Networked Sensor Systems* (pp. 227-238). ACM Press. doi: 10.1145/1031495.1031522

Zhang, X., Schiffman, J., & Gibbs, S. (2009). Securing elastic applications on mobile devices for cloud computing. In *Proceedings of ACM Workshop on Cloud Computing Security* (pp. 127-134). Chicago: ACM.

Zhang, Y., Lou, W., & Fang. (2007). A secure incentive protocol for mobile ad hoc networks. *ACM Wireless Networks, 13*(5), 569-582.

Zhang, H., & Hou, J. C. (2005). Maintaining sensing coverage and connectivity in large sensor networks. *Wireless Ad hoc and Sensor Networks. International Journal (Toronto, Ont.), 2*(1-2), 89–123.

Zhang, W., Wang, G., Xing, Z., & Wittenburg, L. (2005). Distributed stochastic search and distributed breakout: Properties, comparison and applications to constraint optimization problems in sensor networks. *Journal of Artificial Intelligence, 161*(1-2).

Zhang, X., & Jacob, L. (2004). MZRP: An extension of the zone routing protocol for multicasting in MANETs. *Journal of Information Science and Engineering, 20*(3), 535–551.

Zhang, Y., & Fang. (2006). ARSA: An attack-resilient security architecture for multihop wireless mesh network. *IEEE Journal on Selected Areas in Communications, 24*(10), 1916–1928. doi:10.1109/JSAC.2006.877223

Zhang, Y., Yang, L. T., & Chen, J. (Eds.). (2010). *RFID and sensor networks: Architectures, protocols, security and integrations*. New York: Taylor and Francis Group, LLC. doi:10.3837/tiis.2010.06.004

Zhao, M., & Yang, Y. (2009). Bounded relay hop mobile data gathering in wireless sensor networks. In *Proceedings of the 6th International Conference on Mobile Ad hoc and Sensor Systems* (pp. 373-382). Macau, China: IEEE.

Zhen, B., Li, H.-B., & Kohno, A. R. (2009). Networking issues in medical implant communications. *International Journal of Multimedia and Ubiquitous Engineering, 4*(1).

Zheng, Y. (2013). Public key cryptography for mobile cloud. In C. Boyd, & L. Simpson (Eds.), *Information security and privacy*. Berlin: Springer. doi:10.1007/978-3-642-39059-3_30

Zhong, T., Xu, B., & Wolfson, O. (2008). Disseminating real-time traffic information in vehicular ad-hoc networks. In *Proceedings of Intelligent Vehicles Symposium* (pp. 1056-1067). Eindhoven, The Netherlands: IEEE.

Zhong, M., & Cassandras, C. G. (2011). Distributed coverage control and data collection with mobile sensor networks. *IEEE Transactions on Automatic Control, 56*(10), 2445–2455. doi:10.1109/TAC.2011.2163860

Zhon, J., Geng, Weng, & Li. (2012). A cross-layers service discovery protocol for MANET. *Journal of Computer Information Systems, 8*(12), 5085–6092.

Zhou, J., & Mitchell, K. (2009). A scalable delay based analytical framework for CSMA/CA wireless mesh networks. *Computer Networks*.

Zhu, C., Nadeem, T., & Agre, J. R. (2006). Enhancing 802.11 wireless networks with directional antenna and multiple receivers. In *Proceedings of Military Communications Conference, 2006*. IEEE.

Ziolkowski, R. W. (2003). Design, fabrication, and testing of double negative metamaterials. *IEEE Transactions on Antennas and Propagation, 51*(16), 1516–1529. doi:10.1109/TAP.2003.813622

About the Contributors

Mohammad A. Matin is currently working at the Department of Electrical and Electronic Engineering, Institut Teknologi Brunei (ITB), Brunei Darussalam as an Associate Professor. Before joining ITB, he was with the department of Electrical Engineering and Computer Science, North South University as an Associate Professor. He obtained his BSc. degree in Electrical and Electronic Engineering from BUET (Bangladesh), MSc degree in digital communication from Loughborough University, UK, and PhD degree in wireless communication from Newcastle University, UK. He has taught several courses in communications, electronics and signal processing at KUET, Khulna University, and BRAC University during his career. Dr. Matin was a visiting academic staff at the National University of Malaysia (UKM), University of Malaya (UM), etc. He has published over 60 refereed journals and conference papers. He is the author of six academic books and seven book chapters. He has presented invited talks in Bangladesh, and Malaysia and has served as a member of the program committee for more than 50 international conferences like ICCSIT'09, IDCS'09, ICCSN'10, ICCSIT'10, ICCSN'11 etc. He also serves as a referee of a few renowned journals, keynote speaker and technical session chair of few international conferences like MIC-CPE 2008, ICCIT 2008, ICMMT 2010, ICCIT 2010, IEEE GLOBECOM 2010, etc. He is currently serving as a member of editorial board of several international journals such as IET Wireless Sensor Systems (IET-WSS), Journal of Electrical and Computer Engineering (JECE), Hindawi Publishing Corporation, IJCTE, etc. and Guest Editor of special issue of IJCNIS. Dr. Matin is a member of IEEE, IEEE Communications Society (IEEE ComSoc), and several other international organizations. He served as a counselor of IEEE North South University (2008-2011), and secretary of IEEE Communication Society, Bangladesh Chapter (2010-2011). He has received a number of Prizes and Scholarships including the Best student prize (Loughborough University), Commonwealth Scholarship, and Overseas Research Scholarship (ORS). He has been fortunate enough to work in WFS Project with Wireless Fibre Sytems Ltd, UK as an expert. His current research interests include UWB communication, wireless sensor networks, cognitive radio, EM modeling, and antenna engineering.

* * *

Qasim Zeeshan Ahmed received the B.Eng. degree in Electrical Engineering from the National University of Sciences and Technology (NUST), Rawalpindi, Pakistan, in 2001, MSc degree from the University of Southern California (USC), Los-Angeles, USA in 2005, and his Ph.D. degree from the University of Southampton, UK in 2009, respectively. From November 2009 to June 2011, he was an Assistant Professor at National University of Computer and Emerging Sciences (NU-FAST), Islamabad,

Pakistan. Since June 2011, he has been with the King Abdullah University of Science and Technology (KAUST), KSA, where he is currently a post-doctoral Fellow with the Computer, Electrical and Mathematical Sciences, and Engineering division. He has published more than 20 IEEE research papers in journals and conference proceedings and his research interests include mainly low-complexity UWB transceiver design, cooperative communications, adaptive signal processing, and spread-spectrum communications.

Nurul 'Ain Amirrudin received the B.Eng. degree in electrical engineering and telecommunications from Universiti Teknologi Malaysia, Malaysia, in 2008. She joined Telekom Malaysia Berhad as an Assistant Manager of Product Development and Management starting July 2008 until the end of 2011. She is currently pursuing a Ph.D. degree in the UTM-MIMOS Center of Excellence for Communication, Faculty of Electrical Engineering, Universiti Teknologi Malaysia. Her research interests include mobility management, mobility prediction, and their applications in Long Term Evolution.

Sharifah H. S. Ariffin received her Ph.D. in 2006 from Queen Mary University of London,, received her Master degree in Mobility Management in Wireless Telecommunication (2001) from Universiti Teknologi Malaysia, and her B.Eng (Hons) in Electronic and Communication Engineering from University of North London in 1997. She is currently an Associate Professor in the Faculty of Electrical Engineering, Universiti Teknologi Malaysia . Her research interests include Wireless Sensor Network, IPv6 network and mobile computing system, handoff management in WiMax, low rate transmission protocol using IPv6-6loWPAN, network modelling and performance, and priority scheduling in packet network.

Uma Bhattacharya earned her Ph.D. in Computer Science in 1995 from University of Calcutta, India. She did her post-doctoral work in Windsor University, Canada in 1998. She was also in UK as Commonwealth Fellow in 2002-2003. In 2005, She was awarded UK-India networking Fellowship to do research work in Northumbria University, UK. Now, she is working as a Professor in the Department of Computer Science and Technology, Bengal Engineering and Science University, Shibpur, Howrah, India. Her present interest of work is based on optical networks, wireless networks, and mobile computing.

Matthew Butler is currently a Ph.D. candidate in the Tandy School of Computer Science. He graduated with his Bachelor of Science in 2008 and Masters of Science in 2011, both in Computer Science, from the Tandy School of Computer Science. His research interests are security engineering, spatial access control, and embedded automotive communication networks.

Rituparna Chaki completed her Ph. D. in Computer Science in 2003 from Jadavpur University, Kolkata, India. She is presently working as an Associate Professor in AKCSIT, University of Calcutta. Before joining AKCSIT, she had worked for seven years in the Department of Computer Science and Engineering in West Bengal University of Technology (WBUT). She had also an experience of nine years as an executive in the systems department of Joint Plant Committee, Kolkata. She has more than 50 research articles in peer-reviewed journals and books. She is presently working in the domain of wireless sensor networks, cloud computing, and the Internet of Things.

Ali Diab received the B.E. degree in Electronic Engineering from Damascus University, Damascus, Syria, in 1999 and a diploma in Computer Science and Automation from the same university in 2000. After that, he obtained his Dr.-Ing. title from Ilmenau University of Technology, Faculty of Computer Science in 2010. His dissertation focused on mobility management in IP-based networks. Currently, he is pursuing his postdoctoral degree at Ilmenau University of Technology on the topic "Self-Organized Future Mobile Communication Networks".

Hussein S. Eissa received his B.Sc. and M.Sc. degrees from Electronics and Communications Department, Faculty of Engineering, Cairo University at 1993 and 1996. Dr. Eissa earned his Ph.D degree from Electronics and Communications Department, Faculty of Engineering, Cairo University in cooperation with Electrical Engineering Department, University of Pennsylvania, Philadelphia, USA in 2000. Dr. Eissa received an international certificate in business and management from IESES business school, University of Navarra, Spain at 2004. Dr. Eissa is an Associate Professor at Computers and Systems Department, Electronics Research Institute. Dr. Eissa has published 25 papers in the computer networking area. Dr. Eissa is the Director of Information Systems and Crisis Management Department at the Ministry of Communications and Information Technology. Dr. Eissa had managed IS projects and international agreements with budgets exceeding 120 million USD.

George Eleftherakis is a Senior Lecturer and Research Coordinator of the CS department at CITY College Thessaloniki, which is an International Faculty of the University of Sheffield. He also leads the Information and Communication Technologies Research Track of the South Eastern European Research Centre (SEERC). He is a Senior Member of the Association of Computing Machinery (ACM), and a member of the Council of European Chapter Leaders (CECL) of ACM. He is also serving the last years as a member of the administration board of the Greek Computer Society. He holds a BSc in Physics (University of Ioannina, Greece), and an MSc and a PhD in Computer Science (University of Sheffield, UK). His main research work is in the area of Formal Methods, Biologically Inspired Computing, Complex Systems, Emergence and self-adaptive, self-organizing systems, Multi-Agent Systems, Education, and Information Security. He gave more than 15 invited talks and published more than 60 papers in International Conferences and Journals and edited six books. He organized, chaired, and joined scientific committees of several international conferences and journals.

Sherine Abd El-kader has her MSc, and PhD degrees from the Electronics and Communications Department and Computers Department, Faculty of Engineering, Cairo University, at 1998, and 2003. Dr. Abd El-kader is an Associate Professor, Computers and Systems Department at the Electronics Research Institute (ERI). She is currently supervising 3 PhD students, and 10 MSc students. Dr. Abd El-kader has published more than 25 papers, 4 book chapters in computer networking area. She is working in many computer networking hot topics such as; Wi-MAX, Wi-Fi, IP Mobility, QoS, Wireless sensors Networks, Ad-Hoc Networking, realtime traffics, Bluetooth, and IPv6. She was an Associate Professor at Faculty of Engineering, Akhbar El Yom Academy from 2007 till 2009. She is also a technical reviewer for many international Journals. She is heading the Internet and Networking unit at ERI from 2003 until now. She is also heading the Information and Decision Making Support Center at ERI from 2009 until now. She is supervising many automation and web projects for ERI. She is supervising many Graduate Projects from 2006 until now. She was also a technical member at both the ERI projects committee and at the telecommunication networks committee, Egyptian Organization for Standardization and Quality from February 2007 until 2011.

Norsheila Fisal received her B.Sc. in Electronic Communication from the University of Salford, Manchester, U.K. in 1984. M.Sc. degree in Telecommunication Technology, and PhD degree in Data Communication from the University of Aston, Birmingham, U.K. in 1986 and 1993, respectively. Currently, she is a Professor with the Faculty of Electrical Engineering, Universiti Technologi Malaysia leading as a Director of UTM MIMOS CoE in Telecommunication Technology and head of Telematic Research Group (TRG). She is actively involved in research focusing on work related to broadband networking in wired and wireless network, multimedia communication, and teletraffic engineering. Her current research interests are in Wireless Sensor Networks, Wireless Mesh and Relay Networks, Cognitive Radio Networks, LTE Advanced Network, and WiMaX Network.

N. Effiyana Ghazali received her B. Eng. (Hons) Electrical (Telecommunications) from Universiti Teknologi Malaysia in 2007, Master of Engineering in Electrical and Computer Science from Shibaura Institute of Technology, Japan in 2010, and Master of Engineering in Electrical (Electronics and Telecommunications) from Universiti Teknologi Malaysia in 2011. Currently, she is a Ph.D candidate, and her research interests are Long Term Evolution (LTE), WiMAX, WiFi, Network Mobility, IPv6 network, Handover Management in Proxy Mobile IPv6, Mobility Management, and Mobile Computing.

Weisi Guo received his M.Eng., M.A. and Ph.D. degrees from the University of Cambridge. He is currently an Assistant Professor and Co-Director of Cities Research Theme at the School of Engineering, University of Warwick. He is the author of the VCEsim LTE System Simulator, and his research interests are in the areas of heterogeneous networks, smart cities, self-organization, energy-efficiency, nano-communications, and cooperative communications.

Andrei Gurtov received his M.Sc (2000) and Ph.D. (2004) degrees in Computer Science from the University of Helsinki, Finland. He is presently a visiting scholar at the International Computer Science Institute (ICSI), Berkeley. He was a Professor at University of Oulu in the area of Wireless Internet in 2010-12. He is also a Principal Scientist leading the Networking Research group at the Helsinki Institute for Information Technology HIIT. Previously, he worked at TeliaSonera, Ericsson NomadicLab, and University of Helsinki. Dr. Gurtov is a co-author of over 130 publications including two books, research papers, patents, and IETF RFCs. He is a senior member of IEEE.

John Hale is a Professor in the Tandy School of Computer Science and a faculty researcher in the Institute for Information Security at The University of Tulsa. He received his Bachelor of Science in 1990, Master of Science in 1992 and doctorate degree in 1997, all in computer science from the University of Tulsa. Dr. Hale has overseen the development of one of the premier information assurance curricula in the nation while at iSec. In 2000, he earned a prestigious National Science Foundation CAREER award for his education and research initiatives at iSec. His research interests include cyber attack modeling, analysis and visualization, enterprise security management, secure operating systems, distributed system verification, and policy coordination.

Sofiane Hamrioui received the engineering degree in computer science option parallel and distributed systems in 2004 and the magister degree in computer science in 2007 at the Mouloud Mammeri University of Tizi Ouzou, Algeria. Since 2008, he was a teacher in the Department of Computer Science,

University of Science and Technology Houari Boumediene, Algiers, Algeria. He has been enrolled in doctoral thesis since 2008 and it looks as part of its research issues in mobile and ad hoc wireless networks as like interactions between protocols, performance protocols, quality of services, and saving energy.

Peter J. Hawrylak is an Assistant Professor in the Electrical Engineering department at University of Tulsa (TU), chair of the AIM RFID Experts Group (REG), and chair of the Healthcare Initiative (HCI) sub-group of the AIM REG. Dr. Hawrylak is a member of The University of Tulsa's Institute for Information Security (iSec), which is a NSA (U.S. National Security Agency) Center of Excellence. Peter has seven (7) issued patents in the RFID space and numerous academic publications. Peter's research interests are in the areas embedded system security, radio frequency identification (RFID), the Internet of Things, embedded systems, and low power wireless systems. He is Associate Editor of the International Journal of Radio Frequency Identification Technology and Applications (IJRFITA), published by InderScience Publishers, which focuses on the application and development of RFID technology.

Noman Islam received B.S. in Computer Science from University of Karachi, Pakistan and M.S. in Computer Science from National University of Computer and Emerging Sciences, Karachi, Pakistan in 2002 and 2006, respectively. Since 2006, he was associated with National University of Computer and Emerging Sciences, Karachi, Pakistan as a Research fellow and completed his Ph.D. in 2013. He has produced more than twenty publications in various international conferences, journals and books. His current research interests include Mobile Ad hoc Network, Ubiquitous Computing, Semantic Web and Multi Agent Systems.

Nor Syahidatul Nadiah Ismail received her Diploma in Electrical (Communication) and a Bachelor of Electrical Engineering (Telecommunication) from Universiti Teknologi Malaysia, Skudai, Johor in 2006 and 2009 respectively. Currently, she is pursuing a PhD degree at Universiti Teknologi Malaysia. Her research interests include MAC layer design in Wireless Sensor Networks (WSNs) for multimedia applications.

J. G. Joshi received his graduate degree in Electronics and Telecommunication Engineering from Amravati University, India and post graduate degree from Birla Institute of Technology and Science (BITS), Pilani, Rajasthan, India in 1994 and 1996 respectively. He is pursuing a Ph.D. under AICTE, India sponsored Ph. D. QIP (POLY) programme. He is presently serving as Lecturer in Electronics (Senior Scale) at Department of Electronics and Telecommunication Engineering, Government Polytechnic, Nashik (M.S.), India. His total teaching experience is 19 years. He has published two books: (1) Mechatronics (2006, Published by Prentice Hall of India; New Delhi, India) (2) Electronic Measurement and Instrumentation Systems (2001, Khanna Book Publishing Co; New Delhi, India). His book chapter on "Some important aspects to enhance the quality of the technical education system for better industry-institute-interaction" has been published by IGI Global, USA in June 2013. He has 70 technical research papers to his credit. His research interests include metamaterial and microstrip patch antennas, wearable antennas (for Wi-Fi, Wi-Max, WLAN, BAN and public safety band), mechatronics, and instrumentation systems. He is a member of Institution of Electronics and Telecommunication Engineers (IETE), New Delhi, India, Associate Member of Institution of Engineers (India), Life Member of Indian Society for Technical Education, New Delhi, India and Life Member of Instrument Society of India.

Govind R Kadambi is a Professor and Dean (Academic) at M.S. Ramaiah School of Advanced Studies, Bangalore, India. He received a B.E. degree in Electronics and Communications from the University of Mysore, India, and M.S. and Ph.D. degrees from the Indian Institute of Technology (IITM), Madras. He is an Inventor/Co-Inventor of 21 US patents. He has authored/coauthored more than 50 research publications in peer reviewed International Journals and Conferences. He also served as a reviewer for IEEE Transactions on Antennas and Propagation. He was Session Chair and invited speaker at several international conferences. His current research interests are mainly focused on Antennas, Computational Electromagnetics, Digital Beamforming, and Signal Processing for Wireless Communication, Mobile AdHoc Networks, and Adaptive Techniques in Communication Engineering. He has been associated with various R&D Labs in India and abroad. He has been the Principal Investigator and Co-Principal Investigator for a number of Sponsored research projects. He is listed in Marquis Who is Who in the World, Marquis Who is Who in Science and Engineering, Marquis Who is Who in Asia, and Dictionary of International Biography, Cambridge, England.

Thomas Lagkas received the BSc degree (with honors) in computer science from the Department of Informatics, Aristotle University, Thessaloniki, Greece. He received the PhD degree on "Wireless Communication Networks" from the same department, in 2006. During his PhD studies, he was awarded the PhD candidates' scholarship by the Research Committee of the Aristotle University. He is a Lecturer at the Computer Science Department, International Faculty of the University of Sheffield, City College. He has been an adjunct Lecturer at the Department of Informatics and Telecommunications Engineering, University of Western Macedonia, Greece, since 2007. He has also been a Laboratory Associate at the Technological Educational Institute of Thessaloniki since 2004 and a Scientific Associate since 2008. Dr. Lagkas has been awarded the postdoctoral research fellowship by the State Scholarships Foundation of Greece. His interests are in the areas of wireless communication networks with relevant publications at a number of widely recognized international scientific journals and conferences.

Mustapha Lalam received the Master's degree in Computer Architecture from the High School of Computer Science, Algiers, Algeria, in 1980 and the PhD degree in Computer Science from University of Toulouse, France, in 1990. He joined the University of Tizi Ouzou, Algeria in 1993, where he is now a professor in the Computer Science Department at the University of Tizi Ouzou. He has been involved in research and Development of Computer Architecture, Distributed Systems and Mobility management for Wireless Mobile Computing and Communications.

Nurul Mu`Azzah Binti Abdul Latiff received B.Eng (Hons) in Electrical Engineering in Telecommunication at Universiti Teknologi Malaysia (UTM) in 2001 and obtained her MSc Geoge Washington University USA in1988. She obtained MEE and Phd from School of Electrical, Electronic and Computer Engineering, Newcastle University, United Kingdom in 2003 and 2008 respectively. She is currently a lecturer in the Faculty of Electrical Engineering, Universiti Teknologi Malaysia, Malaysia. Her specialization filed includes routing in mobile ad hoc networks and wireless sensor networks, network optimization, clustering algorithms, and evolutionary optimization.

Jaime Lioret received his M.Sc. in Physics in 1997 from University of Valencia. He finished a post-graduate Master in Corporative networks and Systems Integration from the Department of Communications of the Polytechnic University of Valencia in 1999. He received his M.Sc. in electronic Engineering in 2003 from University of Valencia and his Ph.D. in telecommunication engineering (Dr. Ing.) from the Polytechnic University of Valencia in 2006. Until 2008, he had more than 55 scientific papers published in national and international conferences, he had more than 25 educational papers and he had more than 25 papers published in international journals (several of them with Journal Citation Report).

Madhusanka Liyanage received the B.Sc. degree in electronics and telecommunication engineering from the University of Moratuwa, Moratuwa, Sri Lanka, in 2009, the M.Eng. degree from the Asian Institute of Technology, Bangkok, Thailand, in 2011 and the M.Sc. degree from University of Nice Sophia Antipolis, Nice, France in 2011. He is currently a Doctoral Student at the Department of Communications Engineering, University of Oulu, Oulu, Finland. In 2011-2012, he was a research scientist at I3S Laboratory and Inria, Shopia Antipolis, France. His research interests are mobile networks and virtual network security. He is a student Member of IEEE and ICT.

Pascal Lorenz received a PhD degree from the University of Nancy, France. Between 1990 and 1995 he was a research engineer at WorldFIP Europe and at Alcatel-Alsthom. He is a professor at the University of Haute Alsace and responsible of the Network and Telecommunication Research Group. His research interests include QoS, Wireless Networks and High-speed Networks. He is a senior member of the IEEE, member of many international program committees, and has served as a guest editor for a number of journals including Telecommunications Systems, IEEE communications Magazine and LNCS.

N. N. Nik Abd Malik graduated with B. Eng (Electrical-Telecommunication), M. Eng. (Radio Frequency and Microwave Communications), and PhD (Electrical Engineering) from Universiti Teknologi Malaysia (UTM), Malaysia; University of Queensland, Australia, and Universiti Teknologi Malaysia (UTM), Malaysia in 2003, 2005, and 2013, respectively. She was a research and development electrical engineer in Motorola Solutions Penang, Malaysia in 2004. She has been a lecturer with the Faculty of Electrical Engineering, UTM since 2005.

Yogesh Kumar Meena received the Integrated Masters (BTech and MTech) in ABV-Information Technology from Indian Institute of Information Technology and Management (ABVIIITM) Gwalior, India, in 2010. In June 2010, he joined the Information Technology Department at Sharda Group of Institution, Agra, India as an Assistant Professor. He has published a number of papers in various national and international journals/conferences. He is a member of the IEEE, IETE, AICSIT and MIR lab. Meena is a reviewer of IEEE and Springer journals. He was given the Excellent Award in Faculty Development Program, organized by Sharda Group of Institutions, Agra, India.

Natarajan Meghanathan is a tenured Associate Professor of Computer Science at Jackson State University, Jackson, MS. He graduated with a Ph.D. in Computer Science from The University of Texas at Dallas in May 2005. Dr. Meghanathan has published more than 150 peer-reviewed articles (more than half of them being journal publications). He has also received federal education and research grants from the U.S. National Science Foundation, Army Research Lab and Air Force Research Lab. Dr. Meghanathan has been serving in the editorial board of several international journals and in the Technical Program Committees and Organization Committees of several international conferences. His research interests are Wireless Ad hoc Networks and Sensor Networks, Graph Theory, Network and Software Security, Bioinformatics, and Computational Biology.

Andreas Mitschele-Thiel is a full professor at the Ilmenau University of Technology, Germany, and head of the Integrated Communication Systems lab. In addition, he is the head of the International Graduate School on Mobile Communications of the University. 2005 to 2009, he also served as Dean for the Faculty for Computer Science and Automation. In addition, he is co-founder and scientific director of two research spin-offs of the university, Cuculus and IDEO Labs. He received a Diploma in Computer Engineering from the Fachhochschule Esslingen in 1985, a M.S. in Computer and Information Science from Ohio State University in 1989 and a Doctoral degree in Computer Science from the University of Erlangen in 1994. He completed his habilitation in Computer Science at the University of Erlangen in 2000. Andreas Mitschele-Thiel has held various positions in development, research, and management at Alcatel and Lucent Bell Labs. His research focuses on the engineering of telecommunication systems. Special interests are in mobile communication networks, especially self-organization in next generation mobile networks, and IP-based mobile communication systems.

Basma M. Mohammad is an assistant researcher with the Computers and Systems Department at the Electronics Research Institute (ERI) in Egypt. In May 2005, she completed her B.S. in Computers and Control, Science, Faculty of Engineering, Zagazig University. During the 2005-2006 year, she joined the Professional Training Project (Network Management and Infrastructure (Tivoli) Track), which is organized by the Egyptian Ministry of Electricity and Information Technology. In 2007, she occupied the position of research assistant at Electronics Research Institute. Since 2011, she occupied her current position. Her publications exceeds six, she also has experience in the field of patents. Her current research interests focus on networks, especially on Wireless Sensor Networks (WSN). Presently, she is studying for Ph.D. degree.

Philip D. Mumford received his BS in Electrical Engineering from the University of Cincinnati in 1983, MS in Physics from Wright State University in 1991, MS in Electrical Engineering from the University of Dayton in 1992, and PhD from the University of Cincinnati in 1997. His career began with advancing research in microwave power sources including both solid state transistors and vacuum tubes and progressed to development of low power analogue to digital converters and integrated panel architecture radar aperture systems. He is currently a Project Engineer in the Integrated Electronic and Net-Centric Warfare Division of the Air Force Research Laboratory's Sensors Directorate, Wright Patterson AFB, OH specialising in electronic warfare applications. He has over 28 years experience in RF devices, components, and subsystem integration projects.

Mukundan K N is a Senior Engineer at Broadcom India, Bangalore. He has received a B.E. degree in Electronics and Communications from the National Institute of Engineering, University of Mysore, India, and the M.Sc [Eng.] from Coventry University, UK. He has industrial experience of 10 years in VLSI verification and has research interest in wireless networks.

Shyam S. Pattnaik received Ph.D. Degree in Engineering (Electronics and Telecommunication Engineering) from Sambalpur University, India in1992. He is presently serving as a Professor and Head of Educational Television Centre of National Institute of Technical Teachers Training and Research (NITTTR), an autonomous institute of Ministry of Human Resource Development, Government of India, Chandigarh. He is a recipient of National Scholarship, BOYSCAST Fellowship, SERC visiting Fellowship, INSA visiting Fellowship, UGC Visiting Fellowship, Best Paper award, etc. He is a fellow of IETE, Senior member of IEEE, Member of IET (UK), life member of ISTE, and has been listed in the Who's Who in the world. He has 277 technical research papers to his credit. He has conducted number of conferences and seminars. His areas of interest are soft computing and information fusion and their application to bio-medical imaging, antenna design, metamaterial antennas, and video processing. 10 Ph.D. students and 48 M.E. students have completed their theses under the guidance of Prof. (Dr.) S.S. Pattnaik. He also worked in the department of Electrical Engineering, University of Utah, USA under Prof. Om. P. Gandhi.

Steven Reed is currently a graduate student of Computer Science in the Tandy School of Computer Science. He graduated with his Bachelor of Science in Computer Science in 2013 from the Tandy School of Computer Science. His research interests are spatial access control, computer graphics, and image processing.

Zubair Ahmed Shaikh received M.S. and Ph.D. degrees in Computer Science from Polytechnic University, New York, USA, in 1991 and 1994, respectively. He did his B.E. (Computer Systems) in 1989 from Mehran University of Engineering and Technology, Jamshoro, Pakistan. He is currently serving as Professor and Director at National University of Computer and Emerging Sciences, Karachi, Pakistan. He has published more than 70 papers in international conferences and journals. His research interest includes Ubiquitous Computing, Artificial Intelligence, Social Network, Human Computer Interaction, Wireless Sensor Networks, and Data Provenance.

Rinki Sharma is working as Assistant Professor at M.S.Ramaiah School of Advanced Studies and is a research scholar at Coventry University, U.K. in wireless networks. She has 8 years of experience in teaching and research. Her main areas of interest are Computer Networks, Network Programming, Wireless Networks and Protocols, and Mobile Ad-hoc Networks. She has authored/coauthored around 15 research publications and has filed 1 US Patent and 2 Indian Patents. She has delivered many trainings and lectures in esteemed corporate and educational institutes.

Ditipriya Sinha, Ph. D. student in Bengal Engineering and Science University, has submitted her Ph. D. dissertation under the joint guidance of Dr. Uma Bhattacharya and Dr. Rituparna Chaki. She is working as Assistant Professor in Calcutta Institute of Engineering and Management, Kolkata, India. She has more than 10 research articles in peer-reviewed journals, books and conferences. She is presently working in the domain of Mobile Ad-hoc Network.

Miroslav Skoric, YT7MPB, has been a licensed radio amateur since 1989. During two decades he administered various types of amateur radio bulletin board systems (based on MS DOS™, Windows™ and Linux platforms) with VHF/HF radio frequency and Internet inputs/outputs. He voluntarily served as the information manager and secretary in the amateur radio union of Vojvodina province in Serbia, where he compiled technical and scientific information for broadcasting via local amateur radio frequencies. His teaching experience includes classes in a local high-school amateur radio club, tutorials on the amateur radio in engineering education, visiting lectures, and paper presentations in domestic and international conferences. He authored three book chapters, several magazine, and journal articles related to amateur radio, as well as a dedicated web page http://tldp.org/HOWTO/FBB.html, which includes user manuals for amateur radio e-mail servers. He is a member of IEEE (Computer Society, Communications Society, and Education Society), ACM, NIAR, and IAENG.

Aditya Trivedi is a Professor in the Information and Communication Technology (ICT) Department at ABV Indian Institute of Information Technology and Management, Gwalior, India. He has about 20 years of teaching experience. He has published around 60 papers in various national and international journals/conferences. He is a fellow of the Institution of Electronics and Telecommunication Engineers (IETE). In 2007, he was given the IETEs K.S. Krishnan Memorial Award for best system-oriented paper.

Yuri Vershinin is senior lecturer and director of the Intelligent Transport Systems and Telematics (ITS&T) Applied Research Group at Coventry University, UK. He joined Coventry University as Senior Lecturer in late 2000 after his research at Aston University, following his industrial career. His industrial work experience included digital and analog systems design, design and implementation of electronic systems for automation and control, and microprocessor control systems. His recent research has been carried out in the area of adaptive control systems, identification, decentralized control, Kalman filtering, and optimal control. He has worked in machine vision, image processing, and computer graphics. At Coventry University, he has formed the Intelligent Transport Systems and Telematics (ITS&T) Applied Research Group in order to work on practical tasks related to advanced automotive systems, Controller Area Network (CAN-bus), vehicle diagnostics and safety, vehicle position and sensors monitoring based on GPS, wireless data-transmission, modelling and simulation of road traffic networks for optimization of the traffic flow, traffic accidents simulation, and reduction of fuel consumption and exhaust emissions. He has organized the Special Edition "Advanced Control Systems in Automotive Applications" of the *International Journal of Modelling, Identification and Control* (IJMIC). He was the Co-Chair of ICEEE and the Invited Speaker on ICME of the World Congress on Engineering (WCE-2012), Imperial College, London, UK. Also, he has provided the Keynote Speech on the International Multi-Conference of Engineers and Computer Scientists (IMECS-2013), International Conference on Electrical Engineering (ICEE) in Hong Kong, March 2013.

Siyi Wang received his B.Eng. in Communication Engineering, B.A. in English Language and Literature from Shanghai University (China) in 2006, and M.Sc(Eng) (with distinction) in Electrical Engineering from the University of Leeds (UK) in 2007. From 2010 to 2012, he was a research member of Virtual Centre of Excellence (VCE) based in the University of Sheffield (UK). He is currently studying for a Ph.D in the Institute for Telecommunications Research at the University of South Australia. The research for his Ph.D investigates communicating data in harsh environments via the diffusion of nanoparticles and is funded by University President's Scholarships. His research interests include: molecular communications, indoor-outdoor network interaction, small cell deployment, device-to-device (D2D) communications, machine learning, stochastic geometry, theoretical frameworks for complex networks and urban informatics.

Lie-Liang Yang (M'98–SM'02) received the B.Eng. degree in communications engineering from Shanghai Tiedao University, Shanghai, China, in 1988 and the M.Eng. and Ph.D. degrees in Communications and Electronics from Northern (Beijing) Jiaotong University, Beijing, China, in 1991 and 1997, respectively. From June 1997 to December 1997, he was a Visiting Scientist with the Institute of Radio Engineering and Electronics, Academy of Sciences of the Czech Republic, Prague, Czech Republic. Since December 1997, he has been with the University of Southampton, Southampton, U.K., where he is currently a Professor with the School of Electronics and Computer Science. He has published more than 240 research papers in journals and conference proceedings, has been the author or a coauthor of three books, and has published several book chapters. Details about his publications can be found at http://www-mobile.ecs.soton.ac.uk/lly/. He is currently an Associate Editor for the *Journal of Communications and Networks* and the *Security and Communication Networks Journal*. His research interests include wireless communications, networking, and signal processing. Dr. Yang is currently an Associate Editor for the IEEE Transactions on Vehicular Technology.

Mika Ylianttila received his Doctoral Degree on Communications Engineering at the University of Oulu in 2005. He has worked as a researcher and professor at the Department of Electrical and Information Engineering. He is the director of the Center for Internet Excellence (CIE) research and innovation unit. He is also docent at the Department of Computer Science and Engineering. He was appointed as a part-time professor at the Department of Communications Engineering for a three-year term starting January 1, 2013. The field of the professorship is broadband communications networks and systems, especially wireless Internet technologies. He has published more than 80 peer-reviewed articles on networking, decentralized (peer-to-peer) systems, mobility management, and content distribution. Based on Google Scholar, his research has impacted more than 1500 citations, and his h-index is 19. He is a Senior Member of IEEE and Editor for *Wireless Networks* journal.

Farizah Yunus received her Diploma in Electrical (Communication) and a Bachelor of Electrical Engineering (Telecommunication) from Universiti Teknologi Malaysia, Skudai, Johor in 2006 and 2009 respectively. Currently she is pursuing a PhD degree at Universiti Teknologi Malaysia. Her research interests include transport protocol in Wireless Sensor Networks (WSNs) for multimedia applications.

Sharifah Kamilah Binti Syed Yusof received her BSc (cum laude) in Electrical Engineering from Geoge Washington University USA in 1988 and obtained her MEE and PhD in 1994 and 2006 respectively from Universiti Teknologi Malaysia. She is currently Associate Professor with the Faculty of Electrical Engineering, Universiti Teknologi Malaysia, Malaysia. Her research interests include OFDMA based system, Software Defined Radio, and Cognitive Radio.

Mohammed M. Zahra received the B. S. and M. S. degrees from Al-Azhar University, Cairo, Egypt, in 1979 and 1987, respectively, all in electrical engineering. He received the Ph. D. degree from AGH, Krakow, Poland in 1993. His doctoral research focused on bandwidth allocation techniques in ATM networks. He was an Assistant Professor of Communication Networks from 1993 to 2006 and Associate Professor from 2006 to present in Al-Azhar University. He is an IEEE member.

Index